# Impact Cratering

# Impact Cratering
## Processes and Products

Edited by

**Gordon R. Osinski**
Western University London, Ontario

**Elisabetta Pierazzo**
Planetary Science Institute, Tuscon, Arizona

**WILEY-BLACKWELL**

A John Wiley & Sons, Ltd., Publication

*Library of Congress Cataloging-in-Publication Data*
Impact cratering : processes and products / edited by Gordon R. Osinski and Elisabetta Pierazzo.
     pages ; cm
  Includes bibliographical references and index.
   ISBN 978-1-4051-9829-5 (cloth)
 1. Impact craters.   2. Cratering.   I. Osinski, Gordon R., editor of compilation.   II. Pierazzo, Elisabetta, editor of compilation.
  QE612.I47 2013
  551.3'97–dc23
                       2012018762

*Main image:* This Mars Global Surveyor image shows a cluster of impact craters in northwest Arabia Terra. NASA/JPL/Malin Space Science Systems.

*Thumbnails. Left:* The Haughton impact structure in the Canadian High Arctic. It is one of the best-exposed impact structures in the world and is estimated to be 39 million years old. Landsat 7 satellite image. NASA's Goddard Space Flight Center/USGS.

*Centre:* Meteor crater near the Route 66, Arizona, US. ©iStockphoto.com/Helena Lovincic.

*Right:* The south pole of Saturn's moon Enceladus. From afar, Enceladus exhibits a bizarre mixture of softened craters and complex, fractured terrains. NASA/JPL/Space Science Institute.

Cover Design by Steve Thompson

Set in 9/11.5 pt Minion by Toppan Best-set Premedia Limited

Printed and bound by CPI Group (UK) Ltd, Croydon, CR0 4YY

C9781405198295_120124

# *Dedication*

This book is dedicated to my friend, colleague, and co-editor, Elisabetta Pierazzo, known to everyone as Betty.

**Elisabetta (Betty) Pierazzo**
**1963–2011**

Betty was recognized internationally as an expert on impact cratering. She was a Senior Scientist at the Planetary Science Institute in Tucson, Arizona. Betty attended graduate school at the Department of Planetary Sciences at the University of Arizona, receiving her Ph.D. in 1997 and was awarded the Gerard P. Kuiper Memorial Award for this research. After a few years as a Research Associate at the University of Arizona, she joined the Planetary Science Institute as a Research Scientist in 2002. It is not an exaggeration to say that Betty played an integral role in the successful expansion of the Planetary Science Institute and making it the renowned research organization that it is today.

Betty was an exceptional scientist and she made numerous important research contributions throughout her career. Amongst her many achievements, she generated the first robust models of melt production during impact events and of oblique impact events. In more recent years Betty increasingly focused on furthering our understanding of the environmental effects of impact events. Another area of research that Betty played a pioneering role in was in modeling the astrobiological consequences of impact events. This ranged from investigating the delivery of organics to planets to the generation of hydrothermal systems by impacts that may have been favorable sites for life on Mars. Betty's community-minded spirit also led her to lead an international effort to benchmark and validate the various different numerical codes used in the impact cratering community.

This sense of bringing together the community is also reflected in Betty's leadership in developing the *Bridging the Gap* series of conferences. I remember attending the first of these conferences in 2003 as a graduate student and being incredibly impressed by the level of community participation and the time dedicated to discussion and the highlighting of major research problems facing the community. The Lunar and Planetary Institute Contribution (#1162) that arose from this conference stands as a testament to Betty's leadership.

In addition to her research, Betty was passionate about education and public outreach. One of her earliest projects was the *Explorers Guide to Impact Craters* program. Betty hired me to help develop this program during my time as a Research Associate at the University of Arizona. One of my fondest memories of Betty was the trip we took together to Meteor Crater with Frank Chuang, where we did the filming for the virtual tour of this crater. Since that time, this program has grown, reached thousands of school children, and served as a model for other programs at the Planetary Science Institute.

Finally, any dedication to Betty would be incomplete without mentioning Betty's lust for life. Whether it was interacting with colleagues on field trips to impact craters around the world, on the pitch playing soccer, or during the many social gatherings at her house in Tucson, Betty always had a smile on her face that invoked the same sense of enthusiasm and joy in everyone she met. Her approach to life made her an ideal mentor for all budding scientists, but especially for women. She proved by her actions that there were many opportunities available to all hard working scientists. She prioritized the happiness of herself and her family, and by doing so ensured her success in work and life, leading the way for others to do the same. The impact community has lost an invaluable member; we all miss her.

Gordon Osinski

# Contents

*Plate section can be found between pages 162–163*

COMPANION WEBSITE:
This book has a companion website:
www.wiley.com/go/osinski/impactcratering
with Figures and Tables from the book

# *Preface*

Despite first being described on the Moon by Galileo Galilei in the seventeenth century, it was not until the late nineteenth century that the American geologist Grove Gilbert proposed an impact origin for these lunar 'craters'. This is long after the birth of modern geology. It was not until 1906 that the first impact crater on Earth – Meteor or Barringer Crater – was recognized by Daniel Barringer in northern Arizona. By the 1930s, several small craters on Earth were suspected as being of impact origin, based on their association with fragments of meteorites. Robert Dietz established the first reliable geological criterion for the identification of impact structures, in the absence of meteorites, in 1947, with the recognition of shatter cones. The increased recognition of impact sites resulting from this discovery, together with the impetus provided by the Apollo landings on the Moon, led to a more complete understanding of the formation of impact craters in the 1960s thanks to the pioneering work of a small group of scientists, including Ralph Baldwin, Carlyle Beals, Robert Dietz, Bevan French, Eugene Shoemaker and others. An improved understanding of the impact cratering process continued during the 1970s and 1980s, with the recognition of several shock metamorphic criteria and dozens more terrestrial impact sites. Finally, in the 1990s, two events finally resulted in impact cratering entering the geological mainstream. Namely, the spectacular impact of 21 fragments of the comet Shoemaker–Levy 9 into Jupiter over 6 days in July 1994 and the discovery of the approximately 200 km diameter Chicxulub impact structure, Mexico, with its link to the Cretaceous–Palaeogene mass extinction event. As has been argued by others, impact cratering can be considered geology's latest revolution.

The initial idea for this book dates back to my time as a graduate student, in the late 1990s. As a student new to the field of impact cratering, I was struck by the general absence of books on this topic. The few that were in circulation were mostly either conference volumes on particular sub-topics or were out of print (e.g. *Impact Cratering: A Geologic Process*, published by H. Jay Melosh in 1989). I also became aware of the general lack of appreciation in the scientific community as to the importance of impact cratering as a geological process, particularly on the Earth. Three events then occurred in quick succession that led to the development of this book. First, in 2007, I co-organized the second *Bridging the Gap* conference in Montreal with Betty Pierazzo and Robbie Herrick. In January 2008, shortly after moving to Western University, I developed a graduate course on *Impact Cratering: Processes and Products* for the first time. The structure of this book is modelled on this course. Then, in late 2008, I was contacted by Ian Francis from Wiley–Blackwell following the publication of a general article entitled 'Meteorite impact structures: the good and the bad' in *Geology Today*. Ian proposed the concept of a book on impact cratering. With some hesitation – as I was only in my second year of being a professor and not even in a tenure track position – I agreed. Soon afterwards I contacted Betty Pierazzo with the idea of continuing the spirit of the *Bridging the Gap* theme of bringing together scientists from various disciplines who study the impact cratering process, its products and effects. Betty gladly agreed and this book is testament to Betty's knowledge, integrity and professionalism.

Sadly, Betty passed away in May 2011, after a battle with a rare form of cancer. A dedication to Betty can be found on the preceding pages. This book would not have been possible without Betty. I am greatly indebted to Richard Grieve for stepping in to assist with editing during Betty's illness and subsequent passing. I would like to thank Richard and all the authors and co-authors of the various chapters in this book for their contributions and taking time out of their busy schedules to make their contributions happen. In addition to various authors and co-authors, who provided peer reviews of other chapters, I would like to thank Claire Belcher, Veronica Bray, Michael Dence, J. Wright Horton, Fred Jourdan, David King, Lucy Thompson, Mark Pilkington, Uwe Reimold, Ralf-Thomas Schmidt, John Spray, Axel Wittman and Michael Zanetti for providing thoughtful and constructive reviews of various chapters. Lastly, I would like to thank the publishing team at Wiley–Blackwell for their unwavering support and patience throughout this process.

Dr Gordon R. Osinski
Western University
London, Ontario

# List of contributors

**Natalia Artemieva** Institute for Dynamics of Geospheres, Russian Academy of Science, Leninsky pr. 38, blg.1, Moscow 119334, Russia.

**Anna Chanou** Departments of Earth Sciences/Physics and Astronomy, Western University, 1151 Richmond Street, London, ON, N6A 5B7, Canada.

**Philippe Claeys** Earth System Science, Vrije Universiteit Brussel, Pleinlaan 2, BE-1050 Brussels, Belgium.

**Charles S. Cockell** School of Physics and Astronomy, University of Edinburgh, Edinburgh, EH9 3JZ, UK. Email: c.s.cockell@ed. ac.uk.

**Gareth S. Collins** IARC, Department of Earth Science and Engineering, Imperial College London, London, SW7 2AZ, UK. Email: g.collins@imperial.ac.uk.

**Ludovic Ferrière** Natural History Museum, Burgring 7, A-1010 Vienna, Austria. Departments of Earth Sciences/Physics and Astronomy, University of Western Ontario, 1151 Richmond Street, London, ON, N6A 5B7, Canada. Email: ludovic.ferriere@nhm-wien.ac.at.

**Steven Goderis** Earth System Science, Vrije Universiteit Brussel, Pleinlaan 2, BE-1050 Brussels, Belgium. Department of Analytical Chemistry, Universiteit Gent, Krijgslaan 281 – S12, BE-9000 Ghent, Belgium. Email: steven.goderis@vub.ac.be.

**Richard A. F. Grieve** Earth Sciences Sector, Natural Resources Canada, Ottawa, Ontario, K1A 0E8, Canada, and Departments of Earth Sciences/Physics and Astronomy, Western University, 1151 Richmond Street, London, ON, N6A 5B7, Canada. Email: rgrieve@ nrcan.gc.ca.

**Justin J. Hagerty** United States Geological Survey, Astrogeology Science Center, 2255 N. Gemini Drive, Flagstaff, AZ 86001, USA. Email: jhagerty@usgs.gov.

**Simon P. Kelley** Centre for Earth, Planetary, Space and Astronomical Research, Department of Earth and Environmental Sciences, Open University, Milton Keynes, MK7 6AA, UK. Email: s.p.kelley@open.ac.uk.

**Thomas Kenkmann** Institut für Geowissenschaften – Geologie, Albert-Ludwigs-Universität Freiburg, Albertstrasse 23-B, D-79104 Freiburg, Germany. Email: thomas.kenkmann@geologie.uni-freiburg.de.

**Kalle Kirsimäe** Department of Geology, University of Tartu, Ravila 14a, 50411 Tartu, Estonia. Email: Kalle.Kirsimae@ut.ee.

**Cassandra Marion** Departments of Earth Sciences/Physics and Astronomy, Western University, 1151 Richmond Street, London, ON, N6A 5B7, Canada.

**H. Jay Melosh** Earth, Atmospheric and Planetary Sciences Department, 550 Stadium Mall Drive, Purdue University, West Lafayette, IN 47907, USA. Email: jmelosh@purdue.edu.

**Patrick Michel** University of Nice–Sophia Antipolis, CNRS, Côte d'Azur Observatory, BP 4229, 06304 Nice Cedex 4, France. Email: michelp@oca.eu.

**Saumitra Misra** School of Geological Sciences, University of Kwazulu-Natal, Private Bag X54001, Durban 4000, South Africa. Email: misrasaumitra@gmail.com.

**Alessandro Morbidelli** University of Nice–Sophia Antipolis, CNRS, Côte d'Azur Observatory, BP 4229, 06304 Nice Cedex 4, France.

**Joanna Morgan** Department of Earth Science and Engineering, Imperial College London, South Kensington Campus, London, SW7 2AZ, UK. Email: j.v.morgan@imperial.ac.uk.

**Horton E. Newsom** Institute of Meteoritics and Department of Earth and Planetary Sciences MSC03-2050, University of New Mexico, Albuquerque, NM 87131, USA. Email: newsom@unm. edu.

**Gordon R. Osinski** Departments of Earth Sciences/Physics and Astronomy, Western University, 1151 Richmond Street, London, ON, N6A 5B7, Canada. Email: gosinski@uwo.ca.

**François Paquay** Department of Geology and Geophysics, University of Hawaii at Manoa, Honolulu, HI, USA.

**Elisabetta Pierazzo** (Deceased)  Formerly of Planetary Science Institute, 1700 E. Fort Lowell Road, Suite 106, Tucson, AZ 85719, USA.

**Michael S. Ramsey**  Department of Geology and Planetary Science, University of Pittsburgh, 4107 O'Hara Street, SRCC, Pittsburgh, PA 15260-3332, USA.

**Mario Rebolledo-Vieyra**  Centro de Investigación Científica de Yucatán, A.C., CICY, Calle 43 No 130, 97200 Mérida, Mexico.

**Sarah C. Sherlock**  Centre for Earth, Planetary, Space and Astronomical Research, Department of Earth and Environmental Sciences, Open University, Milton Keynes, MK7 6AA, UK.

**Ann M. Therriault**  Earth Sciences Sector, Natural Resources Canada, Ottawa, Ontario, K1A 0E8, Canada.

**Livio L. Tornabene**  Departments of Earth Sciences/Physics and Astronomy, Western University, 1151 Richmond Street, London, ON, N6A 5B7, Canada.

**Mary A. Voytek**  US Geological Survey, MS 430, 12201 Sunrise Valley Drive, Reston, VA 20192, USA.

**Shawn P. Wright**  Institute of Meteoritics and Department of Earth and Planetary Sciences MSC03 2050, University of New Mexico, Albuquerque, NM 87131, USA. Email: spw0007@auburn. edu.

**Kai Wünnemann**  Museum für Naturkunde, Leibniz-Institut an der Humboldt-Universität Berlin, Invalidenstrasse 43, 10115 Berlin, Germany.

# ONE

# Impact cratering: processes and products

**Gordon R. Osinski*** and **Elisabetta Pierazzo**[†]

*Departments of Earth Sciences/Physics and Astronomy, Western University, 1151 Richmond Street, London, ON, N6A 5B7, Canada

[†]Planetary Science Institute, 1700 E. Fort Lowell Road, Suite 106, Tucson, AZ 85719, USA

## 1.1 Introduction

Over the past couple of decades, it is has become widely recognized that impact cratering is a ubiquitous geological process that affects all planetary objects with a solid surface. Indeed, meteorite impact structures are one of the most common geological landforms on all the rocky terrestrial planets, except Earth, and many of the rocky and icy moons of Saturn and Jupiter. A unique result of the impact cratering process is that material from depth is brought to the surface in the form of ejecta deposits and central uplifts. Impact craters, therefore, provide unique windows into the subsurface on planetary bodies where drilling more than a few metres is not a viable scenario for the foreseeable future. On many planetary bodies where planetary-scale regoliths can develop through micrometeorite bombardment, aeolian or cryogenic processes, the crater walls of fresh impact craters also provide unique sites where *in situ* outcrops can be found. It should not be surprising, therefore, that impact craters have been, and remain, high-priority targets for planetary exploration missions to the Moon, Mars and elsewhere.

The impact record on Earth remains invaluable for our understanding of impact processes, for it is the only source of ground-truth data on the three-dimensional structural and lithological character of impact craters. However, Earth suffers from active erosion, volcanic resurfacing and tectonic activity, which continually erase impact structures from the rock record. Despite this, 181 confirmed impact structures have been documented to date, with several more 'new' impact sites being recognized each year (Earth Impact Database, 2012). Although we lack ground truth, apart from a few lunar and Martian sites visited by human and robotic explorers, the results of planetary exploration missions continue to provide a wealth of new high-resolution data about the surface expression of impact craters. The driving paradigm is that impact cratering is governed by physics and the fundamental processes are the same regardless of the planetary target (Melosh, 1989). However, variations in planetary conditions permit the investigation of how different properties lead to slightly different end results. The Moon represents an end-member case with respect to the terrestrial planets. Low planetary gravity and lack of atmosphere result in cratering efficiency, for a given impact, that is higher than on the other terrestrial planets (Stöffler *et al.*, 2006). The relatively simple target geology combined with the lack of post-impact modification by aqueous and aeolian processes makes the Moon an ideal natural laboratory for studying crater morphology and morphometry. Mercury is similar to the Moon, except for a higher impact velocity, and new data from the *MESSENGER* spacecraft (see Solomon *et al.* (2011) and references therein) are providing a wealth of new information on the mercurian impact cratering record (Strom *et al.*, 2008). Venus is almost the antithesis of the Moon and Mercury. The relatively high planetary gravity and thick atmosphere reduce cratering efficiency for a given impact relative to these bodies (Schultz, 1993). Hotter surface and subsurface temperatures affect numerous aspects of the cratering process on Venus, the most spectacular outcome of which is the production of vast impact melt flows (Grieve and Cintala, 1995). The final terrestrial planet, Mars, has a thinner atmosphere, a more complex geology, including the presence of volatiles, and more endogenic geological processes to modify craters (Carr, 2006). It is more Earth-like in this respect, which comes with the associated complications, but its impact cratering record is vastly better preserved and exposed than on Earth (Strom *et al.*, 1992).

Notwithstanding the prior discussion of the ubiquity of impact craters throughout the Solar System, it is important to recognize that, despite being first observed on the Moon by Galileo Galilei in 1609, it was not until the 1960s and 1970s that the importance of impact cratering as a geological process began to be recognized. In 1893, the American geologist Grove Gilbert proposed an impact origin for these lunar craters, but it was not until the 1900s that the first impact crater was recognized on Earth: Meteor or Barringer Crater in Arizona (Barringer,

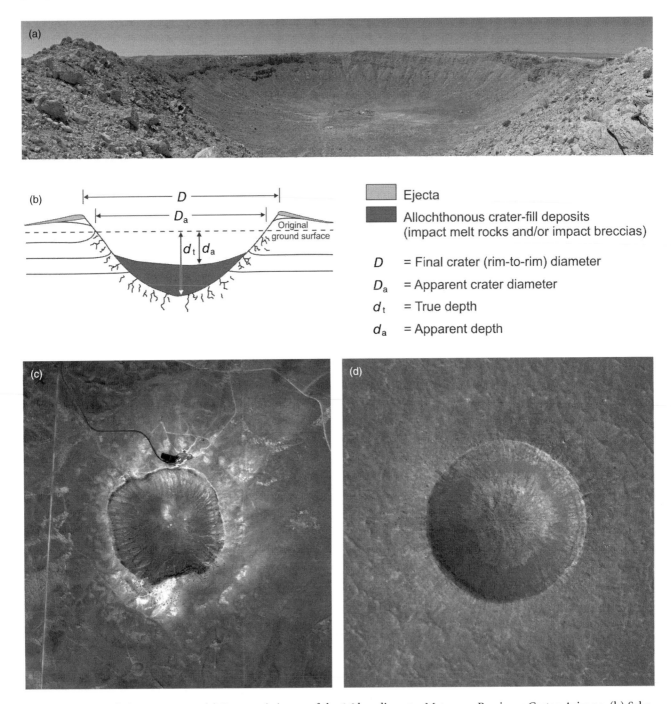

**Figure 1.1**  Simple impact craters. (a) Panoramic image of the 1.2 km diameter Meteor or Barringer Crater, Arizona. (b) Schematic cross-section through a simple terrestrial impact crater. Fresh examples display an overturned flap of near-surface target rocks overlain by ejecta. The bowl-shaped cavity is partially filled with allochthonous unshocked and shocked target material. (c) A 2 m per pixel true-colour image of Barringer Crater taken by WorldView2 (north is up). Image courtesy of Livio L. Tornabene and John Grant. (d) Portion of Lunar Reconaissance Orbiter Camera (LROC) image M122129845 of the 2.2 km diameter Linné Crater on the Moon (NASA/GSFC/Arizona State University). (See Colour Plate 1)

1905; Fig. 1.1a). In the decades that followed, there remained little awareness in the geological community of the importance of impact cratering and there was a general view that impact events were not important for Earth evolution. Indeed, even G. Gilbert, himself, initially disputed the impact origin of Barringer Crater. It was not until the recognition of shock metamorphic criteria (French and Short, 1968; see Chapter 8), which resulted in the increased recognition of terrestrial impact sites, together with the impetus provided by the Apollo landings on the Moon in the 1960s and 1970s, that increasing awareness and a

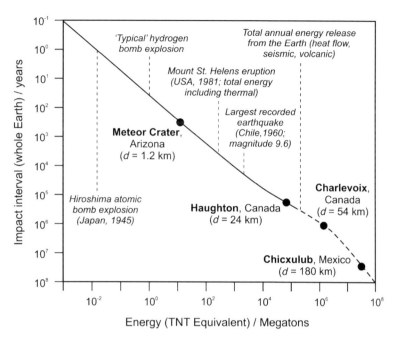

**Figure 1.2**  A comparison of the energy released during impact events with endogenous geological processes and man-made explosions. Note that only the frequency of impact events is shown. The vertical axis represents the frequency of impact events expressed as the estimated interval in years for a particular size of event. For example, an impact event of the size that formed Barringer Crater is expected once every 1900 years. Data from French (1998).

more complete appreciation of the formation of impact craters commenced.

Discussion of the importance of meteorite impacts for Earth evolution finally entered the geological mainstream in 1980, with evidence for a major impact as the cause of the mass extinction event at the Cretaceous–Paleogene (K–Pg) boundary 65 Ma ago (Alvarez *et al.*, 1980). The actual impact site, the approximately 180 km diameter Chicxulub crater, was subsequently identified in 1991, buried beneath approximately 1 km of sediments in the Yucatan Peninsula, Mexico (Hildebrand *et al.*, 1991). The spectacular impact of comet Shoemaker–Levy 9 into Jupiter in July 1994 reminded us that impact cratering is a process that continues to the present day. The result is that it is now apparent that meteorite impact events have played an important role throughout Earth's history, shaping the geological landscape, affecting the evolution of life and producing economic benefits. As summarized in Chapter 2, the evolutions of the terrestrial planets and the Earth's moon have been strongly affected by changes in the population of impactors and in the impact cratering rate through time in the inner Solar System.

To summarize, our understanding of the impact cratering process has come a long way in the past century, but several fundamental aspects of the processes and products of crater formation remain poorly understood. One of the major reasons for this is that, unlike many other geological processes, there have been no historical examples of hypervelocity impact events (French, 1998). This is, of course, fortunate, as impacts release energies far in excess of even the most devastating endogenous geological events (Fig. 1.2). Our understanding is also hindered

by the major differences between impact events and other geological processes, including (1) the extreme physical conditions (Fig. 1.3), (2) the concentrated nature of the energy release at a single point on the Earth's surface, (3) the virtually instantaneous nature of the impact process (e.g. seconds to minutes) and (4) the strain rates involved ($\sim 10^4 - 10^6\,\mathrm{s}^{-1}$ for impacts versus $10^{-3} - 10^{-6}\,\mathrm{s}^{-1}$ for endogenous tectonic and metamorphic processes) (French, 1998). Impact events, therefore, are unlike any other geological process, and the goal of this chapter, and this book, is to provide a modern up-to-date synthesis as to our current understanding of the processes and products of impact cratering.

## 1.2  Formation of hypervelocity impact craters

The formation of hypervelocity[1] impact craters has been divided, somewhat arbitrarily, into three main stages (Gault *et al.*, 1968): (1) contact and compression; (2) excavation; and (3) modification (Fig. 1.4). These are described below. A further stage of 'hydrothermal and chemical alteration' has also sometimes been

---

[1] *Hypervelocity* impact occurs when a cosmic projectile is large enough (typically >50 m for a stony object and >20 m for an iron body) to pass through the atmosphere with little or no deceleration and so strike at virtually its original cosmic velocity (>11 km s$^{-1}$; French, 1998). This produces high-pressure shock waves in the target. Smaller projectiles lose most of their original kinetic energy in the atmosphere and produce small metre-size 'penetration craters', without the production of shock waves.

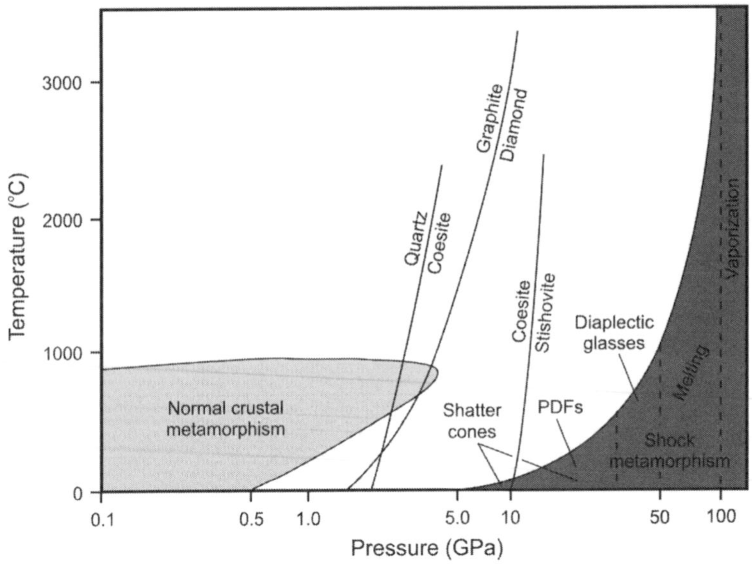

**Figure 1.3**  Pressure–temperature (*P–T*) plot showing comparative conditions for shock metamorphism and 'normal' crustal metamorphism. Note that the pressure axis is logarithmic. The approximate *P–T* conditions needed to produce specific shock effects are indicated by vertical dashed lines below the exponential curve that encompasses the field of shock metamorphism. Modified from French (1998).

included as a separate, final stage in the cratering process (Kieffer and Simonds, 1980), and is also described below.

### 1.2.1  Contact and compression

The first stage of an impact event begins when the projectile, be it an asteroid or comet, contacts the surface of the target (Fig. 1.4) – see Chapter 3 for details. Modelling of the impact process suggests that the projectile penetrates no more than one to two times its diameter (Kieffer and Simonds, 1980; O'Keefe and Ahrens, 1982). The pressures at the point of impact are typically several thousand times the Earth's normal atmospheric pressure (i.e. >100 GPa) (Shoemaker, 1960). The intense kinetic energy of the projectile is transferred into the target in the form of shock waves that occur at the boundary between the compressed and uncompressed target material (Melosh, 1989). These shock waves, travelling faster than the speed of sound, propagate both into the target sequence and back into the projectile itself. When this reflected shock wave reaches the 'free' upper surface of the projectile, it is reflected back into the projectile as a rarefaction or tensional wave (Ahrens and O'Keefe, 1972). The passage of this rarefaction wave through the projectile causes it to unload from high shock pressures, resulting in the complete melting and/or vaporization of the projectile itself (Gault *et al.*, 1968; Melosh, 1989). The increase in internal energy accompanying shock compression and subsequent rarefaction also results in the shock metamorphism (see Chapter 8), melting (see Chapter 9) and/or vaporization of a volume of target material close to the point of impact (Ahrens and O'Keefe, 1972; Grieve *et al.*, 1977). The point at which the projectile is completely unloaded is generally taken as the end of the contact and compression stage (Melosh, 1989; Chapter 3).

### 1.2.2  Excavation stage

The transition from the initial contact and compression stage into the excavation stage is a continuum. It is during this stage that the actual impact crater is opened up by complex interactions between the expanding shock wave and the original ground surface (Melosh, 1989) – see Chapter 4 for details. The projectile itself plays no role in the excavation of the crater, having been unloaded, melted and/or vaporized during the initial contact and compression stage.

During the excavation stage, the roughly hemispherical shock wave propagates out into the target sequence (Fig. 1.4). The centre of this hemisphere will be at some depth in the target sequence (essentially the depth of penetration of the projectile). The passage of the shock wave causes the target material to be set in motion, with an initial outward radial trajectory. At the same time, shock waves that initially travelled upwards intersect the ground surface and generate rarefaction waves that propagate back downwards into the target sequence (Melosh, 1989). In the near-surface region an 'interference zone' is formed in which the maximum recorded pressure is reduced due to interference between the rarefaction and shock waves (Melosh, 1989).

The combination of the outward-directed shock waves and the downward-directed rarefaction waves produces an 'excavation flow-field' and generates a so-called 'transient cavity' (Fig. 1.4 and Fig. 1.5) (Dence, 1968; Grieve and Cintala, 1981). The different trajectories of material in different regions of the excavation flow field result in the partitioning of the transient cavity into an upper 'excavated zone' and a lower 'displaced zone' (Fig. 1.5). Material in the upper zone is ejected ballistically beyond the transient cavity rim to form the continuous ejecta blanket (Oberbeck, 1975) – see Chapter 4. Experiments and theoretical

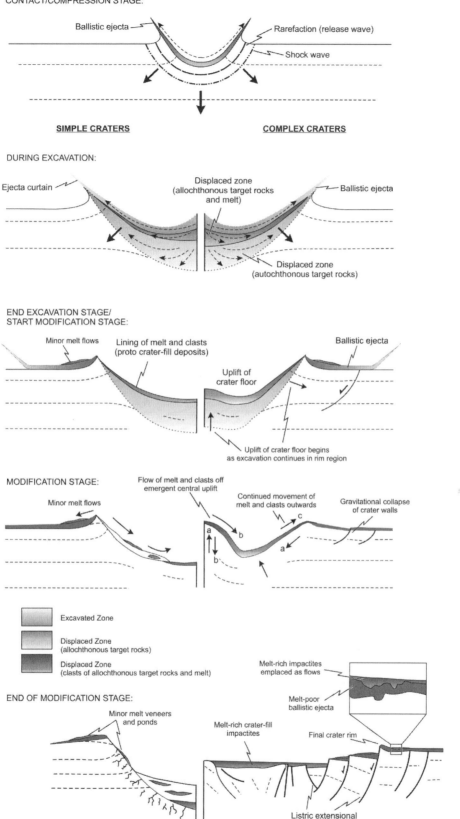

**Figure 1.4** Series of schematic cross-sections depicting the three main stages in the formation of impact craters. This multi-stage model accounts for melt emplacement in both simple (left panel) and complex craters (right panel). For the modification stage section, the arrows represent different time steps, labelled 'a' to 'c'. Initially, the gravitational collapse of crater walls and central uplift (a) results in generally inwards movement of material. Later, melt and clasts flow off the central uplift (b). Then, there is continued movement of melt and clasts outwards once crater wall collapse has largely ceased (c). Modified from Osinski *et al.* (2011). (See Colour Plate 2)

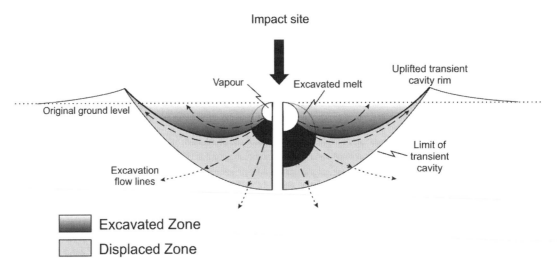

**Figure 1.5**    Theoretical cross-section through a transient cavity showing the locations of impact metamorphosed target lithologies. Excavation flow lines (dashed lines) open up the crater and result in excavation of material from the upper one-third to
one-half the depth of the transient cavity. Modified after Grieve (1987) and Melosh (1989).

considerations of the excavation flow suggest that the excavated
zone comprises material only from the upper one-third to one-
half the depth of the transient cavity (Stöffler *et al.*, 1975). It is
clear that the excavation flow lines transect the hemispherical
pressure contours, so that ejecta will contain material from a
range of different shock levels, including shock-melted target
lithologies. In simple craters (see Section 1.3.1), the final crater
rim approximates the transient cavity rim (Fig. 1.1). In complex
craters (see Section 1.3.2), however, the transient cavity rim is
typically destroyed during the modification stage, such that ejecta
deposits occur in the crater rim region interior to the final crater
rim (Fig. 1.6).

Ejecta deposits represent one of the most distinctive features of
impact craters on planetary bodies (other than Earth), where they
tend to be preserved. It is notable that the continuous ejecta
deposits vary considerably in terms of morphology on different
planetary bodies. For example, on Mars, many ejecta blankets
have a fluidized appearance that has been ascribed due to the
effect of volatiles in the subsurface (Barlow, 2005). Indeed, the
volatile content and cohesiveness of the uppermost target rocks
will significantly affect the runout distance of the ballistically
emplaced continuous ejecta blanket, with impact angle also influ-
encing the overall geometry of the deposits (e.g. the production
of the characteristic butterfly pattern seen in very oblique impacts)
(Osinski *et al.*, 2011). In terms of the depth of excavation $d_e$, few
craters on Earth preserve ejecta deposits and/or have the distinct
pre-impact stratigraphy necessary for determining depth of materials. Based on stratigraphic considerations, $d_e$ at Barringer Crater
is greater than $0.08D$ (Shoemaker, 1963), where $D$ is the final rim
diameter. For complex craters and basins, the depth and diameter
must be referred back to the 'unmodified' transient cavity to
reliably estimate the depth of excavation (Melosh, 1989). The
maximum $d_e$ of material in the ballistic ejecta deposits of the
Haughton and Ries structures, the only terrestrial complex
structures where reliable data are available, yield identical values

of $0.035D_a$, where $D_a$ is the apparent crater diameter (Osinski
*et al.*, 2011). If the initial final rim diameter $D$ is used, which is
the parameter measured in planetary craters, a value $0.05D$ is
obtained for Haughton.

Based on experiments, it was generally assumed that material
in the displaced zone remains within the transient cavity (Stöffler
*et al.*, 1975); however, observations from impact craters on all
the terrestrial planets suggest that some of the melt-rich material
from this displaced zone is transported outside the transient
cavity rim during a second episode of ejecta emplacement
(Osinski *et al.*, 2011). This emplacement of more melt-rich,
ground-hugging flows – the 'surface melt flow' phase – occurs
during the terminal stages of crater excavation and the modification stage of crater formation (see Chapter 4). Ejecta deposited
during the surface melt flow stage are influenced by several
factors, most importantly planetary gravity, surface temperature
and the physical properties of the target rocks. Topography
and angle of impact play important roles in determining the
final distribution of surface melt flow ejecta deposits, with
respect to the source crater (Osinski *et al.*, 2011). A critical
consideration is that the upper layer of ejecta reflects the composition and depth of the displaced zone of the transient cavity
(Fig. 1.4). At Haughton, this value is a minimum of $0.08D_a$
or $0.12D$.

A portion of the melt and rock debris that originates beneath
the point of impact remains in the transient cavity (Grieve *et al.*,
1977). This material is also deflected upwards and outwards parallel to the base of the cavity, but must travel further and possesses
less energy, so that ejection is not possible. This material forms
the crater-fill impactites within impact craters (see Chapter 7 for
an overview of impactites). Eventually, a point is reached at which
the motions associated with the passages of the shock and rarefaction waves can no longer excavate or displace target rock and
melt (French, 1998). At the end of the excavation stage, a mixture
of melt and rock debris forms a lining to the transient cavity.

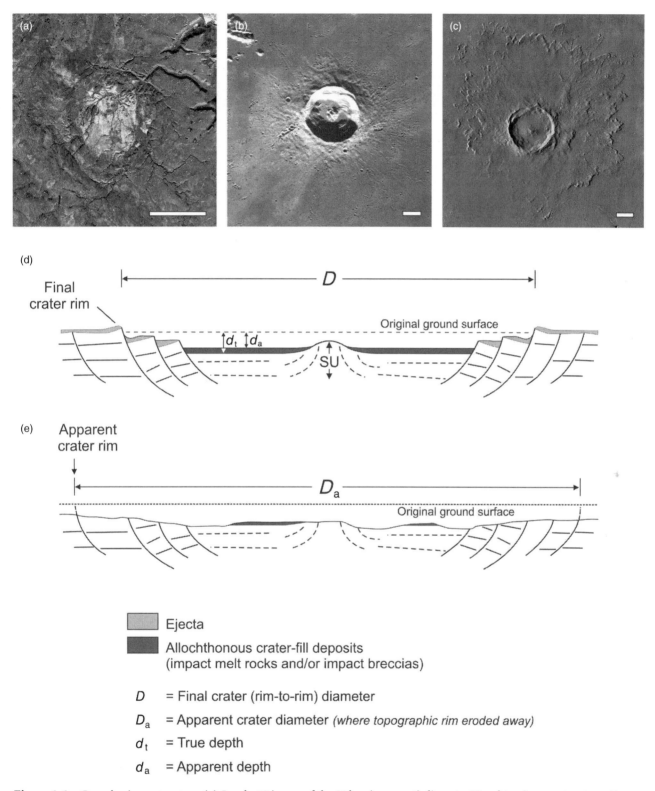

**Figure 1.6** Complex impact craters. (a) *Landsat 7* image of the 23 km (apparent) diameter Haughton impact structure, Devon Island, Canada. (b) Portion of *Apollo 17* metric image AS17-M-2923 showing the 27 km diameter Euler Crater on the Moon. Note the well-developed central peak. (c) Thermal Emission Imaging System (THEMIS) visible mosaic of the 29 km diameter Tooting Crater on Mars (NASA). Note the well-developed central peak and layered ejecta blanket. Scale bars for (a) to (c) are 10 km. (d) Schematic cross-section showing the principal features of a complex impact crater. Note the structurally complicated rim, a down-faulted annular trough and a structurally uplifted central area (SU). (e) Schematic cross-section showing an eroded version of the fresh complex crater in (d). Note that, in this case, only the apparent crater diameter can typically be defined. (See Colour Plate 3)

### 1.2.3    Modification stage

The effects of the modification stage are governed by the size of the transient cavity and the properties of the target rock lithologies (Melosh and Ivanov, 1999) – see Chapter 5 for an overview. For crater diameters less than 2–4 km on Earth, the transient cavity undergoes only minor modification, resulting in the formation of a simple bowl-shaped crater (Fig. 1.1). However, above a certain size threshold the transient cavity is unstable and undergoes modification by gravitational forces, producing a so-called complex impact crater (Fig. 1.4 and Fig. 1.6; Dence, 1965) – see Chapter 5. Uplift of the transient crater floor occurs, leading to the development of a central uplift (Fig. 1.4 and Fig. 1.6). This results in an inward and upward movement of material within the transient cavity. Subsequently, the initially steep walls of the transient crater collapse under gravitational forces (Fig. 1.4). This induces an inward and downward movement of large (~100 m to kilometre-scale) fault-bounded blocks. The diameter at which the transition occurs from simple to complex craters on Earth occurs at approximately 2 km for craters developed in sedimentary targets and approximately 4 km for those in crystalline lithologies. This transition diameter is dependent on the strength of the gravitational field of the parent body and increases with decreasing acceleration of gravity (Melosh, 1989). Thus, the transition from simple to complex craters occurs at approximately 5–10 km on Mars and at approximately 15–27 km on the Moon (Pike, 1980).

It is generally considered that the modification stage commences after the crater has been fully excavated (Melosh and Ivanov, 1999). However, numerical models suggest that the maximum depth of the transient cavity is attained before the maximum diameter is reached (e.g. Kenkmann and Ivanov, 2000). Thus, uplift of the crater floor may commence before the maximum diameter has been reached. As French (1998) notes, the modification stage has no clearly marked end. Processes that are intimately related to complex crater formation, such as the uplift of the crater floor and collapse of the walls (Chapter 5), merge into more familiar endogenous geological processes, such as mass movement, erosion and so on.

### 1.2.4    Post-impact hydrothermal activity

Evidence for impact-generated hydrothermal systems has been recognized at over 70 impact craters on Earth (Naumov, 2005; Osinski *et al.*, 2012), from the approximately 1.8 km diameter Lonar Lake structure, India (Hagerty and Newsom, 2003), to the approximately 250 km diameter Sudbury structure, Canada (Ames and Farrow, 2007). Based on these data, it seems highly probable that any hypervelocity impact capable of forming a complex crater will generate a hydrothermal system, as long as sufficient $H_2O$ is present (see Chapter 6 for an overview). Thus, the recognition of impact-associated hydrothermal deposits is important in understanding the evolution of impact craters through time. There are three main potential sources of heat for creating impact-generated hydrothermal systems (Osinski *et al.*, 2005a): (a) impact melt rocks and impact melt-bearing breccias; (b) physically elevated geothermal gradients in central

uplifts; and (c) thermal energy deposited in central uplifts due to the passage of the shock wave. Interaction of these hot rocks with groundwater and surface water can lead to the development of a hydrothermal system. The circulation of hydrothermal fluids through impact craters can lead to substantial alteration and mineralization. It has been shown that there are six main locations within and around an impact crater where impact-generated hydrothermal deposits can form (Fig. 1.7): (1) crater-fill impact melt rocks and melt-bearing breccias; (2) interior of central uplifts; (3) outer margin of central uplifts; (4) impact ejecta deposits; (5) crater rim region; and (6) post-impact crater lake sediments.

## 1.3    Morphology and morphometry of impact craters

### 1.3.1    Simple craters

Impact craters are subdivided into two main groups based on morphology: simple and complex. Simple craters comprise a bowl-shaped depression (Fig. 1.1). When fresh, they possess an uplifted rim and are filled with an allochthonous breccia lens that comprises largely unshocked target material, possibly mixed with impact melt-bearing lithologies (Fig. 1.1b; Shoemaker, 1960). The overall low shock level of material in the breccia lens suggests that it formed due to slumping of the transient cavity walls, and is not 'fallback' material (Grieve and Cintala, 1981). Simple craters typically have depth-to-diameter ratios of approximately 1:5 to 1:7 (Melosh, 1989). It is important, however, to make the distinction between the true and apparent crater (Fig. 1.1b). Morphometric data from eight simple impact structures (i.e. Barringer, Brent, Lonar, West Hawk, Aouelloul, Tenoumer, Mauritania and Wolfe Creek) define the empirical relationships: $d_a = D^{1.06}$ and $d_t = 0.28D^{1.02}$, where $d_a$ is the depth of the apparent crater, $d_t$ is the depth of the true crater and $D$ is the rim diameter of the structure (Grieve and Pilkington, 1996). As diameter increases, so-called 'transitional craters' form. Such craters have not been recognized on Earth, but on the Moon and Mars, where they are abundant, spacecraft observations show that, while they lack a central peak, they possess some of the other characteristics of complex craters (see below), such as a shallower profile and terraced crater rim. As such, they are neither simple nor complex and the exact mechanism(s) responsible for their appearance remain poorly understood.

### 1.3.2    Complex craters

Observations of lunar craters first revealed that, as diameter increases yet further, a topographic high forms in the centre of a transitional crater, signifying the progression to a so-called complex impact crater. Such craters generally have a structurally complicated rim, a down-faulted annular trough and an uplifted central area (Fig. 1.6). These features form as a result of gravitational adjustments of the initial crater during the modification stage of impact crater formation (Chapter 5). Owing to these late-stage adjustments, complex impact craters are shallower than simple craters, with depth-to-diameter ratios of

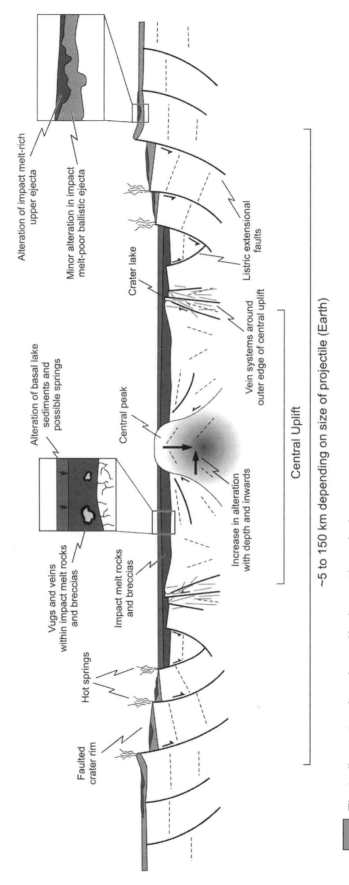

Alteration of impact melt-rich
upper ejecta

Minor alteration in impact
melt-poor ballistic ejecta

Crater lake

Listric extensional
faults

Alteration of basal lake
sediments and
possible springs

Central peak

Vein systems around
outer edge of central uplift

Central Uplift

Vugs and veins
within impact melt rocks
and breccias

Impact melt rocks
and breccias

Increase in alteration
with depth and inwards

~5 to 150 km depending on size of projectile (Earth)

Hot springs

Faulted
crater rim

Ejecta (impact melt rocks and/or impact breccias)

Crater-fill deposits (impact melt rocks and/or impact breccias)

**Figure 1.7**    Distribution of hydrothermal deposits within and around a typical complex impact crater. Modified from Osinski *et al.* (2012). (See Colour Plate 4)

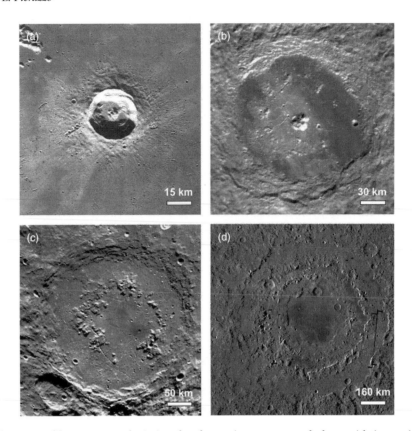

**Figure 1.8**  Series of images of lunar craters depicting the change in crater morphology with increasing crater size. (a) The 27 km diameter Euler Crater possesses a well-developed central peak. Portion of *Apollo 17* metric image AS17-M-2923 (NASA). (b) The 165 km diameter Compton Crater is one of the rare class of central-peak basin craters on the Moon. Clementine mosaic from USGS Map-A-Planet. (c) Clementine mosaic of the 320 km diameter Schrödinger impact crater, which displays a peak ring basin morphology (NASA). (d) The approximately 950 km diameter Orientale Basin is the youngest multi-ring basin on the Moon (NASA/GSFC/Arizona State University).

approximately 1 : 10 to 1 : 20 (Melosh, 1989). The so-called annular trough in complex craters is filled with a variety of impact-generated lithologies (impactites) that will be introduced in Section 1.4.

A unique result of complex crater formation is that material from depth is brought to the surface. As noted above, for many impact sites, these 'central uplifts' provide the only samples of the deep subsurface. This is particularly important on other planetary bodies, but even on Earth they provide vital clues as to the structure of the crust. For example, the central uplift of the approximately 250 km diameter Vredefort impact structure, South Africa, provides a unique profile down to the lower crust (Tredoux *et al.*, 1999). On the Moon and other planets, where post-impact modification of craters is generally minimal, there is a progression with increasing crater size from central peak, central-peak basin (i.e. a fragmentary ring of peaks surrounding a central peak), to peak-ring basins (i.e. a well-developed ring of peaks but no central peak) (Fig. 1.8; Stöffler *et al.*, 2006). On Earth, erosion has modified the surface morphology of all impact craters and it is, therefore, typically not possible to ascertain the original morphology. As such, the term *central uplift* is preferred. Related to this is the fact that a number of relatively young (i.e. only slightly eroded)

terrestrial complex structures (e.g. Haughton (Fig. 1.6a), Canada; Ries, Germany; Zhamanshin, Kazakhstan) lack an emergent central peak (Grieve and Therriault, 2004). These structures are in mixed targets of sediments overlying crystalline basement, with the lack of a peak most likely due to target strength effects. This highlights the problems with making direct comparisons between impact craters on Earth and those on other planetary bodies.

Based on observations of 24 impact craters developed in sedimentary rocks on Earth, the structural uplift of the target rocks in the centre of the crater (Fig. 1.6d) was determined to be $0.086D^{1.03}$, where $D$ is the crater 'diameter' (Grieve and Pilkington, 1996). According to this estimate, a good working hypothesis is that the observed structural uplift is approximately one-tenth of the rim diameter at terrestrial complex impact structures. It is important to note that no data exist on the amount of structural uplift in craters developed in crystalline targets for the obvious reason that stratigraphic markers, upon which this calculation relies upon, are lacking. Despite its widespread application, there is also currently no data to support the hypothesis that this formula for structural uplift holds for craters on other planetary bodies, at least in its current form.

A key descriptor for complex craters is 'diameter'. As noted above, defining the size, or diameter, of a crater is critical for estimating stratigraphic uplift, in addition to energy scaling and numerical modelling of the cratering process. Unfortunately, there is considerable confusion about crater sizes within the literature. This arises largely from the fact that most craters on Earth are eroded to some degree, whereas most craters on other planetary bodies are relatively well preserved. For a discussion of what crater diameter represents, the reader is referred to Turtle *et al.* (2005) and the nomenclature recommended here comes from this synthesis paper. In short, the *rim (or final crater) diameter* is defined as the diameter of the topographic rim that rises above the surface for simple craters, or above the outermost slump block not concealed by ejecta for complex craters (Fig. 1.6d). This is relatively easy to measure on most planetary bodies, where the topographic rim is usually preserved due to low rates of erosion (e.g. Fig. 1.6b,c). On Earth, however, such pristine craters are rare and the rim region is typically eroded away (e.g. Fig. 1.6a). The *apparent crater diameter*, in contrast, is defined as the diameter of the outermost ring of (semi-) continuous concentric normal faults (Fig. 1.6e). For the majority of impact structures on Earth this will be the only measurable diameter. It is not always clear how the apparent diameter is related to the rim diameter, although one would expect the rim diameter to be smaller than the apparent crater diameter. This is consistent with observations at the Haughton impact structure, where an apparent crater diameter of 23 km and a rim diameter of 16 km have been reported (Osinski and Spray, 2005).

Returning to the previous discussion on stratigraphic uplift and its application to planets other than Earth, *D* in Grieve and Pilkington's (1996) formula actually is predominantly based on *apparent* crater diameter estimates (R. A. F. Grieve, personal communication, 2012) and not *rim* diameter estimates, further complicating the discussion about its application to other planets.

### 1.3.3  Multi-ring basins

The largest impact 'craters' in the Solar System are typically surrounded by one or more concentric scarps or fractures and are known as multi-ring basins (Fig. 1.8d). Multi-ring basins are best studied on the Moon and Callisto, where a large number exist, although these structures remain the least understood crater morphology. There are two basic morphological types (e.g. Melosh and McKinnon, 1978). The first type, as exemplified by the Orientale basin on the Moon, exhibits a few to several inward-facing scarps with gentle outward slopes. The second type exhibit tens to hundreds of closely spaced rings consisting of a graben or outward-facing scarps surrounding a central, flat basin (e.g. Valhalla, Callisto). An important observation is that very few multi-ring basins have been documented on Ganymede, despite the obvious similarities with Callisto, and there is no clear evidence for their existence on Mercury, Mars or Venus (Melosh, 1989). In this respect, it is critical to understand that just because an impact crater is very 'large' (e.g. Hellas, Mars; South Pole-Aitken, Moon), this does not necessarily mean that a structure is a multi-ring basin; to be categorized as such, multiple rings must be clearly

observable. It is also important to note that the rings that define multi-ring craters are distinct from the peak rings described in the previous section. In particular, it is thought that rings characteristic of multi-ring craters form outside the final crater. Several mechanisms have been proposed to account for the formation of multi-ring basins, but no agreement exists in the literature to date – see Melosh (1989) for a discussion. Melosh (1989) preferred the so-called ring tectonic theory, where the thickness of the lithosphere plays a dominant role in determining whether or not a ring forms. More recently, a nested melt-cavity model has been proposed to account for transition from complex craters to multi-ringed basins on the Moon (Head, 2010).

Complications arise, as external rings have been documented around much smaller impact structures, such as the proposed (but not confirmed) approximately 20 km diameter Silverpit structure in the North Sea (Fig. 1.9; Stewart and Allen, 2002). Numerical modelling suggests this morphology formed due to an impact into a layer of brittle chalk overlying weak shales (Stewart and Allen, 2002; Collins *et al.*, 2003). Whether multi-ring basins exist on the Earth also remains a topic of debate. Of the three largest structures on Earth (Chicxulub, Sudbury and Vredefort), Chicxulub is the best-preserved large terrestrial impact structure, due to burial. As such, however, the definition of its morphological elements depends on the interpretation of geophysical data. It has an interior topographic 'peak-ring', a terraced rim area and exterior ring faults and, therefore, appears to correspond to the definition of a multi-ring basin, as on the Moon (Grieve *et al.*, 2008).

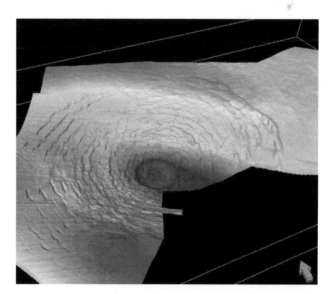

**Figure 1.9**  Perspective view of the top chalk surface at the Silverpit structure, North Sea, UK, a suspected meteorite impact structure. The central crater is 2.4 km wide and is surrounded by a series of concentric faults, which extend to a radial distance of approximately 10 km from the crater centre. False colours indicate depth (yellow: shallow; purple: deep). Image courtesy of Phil Allen and Simon Stewart. (See Colour Plate 5)

## 1.4    Impactites

In terms of the *products* of meteorite impact events, the above considerations of the impact cratering *process* reveal that pressures and temperatures that can vaporize, melt, shock metamorphose[2] and/or deform a substantial volume of the target sequence can be generated. The transport and mixing of impact-metamorphosed[3] rocks and minerals during the excavation and formation of impact craters produces a wide variety of distinctive impactites that can be found within and around impact craters (see Fig. 1.10; 'rock affected by impact metamorphism') (Stöffler and Grieve 2007) – see Chapter 7. Much of our knowledge of impactites comes from impact craters on Earth and, to a lesser extent the Moon, where large numbers and volumes of samples from known locations are available for study.

The transient compression, decompression and heating of the target rocks lead to shock metamorphic effects (see Chapter 8 for an overview), which record pressures, temperatures and strain rates well beyond those produced in terrestrial regional or contact metamorphism (Fig. 1.8 and Fig. 1.9). Given the highly transient nature of shock metamorphic processes, disequilibrium and metastable equilibrium are the norm. The only megascopic shock features are shatter cones, which are distinctive, striated and horse-tailed conical fractures ranging in size from millimetres to tens of metres (Fig. 1.11a). The most-documented shock metamorphic feature is the occurrence of so-called planar deformation features, particularly in quartz (Fig. 1.11b), although they do occur in other minerals (e.g. feldspar and zircon). When fresh, the planar deformation features are parallel planes of glass, with specific crystallographic orientations as a function of shock pressures of approximately 10–35 GPa. At higher pressures, the shock wave can destroy the internal crystallographic order of feldspars and quartz and convert them to solid-state glasses, which still have the original crystal shapes. These are 'diaplectic' glasses (Fig. 1.11c,d), with the required pressures being 30–45 GPa for plagioclase feldspar (also known as maskelynite) and 35–50 GPa for quartz. The extremely rapid compression and then decompression also produces metastable polymorphs, including coesite and stishovite from quartz and diamond and lonsdaleite from graphite (Chapter 8).

### 1.4.1    Classification of impactites

As part of the IUGS Subcommission on the Systematics of Metamorphic Rocks, a study group formulated a series of recommendations for the classification of impactites (Stöffler and

Grieve, 2007). This group suggested that impactites from a single impact should be classified into three major groups irrespective of their geological setting: (1) *shocked rocks*, which are non-brecciated, melt-free rocks displaying unequivocal effects of shock metamorphism; (2) *impact melt rocks* (Fig. 1.10a–c), which can be further subclassified according to their clast content (i.e. clast-free, -poor or -rich) and/or degree of crystallinity (i.e. glassy, hypocrystalline or holocrystalline); (3) *impact breccias* (Fig. 1.10d,e), which can be further classified according to the degree of mixing of various target lithologies and their content of melt particles (e.g. lithic breccias and 'suevites').

It is apparent from the literature that substantial problems exist with the current IUGS nomenclature of impactites, particularly those including impact melt products (see Chapters 7 and 9 for detailed discussions). This is due to several reasons, including the erosional degradation of many impact structures on Earth such that outcrops of impact melt-bearing lithologies preserving their entire original context are relatively rare (Grieve *et al.*, 1977). Other complicating factors are introduced due to inconsistent nomenclature and unqualified use of terms (such as 'suevite' – Fig. 1.10d) for several types of impactites with somewhat different genesis; for example, impactites with glass contents ranging up to approximately 90 vol.% have been termed suevites at the Popigai impact structure (Masaitis, 1999). It is also important to note that the framework for the IUGS classification scheme was developed in the 1990s and remained little changed up to its publication in 2007, despite several major discoveries and advancements in our understanding of impactites. In particular, in recent years, the effect(s) of target lithology on various aspects of the impact cratering process, in particular the generation and emplacement of impactites, has emerged as a major research topic (Osinski *et al.*, 2008a).

### 1.4.2    Impact melt-bearing impactites

The production of impact melt rocks and glasses is a diagnostic feature of hypervelocity impact, and their presence, distribution and characteristics have provided valuable information on the cratering process (Dence *et al.*, 1977; Grieve *et al.*, 1977; Grieve and Cintala, 1992) – see Chapter 9. Within complex impact structures formed entirely in crystalline targets, coherent impact melt rocks or 'sheets' are formed. These rocks can display classic igneous structures (e.g. columnar jointing) and textures (Fig. 1.10a–c). Impact craters formed in 'mixed' targets (e.g. crystalline basement overlain by sedimentary rocks) display a wide range of impact-generated lithologies, the majority of which were typically classed as 'suevites' (Fig. 1.10d; Stöffler *et al.*, 1977; Masaitis, 1999); the original definition of a suevite is a polymict impact breccia with a clastic matrix/groundmass containing fragments and shards of impact glass and shocked mineral and lithic clasts (Stöffler *et al.*, 1977). Minor bodies of coherent impact melt rocks are also sometimes observed, often as lenses and irregular bodies within larger bodies of suevite (e.g. Masaitis, 1999). In impact structures formed in predominantly sedimentary targets, impact melt rocks were not generally recognized, with the resultant crater-fill deposits historically referred to as clastic, fragmental or sedimentary breccias (Masaitis *et al.*, 1980; Redeker and Stöffler 1988; Masaitis 1999). These observations led to the conclusion

---

[2] *Shock metamorphism* is defined as the metamorphism of rocks and minerals caused by shock wave compression and decompression due to impact of a solid body or due to the detonation of high-energy chemical or nuclear explosives (Stöffler and Grieve, 2007).

[3] *Impact metamorphism* is essentially the same as shock metamorphism except that it also encompasses the melting and vaporization of target rocks (Stöffler and Grieve, 2007).

**Figure 1.10**   Field images of impactites. (a) Oblique aerial view of the approximately 80 m high cliffs of impact melt rock at the Mistastin impact structure, Labrador, Canada. Photograph courtesy of Derek Wilton. (b) Close-up view of massive fine-grained (aphanitic) impact melt rock from the Discovery Hill locality, Mistastin impact structure. Camera case for scale. (c) Coarse-grained granophyre impact melt rock from the Sudbury Igneous Complex, Canada. Rock hammer for scale. (d) Impact melt-bearing breccia from the Mistastin impact structure. Note the fine-grained groundmass and macroscopic flow-textured silicate glass bodies (large black fragments). Steep Creek locality. Marker/pen for scale. (e) Polymict lithic impact breccias from the Wengenhousen quarry, Ries impact structure, Germany. Rock hammer for scale. (f) Carbonate melt-bearing clast-rich impact melt rocks from the Haughton impact structure. Penknife for scale. This lithology was originally interpreted as a clastic or fragmental breccia (Redeker and Stöffler, 1988), but was subsequently shown to be an impact melt-bearing impactite (Osinski and Spray, 2001; Osinski *et al.*, 2005b). (See Colour Plate 6)

**Figure 1.11** Shock metamorphic effects in rocks and minerals. (a) Shatter cones in limestone from the Haughton impact structure. Penknife for scale. (b) Planar deformation features in quartz. Image courtesy of L. Ferrière. (c) and (d) Plane- and cross-polarized light photomicrographs, respectively, of diaplectic quartz glass from the Haughton impact structure, Canada. Note the original grain shape of the sandstone quartz grains has been preserved, which is diagnostic for diaplectic glass, but not whole rock glasses.

that no, or only minor, impact melt volumes are apparently present in impact structures formed in predominantly sedimentary targets. However, more recent work suggests that impact melting is more common in sedimentary targets than has, hitherto, been believed and that impact melt rocks are produced (Fig. 1.10f; Osinski *et al.*, 2008b). These observations are generally consistent with numerical modelling studies (Wünnemann *et al.*, 2008), which also suggests that the volume of melt produced by impacts into dry porous sedimentary rocks should be greater than that produced by impacts in a crystalline target. Thus, it seems that the basic products are genetically equivalent regardless of target lithology, but they just appear different. That is, it is the textural, chemical and physical properties of the products that vary (Osinski *et al.* 2008a,b); for example, compare Fig. 1.10f with Fig. 1.10b.

## 1.5   Recognition of impact craters

Several criteria may be used to recognize hypervelocity impact structures, including the presence of a crater form and/or unusual rocks, such as breccias, melt rocks and pseudotachylite; however, on their own, these indicators do not provide definitive evidence for a meteorite impact structure. Geophysics can also provide clues (see Chapter 14), and a geophysical anomaly is often the first indicator of the existence of buried structures. The most common geophysical anomaly is a localized low in the regional gravity field, due to lowering of rock density from brecciation and fracturing (Pilkington and Grieve, 1992). Larger complex impact structures tend to have a central, relative gravity high, which can extend out to approximately half the diameter of the structure. In terms of magnetics, the most common expression is a magnetic low, with the disruption of any regional trends in the magnetic field. This is due to an overall lowering of magnetic susceptibility and the randomizing of pre-impact lithologic trends in the target rocks (Pilkington and Grieve, 1992). Seismic velocities are reduced at impact structures, due to fracturing, and reflection seismic images are extremely useful in characterizing buried structures in sedimentary targets. There is, however, no geophysical anomaly that can provide definitive evidence for a meteorite impact structure.

The general consensus within the impact community is that unequivocal evidence for hypervelocity impact takes the form of shock metamorphic indicators (French and Koeberl, 2010), either megascopic (e.g. shatter cones) or microscopic (e.g. planar deformation features, diaplectic glass) (Fig. 1.11), and the presence of high-pressure polymorphs (e.g. coesite and stishovite) – see Chapter 8 for an overview of shock metamorphism. Unfortunately, this requires investigation and preservation of suitable rocks within a suspected structure. However, this is often not possible for eroded and/or buried structures and/or structures presently in the marine environment (e.g. the Eltanin feature in the South Pacific; Kyte *et al.*, 1988), even though there is strong evidence for an impact origin.

A prime example is the controversy surrounding the Silverpit structure in the North Sea. Stewart and Allen (2002) originally proposed that this structure was an impact crater based on high-resolution three-dimensional (3D) seismic data (Fig. 1.9); and despite some opposition (Thomson *et al.*, 2005), most impact workers accept this; however, without drilling to retrieve samples, this structure is currently relegated to the list of 'possible' impact structures. This is unfortunate, as the seismic dataset for this structure surpasses that available for any known impact structure and may provide important insights into complex crater formation. In order to try to address this issue, Stewart (2003) proposed a framework for the identification of impact structures based on 3D seismic data, but this has received little attention to date within the impact community.

## 1.6　Destructive effects of impact events

In 1980, Luis and Walter Alvarez and colleagues published a paper in *Science* outlining evidence for an extraterrestrial origin for the most recent of the 'big five' mass extinctions: the Cretaceous–Tertiary (now the Cretaceous–Palaeogene) mass extinction event at approximately 65 Ma (Alvarez *et al.*, 1980). A decade later, the source crater – the approximately 180 km diameter Chicxulub impact structure – was found lying beneath approximately 1 km of sediment below and half offshore the present-day Yucatan Peninsula, Mexico (Hildebrand *et al.*, 1991). As outlined in Chapter 10, the Chicxulub impact caused severe environmental effects that ranged from local to global and that lasted from seconds to tens of thousands of years. The local and regional effects of the impact event include the air blast and heat from the impact explosion, tsunamis and earthquakes. Global effects included forest fires ignited by impact ejecta re-entering the Earth's atmosphere, injection of huge amounts of dust in the upper atmosphere, which may have inhibited photosynthesis

for as much as 2 months, and the production of vast quantities of $N_2O$ from the shock heating of the atmosphere (Chapter 10). However, one of the most important findings has been that, in terms of global effects, the severity of the Chicxulub impact was due, in part, to the composition of the target rocks: approximately 3 km of carbonates and evaporites overlying crystalline basement.

While it was initially thought that the vaporization and decomposition of carbonates – producing CaO and releasing $CO_2$, resulting in global warming – was important (O'Keefe and Ahrens, 1989), it appears that the most destructive effect(s) came from the release of sulfur species from the evaporite target rocks (Pope *et al.*, 1997). We know from studies of sulfur-rich volcanic eruptions, such as Mount Pinatubo in 1992, that sulfur aerosols can significantly reduce the amount of sunlight that reaches the Earth's surface, resulting in short-term global cooling. Estimates for Chicxulub suggest as much as a 15 °C decrease in the average global temperatures, which when coupled with the other effects of the impact event would have resulted in severe environmental consequences (see Chapter 10 for an overview).

## 1.7　Beneficial effects of impact events

### 1.7.1　Microbiological effects

As noted in Section 1.6, ever since the proposal of a link between meteorite impacts and mass extinctions, the deleterious effects of impact events have received much attention (Schulte *et al.*, 2010). However, research conducted over the past few years indicates that, although meteorite impacts are indeed destructive, catastrophic events, there are several potential beneficial effects, particularly in terms of providing new habitats for microbial communities (Cockell and Lee, 2002) – see Chapter 11 for an overview. This may have important implications for understanding the origin and evolution of life on Earth and other planets such as Mars.

One of the most important beneficial effects is the generation of a hydrothermal system within an impact crater immediately following its formation. As noted in Section 1.2.4, recent work suggests that impact-associated hydrothermal systems will form following impacts into any $H_2O$-bearing solid planetary body, with exceptions for small impacts and those in extremely arid regions (Naumov, 2005; Osinski *et al.*, 2012) – see Chapter 6. Numerical models of these hydrothermal systems suggest that they may last several million years for large, 100 km size, impact structures (Abramov and Kring, 2004, 2007). This may have important astrobiological implications, as many researchers believe that hydrothermal systems in general might have provided habitats or 'cradles' for the origin and evolution of early life on Earth (Farmer, 2000) and possibly other planets, such as Mars. Excitingly, the first clear evidence for impact-generated hydrothermal systems on Mars has recently been discovered (Marzo *et al.*, 2010).

Other potential habitats exclusive to impact craters include impact-processed crystalline rocks (Cockell *et al.*, 2003), which have increased porosity and translucence compared with un-shocked materials, improving microbial colonization, impact-generated glasses (Sapers *et al.*, 2010), and impact crater lakes, which form protected sedimentary basins that can provide protective environments and increased preservation potential of fossils and organic material (Cockell and Lee, 2002).

### 1.7.2   Economic effects

One of the less well-known aspects of meteorite impact craters, at least in the general scientific community, is the potential association of economic mineral and hydrocarbon deposits, and thus their suitability as exploration targets (see Chapter 12 for an overview). This is exemplified by the large, approximately 200–250 km diameter Sudbury (Canada) and Vredefort (South Africa) impact structures, which host some of the world's largest and most profitable mining camps (Grieve, 2005). As outlined in Chapter 12, economic resources associated with impact craters can be classified as either pre-, syn- or epigenetic with respect to the impact event. At Vredefort, the impact event led to the preservation of pre-impact (progenetic) gold and uranium deposits in the Witwatersrand Basin and their subsequent mobilization and concentration during impact-induced hydrothermal alteration, producing the world's richest gold province. In contrast, at Sudbury, the world's largest nickel–copper ore deposits occur at the base of the impact melt sheet and in radial dikes. These ore deposits are syngenetic and formed through the separation of immiscible sulfide liquids from the silicate impact melt. Subsequent post-impact hydrothermal activity also led to the formation of copper–platinum group element-rich and zinc–copper–lead economic ore deposits at Sudbury (Ames and Farrow, 2007). Economic ore deposits also occur at a number of other smaller terrestrial impact structures, and the lack of detailed studies of many impact sites leaves room for further discoveries.

In addition to economic metalliferous ore deposits, several meteorite impact structures have been exploited for hydrocarbons (Donofrio, 1998). The fracturing and faulting of rocks in central uplifts and faulted crater rims, results in enhanced porosity and permeability, providing valuable reservoirs for oil and gas, even in rocks such as granites that are typically not suitable hydrocarbon reservoirs (e.g. Ames structure, USA; Chapter 12). Post-impact sedimentary crater-fill deposits can also generate suitable source rocks.

### 1.8   When a crater does not exist: other evidence for impact events

The subject of this book is impact cratering, which implies that the emphasis is on *cratering*. However, it is important to note that not all impact events result in the formation of an impact crater. This is obvious for impacts into the Jovian planets, such as the collision of comet Shoemaker–Levy 9 with Jupiter in 1992, but perhaps less so for Earth. The documentation of spherule beds and tektites is a topic that is discussed in chapters on impact ejecta (Chapter 4) and impact melting (Chapter 9). While many of these

occurrences have been linked to source craters, the majority, particularly spherule beds in Archaean-age rocks, have not (Simonson and Glass, 2004). These distal ejecta deposits do, however, provide important information regarding the impact cratering process and should not be overlooked. These Archaean-age spherule beds found in South Africa and Australia are also the most ancient part of the impact record on Earth.

In addition to distal ejecta deposits, there is an increasing discovery of occurrences of natural glasses around the world that are neither spherules nor tektites (e.g. Fig. 1.12). These glasses are either confirmed or suspected as being of impact origin but for which no source crater has been recognized. Some of these glass occurrences are well known and widely accepted as being of impact origin; for example, Libyan Desert Glass (Weeks *et al.*, 1984), Darwin Glass (Meisel *et al.*, 1990), urengoites or South Ural Glass (Deutsch *et al.*, 1997) and Dakhleh Glass (Osinski *et al.*, 2007; Fig. 1.12). Others remain more enigmatic (Haines *et al.*, 2001; Schultz *et al.*, 2006). Several of these occurrences have been ascribed to large aerial bursts or airbursts.

The 1908 Tunguska event represents the largest recorded example of an airburst event on Earth to date (Vasilyev, 1998), with estimated magnitude estimates ranging from 3–5 Mt up to approximately 10–40 Mt. Theoretical calculations coupled with ground- and satellite-based observations of airbursts suggest that the Earth is struck annually by objects of energy 2–10 kt with Tunguska-size events occurring once every 1000 years (Brown *et al.*, 2002). Recent numerical modelling suggests that substantial amounts of glass can be formed by radiative/convective heating of the surface during greater than 100 Mt low-altitude airbursts (Boslough, 2006). When coupled with the observations of natural glasses described above, there is, therefore, growing evidence to suggest that airbursts – and the glass produced by such events – should occur more frequently than has been previously recognized in the geological record.

### 1.9   Concluding remarks

The recognition of impact cratering as a fundamental geological process represents a revolution in Earth and planetary sciences. The study of impact craters and related phenomena is relatively young when compared with other fields of geological study. It is clear that the formation of meteorite impact structures is unlike any other geological process; however, this should not hinder their study. Far from it: coming to terms with understanding a geological process that takes place in only a few seconds to minutes, with energies that can be greater than the total annual internal energy release from the Earth, provides a stimulating framework for research and teaching. What is more, the study of impact craters requires a multi- and inter-disciplinary approach and must take into account observations from throughout the Solar System. Unlike many areas in the geological sciences, there is, therefore, still considerable potential for new and exciting contributions and areas of study.

As outlined above and in other chapters in this book, basic processes such as the mechanics of complex crater formation

**Figure 1.12**  Field images of Dakhleh Glass, the potential product of an airburst event (Osinski *et al.*, 2007; Osinski *et al.*, 2008c). (a) Area of abundant Dakhleh Glass lagged on the surface of Pleistocene lacustrine sediments; the arrows point to some large Dakhleh Glass specimens. (b) Upper surfaces of many large Dakhleh Glass lag samples appear to be in place and are highly vesiculated. This contrasts with the smooth, irregular lower surfaces. (c) In cross-section, it is clear that there is an increase in the number of vesicles towards the upper surface. Together, these features are indicative of ponding of melt and volatile loss through vesiculation. (d) Highly vesicular pumice-like Dakhleh Glass sample. 7 cm lens cap for scale. (See Colour Plate 7)

(see Chapter 5) and the production of vapour plumes and ejecta deposits (see Chapters 3 and 4) are still not fully understood. Major questions concerning the effect of target properties (e.g. volatiles, porosity, layering) on the impact cratering process and the environmental effects of impact events still remain to be resolved. As the exploration of our Solar System continues, furthering our understanding of impact cratering will become even more important.

## References

Abramov, O. and Kring, D.A. (2004) Numerical modeling of an impact-induced hydrothermal system at the Sudbury crater. *Journal of Geophysical Research*, 109, E10007. DOI: 10.1029/2003JE002213.

Abramov, O. and Kring, D.A. (2007) Numerical modeling of impact-induced hydrothermal activity at the Chicxulub crater. *Meteoritics and Planetary Science*, 42, 93–112.

Ahrens, T.J. and O'Keefe, J.D. (1972) Shock melting and vaporization of Lunar rocks and minerals. *Moon*, 4, 214–249.

Alvarez, L.W., Alvarez, W., Asaro, F. and Michel, H.V. (1980) Extraterrestrial cause for the Cretaceous/Tertiary extinction. *Science*, 208, 1095–1108.

Ames, D.E. and Farrow, C.E.G. (2007) Metallogeny of the Sudbury mining camp, Ontario, in *Mineral Deposits of Canada: A Synthesis of Major Deposit-Types, District Metallogeny, the Evolution of Geologic Provinces, and Exploration Methods* (ed. W.D. Goodfellow). Geological Association of Canada, Mineral Deposits Division, Special Publication 5, Geological Association of Canada, pp. 329–350.

Barlow, N.G. (2005) A review of Martian impact crater ejecta structrues and their implications for target properties, in *Large Meteorite Impacts III* (eds T. Kenkmann, F. Hörz and A. Deutsch), Geological Society of America Special Paper 384, Geological Society of America, Boulder, CO, pp. 433–442.

Barringer, D.M. (1905) Coon Mountain and its crater. *Proceedings of the Academy of Natural Sciences of Philadelphia*, 57, 861–886.

Boslough, M.B. (2006) Numerical modeling of aerial bursts and ablation melting of Libyan Desert Glass. *Geological Society of America Abstracts with Programs*, 38, 121.

Brown, P., Spalding, R.E., Revelle, D.O. *et al.* (2002) The flux of small near-Earth objects colliding with the Earth. *Nature*, 420, 294–296.

Carr, M.H. (2006) *The Surface of Mars*, Cambridge University Press, Cambridge.

Cockell, C.S. and Lee, P. (2002) The biology of impact craters – a review. *Biological Reviews*, 77, 279–310.

Cockell C.S., Osinski G.R. and Lee P. (2003) The impact crater as a habitat: effects of impact alteration of target materials. *Astrobiology*, 3, 181–191.

Collins, G.S., Ivanov, B.A., Turtle, E.P. and Melosh, H.J. (2003) Numerical simulations of Silverpit crater collapse. Third International Conference on Large Meteorite Impacts. http://amcg. ese.ic.ac.uk/~gsc/publications/abstracts/lic03.pdf (accessed 15 May 2012).

Dence, M.R. (1965) The extraterrestrial origin of Canadian craters. *Annals of the New York Academy of Science*, 123, 941–969.

Dence, M.R. (1968) Shock zoning at Canadian craters: petrography and structural implications, in *Shock Metamorphism of Natural Materials* (eds B.M. French and N.M. Short), Mono Book Corp., Baltimore, MD, pp. 169–184.

Dence, M.R., Grieve, R.A.F. and Robertson, P.B. (1977) Terrestrial impact structures; principal characteristics and energy considerations, in *Impact and Explosion Cratering* (eds D.J. Roddy, R.O. Pepin and R.B. Merrill), Pergamon Press, New York, NY, pp. 247–275.

Deutsch, A., Ostermann, M. and Masaitis, V.L. (1997) Geochemistry and neodymium–strontium isotope signature of tektite-like objects from Siberia (urengoites, South-Ural glass). *Meteoritics and Planetary Science*, 32, 679–686.

Donofrio, R.R. (1998) North American impact structures hold giant field potential. *Oil and Gas Journal*, 96, 69–83.

Earth Impact Database (2012) http://www.unb.ca/passc/Impact Database (accessed 1 January 2012).

Farmer, J.D. (2000) Hydrothermal systems: doorways to early biosphere evolution. *GSA Today*, 10, 1–9.

French, B.M. (1998) *Traces of Catastrophe: A Handbook of Shock-Metamorphic Effects in Terrestrial Meteorite Impact Structures*, LPI Contribution No. 954. Lunar and Planetary Institute, Houston, TX.

French, B.M. and Koeberl, C. (2010) The convincing identification of terrestrial meteorite impact structures: what works, what doesn't, and why. *Earth-Science Reviews*, 98, 123–170.

French, B.M. and Short, N.M. (1968) *Shock Metamorphism of Natural Materials*, Mono Book Corp., Baltimore, MD.

Gault, D.E., Quaide, W.L. and Oberbeck, V.R. (1968) Impact cratering mechanics and structures, in *Shock Metamorphism of Natural Materials* (eds B.M. French and N.M. Short), Mono Book Corp., Baltimore, MD, pp. 87–99.

Grieve, R.A.F. (1987) Terrestrial impact structures. *Annual Review of Earth and Planetary Science*, 15, 245–270.

Grieve, R.A.F. (2005) Economic natural resource deposit at terrestrial impact structures, in *Mineral Deposits and Earth Evolution* (eds I. McDonald, A.I. Boyce, I.B. Butler *et al.*). Geological Society of London Special Publication 248, Geological Society, London, pp. 1–29.

Grieve, R.A.F., Dence, M.R. and Robertson, P.B. (1977) Cratering processes: as interpreted from the occurrences of impact melts, in *Impact and Explosion Cratering* (eds D.J. Roddy, R.O. Pepin and R.B. Merrill), Pergamon Press, New York, NY, pp. 791–814.

Grieve, R.A.F. and Cintala, M.J. (1981) A method for estimating the initial impact conditions of terrestrial cratering events, exemplified by its application to Brent crater, Ontario. *Proceedings of the Lunar and Planetary Science Conference*, 12B, 1607–1621.

Grieve, R.A.F. and Cintala, M.J. (1992) An analysis of differential impact melt-crater scaling and implications for the terrestrial impact record. *Meteoritics*, 27, 526–538.

Grieve, R.A.F. and Cintala, M.J. (1995) Impact melting on Venus: some considerations for the nature of the cratering record. *Icarus*, 114, 68–79.

Grieve, R.A.F. and Pilkington, M. (1996) The signature of terrestrial impacts. *Journal of Australian Geology and Geophysics*, 16, 399–420.

Grieve, R.A.F. and Therriault, A. (2004) Observations at terrestrial impact structures: their utility in constraining crater formation. *Meteoritics and Planetary Science*, 39, 199–216.

Grieve, R.A.F., Reimold, W.U., Morgan, J.V. *et al.* (2008) Observations and interpretations at Vredefort, Sudbury and Chicxulub: toward an empirical model of terrestrial basin formation. *Meteoritics and Planetary Science*, 43, 855–882.

Hagerty, J.J. and Newsom, H.E. (2003) Hydrothermal alteration at the Lonar Lake impact structure, India: implications for impact cratering on Mars. *Meteoritics and Planetary Science*, 38, 365–381.

Haines, P.W., Jenkins, R.J.F. and Kelley, S.P. (2001) Pleistocene glass in the Australian desert: the case for an impact origin. *Geology*, 29, 899–902.

Head, J.W. (2010) Transition from complex craters to multi-ringed basins on terrestrial planetary bodies: scale-dependent role of the expanding melt cavity and progressive interaction with the displaced zone. *Geophysical Research Letters*, 37, L02203.

Hildebrand, A.R., Penfield, G.T., Kring, D.A. *et al.* (1991) Chicxulub crater: a possible Cretaceous/Tertiary boundary impact crater on the Yucatan Peninsula, Mexico. *Geology*, 19, 867–871.

Kenkmann, T. and Ivanov, B.A. (2000) Low-angle faulting in the basement of complex impact craters; numerical modelling and field observations in the Rochechouart Structure, France, in *Impacts and the Early Earth* (eds I. Gilmour and C. Koeberl), Lecture Notes in Earth Sciences 91, Springer, Berlin, pp. 279–308.

Kieffer, S.W. and Simonds, C.H. (1980) The role of volatiles and lithology in the impact cratering process. *Reviews of Geophysics and Space Physics*, 18, 143–181.

Kyte, F.T., Zhou, L. and Wasson, J.T. (1988) New evidence on the size and possible effects of a Late Pliocene oceanic asteroid impact. *Science*, 241, 63–65.

Marzo, G.A., Davila, A.F., Tornabene, L.L. *et al.* (2010) Evidence for Hesperian impact-induced hydrothermalism on Mars. *Icarus*, 208, 667–683.

Masaitis, V.L., Danilin, A.N., Maschak, M.S. *et al.* (1980) *The Geology of Astroblemes*. Nedra, Leningrad (in Russian).

Masaitis, V.L. (1999) Impact structures of northeastern Eurasia: the territories of Russia and adjacent countries. *Meteoritics and Planetary Science*, 34, 691–711.

Meisel, T., Koeberl, C. and Ford, R.J. (1990) Geochemistry of Darwin impact glass and target rocks. *Geochimica et Cosmochimica Acta*, 54, 1463–1474.

Melosh, H.J. and Mckinnon, W. (1978) The mechanics of ringed basin formation. *Geophysical Research Letters*, 5, 985–988.

Melosh, H.J. (1989) *Impact Cratering: A Geologic Process*, Oxford University Press, New York, NY.

Melosh, H.J. and Ivanov, B.A. (1999) Impact crater collapse. *Annual Review of Earth and Planetary Science*, 27, 385–415.

Naumov, M.V. (2005) Principal features of impact-generated hydrothermal circulation systems: mineralogical and geochemical evidence. *Geofluids*, 5, 165–184.

O'Keefe, J.D. and Ahrens, T.J. (1982) Cometary and meteorite swarm impact on planetary surfaces. *Journal of Geophysical Research*, 103, 28607.

O'Keefe, J.D. and Ahrens, T.J. (1989) Impact production of $CO_2$ by the Cretaceous/Tertiary extinction bolide and the resultant heating of the Earth. *Nature*, 338, 247–249.

Oberbeck, V.R. (1975) The role of ballistic erosion and sedimentation in lunar stratigraphy. *Reviews of Geophysics and Space Physics*, 13, 337–362.

Osinski, G.R. and Spray, J.G. (2001) Impact-generated carbonate melts: evidence from the Haughton Structure, Canada. *Earth and Planetary Science Letters*, 194, 17–29.

Osinski, G.R. and Spray, J.G. (2005) Tectonics of complex crater formation as revealed by the Haughton impact structure, Devon Island, Canadian High Arctic. *Meteoritics and Planetary Science*, 40, 1813–1834.

Osinski, G.R., Lee, P., Parnell, J. *et al.* (2005a) A case study of impact-induced hydrothermal activity: the Haughton impact structure, Devon Island, Canadian High Arctic. *Meteoritics and Planetary Science*, 40, 1859–1878.

Osinski, G.R., Spray, J.G. and Lee, P. (2005b) Impactites of the Haughton impact structure, Devon Island, Canadian High Arctic. *Meteoritics and Planetary Science*, 40, 1789–1812.

Osinski, G.R., Schwarcz, H.P., Smith, J.R. *et al.* (2007) Evidence for a ~200–100 ka meteorite impact in the Western Desert of Egypt. *Earth and Planetary Science Letters*, 253, 378–388.

Osinski, G.R., Grieve, R.A.F., Collins, G.S. *et al.* (2008a) The effect of target lithology on the products of impact melting. *Meteoritics and Planetary Science*, 43, 1939–1954.

Osinski, G.R., Grieve, R.A.F. and Spray, J.G. (2008b) Impact melting in sedimentary target rocks: an assessment, in *The Sedimentary Record of Meteorite Impacts* (eds K.R. Evans, W. Horton, D.K. King, Jr *et al.*), Geological Society of America Special Publication 437, Geological Society of America, Boulder, CO, pp. 1–18.

Osinski, G.R., Kieniewicz, J.M., Smith, J.R. *et al.* (2008c) The Dakhleh Glass: product of an impact airburst or cratering event in the Western Desert of Egypt? *Meteoritics and Planetary Science*, 43, 2089–2017.

Osinski, G.R., Tornabene, L.L. and Grieve, R.A.F. (2011) Impact ejecta emplacement on the terrestrial planets. *Earth and Planetary Science Letters*, 310, 167–181.

Osinski, G.R., Tornabene, L.L., Banerjee, N.R. *et al.* (2012) Impact-generated hydrothermal systems on Earth and Mars. *Icarus*, in press.

Pike, R.J. (1980) Control of crater morphology by gravity and target type – Mars, Earth, Moon. *Proceedings of the Lunar and Planetary Science Conference*, 11, 2159–2189.

Pilkington, M. and Grieve, R.A.F. (1992) The geophysical signature of terrestrial impact craters. *Reviews of Geophysics*, 30, 161–181.

Pope, K.O., Baines, K.H., Ocampo, A.C. and Ivanov, B.A. (1997) Energy, volatile production, and climate effects of the Chicxulub Cretaceous/Tertiary impact. *Journal of Geophysical Research*, 102, 21645–21664.

Redeker, H.J. and Stöffler, D. (1988) The allochthonous polymict breccia layer of the Haughton impact crater, Devon Island, Canada. *Meteoritics*, 23, 185–196.

Sapers, H.M., Osinski, G.R. and Banerjee, N.R. (2010) Enigmatic tubular textures hosted in impact glasses from the Ries impact structure, Germany. Astrobiology Science Conference 2010: Evolution and Life: Surviving Catastrophes and Extremes on Earth and Beyond. http://www.lpi.usra.edu/meetings/abscicon2010/pdf/5373.pdf (accessed 15 May 2012).

Schulte, P., Alegret, L., Arenillas, I. *et al.* (2010) The Chicxulub asteroid impact and mass extinction at the Cretaceous–Paleogene boundary. *Science*, 327, 1214–1218.

Schultz, P.H. (1993) Impact crater growth in an atmosphere. *International Journal of Impact Engineering*, 14, 659–670.

Schultz, P.H., Rate, M., Hames, W.E. *et al.* (2006) The record of Miocene impacts in the Argentine Pampas. *Meteoritics and Planetary Science*, 41, 749–771.

Shoemaker, E.M. (1960) Penetration mechanics of high velocity meteorites, illustrated by Meteor Crater, Arizona, in Report of the International Geological Congress, XXI Session, Norden. Part XVIII, International Geological Congress, Copenhagen, pp. 418–434.

Shoemaker, E.M. (1963) Impact mechanics at Meteor Crater, Arizona, in *The Moon, Meteorites and Comets* (eds B.M. Middlehurst and G.P. Kuiper), University of Chicago Press, Chicago, IL, pp. 301–336.

Simonson, B.M. and Glass, B.J. (2004) Spherule layers – records of ancient impacts. *Annual Review of Earth and Planetary Science*, 32, 329–361.

Solomon, S.C., Mcnutt Jr, R.L. and Prockter, L.M. (2011) Mercury after the *MESSENGER* flybys: an introduction to the special issue of *Planetary and Space Science*. *Planetary and Space Science*, 59, 1827–1828.

Stewart, S.A. (2003) How will we recognize buried impact craters in terrestrial sedimentary basins? *Geology*, 31, 929–932.

Stewart, S.A. and Allen, P.J. (2002) A 20-km-diameter multi-ringed impact structure in the North Sea. *Nature*, 418, 520–523.

Stöffler, D. and Grieve, R.A.F. (2007) Impactites, in *Metamorphic Rocks* (eds D. Fettes and J. Desmons), Cambridge University Press, Cambridge, pp. 82–92.

Stöffler, D., Gault, D.E., Wedekind, J. and Polkowski, G. (1975) Experimental hypervelocity impact into quartz sand: distribution and shock metamorphism of ejecta. *Journal of Geophysical Research*, 80, 4062–4077.

Stöffler, D., Ewald, U., Ostertag, R. and Reimold, W.U. (1977) Research drilling Nördlingen 1973 (Ries): composition and texture of polymict impact breccias. *Geologica Bavarica*, 75, 163–189.

Stöffler, D., Ryder, G., Ivanov, B.A. *et al.* (2006) Cratering history and lunar chronology. *Reviews in Mineralogy and Geochemistry*, 60, 519–596.

Strom, R.G., Croft, S.K. and Barlow, N.G. (1992) The Martian impact cratering record, in *Mars* (eds H.H. Kieffer, B.M. Jakosky, C.W. Snyder and M.S. Matthews), University of Arizona Press, Tucson, AZ, pp. 383–423.

Strom, R.G., Chapman, C.R., Merline, W.J. *et al.* (2008) Mercury cratering record viewed from *MESSENGER's* first flyby. *Science*, 321, 79–81.

Thomson, K., Owen, P. and Smith, K. (2005) Discussion on the North Sea Silverpit Crater: impact structure or pull-apart basin? *Journal of the Geological Society*, 162, 217–220.

Tredoux, M., Hart, R.J., Carlson, R.W. and Shirey, S.B. (1999) Ultramafic rocks at the center of the Vredefort Structure; further evidence for the crust on edge model. *Geology*, 27, 923–926.

Turtle, E.P., Pierazzo, E., Collins, G.S. *et al.* (2005) Impact structures: what does crater diameter mean? In *Large Meteorite Impacts III* (eds T. Kenkmann, F. Hörz and A. Deutsch), Geological Society of America Special Paper 384, Geological Society of America, Boulder, CO, pp. 1–24.

Vasilyev, N.V. (1998) The Tunguska meteorite problem today. *Planetary and Space Science*, 46, 129–150.

Weeks, R.A., Underwood, J.R. and Giegengack, R. (1984) Libyan desert glass: A review. *Journal of Non-Crystalline Solids*, 67, 593–619.

Wünnemann, K., Collins, G.S. and Osinski, G.R. (2008) Numerical modelling of impact melt production on porous rocks. *Earth and Planetary Science Letters*, 269, 529–538.

# TWO

# *Population of impactors and the impact cratering rate in the inner Solar System*

## Patrick Michel and Alessandro Morbidelli

*Lagrange Laboratory, University of Nice–Sophia Antipolis, CNRS, Côte d'Azur Observatory, BP 4229, 06304 Nice Cedex 4, France*

## 2.1 Introduction

The investigation of the surface histories of the solid planets and satellites is a major objective of planetary exploration. Most solid bodies in the Solar System display a record of accumulated impact cratering on their surfaces. These craters range in size from a few metres or smaller in diameter to hundreds of thousands of kilometres (e.g. giant basins on the Moon, Mars and Mercury). Assuming a constant impact rate with time, the age of a surface is proportional to the number of impacts it has experienced. Thus, if it is possible to estimate the rate of crater production on a surface, then the total number of craters can allow the estimate of the age of this surface. However, the cratering rate on a planet is related to the population of projectiles, in particular their orbital and size distribution. The determination of cratering rates, therefore, requires a good knowledge of the population of potential impactors, which allows the determination of their impact frequencies on planets. Then, a good understanding of the cratering process itself is required to establish the relationship between the diameters of the crater and the projectile, as a function of the projectile's mass, velocity, impact angle and planetary characteristics.

The only absolute chronology of craters up to 3.8 Ga that has been studied in detail so far is the lunar case, which was calibrated by dating of lunar samples brought back by the Apollo missions. The lunar crater-production function is the best investigated among those on terrestrial planet surfaces, as it is based on a large image database at various resolutions. Conversely, we do not yet have any samples of known portions of the Martian surface, and this is the main reason why different models have been developed to estimate the Martian cratering rate, with the goal of establishing a Martian surface chronology. On Earth, the situation is even worse, as the geological activity is high and several mechanisms (e.g. erosion, plate tectonics) erase crater features over time, making the cratering record incomplete (see Chapter 16). Of the terrestrial planets, only the Moon, Mercury and Mars

have heavily cratered surfaces, and all these surfaces have complex crater size distributions. For instance, the crater distributions of Mercury and Mars at diameters less than 40 km are steeper than the lunar distribution due to the obliteration of a fraction of small craters by plains formation (Strom *et al.*, 2005). On Venus, the crater density is an order of magnitude less than on Mars. The reason is twofold: (1) only young craters are present due to resurfacing events, which erased older craters; (2) small craters on this planet are rare because the small impactors cannot penetrate the thick atmosphere. Finally, the crater counting can be affected by the potential presence of secondary and multiple craters (Bierhaus *et al.*, 2005).

The cratering record of the outer Solar System suffers from even further complications. The satellites that retain cratering records are composed of substantial amounts of water ice, rather than rock, and the thermal evolution of the icy satellites has led to substantial and complicated post-formation modification of craters. The size–frequency distributions of craters on the icy satellites are also much more complex than in the inner Solar System. Although impacts are likely dominated by cometary projectiles originating in the Kuiper belt (a disk of icy bodies beyond Neptune that includes the dwarf planets Pluto and Eris among others), the cratering records of the icy satellites may also contain contributions from planetocentric populations, including irregular satellites as well as ejecta from impacts into neighbouring bodies, and possibly other small body reservoirs such as the main asteroid belt and the Trojan population. The relative importance of these different populations may be substantially different between giant planet satellite systems, and even between different satellites orbiting the same giant planet (Zahnle *et al.*, 2003).

Let us assume, for arguments sake, that all craters have been identified and counted on a surface. Then, to determine the characteristics of the population of projectiles, one needs a relationship linking a crater's diameter to an impactor's size. Scaling laws have been established for this purpose, by extrapolating the results

*Impact Cratering: Processes and Products*, First Edition. Edited by Gordon R. Osinski and Elisabetta Pierazzo.

from small-scale impacts in the laboratory to large planetary impacts (e.g. Holsapple, 1993). However, they rely on our still incomplete understanding of the complex cratering process and on many unknown parameters, such as the surface conditions and the physical properties of the impactors (e.g. density, material strength, porosity) as well as the impact velocity and angle. In particular, decoupling the impact velocity from the impactor size and even the impact angle is a great problem, as we can only make a reasonable estimate of the impact energy. Thus, the physics of cratering is a major area of research over a wide parameter space, which needs to be continuously investigated by confronting experiments, observations and numerical models. Consequently, the size distribution of projectiles derived from the distribution of crater diameters and vice versa still contains large error bars.

Dynamical studies associated with observational surveys have provided an alternative way to constrain the size distribution of projectiles. They led to the elaboration of a model of the complete orbital and size distributions of the near-Earth object (NEO) population, which, as will be shown below, is at the origin of most impacts on terrestrial planets and the Moon for the last 3.6 Ga. Traces of collisions during earlier epochs of planetary formation have been lost, and the oldest record that we have (the Moon basins) suggests that there was an intense bombardment, called the Late Heavy Bombardment (LHB), about 3.9 Ga ago. According to current models, the LHB was not provoked by the NEO population, but by a complete destabilization in the four giant planets' orbital history that produced a massive injection of planetesimals from the outer part of the Solar System to the inner part, and from the asteroid main belt. Scenarios at the origin of this LHB will be discussed at the end of this chapter.

One issue that will not be addressed in this chapter concerns cratering equilibrium (or saturation). When surfaces are old enough, the density of impact craters may reach equilibrium conditions; that is, for each new crater formed, a crater of roughly the same size is erased, and crater counts over a given size range level off as a function of time and further bombardments. The question of when equilibrium (also called saturation) conditions occur and what such conditions look like is a long-standing problem in the study of cratered surfaces and has some implications on the determination of the impactor distribution as a function of the cratering size distribution, especially for heavily cratered surfaces. We refer interested readers to a recent study of lunar cratering by Richardson (2009), and references therein, for more details on this topic.

## 2.2   Population of impactors in the inner Solar System

Strong biases exist against the discovery of objects on some types of orbits, due to the limited portion of observable space from the ground. In particular, most observations are made toward opposition (when the observed object is exactly opposite the Sun in the sky as seen from the Earth) and close to the ecliptic (the plane of the Earth's orbit). Consequently, the observed orbital distribution of NEOs is not representative of the real distribution. Since the NEO population constitutes the main population of impactors

in the inner Solar System, characterizing the impact cratering rate on planets caused by this population can only be achieved with a complete knowledge of both the orbital and size distributions of its members.

Two methods have been developed to obtain an estimate of the real NEO population from the observed one. The first method relies entirely on the data from observational surveys and tries to apply a correction for observational biases. This approach has been used on the largest detection sample size obtained by the LINEAR project (Stuart, 2001). However, this direct debiasing method requires using one-dimensional projections of absolute magnitude, semi-major axis $a$, eccentricity $e$ and inclination $i$ in order to reduce the small-number statistics problem. A severe limitation of this approach is that it cannot capture any potential dependency of the distribution of an orbital element on another. For instance, if a difference exists between the inclination distribution at low semi-major axes and that at large semi-major axes, this method cannot capture it. Another way of using observational data has been defined more recently by Harris (2007) and is presented at the end of this section. The second method uses theoretical orbital dynamical constraints in combination with the detections from observational programmes with known biases. This method and its results are summarized in the following paragraphs.

From the results of numerical integrations, it is possible to estimate the steady-state orbital distribution of the NEOs coming from each of the main source regions of these bodies, which have been clearly identified in the last decade – see Morbidelli *et al.* (2002a) for a complete review on this topic. In this approach, the key assumption is that the NEO population is currently in steady state. Such an assumption is supported by the lunar and terrestrial crater records, which suggest that the impact flux has been roughly constant (within a factor of two, see Section 2.5) during the last 3.6 Ga (Shoemaker, 1998). To compute the steady-state orbital distribution of the NEOs coming from a given source, the following method is employed: first, the dynamical evolutions of a statistically significant number of particles, initially placed in the considered NEO source region(s), are numerically integrated. The particles that enter the NEO region are followed through a network of cells in the $(a, e, i)$-space during their entire dynamical lifetime. The mean time spent in each cell (called residence time hereafter) is computed. The resulting residence time distribution shows where the bodies from the source statistically spend their time in the NEO region. As it is well known in statistical mechanics, in a steady-state scenario, the residence time distribution is equivalent to the relative orbital distribution of the NEOs that originated from the source. In other words, we can expect a larger number of NEOs in regions where the residence time of particles is greater.

This dynamical approach has been used with modern numerical integrations (Bottke *et al.*, 2000, 2002a). It allows for the computation of the steady-state orbital distributions of the NEOs coming from three sources: the $\nu_6$ secular resonance, the 3:1 mean motion resonance with Jupiter and the Mars crosser population.

The $\nu_6$ secular resonance occurs when the precession frequency of the asteroid's longitude of perihelion is equal to the mean

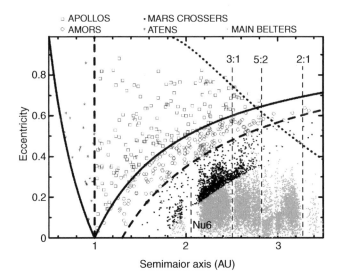

**Figure 2.1** Source regions of the NEO population represented in the (semi-major axis, eccentricity) plane. The three NEO groups, namely Atens, Apollos and Amors, and the Mars-crossers (see Michel *et al.* (2000)) are also indicated, as well as the location of main mean motion resonances with Jupiter (indicated as $3:1$, $5:2$, $2:1$) and the secular resonance $v_6$ (Nu6 in the figure; see text for details). The solid lines represent the Earth-crossing lines (perihelion and aphelion equal to 1 AU), the curved dash line represents the Mars-crossing line and the dotted line at the top right of the plot is the Jupiter-crossing line at the aphelion.

precession frequency of Saturn's longitude of perihelion. This secular resonance essentially marks the inner edge of the main belt. The $3:1$ mean-motion resonance with Jupiter occurs at approximately 2.5 AU from the Sun, where the orbital period of an asteroid is one-third that of Jupiter. Both resonances pump up rapidly the eccentricities of the asteroids and force them to acquire Earth-crossing orbits (e.g. Gladman *et al.*, 1997; Bottke *et al.*, 2002a). The Mars-crossing asteroids have orbits that intersect the orbit of Mars on a few million years' timescale during their secular evolution, but they are not currently NEOs because their perihelion distance is larger than 1.3 AU (Michel *et al.*, 2000). Most of these objects do not come from the main belt through the two main resonances mentioned above, but rather from a network of thin resonances that slowly increase the eccentricity of many asteroids at different locations in the main belt (e.g. Morbidelli and Nesvorny, 1999). The population of Mars-crossers extends up to semi-major axes about 2.8 AU, beyond which it is strongly depleted by the proximity of Jupiter. Figure 2.1 shows the positions of the different sources in the (semi-major axis, eccentricity) plane.

The overall NEO orbital distribution was then constructed as a linear combination of the distributions from these three different sources. Different weights were used for the three different distributions accounting for their relative contributions. The NEO magnitude distribution was then assumed to be source independent. The resulting NEO orbital-magnitude distribution

was then 'virtually' observed by applying to it the observational biases associated with the Spacewatch survey (Jedicke, 1996). This allowed Bottke *et al.* (2000) to determine a good set of weights for the three distributions, which resulted in a satisfactory fit of the orbits and magnitudes of the NEOs discovered or accidentally rediscovered by Spacewatch. To have a better match with the observed population at large semi-major axes (>3 AU), the model has been extended by considering also the steady-state orbital distributions of the NEOs coming from the outer asteroid belt ($a > 2.8$ AU) and from the Jupiter family comets (Bottke *et al.*, 2002a). Besides Spacewatch, this NEO model has been shown to be consistent with the observations of the LINEAR and Catalina surveys (Bottke *et al.*, 2004; Zavodny *et al.*, 2008), once the biases of these surveys are properly taken into account.

An important aspect of this model is that, once the values of the parameters of the model are determined by best-fitting observations of a specific survey, the steady-state orbital-magnitude distribution of the *entire* NEO population is determined. This distribution is valid also in those regions of the orbital space not sampled by any survey, because of extreme observational biases. This underlines the power of the dynamical approach for debiasing the NEO population.

From this model, the total NEO population is estimated to contain about 1200 objects with absolute magnitude $H < 18$ and semi-major axis $a < 7.4$ AU. The absolute magnitude $H$ is defined as the visual magnitude $V$ that an asteroid would have in the sky if observed at 1 AU distance from both the Earth and the Sun, at zero phase angle. It is usually assumed that an object with $H = 18$ as a diameter of 1 km, although the exact relation depends on the albedo of the object (see Section 2.3). In June 2010, approximately 84% of these objects with $H < 18$ have already been discovered. The NEO absolute magnitude distribution is of the type $N(<H) = C \times 10^{(0.35+/-0.02)H}$ in the range $13 < H < 22$, implying $29{,}400 \pm 3600$ NEOs with $H < 22$. Assuming that the albedo distribution is not dependent on $H$, this magnitude distribution implies a power law cumulative size distribution $N(>D) \propto D^q$ with exponent $q = -1.75 \pm 0.1$. This distribution is in perfect agreement with that obtained by Rabinowitz *et al.* (2000), who directly debiased the magnitude distribution observed by the NEAT survey, and slightly shallower than that obtained by Stuart (2001) using the LINEAR database. We will see later that it is also consistent with the crater size distribution on the Moon (exponent around $-2$) when scaling laws are applied to derive the corresponding projectiles' size distribution.

The comparison between the debiased orbital-magnitude distribution of NEOs with $H < 18$ and the observed distributions of discovered objects suggests that most of the undiscovered NEOs have $H > 16$ (<2 km in size) and semi-major axis in the range 1.5–2.5 AU. Given the NEO orbital distribution, and assuming random values for the argument of perihelion and the longitude of node, about 21% of the NEOs turn out to have a minimal orbital intersection distance (MOID) with the Earth smaller than 0.05 AU. The MOID is defined as the minimal distance between the orbits of two objects. By definition, NEOs with MOID < 0.05 AU are classified as potentially hazardous objects (PHOs), and the accurate orbital determination of these bodies is considered a top priority.

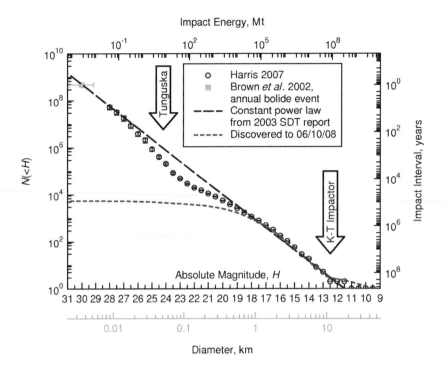

**Figure 2.2** Open blue circles represent the cumulative number of NEOs brighter than a given absolute magnitude *H*, defined as the visual magnitude *V* that an asteroid would have in the sky if observed at 1 AU distance from both the Earth and the Sun, at zero phase angle. A power-law function dashed blue line) is shown for comparison. Ancillary scales give impact interval (right), impact energy in megatons TNT for the mean impact velocity of approximately 20 km s$^{-1}$ (top) and the estimated diameter corresponding to the absolute magnitude *H* (second scale at bottom). Courtesy of A.W. Harris. (See Colour Plate 8)

More recently, Harris (2007) used three overlapping methods to estimate the population of NEOs (disregarding long period comets). In the largest size range, to about *H* = 16, current surveys are essentially complete, so the number of objects discovered up to this magnitude anchors the *H*-distribution of the NEO population at the bright end. In the intermediate range, up to about *H* = 20, the fraction of completeness was estimated from the ratio of re-detections of already known NEOs to the total number of detections for known and new objects in the interval 2005–2006. The re-detection ratio had been bias corrected using a survey simulation model to allow for the fact that NEOs are not equally easy to observe. The resulting size–frequency distribution, represented by blue circles in Fig. 2.2 (see Plate 8), deviates substantially from a power law with fixed exponent in the size range from 1 km down to 10 m. Consequently, the number of NEOs of about 100 m in diameter is almost an order of magnitude smaller than that predicted by extrapolation of a power law. The size of 100 m is of the order of the estimated size of transition from 'rubble pile' (gravitational aggregates) to 'monolithic' bodies, and the 'dip' of the NEO distribution may be related to this transition in physical strength.

## 2.3   Impact frequency of NEOs with the Earth

When a small body collides with the Earth, the corresponding impact energy depends not only on the impact velocity, but also

on the bulk density and size of the object. Therefore, because *H* is related to the diameter by the albedo, it is first necessary to estimate the albedo distribution. The albedo is also used to estimate the body's bulk density.

Two independent approaches, described by Morbidelli *et al.* (2002b) and Stuart and Binzel (2004), have been used to estimate the NEO albedo distribution. The results obtained by these two methods are in very good agreement. In particular, both imply that, on average, the usually assumed conversion *H* = 18 ⇔ *D* = 1 km slightly overestimates the number of kilometre-size objects. There should be approximately 1000 NEOs with *D* > 1 km, against approximately 1200 NEOs with *H* < 18. Once the albedo distribution is determined, a similar procedure is used by Morbidelli *et al.* (2002b) and Stuart and Binzel (2004) to estimate the NEO collision probability with the Earth as a function of collision energy. It is assumed that the density of bright and dark bodies is 2.7 g cm$^{-3}$ and 1.3 g cm$^{-3}$ respectively. These values are taken from spacecraft or radar measurements of a few S-type (bright) and C-type (dark) asteroids. Then, for each set of orbital elements (*a*, *e*, *i*) of the NEO model, the collision probability is computed as described by Bottke *et al.* (1994), assuming that the values of the mean anomalies and of the longitude of perihelia and nodes of both the Earth and the NEOs are uniformly distributed. This assumption fails when NEOs are in a secular or mean motion resonance with a planet, but this occurs not long or frequently enough during their evolution to affect the computation of collision probabilities. The gravitational attraction exerted by the

Earth is also included. There is a remarkable agreement between the final results of Morbidelli *et al.* (2002b) and Stuart and Binzel (2004). The estimates are roughly equivalent to the ones given in Fig. 2.2, which simply assumes a mean value for the collision probability, impact velocity and bulk density of NEOs. In particular, it is found that the Earth should undergo a 1000 Mt (megaton) collision every $50,400 \pm 6400$ years. Such impact energy is, on average, produced by bodies with $H < 20.5$ (size greater than 300 m). The NEOs with $H < 20.5$ discovered so far represent only about 28% of the total population and carry about the same percentage of this collision probability.

## 2.4 Comparison with the impact record on terrestrial planets

### 2.4.1 The Earth

The comparison with the impact record on Earth is difficult because this record is severely incomplete (see Chapter 1). The incompleteness is due not only to the erasure of craters mentioned earlier in this chapter, but also because small projectiles, up to approximately 100 m in diameter, are expected to explode or deliver most of their energy in the atmosphere and, thus, leave little trace, if any, on the ground.

Bland and Artemieva (2006) attempted a detailed modelling of the interaction between bolides and the Earth's atmosphere. From the analysis of the data on explosions in the upper atmosphere, they derived a cumulative size–frequency distribution for upper atmosphere impactors. Then they applied their model results to scale the upper atmosphere curve to a flux at the Earth's surface, elucidating the impact rate of objects less than 1 km diameter on Earth. They found that iron bodies a few metres in diameter, which form craters of about 100 m in diameter, strike the Earth's surface every 500 years. Larger bodies form craters of 0.5 km in diameter every 20,000 years, and craters of 1 km in diameter are formed on the Earth's surface every 50,000 years. Tunguska events (low-level atmospheric disruption of stony bolides with typical size of a few tens of metres) may occur every 500 years. This is within a factor of four of what is shown in Fig. 2.2.

According to Bland and Artemieva (2006), most projectiles smaller than 1 km in diameter should break while crossing the atmosphere. Such fragmentation has been observed recently. On 6 October 2008, a small asteroid of 5 m in diameter was discovered and designated 2008 TC$_3$, about 20 h before it hit the ground. It exploded at 37 km altitude. Although no macroscopic fragments (large enough to not be considered as unrecoverable dust) were expected to survive, a dedicated search along the approach trajectory recovered 47 meteorites, recognized as fragments of a single body with a total mass of 3.95 kg (Jenniskens *et al.*, 2009). These meteorites have been characterized as ureilites, a rare type of achondrite, and are extremely fragile. Another event had already challenged the models: the Carancas impact event, which delivered a meteorite on 15 September 2007. Model predictions suggested that the projectile should not have cratered the ground. In reality, the projectile created a clearly visible crater, hitting the ground with an impact speed larger than the terminal speed that

is roughly the speed most meteoroids reach the surface without getting completely destroyed and leaving meteorites on the ground. Schultz *et al.* (2008) proposed that the meteor did indeed break up into pieces, but shock waves kept them close together and the fragment clump reshaped itself during the fall into a more aerodynamic figure. However, this theory is still being debated. These two events demonstrate that our knowledge regarding the physics involved in the interaction of meteors with the atmosphere is still limited. Moreover, while this knowledge is continuously improving thanks to ongoing studies, the one related to the physical properties of the entering bodies (e.g. their size, density, material strength, degree of porosity) will always remain limited. Consequently, our ability to predict the minimum size of a body capable of reaching the ground will always remain approximate.

### 2.4.2 The other terrestrial planets

In the inner Solar System, the Moon, Mercury and Mars have heavily cratered surfaces. The lunar crater-production function is the best investigated and the most reliable one among those on terrestrial planet surfaces, thanks to a large image database at various resolutions and an absolute dating of the age of some surfaces through the analysis of the samples collected during the Apollo and Luna missions.

It should also be noted that the *Mars Global Surveyor* Mars orbiter camera (MOC), High Resolution Imaging Science Experiment (HiRISE) and Context Camera (CTX) have recently acquired data that can help in establishing the present-day impact cratering rate on the Martian surface, based on the observation of impact craters with diameters in the range 2–150 m created in an area of $21.5 \times 10^6$ km$^2$ between May 1999 and March 2006. Twenty potential new impact sites were first identified using the MOC (Malin *et al.*, 2006). HiRISE confirmed 19 of these sites as pristine impact locations and eventually, as of 7 January 2009, 70 new impact sites were discovered (mostly by CTX) and confirmed by HiRISE (Ivanov *et al.*, 2009). Malin *et al.* (2006) concluded that the values predicted by models that scale the lunar cratering rate to Mars are close to the observed rate. This result supports the extrapolations based on the lunar record to understand the projectile flux on terrestrial planet surfaces and to establish a global chronological scale.

The comparison between the crater record and the impact rate computed from an NEO model requires the use of appropriate scaling laws. Such scaling laws have been derived in a number of studies on the physics of impact cratering on solid bodies. What we exactly mean by the *scaling* of impact events is to apply some relation to predict the outcome of one event from the results of another, or to predict how the outcome depends on the problem parameters. The parameters that are different between the two events are the variables that are 'scaled'. The interested reader may consult the book by Melosh (1989) for a description of the approaches used to derive this scaling. A *scaling law* is usually represented by a formula used to perform the scaling, which relates the size of the projectile to the size of the resulting crater. This law depends strongly on impact velocity and on other parameters, such as the target's gravitational field and the material strength (e.g. Holsapple 1993). A number of questions can be raised about such scaling laws. Clearly, they are the outcome of complex processes involving the balance equations of mass, momentum and energy of continuum mechanics and the constitutive equations of the materials. The impact processes encompass the gamut of pressures from many megabars (where common metals act like a fluid) to near zero (where material strength or other retarding actions limit the final crater growth). Thus, different scaling laws can be derived by using different assumptions and simplifications in the physics of the cratering process – see Holsapple (1993) for a review.

Stuart and Binzel (2004) used three crater scaling laws to compare the rate of crater formation expected from the NEO population that they derived with the observed craters on the Moon. They identified these three scaling laws as Melosh's Pi-scaling (Melosh, 1989), Shoemaker's formula (Shoemaker *et al.*, 1990) and Pierazzo's formula (Pierazzo *et al.*, 1997). In fact, as stated by Melosh (1989), the Pi-group scaling has been defined by Holsapple and Schmidt (1982) on the basis of small-scale laboratory experiments, and Pierazzo's formula is just a different analytical form of these laws, which comes from Grieve and Cintala (1992). In contrast to the Pi-scaling, Shoemaker's formula has been derived from explosion cratering experiments. Such experiments allow the investigation of events of a larger scale than in centrifuge experiments. Nevertheless, all explosion events are still of small scale in comparison with natural impact craters. Therefore, all these laws, in the regime of interest for NEO impacts, are defined by extrapolations of the experimental data.

Stuart and Binzel (2004) found that the following formulae adapted from Shoemaker *et al.* (1990) produce the best match between the NEO population and the observed craters:

$$D_t = 0.01436 \left( W \frac{\rho_i}{\rho_t} \right)^{1/3.4} \left( \frac{g_e}{g_t} \right)^{1/6} (\sin \alpha)^{2/3}$$

$$D_r = 1.56 D_t$$

$$D_f = D_r \quad \text{if } D_r \leq D_*$$

$$D_f = \frac{D_r^{1.18}}{D_*^{0.18}} \quad \text{if } D_r > D_*$$

$$D_* = 4 \text{ km} \left( \frac{g_e}{g_t} \right)$$

where all units are mks; $D_t$, $D_r$ and $D_f$ refer to transient, rim-to-rim and final crater diameters respectively; $\rho$ and $g$ are respectively density and gravity, with subscripts i, t and e indicating impactor, target and Earth respectively; and $\alpha$ ($=\pi/2$ for vertical impacts) is the impact angle. Shoemaker *et al.* (1990) used a simple factor of 1.3 to scale from initial diameters to final, complex crater diameters, and they applied this factor above a transition threshold of $D_*=4$ km on Earth, with $D_*$ scaling inversely with gravity.

With these scaling laws, the models of NEO population predict the formation of $2.73 \times 10^{-14}$ craters larger than 4 km in diameter

per square kilometre per year on the Moon, which compares well with the estimate obtained from crater counting on lunar terrains with known ages (($3.3 \pm 1.7) \times 10^{-14} \text{km}^{-2}\text{yr}^{-1}$; Grieve and Shoemaker, 1994). The size–frequency distribution of the lunar craters that is expected from the NEO models using scaling laws matches well the observed crater size frequency distribution for sizes up to 200 km, but is deficient for larger craters by a factor of approximately three to five (Marchi *et al.*, 2009). The reason for this discrepancy is not clear. It might be due to inappropriate scaling laws for very large impacts. It may also be due to the fact that most large craters were formed in the distant past, when the cratering rate was higher and the projectile size distribution was different (Strom *et al.*, 2005; Marchi *et al.*, 2009). We address this issue in Section 2.6.

## 2.5 Variability of the impact frequency during the last 3 Ga

The impact frequencies indicated in previous sections correspond to average values, assuming a steady-state situation. However, the impact rate in the Solar System may not have been in such a steady state during its entire history. Culler *et al.* (2000) determined the $^{40}\text{Ar}/^{39}\text{Ar}$ formation ages of 155 lunar impact melt beads from a soil sampled during the *Apollo 14* mission. They found that relatively old ages (>3 Ga) and fairly young ones (<500 Ma) are more frequent, which made them conclude that the cratering rate in the inner Solar System decreased by a factor of two to three during the past 3 Ga and then intensified again during the most recent 500 Ma. It should be noted that Hörz (2000) argued that the distribution of bead ages reported by Culler *et al.* (2000) might be consistent with a constant impactor flux. The author pointed out that lunar soils are the products of stochastic processes, and any one soil sample might not faithfully represent the average cratering history through geologic time. Obviously, additional soils need to be investigated using the methods pioneered by Culler *et al.* (2000).

Several independent arguments point toward an intensification of the bombardment flux at 500 Ma and are in favour of the interpretation of lunar data by Culler *et al.* (2000). Approximately two-thirds of L-chondrite meteorites have been heavily shocked and degassed, with $^{39}\text{Ar}/^{40}\text{Ar}$ ages of approximately 470 Ma (Korochantseva *et al.*, 2007) suggesting that the L-chondrite parent body suffered a major impact and catastrophically disrupted at that time – see also Heymann (1967) and Haack *et al.* (1996) and the references therein. The timing of this shock event coincides with the stratigraphic age ($467 \pm 2$ Ma) of the mid-Ordovician strata in southern Sweden where many fossil L-chondrite meteorites have been found (Schmitz *et al.*, 1997, 2003; Greenwood *et al.*, 2007). This suggests that the shocked L-chondrites and the fossil meteorites are related to the same event, namely a catastrophic disruption of a large main-belt asteroid that produced an initially intense meteorite shower on the Earth and that still supplies approximately 30% of the ordinary chondrites falling today. Catastrophic break-ups of large main-belt asteroids leave behind groups of kilometre-sized and larger asteroid fragments, which are then recognized as asteroid families (Hirayama, 1918). Nesvorny *et al.*, (2002, 2007a) first proposed as a possible candidate the prominent Flora family located near the $\nu_6$ secular resonance in the inner asteroid belt. However, a reanalysis of the available data prompted Nesvorny *et al.* (2009) to propose the Gefion family as a plausible alternative.

Family formations events may lead to impact showers on Earth if they occur close enough to powerful resonances that are able to transport some of the fragments onto Earth-crossing orbits on a short timescale (Zappalà *et al.*, 1998). Moreover, if the formation occurs far from an efficient transport route to the Earth, it can still affect the long-term impact rate on our planet by modifying the structure and number of components of the asteroid population in the main belt. As this population ultimately feeds the NEO population, a change in the former causes a change in the *steady-state* distribution of the latter.

Bottke *et al.* (2007) proposed that the formation of the Baptistina family as a result of the break-up of a 170 km-size asteroid at 160 Ma caused another change in the NEO population and may be the main source of NEOs belonging to the C taxonomic class that we see today. They argued that the C-type projectile that caused the Chicxulub crater at the origin of the K/P event at 65 Ma was a former member of this family.

## 2.6 The early cratering history of the Solar System

The Moon and all the terrestrial planets have been resurfaced during a period of intense impact cratering that occurred sometime between 4.5 and 3.8 Ga. In particular, although the interpretation of data is still a subject of debate (e.g. see Hartmann *et al.* (2007) for an alternative view), there is a growing consensus that there was a cataclysmic spike in the cratering rate in the inner solar system at approximately 3.9 Ga, i.e. about approximately 650 Myr after the planets formed (e.g. Hartmann *et al.*, 2000; Ryder *et al.*, 2000; Koeberl 2004, 2006). Several models have been proposed to explain this spike, usually called the Late Heavy Bombardment (LHB) (e.g. Levison *et al.*, 2001, 2004; Chambers and Lissauer 2002). A recent study has proposed a scenario – often called the NICE model (as it originated from a team located at the Côte d'Azur Observatory in Nice, France) – which not only explains the LHB (Gomes *et al.*, 2005), but also other properties of the outer Solar System, such as the current orbital architecture of the giants planets (Tsiganis *et al.*, 2005), the existence and orbital distribution of Jovian Trojans (Morbidelli *et al.*, 2005), the origin of the irregular satellites of the giant planets (Nesvorný *et al.*, 2007b) and the orbital structure of the Kuiper belt (Levison *et al.*, 2008). Thus, the strength of the NICE model is to reproduce many constraints of our Solar System within the framework of a unique scenario.

The main assumptions of this model are that the four giant planets initially had circular orbits, were much closer to each other ($5 < a < 15$ AU) and were surrounded by a massive disk of planetesimals of about 35 Earth masses. As is well known, since the work of Fernández and Ip (1984), dynamical interactions of the planets with this disk caused a slow increase of the orbital separations of the planets. After 500–600 Myr,

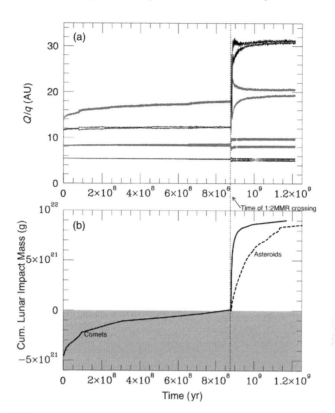

**Figure 2.3** *Top.* Evolutions of the perihelion $q$ and aphelion $Q$ distances of Jupiter (lowest curves), Saturn, Uranus and Neptune until they reach their current orbits (from Gomes *et al.* (2005)). The label 1:2 MMR means that Jupiter and Saturn are in their mutual 1:2 mean motion resonance. When this happens (at 880 Myr in this case), the planetary system is suddenly destabilized. In particular, Uranus and Neptune reach their current orbits. On this plot, note that Neptune is initially at a smaller distance to the Sun than Uranus, but other simulations reproduce the current orbits starting with a more distant Neptune than Uranus. *Bottom.* Cumulative mass of comets (solid curve) and asteroids (dashed curve) at 1 AU from the Sun (Earth's distance). The comet curve is offset so that its value is zero at the time of 1:2 MMR crossing. At this time, there is a drastic increase of material's mass reaching 1 AU over a 100–200 Myr period, in agreement with the magnitude and duration of the LHB from the lunar crater record. Thus, $5 \times 10^{21}$ g of comets was accreted before resonant crossing and $9 \times 10^{21}$ g of cometary material would have struck the Moon during the LHB. Note that, in addition to comets, 95% of the asteroids escape from the main belt and, although it is not obvious from the plot, they eventually dominate the impactor population (see Gomes *et al.* (2005) for details on these calculations).

Jupiter and Saturn crossed their mutual 1:2 mean motion resonance, which destabilized the planetary system as a whole: the orbital eccentricities of these two planets reached their current values, Uranus and Neptune reached their current orbits (Fig. 2.3) and a huge flux of planetesimals was suddenly transported to the orbits of terrestrial planets, from both the asteroid belt and the original trans-Neptunian disk. Simulations show that about $10^{22}$ g of bodies hit the Moon during an approximately 100–200 Myr interval (Fig. 2.3), which is consistent with the magnitude and duration of the LHB from the lunar crater record.

The NICE model has evolved, relative to its first version of 2005. Work has been devoted to remove the arbitrary aspect of the initial conditions of the giant planets used in the original version. The new version of the model starts in a configuration that is consistent with the hydrodynamical phase of the proto-planetary disk; that is, with giant planets that are each in a mean motion resonance with their neighbours. In effect, this is the orbital configuration that the planets should have achieved during their evolutions when they were embedded in a disk of gas (Morbidelli *et al.*, 2007). This configuration also explains why Jupiter and the other giant planets did not migrate to a location closer to the Sun, unlike most of the observed extra-solar planets (Masset and Snellgrove, 2001; Morbidelli and Crida, 2007; Pierens and Nelson, 2008). The evolutions of the giant planets starting from this new configuration are quite analogous to the ones described above. In essence, when the planets are extracted from their resonances, they become unstable. Then close encounters start occurring, which result in an evolution that is very similar to that shown in Fig. 2.3 (Morbidelli *et al.*, 2007; Batygin and Brown, 2010). The epoch of the instability is also late, as in Gomes *et al.* (2005), provided that the planetesimal disk is beyond the orbit of the most distant planet and the self-gravity of the disk is taken into account. It turns out that the epoch of the instability does not depend sensitively on the badly constrained location of the inner edge of the disk, which is a strength of this new version of the NICE model. In terms of small-body dynamics, the old and new versions of the NICE model do not differ significantly. Thus, the results on the LHB quoted above are still valid.

An important piece of information on the origin of the LHB comes from the finding (Strom *et al.*, 2005) that the crater size distribution on the lunar highlands (old terrains, cratered at the time of the LHB) is consistent – via appropriate scaling laws – with the size distribution of the main-belt population. On the other hand, the size distribution of the more recent craters needs impactors whose size distribution is consistent with that of the NEO population – see also Marchi *et al.* (2009).

Today, the size distribution of NEOs is different (steeper) than that of the main belt, because their origin is ultimately related to a non-gravitational, size-dependent process called the Yarkovsky effect. This effect is due to an anisotropy of the thermal re-emission of the light received from the Sun on a small body's surface (Bottke *et al.*, 2002b). It causes a slow drift in semi-major axis of bodies smaller than approximately 20 km, whose rate depends on the body's size. Thanks to this effect, main-belt asteroids are supplied to the resonances that ultimately deliver them to the NEO region. The size dependency of this effect is at the origin of the observed difference between the size distribution of NEOs and main-belt asteroids (Morbidelli and Vokrouhlický, 2003).

If, at the time of the LHB, the impactor population had a size distribution similar to that of main-belt asteroids as claimed by Strom *et al.* (2005), this implies that at this epoch the asteroids had to escape from the main belt via a size-independent (i.e. purely gravitational) process. The only gravitational processes that can eject asteroids from all over the asteroid belt are the sweeping of resonances, caused by the orbital migration of the giant planets, or the passage through the asteroid belt of a massive body, such as a rogue planet. The first scenario is consistent with the NICE model (Gomes *et al.*, 2005). The second scenario would be consistent with an alternative model of the LHB origin developed by Chambers (2007). In this model, there were originally five terrestrial planets. The now missing planet was located between the orbit of Mars and the inner edge of the main asteroid belt. It had a mass a bit smaller than that of the red planet. In several of the simulations by Chambers, this planet becomes unstable after a few 100 Myr. It then crosses briefly the asteroid belt and consequently causes the escape of small bodies from there.

Although Chambers' (2007) model has not been studied in great detail and tested against the large number of constraints of our Solar System, it remains the main competitor of the NICE model. The main difference between these two scenarios is that the NICE model predicts that the LHB occurred also on the giant planets and their satellites, as a result of the injection of trans-Neptunian planetesimals in the inner Solar System; in contrast, the model by Chambers (2007) predicts that only the terrestrial planets have been heavily affected.

Unfortunately, the chronology of the bombardments of the objects of the outer Solar System is not clear. Obviously, the giant planets cannot show traces of impacts. Moreover, most of the surfaces of the icy satellites are young and, therefore, cannot show evidence of anything that occurred in the distant past. Saturn's satellite Iapetus, however, may provide interesting clues. Its surface is old, heavily cratered and supports a dozen impact basins. Thus, it is clear that Iapetus shows that the bombardment rate in the past had to be much heavier than at the current time (Zahnle *et al.*, 2003). The question is: did this heavy bombardment occur early (i.e. soon after the giant planet formation) or late? The ejecta blankets of the basins overlap the equatorial ridge on Iapetus and, therefore, postdate the formation of this ridge. Thus, estimating the age of the ridge becomes fundamental to answering the question of the timing of the bombardment of Iapetus. Until now, the most accepted model for the formation of the ridge is that of Castillo-Rogez *et al.* (2007). This model implies a late formation of the ridge (from 200 to 800 Myr after the formation of the satellite) and, consequently, that the LHB also took place in the outer Solar System (Johnson *et al.*, 2008). This is consistent with the NICE model and goes against that of Chambers. However, future studies of Iapetus might find some mechanisms that can explain an early formation of the ridge, which would then revise our current view.

## 2.7    Conclusions

The impact cratering rates on planets give us some insights into the history of our Solar System. Several lines of evidence point towards an intense bombardment of material on the Moon (and consequently planets) at approximately 3.9 Ga, and dynamical models, such as the NICE model, provide scenarios that explain this so-called LHB. After this period, the flux of impactors declined and stabilized to an almost steady-state regime, with some fluctuations indicated by the formation ages of lunar impact melt beads.

The population at the origin of most of the craters on terrestrial planets produced after the LHB is the NEO population. It has been characterized in terms of its orbital and size distribution, although there are still some unknowns concerning the contribution of cometary bodies. However, the NEO population model and associated impact frequencies on terrestrial surfaces are consistent with the best-characterized crater distribution (i.e. that of the lunar surface) when appropriate scaling laws are applied to convert the impactor size distribution to the distribution of crater diameters. In addition, there is still a discrepancy for the very large craters (>200 km), which may have been formed at earlier epochs or which may require the use of different scaling laws.

Obviously, determining the size distribution of individual craters remains a difficult task, and relating a crater's size to the size of the impactor requires a good understanding of the cratering process. Both problems will still be the subject of many studies. Our knowledge has thus improved in the last decade, thanks to observational and theoretical studies, but it is clear that there are still a lot of uncertainties concerning the impact process, its history in the Solar System and the identification of craters on the surfaces of terrestrial planets.

# References

Batygin, K. and Brown, M.E. (2010) Early dynamical evolution of the Solar System: pinning down the initial condition of the Nice model. *Astrophysical Journal*, 716, 1323–1331.

Bierhaus, E.B., Chapman, C.R. and Merline, W.J. (2005) Secondary craters on Europa and implications for cratered surfaces. *Nature*, 437, 1125–1127.

Bland, P.A. and Artemieva, N.A. (2006) The rate of small impacts on Earth. *Meteoritics and Planetary Science*, 41, 607–631.

Bottke, W.F., Nolan, M.C., Greenberg, R. and Kolvoord, R.A. (1994) Velocity distributions among colliding asteroids. *Icarus*, 107, 255–268.

Bottke, W.F., Jedicke, R., Morbidelli, A. *et al.* (2000) Understanding the distribution of near-Earth asteroids. *Science*, 288, 2190–2194.

Bottke, W.F., Morbidelli, A., Jedicke, R. *et al.* (2002a) Debiased orbital and size distribution of the near Earth objects. *Icarus*, 156, 399–433.

Bottke, W.F., Vokrouhlicky, D., Rubincam, D.P. and Broz, M. (2002b) The effect of Yarkovsky thermal forces on the dynamical evolution of asteroids and meteoroids, in *Asteroid III* (eds W.F. Bottke, A. Cellino, P. Paolicchi and R.P. Binzel), The University of Arizona Press, Tucson, AZ, pp. 395–408.

Bottke, W.F., Morbidelli, A., Jedicke, R. *et al.* (2004) Investigating the near-Earth object population using numerical integration methods and LINEAR data. *Bulletin of the American Astronomical Society*, 36, 1141.

Bottke, W.F., Vokrouhlický, D. and Nesvorný, D. (2007) An asteroid breakup 160 Myr ago as the probable source of the K/T impactor. *Nature*, 449, 48–53.

Castillo-Rogez, J.C., Matson, D.L., Sotin, C. *et al.* (2007) Iapetus' geophysics: rotation rate, shape, and equatorial ridge. *Icarus*, 190, 179–202.

Chambers, J.E. (2007) On the stability of a planet between Mars and the asteroid belt: implications for the Planet V hypothesis. *Icarus*, 189, 386–400.

Chambers, J.E. and Lissauer, J.J. (2002) A new dynamical model for the lunar Late Heavy Bombardment. *Lunar and Planetary Institute Conference Abstracts*, 33, 1093.

Culler, T.S., Becker, T.A., Muller, R.A. and Renne, P.R. (2000) Lunar impact history from $^{40}$Ar/$^{39}$Ar dating of glass spherules. *Science*, 287, 1785–1788.

Fernández, J.A. and Ip, W.-H. (1984) Some dynamical aspects of the accretion of Uranus and Neptune – the exchange of orbital angular momentum with planetesimals. *Icarus*, 58, 109–120.

Gladman, B., Migliorini, F., Morbidelli, A. *et al.* (1997) Dynamical lifetimes of objects injected into asteroid belt resonances. *Science*, 277, 197–201.

Gomes, R., Tsiganis, K., Morbidelli, A. and Levison, H.F. (2005) Origin of the cataclysmic Late Heavy Bombardement period of the terrestrial planets. *Nature*, 465, 466–469.

Greenwood, R.C., Schmitz, B., Bridges, J.C. *et al.* (2007) Disruption of the L chondrite parent body: new oxygen isotope evidence from Ordovician relict chromite grains. *Earth and Planetary Science Letters*, 262, 204–213.

Grieve, R.A.F. and Cintala, M.J. (1992) An analysis of differential impact melt-crater scaling and implications for the terrestrial impact record. *Meteoritics*, 27, 526–538.

Grieve, R.A.F. and Shoemaker, E.M. (1994) The record of past impacts on Earth, in *Hazards Due to Comets and Asteroids* (ed. T. Gehrel), The University of Arizona Press, Tucson, AZ, pp. 417–462.

Haack, H., Farinella, P., Scott, E.R.D. and Keil, K. (1996) Meteoritic, asteroidal, and theoretical constraints on the 500 Ma disruption of the L chondrite parent body. *Icarus*, 119, 182–191.

Harris, A.W. (2007) An update of the population of NEAs and impact risk. *Bulletin of the American Astronomical Society*, 39, 511.

Hartmann, W.K., Ryder, G., Dones, L. and Grinspoon, D. (2000) The time-dependent intense bombardment of the primordial Earth/Moon system, in *Origin of the Earth and Moon* (eds R.M. Canup and K. Righter), The University of Arizona Press, Tucson, AZ, pp. 493–512.

Hartmann, W.K., Quantin, C. and Mangold, N. (2007) Possible long-term decline in impact rates. 2. Lunar impact-melt data regarding impact history. *Icarus*, 186, 11–23.

Heymann, D. (1967) On the origin of hypersthene chondrites: ages and shock effects of black chondrites. *Icarus*, 6, 189–221.

Hirayama, K. (1918) Groups of asteroids probably of common origin. *Astronomical Journal*, 31, 185–188.

Holsapple, K.A. (1993) The scaling of impact processes in planetary sciences. *Annual Reviews of Earth and Planetary Sciences*, 21, 333–373.

Holsapple, K.A. and Schmidt, R.M. (1982) On the scaling of crater dimensions 2. Impact processes. *Journal of Geophysical Research*, 87, 1849–1870.

Hörz, F. (2000) Time-variable cratering rates? *Science*, 288, 2095a.

Ivanov, B.A., Melosh, H.J., McEwen, A.S. and the HiRISE team (2009) Small impact crater clusters in high resolution HiRISE images II. *Lunar and Planetary Institute Conference Abstracts*, 40, 1410.

Jedicke, R. (1996) Detection of near-Earth asteroids based upon their rates of motion. *Astronomical Journal*, 111, 970–983.

Jenniskens, P., Shaddad, M.H., Numan, D. *et al.* (2009) The impact and recovery of asteroid 2008 TC3. *Nature*, 458, 485–488.

Johnson, T.V., Castillo-Rogez, J.C., Matson, D.L. *et al.* (2008) Constraints on outer Solar System chronology. 39th Lunar and Planetary Science Conference (Lunar and Planetary Science XXXIX), League City, TX, March 10–14, LPI Contribution No. 1391, 2314.pdf.

Koeberl, C. (2004) The late heavy bombardment in the inner solar system: is there a connection to Kuiper-belt objects? *Earth, Moon and Planets*, 92, 79–87.

Koeberl, C. (2006) The record of impact processes on the early Earth – a review of the first 2.5 billion years, in *Processes of the Early Earth* (eds W.U. Reimold and R. Gibson), Geological Society of America Special Paper 405, Geological Society of America, Boulder, CO, pp. 1–22.

Korochantseva, E.V., Trieloff, M., Lorenz, C.A. *et al.*, (2007) L-chondrite asteroid breakup tied to Ordovician meteorite shower by multiple isochron $^{40}$Ar–$^{39}$Ar dating. *Meteoritics and Planetary Science*, 42, 113–130.

Levison, H.F., Dones, L., Chapman, C.R. *et al.* (2001) Could the lunar "Late Heavy Bombardment" have been triggered by the formation of Uranus and Neptune? *Icarus*, 151, 286–306.

Levison, H.F., Thommes, E., Duncan M.J. and Dones, L. (2004) A fairy tale about the formation of Uranus and Neptune and the Late Heavy Bombardment, in *Debris Disks and the Formation of Planets* (eds L. Caroff, L.J. Moon, D. Backman and E. Praton), Astronomical Society of the Pacific, ASP Conference Series 324, ASP, San Francisco, CA, pp. 152–167.

Levison, H.F., Morbidelli, A., Vanlaerhoven, C. *et al.* (2008) Origin of the structure of the Kuiper belt during a dynamical instability in the orbits of Uranus and Neptune. *Icarus*, 196, 258–273.

Malin, M.C., Edgett, K.S., Posiolova, L.V. *et al.* (2006) Present-day impact cratering rate and contemporary gully activity on Mars. *Science*, 314, 1573–1577.

Marchi, S., Mottola, S., Cremonese, G. *et al.* (2009) A new chronology for the Moon and Mercury. *Astronomical Journal*, 137, 4936–4948.

Masset, F. and Snellgrove, M. (2001) Reversing type II migration: resonance trapping of a lighter giant protoplanet. *Monthly Notices of the Royal Astronomical Society*, 320, L55–L59.

Melosh, H.J. (1989) *Impact Cratering: A Geological Process*, Oxford University Press, Oxford.

Michel, P., Migliorini, F., Morbidelli, A. and Zappalà, V. (2000) The population of Mars-crossers: classification and dynamical evolution. *Icarus*, 145, 332–347.

Morbidelli, A. and Crida, A. (2007) The dynamics of Jupiter and Saturn in the gaseous protoplanetary disk. *Icarus*, 191, 158–171.

Morbidelli, A. and Nesvorny, D. (1999) Numerous weak resonances drive asteroids toward terrestrial planets orbits. *Icarus*, 139, 295–308.

Morbidelli, A. and Vokrouhlický, D. (2003) The Yarkovsky-driven origin of near-Earth asteroids. *Icarus*, 163, 120–134.

Morbidelli, A., Bottke, W.F., Froeschlé, Ch. and Michel, P. (2002a) Origin and evolution of near-Earth objects, in *Asteroid III* (eds W.F. Bottke, A. Cellino, P. Paolicchi and R.P. Binzel), The University of Arizona Press, Tucson, AZ, pp. 409–422.

Morbidelli, A., Jedicke, R., Bottke, W.F. *et al.* (2002b) From magnitudes to diameters: the albedo distribution of near Earth objects and the Earth collision hazard. *Icarus*, 158, 329–342.

Morbidelli, A., Levison, H.F. Tsiganis, K. and Gomes, R. (2005) Chaotic capture of Jupiter's Trojan asteroids in the early Solar System. *Nature*, 435, 462–465.

Morbidelli, A., Tsiganis, K., Crida, A. *et al.* (2007) Dynamics of the giant planets of the Solar System in the gaseous protoplanetary disk and their relationship to the current orbital architecture. *Astronomical Journal*, 134, 1790–1798.

Nesvorny, D., Morbidelli, A., Vokrouhlicky, D. *et al.* (2002) The Flora family: a case of the dynamically dispersed collisional swarm? *Icarus*, 157, 155–172.

Nesvorny, D., Vokrouhlicky, D., Bottke, W.F. *et al.* (2007a) Express delivery of fossil meteorites from the inner asteroid belt to Sweden. *Icarus*, 188, 400–413.

Nesvorný, D., Vokrouhlický, D. and Morbidelli, A. (2007b) Capture of irregular satellites during planetary encounters. *Astronomical Journal*, 133, 1962–1976.

Nesvorny, D., Vokrouhlicky, D., Morbidelli, A. and Bottke, W.F. (2009) Asteroidal source of L chondrite meteorites. *Icarus*, 200, 698–701.

Pierazzo, E., Vickery, A.M. and Melosh, H.J. (1997) A reevaluation of impact melt production. *Icarus*, 127, 408–423.

Pierens, A. and Nelson, R.P. (2008) Constraints on resonant-trapping for two planets embedded in a protoplanetary disc. *Astronomy and Astrophysics*, 482, 333–340.

Rabinowitz, D., Helin, E., Lawrence, K. and Pravdo, S. (2000) A reduced estimate of the number of kilometre-sized near-Earth asteroids. *Nature*, 403, 165–166.

Richardson, J.E. (2009) Cratering saturation and equilibrium: a new model looks at an old problem. *Icarus*, 204, 697–715.

Ryder, G., Koeberl, C. and Mojsis, S.J. (2000) Heavy bombardment on the Earth ~3.85 Ga: the search for petographic and geochemical evidence, in *Origin of the Earth and Moon* (eds R. Canup and K. Righter), The University of Arizona Press, Tucson, AZ, pp. 475–492.

Schmitz, B., Peucker-Ehrenbrink, B., Lindström, M. and Tassinari, M. (1997) Accretion rates of meteorites and cosmic dust in the Early Ordovician. *Science*, 278, 88–90.

Schmitz, B., Häggström, T. and Tassinari, M. (2003) Sediment-dispersed extraterrestrial chromite traces a major asteroid disruption event. *Science*, 300, 961–964.

Schultz, P.H., Harris, R.S., Tancredi, G. and Ishitsuka, J. (2008) Implications of the Carancas meteorite impact. 39th Lunar and Planetary Science Conference (Lunar and Planetary Science XXXIX), League City, TX, March 10–14, LPI Contribution No. 1391, 2409.pdf.

Shoemaker, E.M. (1998) Impact cratering through geological time. *Journal of the Royal Astronomical Society of Canada*, 92, 297.

Shoemaker, E.M., Wolfe, R.F. and Shoemaker, E.M. (1990) Asteroid and comet flux in the neighborhood of the Earth, in *Global Catastrophes in Earth History: An Interdisciplinary Conference on Impacts, Volcanism, and Mass Mortality* (eds V.L. Sharpton and P.D. Ward) GSA Special Paper 247, Geological Society of America, Boulder, CO, pp. 155–170.

Strom, R.G., Malhotra, R., Ito, T. *et al.* (2005) The origin of planetary impactors in the inner Solar System. *Science*, 309, 1847–1850.

Stuart, J.S. (2001) A near-Earth asteroid population estimate from the LINEAR survey. *Science*, 294, 1691–1693.

Stuart, J.S. and Binzel, R.P. (2004) Bias-corrected population, size distribution, and impact hazard for the near-Earth objects. *Icarus*, 170, 295–311.

Tsiganis, K., Gomes, R., Morbidelli, A. and Levison, H.F. (2005) Origin of the orbital architecture of the giant planets of the Solar System. *Nature*, 435, 459–461.

Zahnle, K., Schenk, P., Levison H.F. and Dones, L. (2003) Cratering rates in the outer Solar System. *Icarus*, 163, 262–289.

Zappalà, V., Cellino, A., Gladman, B.J. *et al.* (1998) Asteroid showers on Earth after family breakup events. *Icarus*, 134, 176–179.

Zavodny, M., Jedicke, R., Beshore, E.C. *et al.* (2008) The orbit and size distribution of small Solar System objects orbiting the Sun interior to the Earth's orbit. *Icarus*, 198, 284–293.

# THREE

# The contact and compression stage of impact cratering

## H. Jay Melosh

*Earth, Atmospheric and Planetary Sciences Department, 550 Stadium Mall Drive, Purdue University, West Lafayette, IN 47907, USA*

## 3.1 Introduction

The impact of an object moving at many kilometres per second on the surface of a planet initiates an orderly sequence of events that eventually produces an impact crater. Although this is really a continuous process, it is convenient to break it up into distinct stages that are each dominated by different physical processes. This division clarifies the description of the overall cratering process, but it should not be forgotten that the different stages really grade into one another and that a perfectly clean separation is not possible. The most commonly used division of the impact cratering process is into contact and compression, excavation and modification. This chapter focuses only on the initial contact and compression stage. Chapters 4 and 5 focus respectively on the subsequent excavation and modification stages.

Contact and compression is the briefest of the three stages, lasting only a few times longer than the time required for the impacting object (referred to hereafter as the 'projectile') to traverse its own diameter $L$. If the impact velocity is denoted by $v_i$, then the duration of this phase $\tau_{cc}$ is given by

$$\tau_{cc} = \frac{L}{v_i} \tag{3.1}$$

In the case of an oblique impact, where the projectile approaches the target surface at an angle $\theta$, the contact time is longer due to the sloping path and is given by

$$\tau_{cc} = \frac{L}{v_i \sin\theta} \tag{3.2}$$

Thus, for a 10 km diameter projectile moving toward the surface at a steep angle at $20\,km\,s^{-1}$, comparable to the one that caused the Cretaceous–Palaeogene mass extinction event 65 million years ago, the contact and compression stage lasts only 0.5 s. During this brief, but important, stage, the projectile first contacts the planet's surface and transfers its energy and momentum to the underlying target rocks (Fig. 3.1). The specific kinetic energy (energy per unit mass, $0.5v_i^2$) possessed by a projectile travelling at even a few kilometres per second is surprisingly large. A.C. Gifford (1924, 1930) first realized that the energy per unit mass of a body travelling at $3\,km\,s^{-1}$ is comparable to that of TNT. Gifford proposed the 'impact-explosion analogy', which draws a close parallel between a high-speed impact and an explosion.

As the projectile plunges into the target, shock waves propagate both into the projectile, compressing and slowing it, and into the target, compressing and accelerating it downward and outward (Fig. 3.1a). At the interface between target and projectile, the material of each body moves at the same velocity. The shock wave in the projectile eventually reaches its back (or top) surface (Fig. 3.1b). At this time, the pressure is released and the surface of the compressed projectile expands upward, while a wave of pressure relief propagates back downward toward the projectile–target interface. The contact and compression stage is considered to end when this relief wave reaches the projectile–target interface. At this time, the projectile has been compressed to high pressure and upon decompression it may be in the liquid or gaseous state due to heat deposited in it during the irreversible compression process. The projectile generally carries off 50% or less of the total initial energy, if the density and compressibility of the projectile and target material do not differ too much. The projectile–target interface at the end of contact and compression is generally less than a projectile diameter below the original surface.

In contrast to conventional chemical processes, which occur at either constant pressure or at constant temperature, the thermodynamic path traversed by highly shocked material starts with a sudden, irreversible compression that causes pressure to jump sharply to more than a few hundred gigapascals, accompanied by temperatures exceeding 10,000 K (see Fig. 3.2, which illustrates the thermodynamic path of shock in $SiO_2$, a particularly well-characterized material that is typical of silicate rocks). For

*Impact Cratering: Processes and Products*, First Edition. Edited by Gordon R. Osinski and Elisabetta Pierazzo.
© 2013 Blackwell Publishing Ltd. Published 2013 by Blackwell Publishing Ltd.

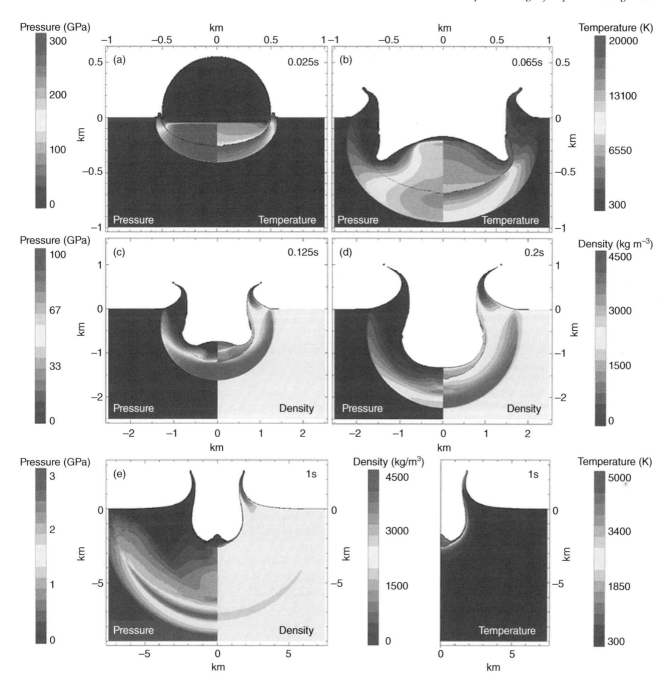

**Figure 3.1**  Contact and compression of the impactor, as it strikes a target surface. Five snapshots from a numerical impact simulation, between 0.025 and 1 s after initial contact, are shown of a 1 km diameter dunite impactor striking a granite target at 18 km s$^{-1}$ (the average impact velocity for asteroidal impactors on Earth). Note the change in the length, pressure and temperature scales between the rows. (a) Impactor just after contact with the target: a small region of high pressure develops along the interface. Both target and impactor are compressed and begin to distort in shape; the rear of the impactor is unaffected by events at the leading edge. (b) The shock wave in the impactor reaches its rear surface and (c) is reflected as a rarefaction (release) wave, unloading the impactor to low pressure. (d) The shock wave in the target propagates outward in an approximately hemispherical geometry, closely followed by the release wave. (e) Behind the detached shock wave, the impact plume (vaporized impactor and proximal target) begins to expand and the excavation flow is initiated. From Collins *et al.* (2012). (See Colour Plate 9)

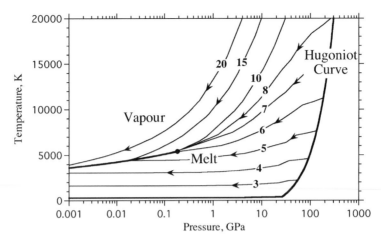

**Figure 3.2**  Thermodynamic paths of the adiabatic release of shocked $SiO_2$ from high pressure, on a log $P$ versus $T$ diagram. The Hugoniot curve indicating the final result of increasingly strong shock compression of quartz is shown as a heavy line, while the thin solid lines are decompression isentropes. The phase curve separating liquid and solid phases is shown as a heavy line and the critical point by a heavy dot. The numbers labelling the release adiabats are the particle velocities in the shocked material in kilometres per second. These velocities can be interpreted as the outcome of an impact experiment between identical materials at twice the particle velocity. Thus, the curve labelled 7 is the release isentrope of a face-on impact between two quartz plates at $14\,km\,s^{-1}$. This isentrope approximately separates states that decompress first to a liquid that boils when it reaches the phase curve from those states so strongly shocked that they decompress as a vapour that then condenses when the isentrope reaches the phase curve. After Melosh (2007).

comparison, the pressure and temperature within the Earth's core is approximately 350 GPa and approximately 5000 K respectively. Under these extreme conditions, atomic bonds are easily broken and chemical rearrangements readily occur. After this brief compression, rarefaction (pressure release) waves from free surfaces surrounding the high-pressure region propagate inward (Fig. 3.1b,c), permitting both the pressure and temperature to drop along a near-adiabatic path back to low pressure (Fig. 3.1c,d), although temperatures may remain high enough to leave the material molten or even vaporized (Fig. 3.1e; Melosh, 1989; Chapter 9). In oblique impacts, the shock wave generated by the impact weakens with decreasing impact angle and becomes asymmetric, with the strongest shock in the downrange direction. As a consequence, while the total volume of highly shocked material decreases with increasing obliquity (and more of the impactor survives), the volume of highly shocked near-surface material is greater in highly oblique impacts (Pierazzo and Melosh, 2000b).

Contact and compression is accompanied by the formation of very high velocity 'jets' of highly shocked material. These jets form where strongly compressed material is close to a free surface; for example, near the circle where a spherical projectile contacts a planar target. The jet velocity depends on the angle between the converging surface of the projectile and target but may exceed the impact velocity by factors as great as five. Jetting was initially regarded as a spectacular but not quantitatively important phenomenon in early impact experiments, where the incandescent streaks of jetted material only amounted to about 10% of the projectile's mass in vertical impacts. However, a study of oblique impacts between planetary-scale bodies (Melosh and Sonnett, 1986) indicates that, in this case, jetting is much more important

and that the entire projectile may participate in a downrange stream of debris that carries much of the original energy and momentum. Oblique impacts are still the least well understood type of impact, and more work needs to be done to clarify the role of jetting early in this process.

The highest pressure attained during contact and compression is almost uniform over a volume roughly comparable to the initial dimensions of the projectile, a volume called the 'isobaric core', reflecting the nearly constant value of the maximum pressure in this region. However, as the shock wave expands away from the impact site the shock pressure declines, as its initial impact energy spreads over an increasingly large volume of rock. The pressure in the shock wave declines as an inverse power of the distance from the impact site, where the power is between two and four, depending on the strength of the shock wave.

The shock wave, with the release wave immediately following, quickly attains the shape of a hemisphere expanding through the target rocks (Fig. 3.1d,e). The high shock pressures are confined to the surface of the hemisphere, as the interior has already decompressed. The shock wave moves very quickly, as fast as or faster than the speed of sound, between about 6 and $10\,km\,s^{-1}$ in most dense rocks. As rocks in the target are overrun by the shock waves, then released to low pressures, mineralogical changes take place in the component minerals. At the highest pressures the rocks may melt or even vaporize upon release (see Chapter 9). As the shock wave weakens, high-pressure polymorphs such as coesite or stishovite arise from quartz in the target rocks, diamonds may be produced from graphite, or maskelynite (diaplectic glass) from plagioclase. Somewhat lower pressures cause pervasive fracturing and 'planar elements' in individual crystals. Still lower pressures create a characteristic cone-in-cone fracture

called 'shatter cones'. These so-called shock metamorphic mineralogical transformations are discussed in Chapter 8.

## 3.2  Maximum pressures during contact and compression

During contact and compression the projectile plunges into the
target, generating strong shock waves as the material of both
objects is compressed. The strength of these shock waves can be
computed from the Hugoniot equations, first derived by P.H.
Hugoniot in a posthumous 1887 memoir, relating quantities in
front of the shock (subscript 0) to quantities behind the shock
(no subscripts):

$$\rho(U - u_p) = \rho_0 U$$
$$P - P_0 = \rho_0 u_p U$$
$$E - E_0 = \frac{1}{2}(P + P_0)\left(\frac{1}{\rho_0} - \frac{1}{\rho}\right)$$

(3.3)

where $P$ is pressure, $\rho$ is density, $u_p$ is particle velocity behind the
shock (the unshocked material is assumed to be at rest), $U$ is
the shock velocity and $E$ is the internal energy per unit mass.
These three equations are respectively equivalent to the conserva-
tions of mass, momentum and energy across the shock front. The
Hugoniot equations hold for all materials, but do not themselves
provide enough information to specify the outcome of an impact.
The Hugoniot equations must be supplemented by a fourth equation, the equation of state, that relates the pressure to the density
and internal energy in each material, $P=P(\rho,E)$. Alternatively, a
relation between shock velocity and particle velocity may be specified, $U=U(u_p)$. As this relation is frequently linear, it often provides the most convenient equation of state in impact processes,
although it is not a full equation of state and does not specify
thermodynamic quantities such as temperature or entropy, so its
use is limited to estimation of shock pressure. Thus:

$$U = c + Su_p$$

(3.4)

where $c$ and $S$ are empirical constants. Table 3.1 lists the measured
values of $c$ and $S$, as well as uncompressed density $\rho_0$ for a variety
of materials.

### 3.2.1  The planar impact approximation

The Hugoniot equations, along with an equation of state, can be
used to compute the maximum pressure, particle velocity, shock
velocity, and so on in an impact (Melosh, 1989). A rough estimate
of these quantities is obtained from the planar impact approximation (sometimes called the impedance-matching solution), which
is valid so long as the lateral dimensions of the projectile are
small compared with the distance the shock has propagated. This
approximation is, thus, valid through most of the contact and
compression stage.

The basic premise of the planar impact approximation is that
two infinitely wide plates of material representative of both the
projectile and target impact one another face-on at a speed equal

**Table 3.1**  Linear shock-particle velocity equation of state parameters

| Material | $\rho_0$ (kg m$^{-3}$) | $c$ (km s$^{-1}$) | $S$ |
|---|---|---|---|
| Aluminium | 2750 | 5.30 | 1.37 |
| Basalt | 2860 | 2.6 | 1.62 |
| Calcite (carbonate) | 2670 | 3.80 | 1.42 |
| Coconino sandstone | 2000 | 1.5 | 1.43 |
| Diabase | 3000 | 4.48 | 1.19 |
| Dry sand | 1600 | 1.7 | 1.31 |
| Granite | 2630 | 3.68 | 1.24 |
| Iron | 7680 | 3.80 | 1.58 |
| Permafrost (water saturated) | 1960 | 2.51 | 1.29 |
| Serpentinite | 2800 | 2.73 | 1.76 |
| Water (25°C) | 9979 | 2.393 | 1.333 |
| Water ice (−15°C) | 915 | 1.317 | 1.526 |

Data are from Melosh (1989: Table AII.2).

to the impact velocity. The assumption of infinitely wide plates
avoids the effects of the lateral edges of either the projectile
or target and permits a one-dimensional approximation to the
impact process, at the cost of ignoring the actual shape of the
projectile or topography in the target. Oblique impacts can also
be approximately treated by replacing the face-on impact velocity
$v_i$ by just the vertical component of the oblique impact velocity
$v_i\sin\theta$, where $\theta$ is the angle between the impact direction and the
plane of the target. Although these simplifications might seem
unrealistic, the planar impact approximation does yield an upper
limit to the pressures expected in an impact event, without the
necessity of performing detailed numerical computations of the
impact. Unfortunately, there is no simple formula even for this
approximation.

The essence of the planar impact approximation is that, after
contact, the pressure in both the target plate and projectile plate
are equal. Furthermore, because the projectile material cannot
interpenetrate the target, the particle velocity of the accelerated
target $u_t$ is identical to that of the decelerated projectile, equal
to $v_i - u_t$ in the projectile's frame of reference (assuming that
non-relativistic velocity summation rules work under these conditions). The Hugoniot equations (eqn 3.3), along with the equations of state of the projectile and target material, are then applied
in the target and projectile frames of reference respectively. As
these individual equations in the target and projectile are linked
by the conditions of equal pressure and no interpenetration at the
interface between the two materials, the set of equations possesses
a unique solution that can be obtained numerically.

A useful algebraic solution to the planar impact approximation
can be obtained, when the materials of the target and projectile
both possess linear shock velocity–particle velocity relations (eqn
3.4). In this case, the expression for the particle velocity in the
target $u_t$ is given by the standard solution to a quadratic equation
(the particle velocity in the projectile is $v_i - u_t$):

$$u_t = \frac{-B + \sqrt{B^2 - 4AC}}{2A}$$

(3.5)

**Table 3.2** Maximum shock pressures in vertical impacts

| Impact velocity (km s$^{-1}$) | Pressure (GPa) | | |
|---|---|---|---|
| | Iron on basalt | Basalt on basalt | Serpentinite (carbonaceous chondrite) on ice |
| 5 | 78 | 48 | 22 |
| 7.5 | 150 | 93 | 44 |
| 10 | 250 | 150 | 73 |
| 15 | 500 | 320 | 160 |
| 30 | 1800 | 1200 | 580 |
| 45 | 3900 | 2500 | 1300 |

where

$$A = \rho_{0t}S_t - \rho_{0p}S_p$$
$$B = \rho_{0t}c_t + \rho_{0p}c_p + 2\rho_{0p}S_p v_i \qquad (3.6)$$
$$C = -\rho_{0p}v_i(c_p + S_p v_i)$$

The subscripts p and t refer to the projectile and target respectively. The above equation can be used in conjunction with the Hugoniot equations and equation of state to obtain any other quantities of interest. Thus, the pressure behind the shock is given by

$$P = \rho_{0t}u_t(c_t + S_t u_t) \qquad (3.7)$$

The pressures in both the target and projectile are the same by construction of the solution. Table 3.2 shows a few typical numbers for maximum shock pressures for typical impact velocities and Solar System materials.

It should be noted that, in the special case that the projectile and target are composed of the same material, the denominator $A$ in eqn 3.5 is zero and the equation apparently has no solution. In reality, the numerator also vanishes in this case, but the ratio is finite and equal to $-C/B$, so that in this important case $u_t = v_i/2$ and the equations are particularly simple. If you plan to implement these equations in a computer program, you must provide a special exception for this case.

### 3.2.2    Energy partition during compression

The initial kinetic energy of the projectile is divided or partitioned among several different reservoirs during contact and compression. This energy is shared between the projectile and target material and between kinetic energy and internal energy in each material. The internal energy may initiate phase changes among these materials, while the kinetic energy may result in rapid ejection of material from the impact site. A detailed accounting of energy partition requires a full numerical computation of the course of an impact. However, the planar impact approximation makes definite predictions of the initial distribution of the projectile's initial kinetic energy that may serve as an approximate guide to the overall flow of energy in an impact. When shock

pressures are high enough to completely vaporize the projectile, a surprisingly large fraction of the energy goes into the early high-speed ejecta that form from the vaporized projectile and a comparable mass of target. Up to half of the total energy of the impact may end up as such high-speed vapour condensates that are eventually widely distributed over the target planet of an impact.

The Hugoniot equations (eqn 3.3) can be solved to show that, when the shocked material is initially at rest and the initial pressure $P_0$ can be neglected, the internal energy increase of shocked material equals its kinetic energy and both are equal to one-half of the particle velocity squared (per unit mass), $E - E_0 = \frac{1}{2}u_p^2$. This relation is true in the target (but not in the projectile, which has a large initial velocity), so the energy partition in the target puts equal amounts of energy into internal and kinetic energies. The total energy in the target, however, also depends on the mass of the material that has been shocked at the time that the projectile is fully engulfed by the shock wave. This occurs at a time $t_c = L/U_t$ after first contact (remember that $L$ is the projectile diameter), at which time the shock wave has reached a depth of $U_t t_c$ in the target. The projectile–target interface has reached a depth of $u_t t_c$ and the rear of the projectile is at a depth of $L - v_i t_c$, which may be either positive (above the original surface) or negative (below it). In the planar impact approximation we examine energy partition per unit area of the (assumed) infinite colliding plates. The kinetic energy of the projectile plate, per unit area, is $\frac{1}{2}r_{0p}Lv_i^2$, while the target's kinetic and internal energies are equal to $\frac{1}{2}r_{0t}U_t t_c u_t^2$. The Hugoniot equations can then be used to derive the *fraction* of the initial kinetic energy partitioned into internal and kinetic energy at the moment that the shock wave reaches the rear of the projectile. This energy will eventually go into opening the crater and heating the target:

$$f_t(KE) = f_t(IE) = \frac{u_p/u_t}{(1 + u_p/u_t)^2} \qquad (3.8)$$

Because the projectile was initially moving downward with respect to the target (at the impact velocity), its kinetic and internal energies are not equal. The fractions that are partitioned into either kinetic or internal energy at the moment that the shock wave just reaches its rear are

$$f_p(KE) = \frac{1}{(1 + u_p/u_t)^2}$$
$$f_p(IE) = \frac{(u_p/u_t)^2}{(1 + u_p/u_t)^2} \qquad (3.9)$$

When both the projectile and target are composed of the same material, $u_t = u_p$ and all four fractions equal 0.25. Half of the total energy is partitioned into the target, while half remains in the projectile. Even when the projectile is iron and the target is basalt, energy is partitioned nearly equally into the projectile and target and all four fractions are nearly the same.

### 3.2.3    Unloading of the projectile

The shock wave moves upward into the projectile, eventually reaching its back (or top) surface (see Fig. 3.1b). At this time, the pressure is released, as the surface of the compressed projectile expands upward, and a pressure relief wave propagates back downward toward the projectile–target interface (see Fig. 3.1c). The contact and compression stage is considered to end when this relief wave reaches the projectile–target interface. During this stage the projectile is compressed to high pressure, often reaching hundreds of gigapascals, and upon decompression it may undergo phase changes caused by heat deposited during this irreversible thermodynamic cycle. Unless the projectile is much denser than the target, it penetrates less than a projectile diameter $L$ into the target during this stage.

The pressure relief that occurs after the shock wave reaches the rear of the projectile is accomplished by a fast-moving adiabatic sound wave in the compressed material that communicates the zero-pressure boundary condition of the free surface to the body of the projectile beneath the surface. Owing to the high pressure in the compressed material, this wave typically moves much faster than the normal sound speed in uncompressed projectile material. Computation of this rarefaction speed typically requires a full equation of state of the material. However, a first estimate of this speed can be made from a Murnaghan-type of equation of state, one that is often used to estimate the sound speed in material deep within the Earth. The Murnaghan equation does not depend on the internal energy of the material and so usually underestimates the actual rarefaction wave speed. On the basis of this equation, the rarefaction speed $c_R$ is given by

$$c_R = \sqrt{(K_0 + nP)/\rho} \qquad (3.10)$$

where the projectile bulk modulus $K_0 = \rho_{0p} c_p^2$, the dimensionless constant $n = 4S_p - 1$ and $\rho$ is the compressed density of the projectile. Note that $c_R$ is the rarefaction wave speed in the rest frame of the compressed material. As this material is moving downward with speed $u_t$, the rarefaction moves downward at the sum of the two wave speeds, $c_R + u_t$ in the target's rest frame. The sum of these two speeds usually exceeds the velocity of the downward-moving shock itself. The compressed projectile, thus, unloads very quickly and the shock wave moving into the target is rapidly weakened by the rarefaction wave catching up with it. The rarefaction wave, travelling at speed $c_R$ with respect to the projectile, must traverse the compressed thickness of the projectile $(\rho_{0p}/\rho)L$ and so the duration $t_R$ of the unloading phase is

$$t_R = \frac{(\rho_{0p}/\rho)L}{c_R} \qquad (3.11)$$

The unloaded projectile material accelerates upward during its release, driven by the strong pressure gradient between the compressed material at depth and the free surface at the rear of the projectile. If the maximum shock pressure is low and the unloaded material is still solid, the 'velocity-doubling rule' applies and the particle velocity of the projectile with respect to the target is $u_t - u_p$. This is nearly zero for a projectile and target of similar composition, but can be either upward or downward depending

on the equations of state of the projectile and target. When the shock pressures are high enough to vaporize the projectile, however, the vaporized projectile plus target material expands upward at a mean velocity nearly equal to the thermal velocity of the vapour (material of this type is visible in Fig. 3.1e). Such high-temperature gases may, thus, leave the impact site at a substantial fraction of the impact velocity itself and either travel large distances away from the impact site or even leave the target planet, if their speed exceeds the planet's escape velocity. Condensates from the vapour produced in high-speed impacts form one of the most widespread of impact sedimentary deposits.

## 3.3 Jetting during contact and compression

Contact and compression is accompanied by the formation of very high velocity 'jets' of highly shocked material. These jets form where strongly compressed material is close to a free surface; for example, near the circle where a spherical projectile contacts a planar target. Figure 3.1a show that two regions of very high pressure develop near the sides of the projectile, as it begins to penetrate the target. These high-pressure zones are immediately adjacent to the free surface, where the pressure is necessarily zero. This creates very large pressure gradients in the immediate vicinity of the intersection between the projectile and target. Elementary fluid mechanics tells us that, unless restrained by other forces, material subjected to a pressure gradient must accelerate by an amount

$$a = \frac{1}{\rho}\frac{dP}{dx} \qquad (3.12)$$

This acceleration can be extreme, when the path length $dx$ is short and, thus, the gradient very high. This acceleration is increased still further, while the projectile is penetrating the target, at a stage before that shown in Fig. 3.1a. At this stage, the curved surface of the spherical projectile intersects the flat surface of the target along an oblique line that moves very rapidly across the surface of the target. Two shock waves that converge obliquely add in a nonlinear fashion to increase the shock pressure locally, well beyond that expected in a face-on impact (Birkhoff *et al.*, 1948; Walsh *et al.*, 1953; Harlow and Pracht, 1966). This process is even more common during oblique impacts, when the oblique convergence angle tends to offset the smaller shock pressures expected from considering only the vertical component of the impact velocity.

The jet velocity depends on the angle between the converging surfaces of the projectile and target, but may exceed the impact velocity by factors as great as five (Kieffer, 1977). Jetting was initially regarded as a spectacular but not quantitatively important phenomenon in early impact experiments, where the incandescent streaks of jetted material only amounted to about 10% of the projectile's mass in vertical impacts. However, the study of Melosh and Sonnett (1986) indicates that in this case jetting is much more important and that the entire projectile may participate in a downrange stream of debris that carries much of the original energy and momentum.

## 3.4   The isobaric core

Numerical computations of impacts that keep track of where maximum pressures are reached during the course of the impact typically show that the maximum shock pressure occurs in a roughly spherical region around the impact point (Croft, 1982; Pierazzo *et al.*, 1997). Outside of this central 'core' region, maximum pressures drop off rapidly with distance from the impact site, as shown in Fig. 3.3. As the pressure is nearly constant in this region, it often goes by the name 'isobaric core'. Models of this kind, thus, argue that the shock pressure around the site of a vertical impact is a function of $d$, the radial distance away from the centre of some point centred beneath the impact:

$$P(d) = P_{max} \left( \frac{r_{ic}}{d} \right)^n \qquad (3.13)$$

where $r_{ic}$ is the radius of the isobaric core and $P_{max}$ is the maximum pressure, which can be determined from the planar impact approximation.

The isobaric core is, thus, the expression of the region most highly shocked during the contact and compression stage. Estimates of the shock pressure and the size and depth of the isobaric core region permit us to compute the volume and provenance of the most highly shocked material during an impact event. Pierazzo *et al.* (1997) numerically studied the size and depth of a

sphere fit to the isobaric core region for impact velocities ranging from 10 to 100 km s$^{-1}$ and for materials including dunite, granite, aluminium and iron in various combinations of projectile and target (Fig. 3.4). All of these data were well fit by a two-parameter equation for the isobaric core radius $r_{ic}$ divided by the projectile radius $R$:

$$\log(r_{ic} / R) = a + b \log v_i$$
$$a = -0.346 \pm 0.034 \qquad (3.14)$$
$$b = 0.211 \pm 0.022$$

where the impact velocity $v_i$ is in kilometres per second. The velocity dependence of the isobaric core radius is rather weak, ranging from about $0.73R$ at 10 km s$^{-1}$ to $1.2R$ at 100 km s$^{-1}$. The depth of the centre of the sphere representing the isobaric core $d_{ic}$ is given by a similar equation:

$$\log(d_{ic} / R) = a + b \log v_i$$
$$a = -0.516 \pm 0.060 \qquad (3.15)$$
$$b = 0.361 \pm 0.038$$

The velocity dependence of the depth is slightly stronger, ranging from about $0.70R$ at 10 km s$^{-1}$ to $1.6R$ at 100 km s$^{-1}$. An isobaric core-like region develops even during highly oblique impacts, although in this case it is elongated along the direction of the projectile's trajectory (see dashed lines in Fig. 3.5).

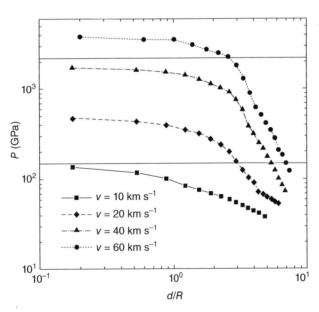

**Figure 3.3**  Shock pressure decays away from the impact point, along a vertical line, for various impact velocities for the impact of a dunite projectile of radius $R$ on a dunite target. Shock pressures corresponding to complete melting and vaporization for dunite are shown as horizontal solid lines. From Pierazzo *et al.* (1997).

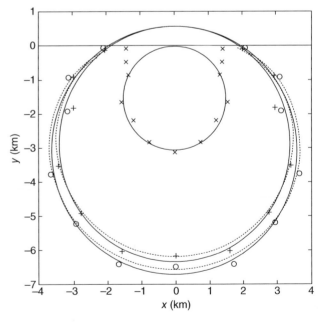

**Figure 3.4**  Melt regions and the isobaric core are shown for a vertical impact of a dunite sphere on a plane dunite target, at a velocity of 20 km s$^{-1}$. Continuous lines are best fits of the data with a circle, dotted lines fit a limacon of Pascal. O: incipient melting; +: complete melting; X: isobaric core. From Pierazzo *et al.* (1997).

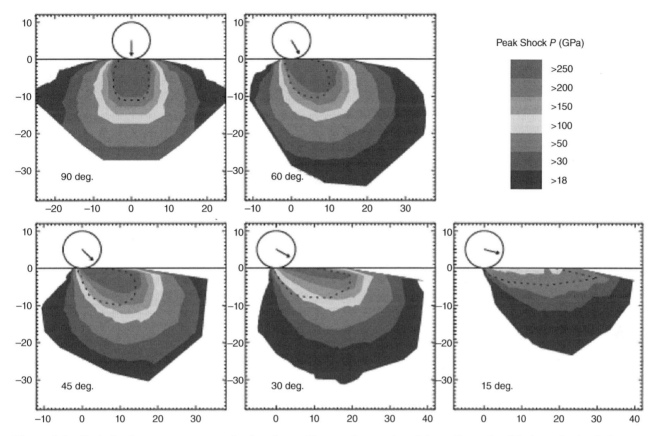

**Figure 3.5**  Peak shock pressure contours in the plane of impact for a series of three-dimensional hydrocode simulations, at various impact angles. Dashed black line represents the isobaric core. The projectile, 10 km in diameter, is shown for scale. Vectors illustrate the direction of impact. From Pierazzo and Melosh (2000b). (See Colour Plate 10)

## 3.5  Oblique impact

Essentially all impacts are oblique to some degree. The probability of an impact occurring at an angle $\theta$ to the surface of a planet within a range $d\theta$ is $2\sin\theta\cos\theta\,d\theta$ per unit area, independent of the gravitational acceleration of the planet (Pierazzo and Melosh, 2000c). The maximum of this probability is at 45° and it is essentially zero for either vertical impacts (90°) or very grazing angles (0°).

Although hypervelocity impacts at nearly all angles produce circular final craters (elliptical craters develop only at grazing angles below about 10° from the surface), this is a consequence of the large expansion of the crater to 10 or 20 times the initial projectile diameter. The effect of oblique convergence is more profound close to the impact site and it is especially important during contact and compression. High-speed phenomena such as jetting, ricochet of the projectile, downrange streaks of highly shocked ejecta and fast ejecta products, such as tektites, are all strongly affected by the obliquity of the impact. The initial coupling of energy into the target is less efficient for oblique impacts, resulting in smaller craters for a given projectile mass and velocity, as well as lower overall shock pressures. In a detailed study of oblique impact, Pierazzo and Melosh (2000b) found that the maximum shock pressure in the isobaric core depends upon

impact angle $\theta$ approximately as $\sin\theta$ down to angles of about 15°, as shown in Fig. 3.5 and Fig. 3.6a. The maximum shock temperature, shown in Fig. 3.6b, has a more complicated dependence on impact angle. The projectile itself sees lower shock pressures for highly oblique impacts and some projectile material located near the top and rear of the projectile may survive nearly intact for low-angle impacts (Pierazzo and Melosh, 2000a), as shown in Fig. 3.7.

The concept of the projectile 'footprint' on the target surface is a useful guide to the geometry of the highly shocked region during contact and compression. The footprint of a spherical projectile is an ellipse of width $L$ equal to the projectile diameter and is elongated along the flight direction by a distance $L/\sin\theta$. This is the area where the projectile directly impacts the target surface and it can reach very high pressures due to this direct impingement of the projectile. Outside of this ellipse, the surface is free and the surface pressures are strictly zero, although strong pressure gradients may accelerate material adjacent to the footprint to high speeds. High pressures are reached in a canoe-shaped region underlying the footprint, shown as a dashed line in Fig. 3.5 for various impact angles. The material in this region also acquires a high downrange velocity component from the tangential velocity component of the initial projectile. At very low angles, the preservation of this downrange component may

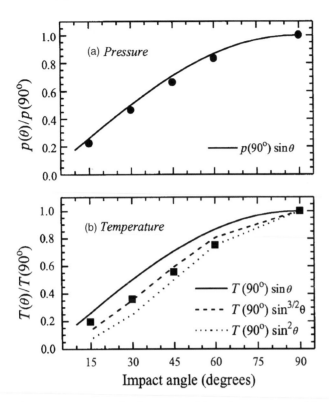

**Figure 3.6**  Mean shock pressure (a) and temperature (b) inside the isobaric core for a three-dimensional hydrocode simulation of a dunite projectile striking a target simulating the Chicxulub crater site at $20 \, \text{km s}^{-1}$. After Pierazzo and Melosh (2000b).

suggest that the projectile ricochets out of the crater, as it acquires a small upward velocity component after it decompresses (Pierazzo and Melosh, 2000a).

## 3.6    The end of contact and compression

After the sudden compression of both the target and projectile, rarefaction waves from the back and sides of the projectile propagate inward and rapidly decompress the shocked material, initiating the rapid and extreme thermodynamic cycle shown in Fig. 3.2. The shock wave moving downward into the target initiates the excavation flow described in Chapter 4, while the shocked and unloaded projectile, which may have undergone radical changes of state, either rides the inside of the opening crater cavity or, if it has been shocked strongly enough to vaporize, expands rapidly out of the nascent crater cavity. Highly shocked projectile material initially expands as a hot supercritical fluid of density comparable to the unshocked material. However, as the pressure and temperature fall, highly shocked projectile and target material may rapidly increase its volume to become a vapour in which radical chemical changes take place. The vapour itself eventually condenses to form liquid droplets that may spread far and wide away from the crater, as it rains out onto the surface of the target planet (Johnson and Melosh, 2012). Although not strictly part of the contact and compression phase, these changes are so strongly coupled to these high-pressure events that they deserve a brief description here.

The physical and chemical changes induced by impacts are most profound close to the point of impact, where the initial shock pressure is highest. In a typical large impact on Earth, the impactor and a roughly equal volume of target rocks are melted and vaporized. The result is a turbulent mixture of vaporized rock, melt droplets, small solid fragments and, if an atmosphere is present, hot atmospheric gases, known variously as the impact plume, the vapour plume or (most dramatically) the 'fireball'. The impact plume rapidly expands from the impact site, carrying much of the impactor material with it, until it equilibrates with the surrounding atmosphere or breaches it completely and extends into space. The evolution and deposition of impact plume material, as well as the concomitant chemical processing of plume constituents, are among the least well understood aspects of the impact process.

Complex molecular solids, of which silicate rocks are a prime example, vaporize in a complex manner. As shown in Fig. 3.8, even the simplest silicate, $SiO_2$, decomposes into molecular fragments as temperature rises. Up to a few thousand degrees above the vaporization temperature (about $3100 \, \text{K}$ at a pressure of $0.1 \, \text{MPa}$), the principal molecules in silica vapour are $SiO$ and $O_2$. At still higher temperatures these diatomic molecules break down further, eventually yielding monatomic Si and O. At still higher temperatures, free electrons appear and the atoms are ionized. The same transitions occur at higher pressures, but they are pushed to higher temperature until, at a sufficiently high pressure and temperature, the critical point is passed, no liquid–vapor transition exists and the very concept of discrete molecular clusters loses its meaning.

Figure 3.2 illustrates the thermodynamic path experienced by shocked silica, $SiO_2$, which is qualitatively similar to the shock behaviour of many other silicate materials. The black curve to the right of this pressure–temperature ($P$–$T$) plot is the Hugoniot curve, the locus of final shock states starting from an initial state of $0.1 \, \text{MPa}$ and $23 \, ^\circ\text{C}$. The thin black lines extending to the left of the Hugoniot curve are release adiabats, which show the history of temperature and pressure of decompressing material. The numbers on these adiabats indicate the particle velocity in kilometres per second behind the shock wave that initiated them. This velocity is approximately one-half of the impact velocity that initiated the decompression.

During its initial release from high pressure, the shocked material is neither a liquid melt nor a gas: the pressure is higher than the critical pressure (approximately $0.19 \, \text{GPa}$ for $SiO_2$; the critical point is represented by a black dot in the figure) and so it behaves as a single-phase hot fluid. The pressure drops below the critical point and the release path intersects the liquid–vapour phase curve (heavier black line). Depending upon the internal energy of the fluid, there are two possible outcomes: either the fluid reaches the phase curve from the liquid side, in which case vapour bubbles appear in the liquid (boiling), or it reaches the phase curve from the vapour side, in which case droplets of liquid melt condense from the gas. In either case, a two-phase mixture suddenly appears. It is then possible for elements to exchange between phases and so initiate chemical changes. Furthermore, the rapid cooling of the mixture of liquid and vapour readily quenches

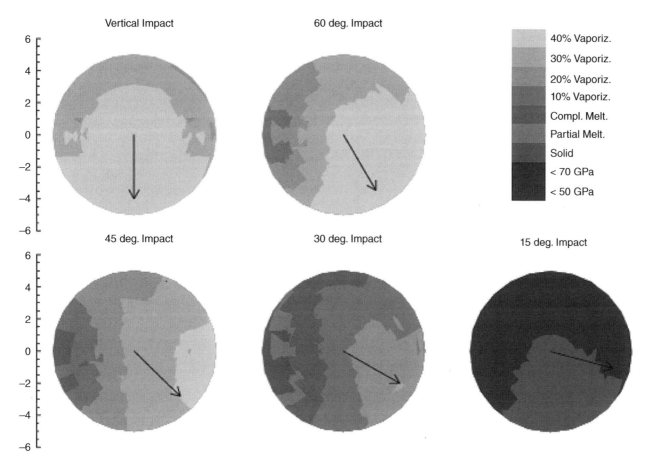

**Figure 3.7** Distribution of melting and vaporization inside a dunite projectile impacting a dunite target at 20 km s$^{-1}$, from three-dimensional hydrocode simulations. The maximum shock pressures corresponding to the various contours are: solid, 135 GPa; partial melting, 149 GPa; complete melting, 186 GPa; 10% vaporization, 224.5 GPa; 20% vaporization, 286.5 GPa; 30% vaporization, 363 GPa; 40% vaporization, 486 GPa; 50% vaporization, 437.2 GPa. From Pierazzo and Melosh (2000a). (See Colour Plate 11)

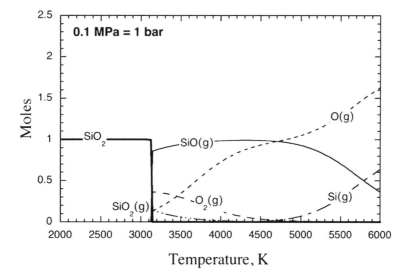

**Figure 3.8** Species present in the vapour phase of SiO$_2$, as a function of temperature, for a pressure of 0.1 MPa (1 bar). The vapour phase is predominantly SiO plus O$_2$, just above the vaporization temperature, with an admixture of about 20 mol% SiO$_2$. At higher temperatures, these molecular clusters break down into atoms, so that the high-temperature limit is a mixture of monatomic Si and O gases. Adapted from Melosh (2007).

high-temperature chemical equilibria and so preserves the high-temperature assemblage of elements.

The most characteristic chemical change in impacts is loss of oxygen or reduction of the melt phase, with a complementary oxidation of the vapour phase (Sheffer, 2007). On the Earth's surface, most rocks contain substantial amounts of oxidized iron, $Fe^{3+}$. However, the iron in the melted or condensed products of impact (e.g. tektites) is dominated by reduced iron, $Fe^{2+}$. On the Moon, where the pre-impact rocks contain mostly $Fe^{2+}$, impact-melted agglutinates contain large amounts of nanophase iron oxide, FeO. The chemical reactions involved are

$$Fe_2O_3 \,(melt) \rightarrow 2FeO\,(melt) + \tfrac{1}{2}O_2 \uparrow$$
$$FeO\,(melt) \rightarrow Fe\,(liquid) + \tfrac{1}{2}O_2 \uparrow$$

where the $Fe_2O_3$ or FeO are taken to be components of a silicate solution. Although the energy of the Fe–O bond might suggest that the above reactions should proceed to the left, the high entropy of the hot vapour phase drives them to the right.

Besides their enrichment in reduced iron (which, incidentally, explains the black or green colours typical of tektite glass), tektites are also strongly depleted in water, containing only about 10–100 ppm by weight of water (O'Keefe, 1964). This is in strong contrast to volcanic glasses, which in some cases may contain up to several weight per cent water. How tektites could lose their initial water over the short interval in which they are heated has been posed as a major mystery, but when it is realized how strongly they are heated, the rapid loss of water (as well as volatile elements such as sodium and potassium) no longer appears so unlikely (Melosh and Artemieva, 2004).

These are only a few of the problems that remain to be understood about the aftermath of strong shock compression of solids during the contact and compression stage of impact cratering. Ongoing and future research will hopefully clarify many of these issues in the years to come and suggest new avenues of investigation.

## References

Birkhoff, G., MacDougall, D.P., Pugh, E.M. and Taylor, G. (1948) Explosives with lined cavities. *Journal of Applied Physics*, 19, 563–582.

Collins, G.S., Melosh, H.J. and Osinski, G.R. (2012) The impact-cratering process. *Elements*, 7, 25–30.

Croft, S.K. (1982) A first-order estimate of shock heating and vaporization in oceanic impacts, in *Geological Implications of Impacts of Large Asteroids and Comets on the Earth* (eds L.T. Silver and P.H. Schultz), Geological Society of America Special Paper 190, The Geological Society of America, Boulder, CO, pp. 143–152.

Gifford, A.C. (1924) The mountians of the Moon. *New Zealand Journal of Science and Technology*, 7, 129–142.

Gifford, A.C. (1930) The origin of the surface features of the Moon. *New Zealand Journal of Science and Technology*, 11, 319–327.

Harlow, F.H. and Pracht, W.E. (1966) Formation and penetration of high-speed collapse jets. *Physics of Fluids*, 9, 1951–1956.

Johnson, B.C. and Melosh, H.J. (2012) Formation of spherules in impact produced vapor plumes. *Icarus*, 217, 416–430.

Kieffer, S.W. (1977) Impact conditions required for formation of melt by jetting in silicates, in *Impact and Explosion Cratering* (eds D.J. Roddy, R.O. Pepin and R.B. Merril), Pergamon Press, New York, NY, pp. 751–769.

Melosh, H.J. (1989) *Impact Cratering*, Oxford University Press, New York, NY.

Melosh, H.J. (2007) A hydrocode equation of state for $SiO_2$. *Meteoritics and Planetary Science*, 42, 2079–2098.

Melosh, H.J. and Artemieva, N. (2004) How does tektite glass lose its water? 35th Lunar and Planetary Science Conference, March 15–19, League City, TX, abstract no. 1723.

Melosh, H.J. and Sonnett, C.P. (1986) When worlds collide: jetted vapor plumes and the moon's origin, in *Origin of the Moon* (eds W.K. Hartmann, R.J. Phillips and G.J. Taylor), Lunar and Planetary Institute, Houston, TX, pp. 621–642.

O'Keefe, J.A. (1964) Water in tektite glass. *Journal of Geophysical Research*, 69, 3701–3707.

Pierazzo, E. and Melosh, H.J. (2000a) Hydrocode modeling of oblique impacts: the fate of the projectile. *Meteoritics and Planetary Science*, 35, 117–130.

Pierazzo, E. and Melosh, H.J. (2000b) Melt production in oblique impacts. *Icarus*, 145, 252–261.

Pierazzo, E. and Melosh, H.J. (2000c) Understanding oblique impacts from experiments, observations and modeling. *Annual Review of Earth and Planetary Sciences*, 28, 141–167.

Pierazzo, E., Vickery, A.M. and Melosh, H.J. (1997) A reevaluation of impact melt production. *Icarus*, 127, 408–423.

Sheffer, A.A. (2007) Chemical reduction of silicates by meteorite impacts and lightening strikes. PhD Thesis, University of Arizona.

Walsh, J.M., Shreffler, R.G. and Willig, F.J. (1953) Limiting conditions for jet formation in high velocity collisions. *Journal of Applied Physics*, 24, 349–359.

# FOUR

# Excavation and impact ejecta emplacement

## Gordon R. Osinski*, Richard A. F. Grieve*,† and Livio L. Tornabene*

*Departments of Earth Sciences/Physics and Astronomy, Western University, 1151 Richmond Street, London, ON, N6A 5B7, Canada
†Earth Sciences Sector, Natural Resources Canada, Ottawa, ON, K1A 0E4, Canada

## 4.1 Introduction

The excavation stage of crater formation encompasses the opening up and enlargement of an initial bowl-shaped cavity following the initial contact and compression stage (see Chapter 3). This initial bowl-shaped cavity, the so-called 'transient cavity', is modified to varying degrees during the subsequent modification stage (see Chapter 5). It is during the excavation stage that one of the most characteristic, but poorly understood, features of meteorite impact craters is formed: namely, ejecta deposits. Impact ejecta deposits can be defined as any target materials, regardless of their physical state, that are transported beyond the rim of the transient cavity formed directly by the cratering flow-field (Fig. 4.1 and Fig. 4.2). In simple craters, the final crater rim approximates the transient cavity rim (Fig. 4.1a). In complex craters, however, the transient cavity rim is typically destroyed during the modification stage, such that ejecta deposits occur in the crater rim region interior to the final crater rim (Fig. 4.1b). Proximal impact ejecta deposits are found in the immediate vicinity of an impact crater (less than five crater radii from the point of impact), whereas distal ejecta deposits are found distant from the crater (greater than five crater radii) and may be dispersed globally depending on the magnitude of the impact event (e.g. the global ejecta layer at the K–Pg boundary, which is associated with the Chicxulub impact in Mexico at approximately 65 Ma). The goal of this chapter is to provide an overview of the excavation stage of crater formation and to summarize and discuss observations of impact ejecta deposits from the terrestrial planets.

## 4.2 Excavation

The transition from the initial contact and compression stage (see Chapter 3) into the excavation stage is a continuum. It is during this stage that the transient cavity is opened up by complex interactions between the expanding shock wave and the original

ground surface (Melosh, 1989). The projectile itself plays little role in the excavation of the crater. It is typically unloaded, melted and vaporized during the initial contact and compression stage so that no physical evidence of the projectile remains. Exceptions do occur, however, such that unmolten projectile material can be found, particularly in small impacts producing simple craters (Shoemaker, 1963; Folco et al., 2011) and, more rarely, in larger impacts (Hart et al., 2002).

During the excavation stage, a roughly hemispherical shock wave propagates out into the target sequence (Fig. 4.3). The centre of this hemisphere will be at some depth in the target sequence (essentially the depth of penetration of the projectile). As this shock wave expands it decays in strength, degrading to a plastic wave and finally an elastic wave. The shock wave is detached, corresponding to a 'shock front', and maintains an approximately constant thickness as it weakens (Melosh, 1989). The passage of the shock wave causes the target material to be set in motion, with an outward and downward radial trajectory. At the same time, shock waves that initially travelled upwards intersect the ground surface and generate rarefaction waves that propagate back downwards into the target sequence. In the near-surface region, an 'interference zone' is formed in which the maximum-recorded pressure is reduced due to interference between the rarefaction and shock waves (Fig. 4.3). Material derived from this region is different to the main body of ejecta (see below) for four main reasons (Melosh, 1989): (1) it is lightly shocked compared with deeper lying material at the same radial distance; (2) it is ejected at high velocities; (3) it is ejected early in the cratering process; and (4) it forms 'spall plates' that are the largest and least shocked fragments thrown out at any given velocity.

The material motions associated with the rarefaction wave are different (except immediately under the point of impact) from those induced by the shock wave. As a result, material motions induced by the combination of the shock and the rarefaction waves produce an 'excavation flow' or 'cratering flow-field' and generate the transient cavity (Fig. 4.2; Dence, 1968; Grieve and

*Impact Cratering: Processes and Products*, First Edition. Edited by Gordon R. Osinski and Elisabetta Pierazzo.
© 2013 Blackwell Publishing Ltd. Published 2013 by Blackwell Publishing Ltd.

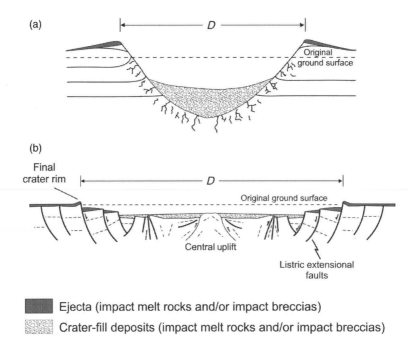

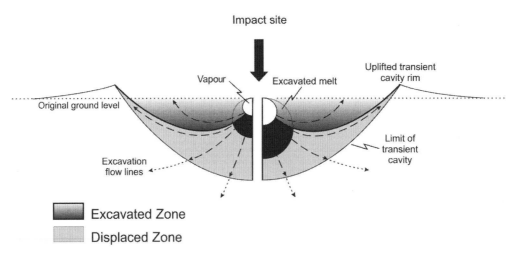

**Figure 4.1**    Typical schematic cross-sections of a simple (a) and complex crater (b). Note that ballistic ejecta is present inside the crater rim in complex craters because it originates from the transient cavity, which is largely destroyed during crater collapse. *D* is the rim (or final crater) diameter, which is defined as the diameter of the topographic rim that rises above the surface for simple craters, or above the outermost slump block not concealed by ejecta for complex craters (Turtle *et al.*, 2005).

**Figure 4.2**    Theoretical cross-section through a transient cavity showing the locations of impact metamorphosed target lithologies. Modified from Melosh (1989).

Cintala, 1981; Melosh, 1989), regardless of the size and on what planetary body the impact takes place. The different trajectories of material in different regions of the excavation flow-field result in the partitioning of the transient cavity into an upper 'excavated zone' and a lower 'displaced zone' (Fig. 4.2). Target materials within the upper zone are ejected beyond the transient cavity rim to form the continuous ejecta blanket (see Section 4.4). It is clear that the excavation flow lines transect the hemispherical pressure contours, so that ejecta will contain material from a range of different shock levels, including shock-melted target lithologies.

Target materials within the displaced zone are accelerated initially downward and outward to form the base of the expanding cavity (Stöffler *et al.*, 1975; Grieve *et al.*, 1977). The bulk of this displaced zone comprises target rocks that are shocked to relatively low to intermediate shock levels and they ultimately come to form the parautochthonous rocks of the true crater floor in simple and central uplift structures, in the case of complex craters. In addition, some allochthonous, highly shocked and melted materials are driven down into the transient cavity and along curved paths parallel to the expanding and displaced floor of the transient

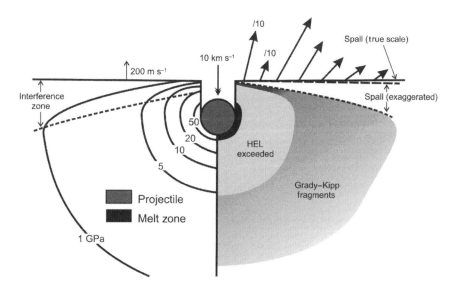

**Figure 4.3** Schematic diagram showing the decrease in pressure away from the point of impact. Beneath the point of impact, pressure contours are approximately hemispherical and grade outwards and downwards from vapour (not shown), melt, then to the limit of crushing at the Hugoniot elastic limit (HEL). Spallation occurs in the near-surface zone. Below the spall zone, tensile stresses fragment the target into Grady–Kipp fragments to considerable depths below the impact site. Modified from Melosh (1989).

cavity. The bulk of these melt-rich materials do not leave the transient cavity and, ultimately, form the allochthonous crater-fill deposits in simple and complex impact structures (Grieve *et al.*, 1977; Melosh, 1989). It has also been recently proposed that some of this material flows outwards during the final stages of crater formation to form patchy or continuous secondary ejecta layers (see Section 4.5). Eventually, the velocity of the cratering flow-field attenuates to a point when it can no longer excavate or displace target rock and melt. For large hypervelocity impacts, including all those on Earth, gravity controls this limit so that excavation stops when insufficient energy remains to lift the overlying material against the force of its own weight. At the end of the excavation stage, a mixture of melt and rock debris forms a lining to the transient cavity.

The discussion so far represents an ideal case; that is, a vertical impact into a dense, homogeneous target. Complications are introduced due to several factors. Data from experiments suggest that, at impact angles as high as 45°, the subsurface flow-fields are significantly different from those created by vertical impacts (Anderson *et al.*, 2004). Most notably, the flow-field centre is offset uprange from the geometric centre of the final crater, which results in changes in the relative distance from the flow-field centre for different portions of the ejecta curtain (Anderson *et al.*, 2004). Particles ejected into the uprange portion of the ejecta curtain originate closer to the initial flow-field centre, whereas downrange particles are ejected farther from the initial flow-field centre.

The presence of layering and pre-existing structures in the target has also been shown to affect the excavation flow-field. Relatively few studies have been carried out regarding the effects of layering on the geometry of the excavation flow-field and final craters. The best-studied example stems from studies of lunar craters, where an unconsolidated fragmental regolith can overlie

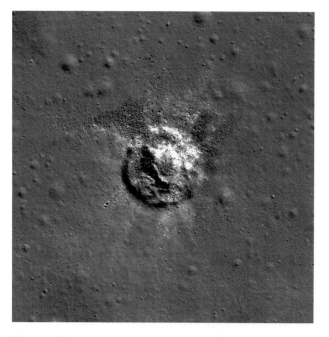

**Figure 4.4** *Lunar Reconnaissance Orbiter* narrow angle camera (NAC) image of a small lunar impact crater showing a characteristic inner bench (51.6 °N, 350.7 °E). Image is 550 m across. LROC NAC image M137610258L (NASA/GSFC/ Arizona State University).

coherent bedrock, typically mare lava flows. Small, kilometre-sized craters in such locations are often seen to possess a characteristic bench or benches on their interior walls (Fig. 4.4). Experiments suggest that this morphology is the end-member in a progression of morphologies that begin with a central mound

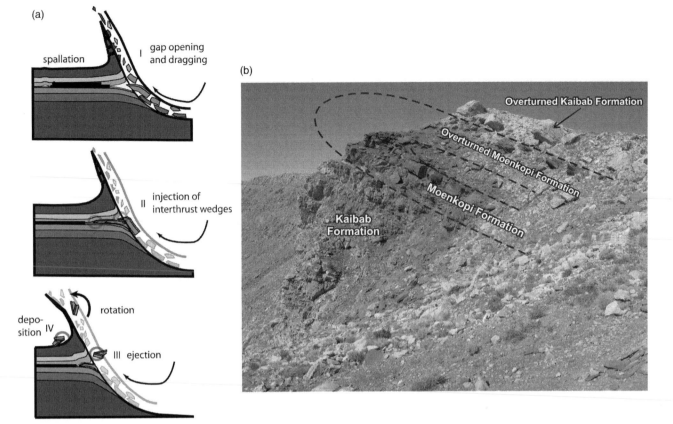

**Figure 4.5**    (a) Model for the formation of interthrust wedges at Barringer Crater by Poelchau *et al.* (2009). After spallation induces horizonatal zones of weakness, small gaps are formed during the excavation stage, while thrust ramps are formed in the sedimentary layers. Wedges of rock are subsequently thrust outward into the crater wall, causing warping of the overlying beds. (b) Field image showing the uplifted and overturned 'flap' at Barringer Crater. (See Colour Plate 12)

and then a flat floor as the size of the crater increases relative to the thickness of the regolith layer (Quaide and Oberbeck, 1968).

A final example of complications introduced in nature comes from studies at Barringer (Meteor) Crater, Arizona. Despite being a prototypical simple crater (Grieve and Garvin, 1984), Barringer is characteristically squarish in shape (Fig. 1.1c). This has been explained by the presence of a prominent set of pre-existing joints along which preferential excavation occurred (Fig. 4.5). These joints were activated as 'tear' faults during the excavation stage and also resulted in vertical displacement of bedrock on either side (Shoemaker, 1963; Poelchau *et al.*, 2009).

### 4.3    Impact plume

During hypervelocity impact, a portion of the kinetic energy of the impactor is converted, irreversibly, into heat. This can result in the melting of significant volumes of target rock, producing characteristic impact melt rocks and melt-bearing breccias (see Chapter 9 for a description and discussion of these products). Closer to the point of impact, the impactor and a portion of the target may vaporize after release from high pressure. This plume of vapour expands outwards in an approximately adiabatic

fashion at velocities comparable to the impact velocity. We prefer the term 'impact plume' (cf. Collins *et al.*, 2011) rather than 'vapour plume' or 'fireball', as it more adequately reflects the intimate mixture of different phases that are present. As noted by Collins *et al.* (2011), the evolution and deposition of impact plume material, as well as the concomitant chemical processing of plume constituents, are among the least well understood aspects of the impact process, and yet it is this impact product that can have the most devastating effect on the local and global environment.

An understanding of the impact plume is important for distal ejecta deposits (see Section 4.7). A characteristic of distal ejecta deposits is the presence of spherules (see Chapter 9 for a complete description of spherules and related tektites). Recent modelling suggests that there is a distinction between 'melt droplets' and 'vapour condensate spherules' (Johnson and Melosh, 2011). The latter condense from the vapour phase in the impact plume; they are more energetic and admixed with projectile material. It has been suggested that it is this type of spherule that is globally distributed to form spherule beds. Melt droplet spherules (and tektites), on the other hand, originate from less highly shocked target rocks and are less widely distributed (Johnson and Melosh, 2011). Impact plumes may also entrain small solid fragments and, if an atmosphere is present, hot atmospheric gases.

## 4.4 Generation of continuous ejecta blankets

Fresh impact craters on all the terrestrial planets are typically surrounded by a 'continuous ejecta blanket' that extends approximately one to two crater radii beyond the crater rim (Melosh, 1989). This continuous ejecta blanket is thickest at the topographic crater rim. Beyond this, the deposits are typically thin and patchy. Once thought to comprise entirely brecciated, ejected debris, it was shown during early studies at Barringer Crater that approximately half the height of crater rim is due to structural uplift of the underlying target rocks (Shoemaker, 1963). This structural uplift occurs due to horizontal compressive forces during the outwards-directed growth of the transient cavity, resulting in the formation of 'interthrust' wedges (Poelchau et al., 2009; Fig. 4.5a). Overlying these structurally uplifted target rocks lies material from the excavated zone of the transient cavity. At Barringer Crater and other simple craters, a so-called 'overturned flap' of ejecta is recognized (Shoemaker, 1963; Maloof et al., 2010), which represents material ejected with such low velocities that the original target stratigraphy is preserved, albeit inverted in places (Fig. 4.5b).

It is widely accepted that the initial emplacement of a continuous ejecta blanket around impact craters on airless bodies, such as the Moon and Mercury, is via the process of ballistic sedimentation (Oberbeck, 1975). In this model, ejecta is ejected from the crater with some initial velocity and follows a near-parabolic flight path (Fig. 4.6). It then falls back to the surface, striking with the same velocity that it possessed upon ejection; hence 'ballistic' ejecta. Innermost ejecta is launched first and with highest velocity,

whereas the outermost ejecta is launched later with lower velocities and so lands closer to crater rim. Experiments suggest that ejection angles initially are high (~55°) and then decrease (to ~45°) up to approximately halfway through crater growth (Anderson et al., 2004). Various factors can affect the velocity of this initial ejecta before it is deposited, with impact velocity and angle having the largest effect (Housen and Holsapple, 2011). Upon landing, secondary cratering and the incorporation of local material (secondary ejecta) in the primary ejecta and subsequent radial flow results in considerable modification and erosion of the local external substrate.

Studies of the continuous ejecta blanket (Bunte Breccia) at the Ries impact structure strongly support the importance of ballistic sedimentation during ejecta emplacement on Earth (Hörz et al., 1983; Fig. 4.7a–c). An important observation is that the Bunte Breccia consists of two main components: (1) primary ejecta excavated from the initial transient cavity (~31 vol.%) and (2) local material or 'secondary ejecta' (~69 vol.%). The incorporation of large amounts of secondary ejecta (Hörz et al. 1983), deformation (Kenkmann and Ivanov, 2006) and radially oriented striations on the pre-impact target surface, very poor sorting (clasts from millimetres to kilometres in size) and overall low shock level of Bunte Breccia deposits are evidence that, after initial ballistic ejection, the ejecta moved radially outwards as some form of ground-flow (Fig. 4.7a,b). This is consistent with observations at the similarly sized 23 km diameter Haughton structure, Canada (Osinski et al., 2005) and at the smaller simple 1.2 km diameter Barringer Crater, USA (Grant and Schultz, 1993), and 1.8 km diameter Lonar Crater, India (Maloof et al., 2010; Table 4.1).

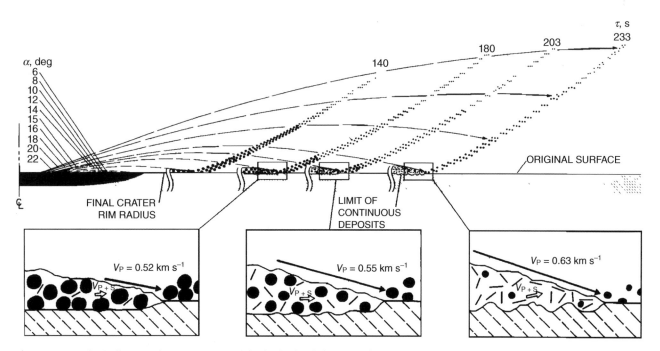

**Figure 4.6** The ballistic sedimentation model of Oberbeck (1975). The ejecta curtain is thickest at its base, where the largest particles with the slowest velocities are also concentrated. Particles highest in the ejecta curtain possess the highest velocities and the finest grain sizes and were launched early in the cratering process. The ejecta curtain sweeps outwards from the crater rim as time progresses.

**Figure 4.7** Impact ejecta of the Ries impact structure, Germany. (a) Schematic cross-section across the Ries impact structure indicating the nature and location of various impactites. Modified from (Schmidt-Kaler, 1978). (b) Large blocks of Malm limestone within the Bunte Breccia at the Gundelsheim Quarry, 7.5 km outside the NE crater rim. The poor sorting, low shock level and modified pre-impact target surface (smooth area at bottom of image) are consistent with ballistic sedimentation and subsequent radial flow (Hörz *et al.*, 1983). (c) The contact between the Bunte Breccia and the underlying limestone displays characteristic striations ('Schliff-Fläche'). These striations demonstrate that the Bunte Breccia, after it was ejected out of the crater on ballistic trajectories, was deposited onto the surface and continued to flow for considerable distances. (d) Image of the Aumühle quarry showing the relationship between suevite (light grey/green) and underlying Bunte Breccia (dark brown/red). Note the sharp contact between the suevite and Bunte Breccia. The height to the top of the outcrop is ~9.5 m. (e) The suevite has clearly filled in a depression in the underlying Bunte Breccia. This is incompatible with a 'fallout' airborne mode of deposition as proposed by some workers (Stöffler, 1977). The infilling suggests a topographic control and is a characteristic of pyroclastic and lava flows (Fisher and Schmincke, 1984). (See Colour Plate 13)

**Table 4.1** Compilation of preserved ejecta deposits around impact structures on Earth[a]

| Crater | Buried | Apparent crater diameter $D_a$ (km) | Rim diameter $D$ (km) | Target stratigraphy[b] | Ejecta description |
|---|---|---|---|---|---|
| Barringer | N | N/A | 1.2 | Sst, Slt, Lst, Dol | Continuous ballistic ejecta blanket with evidence for ground-hugging flow following ballistic emplacement (Shoemaker, 1963), which produced flow lobes (Grant and Schultz, 1993). Small millimetre- to centimetre-sized glassy beads occur as lag deposits overlying the continuous ejecta blanket. |
| Bigach | N | 8 | ? | Sst, Slt, Vol | Ejecta poorly exposed and studied. They are polymict breccias with blocks up to ~20 m across (Masaitis, 1999); it is not clear if any melt or shock effects are present. |
| Boltysh | Y | 24 | ? | Gr, Gn | Ejecta deposits very poorly exposed and eroded in close proximity to the crater rim. In the Tyasmin River valley ~6–8 km outside the crater rim, low-shock, melt-free 'monomict' breccias are overlain by polymict breccias (Gurov et al., 2003). Both breccias comprise crystalline rocks; polymict breccias contain more highly shocked material. The polymict breccias are described as 'lithic breccias', but they also are reported as containing altered melt particles (Gurov et al., 2003). |
| Bosumtwi | N | ? | 10.5 | MSed | Patchy impact melt-bearing breccias have been documented from an area ~1.5 km² and range up to ~15 m thick (Boamah and Koeberl, 2006). In the north of the crater, these breccias are underlain by polymict lithic impact breccias (Koeberl and Reimold, 2005). |
| Chicxulub | Y | N/A | 180 | Gr, Gn overlain by ~3 km of Lst, Dol, Evap | The Chicxulub ejecta deposits vary with distance from the crater. Close to the crater, the UNAM-7 drill core (located 126 km from the crater centre) shows a two-layer stratigraphy with melt-free to poor lithic breccias and megabreccias derived from the sedimentary cover overlain by melt-rich impact melt-bearing breccias (suevites) with abundant crystalline basement clasts (Salge, 2007). The lower sedimentary breccias are interpreted as ballistic ejecta and have been compared with the Bunte Breccia at the Ries structure (Salge, 2007). At greater distances, the ballistic ejecta comprises largely locally derived secondary ejecta (Schönian et al., 2004). |
| Haughton | N | 23 | 16 | Gn and Gr overlain by 1.9 km Lst, Dol, minor Evap, Sst, Sh | Remnants of the ejecta blanket are preserved in the SW of the Haughton impact structure. There, a two-layer sequence of pale yellow–brown melt-poor impact breccias and megablocks overlain by pale grey clast-rich impact melt rocks are preserved (Osinski et al., 2005). The former are derived from depths of >200 m to <760 m and are interpreted as the continuous ballistic ejecta blanket (cf. the Bunte Breccia at the Ries structure). The pale grey impact melt rocks are derived from deeper levels. |
| Lonar | N | N/A | 1.9 | Bs | Continuous ballistic ejecta blanket; evidence for ground-hugging flow following ballistic emplacement (Maloof et al., 2010). Possible at transition of 'simple ballistic emplacement and ballistic sedimentation' (Maloof et al., 2010). |

*(Continued)*

**Table 4.1**   (*Continued*)

| Crater | Buried | Apparent crater diameter $D_a$ (km) | Rim diameter $D$ (km) | Target stratigraphy[b] | Ejecta description |
|---|---|---|---|---|---|
| Mistastin | N | 28 | ? | An, Gr, Mn | Ejecta deposits are preserved in the crater rim region where a complex series of melt-free and -poor lithic impact breccias are overlain by impact melt-bearing breccias and coherent silicate impact melt rocks (Mader *et al.*, 2011). |
| New Quebec (Pingualuit) | N | N/A | 3.4 | Gn | Ejecta largely eroded. Isolated melt samples are found as float beyond the N rim. The glacial direction is to the SE (Bouchard and Saarnisto, 1989); therefore, it is highly improbable that the melt samples originated from inside the crater cavity. They are interpreted as melt 'splashes'. |
| Obolon | Y | 18(poorly constrained) | ? | Gn and Gr overlain by ~300 m of Sst, Clt, Lst | A borehole (#467) close to the crater rim contains a two-layer ejecta sequence: a 22.5 m thick series of melt-free lithic breccias, comprising clays with glide planes, are overlain by a 14 m thick sequence of impact melt-bearing breccias with clasts from the crystalline basement (Gurov *et al.*, 2009). |
| Ragozinka | N | 9 | ? | Folded Lst, Bs, Sst, Sh overlain by 100–200 m of Sst, Slt | Ejecta very poorly exposed and preserved. An 'outlier' near the village of Vostochnyi comprises a series of low-shock polymict impact breccias with clasts up to ~7 m in size (Vishnevsky and Lagutenko, 1986). Lower contacts are not exposed and the upper surface is erosional. |
| Ries | N | 24 | ? | Gn, Gr overlain by 500–800 m of Sst, Sh, Lst | A two-layer sequence of ejecta is preserved at the Ries structure with melt-free to poor Bunte Breccia overlain by impact melt-bearing breccias (suevites) (von Engelhardt, 1990). The Bunte Breccia comprises largely of sedimentary rocks derived from depths of <850 m in the target stratigraphy, whereas the suevites are largely derived from the deeper levels in the target. An occurrence of impact melt rock at Polsingen lies in the same stratigraphic position as the suevites (i.e. overlying Bunte Breccia; Osinski, 2004). |
| Ritland | N | N/A | 2.7 | Gn overlain by Sh | Part of the ejecta blanket is preserved in the east of the crater and is continuous to ~5 km (Kalleson *et al.*, 2012). Lithic breccia comprising clasts of gneiss and shale up to 5 m across. Shatter cones and planar deformation features present. |
| Tenoumer | N | N/A | 1.9 | Gn and Gr overlain by 20–30 m of Lst | The continuous ejecta blanket is partially exposed and comprises millimetre- to metre-size blocks of the crystalline basement. 'Impact melt rock occurs in patches outside the crater' overlying the continuous ejecta blanket, predominantly to the E, NW and SW (Pratesi *et al.*, 2005). |

[a]Modified and updated from Osinski *et al.* (2011).

[b]An: anorthosite; Bs: basalt; Dol: dolomite; Evap: evaporite; Gr: granite; Gn: gneiss; Lst: limestone; Mn: mangerite; MSed: metasedimentary rocks; Sh: shale; Slt: siltstone; Sst: sandstone; Vol: mixed extrusive volcanic rocks.

It is apparent that target lithology plays an important role in the formation of continuous ejecta deposits. At the Ries structure, the volatile content and cohesiveness (e.g. resistant limestone bedrock versus unconsolidated clays and sands) of the uppermost target outside the transient cavity governed the maximum radial extent of ground-hugging flow following ballistic deposition (Hörz *et al.*, 1983). In other words, the presence of volatile-rich and/or unconsolidated/weak surficial materials will result in greater 'fluidization' of secondary ejecta, leading to increased runout. This may have important implications for the formation of so-called fluidized or layered ejecta structures on Mars (Osinski, 2006). The angle of impact also affects the distribution of ballistic ejecta with respect to the source crater, with the preferential downrange concentration of ejecta at angles less than 60°, the development of a 'forbidden zone' uprange of the crater at angles of less than 45° and, finally, at very low angles of less than 20°, the development of a second 'forbidden zone' downrange of the crater, leading to the characteristic 'butterfly' pattern of ejecta deposits (Gault and Wedekind, 1978).

Similar observations have been made at the Chicxulub impact structure, Mexico. In Belize, the outer portion of the continuous ejecta blanket, termed the Albion Formation, comprises a basal spheroid bed and an upper so-called 'diamictite' bed (Pope *et al.*, 1999), which preserve features such as cross-bedding and internal shear planes, indicative of lateral flow outwards from the crater centre (Pope *et al.*, 1999; Kenkmann and Schönian, 2006). Kenkmann and Schönian (2006) proposed the following depositional model: following ballistic deposition at much less than three crater radii, ground-hugging flow occurred driven by the water content of the flow itself. At distances of greater than 3.5 crater radii, the incorporation of local clays further fluidized the flow and allowed it to continue moving for greater distances than would have been possible if the substrate was resistant bedrock. Thus, substrate lithology played a key role in fluidizing the ejecta deposits.

## 4.5 Rayed craters

Rayed craters are an unusual and poorly understood class of craters with a distinctive ejecta morphology (Fig. 4.8). First recognized on the Moon and other airless bodies (Melosh, 1989), they have more recently been discovered on Mars (McEwen *et al.*, 2005; Tornabene *et al.*, 2006). Crater rays can extend several hundred or even thousands of kilometres across a planetary surface. This forms an important distinction to radial lineations and fabrics that occur within, and in close proximity to, continuous ejecta blankets on several terrestrial planets. On the Moon, crater rays comprise bright high-albedo material with the most well known example being Tycho. There appear to be two main types lunar of rays (Hawke *et al.*, 2004). The first are immature rays that are bright because of immature soils and the second are mature compositional rays that are bright because of compositional contrasts. On Mars, crater rays were only recognized following the acquisition of thermal infrared images by the *Mars Odyssey* Thermal Emission Imaging System (THEMIS; Tornabene *et al.*, 2006).

The formation of crater rays remains enigmatic. Hawke *et al.* (2004) suggest that the material to form rays can potentially come from four sources: (1) the emplacement of immature primary ejecta; (2) the deposition of immature local material from secondary craters; (3) the action of debris surges downrange of secondary clusters; and (4) the presence of immature interior walls of secondary impact craters. Tornabene *et al.* (2006) suggest that crater rays may form from high-velocity ejecta ejected at low ejection angles (i.e. material from the spallation zone) during the initial ballistic stage. The scarcity of rayed craters on Mars has also been used to infer that specific conditions are required to form this ejecta morphology, with highly competent

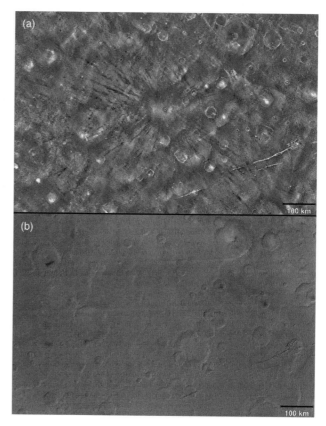

**Figure 4.8** THEMIS and MOC images of the 6.9 km diameter Gratteri Crater, the fourth largest, but most well expressed thermal rayed crater system on Mars (see Tornabene *et al.* (2006) and McEwen *et al.* (2010) for more details on other rayed Martian craters). North is up. (a) A THEMIS night-time thermal infrared brightness temperature (nTIR) mosaic of the Memonia Fossae region centred at approximately 199.9°E, 17.7°S on Gratteri Crater. The darker (cooler) rays are dominated by fine-grained materials (e.g. dust) that show up as cooler deposits in THEMIS nTIR images extending up to ~600 km from the crater rim or ~170 radii. High-resolution images (e.g.,MOC narrow-angle or HiRISE images) of the rays show that the rays consist of densely overlapping secondary craters and their overlapping ejecta. (b) An MOC wide-angle camera mosaic of the exact same region in visible wavelengths. Unlike other rayed crater systems elsewhere in the Solar System (e.g. the Moon), Martian varieties scarcely show any contrast in albedo (a slight contrast can be observed to the northwest of Gratteri's primary cavity – cf. top and bottom images). This explains how these craters went undetected on Mars until high-resolution nTIR images were acquired. Images: JPL/NASA/ASU.

layered (volcanic) targets being the preferred explanation (Tornabene *et al.*, 2006). An interesting association is that these rayed Martian craters produced several hundreds of thousands of secondary craters in addition to the rays themselves (McEwen *et al.*, 2005).

### 4.6    Generation of multiple ejecta layers

#### 4.6.1    Observations

A recent synthesis of observations from all the terrestrial planets suggests that many impact craters display more than one layer of ejecta (Osinski *et al.*, 2011; Table 4.1). These layers may be patchy, impact melt-rich deposits and they occur inside and outside the rim overlying the continuous ejecta blanket at both simple and complex craters. A brief overview of the observations presented by Osinski *et al.* (2011) is given below.

On the Moon, what is generally interpreted to be impact melt ponds on the rim terraces of complex lunar craters and overlying parts of the continuous ejecta blanket have been documented since the 1970s (Howard and Wilshire, 1975; Hawke and Head, 1977; Fig. 4.9a,b). Recent images returned by the Lunar Reconnaissance Orbiter Camera (LROC) show intricate surface textures and morphologies, such as channels and arcuate cracks and ridges, indicative of flow, strongly supporting the impact melt origin (Fig. 4.9c–e). On Venus, spectacular melt outflows have been documented exterior to many craters (Asimow and Wood, 1992). Their increased size with respect to the Moon has been explained by the increased relative efficiency of impact melting on Venus, due to its high gravity and hot surface (Grieve and Cintala, 1997). In addition, any entrained clastic debris in the impact melt is hotter than on the Moon, resulting in higher thermal equilibrium temperatures, lower viscosities and longer cooling time for the impact melt deposits (Grieve and Cintala,

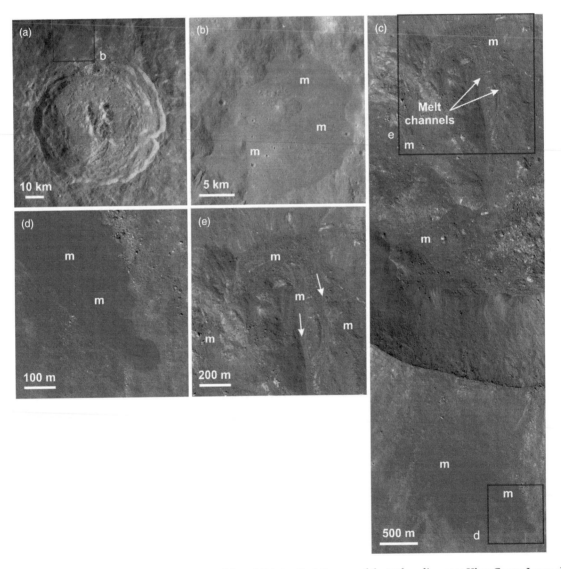

**Figure 4.9**    Observations of complex lunar craters. (a) and (b) *Apollo 16* image of the 76 km diameter King Crater Large showing a large impact melt ('m') pond. Portions of *Apollo 16* image 1580 (NASA). (c) Portion of LROC NAC image pair (M106209806RE) of melt within the interior and overlying the continuous ejecta blanket at the 22 km diameter Giordano Bruno Crater (NASA/GSFC/ASU). (d) A close-up from (c) showing melt overlying the lighter blocky ballistic ejecta, with evidence for flow from the top left to the bottom right. (e) A close-up from (c) showing impact melt flows within the interior of the crater.

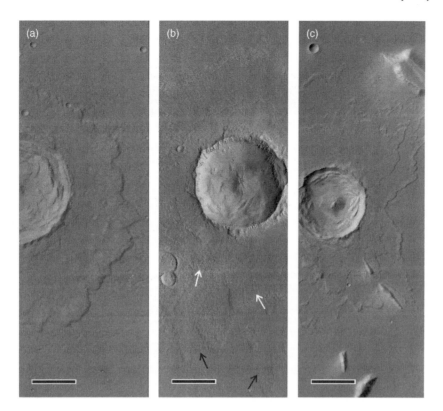

**Figure 4.10** Typical layered ejecta morphologies on Mars. (a) A typical SLE structure with a well-defined outer rampart. CTX image P17_007740_1925_XN_12N270W. (c) A relatively fresh DLE structure. The white arrows indicate the outer margin of the inner layer of ejecta and the black arrows indicated the outer margin of the outer layer. Note the distinctive striae present on the ejecta deposits, the origin of which is still actively debated. CTX image P16_007462_2133_XN_33N241W. (c) A spectacular MLE structure displaying multiple highly lobate ejecta lobes. CTX image P20_008792_1980_XN_18N199W. North is up. Image credits: NASA/JPL/MSSS. All scale bars are 10 km.

1995). It is notable that impact melt outflows are most common in craters resulting from oblique impacts and in larger structures (Chadwick and Schaber, 1993).

On Mars, approximately one-third of all Martian craters of 5 km and greater in diameter possess discernable ejecta blankets, with over 90% possessing so-called layered ejecta that display single (SLE; 86%), double (DLE; 9%) or multiple (MLE; 5%) layer morphologies (Barlow, 2005) (Fig. 4.10). So-called ramparts or ridges, the origins of which are debated, are often seen at the outer edge(s) of the ejecta layers. It is widely accepted that layered ejecta deposits were highly fluidized at the time of their emplacement and occurred as relatively thin ground-hugging flows (Carr *et al.*, 1977). Recent high-resolution imagery of relatively pristine Martian impact craters provides some important new constraints and observations. Some Martian craters are very lunar-like in terms of crater interior and ejecta morphology. The Pangboche Crater is particularly interesting as it is located at approximately 21 km elevation, on the flank of Olympus Mons (Fig. 4.11a), such that it is not possible for volatiles to be present either in the atmosphere or the target rocks. Recent imagery also reveals the presence of impact melt deposits forming large bodies and/or ponds on crater floors (Fig. 4.11a,b), terraces (Fig. 4.11a,b) and overlying continuous ejecta blankets (Fig. 4.11b–d).

The Earth provides the only ground-truth data for the lithological and structural character of impact structures. A review of ejecta at terrestrial impact structures suggests the presence in the proximal ejecta of a low-shock, lithic breccia overlain by a melt-bearing deposit (Table 4.1). The range of target rocks involved suggests that this trait is not due to the effect of volatiles, layering or other effects of target lithology on the impact cratering process. Some of the best-preserved and exposed ejecta deposits on Earth occur at the Ries structure, Germany (von Engelhardt, 1990). The Ries structure clearly displays a distinctive two-layer ejecta configuration (Fig. 4.7a,d,e) with a series of impact melt-bearing breccias ('suevites') and minor impact melt rocks overlying the continuous ejecta blanket (Bunte Breccia). The sharp contact between the Ries ejecta layers (Fig. 4.7d,e) indicates that there is a clear temporal hiatus between emplacement of the ballistic Bunte Breccia and the overlying suevites/impact melt deposits (Hörz, 1982).

In summary, observations at craters on the terrestrial planets suggest that impact melt-bearing deposits occur inside and outside the rim overlying the continuous ejecta blanket at both simple and complex craters. If ballistic sedimentation followed by radial flow accounts for the emplacement of the continuous ballistic ejecta, it begs the question as to the origin, timing and

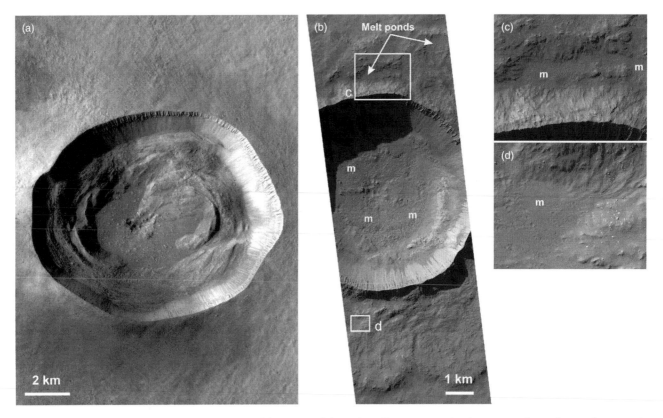

**Figure 4.11**    Images of Martian impact craters. (a) Image of the 11 km diameter Pangboche Crater, situated near the summit of Olympus Mons. Portion of CTX image P02_001643_1974_XN_17N133W (NASA/JPL/MSSS). (b) HiRISE image PSP_008135_1520 (NASA/JPL/UA) showing a fresh unnamed 7 km diameter impact crater in Hesperia Planum (108.9 °E, 27.9 °S) and providing context for (c) and (d). This crater displays characteristic pitted materials likely representing impact melts (Tornabene *et al.*, 2007) ('m') in the crater interior and exterior (observed as pitted and ponded deposits on the ejecta blanket). Without HiRISE imagery, this crater would be misclassified as an SLE crater. (c) and (d) The pitted ponds form in topographic depressions and there is clear evidence for inflow and some remobilization of these materials after their initial emplacement in various locations near the rim and off the distal rampart.

emplacement mechanism(s) of this overlying melt-rich ejecta. Based on these observations and problems, a working hypothesis of a multistage emplacement process has recently been proposed by Osinski *et al.* (2011) and much of the subsequent sections draws on this work. As with the crater-forming process in general – and its subdivision into contact and compression, excavation and modification – ejecta emplacement occurs as a continuum and that the different emplacement processes described below may overlap in time. This will be important in larger craters, where models show that uplift of the transient cavity floor commences before outward growth and excavation of the transient cavity ceases in the rim region (Stöffler *et al.*, 1975; Kenkmann and Ivanov, 2000).

### 4.6.2    Initial impact melt production and early emplacement

It is clear that the ballistic sedimentation model can account for the formation of continuous ejecta blankets (Fig. 4.6). However, the observation of thin melt veneers around some simple lunar and terrestrial craters (e.g. at Tenoumer; Table 4.1) indicates that

some of the melt-rich materials from the displaced zone of the transient cavity are driven up and over the transient cavity walls and rim region (Fig. 4.12). Experiments suggest that this process will be particularly important for oblique impacts, where the subsurface flow-field is displaced downrange (Anderson *et al.*, 2004). For small craters, except in exceptional circumstances (e.g. if a crater rim is breached), the bulk of the melted material lining the transient cavity, however, remains within the cavity and moves inward as the transient cavity walls collapse inward, where they become intercalated with the brecciated cavity wall materials to form the internal breccia lens partially filling simple craters (Grieve and Cintala, 1981). It was suggested that two main parameters will affect this phase of ejecta emplacement: the angle of impact and the initial volume of impact melt generated (Osinski *et al.*, 2011).

It is important to note that the onset of melting porous and volatile-bearing targets (typically, but not necessarily, rocks) is much lower (e.g. ~20 GPa in porous sandstones; Kieffer *et al.*, 1976). For $H_2O$-bearing planetary bodies, ice, and in some cases liquid water, must also be considered; recent calculations suggest that $H_2O$ ice will undergo complete melting at approximately

DURING EXCAVATION:

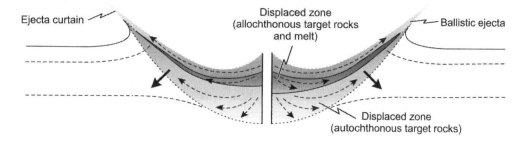

Ejecta curtain

Displaced zone
(allochthonous target rocks
and melt)

Ballistic ejecta

Displaced zone
(autochthonous target rocks)

END EXCAVATION STAGE/
START MODIFICATION STAGE:

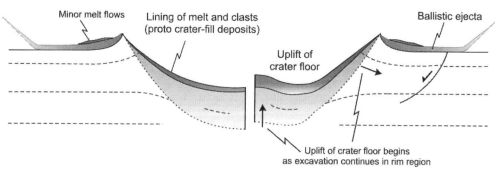

Minor melt flows

Lining of melt and clasts
(proto crater-fill deposits)

Ballistic ejecta

Uplift of
crater floor

Uplift of crater floor begins
as excavation continues in rim region

MODIFICATION STAGE:

Flow of melt and clasts off
emergent central uplift

Continued movement of
melt and clasts outwards

Gravitational collapse
of crater walls

Minor melt flows

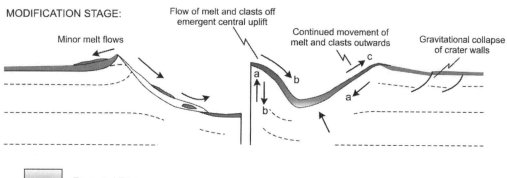

Excavated Zone

Displaced Zone
(allochthonous target rocks)

Displaced Zone
(clasts of allochthonous target rocks and melt)

END OF MODIFICATION STAGE:

Minor melt veneers
and ponds

Melt-rich crater-fill
impactites

Melt-rich impactites
emplaced as flows

Melt-poor
ballistic ejecta

Final crater rim

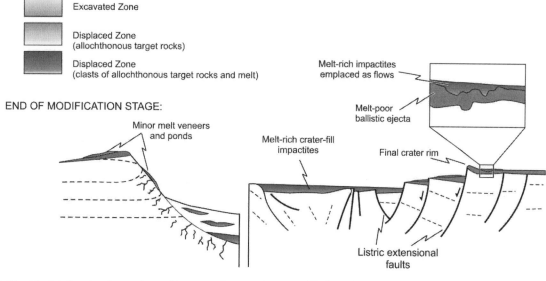

Listric extensional
faults

**Figure 4.12** Model for the formation of impact ejecta and crater fill deposits (modified from Osinski *et al.* (2011)). This multistage model accounts for melt emplacement in both simple (left panel) and complex craters (right panel), as described in the text. It should be noted that this is for a 'typical' impact event. As discussed in the text, the relative timing and important of the different processes can vary, particularly for more oblique impacts. It should be noted that, in the modification-stage section, the arrows represent different time steps, labelled 'a' to 'c'. Initially, the gravitational collapse of crater walls and central uplift (a) results in generally inwards movement of material. Later, melt and clasts flow off the central uplift (b). Then, there is continued movement of melt and clasts outwards once crater wall collapse has largely ceased (c). (See Colour Plate 14)

2–4 GPa depending on temperature (Stewart and Ahrens, 2005). Thus, much more melt will be generated for the same pressures and temperatures than with a dry crystalline target. This could result in proportionally more melt in the displaced melt-rich zone of the transient crater and, thus, a more fluidized and voluminous overlying melt-rich ejecta layer.

### 4.6.3   Late-stage melt emplacement – the surface melt flow phase

Notwithstanding the minor emplacement of melt outside of crater rims by the cratering flow-field discussed in Section 4.6.2, the observations of relatively extensive impact melt-rich deposits overlying ballistic ejecta deposits and the sharp contact between these units imply the general late-stage emplacement of melt-rich ejecta. In larger complex craters, where these melt-rich deposits are most abundant, central structural uplift occurs (Fig. 4.12). Field observations and numerical models suggest that central uplifts partially collapse to varying degrees depending on crater size and target properties (Collins *et al.*, 2002; Osinski and Spray, 2005). In some cases, this uplift may initially overshoot the original target surface and then collapse (Collins *et al.*, 2002). It has been suggested, therefore, that cavity modification, in particular uplift, thus imparts an additional outward momentum to the melt- and clast-rich lining of the transient cavity during the modification stage, resulting in flow towards and over the collapsing crater rim and onto the proximal ballistic ejecta blanket, forming a second thinner and potentially discontinuous layer of non-ballistic ejecta (Fig. 4.12; Osinski *et al.*, 2011). This offers an explanation for the above observations of the presence of melt-rich materials stranded on top of the central peaks and terraces of complex lunar, Martian and terrestrial craters and impact melt ponds within, and without, the crater rim region of all terrestrial planets. This mechanism will also be important for oblique impacts, where field (Scherler *et al.*, 2006) and numerical modelling (Ivanov and Artemieva, 2002; Shuvalov, 2003) studies suggest that the horizontal momentum of the impactor is preserved into the final stages of crater formation such that there is an uprange initiation of crater rim collapse and the migration of the uplifting crater floor downrange.

For oblique impacts, an additional mechanism may be the emplacement of external melt deposits in a process more akin to what is believed to occur at simple craters; that is, by the late stages of the cratering flow-field. In this case, the initial direction of the flows is preferentially downrange. Indeed, some of the most notable melt outflows from complex craters on Venus are associated with craters formed from oblique impacts (Chadwick and Schaber, 1993). Model calculations indicate that, although the volume of melt is lower in oblique impacts, the fraction of melt that is retained in the transient cavity is also less (Ivanov and Artemieva, 2001). They also show that the cratering flow-field is asymmetric, with higher downrange velocities but that subsequent cavity modification is much more symmetric (Ivanov and Artemieva, 2001). Thus, it would appear that preferential initial direction of external melt flows in oblique impacts may be related to asymmetries in the cratering flow-field in addition to subsequent modification processes.

It is important to note that the second, upper layer may be thin and discontinuous and not easily observable from spacecraft. An example of this is that many SLE craters on Mars, with high-resolution imagery, can now be seen to have patchy melt-rich deposits (Fig. 4.10b–d; Tornabene *et al.*, 2007). This is also consistent with observations from terrestrial impact structures (Table 4.1). Emplacement of these upper ejecta layers as surface melt-rich flows is consistent with observations of the proximal surficial 'suevites' at the Ries impact structure (Newsom *et al.*, 1986; Bringemeier, 1994; Osinski *et al.*, 2004). Indeed, the origin of the surficial suevite as the result of fallout from an ejecta plume over the crater is also not supported by recent numerical models (Artemieva *et al.*, 2009).

Finally, it should be noted that other factors can govern the final resting place of these late-stage melt-rich deposits, the most important of which is the local topography of the target region. Impact melt-rich materials can continue to flow for an extended period of time following impact and, thus, will tend to follow topography and collect in topographic lows (Fig. 4.9a,b).

## 4.7   Distal impact ejecta

Ever since the ground-breaking breaking papers by Alvarez *et al.* (1980) and Bohor *et al.* (1987) in which evidence (chemically and physical respectively) was presented for a connection between the K–Pg boundary layer and a possible meteorite impact event, there has been considerable interest in so-called distal ejecta; that is, ejecta deposited at greater than five crater radii from the impact site. Distal ejecta deposits on Earth, collectively termed air fall beds, typically comprise two main types: strewn fields of glassy tektites and microtektites, and spherule beds comprising (formerly) glassy impact spherules and fragments of shocked target rocks. Of the four known terrestrial tektite-strewn fields, all but one have been linked to source craters. Such fields are not globally distributed. In recent years, the discovery of spherule layers dating from the Phanerozoic to the Cenozoic has continued unabated (Simonson and Glass, 2004). These beds vary in thickness from a few millimetres to several tens of centimetres and some may have been globally distributed. The characteristics of tektites and spherules, both products of impact melting, are discussed in Chapter 9.

It is typically assumed that distal ejecta gradually settle out from the atmosphere as individual particles. However, more recent modelling work suggests that distal impact ejecta falling into the atmosphere may clump together into density currents that flow to the ground much more rapidly than might be expected for single particles themselves (Goldin and Melosh, 2009). This model provides an interesting alternative explanation for the presence of sedimentary structures such as cross-bedding, in some of the spherule layers that have previously been interpreted as being deposited by impact tsunamis. A further outcome of this modelling work is that this clumping of particles results in a thermal self-shielding effect of settling spherules, which may have prevented widespread wildfire ignition following the Chicxulub impact event, requiring the global wildfire model to be re-evaluated (see Chapter 10 for further discussion of the environmental effects of meteorites impact events).

The most intensively studied terrestrial distal ejecta deposits are the sediments associated with the K–Pg boundary layer. Impact events the size of Chicxulub have sufficient energy to distribute ejecta materials around the entire Earth. The size of the spherules and shocked minerals at sites around the globe are inversely proportional to their distance from Chicxulub (Morgan *et al.*, 2006). Ejecta deposits from Chicxulub are known worldwide and have been identified at over 350 locations. Although these deposits have been modified, they show a distinct pattern related to their radial distance from their source crater, Chicxulub on the Yucatan peninsula, Mexico. Within Chicxulub, limited drilling data suggest that they are several hundreds of metres thick and analogies have been drawn with the Bunte Breccia and overlying 'suevite' breccia at the much smaller Ries structure in Germany (Sharpton *et al.*, 1996). The Chicxulub impact triggered intense seismic activity and tsunamis in the Gulf of Mexico and the Caribbean, which collapsed the continental shelf locally, producing breccias in some areas that were to become reservoir rocks for hydrocarbon deposits (see Chapter 12) and thick turbidite clastic sequences (Arenillas *et al.*, 2006). Close to Chicxulub (<500 km), these tens of metres thickness clastic beds are overlain by reworked clastic beds that also contain shocked minerals, which are topped by an enrichment in platinum group elements (PGEs) indicative of projectile material (see Chapter 15). At distances of 500–1000 km from Chicxulub, the K–Pg boundary sediments are characterized by centimetre- to metre-thick spherule beds, overlain by high-energy clastic beds. At greater distances (1000–5000 km), the K–Pg boundary sediments consist of two layers: a lower layer up to 10 cm thick, consisting of spherules, and a thinner (up to 0.5 cm) upper layer enriched in PGEs, shocked minerals and Ni-rich spinels (Schulte *et al.*, 2009). In even more distal marine settings, the K–Pg ejecta consist of a few millimetres thickness layer (now clay), with enriched PGE content, shocked minerals and what were once spherules (Smit, 1999).

## 4.8  Depth of excavation

Various estimates for the depth of excavation $d_e$ are in the literature (Melosh, 1989); however, very little ground-truth data exists to estimate this depth, as few terrestrial craters preserve ejecta deposits and/or have the distinct pre-impact stratigraphy necessary for determining depth of materials (Table 4.1). An estimate of the depth of excavation is, of course, critical for determining the original provenance of planetary surface samples, which are not *in situ*, by virtue of an impact event. Based on stratigraphic considerations, $d_e$ at Barringer Crater is greater than $0.08D$ (Shoemaker, 1963), where $D$ is the final rim diameter. Importantly, in large complex craters and basins the depth and diameter must be referred back to the 'unmodified' transient cavity to reliably estimate the depth of excavation (Melosh, 1989). This is inherently difficult, as the transient cavity is essentially destroyed during the modification stage of complex crater formation. The Haughton and Ries structures represent the only terrestrial complex structures where reliable data are available. There, the maximum $d_e$ of material in the ballistic ejecta deposits yield identical values of $0.035D_a$ (Table 4.1), where $D_a$ is the apparent crater diameter (Osinski *et al.*, 2011). If the final rim diameter $D$ is used, which

is the parameter measured in planetary craters, a value $0.05D$ is obtained for Haughton (see Chapter 1 for definitions of final and apparent crater diameter). A critical consideration is that the upper layer of ejecta (and the crater-fill deposits) reflects the composition and depth of the displaced zone of the transient cavity (Fig. 4.2). At Haughton, this value is a minimum of $0.08D_a$ or $0.12D$ (Table 4.1). If sampling deeper seated lithologies is the goal of future planetary sampling missions, these melt ponds could be prime exploration targets (Osinski *et al.*, 2011).

## References

Alvarez, L.W., Alvarez, W., Asaro, F. and Michel, H.V. (1980) Extraterrestrial cause for the Cretaceous/Tertiary extinction. *Science*, 208, 1095–1108.

Anderson, J.L.B., Schultz, P.H. and Heineck, J.T. (2004) Experimental ejection angles for oblique impacts: implications for the subsurface flow-field. *Meteoritics and Planetary Science*, 39, 303–320.

Arenillas, I., Arz, J.A., Grajales-Nishimura, J.M. *et al.* (2006) Chicxulub impact event is Cretaceous/Paleogene boundary in age: new micropaleontological evidence. *Earth and Planetary Science Letters*, 249, 241–257.

Artemieva, N.A., Wünnemann, K., Meyer, C. *et al.* (2009) Ries Crater and suevite revisited: Part II modelling. 40th Lunar and Planetary Science Conference, 1526.pdf.

Asimow, P.D. and Wood, J.A. (1992) Fluid outflows from Venus impact craters: analysis from *Magellan* data. *Journal of Geophysical Research*, 97, 13643–13665.

Barlow, N.G. (2005) A review of Martian impact crater ejecta structrues and their implications for target properties, in *Large Meteorite Impacts III* (eds T. Kenkmann, F. Hörz and A. Deutsch), Geological Society of America Special Paper 384, Geological Society of America, Boulder, CO, pp. 433–442.

Boamah, D. and Koeberl, C. (2006) Petrographic studies of 'fallout' suevite from outside the Bosumtwi impact structure, Ghana. *Meteoritics and Planetary Science*, 41, 1761–1774.

Bohor, B.F., Modreski, P.J. and Foord, E.E. (1987) Shocked quartz in the Cretaceous–Tertiary boundary clays: evidence for a global distribution. *Science*, 236, 705–709.

Bouchard, M.A. and Saarnisto, M. (1989) Déglaciation et paléodrainages du cratére du Nouveau-Québec, in *L'Histoire naturelle du Cratére du Nouveau-Québec, Collection Environnement et Géologie* (eds M.A. Bouchard and S. Péloquin), Université de Montréal, Montreal, pp. 165–189.

Bringemeier, D. (1994) Petrofabric examination of the main suevite of the Otting Quarry, Nordlinger Ries, Germany. *Meteoritics and Planetary Science*, 29, 417–422.

Carr, M.H., Crumpler, L.S., Cutts, J.A. *et al.* (1977) Martian impact craters and emplacement of ejecta by surface flow. *Journal of Geophysical Research*, 82, 4055–4065.

Chadwick, D.J. and Schaber, G.G. (1993) Impact crater outflows on Venus: morphology and emplacement mechanisms. *Journal of Geophysical Research*, 98, 20891–20902.

Collins, G.S., Melosh, H.J., Morgan, J.V. and Warner, M.R. (2002) Hydrocode simulations of Chicxulub Crater collapse and peak-ring formation. *Icarus*, 157, 24–33.

Collins, G.S., Melosh, H.J. and Osinski, G.R. (2011) The impact cratering process. *Elements*, 8, 25–30.

Dence, M.R. (1968) Shock zoning at Canadian craters: petrography and structural implications, in *Shock Metamorphism of Natural*

*Materials* (eds B.M. French and N.M. Short), Mono Book Corp., Baltimore, MD, pp. 169–184.

Fisher, R.V. and Schmincke, H.U. (1984) *Pyroclastic Rocks*, Springer-Verlag, Berlin.

Folco, L., Di Martino, M., El Barkooky, A. *et al.* (2011) Kamil Crater (Egypt): ground truth for small-scale meteorite impacts on Earth. *Geology*, 39, 179–182.

Gault, D.E. and Wedekind, J.A. (1978) Experimental studies of oblique impacts. *Proceedings of the Lunar and Planetary Science Conference*, 9, 3843–3875.

Goldin, T.J. and Melosh, H.J. (2009) Self-shielding of thermal radiation by Chicxulub impact ejecta: firestorm or fizzle? *Geology*, 37, 1135–1138.

Grant, J.A. and Schultz, P.H. (1993) Degradation of selected terrestrial and Martian impact craters. *Journal of Geophysical Research*, 98, 11025–11042.

Grieve, R.A.F. and Cintala, M.J. (1981) A method for estimating the initial impact conditions of terrestrial cratering events, exemplified by its application to Brent crater, Ontario. *Proceedings of the Lunar and Planetary Science Conference*, 12, 1607–1621.

Grieve, R.A.F. and Garvin, J.B. (1984) A geometric model for excavation and modification at terrestrial simple impact craters. *Journal of Geophysical Research*, 89, 11561–11572.

Grieve, R.A.F. and Cintala, M.J. (1995) Impact melting on Venus: some considerations for the nature of the cratering record. *Icarus*, 114, 68–79.

Grieve, R.A.F. and Cintala, M.J. (1997) Planetary differences in impact melting. *Advances in Space Research*, 20, 1551–1560.

Grieve, R.A.F., Dence, M.R. and Robertson, P.B. (1977) Cratering processes: as interpreted from the occurrences of impact melts, in *Impact and Explosion Cratering* (eds D.J. Roddy, R.O. Pepin and R.B. Merrill), Pergamon Press, New York, NY, pp. 791–814.

Gurov, E.P., Kelley, S.P. and Koeberl, C. (2003) Ejecta of the Boltysh impact crater in the Ukrainian Shield, in *Impact Markers in the Stratigraphic Record, Impact Studies*, vol. 3 (eds C. Koeberl and F.C. Martinez-Ruiz), Springer, Heidelberg, pp. 179–202.

Gurov, E., Gurova, E., Chernenko, Y. and Yamnichenko, A. (2009) The Obolon impact structure, Ukraine, and its ejecta deposits. *Meteoritics and Planetary Science*, 44, 389–404.

Hart, R.J., Cloete, M.C., McDonald, I. and Andreoli, M.C. (2002) Siderophile-rich inclusions from the Morokweng impact melt sheet, South Africa: possible fragments of a chondritic meteorite. *Earth and Planetary Science Letters*, 198, 49–62.

Hawke, B.R. and Head, J.W. (1977) Impact melt on lunar crater rims, in *Impact and Explosion Cratering* (eds D.J. Roddy, R.O. Pepin and R.B. Merrill), Pergamon Press, New York, NY, pp. 815–841.

Hawke, B.R., Blewett, D.T., Lucey, P.G. *et al.* (2004) The origin of lunar crater rays. *Icarus*, 170, 1–16.

Hörz, F. (1982) Ejecta of the Ries Crater, Germany, in *Geological Implications of Impacts of Large Asteroids and Comets on the Earth* (eds L.T. Silver and P.H. Schultz), Geological Society of America Special Paper 190, Geological Society of America, Boulder, CO, pp. 39–55.

Hörz, F., Ostertag, R. and Rainey, D.A. (1983) Bunte breccia of the Ries: continuous deposits of large impact craters. *Reviews of Geophysics and Space Physics*, 21, 1667–1725.

Housen, K.R. and Holsapple, K.A. (2011) Ejecta from impact craters. *Icarus*, 211, 856–875.

Howard, K.A. and Wilshire, H.G. (1975) Flows of impact melt at lunar craters. *Journal of Research of the U.S. Geological Survey*, 3, 237–251.

Ivanov, B.A. and Artemieva, N. (2001) Transient cavity scaling for oblique impacts. 32nd Lunar and Planetary Science Conference, abstract no. 1327.

Ivanov, B.A. and Artemieva, N. (2002) Numerical modeling of the formation of large impact craters, in *Catastrophic Events and Mass Extinctions: Impacts and Beyond* (eds C. Koeberl and K.G. MacLeod), Geological Society of America Special Paper 356, Geological Society of America, Boulder, CO, pp. 619–630.

Johnson, B.C. and Melosh, H.J. (2011) Formation of spherules in impact produced vapor plumes. *Icarus*, 217, 416–430.

Kalleson, E., Riis, F., Setsa, R. and Dypvik, H. (2012) Ejecta distribution and stratigraphy – field evidence from the Ritland impact structure. 43rd Lunar and Planetary Science Conference, abstract no. 1351.

Kenkmann, T. and Ivanov, B.A. (2000) Low-angle faulting in the basement of complex impact craters; numerical modelling and field observations in the Rochechouart Structure, France, in *Impacts and the Early Earth* (eds I. Gilmour and C. Koeberl), Lecture Notes in Earth Sciences 91, Springer, Berlin, pp. 279–308.

Kenkmann, T. and Ivanov, B.A. (2006) Target delamination by spallation and ejecta dragging: an example from the Ries crater's periphery. *Earth and Planetary Science Letters*, 252, 15–29.

Kenkmann, T. and Schönian, F. (2006) Ries and Chicxulub: impact craters on Earth provide insights for Martian ejecta blankets. *Meteoritics and Planetary Science*, 41, 1587–1603.

Kieffer, S.W., Phakey, P.P. and Christie, J.M. (1976) Shock processes in porous quartzite: transmission electron microscope observations and theory. *Contributions to Mineralogy and Petrology*, 59, 41–93.

Koeberl, C. and Reimold, W.U. (2005) Bosumtwi impact crater: an updated and revised geological map, with explanations. *Jahrbuch der Geologischen Bundesanstalt, Wien (Yearbook of the Austrian Geological Survey)*, 145, 31–70.

Mader, M., Osinski, G.R. and Marion, C. (2011) Impact ejecta at the Mistastin Lake impact structure, Labrador, Canada. 42nd Lunar and Planetary Science Conference, March 7–11, The Woodlands, TX, LPI Contribution No. 1608, p. 2505.

Maloof, A.C., Stewart, S.T., Weiss, B.P. *et al.* (2010) Geology of Lonar Crater, India. *Geological Society of America Bulletin*, 122, 109–126.

Masaitis, V.L. (1999) Impact structures of northeastern Eurasia: the territories of Russia and adjacent countries. *Meteoritics and Planetary Science*, 34, 691–711.

McEwen, A.S., Preblich, B.S., Turtle, E.P. *et al.* (2005) The rayed crater Zunil and interpretations of small impact craters on Mars. *Icarus*, 176, 351–381.

McEwen, A.S., Banks, M.E., Baugh, N. *et al.* (2010) The High Resolution Imaging Science Experiment (HiRISE) during MRO's Primary Science Phase (PSP). *Icarus*, 205, 2–37. DOI:10.1016/j.icarus.2009.04.023.

Melosh, H.J. (1989) *Impact Cratering: A Geologic Process*, Oxford University Press, New York, NY.

Morgan, J., Lana, C., Kearsley, A. *et al.* (2006) Analyses of shocked quartz at the global K–P boundary indicate an origin from a single, high-angle, oblique impact at Chicxulub. *Earth and Planetary Science Letters*, 251, 264–279.

Newsom, H.E., Graup, G., Sewards, T. and Keil, K. (1986) Fluidization and hydrothermal alteration of the suevite deposit in the Ries Crater, West Germany, and implications for Mars. *Journal of Geophysical Research, B: Solid Earth and Planets*, 91, 239–251.

Oberbeck, V.R. (1975) The role of ballistic erosion and sedimentation in lunar stratigraphy. *Reviews of Geophysics and Space Physics*, 13, 337–362.

Osinski, G.R. (2004) Impact melt rocks from the Ries impact structure, Germany: an origin as impact melt flows? *Earth and Planetary Science Letters*, 226, 529–543.

Osinski, G.R. (2006) Effect of volatiles and target lithology on the generation and emplacement of impact crater fill and ejecta deposits on Mars. *Meteoritics and Planetary Science*, 41, 1571–1586.

Osinski, G.R. and Spray, J.G. (2005) Tectonics of complex crater formation as revealed by the Haughton impact structure, Devon Island, Canadian High Arctic. *Meteoritics and Planetary Science*, 40, 1813–1834.

Osinski, G.R., Grieve, R.A.F. and Spray, J.G. (2004) The nature of the groundmass of surficial suevites from the Ries impact structure, Germany, and constraints on its origin. *Meteoritics and Planetary Science*, 39, 1655–1684.

Osinski, G.R., Spray, J.G. and Lee, P. (2005) Impactites of the Haughton impact structure, Devon Island, Canadian High Arctic. *Meteoritics and Planetary Science*, 40, 1789–1812.

Osinski, G.R., Tornabene, L.L. and Grieve, R.A.F. (2011) Impact ejecta emplacement on the terrestrial planets. *Earth and Planetary Science Letters*, 310, 167–181.

Poelchau, M.H., Kenkmann, T. and Kring, D.A. (2009) Rim uplift and crater shape in Meteor Crater: effects of target heterogeneities and trajectory obliquity. *Journal of Geophysical Research, Planets*, 114, E01006. DOI: 10.1029/2008JE003235.

Pope, K.O., Ocampo, A.C., Fischer, A.G. *et al.* (1999) Chicxulub impact ejecta from Albion Island, Belize. *Earth and Planetary Science Letters*, 170, 351–364.

Pratesi, G., Morelli, M., Rossi, A.P. and Ori, G.G. (2005) Chemical compositions of impact melt breccias and target rocks from the Tenoumer impact crater, Mauritania. *Meteoritics and Planetary Science*, 40, 1653–1672.

Quaide, W.L. and Oberbeck, V.R. (1968) Thickness determinations of the lunar surface layer from lunar impact craters. *Journal of Geophysical Research*, 73, 5247–5270.

Salge, T. (2007) The ejecta blanket of the Chicxulub impact crater, Yucatán, Mexico: petrographic and chemical studies of the K–P section of El Guayal and UNAM boreholes. Humboldt University, Berlin.

Scherler, D., Kenkmann, T. and Jahn, A. (2006) Structural record of an oblique impact. *Earth and Planetary Science Letters*, 248, 43–53.

Schmidt-Kaler, H. (1978) General geologic map of the Ries meteorite crater of southern Germany 1 : 100 000.

Schönian, F., Stöffler, D. and Kenkmann, T. (2004) The fluidized Chicxulub ejecta blanket, Mexico: implications for Mars. 35th Lunar and Planetary Science Conference, abstract no. 1848.

Schulte, P., Deutsch, A., Salge, T. *et al.* (2009) A dual-layer Chicxulub ejecta sequence with shocked carbonates from the Cretaceous–Paleogene (K–Pg) boundary, Demerara Rise, western Atlantic. *Geochimica et Cosmochimica Acta*, 73, 1180–1204.

Sharpton, V.L., Marin, L.E., Carney, J.L. *et al.* (1996) A model of the Chicxulub impact basin based on evaluation of geophysical data, well logs, and drill core samples, in *The Cretaceous–Tertiary Event and Other Catastrophes in Earth History* (eds G. Ryder, D. Fastovsky and S. Gartner), Geological Society of America Special Paper 307, Geological Society of America, Boulder, CO, pp. 55–74.

Shoemaker, E.M. (1963) Impact mechanics at Meteor Crater, Arizona, in *The Moon, Meteorites and Comets* (eds B.M. Middlehurst and G.P. Kuiper), University of Chicago Press, Chicago, IL, pp. 301–336.

Shuvalov, V. (2003) Cratering process after oblique impacts. Third International Conference on Large Meteorite Impacts, no. 4130.

Simonson, B.M. and Glass, B.J. (2004) Spherule layers – records of ancient impacts. *Annual Review of Earth and Planetary Science*, 32, 329–361.

Smit, J. (1999) The global stratigraphy of the Cretaceous–Tertiary boundary impact ejecta. *Annual Review of Earth and Planetary Sciences*, 27, 75–113.

Stewart, S.T. and Ahrens, T.J. (2005) Shock properties of $H_2O$ ice. *Journal of Geophysical Research*, 110, E03005. DOI:10.1029/2004JE002305.

Stöffler, D. (1977) Research drilling Nördlingen 1973: polymict breccias, crater basement, and cratering model of the Ries impact structure. *Geologica Bavarica*, 75, 443–458.

Stöffler, D., Gault, D.E., Wedekind, J. and Polkowski, G. (1975) Experimental hypervelocity impact into quartz sand: distribution and shock metamorphism of ejecta. *Journal of Geophysical Research*, 80, 4062–4077.

Tornabene, L.L., Moersch, J., McSween, H.Y.J. *et al.* (2006) Identification of large (2–10 km) rayed craters on Mars in THEMIS thermal infrared images: implications for possible Martian meteorite source regions *Journal of Geophysical Research*, 111, E10006. DOI: 10.1029/2005JE002600.

Tornabene, L.L., Mcewan, A.S., Osinski, G.R. *et al.* (2007) Impact melting and the role of sub-surface volatiles: implications for the formation of valley networks and phyllosilicate-rich lithologies on early Mars. Seventh International Conference on Mars, 3288.pdf.

Turtle, E.P., Pierazzo, E., Collins, G.S. *et al.* (2005) Impact structures: what does crater diameter mean? in *Large Meteorite Impacts III* (eds T. Kenkmann, F. Hörz and A. Deutsch), Geological Society of America Special Paper 384, Geological Society of America, Boulder, CO, pp. 1–24.

Vishnevsky, S.A. and Lagutenko, V.N. (1986) The Ragozinka astrobleme: an Eocene crater in central Urals. *Akademii Nauk USSR*, 14, 1–42 (in Russian).

Von Engelhardt, W. (1990) Distribution, petrography and shock metamorphism of the ejecta of the Ries crater in Germany – a review. *Tectonophysics*, 171, 259–273.

# FIVE

# *The modification stage of crater formation*

**Thomas Kenkmann\*, Gareth S. Collins[†] and Kai Wünnemann[‡]**

*\*Institut für Geowissenschaften – Geologie, Albert-Ludwigs-Universität Freiburg, Albertstrasse 23-B, D-79104 Freiburg, Germany*
*[†]IARC, Department of Earth Science and Engineering, Imperial College London, London, SW7 2AZ, UK*
*[‡]Museum für Naturkunde, Leibniz-Institut an der Humboldt-Universität Berlin, Invalidenstrasse 43, 10115 Berlin, Germany*

## 5.1 Introduction

In order to achieve a better understanding of the formation of impact craters, Gault *et al.* (1968) were the first to suggest the subdivision of the impact process into different stages, characteristic for specific processes: (1) the contact and compression stage; (2) the excavation stage; and (3) the modification stage. This chapter focuses on the final, modification stage of impact cratering, which comprises processes that change the shape of the transient cavity, produced during the excavation stage, and result in the final crater form.

The initial impact of a body with a planetary surface generates a shock wave at the impactor–target interface (see Chapter 3). As a result, the impactor and a similar volume of target rocks are compressed to very high density, which raises the pressure and temperature of these materials. The amplitude of the excursion decays with distance travelled by the shock wave. Behind the shock wave, a rarefaction wave generated by reflection of the shock wave at the free surface (i.e. the rear of the projectile and the air–target interface), releases the compressed material from its high-pressure state. Shock wave compression is irreversible, whereas decompression is reversible; hence, the passage of the shock wave results in a net increase in temperature (internal energy and entropy) and particle velocity in the rocks (Melosh, 1989; Chapter 3). The residual velocity component in the target rocks after unloading from shock pressure is the most important aspect that distinguishes hypervelocity impacts from the low-velocity impact of a projectile. The value of this final velocity component typically corresponds to one-fifth of the peak particle velocity during shock wave compression (Turtle *et al.*, 2005) and, therefore, it is large enough to induce material flow, the so-called excavation flow (see Chapter 4). This is directed outwards, away from the point of impact, which leads to the actual opening of a deep bowl-shaped cavity. Eventually, the excavation flow is halted and the cavity stops growing, when insufficient kinetic energy remains to displace the target

against its own weight (gravity-dominated cratering) or against the cohesive strength of the target material (strength-dominated cratering). The resulting cavity at the end of the excavation stage is called the transient cavity (Fig. 5.1a) and is typically several (10–20) times the size of the projectile in diameter (Chapter 4). The cross-sectional shape of the transient crater is often assumed to be a paraboloid with a depth/diameter ratio of roughly 1/3.

While the excavation flow describes the motion of target material away from the impact centre that leads to the opening of the transient cavity, the modification flow is the reverse of this and acts to close the transient cavity. Gravity is the principal force that drives the collapse of the transient cavity. Depending on the degree of modification, the final crater is classified into either a simple or complex morphology (Dence, 1965; Fig. 5.1b,c, Fig. 5.2, and Fig. 5.3). Simple and complex impact craters have fundamental differences in morphology and structure, as detailed below.

It is important to note that the stages of crater formation grade into one another and, in particular, the transition from excavation to modification stage is difficult to define at a specific instant in time in the course of crater formation. It has been demonstrated by numerical modelling that the excavation flow does not come to a halt simultaneously all along the transient cavity surface before it reverses its direction (Turtle *et al.*, 2005). Instead, the transient cavity continues to grow horizontally after the vertical deepening has stopped (Fig. 5.4). Thus, the processes of the excavation flow and crater collapse overlap each other. This overlap increases the more the impact process is governed by gravity rather than by cohesive target strength. Moreover, the temporal overlap of the excavation and modification stage increases with increasing impact obliquity, because uprange crater collapse begins well before collapse in the downrange direction (Elbeshausen *et al.*, 2009). In highly oblique impacts of less than 20°, or so, the classical transient cavity model is no longer applicable.

(a) transient crater    $d_t/D_t \sim 0.33$

$D_t$ : transient cavity diameter

$d_t$ : transient cavity depth

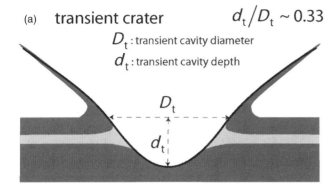

(b) simple crater

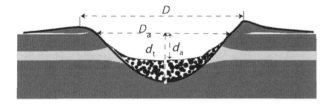

(c) complex crater

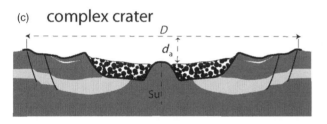

**Figure 5.1** Schematic cross-section through a transient crater (a), a simple crater (b) and a complex impact crater (c). The depth-to-diameter ratio decreases from roughly 0.33 at (a) to 0.1 or even less for complex craters (c).

The concept of the transient cavity is of particular importance in order to estimate the energy required to form a crater of a given size. As the size of the transient cavity is the best measure of impact energy (Schmidt and Housen, 1987), the reconstruction of its extent is crucial and poses a challenging task, particularly for highly modified complex crater structures.

The beginning of the modification stage is a matter of definition. Most commonly, it is defined as when the transient cavity reaches its largest horizontal extent, at the level of the target surface. That is, when the direction of motion at the transient cavity rim switches from outward to inward. The end of the modification stage is reached when significant motion of the target has abated. The duration of crater modification depends on impact energy and strength properties of the target material; its duration increases with impact energy and weakness of the target and can take 15–20 min or so for 100 km-sized impact craters with a weak target. Long-term lithostatic crustal relaxations that are triggered by mass deficits of large impact craters can persist over millions of years and are accommodated by plastic flow at mid to lower crustal levels or even in the underlying mantle. These processes are not detailed further here.

The goal of this chapter is to review recent advances in understanding the process of crater modification. Melosh's classic textbook *Impact Cratering: A Geologic Process* with its chapter 'Cratering mechanics: the modification stage' provides a timeless introduction to processes of crater modification. Here, we briefly recapitulate major concepts and working hypotheses that are more comprehensively presented in Melosh (1989) and, more recently, in Melosh and Ivanov (1999). Our main focus is on the advances in the past 10–20 years since Melosh's textbook was published. Significant improvements have come from numerical simulation of crater collapse phenomena. On the one hand, this was made possible by modifications to the hydrocodes in use that are now capable to consider, for example, three-dimensional

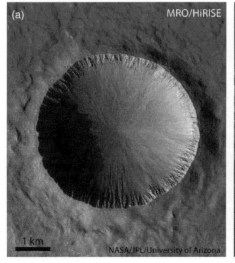

**Figure 5.2** (a) HiRISE image of an unnamed simple crater on Mars (38.7 °N/316.1 °E) displaying an elevated crater rim and steeply dipping upper cavity walls. The mid and lower parts of the wall are covered by talus deposits. Image: NASA/JPL/University of Arizona. (b) *Kaguya/SELENE* image (S0000001616_1906) of the complex impact crater Aristarchus on the Moon, showing a central peak, a flat crater floor with isolated hummocks and an extensive slump terrace zone. Note the different scale bars in the two images.

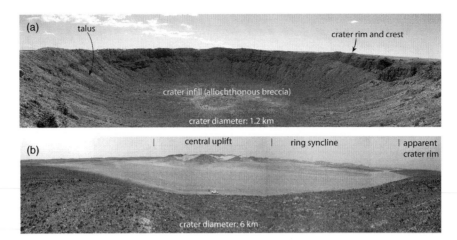

**Figure 5.3**  Composed panorama photographs of a simple and a complex terrestrial impact crater. (a) The simple crater Barringer (Meteor), AZ, USA is 1.2 km in diameter. Autochthonous target rocks outcrop in the upper part of the cavity wall. These rocks dip outward and are truncated by numerous radial faults. Target rocks are overlain by the proximal ejecta blanket that forms the crest of the crater. The proximal ejecta blanket contains overturned strata whose stratigraphic succession is inverted. (b) The complex crater Jebel Waqf as Suwwan, Jordan, measures 6 km from rim to rim (apparent crater rim). The ring syncline is covered by Wadi deposits, the 1 km central uplift is exposed in the centre. (See Colour Plate 15)

geometries, multi-material targets, strength and strain localization. On the other hand, advancements in computational power were a prerequisite for these developments. Significant improvements have also come from structural and kinematic analysis of the subsurface of impact craters. These data provide the ground truth to reconstruct the deformation history of crater formation and there is no review of this topic available currently. Research on impact crater collapse was also stimulated by new spacecraft missions, for example, to Mars and to the icy satellites of Jupiter and Saturn. High-resolution remote-sensing images of complex extraterrestrial craters have provided new constraints for crater formation for a variety of target materials and other boundary conditions, such as surface gravity.

## 5.2  Morphology and morphometry of simple and complex impact craters

The first remote-sensing images of the lunar surface taken by the Lunar Orbiters, in preparation for the Apollo missions, revealed a heavily cratered landscape, where crater morphology varied as a function of size. Crater morphology falls into two classes: simple and complex craters (Dence, 1965). The latter is generally further subdivided into central peak, peak-ring craters, and multi-ring basins. This section provides a brief description of the different crater morphologies and how the size–morphology progression depends, to a first order, on the gravity of the target body.

### 5.2.1  Simple crater morphology

Simple craters are among the most frequent morphological features on planetary surfaces in the solar system. The principal shape of a simple crater consists of a bowl-shaped depression and

a raised crater rim (Fig. 5.1b, Fig. 5.2a). The 1.2 km diameter Barringer Crater (or Meteor Crater), AZ, USA (Fig. 5.3a), is the prime example of a young, well-preserved and well-documented simple impact crater on Earth. Simple craters have well-defined raised rims whose crest line defines the final (rim-to-rim) crater diameter $D$ (Fig. 5.2a). The crater rim is the result of two processes: (1) uplift of the crater wall by the outward and upward movement of target materials by the cratering flow-field and by injection of material during excavation (Roddy, 1979); (2) deposition of the continuous ejecta blanket onto the original target surface. The ejecta near the crater rim consist of near-surface rocks that form a coherent overturned flap, with an inverted stratigraphy (see Chapters 4 and 18). If the raised rim has eroded away, the crater diameter at the present level of erosion is referred to as 'apparent' and denoted $D_a$ (Fig. 5.1b).

The depth/diameter ratio of simple craters is approximately 1/5; thus, they are significantly shallower than the form of the transient cavity (Table 5.1). Bore holes that have been drilled, for example, into the simple crater Brent, Canada, or Barringer, provided a conclusive explanation for this observation. The floor of the morphologically visible apparent crater (depth $d_a$) is underlain by a lens of allochthonous unshocked and shocked target-rock breccias (Grieve, 1987). The lower part of the autochthonous crater wall is often covered by talus deposits.

### 5.2.2  Complex crater morphology

Complex impact craters are larger than simple craters and have a smaller depth/diameter ratio (Fig. 5.1c, Table 5.1). The most remarkable difference between simple and complex impact craters is that complex impact craters contain an uplifted crater floor, which either forms a central uplift (Fig. 5.2b and Fig. 5.3b), a patchy distribution of hills and hummocks, a peak ring or a pan flat crater floor. These uplifted rocks are shocked, brecciated and

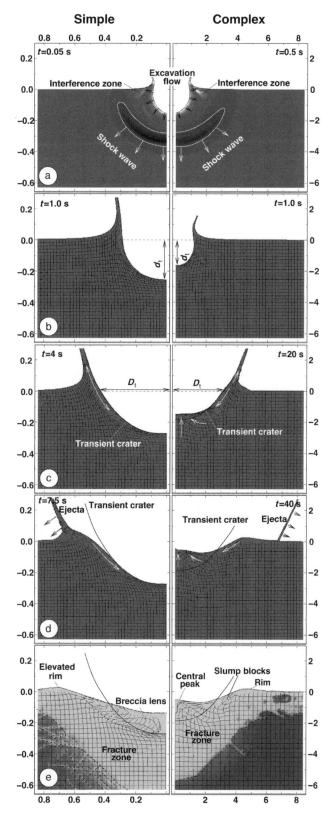

**Figure 5.4** Numerical simulation of crater formation using the iSale hydrocode. The left panel shows five stages of formation of a simple crater; the right panel illustrates a more severe crater modification that results in a complex crater morphology with a central peak. Note that the scale bar in the right panel is 10 times that of the left panel. (a) The hemispherically expanding shock wave induces an excavation flow that opens a rapidly growing transient cavity. (b) The cavity grows continuously, while the shock wave has already left the model boundaries. (c) The parabolically shaped transient cavity reaches its full extent. The floor of the cavity starts to rise in the right panel. (d) The material flow within the cavity is inward directed. In the left model, the inward motion only affects the uppermost transient cavity subsurface. In complex craters (right panels) the transient cavity floor is lifted upwards by buoyancy forces, while the cavity walls move inside the cavity. (e) Extent of the fractured zone beneath the final simple and complex impact craters. Note the lack and presence of a central uplift in the left (simple crater) panels.

**Table 5.1**  Impact crater metrics (see also Chapter 20)

| Crater structure on Earth | Rim diameter $D$ | Data base | Reference |
|---|---|---|---|
| *Transient cavity* | | | |
| depth $d$ | $d = 0.37D$ | | Melosh (1989) |
| *Simple impact craters* | | | |
| apparent depth $d_a$ | $d_a = 0.14D^{1.02}$ | 18 | Pike (1980) |
| | $d_a = 0.13D^{1.06}$ | 7 | Grieve and Pilkington (1996) |
| true depth $d_t$ | $d_t = 0.28D^{1.02}$ | 7 | Grieve and Pilkington (1996) |
| *Complex impact craters* | | | |
| apparent depth $d_a$ | $d_a = 0.27D^{0.16 \pm 0.11}$ | 11 | Pike (1980) |
| stratigraphic uplift su | $su = 0.09D^{1.03}$ | 24 | Grieve and Pilkington (1996) |
| | $su = 0.1D$ | | Melosh and Ivanov (1999) |

heavily deformed. The stratigraphic uplift of the central crater floor systematically increases with increasing final crater size (Table 5.1). Complex impact structures on planetary bodies with volatiles in the subsurface, such as Mars, Callisto and Ganymede, additionally may contain circular to elliptical depressions in their centres. In *floor pit craters* the central pit is located directly on the crater floor; in *summit pit craters* it is situated on top of a central peak (Barlow, 2010). Miscellaneous formation models have been proposed for the occurrences of central pits; for example, central pit formation by vaporization of subsurface volatiles and explosive release of the resulting gases during crater formation (Senft and Stewart, 2009; Barlow, 2010).

The crater floor surrounding a central uplift often appears flat (Fig. 5.2b and Fig. 5.3b). In this region the autochthonous crater floor is not exposed, but is overlain by allochthonous materials, with various shock stages and impact melt rock. The crater rim region of complex impact craters often appears stepped and is subdivided into terraces separated by scarps (Fig. 5.2b). Terraces are narrow close to the crater centre, but increase in width as the rim is approached. The widest, best-defined and last-formed terrace normally occurs just below the crater rim (Pearce and Melosh, 1986; Leith and McKinnon, 1991). The width of the outer terraces $W$ also increases with final crater diameter. Pearce and Melosh (1986) found the empirical relationship between terrace width $W$ and final diameter of $D$: $W \approx 0.09D^{0.87}$. For complex craters it is more difficult to define the final rim accurately because the crater is enlarged by a factor of 1.5–2.0 with respect to the transient cavity (Grieve *et al.*, 1981) and the deformation of the outer crater region ceases gradationally. The continuous ejecta blanket extends from the overturned flap at the collapsed rim of the transient crater outwards and, thus, overlies the terraces and crater rim and does not contribute significantly to the topographical accentuation of the rim (Fig. 5.1c) – see Chapter 4. Nevertheless, the crater rim of pristine extraterrestrial craters can still be delineated by a morphological crest line. Topographical features at the rim are often caused by headscarps of segments of circumferential normal faults.

### 5.2.3  Crater morphology as a function of size

The concept of classifying crater morphology into simple and complex holds true for all bodies in the solar system; however, the crater diameter at which the simple-to-complex transition occurs (and other morphological transitions) varies between planetary bodies depending on surface gravity and target strength. The largest known simple crater in the solar system is approximately 90 km in diameter (Melosh and Ivanov, 1999) and located on Amalthea, the small icy satellite of Jupiter, with a mean radius of only 83 km and an average surface gravity of only 0.02 m s$^{-2}$. On the Moon (1737 km radius, 1.62 m s$^{-2}$ surface gravity), large impact craters with simple geometries have diameters of up to about 16 km; and on Mars (3396 km radius; 3.69 m s$^{-2}$), simple craters reach maximum diameters of approximately 8 km on average. On Earth (6378 km radius, 9.81 m s$^{-2}$), the largest simple craters were formed in crystalline targets and have diameters of up to 4 km (Brent, Canada). In sedimentary targets, the size limit for simple craters on Earth is much less (2–3 km diameter). For example, the 2 km BP impact structure in Libya, which formed in sediments, has a complex morphology.

The comparative planetological analysis of the simple-to-complex transition of impact craters, including studies of craters on Venus and Mercury, shows that the transition diameter is inversely proportional to the surface gravity (Pike, 1988). This indicates that gravity is the main driving force for crater modification. However, the size–morphology progression is also controlled by the strength as the mechanical property of rocks working against the modification of the transient crater. Craters formed in layered sedimentary rocks or in ice are highly susceptible to crater collapse, whereas craters in dense crystalline targets are relatively resistant to collapse.

### 5.3  Kinematics of crater collapse

The excavated transient cavity is negatively buoyant: just as a mountain exerts a downward force on the crust due to its positive mass, a cavity creates an upward force on the crust due to the missing mass. This gravitational force acting to close the crater is resisted by the strength of the target materials (Fig. 5.4 and Fig. 5.5). If the buoyancy force is not strong enough (the transient crater is too small), then strength keeps the material from moving up and inwards to close the cavity and a simple crater is formed (Fig. 5.4, left panel). Alternatively, if the size of the transient cavity exceeds a certain threshold, then the entire cavity

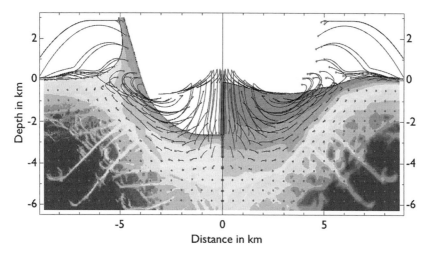

**Figure 5.5** The numerical model (iSale hydrocode) of the impact of a 1 km diameter projectile at 12 km s$^{-1}$ illustrates the motion of tracer particles during transient cavity modification. Red dots (see Plate 16) mark the position of tracers at the moment when the transient cavity has reached its greatest depth. The black lines indicate the subsequent flow path of tracers during crater collapse. The left side shows crater shape and distribution of total plastic strain (blue: low; red: high) at the onset of crater collapse, the right side shows the final crater shape and the final distribution of total plastic strain. For further explanation, see the text. (See Colour Plate 16)

collapses, first at the deepest point of the cavity in the very centre, where the upward direct buoyancy force is strongest and a complex crater is formed (Fig. 5.4, right panel). The relative balance between both mechanisms (the buoyancy acting to close the crater and the material strength acting to stabilize the cavity) determines whether a simple or complex crater is formed.

### 5.3.1 Kinematics of simple crater formation

Simple craters are similar in shape to the transient cavity. Differences are attributed to mass movement of the transient cavity rim and of brecciated material that decorated the surface of the cavity. The evolution of a simple crater is illustrated by a snapshot series from a numerical model of the formation of a 1.5 km diameter crater on Earth in Fig. 5.4b–e (left panels). Observational evidence shows that, depending on target lithology, its state of fragmentation and the presence of volatiles, different types of gravity-driven mass movements may occur, including rolling of blocks, gliding, slumping, debris and mudflow. They initiate along the steepest parts of the crater walls near the slope top end and lead to an increase in cavity diameter by 10–20%. The increase in diameter plus the infill of the cavity explains the decrease of the depth/diameter ratio from 1/3 of the transient cavity to 1/5 of the final simple crater. Depositional fans develop downslope. The fans may have steep fan apices in the upper part like alluvial fans that point to the source region, where wall collapse has occurred. At mid-fan level, adjacent flows converge and form a continuous deposit whose slope depends on the angle of repose of the material. At the cavity bottom, mass flows may collide in frontal or oblique manner with each other. Wall collapse and subsequent downslope deposition can mix with shocked rock fragments or impact melt that lined the transient cavity. Both lithologies may be superimposed by a relatively thin cover of impact breccia.

### 5.3.2 Kinematics of complex crater formation

In contrast to simple craters, the morphology of complex craters differs significantly from the shape of the transient cavity. The strong modification is the result of extensive gravity-driven collapse that is illustrated by a series of snapshots from a numerical model of the formation of a 10 km diameter crater on Earth shown in Fig. 5.4b–e (right panels). The collapse occurs first at the deepest point of the transient cavity. The cavity floor starts to rise, causing a rotational flow field underneath the cavity. The greatest total uplift and uplift rate exist in the very centre, pushing the cavity floor upward (Fig. 5.5). Material underneath the cavity that is located slightly off-centre is uplifted to a lesser degree and simultaneously moves towards the centre. The upward and inward flow creates a mass deficit in the subsurface beneath the cavity rim, which ultimately results in a down-sagging of the steep crater walls and an enlargement of the cavity zone involved into the modification flow. Near the cavity surface, material directly moves downslope towards the centre in a manner similar to that described for simple craters (Fig. 5.5). In large impact craters, the gravitationally driven central uplift can grow so fast and high that inertial flow leads to an overshooting of the equilibrium height of the central uplift. The central uplift, itself, becomes gravitationally unstable and collapses under its own weight. In this case, a downward and outward-directed collapse flow occurs. This idealized pattern of motions holds for cavity collapse in totally homogeneous ductile materials that have no tendency to localize deformation.

While the principal pattern of motions appears to be valid for rocky targets, abundant variations exist that can be studied in terrestrial impact craters. Complex craters on Earth have been eroded to different degrees. Thus, these craters provide opportunities to study the deformation inventory at various levels beneath the original crater floor. In addition, the observed differences can also be attributed to variations in a number of boundary conditions, including impact energy and impact angle, target composition, target rheology and pre-impact target structure.

## 5.4   Subsurface structure of complex impact craters

### 5.4.1   Crater rim

It is difficult to compare pristine extraterrestrial craters with the eroded remnants of terrestrial complex craters, where a morphologically visible crater rim is lacking (Turtle *et al.*, 2005). For example, on Earth, the outermost continuous ring of concentric normal faults is often regarded as the crater diameter ('apparent crater' in Turtle *et al.* (2005)). Examples of relatively weakly eroded craters whose rim is defined by concentric fault zones are the 26 km Ries Crater, Germany, or the 180 km Chicxulub impact crater, Mexico. Crater rim faults (Fig. 5.6) typically undergo unconstrained (free-surface) dip-slip and were denoted as superfaults by Spray (1997). As crater rim normal faults tend to dip toward the centre (60° from the horizontal is the most common angle for normal faults), the apparent crater diameter $D_a$ shrinks

with increasing depth of erosion $E$ by $D_a = D - 1.15E$. Thus, to determine the initial final diameter of a terrestrial crater correctly, it is necessary to know the amount of erosion.

Very deeply eroded impact structures are often not defined by concentric normal faults. Instead, circumferential monoclines (Fig. 5.6) or a combination of inward-dipping normal faults and monoclines are common, particularly if the target is a sedimentary and stratified one. Monoclines are open folds in which one of the fold limbs remained fixed and *in situ*. The inner limb of a crater rim monocline usually dips downward towards the crater and the crater rim is defined by the trace of the monocline's hinge. Examples for this type of crater rim are present at Upheaval Dome, USA (Kriens *et al.*, 1999), Serra da Cangalha, Brazil, or Matt Wilson, Australia (Kenkmann and Poelchau, 2009).

Pristine terrestrial craters are rare and are usually buried beneath sediments, where they were immediately protected from erosion. They usually occur in depo-centres like shallow marine, where continuous sedimentation occurs. Thus, these craters belong to the class of marine-target craters. The presence of a (strength-less) water column as an uppermost target layer strongly affects the modification phase of cratering. Resurging water modifies the upper part of the solid target, particularly the crater rim area (for further details see Section 5.7).

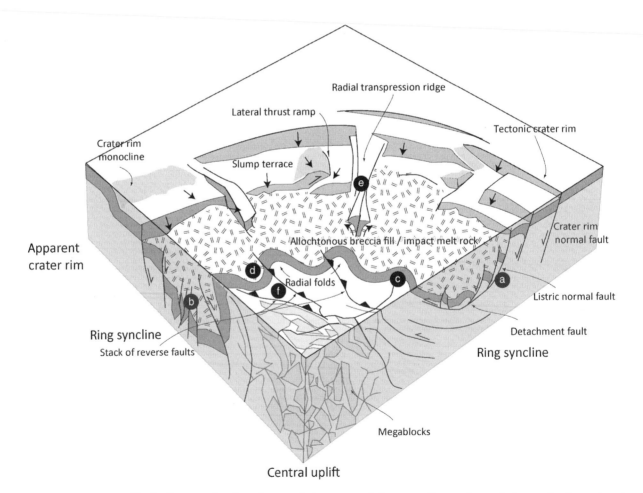

**Figure 5.6**   Schematic block diagram showing characteristic structural features of a partly eroded complex impact crater. Alphabetic characters 'a' to 'f' refer to in photographs in Fig. 5.7. For explanation, see the text. © T. Kenkmann 2011.

## 5.4.2 Ring syncline

The moat between the erosional remnant of the crater rim and the central uplift is termed the ring syncline (Fig. 5.6). Ring synclines are mostly asymmetric in cross-section and often have a steeply dipping or even overturned inner limb that leads over to the central uplift. In almost all known terrestrial complex impact craters, the ring syncline is not a simple synform. It is subdivided into numerous fault-bounded segments or disintegrated into blocks (if fault zones completely frame and isolate a certain rock volume). Between the crater rim and the axis of the ring syncline, normal faults of more-or-less concentric strike are frequent. Normal faulting along non-planar faults is often associated with antithetic or synthetic rotations of the hanging-wall unit. These faults often develop listric shapes, meaning that the fault slope decreases with depth (Fig. 5.6 and Fig. 5.7a). In stratified target rocks they can merge into low-angle detachments at depth to compensate for the inward movement of material during crater collapse (Kriens *et al.*, 1999; Kenkmann *et al.*, 2000; Osinski and Spray, 2005; Fig. 5.6). In this case, bedding planes of the stratified sediments are used as glide planes. Displacements related to the modification stage commonly indicate inward and downward motion within the ring syncline. Owing to the formation of the central uplift, the low-angle normal faults may transform into outward-dipping reverse faults at the inner limb of the ring syncline (Jahn and Riller, 2009). The structural complexity of a ring syncline increases towards the centre because the amount of constriction increases by the motion of rock towards the centre. The resulting convergent particle trajectories can be compensated either by a bulk thickening of inward-sliding masses (tight folding, stacking of rock units along reverse faults (Fig. 5.7b), plastic flow) or by the formation of localized collision zones at the edges of individual mass flows (Kenkmann and von Dalwigk, 2000). In these so-called 'radial transpression zones' material is uplifted to accommodate the converging mass flow and may form morphological ridges (Fig. 5.6). Different modes of uplift are possible, including radial folding (Fig. 5.7c), lateral overthrusting and the formation of positive flower structures (Fig. 5.7d).

## 5.4.3 Central uplift

The highest degree of deformation during the modification stage of cratering occurs within the central uplift. General differences exist between the uplift of crystalline material and stratified sediments. In the absence of appropriate marker beds, impacts into crystalline rock targets often do not allow the motions (during modification) to be reconstructed, in detail. Faulting, however, dominates over folding in these competent rocks. The ratio of the central uplift diameter to the apparent crater diameter increases, with increasing amounts of erosion. In contrast, the observable stratigraphic uplift decreases with erosion. The structural characteristics described in the following were found in central uplifts composed of stratified sediments.

Radial anticlines and synclines are typical for the periphery of central uplifts and the inner ring syncline (Fig. 5.7d,e). They result from the convergent mass flow. They usually plunge outward and cause the serrated appearance of central uplifts in geological maps (e.g. at Decaturville, USA; Offield and Pohn, 1979). The hinge line of these radially striking folds is often bent and plunges more steeply towards the core of the central uplift in such a way that, near the centre, vertical folds or folds with overturned hinges can develop (Fig. 5.6 and Fig. 5.7e). Steeply inclined, vertical or even overturned beds of central uplifts result from this. Moreover, a gradational transition in fold tightness may be detected, with open symmetrical anticlines of the central uplift periphery changing into isoclinal and overturned folds towards the centre. Space incompatibilities of the folds increase with increasing fold tightness. This leads to the initiation of reverse faults in the core of these folds and their rapid propagation into the limbs to finally offset one of the fold limbs from the other. As a consequence, fold limbs become detached into sheet-like blocks, bounded by reverse faults, which 'lean up' against the core of the central uplift. In the core of a central uplift, block sizes decrease further (Kenkmann *et al.*, 2005). In sedimentary targets, the stratigraphic context can be completely broken up. Brecciation and breccia bodies that are at first restricted as fault breccias to the edges of thrust units and blocks can become the dominant rock type in the core of the central uplift. Owing to the immense stratigraphic uplift, very large complex impact structures on Earth, such as the Vredefort Dome, South Africa, show an increase in metamorphic grade toward the centre of the structure. In such cases, an eroded central uplift provides a section through the upper part of the crust (Gibson and Reimold, 2000).

Structural crater modifications may also occur due to post-impact sediment loading (Tsikalas and Faleide, 2007). This has been demonstrated, for example, at the Mjolnir impact structure, Barrents Sea, Norway, where differential compaction during long-term subsidence laterally varies within the crater and caused the formation of a very prominent central high.

## 5.4.4 Peak ring

If the central uplift exceeds a certain size, deformation features can be observed that indicate the gravitational collapse of the central uplift. Among them are: (1) overturning of strata in the central uplift periphery (Morgan *et al.*, 2000; Lana *et al.*, 2003; Jahn and Riller, 2009); (2) normal shear zones in the central area dipping outward and offsetting the uplifted strata (Osinski and Spray, 2005); (3) radial transtension troughs (Kenkmann and von Dalwigk, 2000); and (4) outward kinking and buckle folding of uplifted strata. Eventually, the collapsing central uplift flows outward, thereby superposing the downfaulted rocks of the surrounding ring syncline. The 65 Ma Chicxulub impact crater in Mexico is the most prominent example of a peak-ring crater on Earth. The collapse of the central uplift led to the formation of a rugged peak ring of 40 km radius that stands several hundred metres above the otherwise relatively flat crater basin floor (Brittan *et al.*, 1999). The peak ring most likely consists of heavily brecciated material, as indicated by a gravity and seismic velocity low (Hildebrand and Boynton, 1990; Morgan *et al.*, 2000). The existence of slumped blocks of the annular trough beneath the peak ring was inferred by reflection seismic studies, which demonstrate inward-dipping reflectors, which have been interpreted

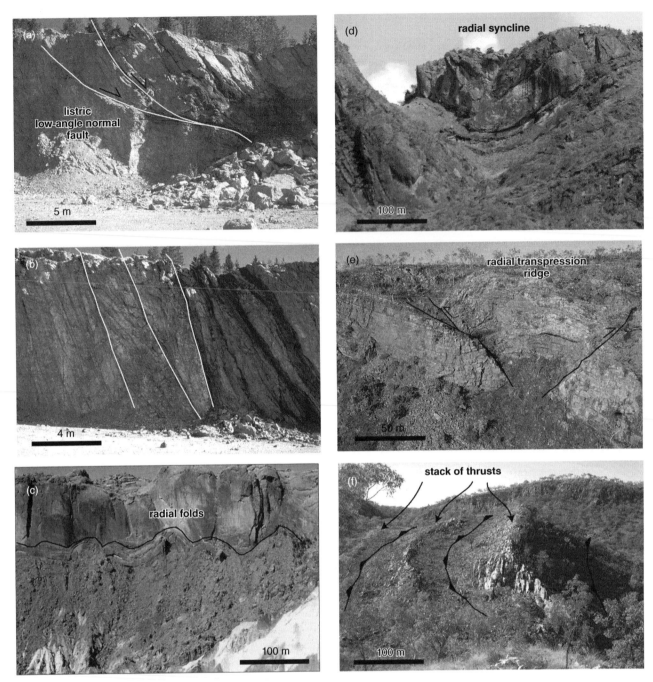

**Figure 5.7**   Field photographs of structural features characteristic for complex terrestrial impact craters. (a) Listric normal faulting observed in the ring syncline of the Siljan impact crater, Sweden (apparent crater diameter: 52–65 km). (Image from Kenkmann and von Dalwigk 2000, with permission.) (b) Steeply dipping stack of Ordovician limestones in the ring syncline of the Siljan impact crater, Sweden. (c) Radial folding in the periphery of the central uplift of Upheaval Dome, UT, USA, observed from the centre of the structure. The fold hinges plunge outward. Note that the inner part of the central uplift is eroded (apparent crater diameter: 5.5 km). (d) Radial syncline within the central uplift of Serra da Cangalha impact crater, Brazil (apparent crater diameter: 14 km). (e) Radial transpression ridge (positive flower structure) in the periphery of the central uplift of the elliptical Matt Wilson impact crater, NT, Australia (apparent crater diameter: 6.3 × 7.5 km). (f) Imbrication of blocks thrusted onto each other in the core of the central uplift of Matt Wilson Crater, NT, Australia. (See Colour Plate 17)

to indicate a simultaneous outward collapse of the central uplift and inward collapse of the transient crater (Brittan *et al.*, 1999). Hydrocode modelling has reproduced this dynamic behaviour, showing that the overturned flap of the transient crater rim moved into the cavity and is finally situated beneath the peak ring (Morgan *et al.*, 2000).

## 5.5 Mechanics of cavity collapse: what makes the target so weak?

The above description of transient cavity collapse is often regarded as phenomenological, rather than physical, because it explains *how* transient cavities collapse, without addressing *why* they collapse (Melosh, 1989). This may seem surprising, given that the phenomenon is supported by illustrations from numerical models of crater formation that, by definition, are governed by the fundamental laws of physics. To match observational constraints from terrestrial and extraterrestrial craters, however, theoretical and numerical models of impact crater formation must assume strength properties for the impacted target that are much weaker than typical for geological materials (Melosh, 1977; McKinnon, 1978). Providing a physical explanation for the apparent transitory low strength of the target is an enduring problem in impact cratering mechanics (Melosh, 1989; Melosh and Ivanov, 1999; Senft and Stewart, 2009).

The paradox of crater collapse is readily appreciated by a simple analysis of stresses surrounding a hemispherical cavity. For an uplift of the transient cavity floor to occur, stresses must exceed the strength $Y$ of the target material. For a constant $Y$, large craters will collapse if their depth exceeds the ratio $Y/\rho g$. Hence, this simple model explains both the presence of a transition from simple crater formation to complex crater formation and the $1/g$ dependence of the crater size at which this transition occurs on different planetary bodies. Rigorous static analysis of cavity slumping has shown that, for substantial rim collapse or floor uplift to occur, the actual effective strength must be less than approximately 3 MPa (Melosh, 1977), with little or no internal friction (McKinnon, 1978). Moreover, the same low strength and friction are required by dynamic models of crater formation (e.g. O'Keefe and Ahrens, 1999; Wünnemann and Ivanov, 2003). These properties are inconsistent with laboratory measurements of the strength of rock, even accounting for the weakening influence of cracks, fissures and faults present in the target rocks on a large scale. The unavoidable conclusion is that the impact process itself must in some way reduce the effective strength of the material that it disturbs.

### 5.5.1 Target disintegration into blocks

Melosh (1989) and Melosh and Ivanov (1999) discuss several ideas proposed to explain the apparent low-strength behaviour of rocks during a large impact event. Here, we present recent geological and geophysical observations that provide constraints on the weakening mechanism and review the weakening mechanisms employed by modern impact models. Detailed geological mapping of eroded impact structures on Earth suggests that the observed deformation depends on the mechanical properties of the target

rocks at their original depth (confining pressure and temperature), as well as their proximity to the crater centre. At many impact structures, geological observations show that the shallow target underneath the transient cavity disintegrates into large 'megablocks' during the cratering process, in particular in the central uplift (Fig. 5.6). These blocks are commonly internally deformed (bent or folded) at the millimetre to decametre scale, rather than being entirely rigid and bounded on all sides by faults. An average block size of approximately 100 m was determined from the Vorotiv Deep Borehole (5374 m) drilled through the central uplift of the 40 km diameter Puchezh-Katunki impact crater in Russia (Ivanov *et al.*, 1996). Mapping at Upheaval Dome, USA (7 km diameter; Kenkmann *et al.*, 2006), and Waqf as Suwwan, Jordan (6 km diameter; Kenkmann *et al.*, 2010), impact structures (Fig. 5.3b) revealed 50–100-m-scale block sizes, with evidence of both a lithological control on block size (smaller blocks were observed in limestone relative to chert) and an increase in block size as a function of distance from the crater centre, which is in accordance with theoretical models of rock fragmentation (Grady and Kipp, 1987). The disintegration of the target into blocks may be responsible for a temporarily weakening of the target (the so-called block oscillation model of Ivanov and Kostuchenko (1997)).

### 5.5.2 Distributed and localized brittle deformation

The intensity of impact deformation increases from the rim to the centre. Growth and collapse of the transient cavity leads to an accumulated strain of approximately 1.0 in the material underneath the crater, decreasing to approximately 0.25 at the crater rim (Collins *et al.*, 2004). Large bulk strains in the centre of an impact crater are often associated with a macroscopically ductile appearance of rocks. This can be achieved by a pervasive cataclastic flow at the grain scale (Kenkmann, 2003) or as a result of brittle faulting with small (millimetre-scale) displacements on a network of numerous closely and uniformly spaced fractures (Kenkmann, 2002). Large bulk strains can also result from highly localized faulting, with large offsets (e.g. Spray, 1997). Distributed brittle deformation, when considered on a macroscale, produces apparent ductile behaviour both within an individual block and within block zones. In general, distributed brittle deformation occurs in the central uplift and crater centre, where blocks are smaller or more internally damaged; large-scale displacements and localized faulting are evident in the crater rim zone, between blocks that are larger and less internally damaged.

These observations suggest that, from a macroscopic point of view, the sub-cavity material (at least in the upper crust) is most appropriately described as a 'rock mass', as it is composed of discrete blocks of rock that are orders of magnitude smaller than the size of the transient cavity. Indeed, pervasive fracturing of the target seems to be an important prerequisite for the formation of impact folds and distributed brittle deformation during collapse. Hence, an obvious weakening mechanism is fracturing and fragmentation induced by the passage of the shock wave. Large-scale, fractured rock masses are considerably weaker than pristine rock samples, which may explain the low effective cohesion of the target. Despite pervasive fractures, however, rock masses typically retain internal friction angles much higher than the few degrees required to explain the observed uplift of transient cavity floors (McKinnon, 1978). Most modern numerical simulations of crater formation account for fracturing of the target and the concomitant reduction in shear strength (e.g. Collins *et al.*, 2004), but still require additional weakening to facilitate uplift of the cavity floor.

### 5.5.3   Localized melting

Another near-ubiquitous characteristic of impact structures is the generation of impact melt (see Chapter 9). The products of impact melting at terrestrial impact structures range from tiny glassy melt fragments within impact breccias to thick sheets of coherent impact melt rock (Grieve, 1991). The volume of impact melt, relative to the volume of the transient cavity, increases with crater size (Grieve, 1991). This differential scaling occurs because gravity controls crater size but has a minor affect on melt production (Grieve and Cintala, 1992). Differential scaling analysis (for impacts into cold targets) suggests that impact melt volume exceeds crater volume for craters larger than about 2000 km on the Moon and about 400 km on Earth (Melosh, 1989; Grieve and Cintala, 1992). In such cases, impact craters will form within a melt pool and collapse completely to a flat surface. In smaller craters, however, melt volume is a tiny fraction of the transient crater volume. Furthermore, impact melt is not distributed uniformly in the material surrounding the crater; rather, the majority is present in a relatively thin sheet lining the final crater floor.

An important secondary component of melt rock at a number of large terrestrial craters, such as Vredefort, South Africa, comes from pseudotachylitic breccias (see Chapter 9). Pseudotachylite is generally regarded as a melt rock formed by frictional melting (e.g. Magloughlin and Spray, 1992). Pseudotachylitic breccias are breccias containing a melted rock matrix that resembles pseudotachylite but where the genetic origin of the melt is unclear (Reimold and Gibson, 2005). Potential mechanisms for producing the melt in these breccias include friction melting (Spray and Thompson, 1995), shock melting (Fiske *et al.*, 1995), decompression melting (Reimold and Gibson, 2005) or drainage from the melt sheet above (Lieger *et al.*, 2009, Riller *et al.*, 2010). Pseudotachylitic breccias are abundant in the central uplift of the Vredefort structure and are particularly pervasive in the central core rocks, which were uplifted furthest, from the greatest depth (Reimold and Gibson, 2006).

The ubiquity of impact melting suggests heat as a weakening mechanism. It is well known that the strength of rock drops substantially as temperature approaches the melting point (e.g. Stesky, 1974). Numerical impact models implicitly include shock heating and all models in use today reduce the strength of the target material to account for the temperature increase (e.g. O'Keefe and Ahrens, 1999). However, only in very large impacts or impacts into planetary surfaces with steep thermal gradients (such as those likely to have been present in the early Solar System) is shock heat an important weakening agent. A possible example of thermal softening as a significant weakening mechanism is the central uplifted core of the 200 km diameter Vredefort impact structure. Lana *et al.* (2003) attributed the apparent lack of megablock formation and the absence of large-displacement localized deformation in these rocks to their relatively large depth of origin and high temperature. They suggest that the large strains

in the core are instead accommodated by many, small (millimetre- to centimetre-scale) displacements along pseudotachylitic breccia veins that are pervasive throughout the central core of the Vredefort structure and may have been caused by shock heating and/or friction melting. The uniform distribution of these small movements gives the impression of large-scale ductile deformation, which may have been aided by the 700–1000 °C temperatures of the originally deep core complex (Gibson *et al.*, 1998).

It has been postulated that shock and frictional melting during the excavation flow might provide sufficient lubrication to lower the strength of block contacts during the later stages of movement (Dence *et al.*, 1977). This idea is supported by field observations of possible frictional melts at several large craters (Spray and Thompson, 1995; Spray 1997). In many cases, however, the significance (in volume and distribution within the crater) of melts that form conclusively as a result of friction is unclear (e.g. Reimold and Gibson, 2005; Lieger *et al.*, 2009). Moreover, although the total displacements and slip rates necessary to generate melt along a fault are easily achieved during crater formation, the volume of friction melt expected is very small (less than a few volume per cent) compared with the volume of collapsing material (Melosh, 2005). Hence, whether or not sufficient friction melt can be formed to lubricate crater collapse remains uncertain.

The relatively low abundance of melt (compared with crater volume) in most terrestrial impact craters (particularly small to mid-sized complex craters) and the predominance of brittle deformation suggests that crater collapse is dominated by the frictional interaction of cold rocks. This might be the internal friction in a breccia, the friction between large blocks or the frictional resistance to slip along discrete fault zones. Hence, the dominant weakening mechanism must temporarily reduce friction in all these settings.

### 5.5.4   Temporary weakening

A relatively well-developed alternative working hypothesis for temporarily reducing friction both within a rock mass and along fault zones is acoustic fluidization (Melosh, 1979, 1996). The premise of acoustic fluidization in the context of an impact is that seismic vibrations generated by the shock wave and scattered within the fractured rock mass surrounding the crater, or within narrow fault zones, result in pressure fluctuations about the ambient overburden pressure. During periods of low pressure, frictional resistance is diminished, leading to slip events in low-pressure zones. The time- and space-averaged effect of this process is that the rock mass behaves rheologically as a viscous fluid. Hence, this mechanism elegantly explains the distributed brittle deformation, which leads to apparent large-scale ductile deformation in cold rocks, as observed in many terrestrial craters. The only prerequisites are that the target is pervasively fractured by the expanding shockwave and that the scattered pressure wave-field behind the shock is of sufficient amplitude and has sufficient longevity to facilitate slip for the duration of transient cavity collapse. As slip movements may generate additional seismic vibrations, providing a positive feedback to the weakening mechanism, the working hypothesis also allows for large displacements along localized fault zones (Melosh, 1996).

Simple parameterizations of the macroscale effect of acoustic fluidization have been used in numerical impact models for many years (e.g. Melosh and Ivanov, 1999), producing dynamic crater formation models in good agreement with observational constraints on cavity collapse and reproducing the general size–morphology progression of craters on crystalline planetary surfaces (e.g. Wünnemann and Ivanov, 2003). The working hypothesis, however, has not been universally accepted. This is, in part, because the transient nature of the pressure fluctuations means that evidence for their existence is not necessarily recorded in the rocks and, in part, because the limited experimental support for the theory implies that some of the model parameters are unconstrained and can be tuned to match observation. Nevertheless, the success of numerical impact models that employ acoustic fluidization in reproducing many of the observed features of large impact craters, with a self-consistent set of model parameters, suggests that the phenomenology of the hypothesis is correct.

Modern numerical impact models are generally successful in predicting the large-scale deformation of target rocks during crater formation. Most models, however, do not reproduce observed zones of highly localized displacement, such as those commonly observed near the crater rim. This is primarily a limitation of resolution. Fault zones in a 10–100-km-scale crater are typically centimetres to metres in width, whereas the minimum cell dimensions in numerical impact models are tens to hundreds of metres in size. Hence, failure along individual faults cannot be modelled explicitly and must be parameterized as an amount of damage spread across an entire computational cell. Moreover, most numerical impact models are two-dimensional, with an axial symmetry; hence, cells represent rings of material and material movement is restricted to the vertical and radial directions – no tangential deformation is permitted. In reality, target motions during crater excavation and collapse have tangential as well as both radial and vertical components (e.g. Kenkmann and von Dalwigk, 2000; Osinski and Spray, 2005). Nevertheless, the inexorable advance of computer power means that two-dimensional numerical impact models today are being performed with close to the resolution required to represent metre-scale strain localization in a 10–100-km-scale crater (Senft and Stewart, 2009). This raises important questions about how existing parametric damage models (e.g. Collins *et al.*, 2004), which assume that fracturing is uniform and isotropic within a cell and operates on a scale below the mesh resolution, should be modified to account for explicit resolution of individual fault zones (Collins *et al.*, 2008b) and the mechanics of faulting (Senft and Stewart, 2009). Inspired by experimental observations of dynamic friction reduction during high-strain, high-strain-rate deformation of rocks (e.g. Di Toro *et al.*, 2004), Senft and Stewart (2009) proposed a parametric strain-rate weakening model that reduces the friction in damaged cells that exceed specified minimum total strain and strain-rate criteria. Despite a dependence of the detail of the results on resolution, the model's success in matching observed features in large craters suggests that temporary weakening of fault zones is sufficient to explain complex crater collapse. The physical mechanism for this weakening, however, is uncertain and may be different in different target lithologies. Potential explanations include friction melting, pore-fluid pressurization, granular flow of fault gouge material, silica gel formation in quartz-rich rocks

acoustic fluidization of fault gouge and flash heating along asperities (Senft and Stewart, 2009 and references therein).

## 5.6 Effects of oblique impact incidences on cavity collapse

Effects of impact obliquity on transient cavity modification and the associated subsurface structure have long been neglected, or even ruled out, and have been investigated only recently. The geometry of the ejecta blanket was often regarded as the only indicator to decipher azimuth and angle for an oblique impact (e.g. Herrick and Hessen, 2006; Chapter 4). The crater outline is in fact insensitive to the impact trajectory and remains circular with the exception for highly oblique impact (<10°). A number of morphological crater features have been cited as diagnostic of oblique impacts, such as a depressed rim with a steepened inner slope uprange, a large central uplift diameter relative to crater diameter or an uprange offset of the central peak (e.g. Schultz and Anderson, 1996). The latter interpretation, however, has been disputed (Ekholm and Melosh, 2001).

Recently, the deep internal structure of the eroded crater floor and central uplift was tested as a tool to derive the impact vector (Scherler *et al.*, 2006; Kenkmann and Poelchau, 2009, Kenkmann *et al.*, 2010). Non-radial structural features possibly indicative for an oblique incidence are: (1) dominance of a thrust direction within a central uplift (Fig. 5.7f); (2) bilateral symmetry of the central uplift, which may be divided into two parts; (3) occurrence of anticlines and synclines parallel to the symmetry axis; (4) normal dipping strata uprange and overturned strata downrange; and (5) normal plunging radial fold axes uprange and overturned plunging axes downrange. (Features (1), (4) and (5) are illustrated in Fig. 5.6 and would indicate an impact from right to left.) The observed deformation features suggest a downrange transport of rock and a central uplift that initiates uprange and migrates downrange as the central uplift and crater grows to its final size. This is in agreement with flow fields inferred from numerical models of oblique impact cratering (Shuvalov and Dypvik, 2004; Elbeshausen *et al.*, 2009). Layered sedimentary rocks with much less resistance to horizontal movement than to vertical movement seem to be particularly susceptible to this type of deformation.

## 5.7 Effects of rheologically complex targets on cavity modification

As discussed in Section 5.5, the termination of transient cavity growth and the degree and nature of subsequent cavity collapse are controlled by gravity and the dynamic strength of the post-shock, fractured target material (Melosh, 1989; Chapter 3). Consequently, variations in density and strength within a target can have a profound effect on crater formation. The standard model of crater formation depicted in Fig. 5.2 and Fig. 5.3 applies to craters formed in a uniform, isotropic crystalline target. However, in many contexts across the Solar System, target volumes are heterogeneous, exhibiting vertical or lateral changes in density, strength and/or rheology.

Vertical heterogeneity typically takes the form of stratigraphic layering. Much of the Earth's surface, for example, is covered by layers of sedimentary rocks and/or a water layer. In many craters, the most remarkable heterogeneity is defined by the transition from layered sedimentary rocks to competent crystalline basement rocks. This is, for example, the case at the Ries impact structure, Germany (Pohl *et al.*, 1977), or at the Chesapeake Bay impact structure, USA (Gohn *et al.*, 2008). A comparative study of three terrestrial craters (El'gygytgyn in Russia, Ries in Germany and Haughton in Canada) of similar size but very different sediment thicknesses demonstrated that important structural differences between the craters were caused by the differences in thickness of the sedimentary cover (Collins *et al.*, 2008a). For example, the presence of a topographic inner ring at Ries, but not at Haughton, appears to be a consequence of the difference in sediment thickness. At Ries, stronger basement rocks are sufficiently close to the surface that they are uplifted and overturned during excavation and remain as an uplifted ring after modification and subsequent erosion. At Haughton, the corresponding zone of the crater shows up- and over-turning of mid-sedimentary sequence strata (Osinski and Spray, 2005). This work also demonstrated that, for constant impact energy, transient and final crater diameters are larger when the thickness of the sedimentary layer is greater.

The effect of a weak layer overlying a strong layer was investigated experimentally as early as the 1960s, to provide insight into the effect of lunar regolith thickness on small impact crater formation on the Moon (Oberbeck and Quaide, 1968). Based on small-scale hypervelocity laboratory impact experiments into layered targets, Oberbeck and Quaide (1968) concluded that, for a sufficient contrast in layer strength, the morphology of the crater was affected by the stronger underlying layer when the ratio of crater diameter to layer thickness exceeded about four. Craters with central mounds, flat floors or a small deep crater nested within a shallow outer crater were produced, depending on the relative strengths of the target layers and the ratio of crater diameter to weak-layer thickness. Oberbeck and Quaide's conclusions were used to correctly estimate lunar regolith thickness before the Apollo missions, using images of small lunar craters.

Target layering exists in many other contexts in the Solar System, due to variations in composition and temperature: ice (water) and sediment layers on Mars; brittle and ductile ice or water layers on the icy satellites; regolith layers on asteroids, comets and other airless bodies; and, at the largest scale, crust over mantle on differentiated planets and satellites.

The rheological stratification of target rocks also influences the geometry and distribution of major impact-induced shear zones. Strain associated with transient cavity collapse of terrestrial impact craters is often concentrated into rheologically soft beds, such as clay or marl layers, which then control further deformation. If the sedimentary target is flat lying, this can result in low-angle normal faulting or detachments (Kriens *et al.*, 1999; Kenkmann *et al.*, 2000; Fig. 5.6). Strain localization along the interface between a strong layer and a weak one could cause a mechanical decoupling of the two layers. This interface could be between basement and sedimentary cover, or within the sedimentary sequence where competent sedimentary rocks, such as

limestone, overlie weaker sedimentary rocks, such as evaporite lithologies.

Oceanic impacts differ in some fundamental ways to impacts on land (Shuvalov and Trubestkaya, 2002; Wünnemann *et al.*, 2010). Most importantly, the water layer changes the coupling of the impactor's kinetic energy to the seafloor. The thicker the water layer is, the smaller is the fraction of the impactor's kinetic energy that is transferred to the seafloor and the smaller the transient cavity excavated. Cratering in the seafloor is completely suppressed if the water depth is approximately six to eight times the impactor diameter (Wünnemann *et al.*, 2010). In addition, strong water currents along the seafloor, due to the collapse of the transient cavity in the water column, can influence and supplement the modification stage of crater formation in several ways. Typical features of submarine impact structures are: (1) chaotic mixing of the uppermost strata at the seafloor; (2) modification of the crater rim (ranging from formation of resurge gullies to complete erosion of the crater rim; Ormö and Lindström, 2000; Lindström *et al.*, 2005); and (3) enhanced uplift of the transient cavity floor, due to the temporary removal of the overburden of the water layer.

The water column's influence on cavity modification depends strongly on the strength properties of the benthic strata and water depth. For example, the 'inverted-sombrero' morphology (a broad, shallow outer basin surrounding a deeper inner basin) characteristic of marine impact craters such as Chesapeake Bay (Gohn *et al.*, 2008) appears to be a consequence of the sedimentary layer being substantially weaker than the underlying basement (because it is poorly lithified or water-saturated, for example). Numerical simulations of these impacts, which include a large contrast in strength between the sedimentary and crystalline layers, give excellent agreement with interpretations of geophysical data from the craters (Collins and Wünnemann, 2005).

The opposite rheological situation exists on the icy satellites, where a strong brittle layer overlies a weaker subsurface layer. Craters on these bodies exhibit the same simple and central peak morphologies seen for craters below approximately 150 km in diameter on the Moon. The largest craters on the icy Galilean satellites, however, display exotic features that have no obvious lunar analogues, such as large central pits and multiple, closely spaced external rings. In addition, the depth-to-diameter ratios of impact craters on Europa, Ganymede and Callisto show two anomalous transitions with diameter, which have been related to transitions in ice rheology with depth (Schenk, 2002). As the gravities of Ganymede and Europa are similar to Earth's moon, the unusual crater morphologies are considered to be due to the mechanical properties of ice or the presence of subsurface liquid layers (e.g. Schenk 2002).

Numerical simulations of cratering on Mars that included a thin subsurface ice layer suggest that the presence of an icy layer can affect the collapse process (Senft and Stewart, 2008). The presence of a weak ice layer caused variations in crater morphometry (depth and rim height) and permitted infill of the crater floor by ice during the late stages of crater formation. Hence, many of the features of Martian craters that distinguish them from craters on other planetary bodies may be explained by the presence of icy layers, including shallow craters with well-preserved ejecta blankets, icy flow-related features, some layered ejecta structures and crater lakes (Senft and Stewart, 2008).

Target surfaces may also contain lateral heterogeneity, such as topography, pre-existing faults, fractures and joints, or tilted stratigraphy. In this case, the heterogeneity can cause asymmetries in the crater planform and result in polygonal crater shapes (Eppler *et al.*, 1983; Öhman *et al.*, 2010). According to current models, polygonal impact craters' straight rim segments reflect the orientations of the target differently depending on whether the crater is simple or complex. The rectangular joint and fissure pattern at Barringer Crater (Meteor Crater), USA, resulted in a more efficient excavation flow along the fissures (Eppler *et al.*, 1983; Poelchau *et al.*, 2009) than between them. In contrast, in polygonal complex impact structures, crater modification obliterates the excavation flow and the straight rim segments are parallel to joints in the target rocks (Eppler *et al.*, 1983).

Asymmetries in crater structure and planform have been attributed to variation in the thickness of target layers across a number of craters. For example, the width and depth of the terrace zone at the Chicxulub crater, Mexico, has been correlated with variation in water depth and sediment thickness (Gulick *et al.*, 2008; Collins *et al.*, 2008a). Dipping stratigraphy also influenced the structural deformation in the collar of the central uplift at Vredefort (Lana *et al.*, 2003) and at Haughton (Osinski and Spray, 2005).

# References

Barlow, N.G. (2010) What we know about Mars from its impact craters. *Geological Society of America Bulletin*, 122, 644–657.

Brittan, J., Morgan, J., Warner, M. and Marin, L. (1999) Near-surface seismic expression of the Chicxulub impact crater, in *Large Meteorite Impacts and Planetary Evolution II* (eds B.O. Dressler and V.L. Sharpton), Geological Society of America Special Paper 339, Geological Society of America, Boulder, CO, pp. 269–279.

Collins, G.S. and Wünnemann, K. (2005) How big was the Chesapeake Bay impact? Insights from numerical modelling. *Geology*, 33, 925–928.

Collins, G.S., Melosh, H.J. and Ivanov, B.A. (2004) Modeling damage and deformation in impact simulations. *Meteoritics and Planetary Science*, 35, 217–231.

Collins, G.S., Kenkmann, T., Osinski, G.R. and Wünnemann, K. (2008a) Mid-sized complex crater formation in mixed crystalline–sedimentary targets: insight from modeling and observation. *Meteoritics and Planetary Science*, 43, 1955–1977.

Collins, G.S., Morgan, J., Barton, P. *et al.* (2008b) Dynamic modeling suggests terrace zone asymmetry in the Chicxulub crater is caused by target heterogeneity. *Earth and Planetary Science Letters*, 270, 221–230. DOI: 10.1016/j.epsl.2008.03.032.

Dence, M.R. (1965) The extraterrestrial origin of Canadian craters. *Annals of the New York Academy of Sciences*, 123, 941–969.

Dence, M.R., Grieve, R.A.F., Robertson, P.B. and Thomas, M.D. (1977) Terrestrial impact structures, in *Impact and Explosion Cratering* (eds J. Roddy, R.O. Pepin and R.B. Merill), Pergamon Press, New York, NY, pp. 247–276.

Di Toro, G., Goldsby, D. and Tullis, T. (2004) Friction falls towards zero in quartz rock as slip velocity approaches seismic rates. *Nature*, 427 (6973), 436–439.

Ekholm, A.G. and Melosh, H.J. (2001) Crater features diagnostic of oblique impacts: the size and position of the central peak. *Geophysical Research Letters*, 28, 623–626. DOI: 10.1029/2000GL011989.

Elbeshausen, D., Wünnemann, K. and Collins, G.S. (2009) Scaling of oblique impacts in frictional targets: implications for crater size and formation mechanisms. *Icarus*, 204, 716–731.

Eppler, D.T., Ehrlich, R., Nummedal, D. and Schultz, P.H. (1983) Sources of shape variation in lunar impact craters. Fourier shape analysis. *Geological Society of America Bulletin*, 94, 274–291.

Fiske, P.S., Nellis, W.J., Lipp, M. *et al.* (1995) Pseudotachylites generated in shock experiments: implications for impact cratering products and processes. *Science*, 270, 281–283.

Gault, D.E., Quaide, W.L. and Oberbeck, V.R. (1968) Impact cratering mechanics and structures, in *Shock Metamorphism in Natural Materials* (eds B.M. French and N.M. Short), Mono Book Corp., Baltimore, MD, pp. 87–99.

Gibson, R.L. and Reimold, W.U. (2000) Deeply exhumed impact structures: a case study of the Vredefort structure, South Africa, in *Impacts and the Early Earth* (eds I. Gilmour and C. Koeberl), Lecture Notes in Earth Sciences, vol. 91, Springer, Berlin, pp. 249–277.

Gibson, R.L., Reimold W.U. and Stevens, G. (1998) Thermal metamorphic signature of an impact event in the Vredefort dome, South Africa. *Geology*, 26, 787–790.

Gohn, G.S., Koeberl, C., Miller, K.G. *et al.* (2008) Deep drilling into the Chesapeake Bay impact structure. *Science*, 320, 1740–1745.

Grady, D.E. and Kipp, M.E. (1987) Dynamic rock fragmentation, in *Fracture Mechanics of Rock* (ed. B.K. Atkinson), Academic Press, San Diego, CA, pp. 429–475.

Grieve, R.A.F. (1987) Terrestrial impact structures. *Annual Review of Earth and Planetary Sciences*, 15, 245–270.

Grieve, R.A.F. (1991) Terrestrial impact: the record in the rocks. *Meteoritics*, 26, 175–194.

Grieve, R.A.F. and Cintala, M.J. (1992) An analysis of differential impact melt-crater scaling and implications for the terrestrial impact record. *Meteoritics*, 27, 526–538.

Grieve, R.A.F. and Pilkington, M. (1996) The signature of terrestrial impacts. *AGSO Journal of Australian Geology and Geophysics*, 16, 399–420.

Grieve, R.A.F., Robertson, P.B. and Dence, M.R. (1981) Constraints on the formation of ring impact structures based on terrestrial data, in *Multi-Ring Basins* (eds P.H. Schultz and R.B. Merill), Pergamon Press, New York, NY, pp. 37–57.

Gulick, S.P.S., Barton, P.J., Christeson, G.L. *et al.* (2008) Importance of pre-impact crustal structure for the asymmetry of the Chicxulub impact crater. *Nature Geoscience*, 1, 131–135.

Herrick, R.R. and Hessen, K.K. (2006) The planforms of low-angle impact craters in the northern hemisphere of Mars. *Meteoritics and Planetary Science*, 41, 1483–1495.

Hildebrand, A.R. and Boynton, W.V. (1990) Proximal Cretaceous–Tertiary boundary impact deposits in the Caribbean. *Science*, 248, 843–847.

Ivanov, B.A. and Kostuchenko, V.N. (1997) Block oscillation model for impact crater collapse (abstract). 28th Lunar and Planetary Science Conference, pp. 631–632.

Ivanov, B.A., Kocharyan, G.G., Kostuchenko, V.N. *et al.* (1996) Puchezh-Katunki impact crater: preliminary data on recovered core block structure (abstract). 27th Lunar and Planetary Science Conference, pp. 589–590.

Jahn, A. and Riller, U. (2009) A 3D model of first-order structural elements of the Vredefort Dome, South Africa – importance for understanding central uplift formation of large impact structures. *Tectonophysics*, 478, 221–229.

Kenkmann, T. (2002) Folding within seconds. *Geology*, 30, 231–234.

Kenkmann, T. (2003) Dike formation, cataclastic flow, and rock fluidization during impact cratering: an example from the Upheaval Dome structure, Utah. *Earth and Planetary Science Letters*, 214, 43–58.

Kenkmann, T. and Poelchau, M.H. (2009) Low-angle collision with Earth: the elliptical impact crater Matt Wilson, NT, Australia. *Geology*, 37, 459–462.

Kenkmann, T. and von Dalwigk, I. (2000) Radial transpression ridges: a new structural feature of complex impact craters. *Meteoritics and Planetary Science*, 35, 1189–1202.

Kenkmann, T., Ivanov, B.A. and Stöffler, D. (2000) Identification of ancient impact structures: low-angle normal faults and related geological features of crater basements, in *Impacts and the Early Earth* (eds I. Gilmour and C. Koeberl), Lecture Notes in Earth Sciences, vol. 91, Springer, Berlin, pp. 279–309.

Kenkmann, T., Jahn, A., Scherler, D. and Ivanov, B.A. (2005) Structure and formation of a central uplift: a case study at the Upheaval Dome impact crater, Utah, in *Large Meteorite Impacts III* (eds T. Kenkmann, F. Hörz and A. Deutsch), Geological Society of America Special Paper 384, Geological Society of America, Boulder, CO, pp. 85–115.

Kenkmann, T., Jahn, A. and Wünnemann, K. (2006) 'Block size' in a complex impact crater inferred from the Upheaval Dome structure, Utah. 37th Lunar and Planetary Science Conference, CD-ROM, no. 1540.

Kenkmann, T., Reimold, W.U., Khirfan, M. *et al.* (2010) The complex impact crater Jebel Waqf as Suwwan in Jordan: effects of target heterogeneity and impact obliquity on central uplift formation, in *Large Meteorite Impacts and Planetary Evolution IV* (eds R.L. Gibson and W.U. Reimold), Geological Society of America Special Paper 465, Geological Society of America, Boulder, CO, pp. 471–487.

Kriens, B.J., Shoemaker, E.M. and Herkenhoff, K.E. (1999) Geology of the Upheaval Dome impact structure, southeast Utah. *Journal of Geophysical Research*, 104, 18867–18887.

Lana, C., Gibson, R.L. and Reimold, W.U. (2003) Impact tectonics in the core of the Vredefort Dome: implication for formation of central uplifts in very large impact structures. *Meteoritics and Planetary Science*, 38, 1093–1107.

Leith, A.C. and McKinnon, W.B. (1991) Terrace width variations in complex Mercurian craters and the transient strength of cratered Mercurian and lunar crust. *Journal of Geophysical Research*, 96, 20923–20931.

Lieger, D., Riller, U. and Gibson, R.L. (2009) Generation of fragment-rich pseudotachylite bodies during central uplift formation in the Vredefort impact structure, South Africa. *Earth and Planetary Science Letters*, 279, 53–64.

Lindström, M., Ormö, J., Sturkell, E. and von Dalwigk, I. (2005) The Lockne crater: revision and reassessment of structure and impact stratigraphy, in *Impact Tectonics: Impact Studies*, (eds C. Koeberl and H. Henkel), Springer, Berlin, pp. 357–388.

McKinnon, W.B. (1978) An investigation into the role of plastic failure in crater modification. 9th Proceedings of Lunar Planetary Science Conference, pp. 3965–3973.

Magloughlin, J.F. and Spray, J.G. (eds) (1992) Frictional melting processes and products in geological materials (special issue). *Tectonophysics*, 202, 197–337.

Melosh, H.J. (1977) Crater modification by gravity: a mechanical analysis of slumping, in *Impact and Explosion Cratering* (eds J. Roddy, R.O. Pepin and R.B. Merill), Pergamon Press, New York, NY, pp. 1245–1260.

Melosh, H.J. (1979) Acoustic fluidization: a new geologic process? *Journal of Geophysical Research*, 84, 7513–7520.

Melosh, H.J. (1989) *Impact Cratering: A Geological Process*, Oxford University Press, New York, NY.

Melosh, H.J. (1996) Dynamical weakening of faults by acoustic fluidization. *Nature*, 379 (6566), 601–606.

Melosh, H.J. (2005) The mechanics of pseudotachylite formation in impact events, in *Impact Tectonics* (eds H. Henkel and C. Koebert), Springer, Berlin, pp. 55–80.

Melosh, H.J. and Ivanov, B.A. (1999) Impact crater collapse. *Annual Review of Earth and Planetary Sciences*, 27, 385–415.

Morgan, J., Warner, M., Collins, G.S. *et al.* (2000) Peak-ring formation in large impact craters: geophysical constraints from Chicxulub. *Earth and Planetary Science Letters*, 183, 347–354.

Oberbeck, V.R. and Quaide, W.L. (1968) Genetic implications of lunar regolith thickness variations. *Icarus*, 9, 446–465.

Offield, T.W. and Pohn, H.A. (1979) Geology of the Decaturville impact structure, Missouri, US Geological Survey Professional Paper, 1042, Department of the Interior, Geological Survey. Reston, VA.

Öhman, T., Aittola, M., Korteniemi, J. *et al.* (2010) Polygonal impact craters in the Solar System: observations and implications, in *Large Meteorite Impacts and Planetary Evolution IV* (eds R.L. Gibson and W.U. Reimold), Geological Society of America Special Paper 465, Geological Society of America, Boulder, CO, pp. 51–65.

O'Keefe, J.D. and Ahrens, T.J. (1999) Complex craters: relationship of stratigraphy and rings to impact conditions. *Journal of Geophysical Research*, 104, 27091–27104.

Ormö, J. and Lindström, M. (2000) When a cosmic impact strikes the sea bed. *Geological Magazine*, 137, 67–80.

Osinski, G.R. and Spray, J.G. (2005) Tectonics of complex crater formation as revealed by the Haughton impact structure, Devon Island, Canadian High Arctic. *Meteoritics and Planetary Science*, 40, 1813–1834.

Pearce, S.J. and Melosh, H.J. (1986) Terrace width variations in complex lunar craters. *Geophysical Research Letters*, 13, 1419–1422.

Pike, R.J. (1980) Formation of complex impact craters: evidence from Mars and other planets. *Icarus*, 43, 1–19.

Pike, R.J. (1988) Geomorphology of impact craters on Mercury, in *Mercury*. (eds F. Vilas, C.R. Chapman and M. Shapley Matthews), University of Arizona Press, Tucson, AZ, pp. 165–273.

Poelchau, M.H., Kenkmann, T. and Kring, D.A. (2009) Rim uplift in simple craters: the effects of target heterogeneities and trajectory obliquity. *Journal of Geophysical Research, Planets*, 114, E01006. DOI: 10.1029/2008JE003235.

Pohl, J., Stöffler, D., Gall, H. and Ernstson, K. (1977) The Ries impact crater, in *Impact and Explosion Cratering* (eds D.J. Roddy, R.O. Pepin and R.B. Merrill), Pergamon Press, New York, NY, pp. 343–404.

Reimold, W.U. and Gibson, R.L. (2005) 'Pseudotachylites' in large impact structures, in *Impact Tectonics. Impact Studies* (eds C. Koeberl and H. Henkel) Springer, Berlin, pp. 1–51.

Reimold, W.U. and Gibson, R.L. (2006) The melt rocks of the Vredefort impact structure – Vredefort Granophyre and pseudotachylitic breccias: implications for impact cratering and the evolution of the Witwatersrand Basin. *Chemie der Erde – Geochemistry*, 66, 1–35.

Riller, U., Lieger, D., Gibson, R.L. *et al.* (2010) Origin of large-volume pseudotachylite in terrestrial impact structures. *Geology*, 38, 619–622.

Roddy, D.J. (1979) Structural deformation at the Flynn Creek impact crater, Tennessee. 10th Lunar and Planetary Science Conference, pp. 2519–2534.

Schenk, P.M. (2002) Thickness constraints on the icy shells of the Galilean satellites from a comparison of crater shapes. *Nature*, 417, 419–421.

Scherler, D., Kenkmann, T. and Jahn, A. (2006) Structural record of an oblique impact. *Earth and Planetary Science Letters*, 248, 28–38.

Schmidt, R.M. and Housen, K.R. (1987) Some recent advances in the scaling of impact and explosion cratering. *International Journal of Impact Engineering*, 5, 543–560.

Schultz, P.H. and Anderson, R.R. (1996) Asymmetry of the Manson impact structure: evidence for impact angle and direction, in *Manson Impact Structure: Anatomy of an Impact Crater* (eds C. Koeberl and R.R. Anderson), Geological Society of America Special Paper 302, Geological Society of America, Boulder, CO, pp. 397–417.

Senft, L.E., and Stewart, S.T. (2008) Impact crater formation in icy layered terrains on Mars. *Meteoritics and Planetary Science*, 43, 1993–2013.

Senft, L.E. and Stewart, S.T. (2009) Dynamic fault weakening and the formation of large impact craters. *Earth and Planetary Science Letters*, 287, 471–482.

Shuvalov, V.V. and Dypvik, H. (2004) Ejecta formation and crater development of the Mjölnir impact. *Meteoritics and Planetary Science*, 39, 467–479.

Shuvalov, V.V. and Trubestkaya, I.A. (2002) Numerical modeling of marine target impacts. *Solar System Research*, 36, 417–430.

Spray, J.G. (1997) Superfaults. *Geology*, 25, 579–582.

Spray, J.G. and Thompson, L.M. (1995) Friction melt distribution in a multi-ring impact basin. *Nature*, 373, 130–132.

Stesky, R. (1974) Friction in faulted rock at high temperature and pressure. *Tectonophysics*, 23, 177–203.

Tsikalas, F. and Faleide, J.I. (2007) Post-impact structural crater modification due to sediment loading: an overlooked process. *Meteoritics and Planetary Science*, 42, 2013–2029.

Turtle, E.P., Pierazzo, E., Collins, G.S. *et al.* (2005) Impact structures: what does crater diameter mean? in *Large Meteorite Impacts III.* (eds T. Kenkmann, F. Hörz and A. Deutsch), Geological Society of America Special Paper 384, Geological Society of America, Boulder, CO, pp. 1–24.

Wünnemann, K. and Ivanov, B.A. (2003) Numerical modelling of impact crater depth–diameter dependence in an acoustically fluidized target. *Planetary and Space Science*, 51, 831–845.

Wünnemann, K., Collins, G.S. and Weiss, R. (2010) Impact of a cosmic body into Earth's ocean and the generation of large tsunami waves: insight from numerical modelling. *Reviews of Geophysics*, 48, RG4006. DOI: 10.1029/2009RG000308.

# SIX

# Impact-induced hydrothermal activity

## Kalle Kirsimäe* and Gordon R. Osinski[†]

*Department of Geology, University of Tartu, Ravila 14a, 50411 Tartu, Estonia
[†]Departments of Earth Sciences/Physics and Astronomy, Western University, 1151 Richmond Street, London, ON, N6A 5B7, Canada

## 6.1 Introduction

Hydrothermal alteration (i.e. interaction of rock-forming minerals with solutions that have temperatures higher than expected from the regional geothermal gradient in the given area) is a common natural phenomenon that occurs in a wide variety of geological settings and rock types (e.g. Cawood and Pirajno, 2008). The formation of a hydrothermal system requires water, fractures/open porosity in the rock and, most importantly, a heat source. As a consequence, hydrothermal phenomena are usually found in close relation to regions with high heat flow, such as volcanic centres, mid-ocean ridges and continental rifts.

Hydrothermal systems in impact structures (defined hereafter as impact-induced hydrothermal systems) result from the strong differential temperatures remaining in the shocked rocks due to the large amount of kinetic energy released into the target by the passage of the shock wave. An additional source of heat available to drive a hydrothermal circulation system results from the uplift of rocks from depth that are hotter due to the geothermal gradient (McCarville and Crossey, 1996). In dry environments, such as the Moon, the heat loss occurs mainly by conduction and radiation transfer. However, if water is present at the crater site, the thermal field remaining in the rocks initiates the convective heat transfer by circulation of water and the emission of steam, forming a hydrothermal system.

The recognition of impact-related hydrothermal alteration dates back to the late 1960s and early 1970s, when hydrothermal alteration products were recognized in impacted rocks and impact glasses at the Ries impact structure, Germany (Förstner, 1967; von Engelhardt, 1972), although it was not until the past decade that the study of impact-induced hydrothermal systems became commonplace. Currently, evidence of impact-induced hydrothermal activity is present in more than 60 terrestrial craters out of the 181 confirmed impact structures (Fig. 6.1; Naumov, 2005). In addition, impact-induced hydrothermal activity has been suggested to occur in extraterrestrial settings, most notably on Mars

(e.g. Newsom, 1980; Allen et al., 1982; Newsom et al., 1986,1996). Recent discoveries have evidenced the hydrous alteration, most probably hydrothermal, in impact structures on Mars (Marzo et al., 2010).

Evidence for hydrothermal activity has been identified in both simple and complex terrestrial craters varying in size from the 1.8 km diameter Lonar Crater in India (Hagerty and Newsom, 2003) to the approximately 200 km diameter Sudbury impact structure in Canada (e.g. Ames et al., 1998). A number of impact-induced hydrothermal systems have been studied in great detail: Chicxulub (e.g. Zürcher and Kring, 2004), Haughton (Osinski et al., 2001,2005), Kärdla (Versh et al., 2005), Lockne (Sturkell et al., 1998), Lonar (Hagerty and Newsom, 2003), Manson (e.g. McCarville and Crossey, 1996), Puchezh-Katunki, Kara and Popigai (Naumov, 2002), Ries (e.g. Osinski, 2005), Siljan (e.g. Komor et al., 1988) and Sudbury (e.g. Farrow and Watkinson, 1992). This chapter introduces impact-induced hydrothermal systems and describes the processes and parameters that control the nature, extent and distribution of such systems within impact craters. The potential for the formation of such systems on Mars and the implications for astrobiology are also discussed.

## 6.2 Formation and development of the post-impact thermal field

### 6.2.1 Impact heating

During the impact process, propagation of the supersonic shock wave into the target and projectile causes pressures and temperatures exceeding greater than 100 GPa and greater than 3000 °C in large impact basins (Melosh, 1989; Chapters 1 and 3). Adiabatic decompression of the compressed projectile and target rocks imparts waste heat that remains in the shocked rocks, and if the pressures during shock wave passage exceed approximately 45 GPa

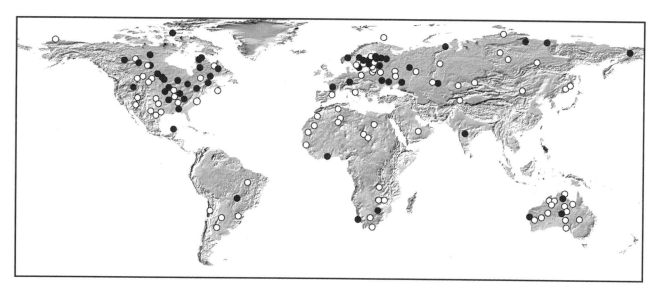

**Figure 6.1** Location of terrestrial impact craters with indications of post-impact hydrothermal activity (filled circles). Modified after Naumov (2002, 2005).

then the rocks may start to melt and/or vaporize (e.g. Ahrens and O'Keefe, 1972; Stöffler, 1972). The temperature field remaining in the crater structure results from the combined action of shock heating and material displacement, the latter mainly due to the uplift of deep-seated and, therefore, warmer rocks in crater centre (i.e. central uplifts; Ivanov and Deutsch, 1999; Turtle *et al.*, 2003; Ivanov, 2004; Chapter 5). The amount of heat deposited during the impact process scales approximately as a function of the energy released into target (O'Keefe and Ahrens, 1977); in large-scale structures, with shock pressures exceeding 40–50 GPa, the passage of the shock wave induces extensive melting and vaporization of the target rocks (Fig. 6.2). The impact melt rocks and melt-rich breccias produced during the impact, especially the continuous melt 'sheet' filling the annular depressions in large structures (e.g. Sudbury, Chicxulub), are significant sources of heat for development of hydrothermal systems (Abramov and Kring, 2004,2007). Two other potential sources of heat for creating impact-generated hydrothermal systems are elevated geothermal gradients in central uplifts and the energy deposited in central uplifts due to the passage of the shock wave. Overall, the temperature increase imparted to the target rocks due to impact can range in the area of the central uplift from a maximum of approximately 100–200 °C in small- to medium-size structures (simple and complex craters of approximately up to 20–30 km in diameter) to greater than 1000 °C in large-scale structures and impact basins, associated with the formation of significant amounts of initially superheated (up to ~2000 °C; Grieve *et al.*, 1977) impact melt (Ivanov, 2004; Fig. 6.2).

### 6.2.2 Formation of the hydrothermal system

The presence and dimensions of the impact melt sheet and/or other melt-rich crater-fill impactites significantly influences the initiation and spatial configuration of impact-induced hydrothermal systems. In large peak-ring structures and multi-ring basins, where the initially impermeable hot melt sheet may cover a significant portion of the crater interior (e.g. the Sudbury and Chicxulub structures) the heat transfer during the first stages of the impact cooling occurs through the conductive and radiative processes. In this case, hydrothermal activity develops at first in the annular trough between the central uplift and final crater rim, with the fluids venting through faults in the crater modification zone (Osinski *et al.*, 2001; Abramov and Kring, 2004,2007). As a consequence, the cooling of these large structures immediately after the impact is mainly controlled by the low permeability of hot rock and/or melt and will be via the least effective heat removal mechanism; that is, conduction.

In contrast, in small- to medium-size craters without significant volumes of impact melt – and particularly those in sedimentary or mixed sedimentary–crystalline targets where little silicate melt is generated – the hydrothermal circulation occurs within and around the central uplift shortly after the impact by gaining the heat from the hot impact melt-bearing crater-fill breccias (Osinski *et al.*, 2001; Osinski, 2005) and the thermal effect of the elevated uplift rocks (McCarville and Crossey, 1996; Jõeleht *et al.*, 2005; Fig. 6.3); circulation also occurs in the faulted crater rim region.

As a result, small-to-medium and large structures are characterized by different cooling histories and, consequently, by different complexities of the resultant hydrothermal systems. Large peak-ring structures will comprise a number of different heat sources, resulting in the development of several independent convection cells, initially in the annular trough, in breccias above the central melt sheet and in the fractured modification zone and likely in the peak-ring area and, then, after solidification and cooling of the melt sheet, in the central uplift area (e.g. Abramov and Kring, 2007). In smaller structures a simple convective cell

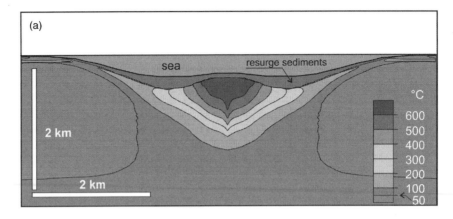

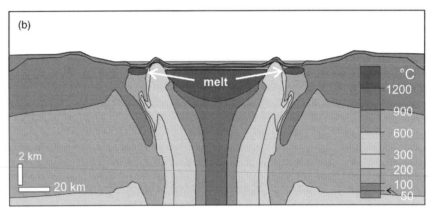

**Figure 6.2** Modelled thermal anomaly remaining in crater rocks immediately after the impact. (a) Kärdla Crater (4 km in diameter); (b) Chicxulub Crater (180 km in diameter). Modified after Jõeleht *et al.* (2005) and Abramov and Kring (2007). (See Colour Plate 18)

system in the most heated central part of the crater will typically develop (e.g. Jõeleht *et al.*, 2005).

Permeability is the key parameter determining the onset of the hydrothermal circulation and the lifetime of the convective systems. Self-sealing of impact-generated hydrothermal systems can reduce permeability as well (Newsom *et al.*, 2001), based on numerous examples from terrestrial hydrothermal systems (Cathles *et al.*, 1997). Fluid inclusion studies of impact-induced hydrothermal mineralization suggest that, regardless of the crater size, the maximum fluid temperatures of fluid inclusions in quartz vary in the same range of 100–350/400 °C (Fig. 6.4; e.g. Komor *et al.*, 1988; Boer *et al.*, 1996; Kirsimäe *et al.*, 2002; Lüders and Rickers, 2004). Similar maximum temperatures are observed in volcanic hydrothermal systems, which show transition from a strongly convecting system to a non-convecting system at approximately 400 °C (e.g. Muraoka *et al.*, 1998) that is attributed to the transition from ductile to brittle deformation in silicate rocks (Fournier, 1991). Nevertheless, the presence of high-temperature secondary phases (such as calcium silicates, garnet, etc.) and geothermobarometry of post-shock thermally influenced target rocks (e.g. McCarville and Crossey, 1996; Gibson *et al.*, 1998; Osinski *et al.*, 2001) suggest that the initial temperatures at the centre of the crater exceeded at least 500–600 °C, resulting in thermal metamorphism and metasomatism of impactites.

### 6.2.3    Lifetime of the hydrothermal system

The lifetimes of impact-induced hydrothermal systems vary significantly with the crater size, depending on the amount of melting and governing mode of heat transport. Abramov and Kring (2004, 2007) estimated lifetimes (defined herein as the time needed for the system to cool below 90 °C within 1 km of the surface everywhere in the model) for the 250 km diameter Sudbury structure in the range 0.2–3.2 Myr and in 180 km diameter Chicxulub in the range 1.5–2.3 Myr for permeabilities of $10^{-4}$ to $10^{-2}$ D (darcy; Fig. 6.5). Such long cooling times in these large structures can be explained by hydrothermal circulation with the most effective convective heat transport taking place only near the surface, whereas the temperatures of deeper parts of the crater are above the brittle–ductile transition and rock is impermeable, thus making conduction the dominant form of heat transport. In addition, in large structures, additional heat is delivered from deeply seated hot rocks to the near surface by circulating fluids (Abramov and Kring, 2004).

In smaller structures with limited melting, cooling is governed mainly by convective heat transport, leading to a more rapid temperature decrease. As an example, in the 4 km diameter Kärdla Crater, cooling to temperatures below 90 °C in the uppermost 1 km takes only 1500–4500 years (Jõeleht *et al.*, 2005), which

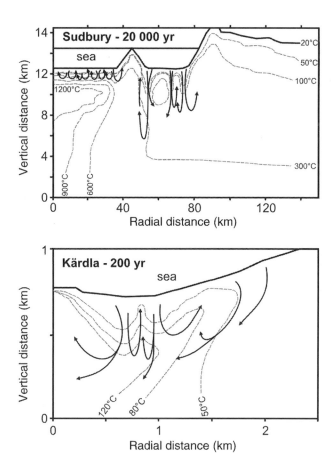

**Figure 6.3** Schematic location of impact-induced hydrothermal convective cells in large-sized (Sudbury) and small-to-medium sized (Kärdla) impact structures. Modified after Abramov and Kring (2004) and Jõeleht *et al.* (2005).

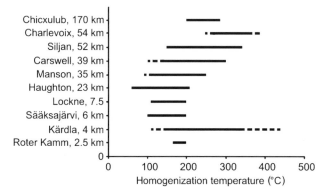

**Figure 6.4** Homogenization temperatures of secondary aqueous quartz fluid inclusions in impact-induced hydrothermal systems. Data sources: Roter Kamm (Koeberl *et al.*, 1989); Kärdla (Kirsimäe *et al.*, 2002); Sääksajärvi (Mutanen, 1979); Lockne (Sturkell *et al.*, 1998); Haughton (Osinski *et al.*, 2005); Manson (Boer *et al.*, 1996); Carswell (Pagel *et al.*, 1985); Siljan (Komor *et al.*, 1988); Charlevoix (Pagel and Poty, 1975); Chicxulub (Lüders and Rickers, 2004).

is two to three orders of magnitude less than in the large structures.

Another important parameter controlling cooling is the phase change of water that can add to the convective fluid circulation by the effect of latent heat of vaporization (Jõeleht *et al.*, 2005). At pressures below the critical point of water (~22 MPa for fresh water) the enthalpy change associated with phase change of water adsorbs the energy needed for vaporization from the heated rocks, leading to effective cooling down to the boiling temperature of the fluid. The vapour flow is limited by the brittle–ductile transition-related permeability properties of the rock discussed above, as well as the hydraulic pressure variation. The latent heat of vaporization is a pressure-dependent parameter, which decreases towards critical pressure. As an example, in marine craters the overload of the deep water column can significantly reduce or completely diminish the vaporization effect (Jõeleht *et al.*, 2006). However, in shallow marine or continental impacts with the limited water supply into the crater depression, vaporization can play an important role during the first stages of the cooling, resulting in degassing pipe structures found in the Ries (Newsom *et al.*, 1986) and Haughton (Osinski *et al.*, 2005) Craters.

## 6.3 Composition and evolution of the hydrothermal fluids and mineralization

### 6.3.1 Composition and properties of hydrothermal fluid

Impact-induced hydrothermal systems can be fed by different types of water sources: seawater, groundwater, meteoric water, deep-formational brines and magmatic fluids from the impact melt sheet are all possible fluid sources (e.g. McCarville and Crossey, 1996; Ames *et al.*, 2004; Zürcher *et al.*, 2005). However, the composition of fluids trapped in hydrothermal minerals suggests that seawater and meteoric water are the most important sources for fluids in impact hydrothermal systems and there is no evidence to date of the involvement of magmatic fluids (Naumov, 2005). In most cases, the fluids entrapped in inclusions are aqueous NaCl (rarely $CaCl_2$ or $CaCl_2$–NaCl–$H_2O$) composition fluids with low to moderate salinity (up to 13% $NaCl_{eq}$) and low gas-phase ($CO_2$) content (e.g. Kirsimäe *et al.*, 2002). However, in a few craters, a higher salinity (14–21% $NaCl_{eq}$ or $CaCl_{2\ eq}$) of the fluids has been reported (Koeberl *et al.*, 1989; Sturkell *et al.*, 1998; Lüders and Rickers, 2004), indicating either different fluid sources (e.g. basinal oil-field saline brine in Chicxulub; Zürcher *et al.*, 2005) or potentially the involvement of the volatiles from thermally decomposed phases such as sulfates and carbonates in sedimentary targets.

The chemical composition of impact-induced hydrothermal fluids depends upon the composition of target rocks present at the impact site as well as on the fluid source. Principally, using the analogy of hydrothermal alteration in young orogenic belts, hydrothermal fluids can be divided according to temperature and activity ratios of aqueous cation species into three types: (1) acid-type alteration with low cation-to-hydrogen ratios in the solution; (2) intermediate alteration subdivided into Ca–Mg-series and K-series with medium cation-to-hydrogen ratios; and

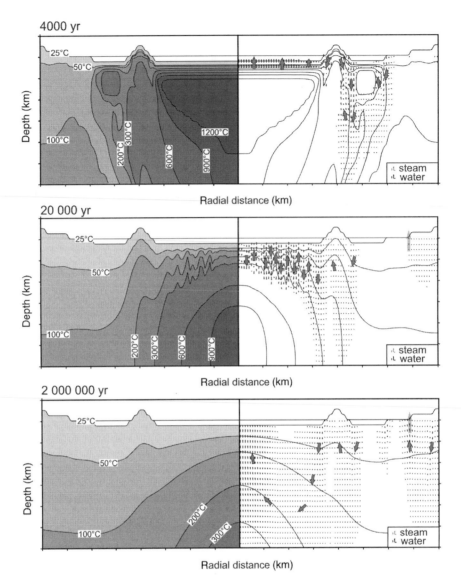

**Figure 6.5**  Cooling of the Chicxulub impact structure. Modified after Abramov and Kring (2007). Numerical model assumes surface permeability $k_0 = 10^{-3}$ D. Block arrows indicate main directions of water movement.

(3) alkaline-type subdivided into Na- and Ca-series with high cation-to-hydrogen ratios. Each alteration type includes mineral zones corresponding to different fluid temperature conditions (Utada, 1980). In this sense the mineralization associated with impact-induced hydrothermal systems will typically correspond to the intermediate type of hydrothermal alteration; whether the Ca–Mg- or K-series dominates will depend on the specifics of an impact: target rock composition (e.g. sedimentary versus crystalline, siliciclastic or carbonate), its size (which governs the magnitude of heating), the nature of the hydrothermal fluids (saline brines, seawater, undersaturated fresh water resources, etc.) and the stage of the development in a hydrothermal system.

### 6.3.2   Evolution of the fluids and mineralization

Naumov (2005) proposed a general scheme for the evolution of the hydrothermal system in impact craters: (1) the reaction

of infiltrating undersaturated water with unstable shocked (alumino-)silicates and impact glasses and the formation of Fe–Mg clay minerals by the anion hydrolysis reaction resulting in weakly alkaline hydrothermal fluids; (2) the dissolution of silicates and, especially, of metastable glasses, which provides a high activity (supersaturation) of silica that creates favourable conditions for the formation of iron smectites and zeolites – these are typically the main mineral phases precipitating in impact-generated hydrothermal systems in silicate target rocks.

Based on mineral associations and the dominant fluid environment regimes, the sequence of hydrothermal alteration can by divided into three stages (e.g. Osinski *et al.*, 2001; Versh *et al.*, 2005): (1) an early stage of vapour-dominated silicate (K-metasomatic) alteration; (2) an intermediate stage of vapour- to liquid-dominated silicate (chlorite–smectite–zeolite) mineralization; and (3) a late stage of liquid-dominated carbonate–sulfide/ iron oxyhydrate mineralization (Fig. 6.6). Commonly, the

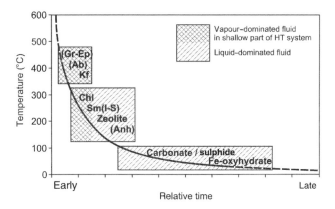

**Figure 6.6** Principal sequence of the mineralization in an impact-induced hydrothermal system. Anh: anhydrite; Gr: garnet; Ep: epidote; Ab: albite; Kf: potassium feldspar; Chl: chlorite; Sm: smectite; I-S: mixed-layer illite–smectite.

hydrothermal alteration type evolves from (Na)K-series during the first two stages into Ca–Mg-series during the last stage of the impact cooling, while in several structures an early calc-silicate thermal metamorphism precedes the fluidal alteration – for example, Chicxulub (Zürcher and Kring, 2004), Manson (McCarville and Crossey, 1996) and Haughton (Osinski *et al.*, 2001).

The mineral paragenetic associations combined with fluid inclusion, geothermobarometric and stable isotope data suggest a continuous temperature decrease through all three evolutionary stages. The homogenization temperatures of quartz fluid inclusions indicate that during the first evolutionary stage the maximum temperatures were as high as 400–500 °C, whereas the majority of aqueous inclusions are trapped at fluid temperatures less than 300 °C (e.g. Kirsimäe *et al.*, 2002), indicating that the first stage of the (K-)metasomatic alteration occurs in a vapour-dominated environment. Owing to efficient heat loss at the water vaporization front, the vapour-dominated stage is typically short (probably a few tens to hundreds of years in smaller structures and thousands of years in larger structures) and progressive cooling of the system results in a transition to mixed two-phase vapour-liquid dominated fluid at temperatures 300–150(100) °C. This second and main stage is associated with the most intensive and widespread/pervasive hydrothermal alteration characterized in silicic targets by argillic alteration and zeolitization of the impactites (Naumov, 2005), and carbonate–sulfate mineralization in carbonate/sulfate-rich sedimentary targets (e.g. Osinski *et al.*, 2001,2005; Osinski, 2005). The third, late stage of hydrothermal mineralization takes place in a liquid-dominated fluid environment at temperatures less than 100 °C and is characterized by precipitation of mainly carbonate–zeolite cements, calcite/gypsum and iron oxyhydrates(sulfides) at temperatures decreasing close to the ambient. The minimum temperature of a hydrothermal solution may be somewhat arbitrary and is limited by kinetic effects of water–rock interaction (Versh, 2006). As the temperature decreases, close to the thermal gradient of the area, any reaction between rock and solution proceeds more slowly and, therefore, the production of secondary minerals is hard to recognize in the rocks due to the short period of interaction.

During the first stages of the alteration the precipitation and replacement of primary feldspars (plagioclase, microcline-type potassium feldspar) with secondary alkali feldspars (albite-, 'pericline'- and 'adularia'-type potassium feldspar; Parsons and Lee, 2009) suggests rather high pH (>8) of the circulating fluid. This is possibly related to the effective removal of evolved gas and strong reactivity of high-temperature solutions with the surrounding rocks that efficiently consumed the available $H^+$ ions through anion hydrolysis of primary silicates. At later stages the pH of the initial fluid becomes lowered, but remains high enough (>7) to promote the clay alteration and carbonate precipitation. Owing to the interactions of aluminosilicate minerals and water, the fluid becomes gradually enriched with respect to Ca–Mg at last stages of cooling from chloritization of mafic minerals and the transformation of plagioclase into alkali feldspars during the first and second stages of alteration. The subsequent precipitation of calcite/dolomite is controlled by the availability of Ca and Mg ions and it can continue through the second stage to the third stage, reaching the environmental conditions close to ambient. However, if carbonate and evaporite lithologies are significant in the target (e.g. Haughton, Chicxulub), the circulating fluids have higher activities of Ca, Mg, Fe, S and $CO_2$, and consequently weakly acidic sulfate-rich mineralizing fluids predominate. Under these conditions the main alteration phases are silica minerals, and calcium–ferriferous-sulfide/sulfate mineral associations (Osinski *et al.*, 2001; Naumov, 2005).

### 6.3.3 Zoned structure of hydrothermal mineralization

Hydrothermal alteration in any setting is typically characterized by the zonal distribution of alteration products that is principally related to mass transfer between minerals and hydrothermal solutions as temperature decreases during circulation of hydrothermal fluids (Helgeson, 1979). The distribution and morphology of alteration zones are influenced by fracture distribution, porosity and permeability of the rocks, and the distance between the high-temperature and low-temperature zones that is a factor of the vertical thermal gradient at a given location (Meunier, 2005). The spatial structure and specific alteration gradients are best described for impact-induced hydrothermal structures in vertical profiles through central uplifts where the thermal gradients are in order of 100 °C km$^{-1}$ (Naumov, 2005).

The vertical zonation of the mineral distribution in the central part of impact-induced hydrothermal systems (Fig. 6.7) is typically characterized by the smectite to chlorite transition and the substitution of calcite by anhydrite downward in the section (Naumov, 2005). This is accompanied by a decrease of the intensity of hydrothermal alteration, which is largely determined by the decrease in permeability and fracturing below the crater floor. In the upper smectite–zeolite zone of alteration, the composition of the smectites and, especially, zeolites varies with depth (Fig. 6.7 and Fig. 6.8). In the 40 km diameter Puzhezh-Katunki impact structure the smectites are progressively substituted: dioctahedral high-alumina varieties (montmorillonite) in upper part of the zone and trioctahedral magnesium–iron smectite (saponites) in the lower parts of the section whereas the tetrahedral alumina content increases with depth in the saponites (Naumov,

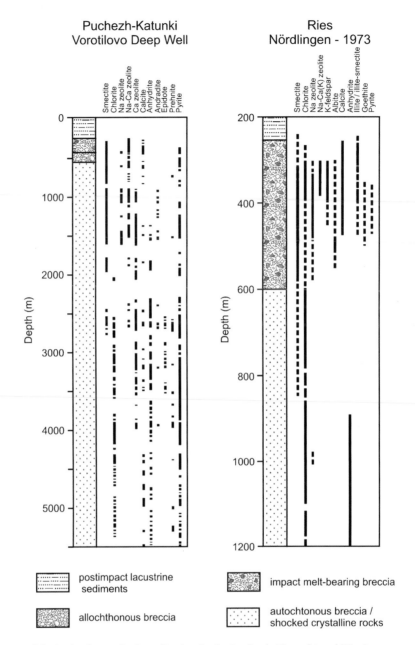

**Figure 6.7**    Distribution of the hydrothermal mineralization in the Puchezh-Katunki and Ries impact structures. Modified after Naumov (2002, 2005) and Osinski (2005).

2002, 2005). In the 24 km diameter Ries impact structure, however, the montmorillonite- and saponite-type smectite occur nearly contemporarily throughout the crater-filling suevite section (Fig. 6.7; Osinski, 2005). Variation of the zeolites in the upper zone is characterized by the successive replacement of sodium zeolites, sodium–calcium zeolites and finally calcium zeolites with increasing depth with accompanying calcite in two uppermost subzones, whereas the zeolites have typically higher Si content that increases with depth compared with their stoichiometric composition (Naumov, 2002, 2005; Osinski, 2005). The lower chlorite–anhydrite zone of the hydrothermal alteration occurs typically in autochthonous brecciated and shocked basement rocks of the crater floor, where saponite-type trioctahedral smectite is replaced by chlorite and/or chlorite smectite phases. The chlorite–anhydrite paragenesis is typically associated with the thermal-metamorphic and/or high-temperature metasomatic calcium–iron silicates (e.g. andradite, ferrosalite, epidote, prehnite and actinolite; Naumov, 2002).

It is important to note that a general depth–temperature-related zonation of the alteration could, however, be significantly disturbed due to the thermal anomalies caused by the distribution of thick melt or melt-rich breccia bodies and permeability/fracturing variability within the crater. Osinski (2005) showed that in the Ries structure the early high-

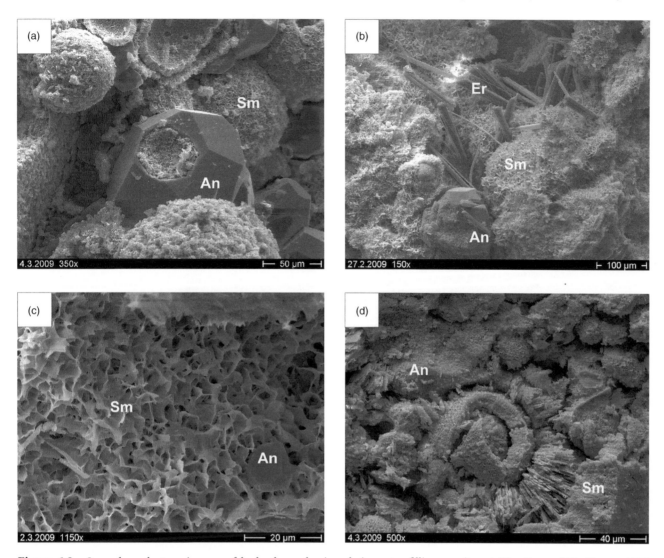

**Figure 6.8** Secondary electron images of hydrothermal minerals in crater-filling suevites at Ries Crater (Nördlingen 1973 drillcore): (a) depth 408.35 m; (b,c) depth 340.30 m; (d) 368.1 m; the hollow smectite spheroid replacing the primary glass vesicle is filled with analcime aggregate. Er: erionite; An: analcime; Sm: smectite.

temperature K-metasomatic alteration and minor albitization, as well as chloritization, is restricted to the upper parts of the suevite layer (Fig. 6.7). Likewise, the vertical distribution of clay minerals has been found locally inverted in some sequences, where chlorites are substituted by saponite- and montmorillonite-type smectites downwards in the section (Naumov, 2005; Osinski, 2005). In the Ries case, the inverted distribution is explained by the thermal anomaly caused by the hot, melt-rich suevite unit in the upper part of the crater suevites with temperatures in excess of approximately 600 °C, with lower temperatures for the underlying melt-poor unit (Miller and Wagner, 1979; von Engelhardt, 1990; Osinski, 2005).

In terms of the lateral variation of alteration, based on mapping of the Haughton impact structure and a review of the existing literature, Osinski *et al.* (2005) suggested that there are five main locations in a complex impact crater where post-impact hydrothermal deposits can form (Fig. 6.9): (1) interior of central uplifts,

as revealed by deep drilling and as discussed above; (2) outer margin of central uplifts – these represent regions of intense and complex faulting and the concentration of alteration in such zones has been documented for the Haughton (Osinski *et al.*, 2005) and Siljan (Hode *et al.*, 2003) impact structures; (3) at the base and edge of the crater-fill impact melt rocks and breccias; (4) in the faulted crater rim region in the form of pipe structures that may represent fossil hydrothermal springs and fumaroles; and (5) within post-impact sedimentary fill deposits – if sedimentation is rapid, hydrothermal activity may still be active, as was the case at the Ries structure, where smectite clays and various zeolite minerals are found throughout the approximately 400 m thick sequence of intra-crater palaeolacustrine sedimentary rocks (Fig. 6.7; Stöffler *et al.*, 1977). In addition, hydrothermal activity has been found in the faulted rim or megablock area outside the transient cavity in the Yax-1 drillcore in the Chicxulub impact basin (e.g. Lüders and Rickers, 2004) and hydrothermal

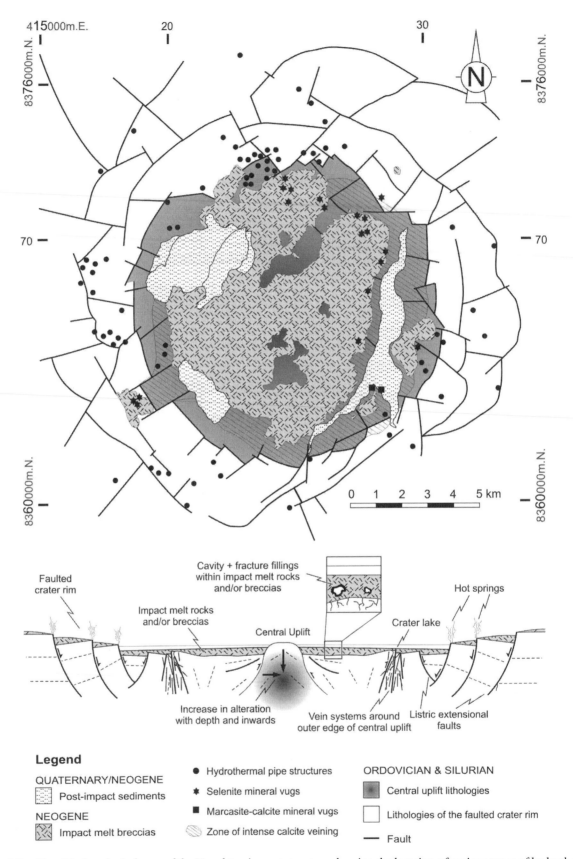

**Figure 6.9**    Simplified geological map of the Haughton impact structure showing the location of various types of hydrothermal deposits (top). At the bottom is a schematic cross-section showing the lateral zonation of the hydrothermal system; it is likely that this applies for other complex impact structures. Modified after Osinski *et al.* (2005).

conditions can be met in thick melt-bearing ejecta around impact basins outside the megablock zone (Salge, 2007).

## 6.4 Implications for extraterrestrial impacts and microbial life

### 6.4.1 Impact-induced hydrothermal systems on Mars?

Impact cratering is a far more common geological process in the Solar System than volcanism. Consequently, the possibility of impact-induced hydrothermal systems in crater structures formed in $H_2O$-bearing targets is potentially much higher than in volcanic hydrothermal systems. Extraterrestrial impact-induced hydrothermal systems, therefore, are interesting targets to study owing to their possible astrobiological implications (see Chapter 11 for a complete discussion of the geomicrobiology of impact craters). Mars has been suggested as the most likely site for preserved evidence of impact-induced hydrothermal systems (e.g. Newsom, 1980; Allen *et al.*, 1982; Newsom *et al.*, 1986,1996; Rathbun and Squyres, 2002; Schulze-Makuch *et al.*, 2007; Marzo *et al.*, 2010). Morphological, chemical and mineralogical evidence from past and ongoing Mars exploration missions suggest that liquid water was present and stable on the surface in the earliest period (Noachian epoch) of the Martian geological history (e.g. Carr, 1996,2006; Bibring *et al.*, 2005; Poulet *et al.*, 2005; Mustard *et al.*, 2008; Ehlmann *et al.*, 2009). Moreover, the presence of substantial amounts of subsurface ground ice in northern latitudes (e.g. Costard and Kargel, 1995; Carr, 1996,2006; Boynton *et al.*, 2002; Head *et al.*, 2003; Soare *et al.*, 2007) implies that the impact-induced hydrothermal activity is also likely to occur following impact events at the present day on Mars.

The results of numerical modelling of impact-induced hydrothermal systems on an early 'wet' Mars (Rathbun and Squyres, 2002; Abramov and Kring, 2005) show their similarity to the hydrothermal systems on Earth, where upwelling water flow in saturated rocks is developed in either the central peak (for 30 km diameter craters) or peak-ring area (for 100 and 180 km diameter craters) (Abramov and Kring, 2005). On the other hand, the simulation of post-impact hydrothermal systems within 45 and 90 km diameter craters on Mars suggests that if the impact were to occur during the post-Noachian time when surface temperatures are below freezing, then discharge would occur only near the centre of the crater, not at the rim (Barnhart *et al.*, 2010).

The modelling of Martian impact-induced hydrothermal systems shows, however, that: first, a crater of a given diameter would be less hot on Mars than on Earth, because less energy is required to create it under lower gravitational force; and second, the early stage of the cooling is characterized by substantial steam emission initially from the surface; but as the near-surface cools, steam production ceases, except for supercritical fluid at high temperatures and pressures (Abramov and Kring, 2005). Consequently, the estimated cooling times of the hydrothermal activity are nearly an order of magnitude less than for similarly sized craters on Earth. Another important aspect of impact structures on Mars is the possible formation of ice-covered crater lakes in large impact sites (<65 km diameter); this would have the effect of retaining liquid water, otherwise unstable at the surface of Mars under present environmental conditions, over thousands of years (Newsom *et al.*, 1996). Cabrol and Grin (1999) suggested that at least 179 Martian craters were occupied by palaeolakes, making them a common phenomenon on Mars.

In contrast to the largely differentiated Earth's crust, possible Martian target rocks are likely to be petrologically primitive mafic rocks that are rich in Mg and Fe, and considerably lower in Si, Al and alkalis. This implies specific alteration mineralogy and alteration sequences. The closest Earth-based analogues are impacts into basalt or amphibolite-facies mafic rocks. These impact-induced systems (e.g. Lonar; Hagerty and Newsom, 2003) are characterized by iron smectite (saponite), corrensite and chlorite-type mineralization at the expense of primary pyroxene–amphibole minerals and volcanic glasses. However, the low rock/water ratio, negligible $O_2$ and high(-er) $CO_2$ fugacity, as well as low atmospheric pressure on Mars, would suggest strong vapour dominance in highly saline fluids and acidic alteration, which is untypical for Earth. This would result in fast self-sealing of fluid conduits by calcium/magnesium/iron carbonate/sulfate precipitation and, second, strong hydrolysis, resulting in abnormally Fe-rich smectite (saponite) and halloysite mineralogies. Thermochemical modelling of alteration mineralogy in a 100 km diameter crater on early Mars (Schwenzer and Kring, 2009) suggests formation of serpentine, chlorite, nontronite and kaolinite as the main hydrous silicates in hydrothermally altered mafic Martian crust.

Recently, Marzo *et al.* (2010) provided evidence of post-Noachian (Hesperian) phyllosilicate formation on Mars in an impact-induced hydrothermal system in the approximately 40 km diameter complex Toro Crater, Syrtis Major Volcanic Plain. Combined geomorphological and spectral data suggest hydrous silicate formation in a post-impact hydrothermal system in and around the central uplift. Morphological features (mounds, vents, veins) are interpreted to be associated with volatile release and liquid flows, whereas the hydrated mineral phases in central uplift and in impact-melt bearing crater-fill deposits are characterized by phrenite, chlorites, smectites and opaline phases. The location and mineralogy of the altered rocks in Toro Crater agrees with the numerical models of the impact-induced hydrothermal systems on Mars presented by Schwenzer and Kring (2009) and Barnhart *et al.* (2010). The distribution of the alteration intensity (i.e. clay abundance) within the terrestrial craters suggests that, in the absence of significant erosion, the supposedly clay-rich zones are not commonly exposed at the surface except at 'pipe' structures and mounds – thought to be expressions of hydrothermal springs or fumaroles – as observed in Haughton and Ries Craters on Earth (Osinski *et al.*, 2001,2005) and in Toro Crater on Mars (Marzo *et al.*, 2010). Nevertheless, signatures of hydrothermal mineralization in the deeper parts of Martian structures could potentially be exposed through erosion or by excavation by later impact events. This may be the case at the approximately 150 km diameter Holden Crater, where a deep canyon, Uzboi Valles, has breached the crater rim and revealed megabreccias that are altered to clays, potentially via impact-induced hydrothermal activity (Tornabene *et al.*, 2009). Also, phyllosilicates have been observed around Martian craters in the northern plains (Carter *et al.*, 2010) and Nili Fossae (e.g. Ehlmann *et al.*, 2009).

### 6.4.2 Microbiological implications of impact-induced hydrothermal systems

The cooling impact structure offers the possibility of intriguing ecological niches for life (e.g. Cockell and Lee, 2002; Cockell et al., 2002). This is exemplified by the present-day preferential microbial colonization of shocked rocks compared with unshocked or low-shocked rocks in the approximately 23 Ma Haughton impact structure, Arctic Canada (Cockell et al., 2002; Chapter 11). With respect to impact-induced hydrothermal systems, the most primitive thermophilic and/or hyperthermophilic microorganisms are of particular interest, mainly due to the exobiological potential of hydrothermal deposits (e.g. Farmer and Des Marais, 1999). Impact hydrothermal systems are also interesting to study with respect to the origin of early life, as they could well have provided conditions supportive of the development of life on the Hadean Earth (Zahnle and Sleep, 1997; Abramov and Mojzsis, 2009), and this cannot be excluded for other planetary bodies as well (Mojzsis and Harrison, 2000).

While volcanic hydrothermal systems have been found to support life on Earth (Corliss et al., 1979) there is no well-documented evidence for life at impact-induced hydrothermal systems to date; although potential evidence includes possible microbial signatures of titanium-rich bacteriomorphs at the Ries structure (Glamoclija et al., 2007) and microbial-like features associated with hydrothermal veins at the 50 km diameter Siljan impact structure, Sweden (Hode et al., 2008). There are several possible reasons for this. First, the cooling of an impact crater occurs geologically in a short time and suitable temperature zones (ecological niches) may change too rapidly in time and space, thus preventing a stable colonization. Second, biosignatures associated with the activity of microbial organisms, which are most adaptive for hydrothermal conditions, are not easy to recognize in fossil hydrothermal deposits. Third, there have been very few detailed studies of biosignatures in impact-induced hydrothermal systems.

The lifetime and structure of the habitable zone is the most critical limiting factor for the colonization of microorganisms. The suitable zones for hyperthermophilic (80–120 °C) and thermophilic (45–80 °C) organisms change significantly in space and volume with time (Abramov and Kring, 2005; Versh et al., 2006). As a general consequence, independent of the size, structure and lifetime of the hydrothermal system, the volume that is habitable for hydrothermal microbial communities (temperature range 45–120 °C) grows rapidly during the first period of the cooling and then fades off with the gradual decrease and deepening of the habitable zone(s) towards the crater centre (Fig. 6.10). In small-sized structures, the cooling of the rim region is quite fast, but in large structures with extensive melt rocks in the rim area the hydrothermal circulation is initiated and remains until the melt sheet has cooled down to 350–400 °C in the centre of the crater. Although in the central peak region the possible microbial colonization is inhibited for about the first thousand years, it is this rock body that retains the optimum temperatures (45–80 °C) needed for thermophiles and/or hyperthermophiles (80–107 °C) for the longest period of time (Abramov and Kring, 2005).

The structure of the possible microbial community within a particular impact-induced hydrothermal system is primarily

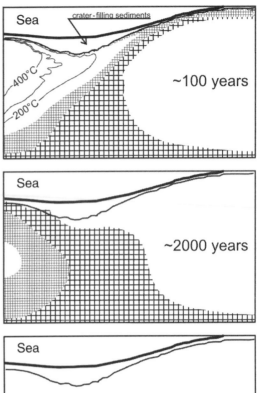

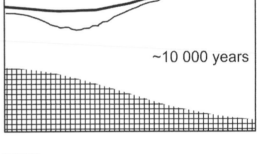

zone (45–80 °C) suitable for thermophilic organisms

zone (80–110 °C) suitable for hyperthermophilic organisms

**Figure 6.10** Modelling simulation of post-impact temperature distribution in Kärdla Crater with optimum growth temperature zones for thermophilic and hyperthermophilic organisms. The area modelled is 2.5 km wide and 1.5 km deep. Modified after Versh et al. (2006).

defined by the environmental conditions, which depend on the temperature and geochemical characteristics of the fluid (i.e. fluid–rock interaction) within the system and also the circumstances of photosynthesis. In terrestrial hydrothermal springs, thermophilic photoautotrophs predominate over other groups, whereas at deep-sea hydrothermal vents the chemosynthetic (chemolithoautotrophic) groups are using the energy from the interaction of reduced compounds in the hydrothermal fluid (sulfur, iron, etc.) and surrounding oxidized seawater (Van Dover, 2000). The terrestrial impact-induced hydrothermal systems, however, are rather typically alkaline and near neutral (pH 6–8) and maintain $E_h$ values of greater than −0.5 V (at least for a neutral environment) in the course of the hydrothermal process. These constrained environmental variables suggest some limitation/specification among possible thermophilic and/or hyperthermophilic microorganisms that could colonize the system. If most of the Archaea and Bacteria are neutrophiles and can easily grow in the very slightly acidic–alkaline pH range (6–8) of the impact-induced environment, then, unlike the volcanic environments that are generally low in oxygen and rich in reductants (Barns and Nierzwicki-Bauer, 1997), impact-induced systems seem to be richer in dissolved oxygen, but not necessarily oxidative, and low in reduced compounds. This would suggest the preference of sulfur-reducing microorganisms in the impact hydrothermal communities, whereas many sulfate reducers are also able to use molecular $H_2$, on which rely most of the chemolithotropic reactions used by microorganisms in hydrothermal vents (McCollom and Schock, 1997).

The microorganisms in impact-induced hydrothermal systems cannot have been very highly specialized, as these systems, particularly in small-sized craters, change rapidly in terms of the space available and the environmental quality. The space habitable for the most primitive hydrothermal microbial communities (temperature range 45–120 °C) grows rapidly during the first period of the impact cooling, reaching the maximum in some thousand years and then fading off in a few tens of thousands of years with the gradual decrease and deepening of the habitable zone(s) towards the crater centre (Versh et al., 2006). This implies that the potential habitats in impact-induced hydrothermal systems would be occupied by organisms capable of adapting to the changing conditions by applying alternative metabolic processes, or organisms adapted to given conditions would migrate inward in accordance with the cooling of rocks.

# References

Abramov, O. and Kring, D.A. (2004) Numerical modeling of an impact-induced hydrothermal system at the Sudbury Crater. *Journal of Geophysical Research*, 109, E10007. DOI: 10.1029/2003JE002213.

Abramov, O. and Kring, D.A. (2005) Impact-induced hydrothermal activity on early Mars. *Journal of Geophysical Research*, 110, E12S09. DOI: 10.1029/2005JE002453.

Abramov, O. and Kring, D.A. (2007) Numerical modeling of impact-induced hydrothermal activity at the Chicxulub Crater. *Meteoritics and Planetary Science*, 42, 93–112.

Abramov, O. and Mojzsis, S.J. (2009) Microbial habitability of the Hadean Earth during the late heavy bombardment. *Nature*, 459, 419–422.

Ahrens, T.J. and O'Keefe, J.D. (1972) Shock melting and vaporization of lunar rocks and minerals. *The Moon*, 4, 214–249.

Allen, C.G., Gooding, J.L. and Keil, K. (1982) Hydrothermally altered impact melt rock and breccia: contributions to the soil of Mars. *Journal of Geophysical Research*, 87, 10083–10101.

Ames, D.E., Watkinson, D.H. and Parrish, R.R. (1998) Dating of a regional hydrothermal system induced by the 1850 Ma Sudbury impact event. *Geology*, 26, 447–450.

Ames, D.E., Kjarsgaard, I.M., Pope, K.O. et al. (2004) Secondary alteration of the impactite and mineralization in the basal Tertiary sequence, Yaxcopoil-1, Chicxulub impact crater, Mexico. *Meteoritics and Planetary Science*, 39, 1145–1167.

Barnhart, C.J., Nimmo, F. and Travis, B.J. (2010) Martian post-impact hydrothermal systems incorporating freezing. *Icarus*, 208, 101–117.

Barns, S.M. and Nierzwicki-Bauer, S.A. (1997) Microbial diversity in ocean, surface and subsurface environments, in *Geomicrobiology: Interactions between Microbes and Minerals* (eds J.F. Banfield and K.H. Nealson), Reviews in Mineralogy 35, Mineralogical Society of America, Washington DC, pp. 35–79.

Bibring, J.-P., Langevin, Y., Gendrin, A. et al. (2005) Mars surface diversity as revealed by the OMEGA/Mars Express observations. *Science*, 307, 1576–1581.

Boer, R.H., Reimold, W.U., Koeberl, C. and Kesler, S.E. (1996) Fluid inclusion studies on drill core samples from the Manson impact crater: evidence for post-impact hydrothermal activity, in *The Manson Impact Structure, Iowa: Anatomy of an Impact Crater* (eds C. Koeberl and R.R. Anderson), Geological Society of America Special Paper 302, Geological Society of America, Boulder, CO, pp. 377–382.

Boynton, W.V., Feldman, W.C., Squyres, S.W. et al. (2002) Distribution of hydrogen in the near surface of Mars: evidence for subsurface ice deposits. *Science*, 297, 81–85.

Cabrol, N.A. and Grin, E.A. (1999) Distribution, classification, and ages of Martian impact crater lakes. *Icarus*, 142, 160–172.

Carr, M.H. (1996) *Water on Mars*, Oxford University Press, New York, NY.

Carr, M.H. (2006) *The Surface of Mars*, Cambridge University Press, Cambridge.

Carter, J., Poulet, F., Bibring, J.-P. and Murchie, S. (2010) Detection of hydrated silicates in crustal outcrops of Mars. *Science*, 328, 1682–1686.

Cathles, L.M, Erendi, A.H.J. and Barrie, T. (1997) How long can a hydrothermal system be sustained by a single intrusive event? *Economic Geology*, 92, 766–771.

Cawood, P. and Pirajno, F. (2008) *Hydrothermal Processes and Mineral Systems*, Springer, Berlin.

Cockell, C.S. and Lee, P. (2002) The biology of impact craters – a review. *Biological Reviews*, 77, 279–310.

Cockell, C.S., Lee, P., Osinski, G. et al. (2002) Impact-induced microbial endolithic habitats. *Meteoritics and Planetary Science*, 37, 1287–1298.

Corliss, J.B., Dymond, J., Gordon, L.I. et al. (1979) Submarine thermal springs on the Galapagos Rift. *Science*, 203, 1073–1083.

Costard, F.M. and Kargel, J.S. (1995) Outwash plains and thermokarst on Mars. *Icarus*, 114, 93–112.

Ehlmann, B.L., Mustard, J.F., Swayze, G.A. et al. (2009) Identification of hydrated silicate minerals on Mars using MRO-CRISM: geologic context near Nili Fossae and implications for aqueous alteration. *Journal of Geophysical Research*, 114, E00D08.

Farmer, J.D. and Des Marais, D.J. (1999) Exploring for a record of ancient Martian life. *Journal of Geophysical Research*, 104, 26977–26995.

Farrow, C.E.G. and Watkinson, D.H. (1992) Alteration and the role of fluids in Ni, Cu and platinum-group element deposition, Sudbury Igneous Complex contact, Onaping–Levack area, Ontario. *Mineralogy and Petrology*, 46, 611–619.

Förstner, U. (1967) Petrographische Untersuchungen des Suevit aus den Bohrungen Deiningen und Wörnitzostheim im Ries von Nördlingen. *Contributions to Mineralogy and Petrology*, 15, 281–308.

Fournier, R.O. (1991) The transition from hydrostatic to greater than hydrostatic fluid pressure in presently active continental hydrothermal systems in crystalline rock. *Geophysical Research Letters*, 18, 955–958.

Gibson, R.L., Reimold, W.U. and Stevens, G. (1998) Thermal metamorphic signature of an impact event in the Vredefort Dome, South Africa. *Geology*, 26, 787–790.

Glamoclija, M., Schieber, J. and Reimold, W.U. (2007) Microbial signatures from impact-induced hydrothermal settings of the Ries Crater, Germany: a preliminary SEM study. XXXVIII Lunar and Planetary Sciences Conference, abstract no. 1989.

Grieve, R.A.F., Dence, M.R. and Robertson, P.B. (1977) Cratering processes: as interpreted from the occurrence of impact melts, in *Impact and Explosion Cratering* (eds D.J. Roddy, R.O. Pepin and R.B. Merrill), Pergamon, New York, NY, pp. 791–814.

Hagerty, J.J. and Newsom, H.E. (2003) Hydrothermal alteration at the Lonar Lake impact structure, India: implications for impact cratering on Mars. *Meteoritics and Planetary Science*, 38, 365–381.

Head, J.W., Mustard, J.F., Kreslavsky, M.A. *et al.* (2003) Recent ice ages on Mars. *Nature*, 426, 797–802.

Helgeson, H.C. (1979) Thermodynamics of hydrothermal systems at elevated temperatures and pressures. *American Journal of Science*, 267, 729–804.

Hode, T., Von Dalwigk, I. and Broman, C. (2003) A hydrothermal system associated with the Siljan impact structure, Sweden – implications for the search for fossil life on Mars. *Astrobiology*, 3, 271–289.

Hode, T., Cady, S.L., Von Dalwigk, I. and Kristiansson, P. (2008) Evidence of ancient microbial life in an impact structure and its implications for astrobiology – a case study, in *From Fossils to Astrobiology: Records of Life on Earth and the search for Extraterrestrial Biosignatures* (eds J. Seckbach and M. Walsh), Springer, Berlin, pp. 249–273.

Ivanov, B.A. (2004) Heating of the lithosphere during meteorite cratering. *Solar System Research*, 38, 266–278.

Ivanov, B.A. and Deutsch, A. (1999) Sudbury impact event: cratering mechanics and thermal history, in *Large Meteorite Impacts and Planetary Evolution II* (eds B. Dressler and R.A.F. Grieve), Geological Society of America Special Paper 339, Geological Society of America, Boulder, CO, pp. 389–397.

Jõeleht, A, Kirsimäe, K., Plado, J. *et al.* (2005) Cooling of the Kärdla impact crater: II. Impact and geothermal modeling. *Meteoritics and Planetary Science*, 40, 21–33.

Jõeleht, A., Kirsimäe, K., Versh, E. and Plado, J. (2006) Rapid post-impact cooling at Kärdla – where are the same conditions met? in *First International Conference on Impact Cratering in the Solar System – Abstract Book* (eds A.F. Chicarro and A.P. Rossi), European Space Agency, ESTEC, Nordwijk, The Netherlands.

Kirsimäe, K., Suuroja, S., Kirs, J. *et al.* (2002) Hornblende alteration and fluid inclusions in Kärdla impact crater, Estonia: evidence for impact-induced hydrothermal activity. *Meteoritics and Planetary Science*, 37, 449–457.

Koeberl, C., Fredriksson, K., Götzinger, M. and Reimold, W.U. (1989) Anomalous quartz from the Roter Kamm impact crater, Namibia: evidence for post-impact activity? *Geochimica et Cosmochimica Acta*, 53, 2113–2118.

Komor, S.C., Valley, J.W. and Brown, P.E. (1988) Fluid inclusion evidence for impact heating at the Siljan Ring, Sweden. *Geology*, 16, 711–715.

Lüders, V. and Rickers, K. (2004) Fluid inclusion evidence for impact-related hydrothermal fluid and hydrocarbon migration in Cretaceous sediments of the ICDP Chicxulub drill core YAX-1. *Meteoritics and Planetary Science*, 39, 1187–1197.

Marzo, G.A., Davila, A.F., Tornabene, L.L. *et al.* (2010) Evidence for Hesperian impact-induced hydrothermalism on Mars. *Icarus*, 208, 667–683.

McCarville, P. and Crossey, L.J. (1996) Post-impact hydrothermal alteration of the Manson impact structure, in *The Manson Impact Structure, Iowa: Anatomy of an Impact Crater* (eds C. Koeberl and R.R. Anderson), Geological Society of America Special Paper 302, Geological Society of America, Boulder, CO, pp. 347–376.

McCollom, T.M. and Schock, E.L. (1997) Geochemical constraints on chemolithoautotrophic metabolism by microorganisms in seafloor hydrothermal system. *Geochimica et Cosmochimica Acta*, 61, 4375–4391.

Melosh, H.J. (1989) *Impact Cratering: A Geologic Process*, Oxford University Press, New York, NY.

Meunier, A. (2005) *Clays*, Springer, Berlin.

Miller, D.S. and Wagner, G.A. (1979) Age and intensity of thermal events by fission track analysis: the Ries impact crater. *Earth and Planetary Science Letters*, 43, 351–358.

Mojzsis, S.J. and Harrison, T.M. (2000) Vestiges of a beginning: clues to the emergent biosphere recorded in the oldest known sedimentary rocks. *GSA Today*, 10, 1–6.

Muraoka, H., Uchida, T., Sasada, M. *et al.* (1998) Deep geothermal resources survey program: igneous, metamorphic and hydrothermal processes in a well encountering 500 °C at 3729 m depth, Kakkonda, Japan. *Geothermics*, 27, 507–534.

Mustard, J.F., Murchie, S.L., Pelkey, S.M. *et al.* (2008) Hydrated silicate minerals on Mars observed by the *Mars Reconnaissance Orbiter* CRISM instrument. *Nature*, 454, 305–309.

Mutanen, T. (1979) Lake Saaksjarvi: an astrobleme after all. *Geology*, 31, 125–130.

Naumov, M.V. (2002) Impact-generated hydrothermal systems: data from Popigai, Kara, and Puchezh-Katunki impact structures, in *Impacts in Precambrian Shields* (eds J. Plado and L.J. Pesonen), Springer-Verlag, Berlin, pp. 71–117.

Naumov, M.V. (2005) Principal features of impact-generated hydrothermal circulation systems: mineralogical and geochemical evidence. *Geofluids*, 5, 165–184.

Newsom, H.E. (1980) Hydrothermal alteration of impact melt sheets with implications for Mars. *Icarus*, 44, 207–216.

Newsom, H.E., Graup, G., Sewards, T. and Keil, K. (1986) Fluidization and hydrothermal alteration of the suevite deposit in Ries Crater, West Germany, and implications for Mars. *Journal of Geophysical Research*, 91, 239–251.

Newsom, H.E., Brittelle, G.E., Hibbitts, C.A. *et al.* (1996) Impact crater lakes on Mars. *Journal of Geophysical Research*, 101 (E6), 14951–14955.

Newsom, H.E., Hagerty, J.J. and Thorsos, I.E. (2001) Location and sampling of aqueous and hydrothermal deposits in Martian impact craters. *Astrobiology*, 1, 71–88.

O'Keefe, J.D. and Ahrens, T.J. (1977) Impact-induced energy partitioning, melting, and vaporization on terrestrial planets. *Proceedings of the Lunar and Planetary Science Conference*, 8, 3357–3374.

Osinski, G.R. (2005) Hydrothermal activity associated with the Ries impact event, Germany. *Geofluids*, 5, 202–220.

Osinski, G.R., Spray, J.G. and Lee, P. (2001) Impact-induced hydrothermal activity within the Haughton impact structure, Arctic Canada; generation of a transient, warm, wet oasis. *Meteoritics and Planetary Science*, 36, 731–45.

Osinski, G.R., Lee, P., Parnell, J. *et al.* (2005) A case study of impact-induced hydrothermal activity: the Haughton impact structure, Devon Island, Canadian High Arctic. *Meteoritics and Planetary Science*, 40, 1859–1878.

Pagel, M. and Poty, B. (1975) Fluid inclusion studies in rocks of the Charlevoix structure (Quebec, Canada). *Fortschritte der Mineralogie*, 52, 479–489.

Pagel, M., Wheatley, K. and Ey, F. (1985) The origin of the Carswell circular structure, in *The Carswell Structure Uranium Deposits (Saskatchewan)* (eds R. Laine, D. Alonso and M. Svab), Geological Association of Canada Special Paper 29, Geological Association of Canada, St John's, Newfoundland, pp. 213–224.

Parsons, I. and Lee, M. (2009) Mutual replacement reactions in alkali feldspars I: microtextures and mechanisms. *Contributions to Mineralogy and Petrology*, 157, 641–661.

Poulet, F., Bibring, J.P., Mustard, J.F. *et al.* (2005) Phyllosilicates on Mars and implications for early Martian climate. *Nature*, 438, 623–627.

Rathbun, S.W. and Squyres, J.A. (2002) Hydrothermal systems associated with Martian impact craters. *Icarus*, 157, 362–372.

Salge, T. (2007) The ejecta blanket of the Chicxulub impact crater, Yucatán, Mexico: petrographic and chemical studies of the K–P section of El Guayal and UNAM boreholes. PhD Dissertation, Humboldt-Universität zu Berlin, Berlin.

Schulze-Makuch, D., Dohm, J.M., Fan, C. *et al.* (2007) Exploration of hydrothermal targets on Mars. *Icarus*, 189, 204–308.

Schwenzer, S.P. and Kring, D.A. (2009) Impact-generated hydrothermal systems capable of forming phyllosilicates on Noachian Mars. *Geology*, 37, 1091–1094.

Soare, R.J., Kargel, J.S., Osinski, G.R. and Costard, F. (2007) Thermokarst processes and the origin of crater-rim gullies in Utopia and western Elysium Planitia. *Icarus*, 191, 95–112.

Stöffler, D. (1972) Deformation and transformation of rock-forming minerals by natural and experimental shock processes: 1. Behavior of minerals under shock compressions. *Fortschritte der Mineralogie*, 49, 50–113.

Stöffler. D., Ewald. U., Ostertag, R. and Reimold, W.U. (1977) Research drilling Nördlingen 1973 (Ries): composition and texture of polymict impact breccias. *Geologica Bavarica*, 75, 163–189.

Sturkell, E.F.F., Broman, C., Forsberg, P. and Torssander, P. (1998) Impact-related hydrothermal activity in the Lockne impact structure, Jämtland, Sweden. *European Journal of Mineralogy*, 10, 589–606.

Tornabene, L.L., Osinski, G.R. and McEwan, A.S. (2009) Parautochthonous megabreccias and possible evidence of impact-induced hydrothermal alteration in Holden Crater, Mars. 40th Lunar and Planetary Science Conference, no. 1766.

Turtle, E.P., Pierazzo, E. and O'Brien, D. (2003) Numerical modelling of impact heating and cooling of the Vredefort impact structure. *Meteoritics and Planetary Science*, 38, 293–303.

Utada, M. (1980) Hydrothermal alteration related to igneous acidity in Cretaceous and Neogene formations of Japan, in *Granitic Magmatism and Related Mineralization* (eds S. Ishihara and S. Takenouchi), Mining Geology Special Issue No. 8, Society of Mining Geologists of Japan, Tokyo, pp. 67–83.

Van Dover, C.L. (2000) *The Ecology of Deep-Sea Hydrothermal Vents*, Princeton University Press, Princeton, NY.

Versh, E. (2006) Development of impact-induced hydrothermal system at Kärdla impact structure. PhD Dissertation, Tartu University, Tartu University Press.

Versh, E., Kirsimäe, K., Jõeleht, A. and Plado, J. (2005) Cooling of the Kärdla impact crater: I. The mineral paragenetic sequence observation. *Meteoritics and Planetary Science*, 40, 3–19.

Versh, E., Kirsimäe, K and Jõeleht, A. (2006) Development of potential ecological niches in impact-induced hydrothermal systems: the small-to-medium size impacts. *Planetary and Space Science*, 54, 1567–1574.

Von Engelhardt, W. (1972) Shock produced rock glasses from the Ries Crater. *Contributions to Mineralogy and Petrology*, 36, 265–292.

Von Engelhardt, W. (1990) Distribution, petrography and shock metamorphism of the ejecta of the Ries Crater in Germany – a review. *Tectonophysics*, 171, 259–273.

Zahnle, K.J. and Sleep, N.H. (1997) Impacts and the early evolution of life, in *Comets and the Origin and Evolution of Life* (eds P.J. Thomas, C.F. Chyba and C.P. McKay), Springer, New York, NY, pp. 175–208.

Zürcher, L. and Kring, D.A. (2004) Hydrothermal alteration in the core of the Yaxcopoil-1 borehole, Chicxulub impact structure, Mexico. *Meteoritics and Planetary Science*, 39, 1199–1221.

Zürcher, L., Kring, D.A., Barton, M.D. *et al.* (2005) Stable isotope record of post-impact fluid activity in the core of the Yaxcopoil-1 borehole, Chicxulub impact structure, Mexico, in *Large Meteorite Impacts III* (eds T. Kenkmann, F. Hörz and A. Deutsch), Geological Society of America Special Paper 384, Geological Society of America, Boulder, CO, pp. 223–238.

# SEVEN

# Impactites: their characteristics and spatial distribution

## Richard A. F. Grieve and Ann M. Therriault

*Earth Sciences Sector, Natural Resources Canada, Ottawa, Ontario, K1A 0E8, Canada*

## 7.1 Introduction

Impactite is the term used for all rocks produced or affected by a hypervelocity impact event. A recent, IUGS-recommended, classification scheme for impactites is given in Stöffler and Grieve (2007). Impactites range from completely reconstituted lithologies, such as impact melt rocks, to fractured target rocks. They generally, but not always, contain evidence of shock metamorphism (Chapter 8). Where evidence of shock metamorphism is lacking, their classification as impactites is generally due to their physical association with an impact structure. The exception to this is when dealing with extraterrestrial materials; for example, meteorites and, in particular, lunar samples (none of which, with the possible exception of an *Apollo 15* basalt sample, were sampled in place (Jolliff *et al.*, 2006)). In these circumstances, a reasoned case can be made that, by analogy, no other geologic process but impact can account for their physical condition.

The combination of shock metamorphism and physical movement of target rocks in an impact event results in phase changes and various degrees of mixing of the lithologies of the original target rocks. This leads to a classification of impactites that is largely based on their lithological character, with consideration given to their original and current locations with respect to both the original target rocks and the parent impact structure (French, 1998). Thus, impactites are grouped according to the extent to which they have been moved from their original pre-impact location by the cratering flow-field and the subsequent modification and collapse of the transient cavity to form the final crater (Chapters 4 and 5). The impactites can be subdivided into autochthonous (formed in place), parautochthonous (moved but appear to be in place) and allochthonous (formed elsewhere and clearly moved to their current location). Allochthonous impactites can be further subdivided into those within and around the final crater (proximal) and those some distance from the final crater (distal). The latter are always ejecta, including air-fall deposits. The classification of impactites and much of the current understanding of their formational processes stems from the observed record of impacts on Earth, which supplies ground-truth data on the lithological and structural character of impact structures in three dimensions (Grieve and Therriault, 2004). Unless subsequently buried, however, thin surface ejecta and air-fall deposits are removed relatively quickly by erosion in the terrestrial environment. This results in an intrinsic weakness in the classification and understanding of distal impactites.

In this chapter we describe the general characteristics of the major impactite groups and consider them with respect to their spatial occurrence at and around simple (Fig. 7.1) and complex impact structures (Fig. 7.2). The principal focus here is on terrestrial impactites, as the reader is most likely to encounter such impactites in the field. There are complications in the nature and significance of some of the more highly shocked terrestrial impactites; for example, impact melt rocks and, particularly, impact melt-bearing breccias or so-called suevitic breccias, which depend on the target rock type: crystalline (igneous or metamorphic), sedimentary and mixed (Osinski *et al.*, 2008). This active area of current research on terrestrial impactites is addressed in Chapter 9 and is subject to evolving interpretations. These complications largely do not occur in current samples of extraterrestrial impactites, as they all derived, at least initially, from crystalline targets. They will, however, likely require consideration if there is a future Mars sample return mission.

## 7.2 Autochthonous impactites

Relatively autochthonous impactites are found in the rim areas and immediately beyond at both simple and complex impact structures. Immediate to the rim area, they may have been uplifted during transient cavity formation but have been returned essentially to their pre-impact location on cavity modification to form the final crater (Chapter 5). Such lithologies broadly retain their pre-impact contact or stratigraphic relations and are sufficiently

---

*Impact Cratering: Processes and Products*, First Edition. Edited by Gordon R. Osinski and Elisabetta Pierazzo.
© 2013 Blackwell Publishing Ltd. Published 2013 by Blackwell Publishing Ltd.

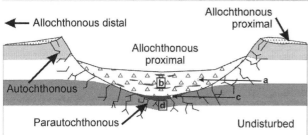

**Figure 7.1** *Top*: *Apollo 10* image (NASA/AS10-29-4324) of the 7 km diameter simple crater Moltke on the Moon. Note bowl-shaped interior, raised rim and exterior rough ejecta deposits. *Bottom*. Schematic cross-section of a simple crater, based on terrestrial data. Large triangles indicate melt-bearing breccia, black areas are more coherent concentrations of impact melt rocks, small triangles are lithic and melt-bearing breccias in ejecta. Letters and boxed letters indicate locations of specific impactite lithologies that are illustrated in later figures.

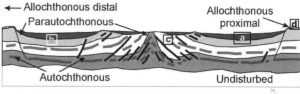

**Figure 7.2** *Top*. *Apollo 17* image (NASA/AS17-2923) of the 28 km diameter complex crater Euler on the Moon. Note slumped material in the rim, flat floor and small central peak. Ejecta facies, including continuous and discontinuous ejecta and secondary craters, are clearly visible. *Bottom*. Schematic cross-section of a complex crater with a small central peak and ring, based on terrestrial data. Black areas are coherent concentrations of impact melt rocks, melt-bearing breccias or lithic breccias filling the crater, depending on the target type. See text for details. Very small triangles are lithic and melt-bearing breccias in ejecta. Boxed letters indicate locations of specific impactite lithologies that are illustrated in later figures.

distant from the point of impact that the shock-wave-induced transient pressures were insufficient to result in diagnostic shock metamorphic features. They are faulted and fractured on local scales, with the degree of brittle deformation increasing as the rim is approached. There are few quantitative studies of the degree of brittle deformation. The relatively young 3.6 Ma El'gygytgyn structure in Siberia is one exception. Here, the measurement of fault and fracture densities indicates impact-related brittle deformation out to approximately three crater radii with their density attenuating at a rate of $r^{-3}$, where $r$ is the radial distance from the centre of this 18 km diameter structure (Gurov *et al.*, 2007). A similar relative distribution of impact-related deformation of autochthonous target rocks was recorded at the Deep Bay structure in Canada (Innes, 1964).

### 7.3 Parautochthonous impactites

Parautochthonous impactites occur in the floor of the so-called true crater (Chapter 18) in simple impact structures. Given that simple impact structures are partially filled with an allochthonous breccia lens (Fig. 7.1), such parautochthonous target rocks are generally encountered through drilling (e.g. Dence *et al.*, 1977)

or, more rarely, at deeply eroded structures (e.g. French *et al.*, 2004; Spooner *et al.*, 2009). These parautochthonous target rocks can exhibit shock metamorphic effects, with the highest recorded shock pressures on the order of 25 GPa (Grieve and Cintala, 1992) and recorded shock levels decay with increasing depth and radial distance from the bottom of the true crater (Dence, 2004). As the parautochthonous target rocks are compressed when they are driven downward during transient cavity formation (Chapter 5), the apparent rate of recorded shock pressure in these rocks is higher than the true (uncompressed) rate (Dence, 2004).

At larger, complex impact structures, the whole-scale modification of the transient cavity results in the parautochthonous target rocks of the transient cavity floor being structurally uplifted to surface or near-surface positions in the final crater-form (Fig. 7.2; Chapter 5). In the terrestrial environment, they are exposed at various levels, depending on degree of subsequent erosion. As with lithologies of the true crater floor in simple structures, these parautochthonous lithologies display shock metamorphic effects, which can record even higher shock levels, up to and including incipient impact melting (Masaitis *et al.*, 1980), due to so-called differential scaling (Grieve and Cintala, 1992). Differential scaling is due to the progressive effect of gravity in reducing the efficiency of crater formation, while not similarly affecting the rate of shock pressure attenuation, as the size of the impact event increases (Melosh, 1989). An indication of this phenomenon is the apparent absence of maskelynite (indicative of recorded shock pressures of approximately 35 GPa (Chapter 8)) in crystalline parautochthonous target rocks of the structural uplift of terrestrial complex impact structures less than 25 km in diameter but its increasing presence at larger structures (Dence, 2004).

Studies at well-exposed structural uplifts at complex impact structures indicate that shock metamorphic effects in the parautochthonous target rocks are concentrically zoned around the centre of the structure, where the highest shock levels are recorded (e.g. Robertson, 1975). Shock studies of drill cores from the central structural uplift of parautochthonous target rocks at such structures as Bosumtwi (Ghana) Clearwater (Canada) and Putchezh-Katunki (Russia) indicate a similar attenuation in recorded shock pressures with depth (Dence *et al.*, 1977; Ivanov *et al.*, 1996; Ferrière *et al.*, 2008). In these structural uplifts, the drill cores indicate that movement, at least in the latter stages of uplift, was between large blocks (tens to hundreds of metres) separated by thin shear zones. This is also indicated by shock attenuation studies, which document reversals in the attenuation of the recorded shock pressure as some block boundaries are crossed (Grieve, 2006). The general block nature of the parautochthonous lithologies in structural uplifts is also indicated by the loss of coherent seismic reflections in the centres of terrestrial complex structures that formed in layered sedimentary targets (Chapter 14).

At the mesoscopic scale, however, outcrops of parautochthonous target rocks in structural uplifts appear remarkably coherent (Dence, 2004), even though they may have travelled several to tens of kilometres from their original pre-impact location during transient cavity formation and subsequent modification to the final crater-form (Chapter 5). Where the parautochthonous target rocks of structural uplifts are exposed by erosion, the lithologies that were originally deepest occur in the centre (Fig. 7.2). In flat-lying sedimentary targets, this is manifested as a suite of concentric parautochthonous lithological units of increasing age, as the centre of the structure is approached; for example, at Gosses Bluff, Australia (Milton *et al.*, 1996). In crystalline targets, this characteristic is less obvious but, in large complex structures, can appear as an increase in the metamorphic grade of the structurally uplifted parautochthonous target rocks as the centre of the structure is approached; for example, at Manicouagan, Canada (Grieve and Head, 1983), and Vredefort, South Africa (Gibson and Reimold, 2005).

Parautochthonous impactites also occur in the down-dropped annular trough surrounding the structural uplift in complex structures (Fig. 7.2). These lithologies are from the upper portion of the target rocks that are outside the transient cavity and are faulted down during cavity modification (Chapter 5). Unlike their counterparts in the structural uplift, they are not shocked, as they are too distant from the point of impact. They are mentioned here only because, in the terrestrial environment, their regional counterparts outside the impact structure may have been removed by erosion and they now exist as isolated outliers in the context of the regional geology. As such, the unexpected presence of such outliers in the regional geology serves to indicate the possibility of the presence of a previously unrecognized eroded complex impact structure.

Some parautochthonous target rocks inside impact structures may be folded and/or brecciated due to late-stage relative movements during crater formation. In such breccias, clasts are of local parautochthonous target rocks and the breccias are most often monomict but can be polymict, depending on the local geology. They are poorly sorted and the clasts tend to be angular, but can be more rounded in weaker lithologies. Such breccias of parautochthonous target rocks develop often as irregular bodies of varying dimension that grade into the local target rocks. If shock features are not present, they are not readily distinguishable from breccias formed by endogenic geologic processes. It is their presence within an identifiable impact structure, with demonstrable shocked materials elsewhere in the structure, which indicates an impact origin.

## 7.4 Allochthonous impactites

### 7.4.1 Proximal varieties

Allochthonous lithologies at impact structures take the form of impact breccias and impact melt rocks. Non-melt-bearing allochthonous breccias are generally termed lithic breccias (French, 1998; Stöffler and Grieve, 2007). They are usually polymict but can be monomict. They make up the bulk of the crater fill at simple structures (Fig. 7.1 and Fig. 7.3a,b; Grieve, 1978). They are relatively less common at larger complex structures in crystalline targets, where relatively more of the target material in the original transient cavity is melted, due to differential scaling. In these cases, the lithic breccias are found in the crater floor (Fig. 7.3c,d), stratigraphically below a coherent sheet of impact melt in crystalline targets. In sedimentary targets, some crater-fill breccias have been simply referred to as allogenic (allochthonous) breccia, with more detailed descriptions (e.g. Masaitis, 1999) indicating that they are equivalent to lithic breccias. Lithic breccias consist of lithic and mineral clasts in a clastic matrix of finer grained material from the same sources. They are poorly sorted and clasts are generally angular to sub-rounded (Fig. 7.3). The clasts may or may not display sub-solidus shock metamorphic features (Chapter 8). As they are allochthonous lithologies, the levels of recorded shock in the clasts are highly variable. Similarly, clast size can be variable and ranges from less than 1 mm to tens or even hundreds of metres. Complications can occur in impactites that formed in marine targets, with resurge deposits being swept back into the crater and, even, further redeposition by water oscillations within the crater (e.g. Ormö *et al.*, 2007; Horton *et al.*, 2008, 2009).

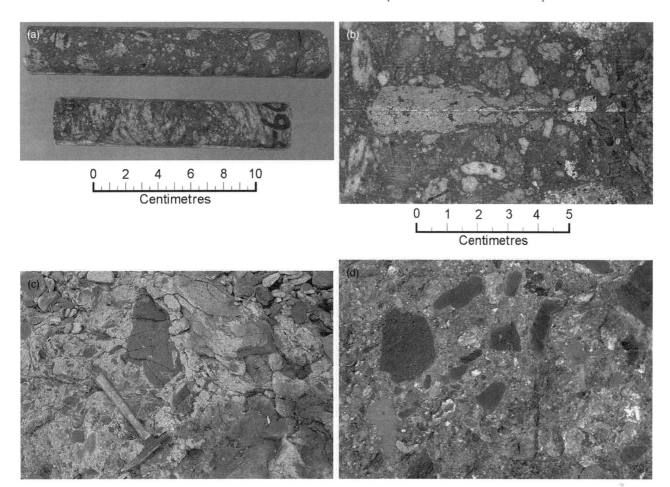

**Figure 7.3**   Various lithic breccias. (a) Monomict lithic breccia from the 3.8 km diameter Brent simple impact structure, Canada. *Bottom.* Core of lithic breccia from the base of the breccia lens, near location 'd' in Fig. 7.1. Note relative paucity of clastic matrix material. *Top.* Core of lithic breccia from slightly higher in the breccia lens, with fewer clasts and more matrix material, near location 'd' in Fig. 7.1. (b) Split core of polymict lithic breccia from Brent. Note more rounded and greater size range of lithic clasts compared with breccias in Fig. 7.3a. Such breccias occur in the main mass of the breccia lens, such as in location 'a' in Fig. 7.1. (c) Outcrop of polymict lithic breccia from the 30 km diameter Slate Islands complex impact structure in Canada. Note angular to sub-rounded, relatively limited variety of 'local' lithic clasts. Hammer for scale. Such breccias occur in the parauto-chthonous rocks of the crater floor beneath the major crater-filling lithologies, near location 'c' in Fig. 7.2. (d) Outcrop of polymict lithic breccia from the Slate Islands. Note greater variety of lithic clasts, compared with Fig. 7.3c. Largest clasts are approximately 10 cm in diameter. Such breccias occur in the parautochthonous rocks of the crater floor beneath the major crater-filling lithologies, near location 'c' in Fig. 7.2.

Allochthonous lithic breccias can also occur as multigenera-tional dikes in the parautochthonous rocks of the floor of complex impact structures (Fig. 7.4a; Lambert, 1981; Bischoff and Oskier-ski, 1987; Dressler and Sharpton, 1997). These dikes are usually polymict and range in width from centimetres to tens of metres. Lithic and mineral clasts tend to be somewhat more rounded than in other lithic breccias and may record sub-solidus shock meta-morphic features. Lambert (1981) identified two dominant types of breccia dikes in the parautochthonous crater floor rocks, based on cross-cutting relationships and differences in texture, structure and composition. He equated the two types with their discrete formation in the compression and modification stages of crater formation (Chapters 3 and 5). In some cases, lithic clasts from lithologies originally higher in the stratigraphic section than the parautochthonous target rocks of the crater floor have been iden-tified in breccia dikes, indicating the downward sense of trans-portation (e.g. Halls and Grieve, 1976).

Although designated as distinct and discrete lithologies, lithic breccias and melt-bearing breccias are both allochthonous units and can grade into each other (Grieve, 1978; Masaitis, 1999; Fer-rière *et al.*, 2007). Melt-bearing breccias, intercalated with lithic breccias, occur as crater-fill products in simple (all target types) and complex craters (in mixed and sedimentary targets; Masaitis, 1999). Melt-bearing breccias can also occur as dikes in the parau-tochthonous target rocks of the crater floor and both flooring and roofing coherent bodies of impact melt rocks at complex struc-tures in crystalline targets. In many cases, these melt-bearing brec-cias have been termed 'suevites' (e.g. Masaitis, 1999; Koeberl *et al.*,

**Figure 7.4**    (a) Outcrop of lithic breccia dike from the 30 km diameter Slate Islands complex impact structure. Note relative paucity of lithic clasts. Such clasts can include lithologies from higher in the stratigraphic sequence. Hammer for scale. Such breccia dikes occur in the parautochthonous rocks of the crater floor beneath the major crater-filling lithologies, near, location 'c' in Fig. 7.2. (b) Outcrop of suevite breccia from the 24 km diameter Ries complex impact structure. Such breccias occur near location 'd' in Fig. 7.2. Dark amorphous to convoluted fluidal clasts (e.g. centre) consist of impact melt glass charged with lithic and mineral clasts. Note wide range in clast sizes. Deutschmark coin (23.5 mm in diameter) for scale. Image courtesy of G. Osinski. (c) Photomicrograph in plane light of suevite breccia from the Ries. Note dark amorphous to convoluted fluidal clasts with inclusions of lithic and mineral clasts and wide range of clasts sizes and shapes. Field of view is 1.0 mm. Image courtesy of G. Osinski.

2007). The term 'suevite' was original coined to describe certain breccias at the Ries impact structure in Germany, where they were described as polymict impact breccias, containing clasts of impact melt glass, shocked (sub-solidus) and unshocked mineral and lithic clasts in a clastic matrix (Fig. 7.4b,c; Pohl *et al.*, 1977). The relatively recent examination of some suevites, particularly from the Ries (in a mixed target) and from within the Haughton structure (in a dominantly sedimentary target) in Canada, using analytical electron microscopy has revealed much more evidence of melting of the sedimentary target rocks in both the clasts and groundmass matrix of the 'suevites' than previously thought from optical studies (Osinski, 2003; Osinski *et al.*, 2004, 2005, 2008). These recent studies are discussed in more detail in Chapter 9. These discoveries have also resulted in a change in the definition of suevite to include the fact that the matrix contains glass particles and that the matrix is better described as 'particulate' rather than 'clastic' (Stöffler and Grieve, 2007).

It is difficult to believe, however, that all the different occurrences of melt-bearing breccias or so-called 'suevite' in impact structures have the same genesis. The current literature record and the use of the term 'suevite' are inconsistent and somewhat confusing. The study and the search for a better understanding of suevite is an area of active research. Until there is a better understanding, it may be best that the term 'suevite' be reserved for the original occurrences at the Ries. Other occurrences should be termed 'suevitic breccias' or, even better, be referred to more generically as melt-bearing breccias. Given the evolution in moving from the optical to electron microscope scale in characterization of the matrix of some of these 'suevitic breccias' and the discovery of evidence for the impact melting of sedimentary lithologies (Chapter 9), it remains to be seen if the past descriptions of the nature of the crater-filling breccias at structures in sedimentary targets, and their current classification, in some cases, as lithic (non-melt-bearing) breccias, will stand the test of time.

Impact melt rocks, as the name indicates, have a matrix of what was melted, and generally mixed, target rock materials. They generally contain both shocked and unshocked mineral and lithic clasts of target material and can be subdivided on the basis of clast content into clast rich, clast poor and clast free. Clasts are, to various degrees, thermally digested and/or reacted out by the melt matrix. The melt matrix itself may be glassy, devitrified glass

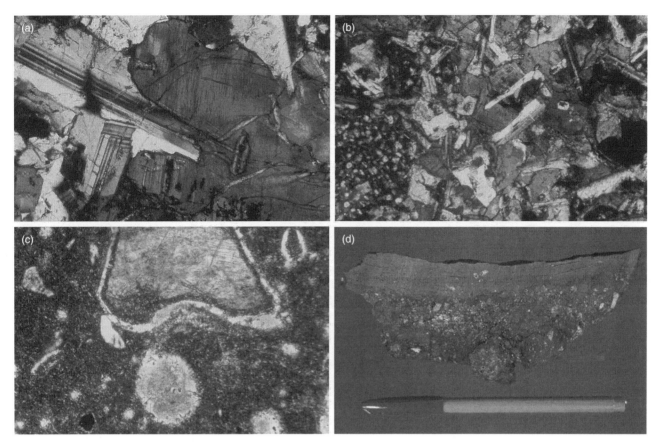

**Figure 7.5** (a) Photomicrograph in crossed polars of relatively 'coarse'-grained clast-poor impact melt rock from the 100 km diameter complex Manicouagan impact structure in Canada. Note apparent absence of lithic and mineral clasts and 'igneous' sub-ophitic texture between plagioclase and pyroxene. Field of view is 2.3 mm. Such melt rocks occur in location 'a' in Fig. 7.2 and location 'a' in Fig. 7.6. (b) Photomicrograph in crossed polars of relatively 'medium'-grained clast-poor impact melt rock from the 28 km diameter complex Mistastin impact structure in Canada. Note small number of thermally altered lithic clasts (e.g. recrystallized anorthosite clast in lower left). Field of view is 1.0 mm. Such melt rocks occur in location 'a' in Fig. 7.2 and location 'b' in Fig. 7.6. (c) Photomicrograph in plane light of relatively 'fine'-grained impact melt rock from Mistastin. Note large number of mineral clasts and fine-grained matrix of crystallites. Large clast (upper centre) is quartz with planar deformation features (Chapter 8) and a reaction corona of high birefringent pyroxene and pale brown high-silica glass. Immediately below are several rounded patches of high-silica glass, where the original quartz clast has been completely reacted out. Field of view is 1.6 mm. Such melt rocks occur in location 'a' in Fig. 7.2 and location 'c' in Fig. 7.6. (d) Slab showing the base of the impact melt sheet at the 28 km diameter complex Mistastin impact structure. Contact is with a melt-bearing breccia containing both lithic and impact melt glass clasts, which some workers would name as 'suevite'. Note, however, the coherent and compact nature of the impact melt clasts compared with those in the Ries suevite illustrated in Fig. 7.4b. Pen for scale. Such melt rocks occur in location 'b' in Fig. 7.2 and location 'd' in Fig. 7.6. (See Colour Plate 19)

or crystalline (fine, medium or coarse grained), depending on the cooling history of the melt rock. Within the same impact structure, impact melt rocks tend to have a common composition, as a chemical mixture of those target rocks melted (e.g. Grieve *et al.*, 1977). They can, however, have a wide range of textures at the microscopic scale (e.g. Floran *et al.*, 1978; Fig. 7.5), as a result of a cooling history determined by geological location within the impact structure.

Impact melt rocks occur as dikes in the parautochthonous rocks of the crater floor of complex structures that formed in crystalline (igneous and metamorphic) targets (Fig. 7.6a,b). However, by far the bulk of impact melt rocks within these complex impact structures in crystalline targets occurs as a

coherent sheet (Fig. 7.6c), overlying parautochthonous target rocks of the crater floor or separated from the floor by an intervening thin layer of breccia (Fig. 7.5d), which contains both impact melt and lithic and mineral clasts of target rocks. Such sheets may display cooling features, such as columnar jointing, to various degrees (Fig. 7.6c), and have well-developed igneous-like matrix textures (Fig. 7.5a). At first appearance, they may look like endogenic volcanic rocks. Apart from their association with an impact structure (which may not have been recognized in the initial discovery of the melt rocks), what distinguishes them from volcanic rocks is that they are single cooling units, often with relatively high clast content and somewhat unusual chemistry.

**Figure 7.6**    (a) Dike of fine-grained impact melt rock in the parautochthonous rocks of the crater floor of the approximately 300 km diameter Vredefort impact basin in South Africa. Such melt rocks occur in location 'c' in Fig. 7.2. Hammer for scale. (b) Dike of fine-grained impact melt rock in the parautochthonous rocks of the crater floor of the approximately 100 km diameter Manicouagan complex impact structure. Note sharp contact with anorthosite host rock and minor xenoliths of anorthosite, which have been dislodged from the wall rock and rafted into the dike. Such melt rocks occur in location 'c' in Fig. 7.2. Hammer for scale. (c) Cliff of impact melt rocks at the approximately 100 km diameter Popigai complex impact structure in Russia. Top of cliff is an erosional surface; base of the melt rocks and their contact with underlying breccias is hidden by scree. Melt rocks have crude columnar jointing. Trees for scale. Boxed letters correspond to the general locations of the impact melt rocks illustrated in Fig. 7.5 and the entire outcrop corresponds to location 'a' in Fig. 7.2. Image courtesy of M. Pilkington.

As a mix of melted target rocks, impact melt rocks do not conform necessarily to the standard subdivision of endogenic volcanic rocks, based on mineralogy and chemistry. In some cases, early discoveries were given 'local' names; for example, 'dellenite' at the Dellen and 'kärnite' at the Lappajärvi structures, in Sweden and Finland respectively (Svensson, 1968a,b). In addition, the coherent impact melt rocks at Popigai, Russia, are referred to as 'tagamite' (Masaitis, 1994). Unfortunately, this term has been extended to other occurrences of coherent impact melt sheets and rocks in the territory of the former Soviet Union (e.g. Masaitis, 1999). The practice of extending locally defined and named impact melt lithologies is discouraged, as it is not descriptive and can lead potentially to unwarranted genetic interpretations (as with the term 'suevite'). The study of impact structures and their lithologies is a relatively recent endeavour that was initially centred in a limited number of countries and, as a result, some confusion has arisen with the nomenclature of impactites. Although the recently IUGS-recommended nomenclature for impactites (Stöffler and Grieve, 2007) required over a decade of discussions among the international community, it is dynamic, as

characterization techniques evolve, and not yet fully accepted nor its use ingrained in the community (Reimold *et al.*, 2008).

Coherent impact melt sheets tend to be more charged with lithic and mineral clasts towards their lower and upper contacts. Cooling of the impact melts is most rapid at these contacts and, as a result, the impact melt rocks are finer grained than in the middle of the impact melt sheets (Fig. 7.5). Longer cooling times in the middle portions of impact melt sheets result in the greater thermal digestion and reaction of clastic material by the melt, with ultimately fewer preserved clasts (Phinney *et al.*, 1978). This local trend in clast distribution with stratigraphic position in impact melts sheets is superimposed on a larger general trend, which dictates the overall initial clast content of impact melt sheets. Namely, larger impacts produce relatively greater volumes of impact melt and, thus, impact melt rocks with less entrained clastic debris than smaller impacts do. The largest known terrestrial impact melt sheet is the Sudbury Igneous Complex (SIC) at the Sudbury structure (Grieve *et al.*, 1991). It had an initial volume on the order of $10^4$ km$^3$ and is close to, but not entirely, clast-free (Therriault *et al.*, 2002). The SIC is up to 3.5 km thick

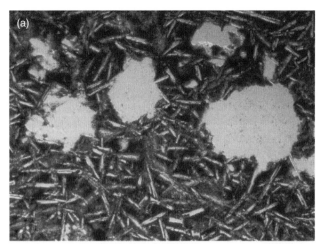

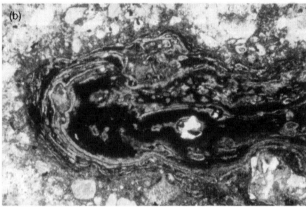

**Figure 7.7**  (a) Photomicrograph in plane light of relatively 'fine'-grained, clast-free impact melt rock from the approximately 3 km New Quebec simple crater in Canada. Texture is a mixture of feldspar and pyroxene crystallites set in a glass matrix. Note also clasts of shocked quartz. Field of view is 2.0 mm. Such melt rocks occur at location 'c' in Fig, 7.1. (b) Photomicrograph in plane light of contorted 'glassy' (now altered) impact melt clast in the breccia lens of the approximately 3.8 km diameter Brent simple crater. Field of view is 2.0 mm. Such impact melt clasts in breccia occur at location 'b' in Fig. 7.1.

and distinguishes itself in the terrestrial impact record by having undergone differentiation due to its relative thickness and long cooling time. The fact that it is differentiated and has so few clasts that exhibit traces of shock metamorphism goes a long way to explain why it took so long for its impact origin to be acknowledged, although Sudbury was first recognized as an impact site in the 1960s (Dietz, 1964; French, 1967).

Samples of impact melt rocks from simple structures are less accessible, as the bulk of the melt rocks are buried in the breccia lens that fills simple structures (Fig. 7.1). Such melt rocks tend to be fine grained to glassy (Fig. 7.7a) and highly charged with clastic material, as befits the relatively small size of simple impact structures and the short cooling time for their impact melts. Within the interior breccia lens, melt rocks can occur as a lens-shaped body near the base of the breccia lens (Fig. 7.1). These melt rocks

are coarser grained and are least charged with clastic material in the middle of the lens. The bulk of the melt rocks, however, are dispersed throughout the breccia lens (Fig. 7.1), as clasts in the breccia (Fig. 7.7b) or as local pods of more concentrated melt (Grieve and Cintala, 1981). Such dispersed melt materials are glassy to very fine grained. As the breccia lens is relatively hot immediately following impact and the terrestrial environment is a relatively wet one, hydrothermal alteration of the melt materials can occur, changing both their texture and their chemistry.

The clast population of impact melt rocks does not necessarily accurately reflect the modal mineralogy of the target rocks, because of the preferential digestion of clasts. For example, it has been noted that, at structures in 'granitic' targets, the fine-grained melt rocks at the base of the melt sheet have both quartz and feldspar clasts but that the relative amount of quartz clasts decreases with stratigraphic height due to thermal reaction with the melt, facilitated by the longer times for the original clasts to react with the melt. In heterogeneous targets, this trend is superimposed upon the fact that the modal clast contents of melt rocks do not necessarily reflect the target material that was melted (McCormick *et al.*, 1989). The volume of the target that was melted is not the same volume of the target that was the source of the clasts (Melosh, 1989). This property of many impact melt rocks has been acknowledged in the study of lunar impact melt rocks and attempting to determine their original provenance. For example, the clast population of so-called Low K Frau Mauro (LKFM) basalts from *Apollo 15*, which are impact melt rocks possibly from the Imbrium Basin (Ryder and Bower, 1977), is not related to the LKFM composition but reflects a troctolitic source. Thus, the clasts provide no evidence to limit the original provenance of LKFM (Spudis *et al.*, 1991).

The above discussion of impact melt lithologies has focused on impact structures formed in crystalline targets. It has been known for some time that impacts into sedimentary targets do not result in coherent sheets of impact melt rocks but rather result in a more complex set of lithologies, including the formation of melt-bearing breccia deposits, which, as noted earlier, are sometimes referred to as 'suevite' (e.g. Masaitis, 1999), in place of coherent impact melt rocks (Kieffer and Simonds, 1980). These breccias occupy the same structural position within the impact structure as the coherent impact melt rocks in crystalline targets and are believed to have a similar genesis (Grieve, 1988; Osinski *et al.*, 2008). Impacts into mixed targets can result in both coherent melt rocks and melt-bearing breccias (Masaitis *et al.*, 1980). The current status of knowledge on impact melt products in various target types is addressed in some detail in Chapter 9.

Impact melt rocks and various breccias (lithic and melt-bearing) occur in the collections of lunar and meteoritic samples (Fig. 7.8). Owing to the nature of the sampling/collection processes, they are, by definition, allochthonous. More emphasis is generally placed on classification under their sample designation, name or meteoritic classification than under their lithological classification as impactites. In general, the descriptive terminology for lunar impactites does not strictly follow that of terrestrial impactites (Stöffler *et al.*, 1980), although the lithologies may be equivalent. The literature contains terms, such as 'fragmental breccia' (Fig. 7.8a), which are not used to describe terrestrial impactites. In some respects, these differences are understandable.

**Figure 7.8** (a) Lunar feldspathic lithic breccia 67016 (NASA/JSC photograph S-75-32783), with angular to sub-rounded clasts, which include anorthosite and previously formed impact melt rocks. Metal cube (upper left) is 1 cm a side. Sample was collected from the rim area of North Ray crater on the *Apollo 16* mission. (b) Lunar vesicular impact melt rock 77135 (NASA photograph S-72-57391). Melt matrix consists mainly of pyroxene, plagioclase and olivine. Lithic clasts are largely recrystallized anorthosite and troctolitic breccia. Cube (lower right) is 1 cm a side. Sample was collected at Station 7 on the *Apollo 17* mission.

Extraterrestrial impactites differ in one important respect from terrestrial examples, in that they can record multiple impact events, a situation so far unrecognized on Earth. Thus, a lunar fragmental breccia can contain clasts of earlier generations of breccia and/or a diverse set of impact melt products as clasts. A number of lunar impact melt rocks (and some terrestrial examples) have also been described as impact melt breccias. They have a very fine-grained to glassy melt matrix and are highly charged with lithic and mineral clasts (French, 1998). This term is no longer recommended and should be replaced with fine-grained, clast-rich impact melt rock (Stöffler and Grieve, 2007). There are

other types of impactite lithologies that occur on the moon and are found in some meteorites. For example, the lunar surface is covered by an impact-generated regolith, which contains a variety of components from rocks to minerals to glasses, and there are also aggregated regolith breccias. The regolith also contains agglutinates, which are themselves also aggregates, but of regolith particles welded by glasses. The abundance of agglutinates is a reflection of regolith maturity, which is a function of the length of time the regolith has been exposed to (micro-) meteorite bombardment (Jolliff *et al.*, 2006).

Pseudotachylite (also spelled pseudotachylyte) was first described from what was later identified as the Vredefort impact structure (Shand, 1916). It has a dark, aphanitic matrix, with lithic and mineral clasts, and forms dikes, veins and anastomosing bodies in the parautochthonous target rocks of the crater floor at large terrestrial impact structures (see Chapter 9; Fig. 7.9). It is, however, not unique to impact and can be found in endogenic fault structures (Sibson, 1975). The term 'impact pseudotachylite' (Stöffler and Grieve, 2007) is used to distinguish pseudotachylite of known impact origin. Spray (1998) suggested that pseudotachylite forms by both shock-wave-induced motions during cavity formation and by frictional melting on cavity modification. While this can explain the occurrence of veins and thin dikes of pseudotachylite, it has difficulties accounting for the volumetrically large and wide (tens to hundreds of metres) bodies. Once frictional melt is formed, the coefficient of friction on the shear fault drops to essentially zero and there is no more friction (heat) generated. Melosh (2005) has proposed that once the frictional melt is formed it is removed from the shear fault and is accumulated in dilational zones, which are physically linked to the shear fault(s). A recent study of pseudotachylite (so-called Sudbury Breccia) at the Sudbury impact structure calls for cataclasis of local target rocks to form the matrix of the breccia and notes two generations of pseudotachylite (Lafrance *et al.*, 2008). This would appear, in part, consistent with some observations at Vredefort, where there is more than one generation of pseudotachylite associated with pre-Vredefort fault structures (Fletcher and Reimold, 1989), which were exploited in both transient cavity formation and subsequent modification (Grieve *et al.*, 2008). Most recently, Lieger *et al.* (2009) and Riller *et al.* (2010) conducted a comprehensive structural analysis in the central core area at Vredefort and found no evidence for shear faults to generate pseudotachylite. They concluded that a tensional fracture system developed during the collapse of the structural uplift (Chapter 5), producing dilation fractures that were later filled by allochthonous melt, possibly from the overlying melt sheet (Riller *et al.*, 2010). As with the current lack of clarity regarding the detailed character and origin(s) of suevite, 'pseudotachylite' may possibly be the result of more than one process and the more appropriate terminology to use may be pseudotachylitic breccia (Reimold, 1995).

### 7.4.2 Distal varieties

The only significant sedimentation process on the Moon is 'impact sedimentation', with impacts that produced the large multi-ring basins, such as Orientale, having the capacity to redistribute lunar crustal material halfway around the moon (Ghent *et al.*, 2008). Apollo missions to the lunar highlands were

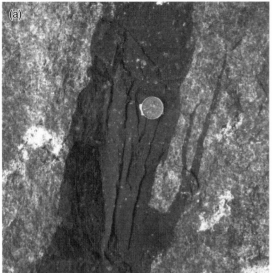

**Figure 7.9** Vein of pseudotachylite in the parautochthonous rocks of the crater floor of the approximately 300 km diameter Vredefort impact basin. Note dark aphanitic matrix with small visible lithic clasts and thin branching veinlets on the right. South African rand coin (19.0 mm in diameter) for scale. Such pseudotachylites occur in location 'c' in Fig. 7.2. (b) Larger pseudotachylite (or pseudotachylitic breccia) veins in parautochthonous rocks of the crater floor of Vredefort. Note incorporation of sub-rounded lithic clasts of the wall rock. Hammer for scale. Such pseudotachylites occur in location 'c' in Fig. 7.2.

specifically designed to sample the ejecta of some of these basins; for example, *Apollo 17* was designed to sample ejecta from the Serenitatis basin (Spudis, 1993). Such source relationships, however, cannot be demonstrated with absolute certainty. In some cases, samples of such lunar impactites as impact melt rocks

have been grouped together on the basis of their similar chemistry and assigned as ejecta from a particular basin (Ryder and Bower, 1977; Spudis, 1993). In other cases, physical association with a particular (small) crater has led to the interpretation that particular lunar samples are ejecta from that crater. For example, samples 14063 and 14064, which are polymict lithic breccias, are interpreted as ejecta from Cone Crater at the *Apollo 14* site (Swann *et al.*, 1977). In planetary studies, ejecta blankets are subdivided, in terms of increasing distance from their source crater, into continuous and discontinuous ejecta and rays, which grade into each other. Secondary craters are generally visible in the zone of discontinuous ejecta and beyond. Such ejecta are also generally described in terms of their surface appearance (e.g. 'hummocky, lineated, textured'). Such descriptions, however, provide no information on the lithological nature of the ejecta facies.

As with other impactites, the terrestrial cratering record is the source of ground-truth data on ejecta. Unfortunately, the high level of surface geologic activity on Earth results in the very poor preservation of ejecta in the terrestrial environment. There are cases of ejecta preserved in the sedimentary record and, in some cases, they can be traced back to their source crater. For example, shocked quartz, altered splash-form melt spherules and accretionary-like lapilli occur in the stratigraphic column in the western Lake Superior region and have been linked to the Sudbury impact event approximately 400–900 km to the east (Addison *et al.*, 2005; Pufahl *et al.*, 2007; Cannon *et al.*, 2010). Such links between shocked materials in the stratigraphic column and a particular impact structure are generally on the basis of 'equivalence' in age and, in some cases, in lithologies and specific minerals in the ejecta and the target area (e.g. Thackrey *et al.*, 2009). There are also cases where impact ejecta cannot be directly linked to a known source impact structure (Montanari and Koeberl, 2000; Koeberl and Martínez-Ruiz, 2003). Unfortunately, discrete occurrences of impact ejecta in the stratigraphic record, which are separated from each other in both space and time, are not ideal for the characterization of impact ejecta as a lithological unit.

Most of the current knowledge of impactites as ejecta comes from observations associated with a limited number of impact structures; in particular, Barringer (or Meteor) Crater, USA (e.g. Kring, 2007), for simple craters and Ries, Germany (e.g. Hörz *et al.*, 1983), and Chicxulub, Mexico (e.g. Pope *et al.*, 2005), for complex craters. Based largely on observations at Barringer, proximal ejecta (continuous ejecta blanket) relatively close to the rim is coarser grained than more distant ejecta and displays an inverted stratigraphy (Osinski *et al.*, 2006). The lower ejecta deposits can be described as lithic breccias, with relatively low shock levels, that are predominantly from relatively shallow depths in the target. The source depth of the ejecta increases with radial range. Due to erosion, there are few examples of melt-bearing lithologies at Barringer but they do exist in lag deposits in the ejecta (Hörz *et al.*, 2002; Osinski *et al.*, 2006).

The relationship between proximal, upper melt-bearing breccias and lower low-shock lithic impact breccias is typified by observations at the Ries (Pohl *et al.*, 1977). Here, melt-bearing breccia (suevite *sensu stricto*; Fig. 7.4b,c), which is dominated by shocked material from relatively deeper sources in the target, overlies a lower lithic breccia (known as Bunte breccia), with low levels of shock. Bunte breccia is dominated by material from relatively higher sources in the target stratigraphy and local surface materials, incorporated due to secondary cratering on landing and subsequent flow across the exterior surface terrain. The amount of local surface material incorporated increases with radial distance from the crater rim (Hörz *et al.*, 1983). The impact of the ballistic ejecta of the continuous blanket and its outward flow are sufficient to cause deformation in the substrate (Kenkmann and Schönian, 2006). These characteristics of proximal ejecta, namely increasing source depth and incorporation of local substrate materials with increasing radial range (Oberbeck, 1975), are tenets in the interpretation of lunar samples (Heiken *et al.*, 1991). However, the contact between the suevite and the underlying Bunte breccia at the Ries is sharp and the concept of the suevite as ballistic ejecta or fallout ejecta from a turbulent suspension in a vapour plume (von Engelhardt, 1990) has been challenged recently by a hypothesis involving ground-hugging, less energetic melt-like flows for the suevite (Osinski *et al.*, 2004; Chapter 4). The two-layer separation of suevite and Bunte breccia can also not be achieved by recent numerical modelling of the Ries impact ejecta (Artemieva *et al.*, 2009).

Chicxulub provides a link between the character of proximal and distal ejecta. Sharpton *et al.* (1996) recognized melt-bearing 'suevite' breccia (interpreted as fallback material) within the impact structure and a proximal ejecta deposit exterior to the structure. The proximal ejecta are lithic breccias, dominated by near surface target-derived materials and recording only low levels of shock. They termed the exterior lithic breccia deposit 'Bunte' breccia. As with 'suevite', the use of 'Bunte' breccia for dominantly lithic breccia ejecta is discouraged as it is not descriptive and the translation of 'Bunte' in English is 'multicoloured', which may apply to the breccia at the Ries but not elsewhere. The initial recognition by Sharpton *et al.* (1996) of these lithologies was from Pemex drill cores. These hydrocarbon-industry holes, however, were not continuously cored and the samples were few and widely spaced. More recent shallow, continuous cored drill holes indicate a melt-bearing breccia atop the lithic breccia ejecta out to greater than 0.6 radius from the crater rim at Chicxulub (Sharpton *et al.*, 1999).

At approximately 3.5 to 4.7 radii from the Chicxulub structure, ejecta in SE Mexico and Belize consists of a several metre thick 'spherule' bed, consisting of altered glass spherules, accretionary-like lapilli and pebble-sized lithic clasts of carbonate, set in a calcite matrix, which is overlain by several metres of lithic breccia, containing minor altered impact melt glass and shocked quartz (Pope *et al.*, 2005; Kenkmann and Schönian, 2006). There is, however, a major difference in interpretation of the origin of these deposits. Kenkmann and Schönian (2006), while attributing the lithic breccia to ejecta from Chicxulub, argue that it is dominated by secondary, local materials and was emplaced as a ground-hugging flow. By contrast, Pope *et al.* (2005) consider that emplacement was from ballistic ejection and collapse of the expanded vapour plume over Chicxulub. While both groups use the non-generic, non-impact term 'diamictite' to describe the lithic breccia deposits, they also both believe that the breccias represent what remains of the distal portion of the continuous ejecta blanket of Chicxulub.

At more distal sites (e.g. at approximately 7.5 to 10 radii in NE Mexico), the Chicxulub ejecta layer consists of a centimetre- to

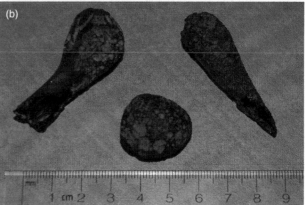

**Figure 7.10**  (a) Ejecta spherules and other forms of partially altered impact glass from Chicxulub, which were recovered from the Cretaceous–Tertiary boundary in Haiti. Tick marks (upper edge) are 1 mm apart. Image courtesy of D. Kring. (b) Spherical and splash form tektites (Indochinites) from Vietnam. Centimetre ruler for scale. The source impact structure for these Australasian tektites is unknown.

metre-thick spherule bed (e.g. Fig. 7.10a) with altered silicate and carbonate phases and accretionary-like lapilli, which have a clear affinity to the Chicxulub target rocks (Schulte and Kontny, 2005). This would favour the interpretation that the spherule beds in SE Mexico and Belize are related to Chicxulub. The overlying lithic breccia deposits, however, are not present, being replaced in NE Mexico by a structure-less sandstone unit (Schulte and Kontny, 2005). This suggests that the absence of the lithic breccia deposits is a primary feature and, in the planetary context of observing ejecta deposits through imagery, the continuous ejecta deposits of Chicxulub extended to between approximately five and seven crater radii from the rim. The observation that the highest energy component of the ejecta (melt-bearing breccia) overlies lithic breccia immediately exterior to Chicxulub's rim (Sharpton *et al.*, 1999) but its equivalent (the spherule bed) lies below lithic breccia at approximately three radii (Pope *et al.*, 2005; Kenkmann and Schönian, 2006) would appear to be significant in terms of changes in the nature and timing of ejecta delivery with distance.

At more distal marine locations (e.g. in the North Atlantic), Chicxulub ejecta consists generally of an approximately 10 cm thick spherule layer, with shocked quartz and platinum group element (PGE) enrichments (Chapter 15) occurring towards the top (e.g. Norris *et al.*, 1999). At terrestrial sites at these distances the ejecta occurs as two distinct layers, with an approximately 2 cm thick lower spherule-rich lower layer, overlain by a thinner spherule-poor upper layer with shocked quartz, Ni-rich spinels and a PGE anomaly (e.g. Izett, 1990; Bohor and Glass, 1995). This dual-layer character indicates distinct delivery mechanisms for different components of the ejecta. At even more distal marine locations (e.g. Europe, Pacific Ocean), Chicxulub ejecta occur as a less than 5 mm single clay layer, with spherules, shocked quartz, Ni-rich spinels and elevated PGE concentrations (Smit, 1999).

The nature of the distal ejecta from Chicxulub has potential implications for the understanding of large-scale impacts on bodies with an appreciable atmosphere (see Chapter 4). The most distal ejecta, with its complement of sub-solidus shocked minerals (low-velocity ejecta), did not reach its destination solely by ballistic ejection. An additional component of non-ballistic atmospheric transportation is required. It has been suggested that redistribution ('skidding') of low-velocity ejecta in the atmosphere may be achieved by atmospheric heating and expansion due to the re-entry of high-velocity ejecta (Artemieva and Morgan, 2009). Similarly, the two-component boundary layer at intermediate distances also suggests two transportation mechanisms. In these cases, shocked and melted target materials could have arrived at the site first through ballistic ejection, to be followed by finer-grained materials, largely condensates, from vaporized projectile, with enhanced PGE content, and fine target components through atmospheric settling (Bohor, 1990; Smit, 1999). There may also be some more proximal effects due to the atmosphere. In the Sudbury ejecta, accretionary lapilli occur only in the more proximal sites. This suggests that there was a turbulent ejecta cloud, which was conducive to the formation of accretionary lapilli that extended out to approximately 650 km, but not to approximately 900 km from the Sudbury impact structure (Pufahl *et al.*, 2007; Cannon *et al.*, 2010).

Ejecta from the Acraman impact structure (apparent diameter 90 km) in Australia has been recorded in the stratigraphic column at distances of 240–540 km from the impact site (Williams and Gostin, 2005). The ejecta consist of mostly sand-sized clasts from the target dacitic volcanic rocks, with rare clasts up to 20 cm in size. Some of these clasts are shocked, containing planar deformation features in quartz and small shatter cones. Altered impact melt spherules and shards are locally abundant in the more proximal sites, which also have an Ir anomaly, indicative of an admixture of projectile material (Gostin *et al.*, 1989). The ejecta bed systematically thins with radial distance from the impact site and its thickness variation corresponds approximately to the McGetchin *et al.* (1973) scaling law for ejecta thickness (Williams and Wallace, 2003). However, it does pinch and swell and is locally missing. Whether the absence of the ejecta layer indicates that it is an example of the impactites of a discontinuous ejecta blanket is not clear, as in some areas the ejecta has been reworked and there is evidence of the actions of impact-induced tsunamis (Wallace *et al.*, 1996).

Tektites are a sub-class of impact melt glasses, with less than 0.02 wt% water, that are associated with more distal ejecta (Koeberl, 1988; Chapter 9). They tend to be spherically symmetric or display splash forms and are several centimetres in size (Fig. 7.10b). They are generally chemically homogeneous, with high silica (68–82% $SiO_2$), and can have inclusions of shocked minerals, high-pressure polymorphs (such as coesite; Chapter 8) and thermal decomposition products (such as baddeleyite from zircon). Tektites have the geochemical signature of the terrestrial upper crust and $^{10}Be$ concentrations consistent with an origin as melted material from the upper few metres of the target stratigraphy (Serefiddin *et al.*, 2007). Gas contents of tektites are very low, indicating solidification at high atmospheric heights (e.g. 40–50 km for a philippinite from the Australasian strewn field (Matsuda *et al.*, 1996)) and, in some cases, there is evidence of remelting on re-entry into the atmosphere (Koeberl, 1994). Tektites are generally considered to be the product of early-stage shock compression of surface materials and their acceleration after decompression (e.g. Stöffler *et al.*, 2002).

There are a number of tektite-strewn fields associated with known impact structures (e.g. Ivory Coast tektites with Bosumtwi, North American tektites with Chesapeake Bay, USA, moldavites with Ries; Stöffler *et al.*, 2002) and strewn fields with no known impact structure (e.g. the Australasian strewn field; Koeberl, 1994), as well as relatively isolated discoveries of tektites (e.g. urengonites; Deutsch *et al.*, 1997). Some of the strewn fields that have been sampled in marine sediments have other discrete sub-solidus shocked materials associated with the tektites (e.g. Glass and Wu, 1993). The marine environment also allows the preservation of micro-tektites, which have sizes on the scale of a millimetre. Some of the spherules associated with Chicxulub ejecta have been referred to as tektites or micro-tektites (e.g. Izett, 1991). As these spherules do not represent near-surface target materials, they do not conform to the definition and character of tektites and the term is being used incorrectly (Montanari and Koeberl, 2000).

## 7.5  Concluding remarks

From a geosciences perspective, the study of impacts and impactites is a recent endeavour. The discovery and study of natural shock metamorphic effects is about 50 years old (e.g. Chao *et al.*, 1960) and, therefore, so is the study of impactites formed as a consequence of impact cratering processes. Significant advances have been made in both characterizing impactites and understanding impact processes, and there is now an IUGS-recommended nomenclature for impactites (Stöffler and Grieve, 2007). The study of impactites, however, is not a static exercise. While the differences between highly shocked impactites in sedimentary or mixed targets and crystalline targets have been documented in the past (e.g. Kieffer and Simonds, 1980), the recent application of electron microscopy techniques has demonstrated commonalities and complications that were unrecognized previously (e.g. Osinski *et al.*, 2008). This has led to the reinterpretation of 'suevite', based on observations in the type area at the Ries, but much remains to be done regarding characterization and documenting the usage of this term in the more global context of terrestrial impact structures. Similarly, the need to study the ejecta of the Chicxulub impact to better constrain potential impact-related 'killing mechanisms' in the global environment 65 million years ago, coupled with advances in computer modelling of ejecta processes, has raised serious questions regarding the robustness of current, and somewhat simplistic, tenets dealing with impact ejecta and ejection processes in large impact events in the presence of an atmosphere. In combination, these recent developments suggest a future focus on better characterization of various melt-bearing breccias and understanding ejecta transportation processes.

## References

Addison, W.D., Brumpton, G.R., Vallini, D.A. *et al.* (2005) Discovery of distal ejecta from the 1850 Ma Sudbury impact event. *Geology*, 33, 193–196.

Artemieva, N. and Morgan, J. (2009) Modeling the formation of the K–Pg boundary layer. *Icarus*, 201, 768–780.

Artemieva, N., Wünnemann, K., Meyer, C. *et al.* (2009) Ries crater and suevite revisited: Part II modeling. 40th Lunar and Planetary Science Conference, CD-ROM, abstract no. 1526.

Bischoff, L. and Oskierski, W. (1987) Fractures, pseudotachylite veins and breccia dikes in the crater floor of the Rochechouart impact structure, SW-France, as an indicator of crater forming processes, in *Research in Terrestrial Impact Structures* (ed. J. Pohl), Friedrich Vieweg and Sohn, Brunswick, pp. 5–29.

Bohor, B.F. (1990) Shocked quartz and more: impact signatures in Cretaceous/Tertiary clays, in *Global Catastrophes in Earth History: An Interdisciplinary Conference on Impacts, Volcanism and Mass Mortality* (eds V.L. Sharpton and P.D. Ward), Geological Society of America Special Paper 247, Geological Society of America, Boulder, CO, pp. 335–342.

Bohor, B.F. and Glass, B.P. (1995) Origin and diagenesis of K/T impact spherules, from Haiti to Wyoming and beyond. *Meteoritics*, 30, 182–198.

Cannon, W.F., Schulz, K.J., Horton Jr, J.W. and Kring, D.A. (2010) The Sudbury impact layer in the Paleoproterozoic iron ranges of northern Michigan, USA. *Geological Society of America Bulletin*, 122, 50–75.

Chao, E.C., Shoemaker, E.M. and Mardsen, B.M. (1960) First natural occurrence of coesite (Arizona). *Science*, 132, 220–222.

Dence, M.R. (2004) Structural evidence from shock metamorphism in simple and complex impact craters: linking observations and theory. *Meteoritics and Planetary Science*, 39, 267–286.

Dence, M.R., Grieve, R.A.F., and Robertson, P.B. (1977) Terrestrial impact structures: principal characteristics and energy considerations, in *Impact and Explosion Cratering* (eds D.J. Roddy, R.O. Pepin and R.B. Merrill), Pergamon Press, New York, NY, pp. 247–275.

Deutsch, A., Ostermann, M. and Masaitis, V.L. (1997) Geochemistry and neodymium–strontium isotope signature of tektite-like objects from Siberia (urengoites, South Ural glass). *Meteoritics and Planetary Science*, 32, 679–686.

Dietz, R.S. (1964) Sudbury structure as an astrobleme. *Journal of Geology*, 72, 412–434.

Dressler, B.O. and Sharpton, V.L. (1997) Breccia formation at a complex impact crater; Slate Islands, Lake Superior, Ontario, Canada. *Tectonophysics*, 275, 285–311.

Ferrière, L., Koeberl, C. and Reimold, W.U. (2007) Drill Core LB-08A, Bosumtwi impact structure, Ghana: petrographic and shock metamorphic studies of material from the central uplift; Bosumtwi crater drilling project. *Meteoritics and Planetary Science*, 42, 611–633.

Ferrière, L., Koeberl, C., Ivanov, B.A. and Reimold, W.U. (2008) Shock metamorphism of Bosumtwi impact crater rocks, shock attenuation, and uplift formation. *Science*, 322, 1678–1681.

Fletcher, P. and Reimold, W.U. (1989) Some notes and speculations on the pseudotachylites in the Witwatersrand Basin and Vredefort Dome. *South African Journal of Geology*, 92, 223–234.

Floran, R.J., Grieve, R.A.F., Phinney, W.C. *et al.* (1978) Manicouagan impact melt, Quebec; 1, stratigraphy, petrology and chemistry. *Journal of Geophysical Research*, 83, 2737–2759.

French, B.M. (1967) Sudbury structure, Ontario; some petrographic evidence for origin by meteorite impact. *Science*, 156, 1094–1098.

French, B.M. (1998) *Traces of a Catastrophe: A Handbook of Shock Metamorphic Effects in Terrestrial Meteorite Impact Structures*, LPI Contribution No. 954, Lunar and Planetary Institute, Houston, TX.

French, B.M., Cordua, W.S. and Plescia, J.B. (2004) The Rock Elm meteorite impact structure, Wisconsin: geology and shock-metamorphic effects in quartz. *Geological Society of America Bulletin*, 116, 200–218.

Gibson, R.L. and Reimold, W.U. (2005) Shock pressure distribution in the Vredefort impact structure, South Africa, in *Large Meteorite Impacts III* (eds T. Kenkmann, F. Hörz and A. Deutsch), Geological Society of America Special Paper 384, Geological Society of America, Boulder, CO, pp. 339–349.

Ghent, R.R., Campbell, B.A., Hawke, B.R. and Campbell, D.B. (2008) Earth-based radar reveal extended deposits of the Moon's Orientale Basin. *Geology*, 36, 343–346.

Glass, B.P. and Wu, J. (1993) Coesite and shocked quartz discovered in the Australasian and North American microtektite layers. *Geology*, 21, 345–438.

Gostin, V.A., Keays, R.R. and Wallace, M.W. (1989) Iridium anomaly from the Acraman impact ejecta horizon: impacts can produce sedimentary iridium peaks. *Nature*, 349, 542–544.

Grieve, R.A.F. (1978) The melt rocks at Brent crater, Ontario, Canada. *Proceedings of the Lunar and Planetary Science Conference*, 9, 2579–2608.

Grieve, R.A.F. (1988) The Haughton impact structure: summary and synthesis of the results of the HISS project. *Meteoritics*, 23, 249–254.

Grieve, R.A.F. (2006) *Impact Structures in Canada*, Geological Association of Canada, St Johns.

Grieve, R.A.F. and Cintala, M.J. (1981) A method for estimating the initial conditions of terrestrial impact events, exemplified by its application to Brent crater, Ontario. *Proceedings of the Lunar and Planetary Science Conference*, 12, 1607–1621.

Grieve, R.A.F. and Cintala, M.J. (1992) An analysis of differential impact melt-crater scaling and implications for the terrestrial impact record. *Meteoritics*, 27, 526–539.

Grieve, R.A.F. and Head, J.W. (1983) The Manicouagan impact structure: an analysis of its original dimensions and form. *Journal of Geophysical Research*, 88 (Suppl. B2), A807–A818.

Grieve, R.A.F. and Therriault, A.M. (2004) Observations at terrestrial impact structures: their utility in constraining crater formation. *Meteoritics and Planetary Science*, 39, 199–216.

Grieve, R.A.F., Dence, M.R. and Robertson, P.B. (1977) Cratering processes: as interpreted from the occurrence of impact melts, in *Impact and Explosion Cratering* (eds D.J. Roddy, R.O. Pepin and R.B. Merrill), Pergamon Press, New York, NY, pp. 791–814.

Grieve, R.A.F., Stöffler, D. and Deutsch, A. (1991) The Sudbury structure: controversial or misunderstood? *Journal of Geophysical Research*, 96, 22753–22764.

Grieve, R.A.F., Reimold, W.U., Morgan, J. *et al.* (2008) Observations and interpretations at Vredefort, Sudbury and Chicxulub: towards an empirical model of terrestrial impact basin formation. *Meteoritics and Planetary Science*, 43, 1–28.

Gurov, E., Koeberl, C. and Yamnichenko, A. (2007) El'gygytgyn impact crater, Russia: structure, tectonics and morphology. *Meteoritics and Planetary Science*, 42, 307–319.

Halls, H.C. and Grieve, R.A.F. (1976) The Slate Islands: a probable meteorite impact structure in Lake Superior. *Canadian Journal of Earth Sciences*, 9, 1301–1309.

Heiken, G.H., Vaniman, D.T. and French, B.M. (eds) (1991) *Lunar Source Book*, Cambridge University Press, Cambridge.

Horton Jr, J.W., Gohn, G.S., Powars, D.S. and Edwards, L.E. (2008) Origin and emplacement of impactites in the Chesapeake Bay impact structure, Virginia, USA, in *The Sedimentary Record of Meteorite Impacts* (eds K.R. Evans, J.W. Horton Jr, D.T. King Jr and J.R. Morrow), Geological Society of America Special Paper 437, Geological Society of America, Boulder, CO, pp. 73–97.

Horton Jr, J.W., Kunk, M.J., Belkin, H.E. *et al.* (2009) Evolution of crystalline target rocks and impactites in the Chesapeake Bay impact structure, ICDP–USGS Eyreville B core, in *The ICDP–USGS Deep Drilling Project in the Chesapeake Bay Impact Structure: Results from the Eyreville Core Holes* (eds G.S. Gohn, C. Koeberl, K.G. Miller and W.U. Reimold), Geological Society of America Special Paper 458, Geological Society of America, Boulder, CO, pp. 277–316.

Hörz, F., Ostertag, R. and Rainey, D.A. (1983) Bunte breccia of the Ries; continuous deposits of large impact craters. *Reviews of Geophysics and Space Physics*, 21, 1667–1725.

Hörz, F., Mittlefehldt, D.W. and See, T.H. (2002) Petrographic studies of the impact melts from Meteor Crater, Arizona, USA. *Meteoritics and Planetary Science*, 37, 501–531.

Innes, M.J.S. (1964) Recent advances in meteorite crater research at the Dominion Observatory, Ottawa, Canada. *Meteoritics*, 2, 219–241.

Ivanov, B.A., Kocharyan, G.G., Kostuchenko, V.N. *et al.* (1996) Puchezh-Katunki impact crater: preliminary data on recovered core block structure. 27th Lunar and Planetary Science Conference, pp. 589–590 (abstract).

Izett, G. (1990) *The Cretaceous/Tertiary Boundary Interval, Raton Basin, Colorado and New Mexico, and its Content of Shock-Metamorphosed Minerals: Evidence Relevant to the K/T Impact Extinction Theory*, Geological Society of America Special Paper 249, Geological Society of America, Boulder, CO.

Izett, G. (1991) Tektites in Cretaceous–Tertiary boundary rocks on Haiti and their bearing on the Alvarez extinction hypothesis. *Journal of Geophysical Research*, 96, 20879–20905.

Jolliff, B.L., Wieczorek, M.A. Shearer, C.K. and Neal, C.R. (eds) (2006) *New Views of the Moon*, Reviews of Mineralogy and Geochemistry 60, Mineralogical Society of America, Chantilly, VA.

Kenkmann, T. and Schönian, F. (2006) Ries and Chicxulub: impact craters on Earth provide insights for Martian ejecta blankets. *Meteoritics and Planetary Science*, 41, 1587–1603.

Kieffer, S.W. and Simonds, C.H. (1980) The role of volatiles and lithology in the impact cratering process. *Reviews of Geophysics and Space Physics*, 18, 143–181.

Koeberl, C. (1988) Water content of tektites and impact glasses and related chemical studies. *Proceedings of the Lunar and Planetary Science Conference*, 18, 403–408.

Koeberl, C. (1994) Tektite origin by hypervelocity asteroidal or cometary impact: target rocks, source craters and mechanisms, in *Large Meteorite Impacts and Planetary Evolution* (eds B.O. Dressler, R.A.F. Grieve and V.L. Sharpton), Geological Society of America Special Paper 293, Geological Society of America, Boulder, CO, pp. 133–151.

Koeberl, C. and Martínez-Ruiz, F. (eds) (2003) *Impact Markers in the Stratigraphic Record*, Springer-Verlag, Berlin.

Koeberl, C., Brandstaetter, F., Glass, B.P. *et al.* (2007) Uppermost impact fallback layer in the Bosumtwi crater (Ghana): mineralogy, geochemistry, and comparison with Ivory Coast tektites; Bosumtwi crater drilling project. *Meteoritics and Planetary Science*, 42, 709–729.

Kring, D.A. (2007) *Guidebook to the Geology of Barringer Meteorite Crater, Arizona (a.k.a. Meteor Crater)*, LPI Contribution No. 1355, Lunar and Planetary Institute, Houston, TX.

Lafrance, B., Legault, D. and Ames, D.E. (2008) The formation of Sudbury breccia in the North Range of the Sudbury impact structure. *Precambrian Research*, 165, 107–119.

Lambert, P. (1981) Breccia dikes; geological constraints on the formation of complex craters, in *Proceedings of Conference on Multi-Ring Basins: Formation and Evolution* (eds R.B. Merrill and P.H. Schultz), Pergamon Press, New York, NY, pp. 59–78.

Lieger, D., Riller, U. and Gibson, R. (2009) Generation of fragment-rich pseudotachylite bodies during central uplift formation at the Vredefort impact structure, South Africa. *Earth and Planetary Science Letters*, 279, 53–64.

Masaitis, V.L. (1994) Impactites from Popigai crater, in *Large Meteorite Impacts and Planetary Evolution* (eds B.O. Dressler, R.A.F. Grieve and V.L. Sharpton), Geological Society of America Special Paper 293, Geological Society of America, Boulder, CO, pp. 152–162.

Masaitis, V.L. (1999) Impact structures of northeastern Eurasia: the territories of Russia and adjacent countries. *Meteoritics and Planetary Science*, 34, 691–711.

Masaitis, V.L., Danilin, A.N., Maschak, M.S. *et al.* (1980) *The Geology of Astroblemes*, Nedra, Leningrad (in Russian).

Matsuda, J., Maruoka, T., Pinti, D.L. and Koeberl, C. (1996) Noble gas study of a phillipinite with an unusually large bubble. *Meteoritics*, 31, 273–277.

McCormick, K.A., Taylor, G.J., Keil, K. *et al.* (1989) Sources of clasts in terrestrial impact melts: clues to the origin of LKFM. *Proceedings of the Lunar and Planetary Science Conference*, 19, 691–696.

McGetchin, T.R., Settle, M. and Head, J.W. (1973) Radial thickness variation in impact crater ejecta: implications for lunar basin deposits. *Earth and Planetary Science Letters*, 20, 226–236.

Melosh, H.J. (1989) *Impact Cratering; a Geologic Process*, Oxford University Press, Oxford.

Melosh, H.J. (2005) The mechanics of pseudotachylite formation in impact events, in *Impact Tectonics* (eds C. Koeberl and H. Henkel), Springer-Verlag, Berlin, pp. 55–80.

Milton, D.J., Glikson, A.Y. and Brett, R. (1996) Gosses Bluff: a latest Jurassic impact structure central Australia. Part 1, geological structure, stratigraphy, and origin. *AGSO Journal of Australian Geology and Geophysics*, 16, 453–486.

Montanari, A. and Koeberl, C. (eds) (2000) *Impact Stratigraphy*, Lecture Notes in Earth Sciences 93, Springer-Verlag. Berlin.

Norris, R.D., Huber, B.T. and Self-Trail, J.M. (1999) Synchroneity of the K–T oceanic mass extinction and meteorite impact: Blake Nose, western North Atlantic. *Geology*, 27, 419–422.

Oberbeck, V.R. (1975) The role of ballistic sedimentation in lunar stratigraphy. *Reviews of Geophysics and Space Physics*, 13, 337–362.

Ormö, J., Sturkell, E. and Lindström, M. (2007) Sedimentological analysis of resurge deposits at the Lockne and Tvären craters: clues to flow dynamics. *Meteoritics and Planetary Science*, 42, 1929–1943.

Osinski, G.R. (2003) Impact glasses in fallout suevites from the Ries impact structure, Germany: an analytical SEM study. *Meteoritics and Planetary Science*, 38, 1641–1667.

Osinski, G.R., Grieve, R.A.F. and Spray J G. (2004) The nature of the groundmass of surficial suevite from the Ries impact structure, Germany, and constraints on its origin. *Meteoritics and Planetary Science*, 39, 1655–1683.

Osinski, G.R., Spray, J. and Lee, P. (2005) Impactites of the Haughton impact structure, Devon Island, Canadian High Arctic. *Meteoritics and Planetary Science*, 40, 1789–1812.

Osinski, G.R., Bunch, T.E. and Wittke, J. (2006) Proximal ejecta at Meteor Crater, Arizona: discovery of impact melt-bearing breccias. 37th Lunar and Planetary Science Conference, CD-ROM, abstract no. 1005.

Osinski, G.R., Grieve, R.A.F., Collins, G.S. *et al.* (2008) The effect of target lithology on impact melting. *Meteoritics and Planetary Science*, 43, 1939–1954.

Phinney, W.C., Simonds, C.H., Cochran, A. and McGee, P.E. (1978) West Clearwater, Quebec impact structure; Part II, petrology. *Proceedings of the Lunar and Planetary Science Conference*, 9, 2659–2693.

Pohl, J., Stöffler, D., Gall, H. and Ernston, K. (1977) The Ries impact crater, in *Impact and Explosion Cratering* (eds D.J. Roddy, R.O. Pepin and R.B. Merrill), Pergamon Press, New York, NY, pp. 343–404.

Pope, K.O., Ocampo, A.C., Fischer, A.G. *et al.* (2005) Chicxulub impact ejecta deposits in southern Quintana Roo, Mexico and southern Belize, in *Large Meteorite Impacts III* (eds T. Kenkmann, F. Hörz and A. Deutsch), Geological Society of America Special Paper 384, Geological Society of America, Boulder, CO, pp. 171–190.

Pufahl, P.K., Hiatt, E.E., Stanley, C.R. *et al.* (2007) Physical and chemical evidence of the 1850 Ma Sudbury impact event in the Baraga Group, Michigan. *Geology*, 35, 827–830.

Reimold, W.U. (1995) Pseudotachylite in impact structures: generation by friction melting and shock brecciation? A review and discussion. *Earth Science Reviews*, 39, 247–265.

Reimold, W.U., Horton Jr, J.W. and Schmitt, R.-T. (2008) Debate about nomenclature – recent problems, in *Large Meteorite Impacts and Planetary Evolution IV*, LPI Contribution No. 1423, Lunar and Planetary Institute, Houston, TX, CD-ROM, abstract no. 3033.

Riller, U., Lieger, D., Gibson, R.L. *et al.* (2010) Origin of large-volume pseudotachylite in terrestrial impact structures. *Geology*, 38, 619–622.

Robertson, P.B. (1975) Zones of shock metamorphism at the Charlevoix impact structure, Quebec. *Geological Society of America Bulletin*, 86, 1630–1638.

Ryder, G. and Bower, J.F. (1977) Petrology of *Apollo 15* black- and white-rocks 15445 and 15455: fragments of the Imbrium melt sheet? *Proceedings of the Lunar and Planetary Science Conference*, 8, 1895–1923.

Schulte, P. and Kontny, A. (2005) Chicxulub impact ejecta from the Cretaceous–Paleogene (K–P) boundary in northeastern Mexico, in *Large Meteorite Impacts III* (eds T. Kenkmann, F. Hörz and A. Deutsch), Geological Society of America Special Paper 384, Geological Society of America, Boulder, CO, pp. 191–221.

Serefiddin, F., Herzog, G.F. and Koeberl, C. (2007) Beryllium-10 concentrations of tektites from Ivory Coast and from central Europe: evidence for near-surface residence of precursor materials. *Geochimica et Cosmochimica Acta*, 71, 1574–1582.

Shand, S.J. (1916) The pseudotachylyte of Parijs (Orange Free State), and its relation to 'trap-shotten gneiss' and 'flinty-crush rock'. *Geological Society of London Quarterly Journal*, 72, 198–221.

Sharpton, V.L., Marin, L.E., Carney, J.L. *et al.* (1996) A model of the Chicxulub impact basin based on evaluation of geophysical data, well logs, and drill core samples, in *The Cretaceous–Tertiary Event and Other Catastrophes in Earth History* (eds G. Ryder, D. Fastovsky and D. Gartner), Geological Society of America Special Paper 307, Geological Society of America, Boulder, CO, pp. 55–74.

Sharpton, V.L., Corrigan, C.M., Marin, L.E. *et al.* (1999) Characterization of impact breccias from the Chicxulub impact basin; implications for excavation and ejecta emplacement. 30th Lunar and Planetary Science Conference, CD-ROM, abstract no. 2179.

Sibson, R.H. (1975) Generation of pseudotachylite by ancient seismic faulting. *Geophysical Journal of the Royal Astronomical Society*, 43, 775–794.

Smit, J. (1999) The global stratigraphy of the Cretaceous–Tertiary boundary impact ejecta. *Annual Reviews of Earth and Planetary Sciences*, 27, 75–113.

Spooner, I., Stevens, G., Morrow, J. *et al.* (2009) Identification of the Bloody Creek structure, a possible bolide impact crater in southwestern Nova Scotia. *Meteoritics and Planetary Science*, 44, 1193–1202.

Spray, J. (1998) Localized shock- and friction-induced melting in response to hypervelocity impact, in *Meteorites: Flux with Time and Impact Effects* (eds M.M. Grady, R. Hutchison, G.J.H. McCall and D.A. Rothery), Geological Society Special Publication 140, Geological Society of London, London, pp. 195–204.

Spudis, P.D. (1993) *The Geology of Multi-Ring Basins: The Moon and Other Planets*, Cambridge University Press, Cambridge.

Spudis, P.D., Ryder, G., Taylor, G.J. *et al.* (1991) Sources of mineral fragments in impact melts 15445 and 15455: towards the origin of low-K Fra Mauro basalt. *Proceedings of the Lunar and Planetary Science Conference*, 21, 151–165.

Stöffler, D. and Grieve, R.A.F. (2007) Impactites, in *Metamorphic Rocks: A Classification and Glossary of Terms, Recommendations of the International Union of Geological Sciences* (eds D. Fettes and J. Desmons), Cambridge University Press, Cambridge, pp. 82–92.

Stöffler, D., Knöll, H.-D., Marvin, U.B. *et al.* (1980) Recommended classification and nomenclature of lunar highland rocks – a committee report, in *Conference on the Lunar Highlands Crust* (eds J.J. Papike and R.B. Merrill), Pergamon Press, New York, NY, pp. 51–70.

Stöffler, D., Artemieva, N.A. and Pierazzo, E. (2002) Modeling the Ries–Steinheim impact event and the formation of the moldavite strewn field. *Meteoritics and Planetary Science*, 37, 1893–1907.

Svensson, N.B. (1968a) The Dellen lakes, a probable meteorite impact in central Sweden. *Geologiska Föreningens i Stockholm Förhandlinger*, 90, 314–315.

Svensson, N.B. (1968b) Lake Lappajärvi, central Finland: a possible meteorite impact structure. *Nature*, 217, 438.

Swann, G.A., Bailey, N.G., Batson, R.M. *et al.* (1977) *Geology of the Apollo 14 Landing Site in the Fra Mauro Highlands*, United States Geological Survey Professional Paper 808, US Government Printing Office, Washington, DC.

Thackrey, S., Walkden, G., Indares, A. *et al.* (2009) The use of heavy mineral correlation for determining the source of impact ejecta: a Manicouagan distal ejecta study. *Earth and Planetary Science Letters*, 285, 163–172.

Therriault, A.M., Fowler, A.D. and Grieve, R.A.F. (2002) The Sudbury igneous complex: a differentiated impact melt sheet. *Economic Geology and the Bulletin of the Society of Economic Geologists*, 97, 1521–1540.

Von Engelhardt, W. (1990) Distribution, petrography and shock metamorphism of the Ries crater in Germany: a review. *Tectonophysics*, 171, 259–273.

Wallace, M.W., Gostin, V.A. and Keays, R.R. (1996) Sedimentology of the Neoproterozoic Acraman impact-ejecta horizon, South Australia. *AGSO Journal of Australian Geology and Geophysics*, 16, 443–457.

Williams, G.E. and Gostin, V.A. (2005) Acraman–Bunyeroo impact event (Ediaaran), South Australia, and environmental consequences; twenty-five years on. *Australian Journal of Earth Sciences*, 52, 607–620.

Williams, G.E. and Wallace, M.W. (2003) The Acraman asteroid impact, South Australia: Magnitude and implications for the late Vendian environment. *Journal of the Geological Society of London*, 160, 545–554.

# EIGHT

# Shock metamorphism

## Ludovic Ferrière[*,†] and Gordon R. Osinski[†]

*Natural History Museum, Burgring 7, A-1010 Vienna, Austria
†Departments of Earth Sciences/Physics and Astronomy, Western University, 1151 Richmond Street, London, ON, N6A 5B7, Canada

## 8.1 Introduction

A requirement for the recognition and confirmation of meteorite impact structures is the presence of shock metamorphic indicators, either megascopic (e.g. shatter cones) or microscopic (e.g. planar deformation features in minerals), or high-pressure polymorphs (e.g. coesite and stishovite) and/or siderophile element (e.g. iridium) or isotopic (osmium) anomalies in specific geological settings. Crater morphology is not a sufficient argument, because a variety of circular features can be formed by completely different geological processes (e.g. volcanism or salt diapirism). The terms 'shock effects' or 'shock-metamorphic effects' cover all types of shock-induced changes, such as the formation of planar microstructures and phase transformations. Impact metamorphism is essentially the same as shock metamorphism, except that it also encompasses the melting, decomposition and vaporization of target rocks (Stöffler and Grieve, 2007). These irreversible changes are produced when rocks are subjected to shock pressures above their Hugoniot elastic limit (HEL). This limit is defined as 'the critical shock pressure at which a solid yields under the uniaxial strain of a plane shock wave' (Stöffler, 1972). The HEL of quartz is in the range of 5–8 GPa and ranges from approximately 1 to 10 GPa for most geological materials (e.g. Stöffler, 1972; Stöffler and Langenhorst, 1994). In nature, at the surface of the Earth, only hypervelocity impacts can generate such high shock pressures.

A great diversity of shock effects in minerals are known and have been abundantly described in the literature during the last 40 years mostly for quartz (e.g. French and Short, 1968; von Engelhardt and Bertsch, 1969; Stöffler, 1972; Stöffler and Langenhorst, 1994; Grieve et al., 1996; French, 1998; Montanari and Koeberl, 2000; Langenhorst, 2002; and references therein) and to some extent for feldspar (e.g. Stöffler, 1967; Robertson, 1975; Ostertag, 1983; Dressler, 1990; Bischoff and Stöffler, 1992; and

references therein), olivine (e.g. Reimold and Stöffler, 1978; Bauer, 1979; Stöffler et al., 1991; Bischoff and Stöffler, 1992; Schmitt, 2000; and references therein), and pyroxene (e.g. Rubin et al., 1997), mostly within meteorites. Less is known about the effects of shock in other minerals, such as the nesosilicates, double-chain inosilicates, phyllosilicates and the carbonate and sulfate minerals in general. Much of our knowledge on shock metamorphism comes from studies of minerals from impact craters formed in dense, non-porous crystalline rocks and from more or less porous meteorites. However, the effects of target lithology on the response of minerals to shock compression remain to be investigated in detail. To date, in the case of sedimentary rocks, only the response of quartz has been investigated in any detail (Table 8.1) – see Kieffer (1971), Kieffer et al. (1976) and Osinski (2007). These observational studies, together with recent numerical simulations (e.g. Wünnemann et al., 2008), have revealed the complicating effects of porosity and volatiles on the response of quartz to impact in sedimentary targets. This will be discussed below, but the main observation is that large differences in shock impedance between solid grains and pore space results in a very heterogeneous distribution of shock wave energy at the microscopic scale. This has the effect of collapsing pore spaces and compressing grain boundaries and producing shock metamorphic effects of very different shock levels within an individual sample. In addition, it is apparent that energy is preferentially transferred into heat and melting of the target lithologies, as opposed to forming shock metamorphic effects in minerals, particularly in poorly or unconsolidated sediments and sedimentary rocks.

This chapter provides an overview of shock metamorphic effects in dense non-porous crystalline rocks and minerals and a comparison with the effects in sedimentary rocks. In addition, two examples of post-shock thermal effects/features, namely toasted quartz and ballen silica, for which occurrence is restricted to impact-derived rocks, are also discussed.

*Impact Cratering: Processes and Products*, First Edition. Edited by Gordon R. Osinski and Elisabetta Pierazzo.
© 2013 Blackwell Publishing Ltd. Published 2013 by Blackwell Publishing Ltd.

**Table 8.1**  Classification of impact metamorphic effects in sandstones. Compiled with data from Barringer Crater (Kieffer, 1971; Kieffer et al., 1976) and the Haughton impact structure (Osinski, 2007)

| Class | Pressure range (GPa)[a] | Temperature range (°C)[a] | Hand specimen observations | Proportion of various SiO$_2$ phases and effects[b] | | | | | Microscopic observations |
|---|---|---|---|---|---|---|---|---|---|
| | | | | Qtz | Dg | Lech | Coesite | Toast | |
| 1a | <3 | <350 | Crude shatter cones | **** | — | — | — | — | Recognizable porosity; no indications of shock |
| 1b | 3–5.5 | <350 | Well-developed shatter cones | **** | — | — | — | — | No recognizable porosity using optical microscopy; fracturing of quartz grains; minor development of PFs |
| 2 | 5.5–10 | 350–950 | Well-developed shatter cones | *** | * | — | * | * | Presence of 'jigsaw' texture; fracturing of individual grains and generation of micro-breccias; symplectic regions along some grain boundaries; PFs; PDFs |
| 3a | 10–20 | >1000 | Difficult to discern individual grains | *** | ** | — | ** | ** | Reduced grain size; multiple sets of PDFs; symplectic regions surrounding majority of quartz grains |
| 3b | 15–20 | >1000 | Difficult to discern individual grains | ** | *** | * | * | ** | Widespread development of diaplectic glass; vesicular SiO$_2$ glass in original pore spaces |
| 4 | 20–30 | >1000 | Few recognizable grains; faint layering white | ** | *** | ** | * | ** | Original texture of sandstone lost; major development of vesicular SiO$_2$ glass; symplectic regions surround all remaining quartz grains |
| 5a | >30 | >1000 | Highly vesicular; white | * | ** | *** | * | * | Isolated remnant quartz grains; almost complete transformation to diaplectic or vesicular SiO$_2$ glass (lechatelierite) |
| 5b | >30 | >1000 | Highly vesicular; white | — | * | *** | * | — | Complete transformation to vesicular SiO$_2$ glass |
| 6 | >30 | >1000 | Dense; grey; translucent | **** | — | — | — | — | Recrystallized SiO$_2$ glass |

[a]Pressures and post-shock temperatures from Kieffer et al. (1976).

[b]Qtz: quartz; Dg: diaplectic quartz glass; Lech: lechatelierite or vesicular SiO$_2$ glass; Toast: toasted quartz; ****: only phase present (>99%); ***: abundant; **: present; *: rare (<10%); —: absent.

## 8.2   Shock metamorphic features

Minerals subjected to shock metamorphism occur in different petrographic assemblages and in different rock types. The full spectrum of diagnostic features may not necessarily be present in all impact structures and is strongly dependent on the lithology and other properties of the target rock(s), and is a function of the magnitude of the hypervelocity impact and of the level of erosion of the crater. The different rocks affected (or produced) by one or more hypervelocity impact(s) are called 'impactites' (see Chapter 7 and reviews by French (1998) and Stöffler and Grieve (2007)). The classification and definition of the various impactites is complex and ongoing; nevertheless, the name of the different rock types described in this chapter follows the classification by Stöffler and Grieve (2007) and is discussed further in Chapter 7.

### 8.2.1   Shatter cones

Shatter cones are the only distinctive shock-deformation feature (i.e. diagnostic evidence of hypervelocity impact) that can be seen with the naked eye (e.g. Dietz, 1960, 1968; French, 1998; French and Koeberl, 2010). These meso- to macro-scale features, consisting of conical striated fracture surfaces, are best developed in fine-grained lithologies (such as limestone; see Fig. 8.1), but can also be observed in coarser grained lithologies, such as granites and gneisses, although they are typically more poorly developed. By definition, shatter cones are 'distinctive curved, striated fractures that typically form partial to complete cones' (French, 1998). The striated surface of shatter cones is either a positive or a negative feature (Fig. 8.1e), with the striations radiating along the surface of the cone. The occurrence of shatter cones has been reported to date for more than half of the currently confirmed impact structures; they usually occur in the central uplifts of complex impact structures, and in some cases isolated fragments/clasts of shatter cones have been found in impact breccias, within or outside the crater (e.g. at the Haughton impact structure, Canada; Osinski and Spray, 2006; see Fig. 8.1b,c,e). The distribution of *in situ* shatter cones at an impact site has also been used as a parameter for estimating the original size of a structure, particularly for old and eroded impact sites, as they occur generally below the crater floor or in the central uplifts of large structures. It is generally accepted that, when restored to their original position prior to the impact, shatter cones' apexes indicate the point of impact. However, shatter cones are generally found as composite groups (rarely as single specimens) of commonly partial to complete cones, with very frequently opposite orientations at the centimetre to decimetre scale. Thus, the use of shatter cone apex orientation to determine the centre of a crater and then its size is likely to yield incorrect results.

The formation of shatter cones, widely accepted as unequivocal proof of a meteorite impact crater, is still not completely resolved (e.g. Dietz, 1960; Johnson and Talbot, 1964; Gash, 1971; Milton, 1977; Baratoux and Melosh, 2003; Sagy *et al.*, 2004; Wieland *et al.*, 2006), but current formation hypotheses suggest that shatter cones originate very early during the impact process. It is generally accepted that shatter cones form at relatively low shock pressures, typically between approximately 2 and 10 GPa (e.g. French, 1998). At the microscopic scale, planar fractures (PFs) and/or planar deformation features (PDFs – see below) occur in minerals in rocks containing shatter cones (e.g. Wieland *et al.*, 2006; Fackelman *et al.*, 2008; Ferrière and Osinski, 2010).

In some cases, shatter cones are not well formed, and it may be hard for non-experts to distinguish them from non-impact features such as cone-in-cone structures, ventifacts or even slickenslides – see discussion in French and Koeberl (2010). Concerning the cone-in-cone structures (Fig. 8.2a), they form uniquely in sedimentary rocks, and, typically, the cone axes are normal to the bedding (see Lugli *et al.* (2005)), whereas shatter cones form in all rock types and at various angles to pre-existing rock. Ventifacts are features commonly found in hot and cold deserts and are the result of preferential abrasion under the effect of the dominant winds (e.g. Higgins, 1956). Wind-abrasion features can resemble shatter cones (Fig. 8.2b); however, they develop only on outcrop surfaces (i.e. they are not penetrative) and, thus, can easily be discriminated from shatter cones. Finally, because the striations in slickensides tend to be parallel, they can also be easily discriminated from shatter cones, in which striations are divergent.

### 8.2.2   Deformation in quartz

#### 8.2.2.1   *Planar microstructures*

Upon shock compression, quartz develops irregular fractures (which are not diagnostic shock effects) and planar microstructures. Planar microstructures in quartz are divided into PFs and PDFs (e.g. French and Short, 1968; von Engelhardt and Bertsch, 1969; Stöffler and Langenhorst, 1994; Grieve *et al.*, 1996; French, 1998). Both types of planar microstructures are crystallographically controlled; therefore, PFs and PDFs are oriented parallel to rational crystallographic planes. The four-digit notation ($hkil$), the so-called Miller–Bravais indices, is used for the indexing of the planes in the crystal. The indices ($hkil$) represent the inverse plane intercepts along the $a_1$, $a_2$, $a_3$ and $c$ axes respectively (e.g. Bloss, 1971).

#### 8.2.2.2   *Planar fractures (PFs)*

PFs are by definition planar, parallel, thin open fissures, generally greater than 3 μm wide and spaced more than 15–20 μm apart (see Fig. 8.3). The spacing between PFs can vary on the order of a few micrometres, but is wider than that of PDFs (Stöffler and Langenhorst, 1994; Grieve *et al.*, 1996; French, 1998; Montanari and Koeberl, 2000; Langenhorst, 2002; Morrow, 2007). They have been documented from both crystalline and sedimentary rocks from a variety of impact structures. The PFs are oriented parallel to rational crystallographic planes, such as (0001) and {10$\bar{1}$1}, and occasionally to {10$\bar{1}$3}.

It is notable that PFs commonly control and/or limit the distribution of adjacent PDF sets, which has been used to suggest that PF formation pre-dates PDF formation in a given quartz crystal grain (e.g. von Engelhardt and Bertsch, 1969; Stöffler and Langenhorst, 1994). Formed at pressures of approximately 5–8 GPa, PFs are not regarded as unambiguous evidence of shock

**Figure 8.1** Photographs of typical shatter cones from different impact structures. (a) Exposure view of shatter cones in quartzite from the Sudbury structure (Canada). Note that shatter cone apices are oriented in one direction on this small outcrop-sized area. (b) A complete cone developed in limestone with apex pointing up and showing the typical divergence of striae away from the cone apex. (c) Two nicely developed partial shatter cones. (d) Hand specimen of shatter cone from the Eagle Butte structure (Canada). (e) Positive (convex) and negative (concave; 'cast') of shatter cone surfaces; note that apices point in opposite directions. (f) Typical horsetailing shatter cone surfaces developed in fine-grained limestone from the Steinheim structure (Germany). Samples illustrated in (b, c, e) are shatter cone clasts from crater-fill impact melt rocks from the Haughton structure (Canada).

**Figure 8.2** Photographs of non-impact-related conical features frequently misidentified as shatter cone. (a) Typical silicified cone-in-cone structures (sedimentary in origin) from the Tafilalt region of Morocco (see Lugli *et al.*, 2005). Photograph courtesy of S. Lugli. (b) Ventifacts (wind abrasion features) developed in quartzitic breccia (sample from the GKCF02 structure, Egypt). Note the presence of a clast at the apex of the cone.

metamorphism, as they also occur, rarely, in quartz grains from non-impact settings (e.g. French, 1998).

### 8.2.2.3  *Feather features*

In the case of some impact structures (in at least 26 according to Poelchau and Kenkmann (2011)) developed within both sedimentary and crystalline target rocks, the occurrence of thinly spaced, short, parallel to subparallel lamellae (Poelchau and Kenkmann, 2011) or incipient PDFs (French *et al.*, 2004) that branch off of PFs (Fig. 8.3c,d) have been observed. This somewhat unusual type of planar microstructure, called 'feather features', appears to be shock related, but its formation mechanism is poorly understood. In a recent publication, Poelchau and Kenkmann (2011) show that these features are crystallographically controlled to a certain degree and they suggest that these microstructures are caused by shearing of planar fractures during shock deformation; however, further investigations are necessary to validate this hypothesis.

### 8.2.2.4  *Planar deformation features (PDFs)*

PDFs in quartz grains, which develop over the pressure range of 5–10 GPa to approximately 35 GPa (see Stöffler and Langenhorst (1994) and French (1998) and references therein), are one of the best criteria for the identification of new impact structures. In contrast to PFs, PDFs are not open fractures. PDFs are typically composed of narrow, individual planes of amorphous material that are less than 2 μm thick, comprising straight, parallel sets spaced 2–10 μm apart (e.g. von Engelhardt and Bertsch, 1969; Stöffler and Langenhorst, 1994). Generally occurring as multiple sets per grain, and typically in more than one crystallographic

orientation, PDFs can be either decorated or non-decorated (Fig. 8.4). The PDFs are generally decorated with tiny fluid inclusions or bubbles, usually less than 2 μm in diameter, which greatly facilitate their detection at the scale of the optical microscope. Decorated PDFs are considered secondary features, in which the decorations form by post-shock annealing and aqueous alteration of non-decorated amorphous PDFs (e.g. Stöffler and Langenhorst, 1994; Grieve *et al.*, 1996; Leroux, 2005).

PDFs are preferentially oriented parallel to rational crystallographic planes, such as $\{10\bar{1}3\}$, $\{10\bar{1}2\}$, $(0001)$, $\{10\bar{1}1\}$, $\{11\bar{2}2\}$, $\{11\bar{2}1\}$, $\{21\bar{3}1\}$, $\{51\bar{6}1\}$, $\{10\bar{1}0\}$ and $\{11\bar{2}0\}$, and more rarely to other planes; the measurement of PDF orientations is possible using transmission electron microscopy (TEM; e.g. Goltrant *et al.*, 1991), as well as with the spindle stage (e.g. Bohor *et al.*, 1987), or using the universal stage technique (e.g. Ferrière *et al.*, 2009a). As specific orientations of PDFs in quartz are formed at different shock pressures (e.g. Hörz, 1968; Müller and Défourneaux, 1968; Huffman and Reimold, 1996), several workers, including Robertson and Grieve (1977), Grieve *et al.* (1990) and Dressler *et al.* (1998), have derived average shock pressure values for a given sample, based on laboratory shock experiments that bracket the pressure ranges associated with the development of individual PDF orientations or assemblages of orientations in quartz grains. For summaries of this method and references, see reviews by Stöffler and Langenhorst (1994) and Grieve *et al.* (1996); however, De Carli *et al.* (2002) and Ferrière *et al.* (2009a) provide a cautionary discussion regarding the potential limitations of this technique.

Because PDFs cannot be clearly resolved under the optical microscope, TEM techniques are required for the characterization of their microstructure (e.g. Kieffer *et al.*, 1976; Goltrant *et al.*, 1991; Trepmann and Spray, 2006). Based on TEM investigations, Kieffer *et al.* (1976) and later Goltrant *et al.* (1991) showed that

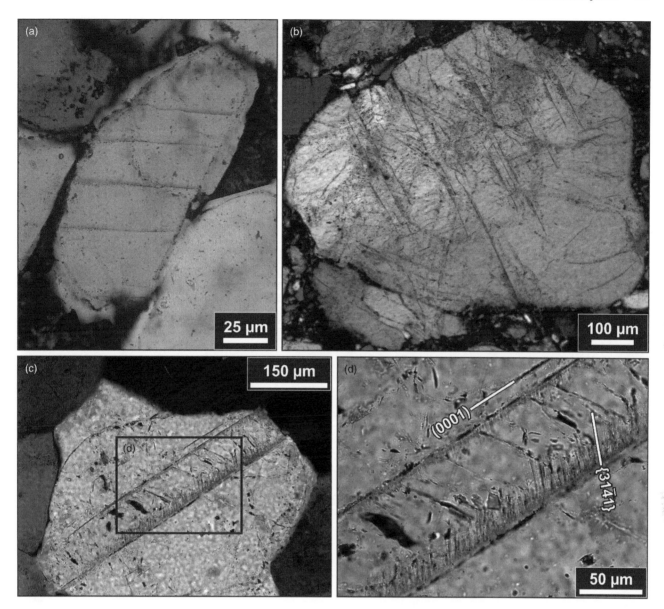

**Figure 8.3** Microphotographs (crossed polars) of quartz grains with PFs. (a) Authigenic quartz grain with one set of typically spaced PFs (sandstone sample from the Libyan Desert Glass strewn field, Egypt). (b) Quartz grain with one prominent set of planar fractures oriented NW–SE; other irregular fractures are also visible in the grain (sample OR-10; Aorounga structure, Chad). (c) Quartz grain showing one set of $c(0001)$ PFs (oriented NE–SW) with feather features that branch off the PF (sandstone sample HMP-04-055; Haughton, Canada). (d) Enlarged part of (c) showing 'feather features' with {31$\bar{4}$1}-equivalent orientation. Directions (e.g. NW–SE) are in relation to an arbitrary 'north' at the top of the image.

PDFs developing parallel to the basal plane (0001) represent mechanical Brazil twins. These Brazil twins are generated at high shear stresses on the basal plane and result from glide motion of the partial dislocations (McLaren *et al.*, 1967).

Planar deformation features form in both crystalline and sedimentary target rocks, although in the latter they have not received much attention and systematic studies are lacking. As such, there are several unanswered questions concerning PDF development in sedimentary rocks. For example, it is clear that PDFs are missing or rare in some craters (e.g. Meteor Crater) but are common in others (e.g. the B.P. and Oasis structures; Grieve *et al.*, 1996). The development of PDFs in sedimentary rocks seems to depend on the grain size of quartz grains, as PDFs are preferentially observed in grains with larger diameters (Grieve *et al.*, 1996). However, a similar grain size effect is reported for leucosome samples from the Ries crater (Walzebuck and von Engelhardt, 1979) and for gneisses from the Charlevoix structure (Trepmann and Spray, 2006). In addition, the orientation of PDFs appears to differ in sedimentary rocks, with some suppression or reduction of some orientations (e.g. {10$\bar{1}$3}) in favour of others (e.g. {10$\bar{1}$1}, {11$\bar{2}$2}) (Robertson, 1980). To our knowledge, unequivocal PDFs have not been recognized at impact sites developed

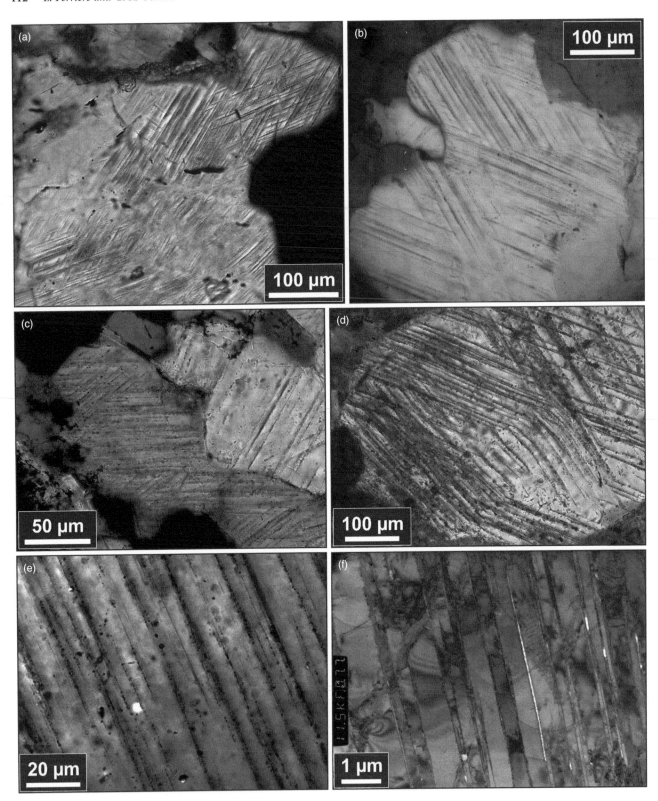

**Figure 8.4** Microphotographs of quartz grains with sets of PDFs. (a) Quartz grain with two relatively non-decorated PDF sets. Quartzite clast in suevite from the Bosumtwi crater (sample KR8-006; depth: 240.36 m). (b) Quartz grain with two PDF sets. Meta-greywacke sample from the Bosumtwi crater (sample KR8-080; depth: 384.54 m). (c) Two sets of decorated PDFs in a quartz grain from a meta-greywacke sample from the Bosumtwi crater (sample KR8-070; depth: 364.12 m). (d) Quartz grain with two PDF sets. Fragmental dike breccia sample from the Manson structure (USA; sample M8-516.3; depth: 157.37 m). (e) One set of decorated (with numerous tiny fluid inclusions) PDFs in a quartz grain from a meta-greywacke sample from the Bosumtwi crater (sample KR8-056; depth: 326.78 m). Microphotographs (a–c) taken in crossed polars and (d, e) in plane-polarized light. (f) TEM bright-field microphotograph of one set of irregularly spaced PDFs in quartz (Bosumtwi sample KR8-006). Note that the light grey network shown in the background corresponds to the carbon net supporting the specimen. (See Colour Plate 20)

in unconsolidated sedimentary materials (e.g. Henbury (Taylor, 1967) and Wabar (Hörz *et al.*, 1989)), although the lack of detailed studies leaves this question open.

### 8.2.2.5   Mosaicism

Mosaicism or mosaic structure is characterized by an irregular or mottled optical extinction pattern, which is distinctly different from undulatory extinction (commonly developed in tectonically deformed quartz). A crystal showing mosaicism comprises several sub-domains with slightly different optical axes (e.g. Dachille *et al.*, 1968; Stöffler, 1972; Stöffler and Langenhorst, 1994; French and Koeberl, 2010), as a result of the distortion (i.e. plastic deformation) of the lattice into small domains that are rotated by low angles against each other. Deformation bands (i.e. bands generally <20 μm across showing extinction directions different from those of the host; also called 'kink bands; e.g. see French and Koeberl (2010)) and/or PFs and PDFs are generally associated with mosaicism (e.g. Stöffler, 1972). Mosaicism can be semi-quantitatively characterized by the X-ray diffraction study of the degree of asterism (i.e. broadening of the normally sharp diffraction spots of lattice planes into elongate spots) in a single crystal, which can be useful for the characterization of shock pressure recorded by minerals (e.g. Hörz and Quaide, 1973). However, no correlation between pressure and degree of mosaicism is valid if minerals were recrystallized during post-shock annealing or subsequent thermal metamorphism (Stöffler, 1972). Even if mosaicism is definitely induced by shock during an impact event, a structure somewhat resembling a mosaic can also be produced by endogenic processes (e.g. Spry, 1969); thus, it cannot be used as a unique diagnostic indicator of shock metamorphism.

### 8.2.2.6   Refractivity, birefringence and density

Optical properties, such as refractivity and birefringence of quartz, have been intensively investigated in the past – see reviews by Stöffler (1974) and Stöffler and Langenhorst (1994). It was shown that, with increasing shock pressure, the birefringence and the refractive index both decrease simultaneously, until the amorphous state (i.e. diaplectic glass) is reached. Similarly, there is a decrease of the density of quartz with increasing shock pressure, from its normal value of $2.650 \pm 0.002\,g\,cm^{-3}$ to values as low as $2.280 \pm 0.002\,g\,cm^{-3}$ for shocked quartz (Langenhorst and Deutsch, 1994). The same authors show that this significant drop in density, between 25 and 35 GPa, depends on the pre-shock temperature, as well as on the orientation of the shock wave relative to the *c*-axis of the quartz crystal.

A unique feature of sandstones is the formation of the characteristic 'jigsaw' texture at pressures less than 10 GPa (Kieffer, 1971). This texture is thought to form due to rotation and shear at quartz grain boundaries, which leaves the interiors of grains relatively undamaged (Kieffer *et al.*, 1976).

## 8.2.3   Deformation in other minerals

Shock-induced deformation occurs in all minerals; deformation that largely depends on the crystal structure and on the composition of the mineral (Stöffler, 1972; Langenhorst, 2002). Two main types of microstructures are formed; namely planar microstructures (i.e. PFs and PDFs) and deformation bands (i.e. kink bands and mechanical twins). Mosaicism, while mainly described for quartz, is also observed in several minerals, such as olivine and pyroxene (e.g. Reimold and Stöffler, 1978; Bauer, 1979; Rubin *et al.*, 1997). These features have been poorly investigated and characterized in minerals other than quartz, with the notable exception of olivine in meteorites (Reimold and Stöffler, 1978; Bauer, 1979; Stöffler *et al.*, 1991; Bischoff and Stöffler, 1992; Schmitt, 2000; and references therein), possibly due to the complexity of such features in other minerals and/or because of the obliteration of these features by secondary alteration.

### 8.2.3.1   Planar microstructures

Aside from quartz, deformation in feldspar is most commonly reported in the literature (e.g. Chao, 1967; Stöffler, 1967, 1972; French, 1998). With increasing shock pressure, fracturing, plastic deformation and PFs, or more frequently PDFs, in both plagioclase and alkali-feldspar occur (Fig. 8.5). Similarly, PDFs have been observed in olivine, pyroxene, amphibole, sillimanite, garnet and apatite (e.g. Stöffler, 1972; French, 1998; Langenhorst, 2002; and references therein). It is notable that PFs and PDFs have not been documented to date in carbonate or sulfate minerals. Stöffler (1970) showed that PFs and PDFs in sillimanite are oriented parallel to bipyramids, prisms and pinacoids. Decorated PDFs have also been observed in feldspar and in amphibole (Stöffler, 1972), and even in zircon grains (e.g. Wittmann *et al.*, 2006 and references therein). It is likely that, as for quartz, PDFs are preferentially oriented parallel to rational crystallographic planes for most rock-forming minerals; however, the lack of studies for minerals other than quartz leaves this question open. In addition, it is unclear as to whether there are clear relationships between shock pressure and specific orientations of resultant PDFs in minerals other than quartz.

Olivine, rarely present in terrestrial impactites, is an important diagnostic mineral used for the classification of shock metamorphism in stony meteorites (e.g. Stöffler *et al.*, 1991; Scott *et al.*, 1992). Olivine with PFs typically parallel to low-index planes is abundant in shocked meteorites (see Langenhorst (2002) and references therein). Importantly, in contrast to quartz, the presence of PFs in olivine is accepted as being indicative of shock, as they are oriented parallel to rational crystallographic planes that do not correspond to the normal cleavage planes of olivine (Langenhorst, 2002).

Zircon grains with shock-induced deformation are also reported in impactites that were subject to shock pressures higher than 20 GPa (e.g. Bohor *et al.*, 1993; Kamo *et al.*, 1996; Wittmann *et al.*, 2006), including mainly planar microdeformation features, such as pervasive micro-cleavage (Fig. 8.6) and dislocation patterns (see Leroux *et al.* (1999)), and granular (or 'strawberry') textures (e.g. Bohor *et al.*, 1993; Gucsik *et al.*, 2004; Wittmann *et al.*, 2006).

### 8.2.3.2   Kind bands

Kink bands are frequently observed in mica, for example in muscovite and in biotite (Cummings, 1965; Chao, 1967; Hörz, 1970;

**Figure 8.5**    Thin-section photomicrographs (crossed polars) of shocked feldspar grains in shatter cone samples from (a) the Keurusselkä (Finland) and (b, c) the Manicouagan (Canada) impact structures. (a) Plagioclase grain with one prominent set of PFs oriented NE–SW (sample KE2; from Jylhänniemi; from Ferrière *et al.* (2010a: fig. 5a)). (b) Plagioclase grain with three sets of planar microstructures. (c) Enlarged part of (b) showing details of the three sets of planar microstructures. Two sets of closely spaced microstructures (possible PDFs?) are visible; one prominent set trending NE–SW is visible on the left part of the photograph; the other possible PDF set is perpendicular to polysynthetic twinning (visible on the right part of the photograph). The third set, most likely PFs, is barely visible on the central part of the photograph (oriented NW–SE). Directions (e.g. NE–SW) are in relation to an arbitrary 'North' at the top of the image.

see Fig. 8.7), but also in other minerals, such as graphite (Stöffler, 1972; El Goresy *et al.*, 2001a). Typically, kink bands form in sheet silicates, without specific orientation relative to the rational crystallographic planes. As kink bands are also observed in minerals from non-impact settings (such as in tectonically deformed rocks) they cannot be used as a diagnostic criterion for the impact origin of a structure. In the case of graphite, in addition to kink bands and with increasing shock pressure, narrowly spaced twin lamellae develop and a partial to complete degradation of the typical prominent bireflection and birefringence is observed (El Goresy *et al.*, 2001a).

### 8.2.3.3 Mechanical twins

Mechanical twins have been observed in a variety of minerals, including pyroxene, amphibole, titanite and ilmenite, and more rarely in plagioclase. These twins appear as sets of parallel bands, submicroscopic to some 10 μm in width (Stöffler, 1972).

Carbonates are known from a variety of impact structures, but relatively little is actually known about shock deformation of calcite and even less for dolomite. One of the few known shock effects is the development of mechanical twins (e.g. Turner *et al.*, 1954; Barber and Wenk, 1973; Robertson and Grieve, 1978; Langenhorst *et al.*, 2002). The first detailed application of twin analysis in calcite has recently been used to quantify shock pressures in calcite from the Serpent Mound impact structure, USA (Schedl, 2006). Given that low shear stresses of approximately 10 MPa are required for twinning in calcite (Schedl, 2006); this technique may be a useful shock indicator for carbonates shocked to relatively low shock levels.

### 8.2.4 Diaplectic glasses

Diaplectic glass forms without melting, by solid-state transformation (De Carli and Jamieson, 1959), generally from framework minerals, such as quartz or feldspar. Other minerals, such as biotite and pyroxene, tend to oxidize or to decompose without forming diaplectic glass. A fundamental diagnostic feature of diaplectic glass is that, although it is amorphous, the pre-shock morphology and texture of the mineral are preserved and flow

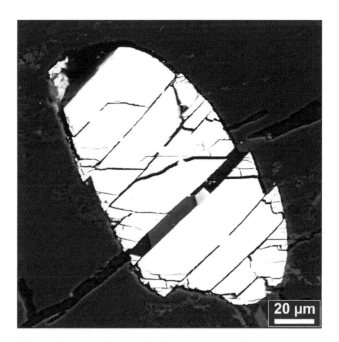

**Figure 8.6**  Backscattered electron image of a zircon grain with one set of planar fractures (impactite sample from the Haughton impact structure, Canada). Image kindly provided by Alaura C. Singleton (London, Canada).

**Figure 8.7**  Microphotographs (crossed polars) of kink-banding in mica. (a) Kink bands in a muscovite grain; suevite sample from the Chesapeake Bay structure (USA; sample CB6-100; depth: 1427.01 m). (b) Kink bands in a biotite grain; polymict impact breccia sample from the Rochechouart structure (France). Note that kink bands in mica are not a diagnostic criterion for shock metamorphism, as they can form also in other (i.e. non-impact) settings (see text). (See Colour Plate 21)

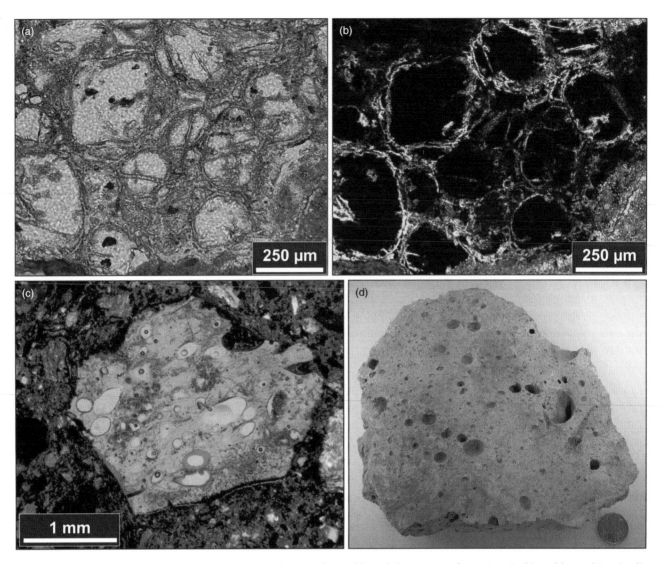

**Figure 8.8**  Photographs of impact metamorphism features formed by solid-state transformation (a, b) and by melting (c, d). (a) Plane-polarized light and (b) crossed-polars microphotographs of diaplectic quartz glass from a shocked sandstone clast in impact breccia from the Haughton impact structure (sample HMP-02-092). Note the occurrence within and between the idiomorphic quartz grains (i.e. now transformed to diaplectic quartz glass) of coesite and some alteration products. (c) Highly vesicular, shard-like, melt glass fragment in impactite from the Bosumtwi crater (sample LB-39a; from outside the crater rim; plane-polarized light). (d) Macrophotograph of a hand specimen of vesicular impact melt rock from the Rochechouart impact structure (France). (See Colour Plate 22)

structures or vesicles are absent (Stöffler and Langenhorst, 1994). Two types of diaplectic glasses are often reported in the literature; namely, diaplectic quartz glass and maskelynite (i.e. diaplectic plagioclase feldspar glass; Tschermak, 1872; Milton and De Carli, 1963; Bunch *et al.*, 1967). The latter has only been documented in crystalline rocks.

Diaplectic glass starts to form in the high-pressure regime, at shock pressures higher than approximately 35 GPa for quartz in the case of dense non-porous crystalline rocks (e.g. Stöffler, 1972; Stöffler and Langenhorst, 1994; French, 1998; and references therein) and at somewhat lower shock pressures, between 28 and 35 GPa, for An-rich feldspar (Stöffler *et al.*, 1986). In

sandstones, diaplectic quartz glass starts to form at pressures as low as approximately 5.5 GPa (Kieffer *et al.*, 1976), and between approximately 10 and 20 GPa almost complete conversion of quartz to diaplectic glass has been observed (Osinski, 2007; Fig. 8.8a,b). Importantly, the pressure limit for complete transformation to diaplectic glass decreases with increasing pre-shock temperature (e.g. Langenhorst and Deutsch, 1994).

Typically, diaplectic glasses have a refractive index and a density that decrease with increasing shock intensity. Thermal annealing experiments have shown that, at temperatures above approximately 1200 °C, diaplectic quartz glass starts to recrystallize (e.g. Rehfeldt-Oskierski, 1986), forming ballen α-cristobalite (see Section 8.3.2 for a discussion of ballen silica). As a result, in some impactites, diaplectic quartz glass is missing, but either ballen α-cristobalite or ballen α-quartz is present.

## 8.2.5 Mineral and whole-rock melt

Melting of individual minerals starts at around 50 GPa (Stöffler, 1972) and at around 60 GPa for the whole-rock in the case of non-porous crystalline rocks, while for sandstones the melting of individual quartz grains starts at pressures as low as approximately 20 GPa and whole-rock melting occurs above approximately 30–35 GPa (Kieffer *et al.*, 1976). Somewhat similar observations as the one by Kieffer *et al.* (1976) on Coconino Sandstone from Meteor Crater were reported recently by Osinski (2007) for sandstone samples from the Haughton impact structure. At both these sites it is apparent that melting is localized in original pore spaces and along grain boundaries.

Mineral and whole-rock melts (Fig. 8.8c,d) have approximately the same composition as the original minerals or mixture of minerals. Impact melt is present in many different forms and settings associated with impact structures, and the reader is referred to Chapter 9 for a detailed overview and discussion of the processes and products of impact melting. Commonly observed in impactites, lechatelierite is an SiO$_2$ melt that forms at very high temperatures (above 1700 °C) without necessarily requiring high shock pressures. Lechatelierite occurs in nature only in fulgurites and in impactites (e.g. Stöffler and Langenhorst, 1994 and references therein).

## 8.2.6 High-pressure polymorphs

High-pressure phases are commonly reported in impactites; for example, coesite and stishovite (from quartz), diamond (from graphite) and reidite (from zircon). However, coesite and diamond are not exclusively formed during shock metamorphism as they are also products of endogenic processes (Schreyer, 1995). Other high-pressure polymorphs, such as jadeite (from plagioclase), majorite (from pyroxene) and wadsleyite and ringwoodite (from olivine), are commonly reported in meteorites (e.g. Ohtani *et al.*, 2004; Fritz and Greshake, 2009; and references therein) – where they are interpreted as products of shock – but these phases have not yet been documented in impactites from terrestrial impact structures. Recently, Stähle *et al.* (2004) presented detailed observations documenting the shock-induced formation of kyanite (Al$_2$SiO$_5$) from sillimanite, arguing that it represents a third shock-induced high-pressure silicate polymorph found in the Ries impact crater, in addition to coesite and stishovite. Finally, two rutile high-pressure phases, namely TiO$_2$-II and akaogiite (El Goresy *et al.*, 2001b, 2010) were discovered in shocked gneisses from the Ries structure; TiO$_2$-II was also found in breccias from the Chesapeake Bay impact structure (Jackson *et al.*, 2006).

### 8.2.6.1 *Coesite and stishovite*

The high-pressure polymorphs of quartz, namely coesite and stishovite, form upon shock compression and are preserved as metastable phases in dense non-porous crystalline rocks that experienced peak pressure ranges between 30 and 60 GPa and

between 12 and 45 GPa respectively (e.g. Stöffler and Langenhorst, 1994). Interestingly, recent investigations (Lakshtanov *et al.*, 2007) have shown that aluminium and possibly hydrogen content in these phases have a large effect on the stability fields of coesite and stishovite. Coesite was named after Loring Coes, who first produced experimentally this high-shock pressure polymorph (Coes, 1953), whereas stishovite was named after Sergei M. Stishov, who first discovered stishovite during high-pressure experiments (Stishov and Popova, 1961). Both polymorphs generally occur in impactites within diaplectic quartz glass (see Fig. 8.9), along grain boundaries or in association with PDFs (Stöffler, 1971; Kieffer *et al.*, 1976, Stähle *et al.*, 2008; and references therein).

In sedimentary rocks, coesite is known to form at pressures as low as approximately 5.5 GPa and is common at pressures above 10 GPa (Kieffer *et al.*, 1976). In sandstones from Meteor Crater and the Haughton impact structure, coesite is typically localized within so-called 'symplectic regions' that are distinguished in transmitted light by their very high relief and the presence of opaque regions. These symplectic regions represent microscopic intergrowths of quartz, coesite and diaplectic glass and form around the rims of sand grains and along fractures (Kieffer, 1971; Kieffer *et al.*, 1976; Osinski, 2007). The generation of coesite (and other shock metamorphic effects) in sandstones at pressures substantially lower than in crystalline rocks has been explained due to the effect of porosity and complex interactions during the collapse of pore spaces (Kieffer *et al.*, 1976). The porosity leads to much higher shock- and post-shock-temperatures within these rocks, especially at pores and fractures, which is preferably the cause for the onset of high-pressure phases in porous rocks at distinctly lower shock pressures. Small polycrystalline aggregates of coesite (up to approximately 20–25 μm in size), colourless to brownish, have been characterized in impactites from a few impact structures, including Meteor Crater (Chao *et al.*, 1960), the Ries (Shoemaker and Chao, 1961), Wabar (Chao *et al.*, 1961), Bosumtwi (Littler *et al.*, 1961) and Haughton (Osinski, 2007). At Meteor Crater and at the Ries, stishovite was also detected. Indeed, the identification of these two polymorphs was used for establishing the impact origin of the Ries crater (Shoemaker and Chao, 1961).

Stishovite has so far only been identified in meteorites and in impact-related rocks (e.g. see the review by Gillet *et al.* (2007)), but post-stishovite is one of the main constituents of the basaltic layer of subducting slabs and it may also occur in the lower mantle and at the core–mantle boundary (e.g. Lakshtanov *et al.*, 2007; Liu *et al.*, 2007). The presence of coesite is frequently reported in kimberlites or in ultra-high-pressure metamorphic rocks, but it can be distinguished easily from coesite in impactites by the paragenesis and by the different geological setting of occurrence. It is important to note that, under static equilibrium conditions, where reaction rates are slower and kinetic factors less important, coesite forms at pressures greater than approximately 2 GPa and stishovite starts to form around 7–8 GPa (Heaney *et al.*, 1994 and references therein).

### 8.2.6.2 *Impact diamonds*

Impact diamonds were first discovered in impactites from the Popigai structure, Siberia (Masaitis *et al.*, 1972), and were later

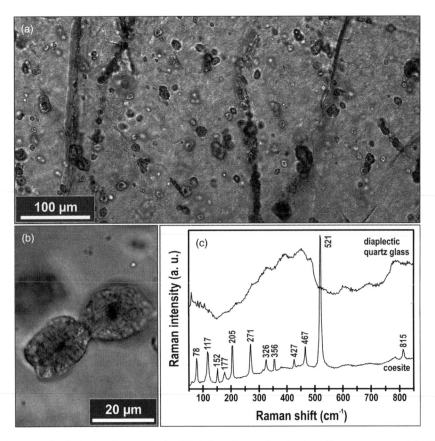

**Figure 8.9**   Microphotographs (a, b; in plane-polarized light) and associated micro-Raman spectra (c) of coesite in diaplectic quartz glass. (a) Coesite aggregates (green to brown clusters) within diaplectic quartz glass from suevite (Bosumtwi crater; sample LB-44B; from outside the crater rim). (b) Enlarged view of two aggregates of coesite from the same sample. (c) Typical micro-Raman spectra of one of the illustrated aggregates of coesite and of the host diaplectic quartz glass.

found at the Ries crater (Rost *et al.*, 1978) and in some other impact structures (see Vishnevsky *et al.* (1997) and El Goresy *et al.* (2001a) and references therein). Impact diamonds are usually polycrystalline and defect rich, and some features of the precursor graphite are inherited (e.g. Koeberl *et al.*, 1997; Langenhorst, 2002). Based on graphite–diamond textural relations, El Goresy *et al.* (2001a) concluded that impact diamonds are formed from graphite by shock-induced solid-state phase transformation during the very short time of the shock compression.

Schmitt *et al.* (2005) discussed two distinct formation mechanisms for impact diamond: (a) diamonds are produced by thermally activated, diffusion-controlled, mechanisms (resulting in the formation of aggregates of small diamond grains of single-crystal structure in paragenesis with shocked graphite or other high-temperature phases of carbon); (b) solid-state martensitic phase transformation of graphite or other carbon phases to diamond (resulting in the formation of diamond paramorphs after graphite or other carbon phases). El Goresy *et al.* (2001a) estimated that impact diamonds from the Ries crater are formed at a peak-shock pressure between 30 and 40 GPa; however, impact diamonds could form at lower pressure, depending on the setting and on the initial crystallinity of the graphite precursor.

Another carbon high-pressure phase with a hexagonal stacking sequence, called lonsdaleite, has been detected using X-ray

diffraction techniques (Frondel and Marvin, 1967; Hanneman *et al.*, 1967); however, it is not yet clear if this phase exists as an independent mineral phase or if it is an effect of stacking faults within impact diamond crystals.

### 8.2.6.3   Reidite

At around 20 GPa, zircon is transformed onto the scheelite-structured high-pressure polymorph reidite. This solid-state transition of zircon to reidite is completed at 52 GPa according to Fiske *et al.* (1994). Reidite was first characterized in zircons from an upper Eocene impact ejecta layer (Glass *et al.*, 2002), and it was named after Alan F. Reid, who first produced experimentally this high-shock pressure polymorph of zircon (Reid and Ringwood, 1969). Subsequently, reidite was also identified in impactites from the Ries impact structure by Gucsik *et al.* (2004). According to Wittmann *et al.* (2006), the transformation of zircon to reidite may be hampered by impurities and pre-impact metamictization. Because reidite is refractory, surviving to a temperature up to approximately 1000 °C (Fiske *et al.*, 1994), and also resistant to alteration, it can be used as an indicator of peak shock pressure in impactites (Glass *et al.*, 2002). Based on the paragenetic relationships of reidite with ballen quartz observed in samples from the Popigai structure, Wittmann *et al.*

(2006) concluded that reidite decomposes to granular-textured zircon at temperatures greater than approximately 1100 °C.

## 8.3   Post-shock thermal features

In several impactites, the presence of post-shock (thermal) features, such as toasted quartz and ballen silica, are also commonly present.

### 8.3.1   Toasted quartz

The term 'toasted' quartz was used first by Short and Gold (1993) in an abstract, for the description of quartz grains showing an orange–brown to greyish–reddish brown colour, and in reference to the perceived aspect of a 'toasted bread' appearance (see Fig. 8.10). Toasted quartz was then studied in more detail by Short and Gold (1996) and Whitehead *et al.* (2002) and has now been documented at many terrestrial impact sites developed in both crystalline and sedimentary target rocks. Toasted quartz is considered to be a post-shock feature, either resulting from 'hydrothermal or other post-shock modification' (Short and Gold, 1996)

or resulting from 'the exsolution of water from glass, primarily along PDFs, during heat-driven recrystallization' (Whitehead *et al.*, 2002). Both studies have also shown that toasted quartz has a higher albedo than untoasted quartz, visible in a hand specimen. In addition, Whitehead *et al.* (2002) noted that 'no compositional origin for the browning is evident'. The brownish appearance of quartz is, according to Whitehead *et al.* (2002), caused by a high proportion of very tiny fluid inclusions that are principally located along decorated PDFs. More recently, shocked quartz with toasted appearance has also been reported from the late Eocene impact ejecta layer at Massignano, Italy (Glass *et al.*, 2004) and within an impact ejecta layer containing Australasian microtektites (Glass and Koeberl, 2006). In a recent abstract, Ferrière *et al.* (2009b) confirm that toasted quartz contains a high proportion of small vesicles and that these vesicles probably enhance scattering of transmitted light. However, Ferrière *et al.* (2009b) note that the vesicles observed in toasted quartz, with highly variable sizes, are likely not related to recrystallization of PDF glass, as previously suggested by Whitehead *et al.* (2002). Using electron microprobe techniques, Ferrière *et al.* (2009b) showed that some trace elements (and inclusions), originally present in quartz, were removed from the quartz structure. They suggest that toasted quartz is formed by vesiculation after pressure release, at high

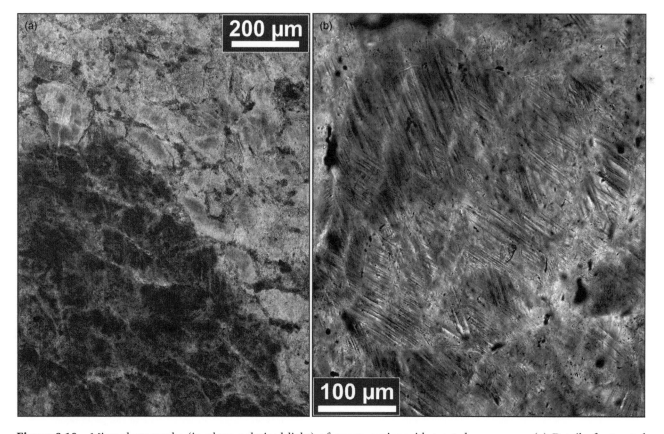

**Figure 8.10**   Microphotographs (in plane-polarized light) of quartz grains with toasted appearance. (a) Detail of a toasted region (lower left side of the photograph) and untoasted region of a sandstone clast; conglomerate sample from the Chesapeake Bay structure (USA; sample CB6-125b; depth: 1522.72 m). (b) Toasted quartz grain with two PDF sets; quartzite clast in suevite from the Bosumtwi crater (sample KR8-006; depth: 240.36 m).

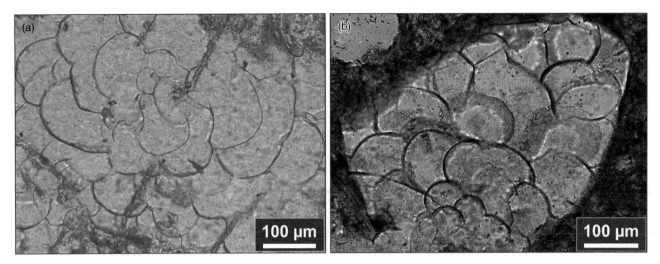

**Figure 8.11**    Microphotographs (in plane-polarized light) of ballen silica. (a) Elongate ovoid (crescent) to roundish α-cristobalite ballen in suevite from the Bosumtwi crater (sample BH1-0790; from Ferrière *et al.* (2009c: fig. 1a). (b) Ballen α-quartz with varied ballen sizes (smaller ones on the lower part of the photograph) in a glassy impact melt rock from the Wanapitei structure (Canada; from Ferrière *et al.* (2010b: fig. 1a).

post-shock temperatures and, thus, represents the beginning of quartz breakdown due to heating.

### 8.3.2    Ballen quartz and cristobalite

Ballen silica, with either an α-quartz or α-cristobalite structure, occurs in impactites as independent clasts or within diaplectic quartz glass or lechatelierite inclusions (see Fig. 8.11). Ballen are more or less spheroidal, in some cases elongate (ovoid), bodies that range in size from approximately 8 to 215 μm; they can intersect or penetrate each other or abut each other (Ferrière *et al.*, 2009c). The occurrence of so-called ballen quartz has been reported from approximately one-fifth of the known terrestrial impact structures, mostly from clasts in impact melt rock and, more rarely, in suevite (e.g. from the Chicxulub crater; Ferrière *et al.*, 2009c). Different types of ballen silica have been described and characterized (Carstens, 1975; Bischoff and Stöffler, 1984; Ferrière *et al.*, 2009c). Several mechanisms have been proposed for the formation of ballen. Most recently, it has been suggested that the formation of ballen can occur via two processes (Ferrière *et al.*, 2009c): (1) impact-triggered solid–solid transition from α-quartz to diaplectic quartz glass, followed by the formation (at high temperature) of ballen of β-cristobalite and/or β-quartz, and finally back-transformation to α-cristobalite and/or α-quartz; or (2) a solid–liquid transition from quartz to lechatelierite followed by nucleation and crystal growth at high temperature. Given that ballen quartz/cristobalite results from back-transformations from shock-induced states, ballen silica cannot be considered direct evidence of shock metamorphism, but represents indirect evidence. For more information, the reader is invited to read the recent detailed observations and review on ballen silica by Ferrière *et al.* (2009c).

It is notable that ballen silica has only been documented in impactites at impact structures developed in entirely crystalline target rocks (e.g. Carstens, 1975; Grieve, 1975; Bischoff and

Stöffler, 1984; Ferrière *et al.*, 2009c) or in mixed crystalline–sedimentary targets but where the crystalline rocks comprise the bulk of the shocked and shock-melted lithologies (e.g. the Ries crater; von Engelhardt, 1972). Indeed, at the Ries crater, ballen silica only occurs in coherent impact melt rocks derived entirely from the crystalline basement (Osinski, 2004), but is not present in the suevite deposits (Osinski *et al.*, 2004), which are derived from a combination of sedimentary and crystalline target rocks. At impact structures formed predominantly in sedimentary rocks, such as the Haughton structure, ballen silica has not been documented to date (Osinski, 2007). This is consistent with the models for ballen silica formation (Ferrière *et al.*, 2009c), which involve high-temperature transitions; sediment-derived impact melts are typically rapidly quenched compared with melts generated from crystalline targets, such that cooling rate may be an important constraint in the formation of ballen silica (Osinski, 2007).

### 8.4    Concluding remarks

After decades of intense investigation of tectosilicates, mostly quartz in terrestrial impactites, the time has arrived to have a closer look at shock effects in nesosilicates, inosilicates and phyllosilicates, and also at carbonate and sulfate minerals in respect to being able to identify impact structures formed in non-silicate target rocks, and also to better evaluate consequences of impact in such types of rock-forming minerals on the environment and atmosphere. We encourage both, field and experimental studies, as we cannot have one without the other if we want to associate proper numbers with observations, even we should be aware that nature is hard to model. Such a type of research is not only needed to expand our knowledge on impact effects on Earth, but also to better understand the formation and evolution of our Solar System.

# References

Baratoux, D. and Melosh, H.J. (2003) The formation of shatter cones by shock wave interference during impacting. *Earth and Planetary Science Letters*, 216, 43–54.

Barber, D.J. and Wenk, H.R. (1973) The microstructure of experimentally deformed limestones. *Journal of Materials Science*, 8, 500–508.

Bauer, J.F. (1979) Experimental shock metamorphism of mono- and polycrystalline olivine: a comparative study. *Proceedings of the Lunar and Planetary Science Conference*, 10, 2573–2596.

Bischoff, A. and Stöffler, D. (1984) Chemical and structural changes induced by thermal annealing of shocked feldspar inclusions in impact melt rocks from Lappajärvi crater, Finland. *Journal of Geophysical Research*, 89, B645–B656.

Bischoff, A. and Stöffler, D. (1992) Shock metamorphism as a fundamental process in the evolution of planetary bodies: information from meteorites. *European Journal of Mineralogy*, 4, 707–755.

Bloss, F.D. (1971) *Crystallography and Crystal Chemistry: An Introduction*, Holt, Rinehart, and Winston, New York, NY.

Bohor, B.F., Modreski, P.J. and Foord, E.E. (1987) Shocked quartz in the Cretaceous–Tertiary boundary clays: evidence for a global distribution. *Science*, 236, 705–708.

Bohor, B.F., Betterton, W.J. and Krogh, T.E. (1993) Impact-shocked zircons: discovery of shock-induced textures reflecting increasing degrees of shock metamorphism. *Earth and Planetary Science Letters*, 119, 419–424.

Bunch, T.E., Cohen, A.J. and Dence, M.R. (1967) Natural terrestrial maskelynite. *The American Mineralogist*, 52, 244–253.

Carstens, H. (1975) Thermal history of impact melt rocks in the Fennoscandian Shield. *Contributions to Mineralogy and Petrology*, 50, 145–155.

Chao, E.C.T. (1967) Shock effects in certain rock-forming minerals. *Science*, 156, 192–202.

Chao, E.C.T., Shoemaker, E.M, and Madsen, B.M. (1960) First natural occurrence of coesite. *Science*, 132, 220–222.

Chao, E.C.T., Fahey, J.J. and Littler, J. (1961) Coesite from Wabar crater, near Al Hadida, Arabia. *Science*, 133, 882–883.

Coes, L. (1953) A new dense crystalline silica. *Science*, 118, 131–132.

Cummings, D. (1965) Kink-bands: shock deformation of biotite resulting from a nuclear explosion. *Science*, 148, 950–952.

Dachille, F., Gigl, P. and Simons, P.Y. (1968) Experimental and analytical studies of crystalline damage useful for the recognition of impact structures, in *Shock Metamorphism of Natural Materials* (eds B.M. French and N.M. Short), Mono Book Corporation, Baltimore, MD, pp. 555–569.

De Carli, P.S. and Jamieson, J.C. (1959) Formation of an amorphous form of quartz under shock conditions. *Journal of Chemical Physics*, 31, 1675–1676.

De Carli, P.S., Bowden, E., Jones, A.P. and Price, G.D. (2002) Laboratory impact experiments versus natural impact events, in *Catastrophic Events and Mass Extinction: Impacts and Beyond* (eds C. Koeberl and K.G. MacLeod), Geological Society of America Special Paper 356, Geological Society of America, Boulder, CO, pp. 595–605.

Dietz, R.S. (1960) Meteorite impact suggested by shatter cones in rock. *Science*, 131, 1781–1784.

Dietz, R.S. (1968) Shatter cones in cryptoexplosion structures, in *Shock Metamorphism of Natural Materials* (eds B.M. French and N.M. Short), Mono Book Corporation, Baltimore, MD, pp. 267–285.

Dressler, B. (1990) Shock metamorphic features and their zoning and orientation in the Precambrian rocks of the Manicouagan structure, Quebec, Canada. *Tectonophysics*, 171, 229–245.

Dressler, B.O., Sharpton, V.L. and Schuraytz, B.C. (1998) Shock metamorphism and shock barometry at a complex impact structure: State Islands, Canada. *Contributions to Mineralogy and Petrology*, 130, 275–287.

El Goresy, A., Gillet, P., Chen, M. et al. (2001a) In situ discovery of shock-induced graphite–diamond phase transition in gneisses from the Ries crater, Germany. *American Mineralogist*, 86, 611–621.

El Goresy, A., Chen, M., Dubrovinsky, L. et al. (2001b) An ultradense polymorph of rutile with seven-coordinated titanium from the Ries crater. *Science*, 293, 1467–1470.

El Goresy, A., Dubrovinsky, L., Gillet, P. et al. (2010) Akaogiite: an ultra-dense polymorph of $TiO_2$ with the baddeleyite-type structure, in shocked garnet gneiss from the Ries crater, Germany. *American Mineralogist*, 95, 892–895.

Fackelman, S.P., Morrow, J.R., Koeberl, C. and McElvain, T.H. (2008) Shatter cone and microscopic shock-alteration evidence for a post-Paleoproterozoic terrestrial impact structure near Santa Fe, New Mexico, USA. *Earth and Planetary Science Letters*, 270, 290–299.

Ferrière, L. and Osinski, G.R. (2010) Shatter cones and associated shock-induced microdeformations in minerals – new investigations and implications for their formation. 41st Lunar and Planetary Science Conference, CD-ROM, abstract no. 1392.

Ferrière, L., Morrow, J.R., Amgaa, T. and Koeberl, C. (2009a) Systematic study of universal-stage measurements of planar deformation features in shocked quartz: implications for statistical significance and representation of results. *Meteoritics and Planetary Science*, 44, 925–940.

Ferrière, L., Koeberl, C., Reimold, W.U. et al. (2009b) The origin of 'toasted' quartz in impactites revisited. 40th Lunar and Planetary Science Conference, CD-ROM, abstract no. 1751.

Ferrière, L., Koeberl, C. and Reimold, W.U. (2009c) Characterization of ballen quartz and cristobalite in impact breccias: new observations and constraints on ballen formation. *European Journal of Mineralogy*, 21, 203–217.

Ferrière, L., Raiskila, S., Osinski, G.R. et al. (2010a) The Keurusselkä impact structure, Finland – impact origin confirmed by characterization of planar deformation features in quartz grains. *Meteoritics and Planetary Science*, 45, 434–446.

Ferrière, L., Koeberl, C., Libowitzky, E. et al. (2010b) Ballen quartz and cristobalite in impactites: new investigations, in *Large Meteorite Impacts and Planetary Evolution IV* (eds R.L. Gibson and W.U. Reimold), Geological Society of America Special Paper 465, Geological Society of America, Boulder, CO, pp. 609–618.

Fiske, P.S., Nellis, W.J. and Sinha, A.K. (1994) Shock-induced phase transitions of $ZrSiO_4$, reversion kinetics, and implications for terrestrial impact craters. *EOS, Transactions, American Geophysical Union*, 75, 416–417 (abstract).

French, B.M. (1998) *Traces of Catastrophe: A Handbook of Shock-Metamorphic Effects in Terrestrial Meteorite Impact Structures*, LPI Contribution No. 954, Lunar and Planetary Institute, Houston, TX.

French, B.M. and Koeberl, C. (2010) The convincing identification of terrestrial meteorite impact structures: what works, what doesn't, and why. *Earth-Science Reviews*, 98, 123–170.

French, B.M. and Short, N.M. (eds) (1968) *Shock Metamorphism of Natural Materials*, Mono Book Corporation, Baltimore, MD.

French, B.M., Cordua, W.S. and Plescia, J.B. (2004) The Rock Elm meteorite impact structure, Wisconsin: geology and shock-metamorphic effects in quartz. *Geological Society of America Bulletin*, 116, 200–218.

Fritz, J. and Greshake, A. (2009) High-pressure phases in an ultramafic rock from Mars. *Earth and Planetary Science Letters*, 288, 619–623.

Frondel, C. and Marvin, U.B. (1967) Lonsdaleite, a hexagonal polymorph of diamond. *Nature*, 214, 587–589.

Gash, P.J.S. (1971) Dynamic mechanism for the formation of shatter cones. *Nature Physical Science*, 230, 32–35.

Gillet, P., El Goresy, A., Beck, P. and Chen, M. (2007) High-pressure mineral assemblages in shocked meteorites and shocked terrestrial rocks: mechanisms of phase transformations and constraints to pressure and temperature histories, in *Advances in High-Pressure Mineralogy* (ed. E. Ohtani), Geological Society of America Special Paper 421, Geological Society of America, Boulder, CO, pp. 57–82.

Glass, B.P. and Koeberl, C. (2006) Australasian microtektites and associated impact ejecta in the South China Sea and the Middle Pleistocene supereruption of Toba. *Meteoritics and Planetary Science*, 41, 305–326.

Glass, B.P., Liu, S. and Leavens, P.B. (2002) Reidite: an impact-produced high-pressure polymorph of zircon found in marine sediments. *American Mineralogist*, 87, 562–565.

Glass, B.P., Liu, S. and Montanari, A. (2004) Impact ejecta in upper Eocene deposits at Massignano, Italy. *Meteoritics and Planetary Science*, 39, 589–597.

Goltrant, O., Cordier, P. and Doukhan, J.-C. (1991) Planar deformation features in shocked quartz; a transmission electron microscopy investigation. *Earth and Planetary Science Letters*, 106, 103–115.

Grieve, R.A.F. (1975) Petrology and chemistry of the impact melt at Mistastin Lake crater, Labrador. *Geological Society of America Bulletin*, 86, 1617–1629.

Grieve, R.A.F., Coderre, J.M., Robertson, P.B. and Alexopoulos, J. (1990) Microscopic planar deformation features in quartz of the Vredefort structure: anomalous but still suggestive of an impact origin. *Tectonophysics*, 171, 185–200.

Grieve, R.A.F., Langenhorst, F. and Stöffler, D. (1996) Shock metamorphism of quartz in nature and experiment: II. Significance in geoscience. *Meteoritics and Planetary Science*, 31, 6–35.

Gucsik, A., Koeberl, C., Brandstätter, F. et al. (2004) Cathodoluminescence, electron microscopy, and Raman spectroscopy of experimentally shock metamorphosed zircon crystals and naturally shocked zircon from the Ries impact crater, in *Cratering in Marine Environments and on Ice* (eds H. Dypvik, P. Claeys and M. Burchell), Springer, Berlin, pp. 281–322.

Hanneman, R.E., Strong, H.M. and Bundy, F.P. (1967) Hexagonal diamonds in meteorites: implications. *Science*, 155, 995–997.

Heaney, P.J., Prewitt, C.T. and Gibbs, G.V. (eds) (1994) *Silica: Physical Behavior, Geochemistry and Materials Applications*, Reviews in Mineralogy 29, Mineralogical Society of America, Washington, DC.

Higgins, C.G. (1956). Formation of small ventifacts. *Journal of Geology*, 64, 506–516.

Hörz, F. (1968) Statistical measurements of deformation structures and refractive indices in experimentally shock loaded quartz, in *Shock Metamorphism of Natural Materials* (eds B.M. French and N.M. Short), Mono Book Corporation, Baltimore, MD, pp. 243–253.

Hörz, F. (1970) Static and dynamic origin of kink bands in micas. *Journal of Geophysical Research*, 75, 965–977.

Hörz, F. and Quaide, W.L. (1973) Debye–Scherrer investigations of experimentally shocked silicates. *The Moon*, 6, 45–82.

Hörz, F., Blanchard, D.P., See, T.H. and Murali, A.V. (1989) Heterogeneous dissemination of projectile materials in the impact melts

from Wabar Crater, Saudi Arabia. *Proceedings of the Lunar and Planetary Science Conference*, 19, 697–709.

Huffman, A.R. and Reimold, W.U. (1996) Experimental constraints on shock-induced microstructures in naturally deformed silicates. *Tectonophysics*, 256, 165–217.

Jackson, J.C., Horton Jr, J.W., Chou, I.-M. and Belkin, H.E. (2006) A shock-induced polymorph of anatase and rutile from the Chesapeake Bay impact structure, Virginia, U.S.A. *American Mineralogist*, 91, 604–608.

Johnson, G.P. and Talbot, R.J. (1964) A theoretical study of the shock wave origin of shatter cones. MS thesis, Air Force Institute of Technology, Wright–Patterson Air Force Base, Dayton, OH.

Kamo, S.L., Reimold, W.U., Krogh, T.E. and Colliston, W.P. (1996) A 2.023 Ga age for the Vredefort impact event and a first report of shock metamorphosed zircons in pseudotachylitic breccias and granophyre. *Earth and Planetary Science Letters*, 144, 369–387.

Kieffer, S.W. (1971) Shock metamorphism of the Coconino sandstone at Meteor Crater, Arizona. *Journal of Geophysical Research*, 76, 5449–5473.

Kieffer, S.W., Phakey, P.P. and Christie, J.M. (1976) Shock processes in porous quartzite: transmission electron microscope observations and theory. *Contributions to Mineralogy and Petrology*, 59, 41–93.

Koeberl, C., Masaitis, V.L., Shafranovsky, G.I. et al. (1997) Diamonds from the Popigai impact structure, Russia. *Geology*, 25, 967–970.

Lakshtanov, D.L., Sinogeikin, S.V., Litasov, K.D. et al. (2007) The post-stishovite phase transition in hydrous alumina-bearing $SiO_2$ in the lower mantle of the Earth. *Proceedings of the National Academy of Sciences of the United States of America*, 104, 13588–13590.

Langenhorst, F. (2002) Shock metamorphism of some minerals: basic introduction and microstructural observations. *Bulletin of the Czech Geological Survey*, 77, 265–282.

Langenhorst, F. and Deutsch, A. (1994) Shock experiments on preheated α- and β-quartz: I. Optical and density data. *Earth and Planetary Science Letters*, 125, 407–420.

Langenhorst, F., Boustie, M., Deutsch, A. et al. (2002) Experimental techniques for the simulation of shock metamorphism: a case study on calcite, in *High-Pressure Shock Compression of Solids V: Shock Chemistry with Applications to Meteorite Impacts* (eds L. Davison, Y. Horie and T. Sekine), Springer, New York, NY, pp. 1–27.

Leroux, H. (2005) Weathering features in shocked quartz from the Ries impact crater, Germany. *Meteoritics and Planetary Science*, 40, 1347–1352.

Leroux, H., Reimold, W.U., Koeberl, C. et al. (1999) Experimental shock deformation in zircon: a transmission electron microscopic study. *Earth and Planetary Science Letters*, 169, 291–301.

Littler, J., Fahey, J.J., Dietz, R.S. and Chao, E.C.T. (1961) Coesite from the Lake Bosumtwi crater, Ashanti, Ghana, in *Abstracts for 1961*, Geological Society of America Special Paper 68, Geological Society of America, Boulder, CO.

Liu, L., Zhang, J., Green II, H.W. et al. (2007) Evidence of former stishovite in metamorphosed sediments, implying subduction to >350 km. *Earth and Planetary Science Letters*, 263, 180–191.

Lugli, S., Reimold, W.U. and Koeberl, C. (2005) Silicified cone-in-cone structures from Erfoud (Morocco): a comparison with impact-generated shatter cones, in *Impact Tectonics (Impact Studies)* (eds C. Koeberl and H. Henkel), Springer-Verlag, Berlin, pp. 81–110.

Masaitis, V.L., Futergendler, S.I. and Gnevushev, M.A. (1972) Diamonds in impactites of the Popigai meteorite crater. *Zapiski Vsesoyuznoge Mineralogicheskogo Obshchestva*, 101, 108–112 (in Russian).

McLaren, A.C., Retchford, J.A., Griggs, D.T. and Christie, J.M. (1967) Transmission electron microscope study of Brazil twins and dislocations experimentally produced in natural quartz. *Physica Status Solidi (b)*, 19, 631–644.

Milton, D.J. (1977) Shatter cones – an outstanding problem in shock mechanics, in *Impact and Explosion Cratering* (eds D.J. Roddy, R.O. Pepin and R.B. Merrill), Pergamon Press, New York, NY, pp. 703–714.

Milton, D.J. and De Carli, P.S. (1963) Maskelynite: formation by explosive shock. *Science*, 140, 670–671.

Montanari, A. and Koeberl, C. (2000) *Impact Stratigraphy: The Italian Record*, Lectures Notes in Earth Sciences 93, Springer Verlag, Heidelberg.

Morrow, J.R. (2007) Shock-metamorphic petrography and micro-Raman spectroscopy of quartz in upper impactite interval, ICDP drill core LB-07A, Bosumtwi impact crater, Ghana. *Meteoritics and Planetary Science*, 42, 591–609.

Müller, W.F. and Défourneaux, M. (1968) Deformationsstrukturen im Quarz als Indikator für Stosswellen: Eine experimentelle Untersuchung an Quartz-Einkristallen. *Zeitschrift für Geophysik*, 34, 483–504.

Ohtani, E., Kimura, Y., Kimura, M. *et al.* (2004) Formation of high-pressure minerals in shocked L6 chondrite Yamato 791384: constraints on shock conditions and parent body size. *Earth and Planetary Science Letters*, 227, 505–515.

Osinski, G.R. (2004) Impact melt rocks from the Ries structure, Germany: an origin as impact melt flows? *Earth and Planetary Science Letters*, 226, 529–543.

Osinski, G.R. (2007) Impact metamorphism of CaCO$_3$-bearing sandstones at the Haughton structure, Canada. *Meteoritics and Planetary Science*, 42, 1945–1960.

Osinski, G.R. and Spray, J.G. (2006) Shatter cones of the Haughton impact structure, Canada. Proceedings of the 1st International Conference on Impact Cratering in the Solar System, European Space Agency Special Publication SP-612 (CD-ROM).

Osinski, G.R., Grieve, R.A.F. and Spray, J.G. (2004) The nature of the groundmass of surficial suevite from the Ries impact structure, Germany, and constraints on its origin. *Meteoritics and Planetary Science*, 39, 1655–1683.

Ostertag, R. (1983) Shock experiments on feldspar crystals. *Journal of Geophysical Research*, 88 (Suppl.), B364–B376.

Poelchau, M.H. and Kenkmann, T. (2011) Feather features: a low-shock-pressure indicator in quartz. *Journal of Geophysical Research*, 116, B02201. DOI: 10.1029/2010JB007803.

Rehfeldt-Oskierski, A. (1986) Stosswellenexperimente an Quarz-Einkristallen und thermisches Verhalten von diaplektischen Quarzgläsern, PhD thesis, University of Münster, Germany.

Reid, A.F. and Ringwood, A.E. (1969) Newly observed high pressure transformations in Mn$_3$O$_4$, CaAl$_2$O$_4$, and ZrSiO$_4$. *Earth and Planetary Science Letters*, 6, 205–208.

Reimold, W.U. and Stöffler, D. (1978) Experimental shock metamorphism of dunite. *Proceedings of the Lunar and Planetary Science Conference*, 9, 2805–2824.

Robertson, P.B. (1975) Experimental shock metamorphism of maximum microcline. *Journal of Geophysical Research*, 80 (14), 1903–1910.

Robertson, P.B. (1980) Anomalous development of planar deformation features in shocked quartz of porous lithologies. 11th Lunar and Planetary Science Conference, pp. 938–940 (abstract).

Robertson, P.B. and Grieve, R.A.F. (1977) Shock attenuation at terrestrial impact structures, in *Impact and Explosion Cratering* (eds D.J. Roddy, R.O. Pepin and R.B. Merrill), Pergamon Press, New York, NY, pp. 687–702.

Robertson, P.B. and Grieve, R.A.F. (1978) The Haughton impact structure. *Meteoritics*, 13, 615–619.

Rost, R., Dolgov, Y.A. and Vishnevsky, S.A. (1978) Gases in inclusions of impact glass in the Ries crater, West Germany and finds of high-pressure carbon polymorphs. *Doklady Akademia Nauk SSSR*, 241, 695–698 (in Russian).

Rubin, A.E., Scott, E.R.D. and Keil, K. (1997) Shock metamorphism of enstatite chondrites. *Geochemica et Cosmochimica Acta*, 61 (4), 847–858.

Sagy, A., Fineberg, J. and Reches, Z. (2004) Shatter cones: branched, rapid fractures formed by shock impact. *Journal of Geophysical Research*, 109, B10209. DOI: 10.1029/2004JB003016.

Schedl, A. (2006) Applications of twin analysis to studying meteorite impact structures. *Earth and Planetary Science Letters*, 244, 530–540.

Schmitt, R.T. (2000) Shock experiments with the H6 chondrite Kernouvé: pressure calibration of microscopic shock effects. *Meteoritics and Planetary Science*, 35, 545–560.

Schmitt, R.T., Lapke, C., Lingemann, C.M. *et al.* (2005) Distribution and origin of impact diamonds in the Ries crater, Germany, in *Large Meteorite Impacts III* (eds T. Kenkmann, F. Hörz and A. Deutsch), Geological Society of America Special Paper 384, Geological Society of America, Boulder, CO, pp. 299–314.

Schreyer, W. (1995) Ultradeep metamorphic rocks: the retrospective view. *Journal of Geophysical Research*, 100, 8353–8366.

Scott, E.R.D., Keil, K. and Stöffler, D. (1992) Shock metamorphism of carbonaceous chondrites. *Geochimica et Cosmochimica Acta*, 56, 4281–4293.

Shoemaker, E.M. and Chao, E.C.T. (1961) New evidence for the impact origin of the Ries Basin, Bavaria, Germany. *Journal of Geophysical Research*, 66, 3371–3378.

Short, N.M. and Gold, D.P. (1993) Petrographic analysis of selected core materials from the Manson (Iowa) impact structure. *Meteoritics and Planetary Science*, 28, A436–A437 (abstract).

Short, N.M. and Gold, D.P. (1996) Petrography of shocked rocks from the central peak at the Manson impact structure, in *The Manson Impact Structure, Iowa: Anatomy of an Impact Crater* (eds C. Koeberl and R.R. Anderson), Geological Society of America Special Paper 302, Geological Society of America, Boulder, CO, pp. 245–265.

Spry, A. (1969) *Metamorphic Textures*, Pergamon Press, New York, NY.

Stähle, V., Altherr, R., Koch, M. and Nasdala, L. (2004) Shock-induced formation of kyanite (Al$_2$SiO$_5$) from sillimanite within a dense metamorphic rock from the Ries crater (Germany). *Contributions to Mineralogy and Petrology*, 148, 150–159.

Stähle, V., Altherr, R., Koch, M. and Nasdala, L. (2008) Shock-induced growth and metastability of stishovite and coesite in lithic clasts from suevite of the Ries impact crater (Germany). *Contributions to Mineralogy and Petrology*, 155, 457–472.

Stishov, S. M. and Popova, S.V. (1961) A new dense modification of silica. *Geokhimiya*, 10, 837–839 (in Russian; English translation in *Geochemistry*, 10, 923–926 (1961)).

Stöffler, D. (1967) Deformation und Umwandlung von Plagioklas durch Stoßwellen in den Gesteinen des Nördlinger Ries. *Contributions to Mineralogy and Petrology*, 16, 51–83.

Stöffler, D. (1970) Shock deformation of sillimanite from the Ries crater, Germany. *Earth and Planetary Science Letters*, 10, 115–120.

Stöffler, D. (1971) Coesite and stishovite in shocked crystalline rocks. *Journal of Geophysical Research*, 76, 5474–5488.

Stöffler, D. (1972) Deformation and transformation of rock-forming minerals by natural and experimental shock processes: I. Behavior of minerals under shock compression. *Fortschritte der Mineralogie*, 49, 50–113.

Stöffler, D. (1974) Deformation and transformation of rock-forming minerals by natural and experimental processes: II. Physical properties of shocked minerals. *Fortschritte der Mineralogie*, 51, 256–289.

Stöffler, D. and Grieve, R.A.F. (2007) Impactites, in *Metamorphic Rocks: A Classification and Glossary of Terms, Recommendations of the International Union of Geological Sciences* (eds D. Fettes and J. Desmons), Cambridge University Press, Cambridge, pp. 82–92 plus glossary.

Stöffler, D. and Langenhorst, F. (1994) Shock metamorphism of quartz in nature and experiment: I. Basic observation and theory. *Meteoritics and Planetary Science*, 29, 155–181.

Stöffler, D., Ostertag, R., Jammes, C. *et al.* (1986) Shock metamorphism and petrography of the Shergotty achondrite. *Geochimica et Cosmochimica Acta*, 50, 889–903.

Stöffler, D., Keil, K. and Scott, E.R.D. (1991) Shock metamorphism of ordinary chondrites. *Geochimica et Cosmochimica Acta*, 55, 3845–3867.

Taylor, S.R. (1967) Composition of meteorite impact glass across the Henbury strewnfield. *Geochimica et Cosmochimica Acta*, 31, 961–962.

Trepmann, C.A. and Spray, J.G. (2006) Shock-induced crystal-plastic deformation and post-shock annealing of quartz: microstructural evidence from crystalline target rocks of the Charlevoix impact structure, Canada. *European Journal of Mineralogy*, 18, 161–173.

Tschermak, G. (1872) Die Meteoriten von Shergotty und Gopalpur. *Sitzungsberichte, Akademie der Wissenschaften, Wien, Mathematisch-Naturwissenschaftliche Klasse*, 65, 122–146.

Turner, F.J., Griggs, D.T. and Heard, H. (1954) Experimental deformation of calcite crystals. *Geological Society of America Bulletin*, 65, 883–934.

Vishnevsky, S.A., Afanasiev, V.P., Argunov, K.P. and Pal'chik, N.A. (1997) *Impact Diamonds: Their Features, Origin and Significance*, Russian Academy of Science Press, Novosibirsk (in Russian and English).

Von Engelhardt, W. (1972) Shock produced rock glasses from the Ries Crater. *Contributions to Mineralogy and Petrology*, 36, 265–292.

Von Engelhardt, W. and Bertsch, W. (1969) Shock induced planar deformation structures in quartz from the Ries Crater, Germany. *Contributions to Mineralogy and Petrology*, 20, 203–234.

Walzebuck, J.P. and von Engelhardt, W. (1979) Shock deformation of quartz influenced by grain size and shock direction: observations on quartz-plagioclase rocks from the basement of the Ries Crater, Germany. *Contributions to Mineralogy and Petrology*, 70, 267–271.

Whitehead, J., Spray, J.G. and Grieve, R.A.F. (2002) Origin of 'toasted' quartz in terrestrial impact structures. *Geology*, 30, 431–434.

Wieland, F., Reimold, W.U. and Gibson, R.L. (2006) New observations on shatter cones in the Vredefort impact structure, South Africa, and evaluation of current hypotheses for shatter cone formation. *Meteoritics and Planetary Science*, 41, 1737–1759.

Wittmann, A., Kenkmann, T., Schmitt, R.T. and Stöffler, D. (2006) Shock-metamorphosed zircon in terrestrial impact craters. *Meteoritics and Planetary Science*, 41, 433–454.

Wünnemann, K., Collins, G.S. and Osinski, G.R. (2008) Numerical modelling of impact melt production in porous rocks. *Earth and Planetary Science Letters*, 269, 529–538.

# NINE

# *Impact melting*

## Gordon R. Osinski, Richard A. F. Grieve, Cassandra Marion and Anna Chanou

*Departments of Earth Sciences/Physics and Astronomy, Western University, 1151 Richmond Street, London, ON, N6A 5B7, Canada*

## 9.1  Introduction

The generation of impact melts and the production of impact melt rocks and glasses is one of the most characteristic features of hypervelocity impact events. Such lithologies generally represent whole-rock melts formed during the rapid shock-induced melting of target rocks directly beneath the impact point. Much of the early work on terrestrial impact structures focused on the analysis and origin of impact melt products. This was driven, in a large part, by the recognition of smooth deposits on the Moon interpreted as impact melt rocks (e.g. Howard, 1974; Howard and Wilshire, 1975; Hawke and Head, 1977) and, then, their subsequent identification in samples brought back by the Apollo astronauts. Early field and analytical studies carried out in the 1960s and 1970s at several Canadian impact structures (e.g. Brent, Manicouagan, Mistastin and the twin Clearwater Lake structures) provided observational information as to the character and distribution of impact-melted material in terrestrial impact structures (Dence, 1971; Grieve, 1975, 1978; Simonds *et al.*, 1978a; Palme *et al.*, 1979). This information, in turn, provided valuable constraints on processes operating during hypervelocity impact events (Grieve *et al.*, 1977). Problems associated with classification and nomenclature have been highlighted in Chapter 7. However, our understanding of impact melting is incomplete, due to several factors. Given their largely surficial nature, the erosional state of terrestrial impact structures means that outcrops of impact melt-bearing lithologies preserving their entire original context are relatively rare (Grieve *et al.*, 1977). In recent years, the effect of target lithology on the processes and products of impact melting has been identified as a major research topic and is addressed below.

## 9.2  Why impact melting occurs

Impact melting, sometimes referred to somewhat erroneously as 'shock melting', occurs upon decompression from high shock pressure and temperatures (Grieve *et al.*, 1977). It is important to note that impact melting is unlike thermal melting, such as experienced in endogenic magmatic systems. Endogenic magmatic processes are governed by thermodynamic relations, such as eutectic compositions and partial melting of progenitor lithologies. Impact melting, on the other hand, is a function of shock pressure and the compressibility of the target rock lithologies and their constituent minerals. It also occurs in the highly dynamic physical environment of impact crater formation, when target lithologies are in high speed and differential motion (see Chapter 4). Mass, momentum and energy are conserved across a shock wave, and the state of material, as it is subject to shock compression, can be defined by so-called Hugoniot equations, which describe the pressure in front and behind the shock wave, the particle velocity of material after the shock wave has passed and the specific internal energy of material in front and behind the shock wave (Melosh, 1989). Geological materials also have a Hugoniot equation of state, which is a locus of a series of discrete shock states that are usually expressed in terms of specific volume and shock pressure. It is important to understand that imposition of a shock wave represents a discontinuous event and the target materials change suddenly across it. Unlike endogenic metamorphism and melting, there are no intermediate states and shock compression is a thermodynamically irreversible process.

Under shock compression, the target materials increase their internal energy, as considerable pressure–volume work is done on the target materials. If the target materials are porous, the pores are closed by relatively low shock pressures. As a result, more pressure–volume work is done on porous geological materials, for a given shock pressure, compared with non-porous target materials (Kieffer *et al.*, 1976; Wünnemann *et al.*, 2008). Decompression from the shocked state is by a rarefaction wave (Chapter 4) and decompression occurs via a release adiabat. As a result, not all the pressure–volume work, resulting from shock compression, is recovered and the pressure–volume work that remains in the target materials is manifested as waste heat. It is this remaining

*Impact Cratering: Processes and Products*, First Edition. Edited by Gordon R. Osinski and Elisabetta Pierazzo.
© 2013 Blackwell Publishing Ltd. Published 2013 by Blackwell Publishing Ltd.

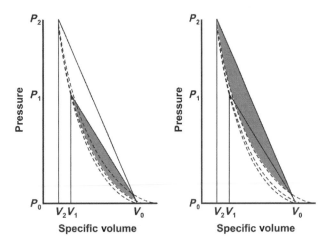

**Figure 9.1** Schematic pressure–specific volume Hugoniot showing two cases of shock and release paths and how the amount of unrecovered $P\Delta V$ work (waste heat) on decompression (expressed as the grey area between the path to the initial shock state and the release adiabat) is greater in the higher shock case.

waste heat on shock decompression that is responsible for impact melting and vaporization phenomena. The initial temperature of the impact melt is a function of the amount of waste heat remaining on decompression. As a result, unlike endogenic melts, impact melts can be and are, as a rule, initially superheated. The paths of compression and decompression due to a shock wave are illustrated schematically in Fig. 9.1. Further details of the physics of these processes can be found in Melosh (1989).

More compressible materials retain more post-shock waste heat, which results in impact melting at lower shock pressures. This characteristic of impact phenomena accounts, for example, for the observation that samples of the lunar regolith are heavily charged with impact-melted materials, compared with associated solid lunar rocks (Heiken *et al.*, 1991). As rocks are made up of a variety of different individual minerals, which also vary in their compressibility, the more compressible minerals retain more waste heat and either thermally decompose or melt at lower shock pressures than their less compressible counterparts, and the first petrographic signs of impact melting are individual or, more often, mixed mineral melts (Lambert and Lange, 1984). The pressure range over which individual minerals melt is not large, a few tens of gigapascals (Stöffler and Grieve, 2007, and references therein), and whole-rock melts are the norm, at least for impacts into crystalline targets (Osinski *et al.*, 2008a). As impact-melted materials are being driven down into the expanding transient cavity (Chapter 4), they have differential particle velocities, depending on the shock pressure they were subject to in the impact event. Thus, such high-velocity turbulent flow results in impact melt bodies of a generally mixed composition, corresponding to the volume of the target that was melted. Depending on their subsequent thermal history (size of the impact event, location in the final crater form), impact melts form a variety of products, which are described in the next section.

## 9.3   Terrestrial impact melt products

### 9.3.1   Overview

Impact melt products have been documented at approximately half of the known terrestrial impact structures. Field studies at these sites reveal that impact melt products are found within and around impact craters in a variety of settings, including up to thousands of kilometres away from the crater in the case of tektites (Fig. 9.2): (1) large kilometre-scale impact melt layers or 'sheets' and/or isolated bodies within the crater interior and rim region; (2) small, metre- to centimetre-scale, glassy (often devitrified or altered) particles, either within impact breccias or on their own, within and without the crater structure; (3) centimetre- to tens of meter-scale injection dikes in the crater floor and walls; and (4) discrete millimetre- to centimetre-scale glassy particles (spherules or tektites and microtektites) distributed regionally to globally.

These impact melt deposits vary from crystalline melt rocks with phaneritic to aphanitic textures, to glassy, often vesiculated, bodies or individual particles (Fig. 9.3). Impact-generated glasses are particularly common and are found in the majority of fresh impact craters, regardless of target lithology. Various types have been referred to, and are defined, in the literature. Stöffler (1984) recognized that an important distinction should be made between 'mineral glasses' and 'rock glasses'. *Rock glasses* represent whole-rock melts and have the composition of a rock or a mixture of minerals (Stöffler, 1984; Fig. 9.4a,b). *Mineral glasses* were then subclassified into two main types by Stöffler (1984). The first are typically called 'diaplectic glasses' in the modern literature, but have also been termed 'thetomorphic glasses' historically (Chao, 1967). Such glasses form without melting, by solid-state transformation, and retain the identical composition and morphologic form of precursor mineral (Fig. 9.4c,d; Stöffler, 1984). They are considered sub-solidus shock metamorphic effects and, because no melting occurs, are not addressed here but are described in detail in Chapter 8. In contrast, so-called *normal glasses* display evidence for flow and vesiculation and clearly represent the product of melting. However, this term does not adequately convey the natural variation seen in these glasses. In particular, it is clear that these 'normal glasses' can form by two different mechanisms: namely, either by melting at grain boundaries of minerals and pores (Fig. 9.4e,f), in which case they represent melts of mixed composition; or they can form via the melting of individual specific minerals with low shock impedance and/or low melting points within the rock matrix (Fig. 9.4g), in which case these are monomineralic melts (Keil *et al.*, 1997) (e.g. so-called 'lechatelierite' or 'melted/fused silica'). This process has been termed 'shock-induced localized melting' by some (Keil *et al.*, 1997) or 'shock-induced selective melting' by others (Schaal and Hörz, 1977). Owing to these ambiguities, it is suggested that the nomenclature in Table 9.1 be used.

A common feature of all impact glasses, except diaplectic varieties, are quench crystallites or microlites (Fig. 9.4h,i), with pyroxene, plagioclase and olivine being the most common. These crystallites typically display skeletal, hollow and acicular forms (Fig. 9.4h,i). These are well-understood quench crystal

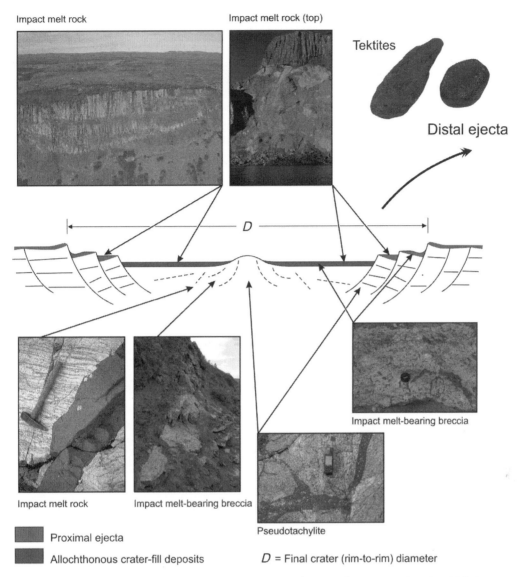

**Figure 9.2**   Schematic diagram of a typical complex impact structure showing the main settings in which impact melt-bearing materials are typically found. (See Colour Plate 23)

morphologies, which indicate rapid crystallization from a melt in response to high degrees of undercooling and supersaturation, and low nucleation densities (Bryan, 1972; Lofgren, 1974; Donaldson, 1976).

### 9.3.2   Effects of target lithology

In terms of the products of impact melting, early studies in North America focused on impact structures developed in dense non-porous crystalline rocks of the Canadian Shield (e.g. the Brent, East and West Clearwater Lake, Manicouagan and Mistastin impact structures; (Grieve, 1975, 1978; Simonds *et al.*, 1978a,b). In contrast, in Europe and the former Soviet Union there was more diversity in the target types, but the bulk of the studies were largely on the description of impact lithologies from structures in mixed targets of sedimentary lithologies overlying crystalline

basement (von Engelhardt *et al.*, 1969; Stöffler, 1977; Masaitis *et al.*, 1980). Sedimentary rocks differ from crystalline rocks in several ways. In particular, they are typically rich in volatiles (mainly $H_2O$, $CO_2$, and $SO_x$), have high porosity and are typically layered; these are the main properties that exert a considerable influence on the results of the impact cratering process (e.g. Kieffer and Simonds, 1980). An early view was that impact melt products were virtually absent at impact structures formed in sedimentary targets, resulting in anomalously low impact melt volumes (Fig. 9.5). Instead, the resultant impactites were historically referred to as clastic, fragmental or sedimentary breccias (Masaitis *et al.*, 1980; Redeker and Stöffler, 1988; Masaitis, 1999).

It was recognized at an early stage that this posed a significant problem, as theoretical calculations suggest that the volume of target material shocked to pressures sufficient for melting is not

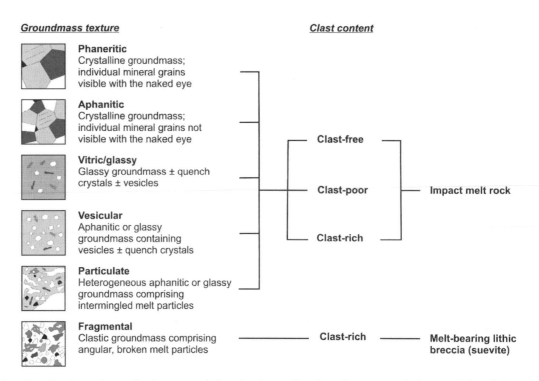

**Figure 9.3** Classification scheme for impact melt-bearing impactites, based on textural characteristic of the groundmass/ matrix, with respect to melt phases, and clast content (Osinski *et al.*, 2008a). Note that clast-rich impact melt rocks have also been termed impact melt breccias in the literature, particularly with respect to lunar impactites; this term is to be avoided following the recommendation of Stöffler and Grieve (2007). In this scheme, the original definition of so-called 'suevite' is used, which is a polymict impact breccia with a clastic matrix/groundmass containing fragments of impact glass and shocked mineral and lithic clasts (Stöffler *et al.*, 1977). Where there is evidence that the groundmass remained molten during and following deposition, but comprises a series of intermingled melts of different composition, the prefix 'particulate' has been suggested. Thus, 'particulate, clast-rich impact melt rock' describes some of the melt-rich crater-fill and ejecta impactites at impact craters developed in heterogeneous mixed sedimentary–crystalline targets, such as the Haughton and Ries impact structures. Fragmental melt-bearing lithic breccias (i.e. impactites conforming to the original definition of suevite) are present at some structures; for example, at Mistastin.

significantly different in sedimentary or crystalline rocks and that both wet and dry sedimentary rocks should yield as much, or even more, volumes of melt on impact than crystalline targets (Kieffer and Simonds, 1980). Faced with this apparent disagreement between observations and theory, Kieffer and Simonds (1980) suggested that this 'anomaly' was due to 'the formation and expansion of enormous quantities of sediment-derived vapour' (e.g. $H_2O$, $CO_2$, $SO_2$), which could result in the 'unusually wide dispersion of shock-melted sedimentary rocks'.

More recent work, focused on field and analytical studies of impactites from complex craters developed in sedimentary targets, suggests that the impact melting *process* is essentially the same in sedimentary and crystalline rocks and that the production of impact melts from sedimentary rocks is the norm, not the exception (see Osinski *et al.* (2008b) for a review; Fig. 9.5). It is also clear that impact melts are generated during marine impacts (see Section 9.3.6). This is in keeping with early theoretical calculations, as well as recent numerical modelling studies (Wünnemann *et al.*, 2008) and improvements in our understanding of phase relations, particularly for carbonates (Ivanov and Deutsch,

2002). What is clear, however, is that the *products* of impact melting in sedimentary rocks do differ substantially in their appearance and characteristics.

### 9.3.3  Impact melt products: crystalline targets

Impacts into entirely dense, non-porous crystalline rocks represent an end-member case, in which no significant amounts of sedimentary rocks are melted. The Canadian Shield hosts several well-known crystalline impact structures, some of which will be discussed in this section. Although the Canadian Shield has been eroded in varying processes through time, most recently by glaciation, much can be studied and learned from preserved impact melt rocks and glasses. Impact melt rocks often display classic igneous structures (e.g. columnar jointing) and textures (Fig. 9.6a). Large coherent bodies of melt cool very quickly at the margins and slower towards the centre of an impact structure due to contact with the crater floor and atmosphere, but also by the incorporation of a large amount of clasts, which can significantly affect the physical properties of the melt (Onorato *et al.*, 1978).

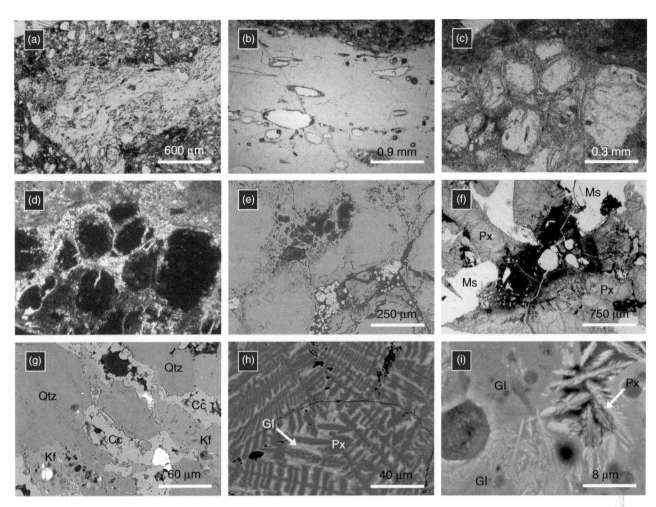

**Figure 9.4** Impact glasses (see Table 9.1 for nomenclature explanations). (a) Whole-rock impact glass clast from impact melt-bearing breccias at the Mistastin impact structure, Labrador, Canada. Note the abundant flow textures and schlieren. (b) Very fresh, colourless, whole-rock impact glass clast from impact melt-bearing breccias at the Ries impact structure, Germany. Plane-(c) and cross-polarized (d) light photomicrographs of diaplectic quartz glass from the Haughton impact structure, Canada. Note the original grain shape of the sandstone quartz grains has been preserved, which is diagnostic for diaplectic glass, but not whole-rock glasses (cf. (a) and (b)). (e) Backscattered electron image of vesiculated interstitial impact melt glass formed around a pore space in sandstone from the Haughton impact structure, Canada. (f) Impact melt pocket (interstitial impact melt glass (Gl)) in a Martian meteorite. Maskelynite (Ms): clear, colourless; pyroxene (Px): brown. There is also a smaller subsidiary melt pocket, branching off the central pocket in the upper right-hand corner. All other opaque phases are oxide assemblages. Image courtesy of Erin Walton. (g) Highly shocked sample in which potassium feldspar (Kf) has melted to form a mineral glass, while quartz (Qtz) is still retained in the structure. Minor calcite (Cc) is also present. Haughton impact structure, Canada. (h) Quench pyroxene crystallites in impact glass from Barringer Crater, Arizona, USA. (i) Quench pyroxene crystallites in impact glass from the Ries impact structure, Germany.

Consequently, the textures formed by silicate impact melts display varying degrees of crystallinity, clast content and vesicularity (Fig. 9.3 and Fig. 9.6).

Impact melt rocks in large complex impact craters can occur as sheets that extend many kilometres laterally and can be several hundred metres thick. The 100 km diameter Manicouagan impact structure (Quebec, Canada) has a preserved impact melt sheet diameter of 55 km and an average thickness of 230 m (Floran *et al.*, 1978; Simonds *et al.*, 1978a). New drill cores and mapping

evidence from Manicouagan has revealed that, although there is a relatively flat surface to the melt sheet, the basal contact of the melt rocks has a significantly variable morphology (from >600 m to >1.4 km thick within a horizontal distance of <1 km; Spray and Thompson, 2008).

Impact melt at the intermediate-sized 28 km Mistastin Lake impact structure in northern Labrador (Canada) occurs in a variety of forms and settings (Grieve, 1975; Marion and Sylvester, 2010). The thickest preserved unit is a massive, dark blue–grey

**Table 9.1**   Summary of the types of glasses formed by hypervelocity impact

| Suggested nomenclature | Stöffler (1984) | Product | Process |
|---|---|---|---|
| Diaplectic glass | Mineral glass | Glass with same composition and retains grain shape | Solid-state transformation |
| Mineral glass | Mineral glass | Glass with same composition as host mineral | Selective melting of individual specific minerals |
| Interstitial impact melt glass | Mineral glass | Glass has a composition that is a mixture of adjacent minerals | Localized melting at grain boundaries of minerals and pores (also termed 'melt pockets' in meteorites) |
| Whole-rock impact glass | Rock glass | Glass has the composition of whole rock(s) | Whole-rock melting |

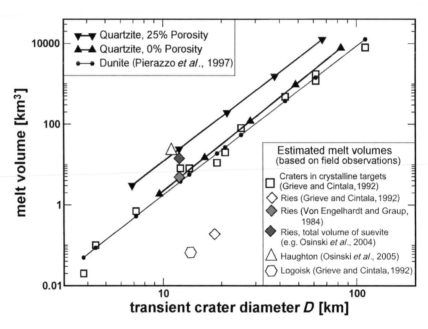

**Figure 9.5**   Comparison between observed melt volumes at various terrestrial craters of different size and hydrocode-generated data for nonporous dunite (Pierazzo *et al.*, 1997) and 0 and 25% porous quartzite (Wünnemann *et al.*, 2008).

melt rock 80 m thick with well-developed columnar jointing (Fig. 9.6a). In general, the melt rocks at Mistastin grade stratigraphically from bottom to top as glassy melt clasts in lithic breccias (Fig. 9.6e,f); as glassy clast-rich impact melt rock; clast-rich glassy to very fine-grained melt rock (Fig. 9.6b); and as clast-poor fine-grained microporphyritic melt (Grieve 1975; Osinski *et al.*, 2008a; Marion and Sylvester, 2010). Vesicular melt rocks can be observed in several of the thinner melt rock units at Mistastin, as well as in basal clast-rich zones. Unfortunately, there are no preserved units of the original upper layers of Mistastin's melt and no drilling has been conducted yet at this site. At the 26 km diameter East Clearwater structure, Quebec, a crater similar in size and structure to Mistastin, core drilling sampled 41 m into the top of the preserved melt sheet. From the base of the core depth, East Clearwater's melt is clast free and fine grained; overlain by a zone of decreasing grain size towards an upper margin; overlain by a 6 m thick vesicular zone; topped by a thin, fine-grained clast-rich zone; and finally capped by a glassy upper unit (Palme *et al.*, 1979). Large coherent melt bodies have not yet been found in any small craters

with crystalline targets, although a 34 m thick, inclusion-rich melt lens has been encountered by drilling in the 3.8 km diameter Brent crater (Grieve, 1978).

Standard crystalline targets form silica-rich melts that can behave quite differently to igneous melts with respect to melting and crystallization (see Section 9.2). These are superheated melts on the order of 1500–2500 °C (Grieve *et al.*, 1977), which are subjected to a vigorous mixing process followed by relatively rapid cooling (Phinney and Simonds, 1977). It is generally accepted that impact melt rocks derived from crystalline targets in intermediate to large impact craters have a relatively homogeneous composition when compared with their original target materials (Dressler and Reimold, 2001). This is the case when comparing bulk rock compositions of major element compositions of the melt with target rocks. For example, Palme *et al.* (1979) describe the main body of melt at the East Clearwater impact structure as a uniform, granodioritic composition derived from three main target rocks (granodiorite, quartz monzonite and granulite). The initial impact melt composition is a mixture

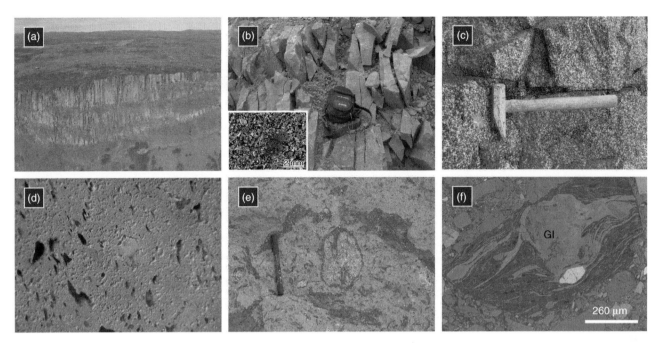

**Figure 9.6** Impact melt-bearing impactites from craters in crystalline targets. (a) Oblique aerial view of the approximately 80 m high cliffs of impact melt rock at the Mistastin impact structure, Labrador, Canada. Photograph courtesy of Derek Wilton. (b) Close-up view of massive fine-grained (aphanitic) impact melt from the Discovery Hill locality, Mistastin impact structure. Camera case for scale. Inset is a cross-polarized light photomicrograph. (c) Coarse-grained granophyre impact melt rock from the Sudbury Igneous Complex, Canada. Rock hammer for scale. (d) Hand sample of vesicular impact melt rock from the Roche-chouart impact structure, France. Image is 8 cm across. (e) Impact melt-bearing breccia from the Mistastin impact structure. Note the fine-grained groundmass and macroscopic flow-textured silicate glass bodies (large black fragments). Steep Creek locality. Marker/pen for scale. (f) Backscattered electron image of suevite showing an angular impact glass clast within a fragmental groundmass. Mistastin impact structure.

of the different target lithologies at the point of impact (see Section 9.2). The melt composition can then be considerably modified by turbulent mixing and incorporation of target rock clasts from underlying breccias as the accelerated melt moves rapidly and turbulently outwards along the growing crater floor from the point of impact. High-temperature impact melt, particularly large volumes of melt early in its formation, has the capacity to incorporate and wholly melt more target material many kilometres away from the original point of impact (Grieve, 1975; Zieg and Marsh, 2005; Marion and Sylvester, 2010). Elements from the melted and vaporized projectile can also contribute to the composition of the impact melt (see Chapter 15 and Section 9.5).

The term homogeneous may be subjective when applied to crystalline melt rocks, which are so often compared with their heterogeneous mixed and sedimentary cousins. A study by Marion and Sylvester (2010) shows that a uniform composition for crystalline melt rocks may not be the norm. At the Mistastin impact structure, modelling of target rock proportions from crystalline-derived impact melt rocks using bulk rock versus matrix analysis from a suite of widely distributed melt rock samples gave notably different results. The initial impact melt composition determined with matrix analyses of a thin melt unit is approximately 73% anorthosite, 20% granodiorite and 7% mangerite, whereas the bulk rock analyses, including dominantly

unmelted to partially melted mangerite clasts, suggest that mangerite was the secondary contributor to the melt, particularly in thick melt units (Fig. 9.7). The Mistastin impact melt rocks are, in fact, heterogeneous: large, metre-scale variations are primarily generated after the entrainment of target rock clasts and as a function of cooling rates, whereas the micrometre-scale variations are due to variations in texture and the homogenizing feature of previous bulk rock analyses. This could be the case in other craters formed in crystalline rocks, where clasts and partial melts incorporated into melt are volumetrically significant and/or differ substantially in composition within the target rocks.

It has been shown that impact melts are largely the result of direct mixing of their target rocks and that, due to the relatively rapid surface cooling of impact melt, fractionation rarely occurs. However, there are two known compositionally layered impact melt rock units in impact structures on Earth: the Sudbury Igneous Complex (SIC) and the Manicouagan melt sheet (both in Canada). As mentioned above, the Manicouagan melt sheet is an average of approximately 230 m thick, but the newly discovered greater than 1100 m thick unit of clast-free melt from a drill core in the central region of the melt sheet reveals a compositional segregation. It appears that melt was thick enough to allow fractionation into an approximately 450 m lower monzodiorite zone, an approximately 180 m quartz monzodiorite transition zone and an approximately 450 m upper quartz monzonite zone (Spray and

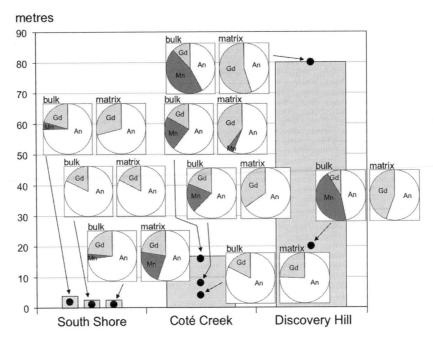

**Figure 9.7**   Modelled proportions of target rocks in bulk and matrix compositions of impact melt rock samples from Mistastin Lake impact structure as a function of melt unit thickness and stratigraphic position. An: anorthosite; Mn: mangerite; Gd: granodiorite. Reprinted from Marion and Sylvester, 2010, with permission from Elsevier.

Thompson, 2008). The bulk of the thinner impact melt at Mani-couagan is similar to the composition of the transition zone. The originally approximately 250 km diameter, 1.85 Ga Sudbury impact structure has a varied stratigraphy and very complicated melt products. The SIC in part, or in its entirety, has been interpreted as impact melt (Grieve *et al.*, 1991). The SIC has been differentiated into a number of coarse-grained phases and units (Fig. 9.6c); from bottom to top it consists of a quartz dioritic/noritic sublayer, basal norite unit, a central quartz gabbro unit (transition zone) and an upper granophyre unit which grades into a more plagioclase-rich granophyre (Therriault *et al.*, 2002).

### 9.3.4   Impact melt products: sedimentary targets

Sedimentary rocks are present in the target sequence of approximately 70% of the world's known impact structures (Osinski *et al.*, 2008b). A large percentage of these craters can be considered 'mixed' targets, comprising varying thickness of sedimentary rocks overlying crystalline basement rocks. These are discussed in Section 9.3.5. When considering impacts into sedimentary rocks, substantial complexities are introduced due to the large number of different types of sedimentary rocks and the dramatic differences in composition between them (e.g. limestone versus sandstone versus gypsum). For a complete synthesis of the processes and products of impact melting in sedimentary targets, the reader is referred to Osinski *et al.* (2008b).

The impact melting of terrigenous clastic sedimentary rocks, dominated by SiO$_2$-rich sandstones but also including conglomerates and various mudrocks (siltstone, claystone, shale), is, perhaps, the best understood. Impact melted sandstones were first

studied in detail at Meteor Crater, Arizona (Kieffer, 1971; Kieffer *et al.*, 1976; Table 9.2). These studies revealed the dramatic effects of porosity, grain characteristics and volatiles on the response of rocks (in this case sandstone) to impact in sedimentary targets. For example, in crystalline rocks, quartz will typically not melt at pressures less than 50–60 GPa (Grieve *et al.*, 1996 and references therein), but whole-rock melting occurred at Meteor Crater at greater than 30–35 GPa (Kieffer, 1971; Kieffer *et al.*, 1976). Impact-melted sandstones have been recognized at a number of terrestrial impact structures in a variety of different stratigraphic settings (Fig. 9.8a; Table 9.2) and impact-melted shales have also been recognized from the Haughton structure (Redeker and Stöffler 1988; Osinski *et al.*, 2005). At the majority of these sites, these impact-melted sandstones exist as isolated clasts within various impactite types. At the Gosses Bluff (Milton *et al.*, 1996) and Goat Paddock (Milton and Macdonald, 2005) structures (both in Australia), however, impact breccias with a groundmass of predominantly silica glass are present. These should, therefore, be classified as clast-rich impact melt rocks (see Chapter 7), and they indicate that considerable amounts of sandstone-derived impact melts can be generated.

Carbonates are present in approximately one-third of the world's impact structures. Historically, it was suggested that carbonates decompose during impact to produce CO$_2$ and oxides (CaO from limestone and MgO from dolomite; Boslough *et al.*, 1982; Agrinier *et al.*, 2001). The phase relations of CaCO$_3$, however, suggest that melting is the dominant response, with decomposition only occurring at very high temperatures, following pressure release and post-shock cooling (Ivanov and Deutsch, 2002). This is consistent with observations, where the only

**Table 9.2** Terrestrial impact structures that preserve evidence for shock-melted sandstones (updated from Osinski *et al.* (2008b))

| Impact site | Impactite setting[a] | | | Reference |
|---|---|---|---|---|
| | Crater-fill | Proximal ejecta | Distal ejecta | |
| Aouelloul | – | CL | – | Koeberl *et al.* (1998) |
| Goat Paddock | CL, GR | – | – | Milton and Macdonald (2005) |
| Gosses Bluff | CL, GR | – | – | Milton *et al.* (1996) |
| Haughton | CL, GR | CL, GR | – | Metzler *et al.* (1988), Redeker and Stöffler (1988), Osinski *et al.* (2005) |
| Meteor Crater | – | CL | – | Kieffer (1971), Kieffer *et al.* (1976) |
| Ries | – | CL, GR | CL | Von Engelhardt *et al.* (1987), Osinski (2003), Osinski *et al.* (2004) |
| Wabar | – | CL | – | Hörz *et al.* (1989) |

[a]CL: occurrence as clasts; GR: occurrence in groundmass (i.e. emplaced in a molten state).

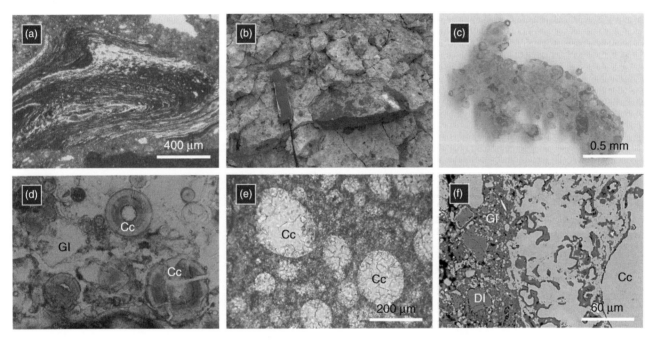

**Figure 9.8** Impact melt-bearing impactites from craters in sedimentary targets. (a) Shock-melted sandstone (whole-rock glass) from the Haughton impact structure, Canada. (b) Field photograph of the carbonate melt-bearing crater-fill clast-rich impact melt rocks from the Haughton impact structure. (c) Impact melt clast from Meteor Crater, Arizona, USA. (d) Plane-polarized light photomicrograph of the glass shown in (c) showing calcite melt spherules. (e) Plane-polarized light photomicrograph of calcite melt spherules. Note that these features are not vesicle fillings, as discussed by Osinski and Spray (2001) and Osinski *et al.* (2005). Haughton impact structure. (f) Backscattered electron image showing shocked dolomite clasts (pale grey) within a predominantly microcrystalline calcite melt groundmass (white). (f) Intermingling of groundmass-forming calcite melt (white) and silicate glass (dark grey). Dolomite clasts (pale grey) are also present. Haughton impact structure.

evidence for the decomposition of carbonates comes from rocks that were juxtaposed with high-temperature silicate impact melt (Spray, 2006; Deutsch and Langenhorst, 2007; Rosa and Martin, 2010). In other words, this decomposition resulted from post-impact thermal metamorphism and not from the impact cratering process *sensu stricto* (Osinski *et al.*, 2008b). Carbonate impact melts have been recognized in a variety of products and settings (Fig. 9.8b,c; Table 9.3). Textural and chemical evidence for the impact melting of carbonates during impact (Osinski *et al.*,

2008b) is provided by: (1) carbonate spherules (Fig. 9.8d,e); (2) liquid immiscible textures (Fig. 9.8f); (3) quench textures; (4) euhedral calcite crystals within impact glass clasts; (5) carbonates intergrown with CaO–MgO-rich silicates; (6) CaO–MgO–$CO_2$-rich glasses; and (7) unusual carbonate chemistry.

In terms of the geochemistry of melt products from craters developed in sedimentary targets, a critical factor is that, in general, impact-generated melts from target rocks at different stratigraphic levels will have widely variable compositions. (Rare

**Table 9.3**  Impact sites where evidence for the melting of carbonates has been reported (updated from Osinski *et al.* (2008b))

| Impact site | Impactite | Setting of carbonates | Textures and geochemistry[a] | | | | | | | References |
|---|---|---|---|---|---|---|---|---|---|---|
| | | | Liq Imm | Quen | Sph | Euh Carb | Ca-Mg-rich silicates | Ca-Mg-CO$_2$-rich glass | Carb Chem | |
| Chicxulub | Proximal ejecta | Individual particles | N | Y | N | N | N | N | n.a. | Jones et al. (2000), Dressler et al. (2004), |
| | Distal ejecta | Individual particles | N | N | Y | N | N | N | n.a. | Kring et al. (2004), Tuchscherer et al. |
| | Proximal ejecta | Embedded in glass clasts | Y | N | N | N | N | N | n.a. | (2004), Nelson and Newsom (2006) |
| | Proximal ejecta | Within groundmass | T | N | N | ? | N | Y | n.a. | |
| | Proximal ejecta | Within groundmass | N | N | N | ? | N | Y | n.a. | |
| Haughton | Crater-fill | Groundmass | Y | N | Y | Y | N | Y | Y | Osinski and Spray (2001), Osinski et al. |
| | Crater-fill | Embedded in glass clasts | Y | N | Y | Y | N | Y | Y | (2005), Osinski (2007) |
| | Proximal ejecta | Groundmass | Y | N | N | N | N | N | N | |
| Lockne | Crater-fill | Embedded in glass clasts | Y | N | Y | N | N | N | Y | Sjöqvist et al. (2012) |
| Meteor Crater | Proximal ejecta | Embedded in glass clasts | N | N | Y | Y | Y | Y | Y | Osinski et al. (2003) |
| Ries | Proximal ejecta | Embedded in glass clasts | Y | N | Y | N | N | N | n.a. | Graup (1999), Osinski (2003), Osinski et al. |
| | Proximal ejecta | Individual particles | N | Y | Y | N | N | N | n.a. | (2004) |
| | Proximal ejecta | Groundmass | Y | N | N | N | N | Y | Y | |
| | Crater-fill | Glass clasts | Y | N | N | N | N | N | Y | |
| Steinheim | Crater-fill | Embedded in glass clasts | Y | Y | Y | N | N | N | Y | Anders et al. (2011) |
| | | Groundmass | Y | Y | Y | N | N | N | Y | |
| Tenoumer | Proximal ejecta | Glass clasts | Y | N | N | N | N | N | n.a. | Pratesi et al. (2005) |

[a]Liq Imm: liquid immiscible textures; Sph: spherule; Quen: quench-textured feathery carbonate; Carb Chem: unusual carbonate chemistry; Y: yes, N: no; n.a.: no analyses presented.

exceptions may occur, although none is currently known, where homogeneous bodies of melt may be formed from impacts into homogeneous sedimentary targets, such as pure limestone.) In these instances, it is highly unlikely that these impact melt packages will ever completely mix or homogenize due to their different physical and chemical properties. Instead, they will remain largely immiscible and retain roughly the same chemistry as the protolith. It is important to note that liquid immiscibility, *sensu stricto*, describes the process whereby an initially homogeneous melt reaches a temperature at which it can no longer exist stably and so it unmixes into two liquids of different composition and density (Roedder, 1978). Evidence for immiscible textures between impact melts of differing composition is common (e.g. Fig. 9.7f); however, these melts likely never homogenized and this is not liquid immiscibility *sensu stricto*. Thus, the term 'carbonate–silicate liquid immiscibility' should be avoided, unless there is unequivocal evidence for the unmixing of an originally homogeneous impact melt (Osinski *et al.*, 2008b). In summary, in sedimentary targets, the difference in composition and temperature of melts derived from different lithologies will result in unmixed and heterogeneous melts in the form of particulate clast-rich impact melt rocks (Fig. 9.3; Osinski *et al.*, 2008a), even though the fundamental formational processes may be essentially the same.

### 9.3.5  Impact melt products: mixed targets

A wide range of impact-generated lithologies have been described for impact craters formed in so-called 'mixed targets' (i.e. crystalline basement overlain by sedimentary rocks). Coherent 'sheets' of silicate impact melt rocks completely lining the crater interior are not observed in these craters, in contrast to craters developed in purely crystalline rocks (see Section 9.3.3). Instead, the dominant product to be described in the literature comprises impact melt-bearing breccias, historically referred to as 'suevites' (Stöffler, 1977; Masaitis, 1999; Fig. 9.9a–c). As noted in Chapter 7, substantial ambiguity surrounds the use of the term 'suevite' and its use should be restricted to the impactites at the Ries impact structure, Germany, where they were first described. In addition, to these impact melt-bearing breccias, minor bodies of coherent impact melt rocks are also sometimes observed, often as lenses and irregular bodies within larger bodies of suevite (Fig. 9.2 and Fig. 9.8d,e). These may reach kilometre-scale proportions in large craters such as Popigai, Russia (Fig. 9.2; Masaitis, 1999).

One prominent feature of the impact melt-bearing breccias from craters formed in mixed targets is the centimetre-sized glassy (when fresh) 'particles' of whole-rock impact melt glass (Fig. 9.9a–c). The geochemistry shows that these glasses are predominantly derived from the crystalline portion of the target

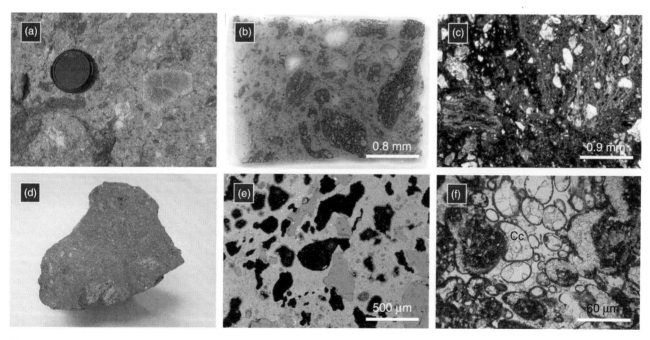

**Figure 9.9**  Impact melt-bearing impactites from craters in mixed sedimentary–crystalline targets. (a) Field photograph of impact melt-bearing breccias from the Ries impact structure, Germany. Note the fine-grained groundmass and macroscopic irregular silicate glass bodies (dark grey to black). Lens cap for scale. (b) Scanned thin section of impact melt-bearing (black) breccias from the Popigai impact structure, Russia. (c) Plane-polarized light photomicrograph of impact melt-bearing breccias from the Ries impact structure. (d) Vesicular, red, silicate impact melt rock from the Polsingen locality, Ries impact structure. (e) Backscattered electron image of (d) showing a groundmass of plagioclase quench crystals (bright), quartz clasts (dark grey) and vesicles (black). (f) Plane-polarized light photomicrograph of groundmass/matrix of surficial suevite showing globules of calcite within a silicate glass–calcite groundmass. Zipplingen locality, Ries impact structure.

sequence (von Engelhardt, 1972; Stähle, 1972). As with impacts into purely sedimentary targets, it was historically believed that impact melts derived from the sedimentary portions of these target sequences were lacking. More recent work conducted with high-resolution microbeam techniques has shown that impact melts derived from the sedimentary proportions of these target sequences are also common (Jones *et al.*, 2000; Osinski *et al.*, 2004; Pratesi *et al.*, 2005). These melt products are, however, typically fine grained and intermixed in the groundmass or within larger bodies of silicate glass (Fig. 9.9f).

### 9.3.6   Impact melt products: marine impacts

Approximately 20% of the world's known impact craters occurred in a marine environment (see review by Dypvik and Jansa (2003)). Many remain in the marine environment and, therefore, are challenging to study. As with subaerial impacts into sedimentary rocks, it was initially thought that impact melt products were rare or lacking from marine impacts. However, Dypvik and Jansa (2003) note that this is not the case and that impact melt rocks and glasses are produced during marine impacts and have been documented at a number of sites. For example, at the 45 km diameter Montagnais structure (offshore Nova Scotia, Canada), which formed in a mixed sedimentary–crystalline target, the crater fill comprises a series of impactites including crystalline impact melt rocks and impact melt-bearing breccias (cf. impacts into similar targets in subaerial environments; Section 9.3.5). Recent work has documented the presence of impact melt-bearing breccias, including mixed silicate–carbonate melt particles, at the smaller 13.5 km Lockne structure, Sweden (Sjöqvist *et al.*, 2012), suggesting that substantial volumes of

impact melt-bearing impactites are indeed generated during marine impact events.

### 9.3.7   Impact melt products in distal ejecta

Two main types of impact melt-bearing impactites occur in distal ejecta deposits: tektites and spherules (Fig. 9.2). Tektites have been defined as 'an impact glass formed at terrestrial impact craters from melt ejected ballistically and deposited sometimes as aerodynamically shaped bodies in a strewn field outside the continuous ejecta blanket' (Stöffler and Grieve, 2007). Tektites are typically black, but also occur as green, brown or grey glassy particles that range in size from the submillimetre to decimetre, though microscopic tektites or 'microtektites' have been found in deep-sea sediments. Tektite glasses are found scattered around the globe in four large localized strewn fields (Fig. 9.10): the Australasian, the Ivory Coast, the central European (Czechoslovakian/Moldavian) and the North American strewn fields (e.g. Glass, 1990 and references therein). Tektites within each strewn field are related to each other through petrological, geochemical and geochronological analysis (Table 9.4 and Table 9.5). The chemistry of tektites generally does not perfectly match the target rocks; and as with larger bodies of impact melt, they are the product of a mixed target rock melt. The $^{10}$Be concentrations in younger tektites, an excellent indicator of depth in sediment stratigraphy, have also shown that the terrestrial target source region must have been in the upper few hundred metres (Koeberl, 1994). Tektites show high chemical resistance, long-term stability against devitrification and high $Fe^{2+}/Fe^{3+}$ ratios (Heide *et al.*, 2001). In addition, tektites are volatile poor and contain less than 0.02% (200 ppm) $H_2O$; the exception is the case of the Muong Nong

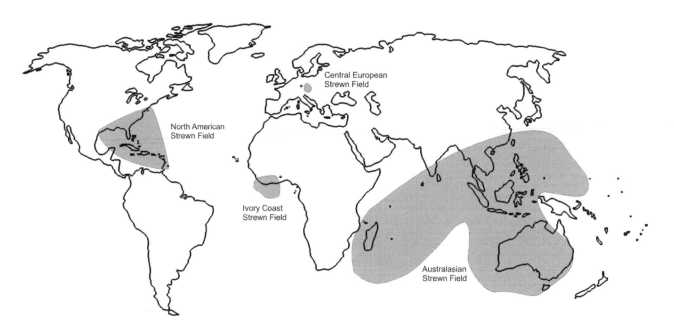

**Figure 9.10**   Global tektite distribution (grey polygons) and their source craters (black dots adjacent strewn fields): North American strewn field – Chesapeake Bay impact structure; Ivory Coast strewn field – Bosumtwi impact structure; Australasian strewn field – no known source crater; and Central European strewn field – Ries and Steinheim impact structures.

**Table 9.4**   Summary of quantitative information on tektites[a]

|  | North American | Central European | Ivory Coast | Australasian |
|---|---|---|---|---|
| Age (Ma) | 35.4 | 15 | 1.09 | 0.77 |
| Area ($10^6$ km$^2$) | 10 | 0.3 | 4 | 50 |
| Total mass ($10^6$ t) | 300–42,000 | >5 | 20 | >2,000 |
| Age (Ga) of source terrain (from Sm–Nd) | 0.7 | 0.9 | 1.9 | 1.11 |
| Sedimentation age (Ma) (from Rb–Sr) | 400 | 15 | 950 | 175 |
| Source crater (location) | Chesapeake Bay (USA) | Ries and Steinheim (Germany) | Bosumtwi (Ghana) | Unknown |
| Source crater diameter (km) | 90 | 24 | 10.5 | 50–100 |

[a]Table modified from Koeberl (1994) and data sources from references therein.

**Table 9.5**   Average range in composition of tektites[a]

| | Composition range (wt%) | | | |
|---|---|---|---|---|
| | North American | Central European | Ivory Coast | Australasian |
| $SiO_2$ | 76.6–81.1 | 78.6–82.6 | 66.17–68–48 | 70.4–77.2 |
| $TiO_2$ | 0.5–0.8 | 0.3–0.4 | 0.54–0.61 | 0.7–0.8 |
| $Al_2O_3$ | 10.9–13.7 | 8.2–11.5 | 16.28–17.72 | 10.8–13.4 |
| FeOTotal | 2.6–4.0 | 1.2–2.2 | 5.84–6.45 | 4.0–6.7 |
| MgO | 0.6–0.7 | 1.4–2.3 | 2.98–4.39 | 1.6–4.2 |
| CaO | 0.5–0.6 | 1.3–3.0 | 1.21–1.52 | 1.5–3.9 |
| $Na_2O$ | 1.1–1.5 | 0.2–0.6 | 1.53–2.08 | 1.2–1.6 |
| $K_2O$ | 2.1–2.4 | 2.5–3.6 | 1.73–2.13 | 1.9–2.4 |
| $H_2O$ | 0.005–0.030 | 0.006–0.017 | 0.002–0.003 | 0.004–0.030 |

[a]Table modified from Heide *et al.* (2001) and references therein; Ivory Coast tektites data are from Koeberl *et al.* (1997) microprobe analysis.

tektites, which have up to 0.03 wt% $H_2O$ (Beran and Koeberl, 1997). Tektites may also contain lechatelierite inclusions, spherules and clasts of other mineral grains, some of which display sub-solidus shock metamorphic features (Glass, 1990).

The discovery of microtektites has helped define and expand the limits of the strewn fields. Microtektites are typically well preserved in the deep-sea sediments until exposed at the surface to strong weathering agents (Glass, 1990; Stöffler and Grieve, 2007). Tektites are typically divided into three groups, based on their exterior morphology: (a) splash form; (b) aerodynamically shaped; and (c) Muong Nong. Splash forms are the most common and are in the shape of teardrops, spheres, dumbbells or bars, etc. and are the result of solidification of rotating liquids in the air or a vacuum. The aerodynamically shaped tektites are essentially splash-form tektites that show evidence of atmospheric ablation in the form of pits, grooves, notches and particularly flanged button shapes, as a result of partial melting followed by quenching as they were ejected outside Earth's atmosphere (Koeberl, 1992). Muong Nong-type tektites are found primarily in Asia and are typically larger and have a blocky shape and layered vesicular texture (Glass, 1990; Koeberl, 1992).

The origin of tektites has been the source of some controversy. A Moon origin was considered early on, but was later rejected on the basis of composition and the fact that no lunar craters correlate in size and age with any tektites (Taylor, 1973; Baldwin,

1981). Numerical modelling of the formation and distribution of tektites suggests that they are produced by high-velocity (>15 km s$^{-1}$) impacts into silica-rich targets with projectile impact angles between 30° and 50° (Artemieva, 2002; Stöffler *et al.*, 2002). Tektites are high-temperature impact melt and vaporized gaseous material originating from the uppermost target surface, which is ejected at high velocity very early in the crater formation process (see Chapters 3 and 4). While tektites are formed primarily of the quenched impact melt ejecta, microtektites may also involve condensation of vaporized target material in the upper atmosphere (Stöffler *et al.*, 2002). Less-oblique impacts eject material at steeper trajectories, whereas very low angle impacts are not as efficient at producing tektites, and would consist of some projectile contamination (Artemieva, 2008).

In addition to tektites and microtektites, occurrences of glassy to microcrystalline spherules are being increasingly recognized in the geological record (e.g. Simonson and Glass, 2004). The spherules are typically small, millimetre-sized, and typically occur in thin, discrete layers, although layers up to several metres thick have been documented. Known examples range from less than 1 Ma to greater than 3.5 Ga in age (Simonson and Glass, 2004) and they represent the only record of meteorite impacts greater than 2.5 Ga. Recent modelling work suggests that spherules can be formed in two distinct scenarios. Johnson and Melosh (2012) propose that the global ejecta layers are formed from condensation of rock vaporized by the impact; these spherules contain a distinct chemical and isotopic signature of the projectile. Other 'spherules' are less widely distributed from the point of impact, lack this projectile signature and represent less-shocked material in the form of impact melt, ejected by the impact. Johnson and Melosh (2012) propose the term 'melt droplets' for such spherules to distinguish them from spherules condensed from vapour.

### 9.3.8    Impact melting during airburst events

In recent years, several occurrences of natural glass that are either confirmed or suspected as being of impact origin have been documented but for which no source crater has been recognized. Some of these glass occurrences are well known and widely accepted as being of impact origin; for example, the Libyan Desert Glass (Weeks *et al.*, 1984), Darwin Glass (Meisel *et al.*, 1990), South Ural Glass (Deutsch *et al.*, 1997) and Dakhleh Glass (Osinski *et al.*, 2007; Osinski *et al.*, 2008c); other occurrences remain more enigmatic (Haines *et al.*, 2001; Schultz *et al.*, 2006). One obvious explanation is that these glasses represent tektites (see Section 9.3.5). Some of these glasses have indeed been interpreted as tektites (Weeks *et al.*, 1984; Deutsch *et al.*, 1997; Glass and Koeberl, 2006), but others do not display the characteristics of tektites and instead resemble impact glasses typically found in the proximal ejecta deposits of impact structures (e.g. the Dakhleh Glass of Egypt and the Darwin and Edoewie Glasses of Australia).

An alternative mechanism for the formation of these glasses is heating due to large airburst events. Wasson (2003) proposed that the Libyan Desert glass and the Muong Nong tektites of southeast Asia, despite their name, may have formed from such events, but it was unclear whether sufficient heating occurs during such events. Recent numerical modelling suggests that substantial amounts of glass can indeed be formed by radiative/convective heating of the surface during greater than 100 Mt low-altitude airbursts (Boslough and Crawford, 2008). This modelling suggests that, during such events, a high-temperature jet descends towards the surface and transfers its kinetic and internal energy to the atmosphere. Above a certain size threshold, this jet will make contact with the Earth's surface and expand radially outwards in the form of a fireball with temperatures exceeding the melting temperatures of most rock-forming minerals (Boslough and Crawford, 2008). The model predicts the following: (1) the formation of glass derived from surface materials; (2) distribution of glass over a large area; (3) the lack of a hypervelocity impact crater; and (4) evidence for ponding and collection of melt. The Dakhleh Glass (Osinski *et al.*, 2007, 2008c) and Muong Nong tektites (Wasson, 2003) display these characteristics and the best explanation for their origin is the melting of surface materials during a large airburst event. Given the recent modelling results of Boslough and Crawford (2008), further work on these glass occurrences is warranted.

### 9.3.9    Pseudotachylite

One final type of impact-generated melt-bearing impactite to be considered is the so-called 'pseudotachylite' (Fig. 9.2). Shand (1916) first introduced the term 'pseudotachylyte' and was the first to identify and describe it in the Vredefort impact structure, South Africa. At that time, the impact origin of the Vredefort structure was unknown. However, it was clear from the start that 'pseudotachylytes' were of non-volcanic origin (Shand, 1916).

Pseudotachylites most typically occur as geometrically planar to sub-planar features dominated by a dark, flinty-looking matrix (although they can also form massive irregular bodies). A main generation zone with irregular injection veins on either side is their typical morphology. The wall rock hosting these injection veins is also known as the reservoir zone. Both the injection veins and the generation zone may usually exhibit sharp boundaries with the hosting wall rock, and they can appear to be layered along the contact rim, due to either selective melting of the wall rock or the presence of chilled margins (Lin, 1998). Pseudotachylites can be fragment rich or fragment poor, and the fragments can vary greatly in size, ranging from sub-centimetre to several metres; and in shape, from well rounded to sub-rounded. The general assumption is that pseudotachylites form by melting of the immediately adjacent wall rock. However, geochemical studies of pseudotachylite matrix (Riller *et al.*, 2010) show that there is not always agreement between compositions.

Two main pseudotachylite types, of distinct morphology and spatial occurrence, can be observed in impact structures. In and around the area of the central uplift, anastomosing vein systems (typically <2 mm wide) of pseudotachylite pervasively occur within the basement rock. Impact pseudotachylites are also observed as massive (hundreds of metres) fragment-rich bodies, related to the impact-generated faults or to major lithological contacts (Martini, 1991; Reimold, 1995; Spray *et al.*, 1998).

Pseudotachylite formation generally requires two basic processes to operate: fragmentation (cataclasis) and melt production. Potential mechanisms responsible are extensive comminution, shock melting, decompression melting and friction melting. Extensive comminution is a prerequisite for friction melting, so these two mechanisms are directly related (Wenk, 1978). Pseudotachylitic vein networks (referred also as A-type (Martini, 1991), or S-type (Spray *et al.*, 1998)) have been interpreted as the product of the interaction between the shock wave and target rocks. The energy released during decompression (immediately after the passage of the shock wave front) may match or even exceed the vaporization temperature of the rock (Martini, 1991; Fiske *et al.*, 1995; Spray *et al.*, 1998). While it is generally agreed that the major formation mechanism behind the large, fault-related, impact pseudotachylites is friction melting, due to the rapid slip displacement of the faults, there is still uncertainty as to the way this process operates. There are several studies of experimentally produced pseudotachylite through rapid slip friction melt (Tsutsumi and Shimamoto, 1997). The limitations of the physical model suggested for this genetic mechanism is well illustrated by Melosh (2005). The physical model can only predict the formation of sub-millimetre veins. It is experimentally observed and physically explained that the amount of melt production is directly related to the amount of displacement of the fault. In natural systems, however, this relationship fails to explain the massive amount of melt in the pseudotachylite bodies. The appearance of melt during rapid slip displacement is expected to lubricate the fault and result in the sudden decrease of the coefficient of friction and the subsequent shut-off of the friction melting process (Melosh, 2005). Thus, friction melting, on its own, is a self-limiting process that cannot explain the massive bodies of pseudotachylite observed in tectonic settings. In order to prolong the effect of friction melting, the melt produced needs to successively be injected elsewhere. Further complications, which highlight our incomplete understanding of the generation of pseudotachylites in impact craters, comes from recent studies based on the different compositions between wall rock and the pseudotachylite matrix and additional structural evidence from the field. The suggestion is that some large-body pseudotachylites may be impact melt derived from the melt sheet that has been injected into tension fractures in crater floors, in the case of large impact structures (Riller *et al.*, 2010). The debate continues with even more recent work suggesting that pseudotachylites in the central uplift of the Vredefort impact structure can be formed *in situ* during shock compression and/or decompression (Mohr-Westheide and Reimold, 2011).

## 9.4　Planetary impact melt products

### 9.4.1　Effect(s) of gravity

An observation made relatively early in the study of impact melt rocks from impacts into crystalline targets was that the volume of impact melt rocks relative to the estimated dimensions of the transient cavity appeared to increase in larger terrestrial impact structures compared with smaller ones (Grieve *et al.*, 1977). It is now recognized that this is a result of the variation of the effects of planetary gravity on impacts of different size (Melosh, 1989). The original observation has since been supplemented with considerably more terrestrial data and the effect is sometimes referred to 'differential scaling' (Grieve and Cintala, 1992). Differential scaling also applies to impacts structures on different planetary bodies, such that impact structures on the Moon have less impact melt rocks than equivalent-sized terrestrial impact structures (Cintala and Grieve, 1994). There are also differences in the amount of impact melt produced due to variations in the composition of the impactor and target and the impact velocity and angle; but, all things being relatively equal, planetary gravity and event size are the main variables in relative melt volume differences.

Gravity acts against the ejection of materials in the formation of a crater. Thus, impact cratering efficiency (that is, the ability to form a transient cavity for a given set of impact conditions) is relatively reduced on a high-gravity planet compared with a low-gravity planet. Gravity, however, has no appreciable effect on the amount of impact melt generated in a given impact on a planet, beyond how it affects the final impact velocity. As a result of the differential effect of the gravity variable in planetary impact processes, relatively more impact melt is produced for a given transient cavity size on higher gravity planets, such as Earth (Fig. 9.5). In addition, as gravity is an acceleration term and, thus, has a time component, the larger the impact event is the longer time gravity also has to act to retard the ejection process and reduce relative cratering efficiency (Melosh, 1989). As a result, larger impacts on a given planet result in the production of relatively more impact

melt in relation to transient cavity size, other variables being equal (Fig. 9.5).

This phenomenon of differential scaling of the effects of gravity also affects the potential amount of clastic debris available to be incorporated in the impact melt as it is driven down into the expanding transient cavity (Melosh, 1989). As such, impact melt rocks on higher gravity planets and/or impact melt rocks produced by larger impacts on a given planet contain relatively less lithic and mineral clasts. It is the initial thermal equilibration between superheated impact melt and any (cold) clastic debris that rapidly lowers the temperature of the superheated melt (Onorato *et al.*, 1978); therefore, impact melts and their entrained clastic debris from impacts on higher gravity planets and/or from larger impacts are hotter and have lower viscosities than those from low gravity planets and/or smaller impacts. Amongst the terrestrial planets, impact melting relative to crater size is most efficient on Venus, as it not only has a relatively high planetary gravity, but also the surface target rocks are already at elevated temperatures (Cintala and Grieve, 1994). This, and the resultant higher temperature and lower viscosity of the impact melts, accounts for the long run-out flows of impact melt that are observed outside some large impact structures on Venus (Asimow and Wood, 1992).

### 9.4.2　Spacecraft data of interpreted impact melts on the terrestrial planets

The increased resolution of cameras and multispectral instruments onboard recent missions to the Moon, Mercury and Mars have provided important new constraints on the characteristics and properties of impact melt deposits. The Moon, in particular, preserves a unique record of the character and distribution of impact melt deposits in and around fresh simple and complex impact craters. Using Lunar Orbiter and Apollo images, several workers recognized the presence of generally smooth and low-albedo 'lava-like' deposits within crater interiors, in the faulted rim (terrace) region and overlying ballistic ejecta deposits (e.g. Howard, 1974; Howard and Wilshire, 1975; Hawke and Head, 1977) (Fig. 9.11a,b). These deposits were interpreted as being impact melts, based on their spatial distribution with respect to individual impact craters, the morphology of the deposits (ponds, leveed channels, flow features, cooling cracks), time of emplacement, and lack of volcanic sources. This interpretation has been further strengthened by images returned by the Lunar Reconnaissance Orbiter Camera (LROC), which show impact-melt forming ponds and veneers within the crater interior, the terraced crater rim region, overlying ballistic ejecta deposits, and draping central uplifts within and around many simple and complex lunar craters (e.g. Osinski *et al.*, 2011; Fig. 9.11c–e). These melt deposits show intricate surface textures and morphologies, such as channels and arcuate cracks and ridges, indicative of flow (e.g. Fig. 9.11c). The recognition of impact melt deposits overlying ballistic ejecta is particularly noteworthy and has implications for the origin and emplacement of impact ejecta deposits (see discussion in Chapter 4).

High-resolution imagery is not currently available for Mercury; however, images recently returned by the *MESSENGER* spacecraft show the presence of what is interpreted as melt ponds around

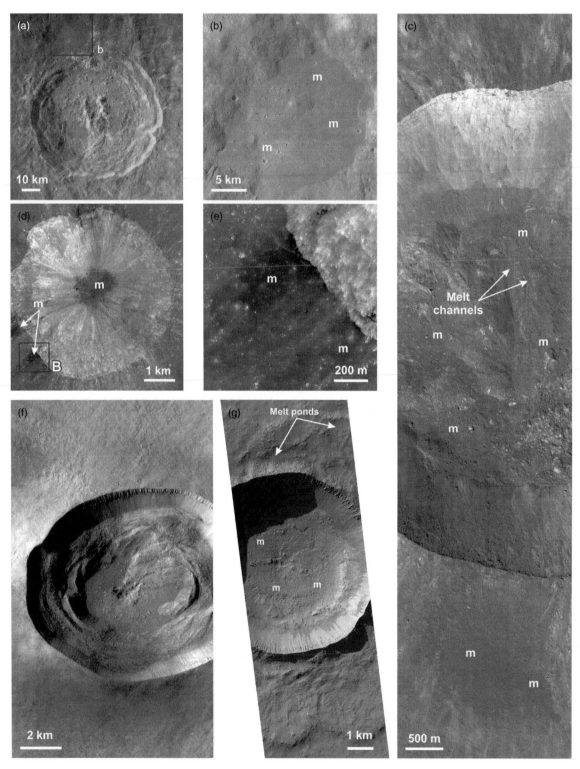

**Figure 9.11**    Observations of lunar and Martian impact craters. (a) and (b) *Apollo 16* image of the 76 km diameter King crater showing a large impact melt ('m') pond. Portion of *Apollo 16* image 1580 (NASA). (c) A portion of LROC NAC image pair (M106209806RE) of melt within the interior and overlying the continuous ejecta blanket at the 22 km diameter Giordano Bruno crater (NASA/GSFC/ASU). (d) and (e) Blocky, ballistic ejecta deposits overlain by dark, patchy impact melt ('m') around a 5 km diameter simple crater. Portions of LROC NAC image M125733619. Images of Martian impact craters (NASA). (f) Image of the 11 km diameter Pangboche crater, situated near the summit of Olympus Mons. Portion of CTX image P02_001643_1974_ XN_17N133W (NASA/JPL/MSSS). (g) HiRISE image PSP_008135_1520 (NASA/JPL/UA) showing a fresh unnamed 7 km diameter impact crater in Hesperia Planum (108.9°E, 27.9°S), Mars. This crater displays characteristic pitted materials likely representing impact melts (Tornabene *et al.*, 2007) ('m') in the crater interior and exterior (observed as pitted and ponded deposits on the ejecta blanket).

several Mercurian impact craters (Prockter *et al.*, 2010). On Venus, spectacular outflows of what are interpreted to be impact melt are found exterior to many Venusian impact craters (Asimow and Wood, 1992). It is recognized that there is considerable diversity in these outflow features. Asimow and Wood (1992) recognized erosive, channel-forming outflows of what was probably impact melt that may consist of a mix of solid and melt. Three-dimensional numerical simulations of the outflows separated them into two types: 'catastrophic' outflows originating at or near the rim, with outflow rates of $10^{10}\,m^3\,s^{-1}$ lasting less than 100 s, and 'gentle' outflows originating in the ejecta, with outflow rates of $10^4\,m^3\,s^{-1}$ lasting more than $10^5\,s$ (Miyamoto and Sasaki, 2000). The emplacements of the catastrophic outflows were attributed directly to the cratering process, while the gentle outflows were attributed to secondary segregation and drainage of melt materials from within previously emplaced ejecta.

Mars is a geologically complex body, more like Earth in terms of the potential range in target rocks. The Martian impact cratering record is diverse, but impact melt deposits were generally not thought to be widely present (Strom *et al.*, 1992). This view has radically changed in recent years, particularly since the availability of high-resolution images provided by cameras onboard the *Mars Reconnaissance Orbiter* (MRO) mission. Deposits consistent with being impact melt form large bodies and/or ponds on crater floors and terraces (Fig. 9.11f; Morris *et al.*, 2010), and overlying continuous ejecta blankets (Fig. 9.11g). These Martian deposits share many similar morphological characteristics with lunar impact melt deposits; in particular, ponds, leveed channels and flow features. One notable difference between lunar and some Martian impact melt deposits is the widespread presence of characteristic pits covering the surface of the latter (Tornabene *et al.*, 2007). These pits are quasi-circular to circular or polygonal to irregular-shaped cavities that are relatively shallow and generally lack raised rims and any sign of boulders or ejected materials deposited around them (Fig. 9.11g). The current working hypothesis is that these pits form due to the interaction of hot impact melt-bearing impactites with volatiles. It is still unknown whether these volatiles are from the impact melt deposits themselves – in which case these deposits could be similar to volatile-rich impact melt-bearing breccias ('suevites'), which often contain so-called degassing pipes (Newsom *et al.*, 1986) – or whether the volatiles are from overlying immediately post-impact sediments. Whatever their origin, it appears that substantial volumes of impact melt are generated and preserved within Martian impact craters.

### 9.4.3 Meteorites and Apollo samples

In keeping with the interpretation of widespread occurrence of impact melt deposits on the Moon based on spacecraft observations (Section 9.4.2), impact melt-bearing impactites are common in the Apollo sample collection and in some lunar meteorites. Impact melt glasses and rocks comprise approximately 30–50% of hand specimens returned from the lunar highlands and approximately 50% of lunar soils and mare material (Heiken *et al.*, 1991). In general, these samples display the same characteristics, and may be classified in the same way, as impact melt-bearing impactites from terrestrial craters developed in crystalline targets (see Section 9.3.3 and Chapter 7). At this time, it is not

known whether impact melt sheets in large lunar impact basins differentiated, as at Sudbury.

As discussed in Chapter 7, complications arise as the nomenclature used to describe some lunar impact melt-bearing impactites does not always correspond to that used for terrestrial impactites. One widely used term is 'impact melt breccias', which has also been applied to terrestrial impactites. Such rocks on the Moon typically have a very fine-grained to glassy melt matrix and contain a large proportion of lithic and mineral clasts. This term is no longer recommended and should be replaced with fine-grained, clast-rich impact melt rock (Fig. 9.3; Stöffler and Grieve, 2007). One impact melt-bearing material that is abundant on the Moon, but which is not known in the terrestrial impact cratering record, is 'agglutinate'. Agglutinates comprise, on average, approximately 25–30% of the lunar soil and form by micrometeorite impact into the lunar 'soil' (Heiken *et al.*, 1991). They are typically very fine grained and comprise soil/regolith particles bonded or welded together. They cool or quench extremely rapidly, so that the impact melt does not homogenize.

Impact melt rocks or melt-bearing breccias are not known in the current Martian meteorite collection; however, so-called impact melt 'pockets' are found in many samples (Fig. 9.4f; e.g. Walton and Herd, 2007). Aphanitic to glassy impact-generated melt rocks and melt-bearing breccias are also present in many meteorites derived from asteroids. The processes are similar, although different physical parameters mean that the relative volumes of impact melt generated on asteroids is much smaller (Keil *et al.*, 1997).

### 9.5 Impactor contamination

In some cases, impact melt rocks contain a small geochemical enrichment in siderophile elements, particularly platinum-group elements (PGEs), relative to the target rocks. This represents the admixture of impactor material, given that several classes of meteoritic bodies are enriched in siderophiles relative to crustal lithologies, which have had their siderophile contents depleted by core formation. In some cases, impactor type identification has been tentative, while in others it has been possible to identify the impactor type down to a particular meteorite class (Palme *et al.*, 1979; Tagle and Hecht, 2006). One problem is that the geochemical contribution from the impactor to the impact melt is generally small (<1%), although unevenly distributed contributions over 5% have been noted (Palme *et al.*, 1979). Thus, there can be problems with any 'indigenous' correction for siderophiles contributed to the melt rock composition from the target lithologies. The problem of the indigenous correction, however, can be effectively eliminated if siderophile ratios are used, in place of relative amounts of particular siderophile elements (e.g. Tagle and Hecht, 2006). Osmium isotope ratios ($^{187}Os/^{188}Os$) are the most sensitive methodology to detect very small impactor contributions to impact melt rock compositions (Koeberl *et al.*, 2002), but they cannot discriminate between different types of impactors. Chromium isotopes can be used to discriminate between impactor types, but require a relatively large contribution (several per cent) of impactor material in the impact melt rock (Koeberl *et al.*, 2002). Full details of the methodologies for the detection of impactor material and a tabulation of the impactor types so far identified can be found in Chapter 15.

## 9.6    Concluding remarks

Impact melting is one of the most characteristic aspects of the cratering process. The products have been studied in terrestrial and lunar crater samples and in meteorites, and remote-sensing data suggest they occur on all the inner Solar System planets. The occurrences and properties of impact melts have provided much information regarding the crater formation process itself. Despite this, there are still considerable gaps in our knowledge of the process and products of impact melting, most notably the effect(s) of target lithology. The genesis of certain types of pseudotachylite and glass occurrence also remains enigmatic, even after decades of detailed investigations.

## References

Agrinier, P., Deutsch, A., Schärer, U. and Martinez, I. (2001) Fast back-reactions of shock-released $CO_2$ from carbonates: an experimental approach. *Geochimica et Cosmochimica Acta*, 65, 2615–2632.

Anders, D., Kegler, P., Buchner, E. and Schmieder, M. (2011) Carbonate melt lithologies from the Steinheim impact crater (SW Germany). 42nd Lunar and Planetary Science Conference, abstract no. 1997.

Artemieva, N. (2002) Tektite origin in oblique impact: numerical modeling, in *Impacts in Precambrian Shields* (eds C. Koeberl and J. Plado), Springer-Verlag, Berlin, pp. 257–276.

Artemieva, N. (2008) Tektites: model versus reality. 39th Lunar and Planetary Science Conference, abstract no. 1651.

Asimow, P.D. and Wood, J.A. (1992) Fluid outflows from Venus impact craters: analysis from *Magellan* data. *Journal of Geophysical Research*, 97, 13643–13665.

Baldwin, R.B. (1981) Tektites: size estimates of their source craters and implications for their origin. *Icarus*, 45, 554–563.

Beran, A. and Koeberl, C. (1997) Water in tektites and impact glasses by Fourier-transformed infrared spectrometry. *Meteoritics and Planetary Science*, 32, 211–216.

Boslough, M.B.E. and Crawford, D.A. (2008) Low-altitude airbursts and the impact threat. *International Journal of Impact Engineering*, 35, 1441–1448. DOI: 10.1016/j.ijimpeng.2008.07.053.

Boslough, M.B., Ahrens, T.J., Vizgirda, J. *et al.* (1982) Shock-induced devolatilization of calcite. *Earth and Planetary Science Letters*, 61, 166–170.

Bryan, W.B. (1972) Morphology of quench crystals in submarine basalts. *Journal of Geophysical Research*, 77, 5812–5819.

Chao, E.C.T. (1967) Shock effects in certain rock-forming minerals. *Science*, 156, 192–202.

Cintala, M.J. and Grieve, R.A.F. (1994) The effects of differential scaling of impact melt and crater dimensions on lunar and terrestrial craters: some brief examples, in *Large Meteorite Impacts and Planetary Evolution* (eds B.O. Dressler and V.L. Sharpton), Geological Society of America Special Paper 293, Geological Society of America, Boulder, CO, pp. 51–59.

Dence, M.R. (1971) Impact melts. *Journal of Geophysical Research*, 76, 5552–5565.

Deutsch, A. and Langenhorst, F. (2007) On the fate of carbonates and anhydrite in impact processes – evidence from the Chicxulub event. *Geologiska Föreningar Förhandingar*, 129, 155–160.

Deutsch, A., Ostermann, M. and Masaitis, V.L. (1997) Geochemistry and neodymium–strontium isotope signature of tektite-like objects from Siberia (urengoites, South-Ural glass). *Meteoritics and Planetary Science*, 32, 679–686.

Donaldson, C.H. (1976) An experimental investigation of olivine morphology. *Contributions to Mineralogy and Petrology*, 57, 187–213.

Dressler, B.O. and Reimold, W.U. (2001) Terrestrial impact melt rocks and glasses. *Earth Science Reviews*, 56, 205–284.

Dressler, B.O., Sharpton, V.L., Schwandt, C.S. and Ames, D.E. (2004) Impactites of the Yaxcopoil-1 drilling site, Chicxulub impact structure: petrography, geochemistry, and depositional environment. *Meteoritics and Planetary Science*, 39, 857–878.

Dypvik, H. and Jansa, L.F. (2003) Sedimentary signatures and processes during marine bolide impacts: a review. *Sedimentary Geology*, 161, 309–337.

Fiske, P.S., Nellis, W.J., Lipp, M. *et al.* (1995) Pseudotachylites generated in shock experiments: implications for impact cratering products and processes. *Science*, 270, 281–283.

Floran, R.J., Grieve, R.A.F., Phinney, W.C. *et al.* (1978) Manicouagan impact melt, Quebec, 1, stratigraphy, petrology, and chemistry. *Journal of Geophysical Research*, 83, 2737–2759.

Glass, B.P. (1990) Tektites and microtektites: key facts and inferences. *Tectonophysics*, 171, 393–404.

Glass, B.P. and Koeberl, C. (2006) Australasian microtektites and associated impact ejecta in the South China Sea and the Middle Pleistocene supereruption of Toba. *Meteoritics and Planetary Science*, 41, 305–326.

Graup, G. (1999) Carbonate–silicate liquid immiscibility upon impact melting: Ries Crater, Germany. *Meteoritics and Planetary Science*, 34, 425–438.

Grieve, R.A.F. (1975) Petrology and chemistry of impact melt at Mistastin Lake crater, Labrador. *Geological Society of America Bulletin*, 86, 1617–1629.

Grieve, R.A.F. (1978) The melt rocks at Brent Crater, Ontario, Canada. *Proceedings of the Lunar and Planetary Science Conference*, 9, 2579–2608.

Grieve, R.A.F. and Cintala, M.J. (1992) An analysis of differential impact melt-crater scaling and implications for the terrestrial impact record. *Meteoritics*, 27, 526–538.

Grieve, R.A.F., Dence, M.R. and Robertson, P.B. (1977) Cratering processes: as interpreted from the occurrences of impact melts, in *Impact and Explosion Cratering* (eds D.J. Roddy, R.O. Pepin and R.B. Merrill), Pergamon Press, New York, NY, pp. 791–814.

Grieve, R.A.F., Stöffler, D. and Deutsch, A. (1991) The Sudbury structure: controversial or misunderstood? *Journal of Geophysical Research*, 96, 22753–22764.

Grieve, R.A.F., Langenhorst, F. and Stöffler, D. (1996) Shock metamorphism of quartz in nature and experiment: II. Significance in geoscience. *Meteoritics and Planetary Science*, 31, 6–35.

Haines, P.W., Jenkins, R.J.F. and Kelley, S.P. (2001) Pleistocene glass in the Australian desert: the case for an impact origin. *Geology*, 29, 899–902.

Hawke, B.R. and Head, J.W. (1977) Impact melt on lunar crater rims, in *Impact and Explosion Cratering* (eds D.J. Roddy, R.O. Pepin and R.B. Merrill), Pergamon Press, New York, NY, pp. 815–841.

Heide, K., Heide, G. and Kloess, G. (2001) Glass chemistry of tektites. *Planetary and Space Science*, 49, 839–844.

Heiken, G., Vaniman, D. and French, B.M. (1991) *Lunar Sourcebook: A User's Guide to the Moon.* Cambridge University Press, New York, NY.

Hörz, F., See, T.H., Murali, A.V. and Blanchard, D.P. (1989) Heterogeneous dissemination of projectile materials in the impact melts from Wabar Crater, Saudi Arabia. *Proceedings of the Lunar and Planetary Science Conference*, 19, 697–709.

Howard, K.A. (1974) Fresh lunar impact craters: review of variations with size. *Proceedings of the Lunar and Planetary Science Conference*, 5, 61–69.

Howard, K.A. and Wilshire, H.G. (1975) Flows of impact melt at lunar craters. *Journal of Research of the U.S. Geological Survey*, 3, 237–251.

Ivanov, B.A. and Deutsch, A. (2002) The phase diagram of $CaCO_3$ in relation to shock compression and decompression. *Physics of the Earth and Planetary Interiors*, 129, 131–143.

Johnson, B.C. and Melosh, H.J. (2012) Formation of spherules in impact produced vapor plumes. *Icarus*, 217, 416–430.

Jones, A.P., Claeys, P. and Heuschkel, S. (2000) Impact melting of carbonates from the Chicxulub Crater, in *Impacts and the Early Earth* (eds I. Gilmour and C. Koeberl), Lecture Notes in Earth Sciences 91, Springer-Verlag, Berlin, pp. 343–361.

Keil, K., Stöffler, D., Love, S.G. and Scott, E.R.D. (1997) Constraints on the role of impact heating and melting in asteroids. *Meteoritics and Planetary Science*, 32, 349–363.

Kieffer, S.W. (1971) Shock metamorphism of the Coconino sandstone at Meteor Crater, Arizona. *Journal of Geophysical Research*, 76, 5449–5473.

Kieffer, S.W. and Simonds, C.H. (1980) The role of volatiles and lithology in the impact cratering process. *Reviews of Geophysics and Space Physics*, 18, 143–181.

Kieffer, S.W., Phakey, P.P. and Christie, J.M. (1976) Shock processes in porous quartzite: transmission electron microscope observations and theory. *Contributions to Mineralogy and Petrology*, 59, 41–93.

Koeberl, C. (1992) Geochemistry and origin of Muong Nong-type tektites. *Geochemica et Cosmochimica Acta*, 56, 1033–1064.

Koeberl, C. (1994) Tektite origin by hypervelocity asteroidal or cometary impact: target rocks, source craters, and mechanisms, in *Large Meteorite Impacts and Planetary Evolution* (eds B.O. Dressler, R.A.F. Grieve and V.L. Sharpton), Geological Society of America Special Paper 293, Geological Society of America, Boulder, CO, pp. 133–152.

Koeberl, C., Bottomley, R., Glass, B.P. and Storzer, D. (1997) Geochemistry and age of Ivory Coast tektites and microtektites. *Geochimica et Cosmochimica Acta*, 61, 1745–1772.

Koeberl, C., Reimold, W.U. and Shirey, S.B. (1998) The Aouelloul crater, Mauritania: on the problem of confirming the impact origin of a small crater. *Meteoritics and Planetary Science*, 33, 513–517.

Koeberl, C., Peucker-Ehrenbrink, B., Reimold, W.U. *et al.* (2002) Comparison of the osmium and chromium isotopic methods for the detection of meteoritic components in impactites: examples from the Morokweng and Vredefort impact structures, South Africa, in *Catastropic Events and Mass Extinctions: Impacts and Beyond* (eds C. Koeberl and K.G. MacLeod), Geological Society of America Special Paper 356, Geological Society of America, Boulder, CO, pp. 607–617.

Kring, D.A., Hörz, F., Zurcher, L. and Urrutia Fucugauchi, J. (2004) Impact lithologies and their emplacement in the Chicxulub impact crater: initial results from the Chicxulub Scientific Drilling Project, Yaxcopoil, Mexico. *Meteoritics and Planetary Science*, 39, 879–897.

Lambert, P. and Lange, M.A. (1984) Glasses produced by shock melting and devolatilization of hydrous silicates. *Journal of Non-Crystalline Solids*, 67, 521–542.

Lin, A. (1998) *Fossil Earthquakes: The Formation and Preservation of Pseudotachylytes*, Lecture Notes in Earth Sciences 111, Spinger, Berlin.

Lofgren, G. (1974) An experimental study of plagioclase crystal morphology: isothermal crystallization. *American Journal of Science*, 274, 243–273.

Marion, C.L. and Sylvester, P.J. (2010) Composition and heterogeneity of anorthositic impact melt at Mistastin Lake crater, Labrador. *Planetary and Space Science*, 58, 552–573.

Martini, J.E.J. (1991) The nature, distribution and genesis of the coesite and stishovite associated with the pseudotachylite of the Vredefort Dome, South Africa. *Earth and Planetary Science Letters*, 103, 285–300.

Masaitis, V.L. (1999) Impact structures of northeastern Eurasia: the territories of Russia and adjacent countries. *Meteoritics and Planetary Science*, 34, 691–711.

Masaitis, V.L., Danilin, A.N., Maschak, M.S. *et al.* (1980) *The Geology of Astroblemes*, Nedra, Leningrad (in Russian).

Meisel, T., Koeberl, C. and Ford, R.J. (1990) Geochemistry of Darwin impact glass and target rocks. *Geochemica et Cosmochimica Acta*, 54, 1463–1474.

Melosh, H.J. (1989) *Impact Cratering: A Geologic Process*, Oxford University Press, New York, NY.

Melosh, H.J. (2005) The mechanics of pseudotachylite formation in impact events, in *Impact Tectonics* (eds C. Koeberl and H. Henkel), Springer, Berlin, pp. 55–80.

Metzler, A., Ostertag, R., Redeker, H.J. and Stöffler, D. (1988) Composition of the crystalline basement and shock metamorphism of crystalline and sedimentary target rocks at the Haughton impact crater, Devon Island, Canada. *Meteoritics*, 23, 197–207.

Milton, D.J. and Macdonald, F.A. (2005) Goat Paddock, Western Australia: an impact crater near the simple–complex transition. *Australian Journal of Earth Sciences*, 52, 689–697.

Milton, D.J., Glikson, A.Y. and Brett, R. (1996) Gosses Bluff – a latest Jurassic impact structure, central Australia. Part, 1, geological structure, stratigraphy, and origin. *Journal of Australian Geology and Geophysics*, 16, 453–486.

Miyamoto, H. and Sasaki, S. (2000) Two differennt styles of crater outflow materials on Venus inferred from numericaal simulations over DEMs. *Icarus*, 145, 533–545.

Mohr-Westheide, T. and Reimold, W.U. (2011) Formation of pseudotachylitic breccias in the central uplifts of very large impact structures: scaling the melt formation. *Meteoritics and Planetary Science*, 46, 543–555.

Morris, A.R., Mouginis-Mark, P.J. and Garbeil, H. (2010) Possible impact melt and debris flows at Tooting Crater, Mars. *Icarus*, 209, 369–389.

Nelson, M.J. and Newsom, H.E. (2006) Yaxcopoil-1 impact melt breccias: silicate melt clasts among dolomite melt and implications for deposition. 37th Lunar and Planetary Science Conference, abstract no. 2081.

Newsom, H.E., Graup, G., Sewards, T. and Keil, K. (1986) Fluidization and hydrothermal alteration of the suevite deposit in the Ries Crater, West Germany, and implications for Mars. *Journal of Geophysical Research, B, Solid Earth and Planets*, 91, 239–251.

Onorato, P.I.K., Uhlmann, D.R. and Simonds, C.H. (1978) The thermal history of the Manicouagan impact melt sheet, Quebec. *Journal of Geophysical Research*, 83, 2789–2798.

Osinski, G.R. (2003) Impact glasses in fallout suevites from the Ries impact structure, Germany: an analytical SEM study. *Meteoritics and Planetary Science*, 38, 1641–1668.

Osinski, G.R. (2007) Impact metamorphism of $CaCO_3$-bearing sandstones at the Haughton structure, Canada. *Meteoritics and Planetary Science*, 42, 1945–1960.

Osinski, G.R. and Spray, J.G. (2001) Impact-generated carbonate melts: evidence from the Haughton Structure, Canada. *Earth and Planetary Science Letters*, 194, 17–29.

Osinski, G.R., Bunch, T.E. and Wittke, J. (2003) Evidence for shock melting of carbonates from Meteor Crater, Arizona. 66th Meteoritical Society Meeting, abstract no. 5070.

Osinski, G.R., Grieve, R.A.F. and Spray, J.G. (2004) The nature of the groundmass of surficial suevites from the Ries impact structure, Germany, and constraints on its origin. *Meteoritics and Planetary Science*, 39, 1655–1684.

Osinski, G.R., Spray, J.G. and Lee, P. (2005) Impactites of the Haughton impact structure, Devon Island, Canadian High Arctic. *Meteoritics and Planetary Science*, 40, 1789–1812.

Osinski, G.R., Schwarcz, H.P., Smith, J.R. *et al.* (2007) Evidence for a ~200–100 ka meteorite impact in the Western Desert of Egypt. *Earth and Planetary Science Letters*, 253, 378–388.

Osinski, G.R., Grieve, R.A.F., Collins, G.S. *et al.* (2008a) The effect of target lithology on the products of impact melting. *Meteoritics and Planetary Science*, 43, 1939–1954.

Osinski, G.R., Grieve, R.A.F. and Spray, J.G. (2008b) Impact melting in sedimentary target rocks: an assessment, in *The Sedimentary Record of Meteorite Impacts* (eds K.R. Evans, W. Horton, D.K. King Jr *et al.*), Geological Society of America Special Publication 437, Geological Society of America, Boulder, CO, pp. 1–18.

Osinski, G.R., Kieniewicz, J.M., Smith, J.R. *et al.* (2008c) The Dakhleh glass: product of an impact airburst or cratering event in the Western Desert of Egypt? *Meteoritics and Planetary Science*, 43, 2089–2017.

Osinski, G.R., Tornabene, L.L. and Grieve, R.A.F. (2011) Impact ejecta emplacement on the terrestrial planets. *Earth and Planetary Science Letters*, 310, 167–181.

Palme, H., Gobel, E. and Grieve, R.A.F. (1979) The distribution of volatile and siderophile elements in the impact melt of East Clearwater (Quebec). *Proceedings of Lunar and Planetary Science Conference*, 10, 2465–2492.

Phinney, W.C. and Simonds, C.H. (1977) Dynamical implications of the petrology and distribution of impact melt rocks, in *Impact and Explosion Cratering* (eds D.J. Roddy, R.O. Pepin and R.B. Merrill), Pergamon Press, New York, NY, pp. 771–790.

Pierazzo, E., Vickery, A.M. and Melosh, H.J. (1997) A reevaluation of impact melt production. *Icarus*, 127, 408–423.

Pratesi, G., Morelli, M., Rossi, A.P. and Ori, G.G. (2005) Chemical compositions of impact melt breccias and target rocks from the Tenoumer impact crater, Mauritania. *Meteoritics and Planetary Science*, 40, 1653–1672.

Prockter, L.M., Ernst, C.M., Denevi, B.W. *et al.* (2010) Evidence for young volcanism on Mercury from the third *MESSENGER* flyby. *Science*, 329, 668–671.

Redeker, H.J. and Stöffler, D. (1988) The allochthonous polymict breccia layer of the Haughton impact crater, Devon Island, Canada. *Meteoritics*, 23, 185–196.

Reimold, W.U. (1995) Pseudotachylite in impact structures – generation by friction melting and shock brecciation? A review and discussion. *Earth Science Reviews*, 39, 247–265.

Riller, U., Lieger, D., Gibson, R.L. *et al.* (2010) Origin of large-volume pseudotachylite in terrestrial impact structures. *Geology*, 38, 619–622.

Roedder, E. (1978) Silicate liquid immiscibility in magmas and in the system $K_2O–FeO–Al_2O_3–SiO_2$: an example of serendipity. *Geochimica et Cosmochimica Acta*, 42, 1597–1617.

Rosa, D.F. and Martin, R.F. (2010) A spurrite-, merwinite- and srebrodolskite-bearing skarn assemblage, West Clearwater Lake impact crater, Northern Quebec. *Canadian Mineralogist*, 48, 1519–1532.

Schaal, R.B. and Hörz, F. (1977) Shock metamorphism of lunar and terrestrial basalts. Proceedings of the 8th Lunar Science Conference, pp. 1697–1729.

Schultz, P.H., Rate, M., Hames, W.E. *et al.* (2006) The record of Miocene impacts in the Argentine Pampas. *Meteoritics and Planetary Science*, 41, 749–771.

Shand, S.J. (1916) The pseudotachylyte of Parijs (Orange Free State), and its relation to 'trap-shotten gneiss' and 'flinty crush-rock'. *Quarterly Journal of the Geological Society of London*, 72, 198–221.

Simonds, C.H., Floran, R.J., McGee, P.E. *et al.* (1978a) Petrogenesis of melt rocks, Manicouagan impact structure, Quebec. *Journal of Geophysical Research*, 83, 2773–2778.

Simonds, C.H., Phinney, W.C., McGee, P.E. and Cochran, A. (1978b) West Clearwater, Quebec impact structure, Part I: field geology, structure and bulk chemistry. *Proceedings of the Lunar and Planetary Science Conference*, 9, 2633–2658.

Simonson, B.M. and Glass, B.J. (2004) Spherule layers – records of ancient impacts. *Annual Review of Earth and Planetary Science*, 32, 329–361.

Sjöqvist, A.S.L., Lindgren, P., Mansfeld, P. *et al.* (2012) Carbonate melt fragments in resurge deposits from the Lockne impact structure, Sweden. 43rd Lunar and Planetary Science Conference, abstract no. 1962.

Spray, J.G. (2006) Ultrametamorphism of impure carbonates beneath the Manicouagan impact melt sheet: evidence for superheating. 37th Lunar and Planetary Science Conference.

Spray, J.G. and Thompson, L.M. (2008) Constraints on central uplift structure from the Manicouagan impact crater. *Meteoritics and Planetary Science*, 43, 2049–2057.

Spray, J.G., Grady, M.M., Hutchinson, R. *et al.* (1998) Localized shock- and friction-induced melting in response to hypervelocity impact, in *Meteorites: Flux with Time and Effects* (eds M.M. Grady, R. Hutchinson, G.J.H. McCall and D.A. Rothery), Geological Society Special Publication 140, Geological Society, London, pp. 195–204.

Stähle, V. (1972) Impact glasses from the suevite of the Nordlinger Ries. *Earth and Planetary Science Letters*, 17, 275–293.

Stöffler, D. (1977) Research drilling Nördlingen, 1973, polymict breccias, crater basement, and cratering model of the Ries impact structure. *Geologica Bavarica*, 75, 443–458.

Stöffler, D. (1984) Glasses formed by hypervelocity impact. *Journal of Non-Crystalline Solids*, 67, 465–502.

Stöffler, D. and Grieve, R.A.F. (2007) Impactites, in *Metamorphic Rocks* (eds D. Fettes and J. Desmons), Cambridge University Press, Cambridge, pp. 82–92.

Stöffler, D., Ewald, U., Ostertag, R. and Reimold, W.U. (1977) Research drilling Nördlingen 1973 (Ries): composition and texture of polymict impact breccias. *Geologica Bavarica*, 75, 163–189.

Stöffler, D., Artemieva, N.A. and Pierazzo, E. (2002) Modeling the Ries–Steinheim impact event and the formation of the moldavite strewn field. *Meteoritics and Planetary Science*, 37, 1893–1907.

Strom, R.G., Croft, S.K. and Barlow, N.G. (1992) The Martian impact cratering record, in *Mars* (eds H.H. Kieffer, B.M. Jakosky, C.W. Snyder and M.S. Matthews), University of Arizona Press, Tucson, AZ, pp. 383–423.

Tagle, R. and Hecht, L. (2006) Geochemical identification of projectiles in impact rocks. *Meteoritics and Planetary Science*, 41, 1721–1735.

Taylor, S.R. (1973) Tektites: a post-Apollo view. *Earth Science Reviews*, 9, 101–123.

Therriault, A.M., Fowler, A.D. and Grieve, R.A.F. (2002) The Sudbury Igneous Complex: a differentiated impact melt sheet. *Economic Geology*, 97, 1521–1540.

Tornabene, L.L., Mcewan, A.S., Osinski, G.R. *et al.* (2007) Impact melting and the role of sub-surface volatiles: implications for the formation of valley networks and phyllosilicate-rich lithologies on early Mars. Seventh International Conference on Mars, abstract no. 3288.

Tsutsumi, A. and Shimamoto, T. (1997) High-velocity frictional properties of gabbro. *Geophysical Research Letters*, 24, 699–702.

Tuchscherer, M.G., Reimold, W.U., Koeberl, C. *et al.* (2004) First petrographic results on impactites from the Yaxcopoil-1 borehole, Chicxulub structure, Mexico. *Meteoritics and Planetary Science*, 39, 899–931.

Von Engelhardt, W. (1972) Shock produced rock glasses from the Ries Crater. *Contributions to Mineralogy and Petrology*, 36, 265–292.

Von Engelhardt, W. and Graup, G. (1984) Suevite of the Ries crater, Germany: source rocks and implications for cratering mechanics. *Geologische Rundschau*, 73, 447–481.

Von Engelhardt, W., Stöffler, D. and Schnieder, W. (1969) Petrologische Untersuchungen im Ries. *Geologica Bavarica*, 61, 229–295.

Von Engelhardt, W., Luft, E., Arndt, J. *et al.* (1987) Origin of moldavites. *Geochimica et Cosmochimica Acta*, 51, 1425–1443.

Walton, E.L. and Herd, C.D.K. (2007) Dynamic crystallization of shock melts in Allan Hills 77005: implications for melt pocket formation in Martian meteorites. *Geochimica et Cosmochimica Acta*, 71, 5267–5285.

Wasson, J.T. (2003) Large aerial bursts: an important class of terrestrial accretionary events. *Astrobiology*, 3, 163–179.

Weeks, R.A., Underwood, J.R. and Giegengack, R. (1984) Libyan desert glass: a review. *Journal of Non-Crystalline Solids*, 67, 593–619.

Wenk, H.R. (1978) Are pseudotachylites products of fracture of fusion? *Geology*, 6, 507–511.

Wünnemann, K., Collins, G.S. and Osinski, G.R. (2008) Numerical modelling of impact melt production on porous rocks. *Earth and Planetary Science Letters*, 269, 529–538.

Zieg, M.J. and Marsh, B.D. (2005) The Sudbury Igneous Complex: viscous emulsion differentiation of a superheated impact melt sheet. *Geological Society of America Bulletin*, 117, 1427–1450.

# TEN

# Environmental effects of impact events

**Elisabetta Pierazzo\* and H. Jay Melosh[†]**

*\*Planetary Science Institute, 1700 E. Fort Lowell Road, Suite 106, Tucson, AZ 85719, USA*
*[†]Earth, and Atmospheric and Planetary Sciences Department, 550 Stadium Mall Drive, Purdue University, West Lafayette, IN 47097, USA*

## 10.1 Introduction

Awareness of the devastating consequences of impact events on the fragile terrestrial ecosystem arose in the 1980s with the revolutionary theory of Alvarez *et al.* (1980) linking the 65 Ma Cretaceous–Palaeogene (K–Pg) mass extinction event with an asteroidal impact – see review by Schulte *et al.* (2010). However, while impact events punctuate the terrestrial geological record, they are not normally associated with mass extinction events. Factors like the magnitude of the impact event and the characteristics of the target region define the impact hazard. The K–Pg boundary, up to now, is the only recognized mass extinction event coinciding with an impact event. On the other hand, although most asteroids and comet impacts will not cause global-scale extinctions, they can still strongly affect the environment on a local or even global scale. Unfortunately, environmental catastrophes associated with such impact events are much more difficult to identify in the geological record. To date, qualitative assessments of impact-related environmental and climatic effects abound, but a comprehensive quantitative investigation is still mostly lacking.

This chapter is designed to provide an overview of the environmental effects of extraterrestrial impact events. It describes the devastating environmental effects of impacts and includes a brief discussion of potential impact-related environmental effects that are favourable to life.

## 10.2 The impact hazard

The impact origin of the first recognized impact structure on Earth, Meteor Crater, was not proven until 1960 (e.g. Shoemaker *et al.*, 1960). It took another two decades before the link of the K–Pg mass extinction to a large impact event (Alvarez *et al.*, 1980) brought the realization of the devastating consequences of impact events on the terrestrial ecosystem. Over the past 50 years,

therefore, impact cratering arose from highly doubtful to a major geological process ubiquitous on solid surfaces of the Solar System.

Geological and dynamical arguments, discussed in detail in Chapter 2, provide the crucial evidence for understanding the past flux of bodies impacting Earth. On the one hand, the lunar record of impacts combined with the dating of lunar samples brought back by the Apollo missions provide an absolute chronology of lunar and terrestrial craters up to approximately 3.9 Ga. On the other hand, dynamical studies associated with observational surveys constrain the impactors' size distribution and, through scaling laws, planetary crater production functions. This approach allows for the modelling of orbital and size distribution of near-Earth object (NEO) populations that make up most of the impactors hitting the surface of inner Solar System bodies.

The support of observational surveys has become a crucial tool in constraining the current impact hazard. The NASA NEOs website (http://neo.jpl.nasa.gov/stats/) indicates that, as of July 2012, the number of discovered NEOs larger than 1 km in diameter is 851, suggesting that approximately 90% of large NEOs have been discovered. Figure 10.1 shows the Harris (2008) estimate of the cumulative population of NEOs versus size compared with observed NEOs. Overall, there is high confidence that all asteroids of 10 km in diameter and above have been discovered, ruling out the hazard of a mass extinction impact event (Harris, 2008). The NEOs discovered so far have a very small probability of hitting Earth in the next century, thus reducing the 'risk' of impact in the near future. Still, well over a hundred objects 1–2 km in diameter are probably looming undiscovered in the Earth's neighbourhood. Smaller and much more abundant NEOs between about 500 m and 1 km in diameter are also believed to pose a significant threat to human civilization.

Comets can be a significant part of the impact hazard. They are generally much fainter than comparable-size asteroids and, thus, much more difficult to discover before they 'turn on' (due to sublimation of volatiles to form the comet's coma and tail)

*Impact Cratering: Processes and Products*, First Edition. Edited by Gordon R. Osinski and Elisabetta Pierazzo.

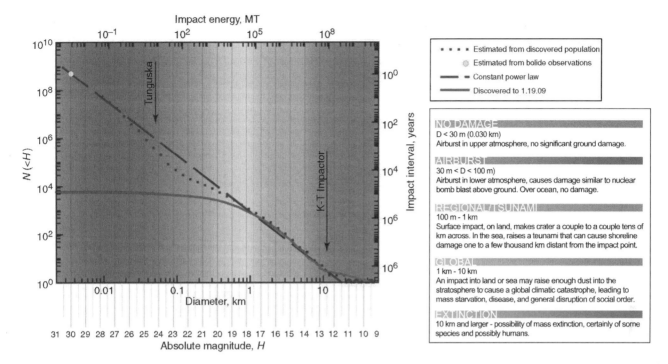

**Figure 10.1** Estimated cumulative population of NEOs with absolute magnitude *H* less than a given magnitude versus size (or impact energy), compared with the discovered NEOs as of January 2009. Absolute magnitude is the brightness at 1 AU from Earth. Object sizes are estimated from *H* using a mean reflectivity value, while impact energy requires additional knowledge of the object's orbital characteristics. The greyscale scheme shows the corresponding type of damage expected if NEOs of a given size were to impact the Earth's surface. A colour version of this figure is available in Pierazzo and Artemieva (2012). Image credit: A.W. Harris, Space Science Institute/Large Synoptic Survey Telescope.

while approaching the inner Solar System. Comets are only a fraction (<10%) of all potential impactors, but they typically have higher impact velocities (15–35 km s$^{-1}$ for short-period comets and 40–70 km s$^{-1}$ for long-period comets; Shoemaker *et al.*, 1994) and correspondingly higher impact energies than similar mass asteroids, thus constituting a significant component of the impact hazard for similar size objects.

Determining the 'hazard' from a particular impactor depends on many unknown factors and utilizes a statistical criterion that uses expected impact frequency and mortality rates to estimate annual risks. While it is important to compare the hazard with other potential natural catastrophes (in terms of 'mortality per year' or 'risk over a lifetime'), an important and unique characteristic of the impact hazard is that a single relatively 'low-risk' global impact catastrophe could lead to the breakdown of civilization. Chapman and Morrison (1994) define a globally catastrophic impact as one that would disrupt global agricultural production and lead, directly or indirectly, to the death of more than one-fourth of the world's population, and the destabilization of modern civilization. They estimate that the threshold for such a global catastrophe from impact events is within energies of $1.5 \times 10^4$ to $10^7$ Mt (objects ~600 m to ~5 km in diameter; 1 Mt = $4.184 \times 10^{15}$ J), while a $10^3$ Mt event (~250 m object) would cause mainly local devastation.

## 10.3 The impact cratering process

For planetary bodies with a substantial atmosphere the impact process begins as the impactor enters and moves through the atmosphere, eventually (if the impactor is large enough) impacting the planetary surface and excavating a crater (see Chapters 3–5).

### 10.3.1 Atmospheric penetration

As an impactor traverses the atmosphere it may decelerate significantly due to atmospheric drag. Because the atmospheric density increases downward, the stagnation pressure (proportional to impactor velocity and atmospheric density: $\rho_a v^2$) at the leading edge of the impactor increases as it penetrates deeper into the atmosphere. When the stagnation pressure becomes larger than the internal strength of the impactor, it begins to break up. Catastrophic fragmentation causes the impactor to deform according to the pressure gradient (between stagnation pressure at the leading edge and essentially zero pressure at the trailing edge), flattening along the direction of motion and spreading in the plane perpendicular to the motion. If the fragmentation begins well above the surface, the impactor may become dispersed over a wide area, increasing the overall drag and deceleration process.

Objects greater than 1 km in diameter do not significantly disperse while penetrating the Earth's atmosphere. Very small impactors, up to a few metres in diameter, will either be completely ablated or aerobraked to free-fall speed while traversing the atmosphere. Objects between tens to hundreds of metres in diameter are entirely fractured during their atmospheric traverse; depending on their size and strength, they may retain sufficient momentum to hit the surface with enough energy to excavate a large crater. Alternatively, they may deposit all of their kinetic energy well above the surface. In this case, the main environmental consequence of the impact event is a strong blast wave in the atmosphere that may or may not strongly affect the surface.

An important consequence of the penetration of the atmosphere by a meteoroid is the generation of atmospheric shock waves that interact with the atmospheric gases, causing a chemical reaction between molecular oxygen and nitrogen to form nitric oxide (NO). It has been estimated that the shock wave generated by the 1908 Tunguska event (a small impactor airburst over Siberia; e.g. Vasilyev (1998) and references therein) produced as much as $3 \times 10^{10}$ kg of NO. The reaction of shock-generated NO with atmospheric oxygen may have provided nitrogen compounds that acted as powerful fertilizers, causing an unprecedented growth of vegetation registered soon after the devastation (Florensky, 1965; Turco *et al.*, 1982).

## 10.3.2   Impact dynamics and local devastation

As the impactor hits the planetary surface, a shock wave propagates through the target and impactor, quickly changing the thermodynamic state of material near the impact point in an irreversible process (see Chapter 3). Material left behind after the passage of the shock wave and rarefaction wave is fractured, shock-heated, shaken and set into motion, which eventually causes the opening of the crater during the excavation stage (see Chapter 4). At the surface, global destruction occurs within the final crater, extending at least one more crater radius beyond that as a result of the large volume of rock debris ejected from the crater and deposited in a continuous blanket.

## 10.4   Shock wave effects

The shock wave generated by an impact event decays as it propagates away from the impact point in a hemispheric shell (see Chapter 3). Therefore, the direct effect of shock waves (both in solids and in the atmosphere) is strongest close to the crater. On the microscopic scale, shock waves affect solids by causing: (1) structural modifications (microstructural changes, glass production, polymorphism); (2) chemical activation (increase in reaction rates and catalytic activity); and (3) immediate chemical reactions that take place at unprecedented rates at or immediately behind the shock front (see Chapters 7–9). These changes can characterize the environmental effects of impact events, such as release of climatically important gases or production of toxic chemicals. On the macroscopic scale, fractured material becomes mixed up during crater excavation, forming various types of impact breccia (see Chapter 7).

### 10.4.1   Air blast

Locally, the air blast, the impact-induced shock wave in the atmosphere, is likely the most destructive impact effect beyond crater formation and ejecta distribution. The intensity of the air blast depends on the impact energy and the atmospheric height at which the energy is deposited, ranging from zero for a surface impact to the airburst altitude for impactors that never reach the surface. The air blast compresses the air to high pressures, generating violent winds immediately behind the high-pressure front. Scaling from nuclear explosion data (Glasstone and Dolan, 1977) has been used to evaluate the damage associated with impact-related air blasts, in terms of peak overpressure (maximum pressure in excess of 1 atm) and maximum wind speed. Glasstone and Dolan (1977) estimated that, for maximum wind velocities over 40 m s$^{-1}$, approximately 30% of trees are blown down (the remainder having branches and leaves blown off), reaching 90% for maximum wind speeds exceeding 62 m s$^{-1}$. Applying these estimates to the Meteor Crater impact event (~20–40 Mt), Kring (1997) estimated that trees would have been flattened by the blast wave over a radial distance of approximately 14–19 km (23–32 crater radii away), with up to 50% casualty rates for human-size mammals up to 9–14 km from the impact. Meteor Crater was formed by a small iron impactor, 30–40 m in size, that began disintegrating in the atmosphere before excavating a crater (Melosh and Collins, 2005; Artemieva and Pierazzo, 2009). Similar-size stony objects, usually much weaker than irons, would normally explode in the atmosphere without cratering the surface (e.g. Chyba *et al.*, 1993).

In an airburst, the main effect at the surface is the air blast. This would generally occur for impactors greater than 20 m in diameter, which can reach the lower part of the atmosphere where they produce megatonne-scale explosions. Smaller objects would generate high-altitude airbursts of much smaller energy, with limited ground effects (Boslough and Crawford, 2008). The best example of a megatonne-scale atmospheric explosion is the Tunguska event (30 June 1908; Vasilyev, 1998). The explosion was caused by a stony object entering Earth's atmosphere over Tunguska, central Siberia. The event has been estimated to have released 5–20 Mt of energy at an altitude between 5 and 15 km (e.g. Chyba *et al.*, 1993; Shuvalov and Artemieva, 2002; Artemieva and Shuvalov, 2007; Boslough and Crawford, 2008). The best known ground effect of this event is a butterfly-shape region of devastated taiga forest of about 40 km × 60 km (>2000 km$^2$) shown in Fig. 10.2a. A city the size of Washington, DC, would be completely destroyed had a similar atmospheric explosion occurred above it. A similar butterfly-shape region of surface winds strong enough to blow down trees was obtained in numerical simulations of a 50 m object entering the Earth's atmosphere at 20 km s$^{-1}$ and 45° from the surface (Fig. 10.2b; Artemieva and Shuvalov, 2007). Other recorded effects associated with the Tunguska event include seismic waves, local magnetic storms and atmospheric optical anomalies such as the bright nights recorded over western Europe for several days after the event (e.g. Vasilyev, 1998). Similar events

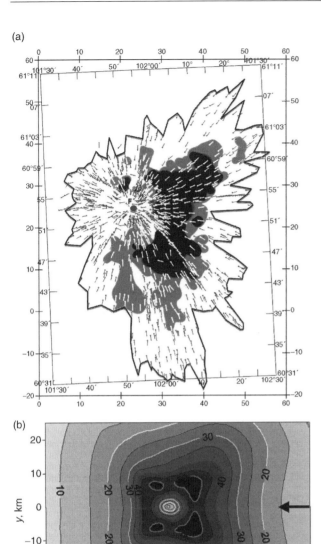

(a)

(b)

**Figure 10.2** (a) Fallen tree directions (small vectors) in the Tunguska airburst. Vectors point to the airburst epicentre (where dead trees with their crowns torn off are still standing, like 'telegraph poles'). Black, grey and white (delimited by the black line) areas correspond to high, medium and low reliability. The external frame represents 'kilometre' coordinates; the inner frame corresponds to geographical coordinates. Image credit: Longo *et al.* (2005). (b) Contours of maximum horizontal velocity (m s$^{-1}$) around the site of the Tunguska explosion. The 20 m s$^{-1}$ contour delimits the area of devastation (roughly 1700 km$^2$), while the 40 m s$^{-1}$ contour delimits totally damaged forest. The region of maximum wind speed is greater than 50 m s$^{-1}$. The black arrow indicates the bolide trajectory. Image credit: N. Artemieva, Planetary Science Institute/ Institute for Dynamics of Geospheres.

would not leave any obvious traces in the geological record, but they do cause significant local devastation.

### 10.4.2 Earthquakes

The propagation of the shock waves along the surface causes violent ground shaking up to several crater radii away. The intensity of the resulting seismic wave scales with the impact energy and decreases with distance from the impact. Impact experiments suggest that about 10$^{-3}$ to 10$^{-5}$ of the impact energy is transferred to seismic energy. Using $E_{seis} = 10^{-4} E_{imp}$ (Schultz and Gault, 1975; Melosh, 1989) in the classic Gutenberg–Richter magnitude–energy relation, the relation between seismic magnitude and impact energy becomes

$$M = 0.67 \log_{10} E_{imp} - 5.87 \qquad (10.1)$$

According to eqn (10.1), the Meteor Crater impact generated an earthquake of magnitude approximately 5.5, while the K–Pg impact (around 10$^8$ Mt) produced an earthquake of magnitude close to 10. Seismic shaking from the very large Chicxulub impact initiated massive submarine landslides that devastated the continental slope many thousands of kilometres from the impact site (Klaus *et al.*, 2000; Norris *et al.*, 2000). It has been proposed that these landslides may have caused massive releases of methane from excavated methane clathrates which then affected the Earth's climate (Day and Maslin, 2005).

Seismic events associated with airbursts or oceanic impacts may require lower seismic efficiencies. In the Tunguska event, the maximum recorded seismic magnitude was about 5; an impact energy of 10 Mt (Toon *et al.*, 1997) implies an efficiency of around 5 × 10$^{-5}$; however, a recent estimate of 4–5 Mt by Boslough and Crawford (2008) lines up the seismic efficiency for the Tunguska airburst event with the assumed 10$^{-4}$ value.

### 10.4.3 Tsunami

The best-known consequence of oceanic impacts is the generation of large tsunami. The importance of waves generated by either explosions at or below sea surface or impact events has received considerable attention over the years (e.g. Van Dorn *et al.*, 1968; Hills *et al.*, 1994; Ward and Asphaug, 2000; Matsui *et al.*, 2002; Melosh, 2003; Weiss *et al.*, 2006), with mixed conclusions. While some studies raise the hazard of impact-generated tsunami (Hills *et al.*, 1994; Ward and Asphaug, 2000), others de-emphasize the overall effect (Van Dorn, 1968; Melosh, 2003).

In modelling impact-generated tsunami, Ward and Asphaug (2000) treated them as solitary waves (no dissipation) and ignored non-linear wave behaviour during crater collapse. Wünnemann *et al.* (2007) showed that, in oceanic impacts, rim waves are the predominant waves in shallow-water impacts but decay rapidly in deep-ocean impacts. Collapse waves, on the other hand, are not a factor in shallow-water impacts, where the crater rim tends to stop water from flowing back into the crater and be affected by crater collapse, but they become the dominant waves in deep-ocean impacts. This affects wave propagation: while rim waves may be treated as solitary waves and decay slowly with distance, collapse waves decay much more rapidly with distance. The critical parameter is $d/H$, the ratio of impactor diameter $d$ and ocean depth $H$. Wünnemann *et al.* (2007) found that the smaller the $d/H$ is the more rapidly the wave amplitude is attenuated with distance. For $d/H < 0.4$, rim waves are quickly dissipated and the main impact-generated wave is a collapse wave. Assuming an ocean depth greater than 3 km (>75% of the ocean floor), this means that impactors of 1 km in diameter and less do not pose as big a threat to coastal populations. A strong hazard from impact-generated tsunami occurs for $d/H > 0.6$, corresponding to impactors of 1.8 km in diameter and above for oceans greater than 3 km deep. The area in between, $0.4 < d/H < 0.6$, is where the effects are most complex, because both rim waves and collapse waves are important.

The consequences of a tsunami reaching coastal regions depends on local conditions, such as distance of the coastline from impact, ocean bathymetry and shore and coastal configuration (offshore slopes), and cannot be addressed easily in a general context (Korycansky and Lynett, 2005, 2007; Wünnemann *et al.*, 2007). Overall, a tsunami constitutes a short-term, mostly localized effect of an oceanic impact (at most the coastal regions delimiting the ocean basin), but it can have potentially devastating effects in highly populated coastal regions, thus creating a significant regional threat for human populations.

## 10.5   Ejecta launch

During crater excavation, material near the impact point is thrown out of the crater on ballistic trajectories (see Chapter 4). This material buries the original surface within about one crater diameter of the rim, while flying fragments of ejected rock constitute a hazard to individual organisms at greater distances. High-speed ejecta of very large impacts may have global effects: heating from the re-entry of fast ejecta debris that encompassed the entire Earth has been indicated as responsible for the immediate devastation caused by the K–Pg impact (Melosh *et al.*, 1990).

### 10.5.1   Proximal ejecta curtain

Most impact ejecta land on the surface at distances reaching several crater diameters away from the impact. Up to about one crater radius away, ejecta are deposited in a continuous blanket, with a thickness varying azimuthally by factors of three to five; the thickest portions often form discrete rays (Chapter 4). Large boulders are usually mixed with finer grained material. Farther out, ejecta distribution becomes patchy, with the surface peppered by a scattered assortment of fine-grained dust and larger pieces of rock, some large enough to produce secondary craters. Data from explosion experiments (McGetchin *et al.*, 1973) suggest that ejecta thickness varies with distance from the explosion/impact point according to the formula

$$\delta = f(R)\left(\frac{r}{R}\right)^{-3\pm0.5} \quad \text{for } r \geq R \quad (10.2)$$

For impact cratering, $f(R)$ is a poorly known function, representing the constraint that the total amount of ejecta is equivalent to the amount of material excavated in the impact (Melosh, 1989).

This formula was applied by Kring (1995) to available proximal K–Pg sites to support a diameter of less than 200 km for the Chicxulub structure.

### 10.5.2   Distal ejecta and thermal effects

During an impact, a significant fraction of the impactor's kinetic energy is converted into thermal energy that melts and vaporizes both the impactor and target. Vaporized/melted material, combined with some fragmented solid material, is then ejected into a rapidly expanding, hot plume (the 'fireball'). The fireball radiates thermal energy that may ignite fires and scorch wildlife within sight. For large impacts, the expansion plume carries material well beyond the Earth's atmosphere. Its re-entry into the upper atmosphere is one of the most famous effects proposed for the K–Pg impact (e.g. Melosh *et al.*, 1990; Covey *et al.*, 1994). Jones and Kodis (1982) reported that the fireball from impacts with energies of $10^3$–$10^4$ Mt (roughly equivalent to impactors 300–500 m in diameter) could reach heights close to 100 km in an undisturbed atmosphere, while the material from impacts with energies above about $2.5 \times 10^4$ Mt (>700 m diameter objects) would breach the atmosphere and be ejected ballistically to shower over the entire Earth. The interaction of ejecta with the air as they fall back through the atmosphere produces strong frictional heating of atmosphere and ejecta particles. This process was directly observed in the 1994 collision of comet Shoemaker/Levy 9 with Jupiter: Almost half of the comet fragments' impact energy was deposited into Jupiter's atmosphere as the ejecta plumes collapsed (Zahnle, 1996).

According to early estimates, the infrared (IR) radiation emitted by hot dust particles stopped upon re-entering the upper atmosphere in the K–Pg impact could have been strong enough to ignite surface biomass (Melosh *et al.*, 1990) and kill any unsheltered organisms (Robertson *et al.*, 2004). The fires, in turn, fill the lower atmosphere with smoke, dust and pyrotoxins. This hypothesis, however, does not find support in detailed analysis of soot identified at various marine and non-marine K–Pg boundary sites (e.g. Wolbach *et al.*, 1985; Belcher *et al.*, 2003, 2005, 2009; Harvey *et al.*, 2008). Soot analyses indicate that morphology and composition are more consistent with soot produced from hydrocarbon combustion (Belcher *et al.*, 2005; Harvey *et al.*, 2008). The proximity of the Chicxulub structure to the Cantarelli oil reservoir, one of the most productive oil fields on Earth, is highly suggestive of above average abundances of organic carbon at the Chicxulub target (Harvey *et al.*, 2008). A recent investigation of the thermal effect of the atmospheric re-entry of K–Pg ejecta using a two-phase fluid flow code with radiation (Goldin and Melosh, 2009) showed that particles, as they settle through the atmosphere and cool, partially shield the surface from the IR radiation originating from new particles entering the upper atmosphere. The resulting surface thermal effect is significantly lower ($\leq 7$ kW m$^{-2}$) and shorter (below solar constant levels in less than 30 min) than originally estimated (>12.5 kW m$^{-2}$ for several hours; Melosh *et al.*, 1990). The presence of sub-micrometre dust at the top of the atmosphere may reflect downward some of the IR radiated upward (Goldin and Melosh, 2009), and increase ground radiation. On the other hand, numerical simulations by Artemieva and Morgan (2009) indicate that a significant fraction of the K–Pg

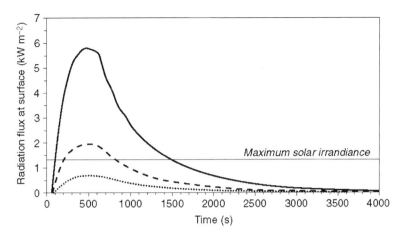

**Figure 10.3** Thermal radiation flux at the Earth's surface for 45° spherules re-entering at 3 km s$^{-1}$ (dotted line), 5 km s$^{-1}$ (dashed line) and 8 km s$^{-1}$ (solid line). Spherules' diameter and mass flux, corresponding to the typical size and mass density of distal K–Pg layers, are the same in all simulations. Image credit: T.J. Goldin, University of Arizona.

distal ejecta may not have been distributed ballistically worldwide, but may have followed a 'postballistic skidding' propagation; this strongly reduces the amount of energy carried by ballistic ejecta. The assumption of a ballistic distribution for all the material in the K–Pg layer requires high ejecta velocities (>7 km s$^{-1}$) and, consequently, very large kinetic energies (Melosh *et al.*, 1990; Toon *et al.*, 1997; Goldin and Melosh, 2009). Reducing the amount of K–Pg layer material emplaced ballistically significantly reduces the initial energy deposited in the atmosphere and, consequently, the intensity of the thermal pulse at the surface, as shown in Fig. 10.3. These new results bring into question the 'widespread wildfires' hypothesis (Melosh *et al.*, 1990) and may cast doubts on the 'land extinction from the heat pulse' hypothesis (Robertson *et al.*, 2004) for the K–Pg impact. Using existing data for humans, Goldin and Melosh (2009) conclude that it is plausible that the Chicxulub-related IR pulse may have been lethal to small thin-skinned animals and caused some degree of dermal damage to larger thick-skinned animals.

Overall, it can be safely stated that wildfires are not a common outcome of extraterrestrial impacts. Even Melosh *et al.* (1990) stated that the K–Pg impact was at the lower limit for the ignition of global wildfires. A detailed study of charcoal associated with various impact craters concluded that small to medium impacts (Tunguska event to Ries crater) do not have enough energy to trigger large wildfires (e.g. Jones and Lim, 2000), while large impacts may trigger wildfires close to the impact, due to the fireball, but generally not as a result of ejecta re-entry.

## 10.6   Long-term atmospheric perturbation

### 10.6.1   Dust

Atmospheric dust injection in large impacts can produce long-term perturbations of the atmosphere's radiative balance. Based on an analogy to nuclear explosion tests, Toon *et al.* (1997) estimated that dust reaching the upper atmosphere (or beyond) in large land impacts from material melted and vaporized in the impact event is about 15 projectile masses for impact velocities of 25 km s$^{-1}$.

In deep-ocean impacts (about twice as likely as land impacts) impactors must be at least 200 m ($H/15$ to $H/20$, where $H$ is the mean ocean depth; Gault and Sonett, 1982; O'Keefe and Ahrens, 1982) to 800 m in diameter ($H/5$; Artemieva and Shuvalov, 2002) to be able to crater the ocean floor. The threshold for significant production of dust in an oceanic impact can be estimated by equating the impactor mass to the mass of water encountered in its motion through the ocean (Zahnle, 1990). For an average ocean depth of 3 km and a typical asteroid density of about 2500 kg m$^{-3}$ this corresponds to asteroids roughly 2–4 km in diameter (for impacts from 90° to 30° from the surface). Thus, in oceanic impacts, the overall atmospheric perturbation from dust injection is negligible for impactors with diameters less than 2 km.

The long-term effect of stratospheric dust from a large impact depends on the recognition that only the stratospheric portion of the fine (sub-micrometre) dust from the impact, a small fraction of the impact-produced dust, can affect the global climate over a significant period of time. Unfortunately, the size distribution of impact-released dust is not well constrained. Data from nuclear tests and laboratory impact studies suggest that approximately 0.1% of the total impact ejecta (~30% of the impactor mass) is less than 1 μm (e.g. Toon *et al.*, 1997; Pope, 2002). The climatic effect of dust injection was investigated in relation to the K–Pg impact with an atmospheric general circulation model (GCM) simulation which included optical effects of a thick dust layer in the lower stratosphere (Covey *et al.*, 1994). The results indicate a strong and 'patchy' cooling on land, with temperature declining by up to approximately 12 °C, and a mild cooling (few degrees) over the oceans, accompanied by a collapse of the hydrologic cycle. Soot is a strong absorber of shortwave radiation; even small amounts of stratospheric soot could prevent solar radiation from reaching the Earth's surface. The evidence for soot found at several marine K–Pg boundary sites (e.g. Wolbach *et al.*, 1985, 2003, and references therein) is consistent with combustion of fossil organic matter present in the target at the impact site (Wolbach *et al.*, 1985; Belcher *et al.*, 2003, 2005, 2009; Harvey *et al.*, 2008). Thus, a significant amount of soot may have been produced upon impact and distributed ballistically with the dust in the K–Pg impact. Once in the stratosphere, soot could have a long atmospheric residence time with dramatic climatic consequences. Furthermore, Mills *et al.* (2008) show that soot injections in the upper atmosphere could have dramatic repercussions on stratospheric ozone.

### 10.6.2   Greenhouse gases

The release of climatically active gases in an impact event depends on the characteristics of the target. Release of $CO_2$ and $SO_x$ have been associated with the presence of a significant sedimentary layer in the K–Pg impact, while large amounts of water vapour are released in oceanic impacts.

#### 10.6.2.1   *Carbon dioxide*

Detailed modelling of the K–Pg impact (Ivanov *et al.*, 1996; Pierazzo *et al.*, 1998) suggest that the amount of $CO_2$ released in the impact may be have increased the end-Cretaceous atmospheric inventory by up to approximately 50% (e.g. Pierazzo *et al.*, 1998). The associated radiative forcing, $1.2–3.4\,W\,m^{-2}$ (Pierazzo *et al.*, 2003), is comparable to the estimated greenhouse gases forcing due to industrialization. The much larger contribution to atmospheric $CO_2$ hypothesized to originate from global wildfires is now mostly ruled out in view of the new results on the intensity of the IR heat pulse (Goldin and Melosh, 2009).

#### 10.6.2.2   *Sulfur oxides*

Production of sulfate aerosols from the release of $SO_x$ and water vapour in the stratosphere is well documented for volcanic eruptions. Sulfate aerosols scatter shortwave radiation and can be strong absorbers of longwave radiation (if $>1\,\mu m$ in diameter), causing a net cooling of the Earth's surface. The potential effects of impact-related sulfate production in the stratosphere were investigated for the K–Pg impact with simple one-dimensional atmospheric models combined with simple coagulation models (e.g. Pope *et al.*, 1997; Pierazzo *et al.*, 2003). Using 200 Gt of $SO_2$ and water vapour, Pope *et al.* (1997) found a significant reduction of solar transmission for about 8–13 years after the impact that would cause continental surface temperatures to approach freezing for several years. Using a similar approach, Pierazzo *et al.* (2003) obtained a slightly shorter duration of the sulfate effect, with a 50% reduction in solar transmission for 4–5 years after the impact.

#### 10.6.2.3   *Water vapour*

Oceanic impacts are about two times more likely than land impacts, since more than 60% of the Earth's surface is covered by deep oceans. The consequence of injecting large amounts of water into the upper atmosphere is an important, yet not well-constrained global effect from oceanic impacts. A K–Pg impact into a deep ocean would inject in the upper atmosphere an

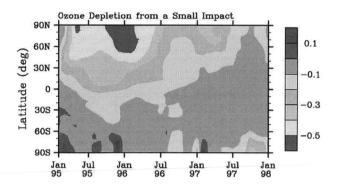

**Figure 10.4**   Zonally averaged monthly mean fractional changes of atmospheric ozone column following the impact of a 1 km asteroid in the central Pacific. Values are calculated with respect to the unperturbed case. Negative contours (ozone depletion) are dashed. Regions where ozone depletion is 50% or more (fractional changes of −0.5) are hatched. Image credit: Pierazzo *et al.* (2010). (See Colour Plate 24)

amount of water equivalent to more than three times the amount of water vapour that an unperturbed middle atmosphere can hold at saturation ($0.2\,g\,cm^{-2}$ integrated over the depth of the atmosphere or ~1000 Gt for the entire planet; Pierazzo, 2005). Water produces free radicals OH and $HO_2$, which participate in the $HO_x$ catalytic cycle that destroys ozone. Furthermore, salts from seawater (average salinity is 35‰) would contribute about 65 Gt of Cl and 3 Gt of S to the upper atmosphere (Pierazzo, 2005). These injections would provide a large perturbation of the atmospheric radiative balance (water is a strong absorber of IR radiation) and chemistry. Recent studies by Birks *et al.* (2007) and Pierazzo *et al.* (2010) show that even smaller impacts (500 m to 1 km in size) would cause a major perturbation of the normally dry upper atmosphere. Combining numerical impact simulations with a whole-atmosphere GCM coupled with an interactive chemistry package, Pierazzo *et al.* (2010) found that mid-latitude oceanic impacts of asteroids 1 km in diameter can produce a multiyear ozone depletion comparable to the mid-1990s ozone hole records, shown in Figure 10.4. At the surface, ozone depletion causes an increase in ultraviolet-B (UV-B, 280–315 nm) radiation that far exceeds levels currently experienced anywhere on the Earth's surface. Important biological repercussions associated with increased surface UV-B irradiance include increased incidence of erythema (skin reddening) and cortical cataracts, decreased plants' height, shoot mass and foliage area, as well as damage to molecular DNA (e.g. UNEP, 2003). Current understanding of the sensitivity of ecosystems to increased UV-B suggests that such an impact would have a long-lasting negative impact on global food production, which, in turn, may affect the sustainability of the current human population.

## 10.7   The response of the Earth system to large impacts

The K–Pg extinction so far provides us with the best information of the response of the Earth system to a large impact event.

The Late Cretaceous climate was slightly cooler than the mid-Cretaceous (up to 8–10 °C), but still much warmer than the present climate, with no evidence of extensive glaciation. Oceanic and continental data suggest equatorial palaeotemperatures similar to the present day (Wilson and Opdyke, 1996; Pearson *et al.*, 2001), and warmer high latitudes (Wolfe and Upchurch, 1987; Huber *et al.*, 1995). Across the K–Pg boundary, plant records indicate a major ecological disruption followed by quick recovery, attributed to a brief low-temperature excursion and change in the hydrologic cycle in the Northern (Wolfe, 1990; Sweet and Braman, 2001) and the Southern Hemispheres (Saito *et al.*, 1986; Vajda *et al.*, 2001). In the end, the early Palaeocene climate (~64 Ma) was very similar to that of the latest Maastrichtian (~67 Ma).

The climate change associated with a K–Pg-type atmospheric perturbation was investigated by Luder *et al.* (2003). They integrated a two-dimensional (2D), zonally averaged dynamic ocean circulation model with surface radiative fluxes modulated by a thick atmospheric dust layer from a one-dimensional radiation balance model. They found that the upper 200 m of the oceans cooled by several degrees in the first year after the impact, while deep-sea temperatures decreased by no more than a few tenths of a degree Celsius. It appears, therefore, that the dust-related climatic change did not affect the overall structure of the ocean circulation, which is the main moderator of the Earth's climate. These results are confirmed by the marine isotopic record, which shows little evidence for either warming or cooling across the K–Pg boundary (e.g. Zachos and Arthur, 1986; Zachos *et al.*, 1989).

The response of terrestrial ecosystems to a large increase in atmospheric $CO_2$ from the K–Pg impact was investigated with a dynamic vegetation–biogeochemistry model (Lomax *et al.*, 2000, 2001). Lomax *et al.* (2001) also included the cooling effect of dust/sulfates by artificially reducing the mean annual temperature by 6 °C for 100 years, and added the potential effects of wildfires by burning 25% of the vegetation carbon ($CO_2$ concentration set to 10 times pre-impact levels). They found that a 4- to 10-fold increase of the atmospheric end-Cretaceous $CO_2$ inventory causes spatially heterogeneous increases in net primary productivity, and a biotic feedback mechanism that would ultimately help climate stabilization. Although the assumed $CO_2$ increase may be a gross overestimate of the real increase from the K–Pg impact event, results are still indicative of the terrestrial ecosystem response to a strong perturbation. These models show that the initial collapse of the Earth's net primary productivity is followed by a total recovery within a decade. In particular, model results suggest that changes in productivity and vegetation biomass were larger at low latitudes, consistent with terrestrial palaeobotanical data.

## 10.8 Environmental impact effects favourable for life

Impact cratering can affect environments in a positive way, creating niches for life to flourish. Studies on this aspect of impact cratering range from the potential delivery of complex organic molecules to planetary surfaces (e.g. Pierazzo and Chyba, 1999, 2006) to the creation of the conditions for a habitat more conducive to life (e.g. Farmer, 2000; Abramov and Kring, 2004, 2005) – see Chapter 11 for an overview.

The investigation of the possibility that comet/asteroid impacts might deliver prebiotic organic molecules to Earth has provided contrasting results. Theoretical and experimental studies indicate that, at the time of the origin of life on Earth, cometary impacts could have delivered significant amounts of certain complex organic material. This is especially important in the case of very low angle impacts, which boost the concentration of important prebiotic molecules in the Earth's oceans (Pierazzo and Chyba, 1999; Blank *et al.*, 2001). On Mars, a low-angle impact would cause most of the projectile material and its precious cargo of prebiotic molecules to reach escape velocity and be lost. Loss of impactor also seems the typical outcome of impacts on other small planetary bodies, such as Europa and the Moon (Pierazzo and Chyba, 2006).

Hydrothermal systems are possible sites for the origin and early evolution of life on Earth (e.g. Farmer, 2000). Evidence of hydrothermal circulation underneath terrestrial structures abounds (see Chapter 6), suggesting that hydrothermal systems could have formed underneath large Martian impact structures as well (Newsom, 1980). Theoretical studies suggest that large impacts can generate large amounts of heat (e.g. Ivanov and Pierazzo, 2011) which could drive substantial hydrothermal activity for hundreds of thousands of years, even under cold climatic conditions, supporting the idea that impact events may have played an important biological role on early Earth and on Mars (Abramov and Kring, 2004, 2005).

## 10.9 Concluding remarks

Impact events have punctuated the Earth's geological past. They bring local if not global devastation, affecting the equilibrium of the terrestrial ecosystem, perturbing the climate and affecting the evolution of life. Smaller, more frequent, impact events affect the environment on a local scale, while larger, rarer impact events have a global reach, affecting the environment and the climate system of the entire planet. Over the last few decades, the study of the K–Pg boundary layer has provided a great deal of information on the impact event that created the Chicxulub structure and its potential for creating a worldwide environmental catastrophe.

Although only the largest (and rarest) of impact events have the capability of producing mass extinction of life, smaller events may still have dramatic consequences. Today's world is strongly affected by the rise of human civilization. Natural resources can barely support the growing world population, and we must rely on the development of new technologies (genetically enhanced crops and livestock, alternative/advanced energy sources) to support our civilization. This brings forward a new view about the consequences for our civilization of asteroidal and cometary impacts on the Earth. Little is yet known about the potential consequences of more probable non-mass-extinction-size impacts on the environment and biosphere. However, smaller impact events may affect the environment and climate enough to cause partial crop failure for a short period of time, with dramatic consequences for human civilization.

## References

Abramov, O. and Kring, D.A. (2004) Numerical modeling of an impact-induced hydrothermal system at the Sudbury crater. *Journal of Geophysical Research* 109, E10007. DOI: 10.1029/2003JE002213.

Abramov, O. and Kring, D.A. (2005) Impact-induced hydrothermal activity on early Mars. *Journal of Geophysical Research* 110, E12S09. DOI: 10.1029/2005JE002453.

Alvarez, L.W., Alvarez, W., Asaro, F. and Michel, H.V. (1980) Extraterrestrial cause for the Cretaceous–Tertiary extinction. *Science*, 208, 1095–1108.

Artemieva, N. and Morgan, J. (2009) Modeling the formation of the K/Pg boundary layer. *Icarus*, 201, 768–780.

Artemieva, N. and Pierazzo, E. (2009) The Canyon Diablo impact event: projectile motion through the atmosphere. *Meteoritics and Planetary Sciences*, 44, 25–45.

Artemieva, N. and Shuvalov, V. (2002) Shock metamorphism on the ocean floor (numerical simulations). *Deep Sea Research*, 49, 959–968.

Artemieva, N. and Shuvalov, V. (2007) 3D effects of Tunguska event on the ground and in the atmosphere. 38th Lunar and Planetary Science Conference, abstract no. 1537.

Belcher, C.M., Collinson, M.E., Sweet, A.R. *et al.* (2003) Fireball passes and nothing burns: the role of thermal radiation in the Cretaceous–Tertiary event; evidence from the charcoal record of North America. *Geology*, 31, 1061–1064.

Belcher, C.M., Collinson, M.E. and Scott, A.C. (2005) Contraints on the thermal energy released from the Chicxulub impactor; new evidence from multi-method charcoal analysis. *Journal of the Geological Society of London*, 162, 591–602.

Belcher, C.M., Finch, P., Collinson, M.E. *et al.* (2009) Geochemical evidence for combustion of hydrocarbons during the K–T impact event. *Proceedings of the National Academy of Sciences of the United States of America*, 106, 4112–4117.

Birks, J.W., Crutzen, P.J. and Roble, R.G. (2007) Frequent ozone depletion resulting from impacts of asteroids and comets, in *Comet/Asteroid Impacts and Human Society* (eds P. Bobrowsky and H. Rickman), Springer, Berlin, pp. 225–245.

Blank, J.G., Miller, G.H., Ahrens, M.J. and Winans, R.E. (2001) Experimental shock chemistry of aqueous amino acid solutions and the cometary delivery of prebiotic compounds. *Origins of Life and Evolution of the Biosphere*, 31, 15–51.

Boslough, M. and Crawford, D.A. (2008) Low-altitude airbursts and the impact threat. *International Journal of Impact Engineering*, 35, 1441–1448.

Chapman, C.R. and Morrison, D. (1994) Impact on the Earth by asteroids and comets: assessing the hazard. *Nature*, 367, 33–40.

Chyba, C.F., Thomas, P.J. and Zahnle, K.J. (1993) The 1908 Tunguska explosion: atmospheric disruption of a stony asteroid. *Nature*, 361, 40–44.

Covey, C., Thompson, S.L., Weissman, P.R. and MacCracken, M.C. (1994) Global climatic effects of atmospheric dust from an asteroid or comet impact on Earth. *Global Planetary Change*, 9, 263–273.

Day, S. and Maslin, M. (2005) Widespread sediment liquefaction and continental slope failure at the K–T boundary: the link between large impacts, gas hydrates and carbon isotope excursions, in *Large Meteorite Impacts III* (eds T. Kenkmann, F. Hörz and A. Deutsch), Geological Society of America Special Paper 384, Geological Society of America, Boulder, CO, pp. 239–258.

Farmer, J. (2000) Hydrothermal systems: doorways to early biosphere evolution. *GSA Today*, 10, 1–9.

Florensky, K.P. (1965) Preliminary results from the 1961 combined Tunguska meteorite expedition. *Meteoritika*, 23, 3–27 (in Russian).

Gault, D.E. and Sonett, C.P. (1982) Laboratory simulation of pelagic asteroidal impact: atmospheric injection, benthic topography, and the surface wave radiation field, in *Geological Implications of Impacts of Large Asteroids and Comets on Earth* (eds L.T. Silver and P.H. Schultz), Geological Society of America Special Paper 190, Geological Society of America, Boulder, CO, pp. 69–92.

Glasstone, S. and Dolan, P.J. (1977) *The Effects of Nuclear Weapons*, 3rd edn, United States Department of Defense and Department of Energy.

Goldin, T.J. and Melosh, H.J. (2009) Self-shielding of thermal radiation by Chicxulub impact ejecta: firestorm or fizzle? *Geology*, 37 (12), 1135–1138.

Harris, A. (2008) What Spaceguard did. *Nature*, 453, 1178–1179.

Harvey, M.C., Brassell, S.C., Belcher, C.M. and Montanari, A. (2008) Combustion of fossil organic matter at the Cretaceous–Paleogene (K–P) boundary. *Geology*, 36, 355–358. DOI: 10.1130/G24646A.1.

Hills, J.G., Nemchinov, I.V., Popov, S.P. and Teterev, A.V. (1994) Tsunami generated by small asteroid impacts, in *Hazards due to Comets and Asteroids* (ed. T. Gehrels), University of Arizona Press, Tucson, AZ, pp. 779–790.

Huber, B.T., Hodell, D.A. and Hamilton, C.P. (1995) Middle–Late Cretaceous climate of the southern high latitudes: stable isotopic evidence for minimal equator-to-pole thermal gradients. *Geological Society of America Bulletin*, 107, 1164–1191.

Ivanov, B.A. and Pierazzo, E. (2011) Impact cratering in $H_2O$-bearing targets on Mars: thermal field under craters as starting conditions for hydrothermal activity. *Meteoritics and Planetary Sciences*, 46, 601–619.

Ivanov, B.A., Badjukov, D.D., Yakovlev, O.I. *et al.* (1996) Degassing of sedimentary rocks due to Chicxulub impact: hydrocode and physical simulations, in *The Cretaceous–Tertiary Event and Other Catastrophes in Earth History* (eds G. Ryder, D.E. Fastovsky and S. Gartner), Geological Society of America Special Paper 307, Geological Society of America, Boulder, CO, pp. 125–139.

Jones, E.M. and Kodis, J.W. (1982) Atmospheric effects of large body impacts: the first few minutes, in *Geological Implications of Impacts of Large Asteroids and Comets on Earth* (eds L.T. Silver and P.H. Schultz), Geological Society of America Special Paper 190, Geological Society of America, Boulder, CO, pp. 175–186.

Jones, T.P. and Lim, B. (2000) Extraterrestrial impacts and wildfires. *Palaeogeography, Palaeoclimatology, Palaeoecology*, 164, 57–66.

Klaus, A.R., Norris, D., Kroon, D. and Smit, J. (2000) Impact-induced mass wasting at the K–T boundary: Blake Nose, western North Atlantic. *Geology*, 28, 319–322.

Korycansky, D.G. and Lynett, P.J. (2005) Offshore breaking of impact tsunami: the Van Dorn effect revisited. *Geophysical Research Letters*, 32, L10608. DOI: 10.1029/2004GL021918.

Korycansky, D.G. and Lynett, P.J. (2007) Run-up from impact tsunami. *Geophysical Journal International*, 170, 1076–1088.

Kring, D.A. (1995) The dimensions of the Chicxulub impact crater and impact melt sheet *Journal of Geophysical Research*, 100, 16979–16986.

Kring, D.A. (1997) Air blast produced by the Meteor Crater impact event and a reconstruction of the affected environment. *Meteoritics and Planetary Science*, 32, 517–530.

Lomax, B.H., Beerling, D.J., Upchurch Jr, G.R. and Otto-Bliesner, B.L. (2000) Terrestrial ecosystem responses to global environmental change across the Cretaceous–Tertiary boundary. *Geophysical Research Letters*, 27, 2149–2152.

Lomax, B., Beerling, D., Upchurch Jr, G. and Otto-Bliesner, B. (2001) Rapid (10-yr) recovery of terrestrial productivity in a simulation study of the terminal Cretaceous impact event. *Earth and Planetary Science Letters*, 192, 137–144.

Longo, G., Di Martino, M., Andreev, G. *et al.* (2005) A new unified catalogue and a new map of the 1908 tree fall in the site of the Tunguska cosmic body explosion. All-Russian Conference: Asteroid–Comet Hazard – 2005, Institute of Applied Astronomy of the Russian Academy of Sciences, St Petersburg, Russia, pp. 222–225.

Luder, T., Benz, W. and Stocker, T.F. (2003) A model for long-term climatic effects of impacts. *Journal of Geophysical Research*, 108, 5074. DOI: 10.1029/2002JE001894.

Matsui, T., Imamura, F., Tajika, E. *et al.* (2002) Generation and propagation of a tsunami from the Cretaceous–Tertiary impact event, in *Catastrophic Events and Mass Extinctions: Impacts and Beyond* (eds C. Koeberl and K.G. MacLeod), Geological Society of America Special Paper 356, Geological Society of America, Boulder, CO, pp. 69–77.

McGetchin, T.R., Settle, M. and Head, J.W. (1973) Radial thickness variation in impact crater ejecta: implications for lunar basin deposits. *Earth and Planetary Science Letters*, 20, 226–236.

Melosh, H.J. (1989) *Impact Cratering: A Geologic Process*, Oxford University Press, New York, NY.

Melosh, H.J. (2003) Impact-generated tsunamis: an over-rated hazard. 34th Lunar and Planetary Science Conference, abstract no. 2013.

Melosh, H.J. and Collins, G.S. (2005) Meteor Crater formed by a low-velocity impact. *Nature*, 434, 157.

Melosh, H.J., Schneider, N.M., Zahnle, K.J. and Latham, D. (1990) Ignition of global wildfires at the K/T boundary. *Nature*, 373, 399–404.

Mills, M.J., Toon, O.B., Turco, R.P. *et al.* (2008) Massive global ozone loss predicted following regional nuclear conflict. *Proceedings of the National Academy of Sciences of the United States of America*, 105, 5307–5312. DOI: 10.1073/pnas.0710058105.

Newsom, H. (1980) Hydrothermal alteration of imapct melt sheets with implications for Mars. *Icarus*, 44 (1), 207–216.

Norris, R.D., Firth, J., Blusztajn, J.S. and Ravizza, G. (2000) Mass failure of the North Atlantic margin triggered by the Cretaceous–Paleogene bolide impact. *Geology*, 28, 1119–1122.

O'Keefe, J.D. and Ahrens, T.J. (1982) Impact mechanisms of large bolides interacting with Earth and their implication to extinction mechanisms, in *Geological Implications of Impacts of Large Asteroids and Comets on Earth* (eds L.T. Silver and P.H. Schultz), Geological Society of America Special Paper 190, Geological Society of America, Boulder, CO, pp. 103–120.

Pearson, P.N., Dicthfield, P.W., Singano, J. *et al.* (2001) Warm tropical sea surface temperatures in the Late Cretaceous and Eocene epochs. *Nature*, 413, 481–487.

Pierazzo, E. (2005) Assessing atmospheric water injections from oceanic impacts. 36th Lunar and Planetary Science Conference, abstract no. 1987.

Pierazzo, E. and Artemieva, N. (2012) Local and global environmental effects of impacts on Earth. *Elements*, 8, 55–60. DOI: 10.2113/gselements.8.1.55. http://users.unimi.it/paleomag/geo2/Pierazzo&Artemieva2012.pdf (last access June 2012).

Pierazzo, E. and Chyba, C.F. (1999) Amino acid survival in large cometary impacts. *Meteoritics and Planetary Sciences*, 34, 909–918.

Pierazzo, E. and Chyba, C.F. (2006) Impact delivery of prebiotic organic matter to planetary surfaces, in *Comets and the Origin and Evolution of Life* (eds P.J. Thomas, R.D. Hicks, C.F. Chyba and C.P. McKay), Springer, Berlin, pp. 137–168.

Pierazzo, E., Kring, D.A. and Melosh, H.J. (1998) Hydrocode simulations of the Chicxulub impact event and the production of climatically active gases. *Journal of Geophysical Research*, 103, 28607–28626.

Pierazzo, E., Hahmann, A.N. and Sloan, L.C. (2003) Chicxulub and climate: effects of stratospheric injections of impact-produced S-bearing gases. *Astrobiology*, 3, 99–118.

Pierazzo, E., Garcia, R.R., Kinnison, D.E. *et al.* (2010) Ozone perturbation from medium size asteroid impacts in the ocean. *Earth and Planetary Science Letters*, 299, 263–272.

Pope, K.O. (2002) Impact dust not the cause of the Cretaceous–Tertiary mass extinction. *Geology*, 30, 99–102.

Pope, K.O., Baines, K.H., Ocampo, A.C. and Ivanov, B.A. (1997) Energy, volatile production, and climatic effects of the Chicxulub Cretaceous/Tertiary impact. *Journal Geophysical Research*, 102, 21645–21664.

Robertson, D.S., McKenna, M.C., Toon, O.B. *et al.* (2004) Survival in the first hours of the Cenozoic. *Geological Society of America Bulletin*, 116, 760–768.

Saito, T., Yamanoi, T. and Kaiho, K. (1986) End-Cretaceous devastation of terrestrial flora in the boreal Far East. *Nature*, 323, 253–255.

Schulte, P., Alegret, L., Arenillas, I. *et al.*, (2010) The Chicxulub asteroid impact and mass extinction at the Cretaceous–Paleogene boundary. *Science*, 327, 1214–1218. DOI: 10.1126/science.1177265.

Schultz, P.H. and Gault, D.E. (1975) Seismic effects from major basin formation on the Moon and Mercury. *The Moon*, 12, 159–177.

Shoemaker, E.M., Chao, E.C.T. and Madsen, B.M. (1960) First natural occurrence of coesite from Meteor Crater, Arizona. *Science*, 132, 220–222.

Shoemaker, E.M., Weissman, P.R. and Shoemaker, C.S. (1994) The flux of periodic comets near Earth, in *Hazards due to Comets and Asteroids* (ed. T. Gerhels), University of Arizona Press, Tucson, AZ, pp. 313–336.

Shuvalov, V.V. and Artemieva, N.A. (2002) Numerical modeling of Tunguska-like impacts. *Planetary and Space Science*, 50, 181–192.

Sweet, A.R. and Braman, D.R. (2001) Cretaceous–Tertiary palynofloral perturbations and extinctions within the Aquilapollenites Phytogeographic Province. *Canadian Journal of Earth Sciences*, 38, 249–269.

Toon, O.B., Zahnle, K., Morrison, D. *et al.* (1997) Environmental perturbations caused by the impacts of asteroids and comets. *Reviews of Geophysics*, 35, 41–78.

Turco, R.P., Toon, O.B., Park, C. *et al.* (1982) An analysis of the physical, chemical, optical and historical impacts of the 1908 Tunguska meteor fall. *Icarus*, 50, 1–52.

UNEP (2003) Environmental effects of ozone depletion and its interactions with climate change: 2002 assessment. *Photochemical and Photobiological Sciences*, 2 (UNEP Special Issue), 1–4.

Vajda, V., Raine, J.I. and Hollis, C.J. (2001) Indication of global deforestation at the Cretaceous–Tertiary boundary by New Zealand fern spike. *Science*, 294, 1700–1702.

Van Dorn, W.G., LeMéhauté, B. and Whang, L.-S. (1968) Handbook of explosion-generated water waves. Vol. 1 – state of the art. TC-130 Final Report Tetra Technologies.

Vasilyev, N.V. (1998) The Tunguska meteorite problem today. *Planetary and Space Science*, 46, 129–150.

Ward, S.N. and Asphaug, E. (2000) Asteroid impact tsunami: a probabilistic hazard assessment. *Icarus*, 145, 64–78.

Weiss, R., Wünnemann, K. and Bahlburg, H. (2006) Numerical modelling of generation, propagation and run-up of tsunamis caused by oceanic impacts: model strategy and technical solutions. *Geophysical Journal International*, 167, 77–88.

Wilson, P.A. and Opdyke, B.N. (1996) Equatorial sea-surface temperatures for the Maastrichtian revealed through remarkable preservation metastable carbonate. *Geology*, 24, 555–558.

Wolbach, W.S., Lewis, R.S. and Anders, E. (1985) Cretaceous extinctions: evidence for wildfires and search for meteoritic materials. *Science*, 230, 167–170.

Wolbach, W.S., Widicus, S. and Kyte, F.T. (2003) A search for soot from global wildfires in central Pacific Cretaceous–Tertiary boundary and other extinction and impact horizon sediments. *Astrobiology*, 3, 91–97.

Wolfe, J.A. (1990) Palaeobotanical evidence for a marked temperature increase following the Cretaceous/Tertiary boundary. *Nature*, 343, 153–156.

Wolfe, J.A. and Upchurch Jr, G.R. (1987) North American nonmarine climates and vegetation during the late Cretaceous. *Palaeogeography, Palaeoclimatology, Palaeoecology*, 61, 33–77.

Wünnemann, K., Weiss, R. and Hofmann, K. (2007) Characteristics of oceanic impact-induced large water waves – re-evaluation of the tsunami hazard. *Meteoritics and Planetary Sciences*, 42, 1893–1903.

Zachos, J.C. and Arthur, M.A. (1986) Paleoceanography of the Cretaceous/Tertiary boundary event: inferences from stable isotopic and other data. *Paleoceanography*, 1, 5–26.

Zachos, J.C., Arthur, M.A. and Dean, W.E. (1989) Geochemical evidence for suppression of pelagic marine productivity at the Cretaceous/Tertiary boundary. *Nature*, 337, 61–64.

Zahnle, K.J. (1990) Atmospheric chemistry by large impacts, in *Global Catastrophes in Earth History* (eds V. Sharpton and P. Ward), Geological Society of America Special Paper 247, Geological Society of America, Boulder, CO, pp. 271–288.

Zahnle, K. (1996) Dynamics and chemsitry of SL9 plumes, in *The Collision of Comet Shoemaker–Levy 9 and Jupiter* (eds K.S. Noll, H.A. Weaver and P.D. Feldman), Cambridge University Press, Cambridge, pp. 183–212.

# ELEVEN

# The geomicrobiology of impact structures

**Charles S. Cockell\*, Gordon R. Osinski[†] and Mary A. Voytek[‡]**

*\*School of Physics and Astronomy, University of Edinburgh, Edinburgh, EH9 3JZ, UK*
*[†]Departments of Earth Sciences/Physics and Astronomy, University of Western Ontario, 1151 Richmond Street, London, ON, N6A 5B7, Canada*
*[‡]US Geological Survey, MS 430, 12201 Sunrise Valley Drive, Reston, VA 20192, USA*

## 11.1 Introduction

Since the time of the origin of life, the Earth has been bombarded by asteroid and comets (French, 2004). Insofar as these energetic and perturbing events would be expected to alter access to nutrients and energy and change the availability of habitats, then one would predict that they would have an important influence on the conditions for life (Cockell *et al.*, 2002). Impact events are much less common on Earth today than they were in the past, particularly during the period of late heavy bombardment, during which microbial life may have emerged. In the early Achaean, environmental perturbations caused by large impacts are thought to have been so great that the oceans may have boiled (Maher and Stevenson, 1988; Sleep *et al.*, 1989), periodically and locally threatening conditions for microorganisms, although not necessarily on a planetary scale (Abramov and Mojzsis, 2009). Although impacts can change conditions on a global scale, most alter the environmental conditions on a local scale (Toon *et al.*, 1997; Kring, 1997, 2003).

The first stage of a meteorite impact is a contact and compression stage during which the impactor makes contact with the ground or water (see Chapter 3), followed by an excavation phase during which a bowl-shaped 'transient' crater cavity is formed (see Chapter 4); for diameters greater than 2–4 km on Earth, subsequent modification can occur, resulting in a central peak and/or peak ring, depending on the magnitude of the event (see Chapter 5; Melosh, 1989). The kinetic energy of such an event is enormous on account of the high velocity of objects intersecting with the Earth's orbit. The mean impact velocity with the Earth is about $21\,\mathrm{km\,s^{-1}}$ (Stuart and Binzel, 2004). The energy is released in large part as heat and the longevity of this thermal excursion depends upon the target lithology, the availability of water and local climate, among other factors (Naumov, 2005; Osinski et al., 2005a; Versh *et al.*, 2005; Abramov and Kring, 2007). The formation of a crater cavity during the excavation and modification stages influences the hydrological cycle, particularly through the formation of long-lived water bodies. About half of the impact craters known on Earth today with surface expressions host some type of intra-crater water body. Lakes formed in the intra-crater cavity of land-based craters can persist for millions of years and are known to host unique microbiotas (Schoeman and Ashton, 1982; Ashton and Schoeman, 1983; Ashton, 1999; Gronlund *et al.*, 1990; Maltais and Vincent, 1997; Cremer and Wagner, 2003; Wani *et al.*, 2006; Surakasi *et al.*, 2007; Joshi *et al.*, 2008). Lakes are eventually subject to breach of the crater rim or infilling of the crater (they are, of course, not relevant for impacts in the marine environment), making them, in most cases, more short-lived than changes in the target geology.

The biological recovery following the perturbation caused by an impactor can be broadly split into three phases, which were first proposed in analogy to successional changes following other ecological disturbances (Cockell and Lee, 2002). The first phase is one of thermal biology, which is in essence any biological alteration associated with the transient thermal pulse delivered by the impactor. In the case of small impactors or environments where cooling is rapid, this phase may be inconsequential. The second phase is a phase of post-impact succession and climax, in which ecosystems become established in or on the various substrates associated with the impact, such as an intra-crater lake or impact-generated breccia materials. The third and final phase is a phase of assimilation, during which the unique ecology of the impact site is lost as the crater is eroded. This phase may never actually be achieved in a planetary lifetime if the crater remains buried and is not subducted. However, small craters can be rapidly eroded by aeolian and aqueous activity.

Insofar as no solar system formation process is known that is completely free of remnant debris, then impact events are a universal phenomenon and would be expected to bring destructive energy to the surface of most terrestrial-type rocky planets (some planets, such as Venus, have atmospheres sufficiently thick to vaporize many impactors, but in the case of Venus the thick $CO_2$ atmosphere is the cause of a runaway greenhouse effect that

*Impact Cratering: Processes and Products*, First Edition. Edited by Gordon R. Osinski and Elisabetta Pierazzo.
© 2013 Blackwell Publishing Ltd. Published 2013 by Blackwell Publishing Ltd.

makes the planet inimical to life). Thus, impact events are of astrobiological importance in understanding the conditions for emergence and persistence of life elsewhere.

In this chapter we will review data on the influence of asteroid and comet impacts on the geomicrobiological conditions at the site of impact. In this review we are primarily concerned with the colonization of altered target materials.

## 11.2   Physical changes

### 11.2.1   Impact fracturing

The most obvious influence of impacts on the local geology is to fracture rocks, increasing their porosity and permeability. Biologically, this might be expected to have two major consequences. First, space for microbial growth is increased by the formation of fractures. However, at least in the deep subsurface, the limitation for microorganisms is generally not the available space, but the presence of sufficient nutrients or redox couples for growth (Horsfield *et al.*, 2007). Nevertheless, impact-induced fracturing can also have the effect of increasing fluid flow through target material, which might increase the availability of nutrients and redox couples after temperatures have become suitable for life. Evidence for impact-induced fracturing in large quantities of target rock is found in the data of electrical conductivity in and around the Siljan impact structure, Sweden, where conductivity diminishes outwards from the centre of the structure (Henkel, 1992). The increasing electrical conductivity towards the centre of the structure is proposed to be caused by the presence of saline fluids in pores and fractures – an indirect demonstration of impact-induced fracturing and the associated increase in fluid paths. In this specific case the age of the crater is $376.8 \pm 1.7$ Ma, making the present-day observations very late post-impact.

The hypothesis that this would influence the conditions for microorganisms is supported by observations. An increase in microbial abundance is associated within a region below a granite megablock in the region corresponding to suevite and fractured schist and pegmatite rock in the deep subsurface of the Chesapeake Bay impact structure. The Chesapeake Bay impact structure was drilled by the International Continental Drilling Program (ICDP) and United States Geological Survey (USGS) in 2005 (Gohn *et al.*, 2006, 2008), during which robust microbiological contamination controls were implemented (Gronstal *et al.*, 2009). The structure is a buried upper Eocene impact structure formed by an impact into the continental shelf of eastern North America. It is the best-preserved example of a large marine impact on Earth (Poag *et al.*, 1992, 2004; Powars *et al.*, 1993; Koeberl *et al.*, 1996; Poag, 1999; Powars and Bruce, 1999). The crater has a diameter of approximately 85 km and now lies in the Atlantic margin of Virginia, USA (Fig. 11.1a; Horton *et al.*, 2005). It is comprised of an approximately 900 m high central uplift surrounded by a central crater (~35 km diameter) which is surrounded by an annular trough. Two coreholes (A and B) were drilled into the structure to a total depth of 1.76 km.

Cell numbers were around $10^6\,\mathrm{g}^{-1}$ in the impact melt-bearing and lithic impact breccias at a depth of 1397–1551 m (Cockell

*et al.*, 2009a; Fig. 11.1b). The upper part of the section from 1397 to 1474 m contained 20–30% impact-melt volume (Horton *et al.*, 2009). Other minerals such as quartz showed shock features, including planar deformation features. The shock features and presence of melt in the upper part of the section shows that the section (or its constituent minerals) was subjected to sterilizing temperatures during impact. The quantity of melt phases suggests an average temperature at the time of deposition of greater than 350 °C (Malinconico *et al.*, 2009), above the upper limit for life (>121 °C steam or >160 °C dry heat). Although there is no evidence to support regional lateral advection, microorganisms could have gradually diffused in from nearby clasts that were not sterilized or may have been carried in by compaction-driven vertical advection from the permeable schist and pegmatite region below (Sanford *et al.*, 2007). Once ambient temperatures within the structure dropped below the upper temperature limit for microbial growth, colonization could occur.

In the lower part of the colonized section in the Chesapeake structure (1551–1766 m), the rocks do not have shock features, but breccia veins and dykes are present. They contain shock-deformed clasts. Fracture networks are associated with the dykes and some narrow veins show micro-fracture networks. These features indicate that deformation and fracture formation occurred, and probably continued, after the impact event. The presence of breccias within the material suggests that dilatancy, or the opening of fractures, occurred during their emplacement before the introduction of a sedimentary cover, which would have contributed to increasing permeability for biological recolonization.

Colonization of the subsurface breccia is reported in the 1.07 Ma Bosumtwi impact structure in which archaeal lipids have been detected (Escala *et al.*, 2008). The authors proposed three possible sources of the lipids. They might be pre-impact in origin, derived from the hydrothermal system in the crater or washed into the lake sediments in the crater after the impact. If the archaeal lipids are post-impact, then the data would be another example of deep subsurface colonization of brecciated impact materials.

The advantages to a biota to be gained from the impact-induced fracturing of rock are particularly evident in surface extreme environments where, if conditions are hostile, the fractures inside rocks provide spaces for microorganisms to grow that are protected from the external conditions. Two cases that can be considered are chasmoendolithic habitats (i.e. habitats within fractures linked to the surface of the rock) and cryptoendolithic habitats (i.e. habitats within the interstices of the rock structure). Insofar as pore spaces within a rock are ultimately connected to the surface, then chasmoendolithic habitats would seem to have little that distinguishes them from cryptoendolithic habitats (Fig. 11.2). However, the former can be understood as growth on the surface of contiguous cracks or fractures that connect the surface to the interior of the rock, whereas the latter can be understood as growth within the pore spaces of the rock.

In the Haughton impact structure located in a polar desert environment in the Canadian High Arctic (Fig. 11.3), both of these habitats, generated by impact, are observed to have provided new microenvironments for microorganisms (Cockell *et al.*, 2002). Figure 11.4 (see also Plate 25) shows phototrophs associated with a chasmoendolithic habitat in shocked dolomite within

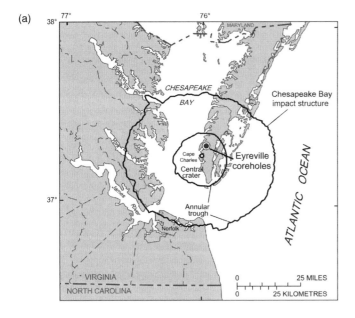

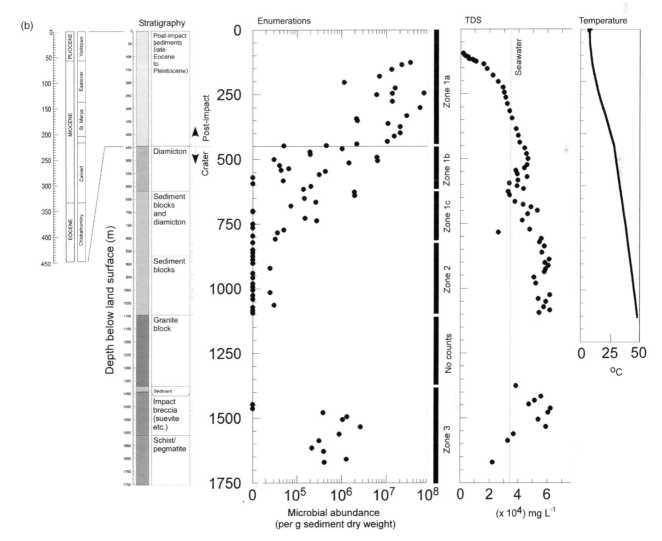

**Figure 11.1** (a) Location of the Chesapeake impact crater and location of drill site (Eryeville) within the structure for the 2005 drilling project (from Gohn *et al.* (2008)). (b) Microbial enumerations (log abundance per gram dry weight) through the Chesapeake Bay impact structure. The stratigraphy, total dissolved solids(TDS) concentrations and temperature profile (to 1100 m) are shown.

the structure (Fig. 11.3). The block is just one example of shocked dolomite blocks that were examined in the crater. These habitats are clearly not unique to impacts, since freeze–thaw, and tectonic activity over the longer term, for instance, fracture rocks. Impact events have the potential to increase the abundance of fractures

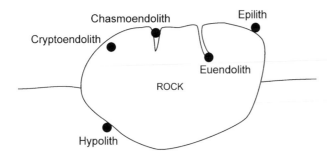

**Figure 11.2** Schematic of nomenclature of organisms that inhabit rocks. Cryptoendoliths inhabit the rock interstices, chasmoendoliths inhabit macroscopic cracks, euendoliths actively bore into rocks, hypoliths live on the underside of rocks and epiliths live on the surface of rocks.

for microorganisms, whether in the surface or subsurface. In the case of the Haughton impact structure, which is approximately 39 million years old, these habitats are clearly not immediately post-impact, but the rocks still retain the fractures and pore spaces generated by impact metamorphism, so that the present-day colonization of the rocks provides valid insights into the immediate effects of impact on lithic habitats.

### 11.2.2  Complex changes in bulk characteristics

An effect more specific to impacts is the heating and pressurization of rocks over very short time scales, leading to gross physical changes in the bulk characteristics of the rock. In the Haughton structure this is manifested most clearly in the cryptoendolithic colonization of shocked gneisses, which are colonized by cyano-bacterial genera that inhabit the pore spaces within shocked material exposed on the carbonate-rich melt rock hills (Cockell *et al.*, 2002, 2003a, 2005). One of the highest abundances of cryptoendolithic habitats is on Anomaly Hill (Fig. 11.3), a local-ized area with a high abundance of heavily shocked clasts (Osinski *et al.*, 2005b). In contrast to the impact-induced chasmoendo-lithic habitats, the shocked gneisses have a highly porous glassy

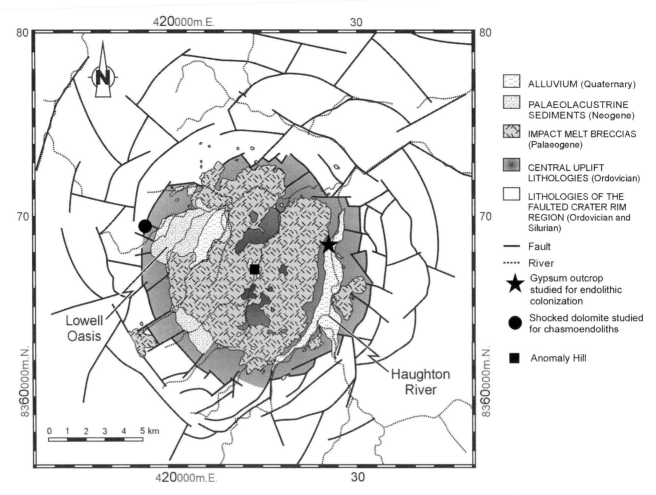

**Figure 11.3**  The Haughton impact structure showing location of some sites whose present-day geomicrobiology is described in this chapter. Modified from Osinski *et al.* (2005a).

**Figure 11.4** Chasmoendolithic colonization of fractured rock. (a) Block of heavily fractured dolomite at location 75°24.397′N, 89°49.894′W. Scale bar 1 m. (b) Close-up showing fractures and colonization along surfaces of fragments of rock removed from the block (centre). (c) Bright-field micrograph of algal and cyanobacterial colonists of the block. (See Colour Plate 25)

structure (Fig. 11.5; Metzler *et al.*, 1988) with sufficient permeability to allow the organisms not merely to invade the rock along any fractures connected to the surface, but also to grow through the spaces within the rock, forming coherent endolithic bands. The increase in porosity provides space for any microorganisms to colonize the rock. The density and porosity of shocked rocks (shocked to >20 GPa) is on the order of $1 \times 10^3\,\mathrm{kg\,m^{-3}}$ and 18.3% respectively. The low-shocked or unshocked rocks have a more typical density and porosity of greater than $2.5 \times 10^3\,\mathrm{kg\,m^{-3}}$ and 9.3% respectively.

Colonization by cyanobacteria is of specific interest because of their requirement for light. The increase in permeability caused

by impact increased the translucence of the rock. Transmission at 680 nm, the chlorophyll *a* absorption maximum, is increased by approximately an order of magnitude in the shocked rocks compared with the unshocked material. The lower limit of the band of colonization is established by the depth at which light is reduced to below the minimum for photosynthesis. In the shocked gneisses at Haughton this was calculated to be 3.6 mm in the rocks studied by Cockell *et al.* (2002), although this will clearly vary within rocks (caused by heterogeneous fracturing and porosity) and with shock level. The cyanobacteria that inhabit the rocks are not unique to the crater; they are found on the surface and underside of other rocks in the Arctic (Cockell and Stokes, 2004, 2006). Thus, the habitat hosts opportunists.

Like other cryptoendolithic habitats, the shocked gneisses at Haughton provide their inhabitants with physical advantages in addition to the availability of light for phototrophs. One of these is the elevated temperature experienced within the microenvironment of the rock (Fig. 11.6b) compared with air temperatures. As the air temperatures in the Arctic are generally low (Fig. 11.6a), then elevated temperatures would be important during the brief growing season. Similar to temperature excursions recorded in hypolithic habitats in the Arctic (Cockell and Stokes, 2006), the interior of the endolithic habitat was predicted to protect against some freeze–thaw events.

These expectations are confirmed experimentally. Temperatures within a gneissic endolithic habitat were measured from 18 to 26 July 2002 (using a rock of dimensions $6 \times 6 \times 7\,\mathrm{cm}^3$). Temperatures were measured using copper–constantan thermocouples attached to a Campbell CR-10X datalogger set to read at 20 s intervals and using a CR10TCF internal temperature reference. Measurements of ground temperatures near the rock were also obtained over a 1 year period at 2 h intervals from 14 August 2001 to 10 July 2002.

During the 9 days of measurements of the rock surface and interior the mean air temperature was 4.51 °C, the mean rock surface temperature was 5.49 °C and the mean temperature in the endolithic habitat was 5.93 °C (Fig. 11.6b). The highest temperature recorded in the endolithic habitat was 21.78 °C at 13:17 on 20 July. The corresponding air temperature was 10.57 °C and the rock surface temperature was 16.11 °C. The coldest temperature recorded in the endolithic habitat was −1.45 °C at 01:24 on 24 July. The corresponding air temperature was −1.54 °C and the rock surface temperature was −1.39 °C. During this period, the endolithic temperature dropped below freezing at 23:08 on 23 July and remained below freezing until 07:50 the following morning. During the period of measurement, the only other time that the air temperature dropped to below freezing was at 08:04 on 18 July. At this time the endolithic temperature was 1.25 °C and it remained above freezing, showing how the endolithic habitat can protect against freeze–thaw. From 14 August 2001 to 10 July 2002 the mean ground temperature at the field site was −20.34 °C and the minimum temperature was −45.35 °C (Fig. 11.6a). Of the 331 days during which measurements were made, 38 days showed a diurnal freeze–thaw cycle.

The second advantage to be obtained in an exposed shocked endolithic habitat is the extended period of liquid water availability, since water from snowmelt at the beginning of the season or brief rain events later in the season is trapped within the pore

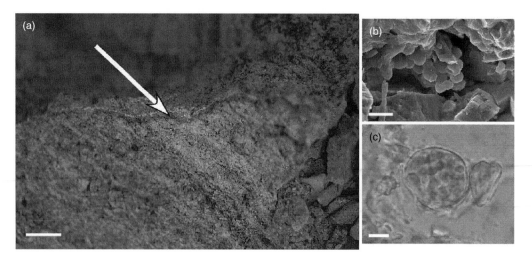

**Figure 11.5**    Cryptoendolithic colonization of impact shocked gneiss. (a) Band of cryptoendolithic colonization of shocked gneiss (scale bar 1 cm). (b) Secondary scanning electron image of cyanobacterial colony within cryptoendolithic zone (scale bar 10 μm). (c) Bright-field micrograph of presumptive *Chroococcidiopsis* sp. colony in cryptoendolithic zone (scale bar 10 μm). Modified from Cockell *et al.* (2002). (See Colour Plate 26)

spaces of the shocked rocks (Fig. 11.7). The interior of the rock shields the organisms from wind-induced desiccation on the surface of the rock. This can be demonstrated with a simple experiment. Two pieces of shocked gneiss with different mass (a: 207.2 g; b: 68.9 g) were collected. At the beginning of the experiment the rocks were immersed in a pan of water for 4 h to simulate immersion in snowmelt and placed back in the field site. They were weighed at 4 h intervals over 8 days from 17 to 25 July 2002. After this experiment, their dry mass was determined after heating the rocks at 105 °C for 48 h in a thermally controlled oven. This value was subtracted from the field measurements to determine the moisture content of the rocks over time.

The rocks could retain water for days after saturation. After artificial immersion in water for 4 h, the larger of the specimens had drawn up 8.4% of its dry mass in water and the smaller specimen 6.1% (Fig. 11.7). Following exposure to field conditions without rain, the larger specimen had retained 3.3% its mass in water after 48 h and 1.4% after 5 days. The smaller specimen (with a smaller surface area, but a larger surface area to volume ratio) had retained 0.9% of its mass in water after 48 h and 0.4% after 5 days. After natural rain events, water uptake was rapid. After the delivery of 1 mm of rain in 1 h (first rain event in Fig. 11.7), the larger specimen had increased its mass in water (expressed as percentage of the dry mass) from 1.3% to 4.8%. The smaller specimen increased its mass from 0.3% to 5.2%. After the second, more continuous, rain event and at the end of the experiment, the large specimen had increased its mass by 11.2% and the smaller specimen by 9.0 %. The extended rain event led to a greater mass gain than immersion of the whole rock in water for 4 h. This is probably caused by air becoming trapped within the rock during immersion, whereas during the rain event a more thorough saturation was achieved.

The third advantage provided by the endolithic habitat is protection against UV radiation. Exposure to UV radiation was evaluated (Rettberg *et al.*, 1999; Cockell *et al.*, 2003b; Rettberg and

Cockell, 2004) using a UV radiation biodosimeter (Puskeppeleit *et al.*, 1992; Quintern *et al.*, 1992, 1994). The experiments were placed in an unshaded location for 10 days (18–28 July 2001). The dosimeters were retrieved ($n=1$ under each rock section). Control dosimeters were run alongside the rock sections and two were retrieved after each of 24, 48, 72 and 120 h.

After 24 h, the control dosimeter spores were all killed, as they were for longer time periods. After 10 days only the thinnest section of gneiss (0.08 ± 0.02 mm) had measurable spore inactivation. The next thickest section (0.13 ± 0.03 mm) had no measurable spore inactivation, in common with all of the other sections. If it is assumed that the dosimeters exposed for 24 h are only just fully exposed (a conservative estimate), then the data show that a 1 mm thick rock section is providing at least a two orders of magnitude reduction in UV exposure. Over the entire summer season in the Arctic, organisms under even 1 mm of gneiss would receive a total UV radiation dose not much greater than that received in 1 day of exposure to unattenuated UV radiation.

Although the gneissic habitats resemble endolithic habitats found in sedimentary rocks such as sandstones and limestones (Friedmann, 1982; Weber *et al.*, 1996; Büdel *et al.*, 2004; Omelon *et al.*, 2006), a significant factor is that they represent the formation of an endolithic habitat in crystalline rocks – substrates which are usually unsuitable for endolithic colonization. There are no other reports of gneissic rocks containing endolithic organisms at the time of writing. This observation is important because it shows that the impact event has not merely increased the abundance of a geomicrobiological habitat, but that it has generated an entirely new, or at least extremely rare, type of cryptoendolithic habitat (Cockell *et al.*, 2002).

From the point of view of post-impact succession in the surface environment, these data show that phototrophs – in analogy to volcanically perturbed environments (Carson and Brown, 1978; Fermani *et al.*, 2007; Herrera *et al.*, 2009) – may be some of

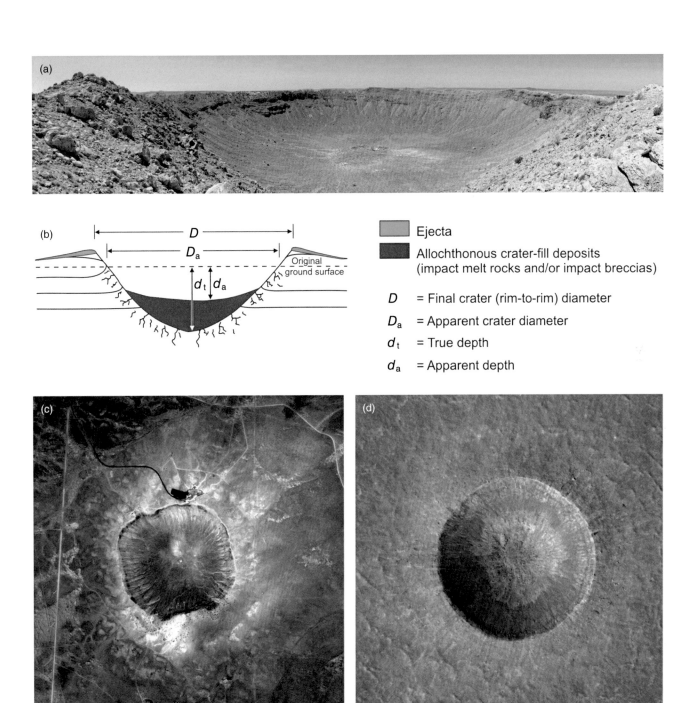

**Plate 1 (Figure 1.1)**  Simple impact craters. (a) Panoramic image of the 1.2 km diameter Meteor or Barringer Crater, Arizona. (b) Schematic cross-section through a simple terrestrial impact crater. Fresh examples display an overturned flap of near-surface target rocks overlain by ejecta. The bowl-shaped cavity is partially filled with allochthonous unshocked and shocked target material. (c) A 2 m per pixel true-colour image of Barringer Crater taken by WorldView2 (north is up). Image courtesy of Livio L. Tornabene and John Grant. (d) Portion of Lunar Reconaissance Orbiter Camera (LROC) image M122129845 of the 2.2 km diameter Linné Crater on the Moon (NASA/GSFC/Arizona State University).

*Impact Cratering: Processes and Products*, First Edition. Edited by Gordon R. Osinski and Elisabetta Pierazzo.
© 2013 Blackwell Publishing Ltd. Published 2013 by Blackwell Publishing Ltd.

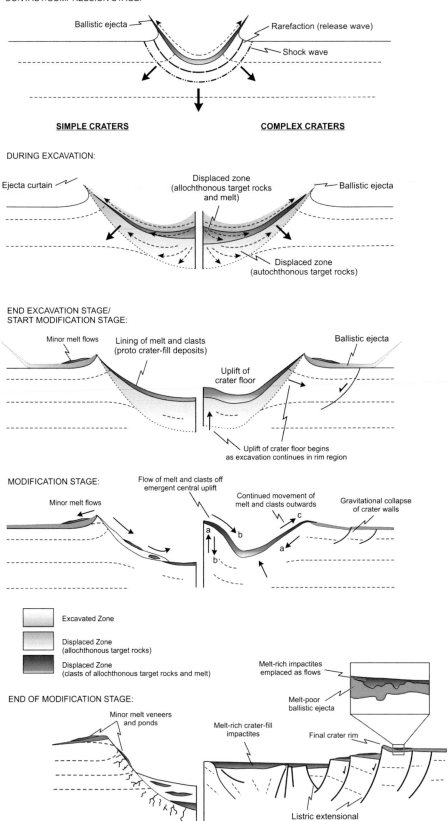

**Plate 2 (Figure 1.4)** Series of schematic cross-sections depicting the three main stages in the formation of impact craters. This multi-stage model accounts for melt emplacement in both simple (left panel) and complex craters (right panel). For the modification stage section, the arrows represent different time steps, labelled 'a' to 'c'. Initially, the gravitational collapse of crater walls and central uplift (a) results in generally inwards movement of material. Later, melt and clasts flow off the central uplift (b). Then, there is continued movement of melt and clasts outwards once crater wall collapse has largely ceased (c). Modified from Osinski *et al.* (2011).

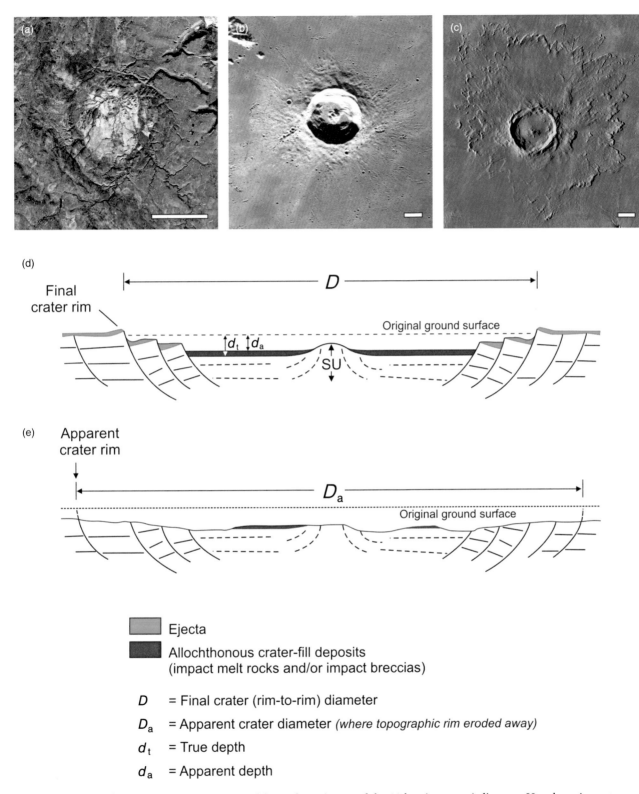

**Plate 3 (Figure 1.6)** Complex impact craters. (a) *Landsat 7* image of the 23 km (apparent) diameter Haughton impact structure, Devon Island, Canada. (b) Portion of *Apollo 17* metric image AS17-M-2923 showing the 27 km diameter Euler Crater on the Moon. Note the well-developed central peak. (c) Thermal Emission Imaging System (THEMIS) visible mosaic of the 29 km diameter Tooting Crater on Mars (NASA). Note the well-developed central peak and layered ejecta blanket. Scale bars for (a) to (c) are 10 km. (d) Schematic cross-section showing the principal features of a complex impact crater. Note the structurally complicated rim, a down-faulted annular trough and a structurally uplifted central area (SU). (e) Schematic cross-section showing an eroded version of the fresh complex crater in (d). Note that, in this case, only the apparent crater diameter can typically be defined.

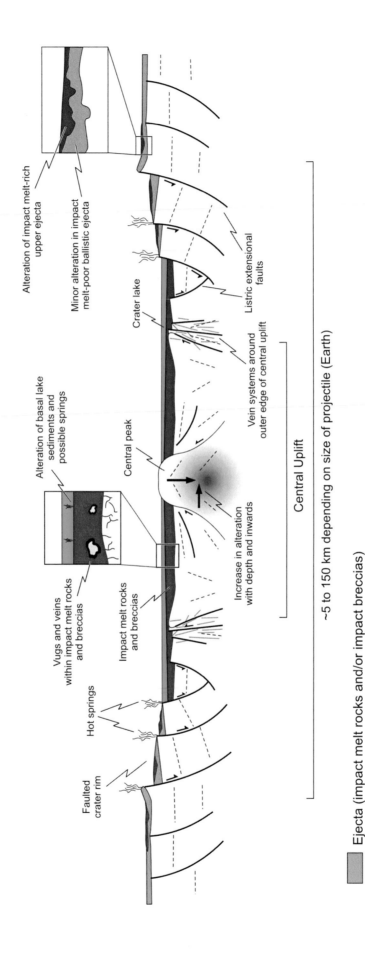

**Plate 4 (Figure 1.7)** Distribution of hydrothermal deposits within and around a typical complex impact crater. Modified from Osinski *et al.* (2012).

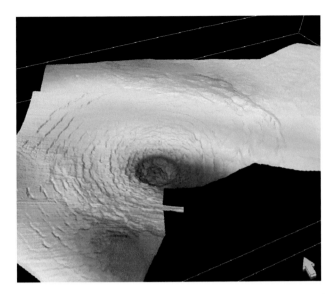

**Plate 5 (Figure 1.9)**  Perspective view of the top chalk surface at the Silverpit structure, North Sea, UK, a suspected meteorite impact structure. The central crater is 2.4 km wide and is surrounded by a series of concentric faults, which extend to a radial distance of approximately 10 km from the crater centre. False colours indicate depth (yellow: shallow; purple: deep). Image courtesy of Phil Allen and Simon Stewart.

**Plate 6 (Figure 1.10)** Field images of impactites. (a) Oblique aerial view of the approximately 80 m high cliffs of impact melt rock at the Mistastin impact structure, Labrador, Canada. Photograph courtesy of Derek Wilton. (b) Close-up view of massive fine-grained (aphanitic) impact melt rock from the Discovery Hill locality, Mistastin impact structure. Camera case for scale. (c) Coarse-grained granophyre impact melt rock from the Sudbury Igneous Complex, Canada. Rock hammer for scale. (d) Impact melt-bearing breccia from the Mistastin impact structure. Note the fine-grained groundmass and macroscopic flow-textured silicate glass bodies (large black fragments). Steep Creek locality. Marker/pen for scale. (e) Polymict lithic impact breccias from the Wengenhousen quarry, Ries impact structure, Germany. Rock hammer for scale. (f) Carbonate melt-bearing clast-rich impact melt rocks from the Haughton impact structure. Penknife for scale. This lithology was originally interpreted as a clastic or fragmental breccia (Redeker and Stöffler, 1988), but was subsequently shown to be an impact melt-bearing impactite (Osinski and Spray, 2001; Osinski *et al.*, 2005b).

**Plate 7 (Figure 1.12)** Field images of Dakhleh Glass, the potential product of an airburst event (Osinski *et al.*, 2007; Osinski *et al.*, 2008c). (a) Area of abundant Dakhleh Glass lagged on the surface of Pleistocene lacustrine sediments; the arrows point to some large Dakhleh Glass specimens. (b) Upper surfaces of many large Dakhleh Glass lag samples appear to be in place and are highly vesiculated. This contrasts with the smooth, irregular lower surfaces. (c) In cross-section, it is clear that there is an increase in the number of vesicles towards the upper surface. Together, these features are indicative of ponding of melt and volatile loss through vesiculation. (d) Highly vesicular pumice-like Dakhleh Glass sample. 7 cm lens cap for scale.

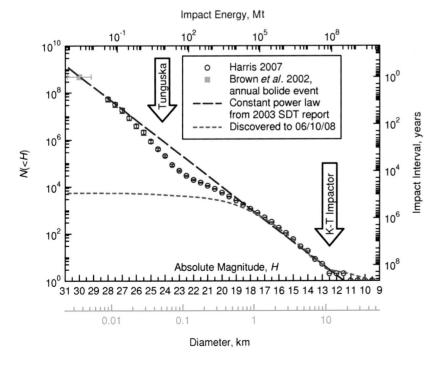

**Plate 8 (Figure 2.2)** Open blue circles represent the cumulative number of NEOs brighter than a given absolute magnitude $H$, defined as the visual magnitude $V$ that an asteroid would have in the sky if observed at 1 AU distance from both the Earth and the Sun, at zero phase angle. A power-law function dashed blue line) is shown for comparison. Ancillary scales give impact interval (right), impact energy in megatons TNT for the mean impact velocity of approximately 20 km s$^{-1}$ (top) and the estimated diameter corresponding to the absolute magnitude $H$ (second scale at bottom). Courtesy of A.W. Harris.

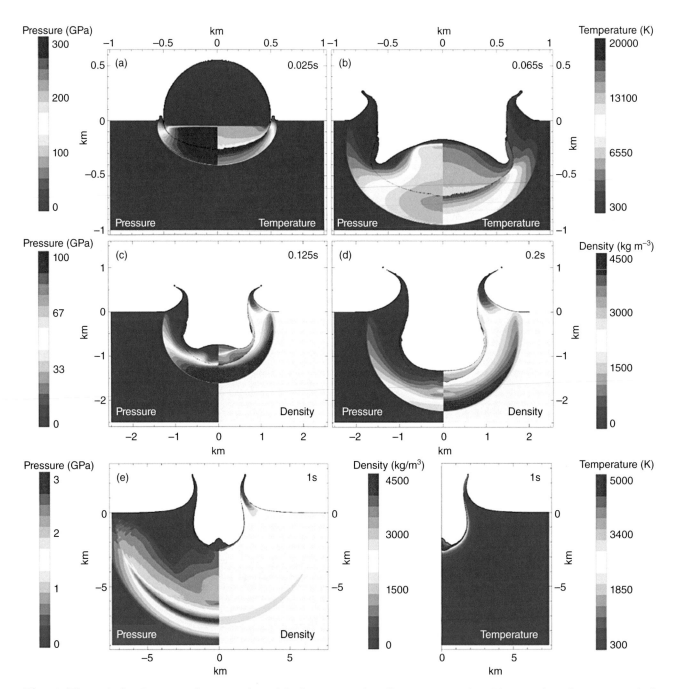

**Plate 9 (Figure 3.1)** Contact and compression of the impactor, as it strikes a target surface. Five snapshots from a numerical impact simulation, between 0.025 and 1 s after initial contact, are shown of a 1 km diameter dunite impactor striking a granite target at 18 km s$^{-1}$ (the average impact velocity for asteroidal impactors on Earth). Note the change in the length, pressure and temperature scales between the rows. (a) Impactor just after contact with the target: a small region of high pressure develops along the interface. Both target and impactor are compressed and begin to distort in shape; the rear of the impactor is unaffected by events at the leading edge. (b) The shock wave in the impactor reaches its rear surface and (c) is reflected as a rarefaction (release) wave, unloading the impactor to low pressure. (d) The shock wave in the target propagates outward in an approximately hemispherical geometry, closely followed by the release wave. (e) Behind the detached shock wave, the impact plume (vaporized impactor and proximal target) begins to expand and the excavation flow is initiated. From Collins *et al.* (2012).

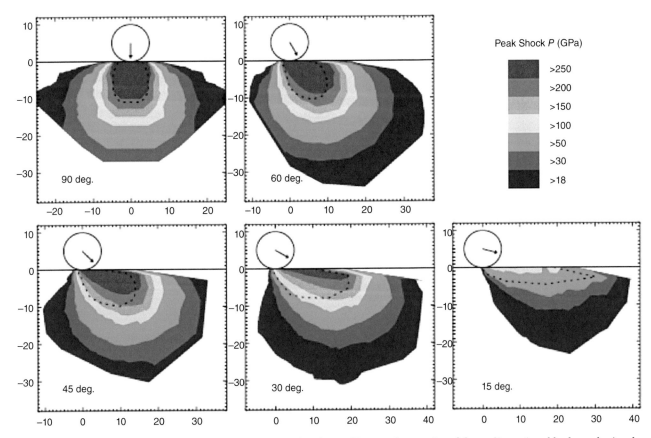

**Plate 10 (Figure 3.5)**   Peak shock pressure contours in the plane of impact for a series of three-dimensional hydrocode simulations, at various impact angles. Dashed black line represents the isobaric core. The projectile, 10 km in diameter, is shown for scale. Vectors illustrate the direction of impact. From Pierazzo and Melosh (2000b).

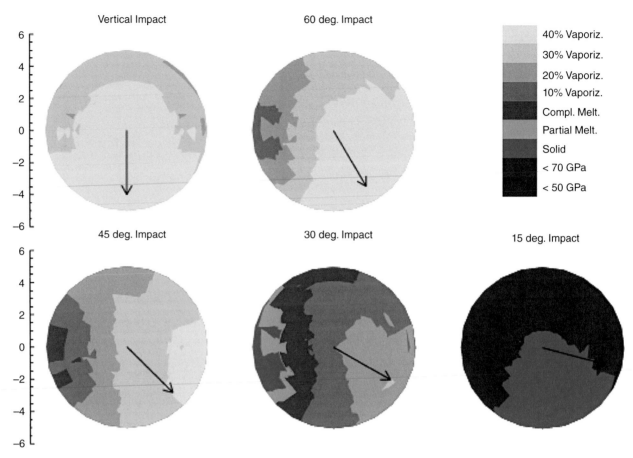

**Plate 11 (Figure 3.7)**    Distribution of melting and vaporization inside a dunite projectile impacting a dunite target at 20 km s$^{-1}$, from three-dimensional hydrocode simulations. The maximum shock pressures corresponding to the various contours are: solid, 135 GPa; partial melting, 149 GPa; complete melting, 186 GPa; 10% vaporization, 224.5 GPa; 20% vaporization, 286.5 GPa; 30% vaporization, 363 GPa; 40% vaporization, 486 GPa; 50% vaporization, 437.2 GPa. From Pierazzo and Melosh (2000a).

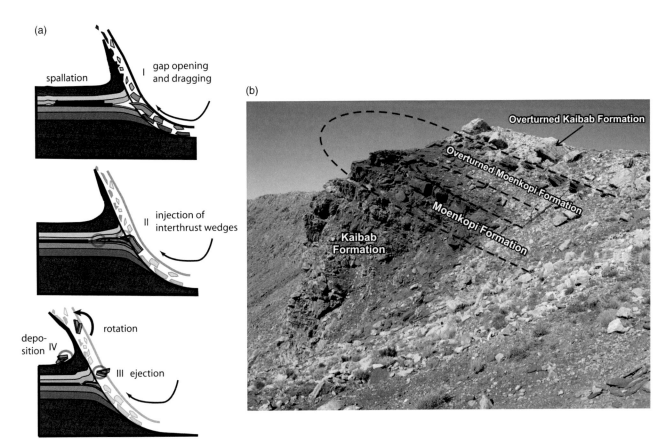

**Plate 12 (Figure 4.5)**  (a) Model for the formation of interthrust wedges at Barringer Crater by Poelchau *et al.* (2009). After spallation induces horizonatal zones of weakness, small gaps are formed during the excavation stage, while thrust ramps are formed in the sedimentary layers. Wedges of rock are subsequently thrust outward into the crater wall, causing warping of the overlying beds. (b) Field image showing the uplifted and overturned 'flap' at Barringer Crater.

**Plate 13 (Figure 4.7)** Impact ejecta of the Ries impact structure, Germany. (a) Schematic cross-section across the Ries impact structure indicating the nature and location of various impactites. Modified from (Schmidt-Kaler, 1978). (b) Large blocks of Malm limestone within the Bunte Breccia at the Gundelsheim Quarry, 7.5 km outside the NE crater rim. The poor sorting, low shock level and modified pre-impact target surface (smooth area at bottom of image) are consistent with ballistic sedimentation and subsequent radial flow (Hörz *et al.*, 1983). (c) The contact between the Bunte Breccia and the underlying limestone displays characteristic striations ('Schliff-Fläche'). These striations demonstrate that the Bunte Breccia, after it was ejected out of the crater on ballistic trajectories, was deposited onto the surface and continued to flow for considerable distances. (d) Image of the Aumühle quarry showing the relationship between suevite (light grey/green) and underlying Bunte Breccia (dark brown/red). Note the sharp contact between the suevite and Bunte Breccia. The height to the top of the outcrop is ~9.5 m. (e) The suevite has clearly filled in a depression in the underlying Bunte Breccia. This is incompatible with a 'fallout' airborne mode of deposition as proposed by some workers (Stöffler, 1977). The infilling suggests a topographic control and is a characteristic of pyroclastic and lava flows (Fisher and Schmincke, 1984).

DURING EXCAVATION:

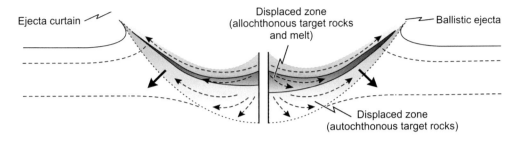

Ejecta curtain

Displaced zone
(allochthonous target rocks
and melt)

Ballistic ejecta

Displaced zone
(autochthonous target rocks)

END EXCAVATION STAGE/
START MODIFICATION STAGE:

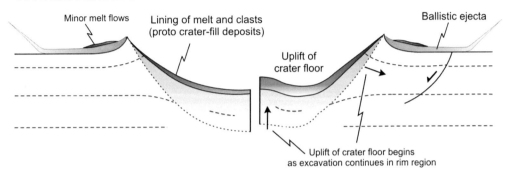

Minor melt flows

Lining of melt and clasts
(proto crater-fill deposits)

Ballistic ejecta

Uplift of
crater floor

Uplift of crater floor begins
as excavation continues in rim region

MODIFICATION STAGE:

Flow of melt and clasts off
emergent central uplift

Continued movement of
melt and clasts outwards

Gravitational collapse
of crater walls

Minor melt flows

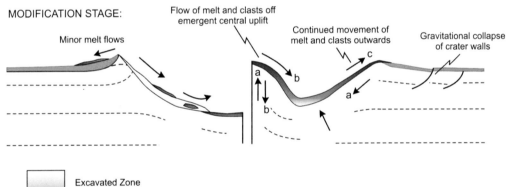

a  b
c
a

b

Excavated Zone

Displaced Zone
(allochthonous target rocks)

Displaced Zone
(clasts of allochthonous target rocks and melt)

END OF MODIFICATION STAGE:

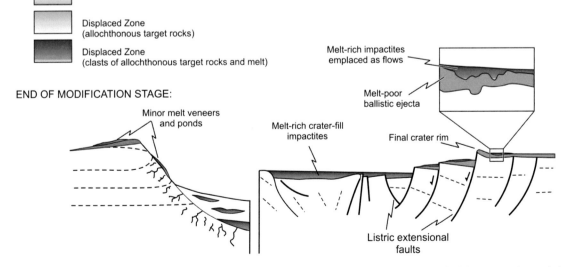

Minor melt veneers
and ponds

Melt-rich crater-fill
impactites

Melt-rich impactites
emplaced as flows

Melt-poor
ballistic ejecta

Final crater rim

Listric extensional
faults

**Plate 14 (Figure 4.12)**   Model for the formation of impact ejecta and crater fill deposits (modified from Osinski *et al.* (2011)). This multistage model accounts for melt emplacement in both simple (left panel) and complex craters (right panel), as described in the text. It should be noted that this is for a 'typical' impact event. As discussed in the text, the relative timing and important of the different processes can vary, particularly for more oblique impacts. It should be noted that, in the modification-stage section, the arrows represent different time steps, labelled 'a' to 'c'. Initially, the gravitational collapse of crater walls and central uplift (a) results in generally inwards movement of material. Later, melt and clasts flow off the central uplift (b). Then, there is continued movement of melt and clasts outwards once crater wall collapse has largely ceased (c).

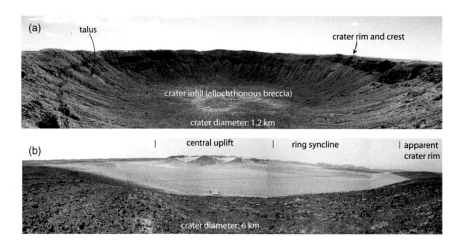

**Plate 15 (Figure 5.3)** Composed panorama photographs of a simple and a complex terrestrial impact crater. (a) The simple crater Barringer (Meteor), AZ, USA is 1.2 km in diameter. Autochthonous target rocks outcrop in the upper part of the cavity wall. These rocks dip outward and are truncated by numerous radial faults. Target rocks are overlain by the proximal ejecta blanket that forms the crest of the crater. The proximal ejecta blanket contains overturned strata whose stratigraphic succession is inverted. (b) The complex crater Jebel Waqf as Suwwan, Jordan, measures 6 km from rim to rim (apparent crater rim). The ring syncline is covered by Wadi deposits, the 1 km central uplift is exposed in the centre.

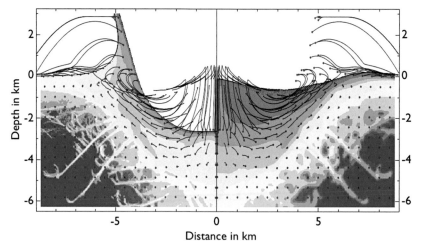

**Plate 16 (Figure 5.5)** The numerical model (iSale hydrocode) of the impact of a 1 km diameter projectile at 12 km s$^{-1}$ illustrates the motion of tracer particles during transient cavity modification. Red dots mark the position of tracers at the moment when the transient cavity has reached its greatest depth. The black lines indicate the subsequent flow path of tracers during crater collapse. The left side shows crater shape and distribution of total plastic strain (blue: low; red: high) at the onset of crater collapse, the right side shows the final crater shape and the final distribution of total plastic strain. For further explanation, see the text.

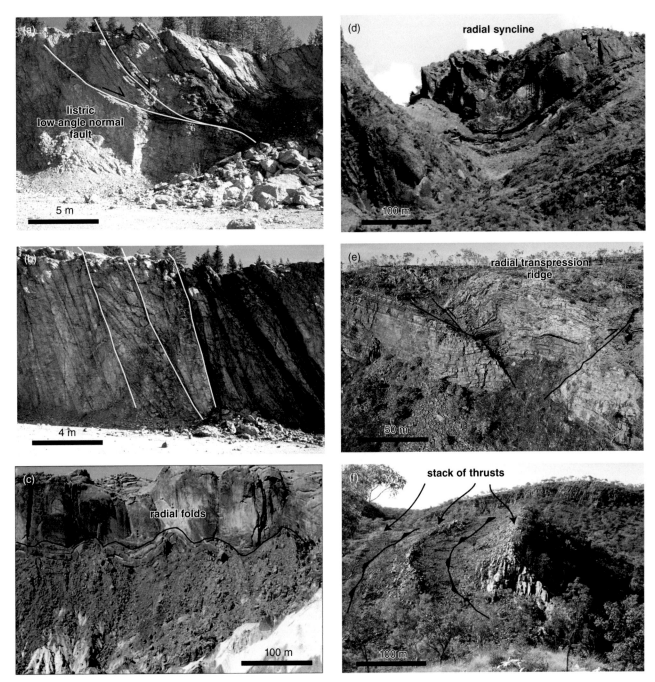

**Plate 17 (Figure 5.7)** Field photographs of structural features characteristic for complex terrestrial impact craters. (a) Listric normal faulting observed in the ring syncline of the Siljan impact crater, Sweden (apparent crater diameter: 52–65 km). (Image from Kenkmann and von Dalwigk 2000, with permission.) (b) Steeply dipping stack of Ordovician limestones in the ring syncline of the Siljan impact crater, Sweden. (c) Radial folding in the periphery of the central uplift of Upheaval Dome, UT, USA, observed from the centre of the structure. The fold hinges plunge outward. Note that the inner part of the central uplift is eroded (apparent crater diameter: 5.5 km). (d) Radial syncline within the central uplift of Serra da Cangalha impact crater, Brazil (apparent crater diameter: 14 km). (e) Radial transpression ridge (positive flower structure) in the periphery of the central uplift of the elliptical Matt Wilson impact crater, NT, Australia (apparent crater diameter: 6.3 × 7.5 km). (f) Imbrication of blocks thrusted onto each other in the core of the central uplift of Matt Wilson Crater, NT, Australia.

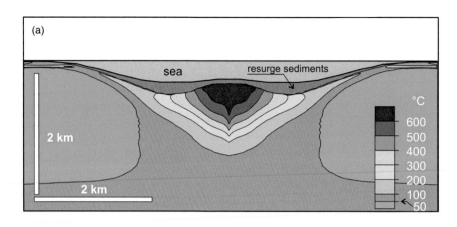

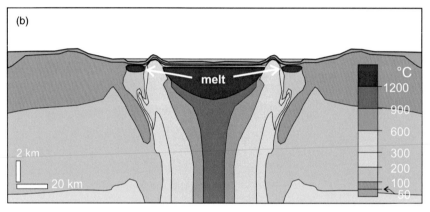

**Plate 18 (Figure 6.2)** Modelled thermal anomaly remaining in crater rocks immediately after the impact. (a) Kärdla Crater (4 km in diameter); (b) Chicxulub Crater (180 km in diameter). Modified after Jõeleht *et al.* (2005) and Abramov and Kring (2007).

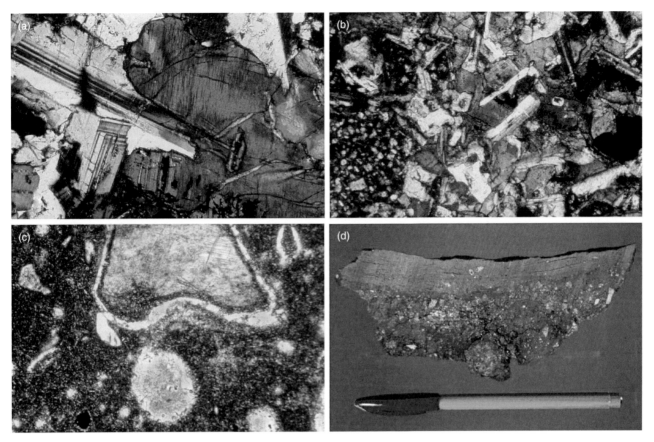

**Plate 19 (Figure 7.5)** (a) Photomicrograph in crossed polars of relatively 'coarse'-grained clast-poor impact melt rock from the 100 km diameter complex Manicouagan impact structure in Canada. Note apparent absence of lithic and mineral clasts and 'igneous' sub-ophitic texture between plagioclase and pyroxene. Field of view is 2.3 mm. Such melt rocks occur in location 'a' in Fig. 7.2 and location 'a' in Fig. 7.6. (b) Photomicrograph in crossed polars of relatively 'medium'-grained clast-poor impact melt rock from the 28 km diameter complex Mistastin impact structure in Canada. Note small number of thermally altered lithic clasts (e.g. recrystallized anorthosite clast in lower left). Field of view is 1.0 mm. Such melt rocks occur in location 'a' in Fig. 7.2 and location 'b' in Fig. 7.6. (c) Photomicrograph in plane light of relatively 'fine'-grained impact melt rock from Mistastin. Note large number of mineral clasts and fine-grained matrix of crystallites. Large clast (upper centre) is quartz with planar deformation features (Chapter 8) and a reaction corona of high birefringent pyroxene and pale brown high-silica glass. Immediately below are several rounded patches of high-silica glass, where the original quartz clast has been completely reacted out. Field of view is 1.6 mm. Such melt rocks occur in location 'a' in Fig. 7.2 and location 'c' in Fig. 7.6. (d) Slab showing the base of the impact melt sheet at the 28 km diameter complex Mistastin impact structure. Contact is with a melt-bearing breccia containing both lithic and impact melt glass clasts, which some workers would name as 'suevite'. Note, however, the coherent and compact nature of the impact melt clasts compared with those in the Ries suevite illustrated in Fig. 7.4b. Pen for scale. Such melt rocks occur in location 'b' in Fig. 7.2 and location 'd' in Fig. 7.6.

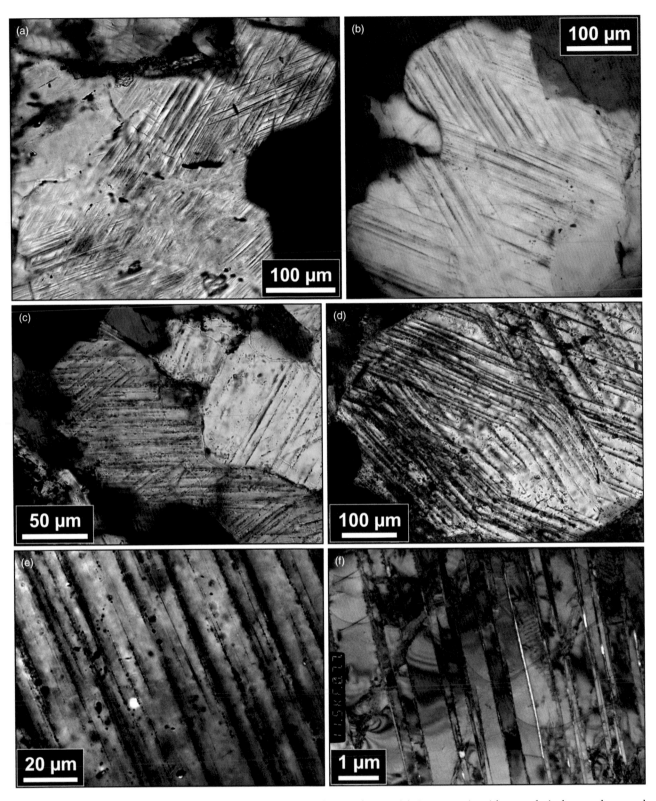

**Plate 20 (Figure 8.4)**   Microphotographs of quartz grains with sets of PDFs. (a) Quartz grain with two relatively non-decorated PDF sets. Quartzite clast in suevite from the Bosumtwi crater (sample KR8-006; depth: 240.36 m). (b) Quartz grain with two PDF sets. Meta-greywacke sample from the Bosumtwi crater (sample KR8-080; depth: 384.54 m). (c) Two sets of decorated PDFs in a quartz grain from a meta-greywacke sample from the Bosumtwi crater (sample KR8-070; depth: 364.12 m). (d) Quartz grain with two PDF sets. Fragmental dike breccia sample from the Manson structure (USA; sample M8-516.3; depth: 157.37 m). (e) One set of decorated (with numerous tiny fluid inclusions) PDFs in a quartz grain from a meta-greywacke sample from the Bosumtwi crater (sample KR8-056; depth: 326.78 m). Microphotographs (a–c) taken in crossed polars and (d, e) in plane-polarized light. (f) TEM bright-field microphotograph of one set of irregularly spaced PDFs in quartz (Bosumtwi sample KR8-006). Note that the light grey network shown in the background corresponds to the carbon net supporting the specimen.

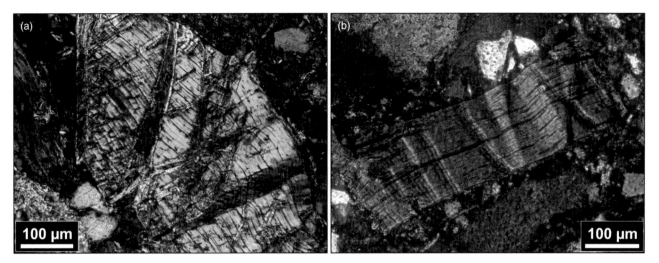

**Plate 21 (Figure 8.7)** Microphotographs (crossed polars) of kink-banding in mica. (a) Kink bands in a muscovite grain; suevite sample from the Chesapeake Bay structure (USA; sample CB6-100; depth: 1427.01 m). (b) Kink bands in a biotite grain; polymict impact breccia sample from the Rochechouart structure (France). Note that kink bands in mica are not a diagnostic criterion for shock metamorphism, as they can form also in other (i.e. non-impact) settings (see text).

**Plate 22 (Figure 8.8)** Photographs of impact metamorphism features formed by solid-state transformation (a, b) and by melting (c, d). (a) Plane-polarized light and (b) crossed-polars microphotographs of diaplectic quartz glass from a shocked sandstone clast in impact breccia from the Haughton impact structure (sample HMP-02-092). Note the occurrence within and between the idiomorphic quartz grains (i.e. now transformed to diaplectic quartz glass) of coesite and some alteration products. (c) Highly vesicular, shard-like, melt glass fragment in impactite from the Bosumtwi crater (sample LB-39a; from outside the crater rim; plane-polarized light). (d) Macrophotograph of a hand specimen of vesicular impact melt rock from the Rochechouart impact structure (France).

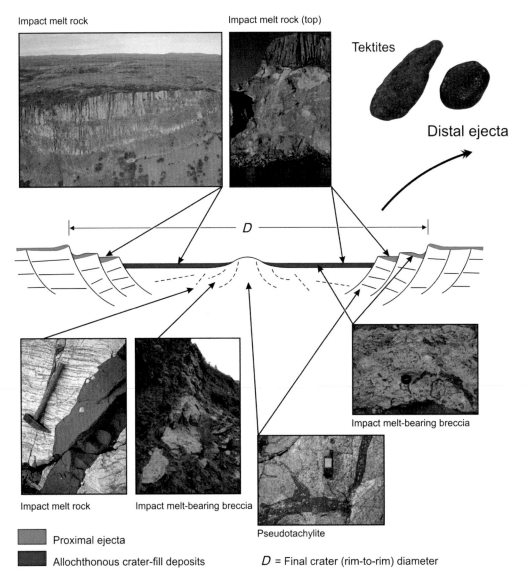

**Plate 23 (Figure 9.2)** Schematic diagram of a typical complex impact structure showing the main settings in which impact melt-bearing materials are typically found.

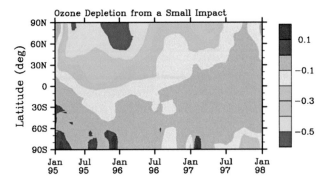

**Plate 24 (Figure 10.4)** Zonally averaged monthly mean fractional changes of atmospheric ozone column following the impact of a 1 km asteroid in the central Pacific. Values are calculated with respect to the unperturbed case. Negative contours (ozone depletion) are dashed. Regions where ozone depletion is 50% or more (fractional changes of −0.5) are hatched. Image credit: Pierazzo *et al.* (2010).

**Plate 25 (Figure 11.4)** Chasmoendolithic colonization of fractured rock. (a) Block of heavily fractured dolomite at location 75°24.397′N, 89°49.894′W. Scale bar 1 m. (b) Close-up showing fractures and colonization along surfaces of fragments of rock removed from the block (centre). (c) Bright-field micrograph of algal and cyanobacterial colonists of the block.

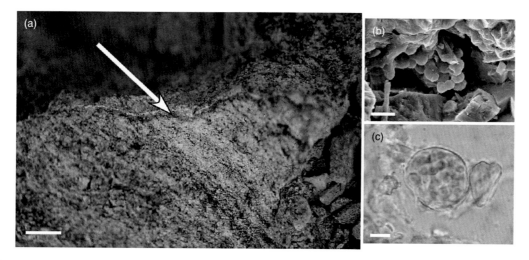

**Plate 26 (Figure 11.5)** Cryptoendolithic colonization of impact shocked gneiss. (a) Band of cryptoendolithic colonization of shocked gneiss (scale bar 1 cm). (b) Secondary scanning electron image of cyanobacterial colony within cryptoendolithic zone (scale bar 10 μm). (c) Bright-field micrograph of presumptive *Chroococcidiopsis* sp. colony in cryptoendolithic zone (scale bar 10 μm). Modified from Cockell *et al.* (2002).

**Plate 27 (Figure 11.9)** Colonization of disrupted gypsum beds in the meltrock hills of the Haughton impact structure. (a) Gypsum outcrop at 75°2385′N, 89°3237′W embedded within the melt rocks of the Haughton impact structure and harbouring evaporitic communities. Note rotation and disruption of bedding planes (white dashed lines) caused by impact (scale bar 1 m). (b) Image of green zone of phototrophs within precipitated gypsum (arrow; scale bar 1 cm). (c) Micrograph of cyanobacterial morphotypes inhabiting the green zone in the gypsum (scale bar 10 μm). (d) SYBR Green I staining of prokaryotes associated with gypsum crystals and cyanobacteria. Presumptive heterotrophs associated with phototrophs are highlighted by an arrow. Phototrophs are highlighted with a dashed arrow. The SYBR Green I image is superposed with the background of the bright-field image to show location of heterotrophs with respect to phototrophs.

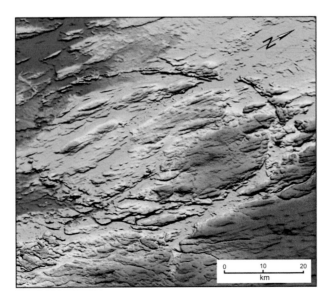

**Plate 28 (Figure 12.1)** Synoptic image of the topography of the Carswell structure, based on the Shuttle Radar Topographic Mission. Reds and yellows are topographic highs and blues are lows.

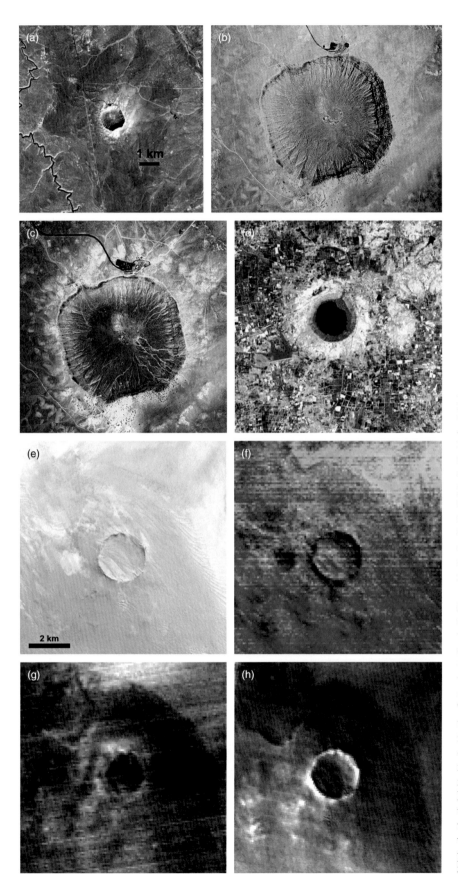

**Plate 29 (Figure 13.2)** (a) 15 m resolution ASTER VNIR of Meteor Crater, Arizona displayed as bands 3–2–1 in red–green–blue, showing very little vegetation save for a small farm with trees east of the crater. (b) True-colour aerial photograph of Meteor Crater with approximately 1 m resolution. (c) False-colour aerial photograph of Meteor Crater stretched to display ejecta lobes of bright-white Coconino and Kaibab ejecta. (d) False-colour ASTER VNIR image of Lonar Crater, India (Chapter 18), displayed as bands 3–2–1 in red–green–blue; vegetation (as red) can be seen circum the crater lake but inside the crater rim. (e)–(h) ASTER VNIR and TIR data of Roter Kamm, Namibia; (e) daytime VNIR colour image showing the pervasive mantling of aeolian sand; (f) daytime TIR decorrelation stretch of TIR bands 14, 12, 10 in R, G, B (respectively) indicating significant variability in the sand composition (Fe-rich silicate sands shown in blue/purple and non-Fe-rich silicates shown in red); (g) night-time temperature image showing less than 4 °C variation from crater rim to the near-rim ejecta; (h) apparent thermal inertia (ATI) image (see text) with high ATI values (bright) indicating potential blocky and less mantled ejecta. See text for explanations of acronyms.

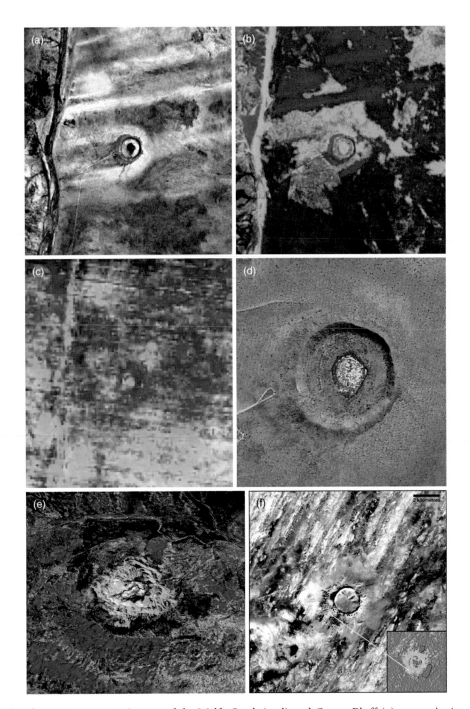

**Plate 30 (Figure 13.3)** Various remote images of the Wolfe Creek (a–d) and Gosses Bluff (e) craters in Australia. (a) ASTER VNIR image with bands 3, 2 and 1 displayed as red, green and blue, with a 2% stretch. Image taken 02-02-2003. (b) Landsat band ratio image with Fe ratio, normalized difference vegetation index and OH-bearing ratio (see text for details) displayed as red, green and blue. Data are stretched 2% to bring out differences. Image from 09-03-1999. (c) ASTER TIR decorrelation stretch with bands 14, 12 and 10 displayed as red, green and blue. Image from 02-02-2003. (d) High-resolution image derived from © Google Earth, 2012. (e) View to northwest of the 23 km diameter Gosses Bluff Crater, with a hue saturation value transformed VNIR with SWIR minimum noise fraction (MNF) bands applied to a VNIR image (15 m resolution) and draped on a DTM with a vertical exaggeration of 3. Green represents bedrock from several Mesozoic sandstone lithologies (Barlow, 1979; Prinz, 1996) uplifted above alluvium shown as blue and purple. (f) ASTER stereo-derived DEM and VNIR data covering the 1.9 km terrestrial impact structure Tenoumer located at 22°55′7″N, 10°24′29″W, in Mauritania, Africa. The 15 m/pixel colour infrared image of Tenoumer reveals what appears to be a two-facies, circumferential deposit around the crater rim reminiscent of ejecta around fresh impact craters on other planetary surfaces. The colour-coded DEM in the inset shows elevated terrain consistent with the inner most facies of these deposits, which strongly supports the idea that some ejecta was preserved around this infilled simple crater. The DEM shows similar to results from Meteor Crater (Garvin *et al.*, 1989), that the wind regime that has dominated this region since the time of crater formation has preferentially muted and eroded the windward (southwest) side. (Note: the two areas outlined in black and in solid purple are areas of spurious data values where difficulties due to surface or atmospheric properties confounded attempts to derive digital elevation for these areas.)

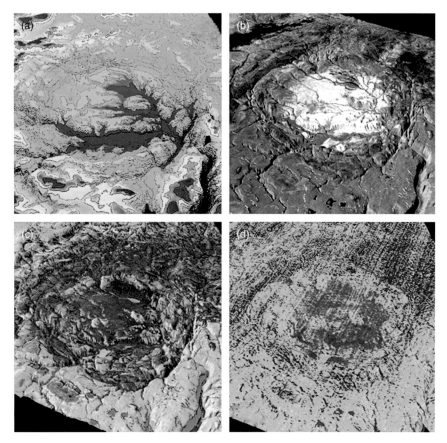

**Plate 31 (Figure 13.4)**   Morphometric and spectral maps (VNIR, SWIR and TIR) of the Haughton impact structure, Devon Island, Nunavut territory, Canada. North is to the bottom right of each subimage. (a) A 25 m colourized DTM of the Haughton impact structure (courtesy of the Natural Resources of Canada http://www.nrcan-rncan.gc.ca/com/index-eng.php), with the vertical exaggeration set to 10. (b) VNIR image from *Landsat 7* ETM+ (30 m/pixel). The impact melt rock of the Haughton Formation is the most apparent lithological unit. This is manifested as a white unit that predominately lines the crater floor, but smaller occurrences can also be observed in the vicinity of the wall/terraces and rim area. (c) ASTER SWIR (1.656–2.400 μm) colour composite of MNF-transformed bands using principle components 2, 1 and 6 in RGB (30 m/pixel). Lithological units are as follows: magenta, dolomite; cyan, limestone; red, gypsum; blue, impact melt rock. (d) ASTER TIR colour-composite using DCS bands 14, 12 and 10 (90 m/pixel; Tornabene *et al.*, 2005). Lithological units are as follows: dark blue–green is dolomite; cyan–light green is limestone; red is gypsum; and magenta–pink is impact melt rock.

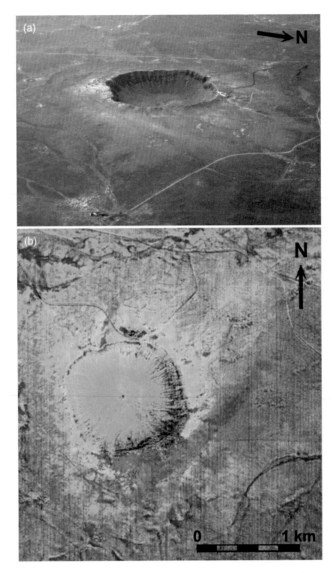

**Plate 32 (Figure 13.8)** (a) Meteor Crater as seen from the east. (b) TIMS image with Coconino Sandstone spectral end-member displayed as red, Kaibab Limestone spectral end-member displayed as green and Moenkopi Formation (siltstone/mudstone in this region) displayed as blue.

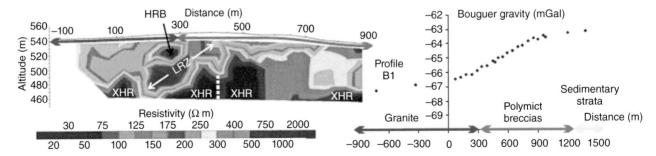

**Plate 33 (Figure 14.1)** Left-hand figure is a tomographic model of electrical resistivity at Araguainha Crater; right-hand figure is Bouguer gravity and its relationship with surface outcrop (from Tong *et al.* (2010)). A low-resistivity zone (LRZ), which corresponds to polymict breccias at surface, dips beneath the granitic core of the central uplift.

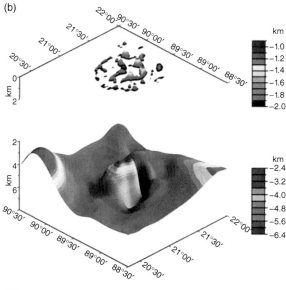

(b)

(d)

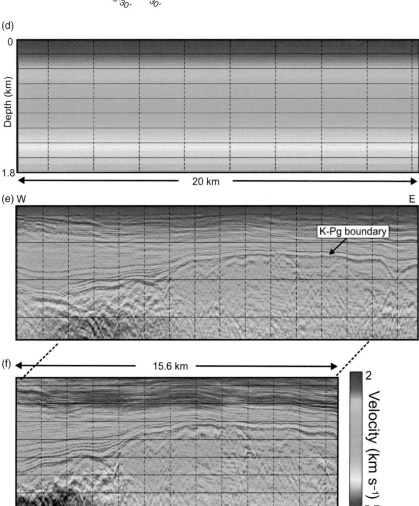

**Plate 34 (Figure 14.2)** (b) Result from the inversion of magnetic data across Chicxulub (from Pilkington and Hildebrand (2000)). The inversion is constrained by imposing the assumption that the long-wavelength magnetic anomaly is produced by magnetized basement rocks and the short-wavelength features by magnetic bodies in the allogenic melt breccias. (d) Starting velocity model for travel-time inversions across the peak ring at Chicxulub (from Morgan *et al.* (2011)). (e) Tomographic velocity model (colour), obtained by inverting travel-time data along seismic profile 10 (acquired in 2005) using the FAST program (Zelt and Barton, 1998). Background is 2D seismic reflection data for the same profile, converted to depth using the tomographic velocity model (Morgan *et al.*, 2011). (f) Tomographic velocity model (colour), obtained by inverting for the full-seismic wavefield along profile 10 (Morgan *et al.*, 2011). Background is 2D seismic reflection data for the same profile, converted to depth using the tomographic velocity model. Model edges and depths below 1.6 km have been omitted as the model is considered to be poorly constrained in these areas.

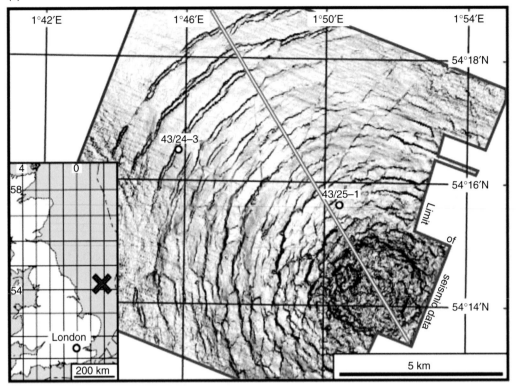

(a)

(b)

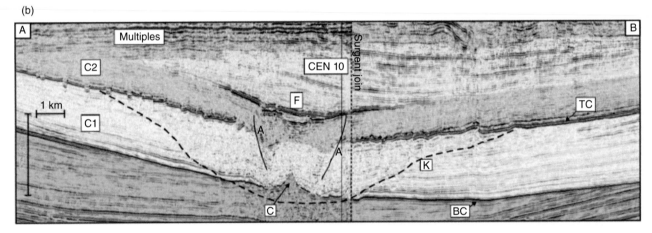

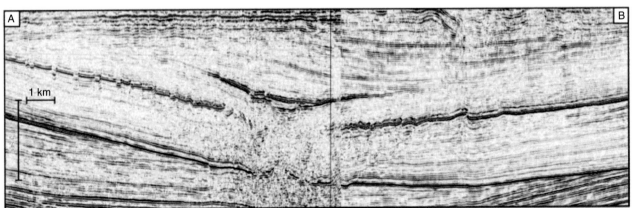

**Plate 35 (Figure 14.3)** (a) The topography of a single horizon at Silverpit (the reflection labelled TC in (b)), that lies a few hundred metres below the seabed in the North Sea, UK – see inset for location (from Stewart and Allen (2002)). Local drill holes (open circles) are used to interpret reflection TC as the top layer of Cretaceous chalk. Offsets in the horizon are observed as continuous rings that are shown to be extensional fault-bounded grabens in (b). (b) A 2D vertical time slice through the 3D seismic reflection volume that crosses through the centre of the Silverpit structure (from Wall *et al.* (2008)). The horizontal bar is 1 km and the vertical bar is 0.5 s two-way travel-time (~1 km). The Silverpit structure forms a depression in Cretaceous and Palaeogene reflections (packages C1 and C2); reflections exhibit a parallel onlap fill onto the bowl-shaped crater floor (package F).

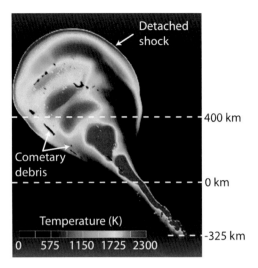

**Plate 36 (Figure 17.2)** Impact plume (fireball) formed by the impact of a 3 km fragment of comet Shoemaker–Levy 9 on Jupiter (simulated using the 3D CTH hydrocode) approximately 69 s after entering Jupiter's atmosphere taken from Crawford *et al.* (1994). This is a cross-sectional view with temperature represented by colour; the vertical scale is altitude above 1 bar (0.1 MPa) atmospheric pressure level. The computational mesh consists of 6.3 million 5-km-sized cubical cells.

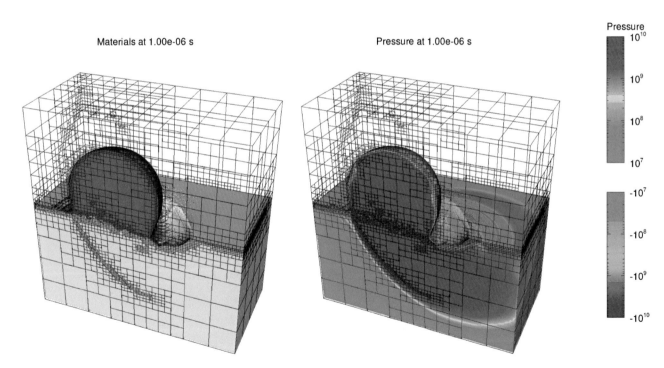

**Plate 37 (Figure 17.5)** Contour plots of material (left) and pressure (Pa, right) from a high-resolution CTH simulation with AMR of a ¼-inch aluminium sphere impacting an aluminium target at 5 km s$^{-1}$ and at an angle of 15 to the horizontal. AMR is a mechanism for dynamically modifying the resolution during a simulation to ensure that computational effort is expended only where necessary. Observe in this example how resolution is focused along material interfaces and around the asymmetric shock wave, where it is most needed for an accurate solution. Note that the smallest mesh-block outlines (shown in black) represent 8 × 8 × 8 computational cells. This image is courtesy of Dave Crawford.

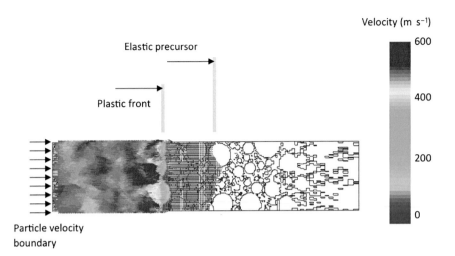

**Plate 38 (Figure 17.8)** A 3D meso-mechanical simulation of shock wave propagation in concrete (Riedel, 2000). The computational domain comprises zones of cement and zones of gravel, for which the EoSs are known, to represent a concrete mixture. A velocity boundary condition is used to generate a plastic wave and an elastic precursor that propagate through the mixture. The calculated shock wave velocity and average material velocity behind the shock can be used to define the Hugoniot curve for the concrete mixture and hence an EoS.

**Plate 39 (Figure 18.4)** Portion of a four-band (RGB + NIR) pan-sharpened (0.6 m resolution) Quickbird image showing Lonar Crater (Maloof *et al.*, 2010). The Ambar Lake ('Little Lonar') is a blue patch surrounded by a circular feature consisting of distal ejecta beneath the label. Distal ejecta are also well exposed in outcrops near the Kalapani Dam and in hand dug water wells southeast of the crater. Image provided courtesy of Dr. Sarah Stewart. (http://www.fas.harvard.edu/~planets/sstewart/resources/Lonar-quickbird-RGB.jpg)

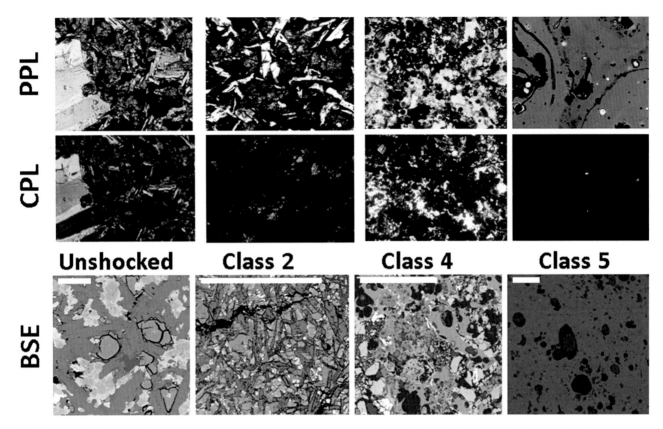

**Plate 40 (Figure 18.13)** Images of Lonar basalts with different stages of shock ranging from unshocked to Class 5. The top row of images is in plane polarized light (PPL), the second row shows the same areas of the thin sections with crossed polarized light (CPL) (optical images are 5× magnification). The third row shows a series of backscattered electron (BSE) images for the different shock levels. Optical images approximately 1.5 mm across; scale bars for the BSE images are 1 mm long. Figure this study.

**Plate 41 (Figure 19.5)** Field photographs of impactites from impact craters developed in sedimentary (Haughton; a,b), mixed sedimentary–crystalline (Ries; b,c), and crystalline (Mistastin; e,f) targets. (a) Field photograph of a well-exposed section of the crater-fill impact melt rocks at the Haughton structure. (b) Close-up view of impact melt rocks showing the clast-rich character and fine-grained microscopic nature of the pale grey groundmass. (c) Surficial impact melt-bearing breccia ('suevite') (light grey) overlying Bunte Breccia (dark grey) at the Aumühle quarry. (d) Close-up view of the impact melt-bearing breccia showing the fine-grained groundmass and macroscopic irregular silicate glass bodies. (e) Oblique aerial view of the approximately 80 m high cliffs of impact melt rock at the Discovery Hill locality, Mistastin impact structure, Labrador. Photograph courtesy of Derek Wilton. (f) Close-up view of massive fine-grained (aphanitic) impact melt from the Discovery Hill locality. Camera case for scale.

(a)

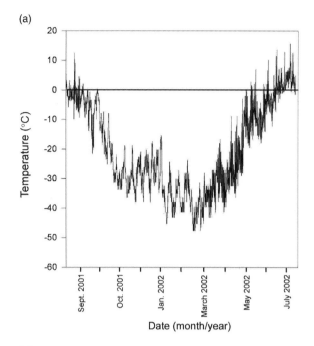

(b)

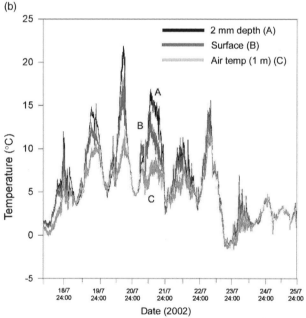

**Figure 11.6** Temperatures within the shocked gneissic endolithic habitat. (a) Soil temperatures measured next to measurements in (b) from 14 August 2001 to 10 July 2002. (b) Temperatures within a cryptoendolithic habitat measured from 18 to 26 July 2002. Upper curve (A) shows temperature recorded at 2 mm depth within endolithic habitat. Middle curve (B) shows temperature on surface of the rock. Lower curve (C) shows air temperature.

the first successful colonists of impact-perturbed environments (Cockell *et al.*, 2002), colonizing impact substrates and providing a source of organic carbon for the heterotrophic populations, consistent with their importance today in the shocked rocks and disrupted evaporite minerals in the Haughton structure. The

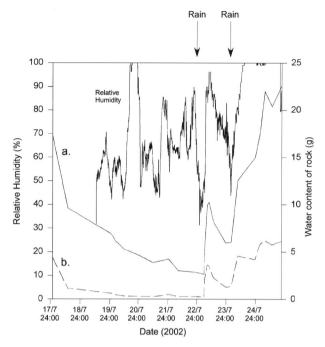

**Figure 11.7** Water availability within the shocked gneissic endolithic habitat. Relative humidity (left axis) from 18 to 25 July 2002. The right axis shows the mass (water) gained by two impact-shocked gneissic rock samples (sample a, 207.2 g; sample b, 68.9 g). On 18 July the experiment begins after immersion of the rocks in water for 4 h and then the rocks dry naturally in the field environment. The first natural rain event is in the early morning of 23 July and lasts for approximately 1 h. The second rain event begins in the early morning of 24 July and continues into 25 July.

communities that inhabit the gneissic rocks today are obviously not immediate post-impact successional communities, and the rocks have been subjected to 39 million years of climatic variation since the impact event (Sherlock *et al.*, 2005). However, the gross changes in permeability are caused by impact metamorphism, so that the present-day rocks can be considered to provide important insights into how impacts alter the colonization of rocky environments.

### 11.2.3 Changes in sedimentary rocks

An obvious corollary question to the above observations is: What is the fate of sedimentary rocks that are usually suitable geomicrobiological habitats for life? Some of the energy from impact must be consumed in physical disruption of the target material, as well as that released as heat, sound and light. For low-porosity rocks this energy will be translated into rock fracturing, vesiculation, melting and so on, as discussed above. Although this will also be the case for sedimentary rocks, in the case of rocks that are initially very porous, some of the energy will be taken up in pore collapse (Kieffer, 1971; Kieffer *et al.*, 1976). Thus, a prediction would be that rocks that are initially very favourable habitats would become less so.

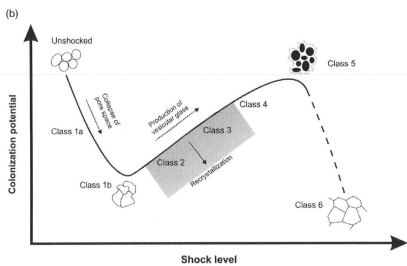

**Figure 11.8**    Colonization of shocked sandstones from the Haughton impact structure. (a) Natural cryptoendolithic coloniza-tion of class 3 shocked sandstone by *Gloeocapsa* sp. Scale bar 1 cm. (b) Schematic of the influence of shock pressures on coloniza-tion potential of sandstones (scale bar 1 cm) (from Cockell and Osinski (2007)).

Field observations of sandstones, again from the Haughton structure (Osinski, 2007), provide a direct comparison with the gneissic habitats (Cockell and Osinski, 2007). It is clear that the effects of impact are not as simple as the hypothesis would predict. Laboratory colonization experiments were used to examine the effects of shock on colonization. The coccoid cyanobacterium *Chroococcidiopsis* and the Gram-positive bacterium *Bacillus sub-tilis* were used to colonize samples of sandstones collected from the Haughton structure and shocked to different levels. The pat-terns of colonization and the ability of the organisms to grow through the rock substrate were examined. Low shock pressures up to just above approximately 5 GPa cause pore collapse as would be expected, rendering the rocks less suitable for endolithic colonization. However, higher shock pressures caused vesicula-tion of the rocks, generating interconnected spaces that allow for the natural colonization of the rocks. Class 3 sandstones shocked to greater than 10 GPa, but less than 20–25 GPa, show natural endolithic colonization by cyanobacteria such as *Gloeocapsa* and *Chroococcidiopsis* (Fig. 11.8a). At shock pressures higher than approximately 30 GPa, solid glass is formed, which becomes again impenetrable to microorganisms. The effects are further compli-cated by the fact that many of these rocks are included in the crater-fill melt sheets and would have been subjected to heating during the cooling of the crater. The long-term heating and annealing of the rocks changes the characteristics of the pore spaces and their interconnectedness (permeability).

These results lead to a schematic of the effects of shock pressure on colonization potential of the Haughton sandstones as shown in Fig. 11.8b. This schematic is simplified and the shock pressures at which these types of transitions occur will depend upon the exact composition of the target lithology. However, it illustrates the point that impacts do not merely fracture or compress rocks, but from the point of view of geomicrobiology they generate a complex suite of rocks with colonization potential that is linked to the exact shock and temperature pressures that have been experienced by the target material. From a geomicrobiological perspective, impacts events must be regarded as quite distinct events in their effects on habitats for life compared with other geological changes that perturb the environment for microorganisms, such as volcanism, tectonism, etc.

### 11.2.4 Other rock types

Microorganisms inhabit other rocks and minerals in addition to crystalline or sedimentary rocks. Another mineral suite of geomicrobiological importance is evaporites. In delivering energy to the site of impact, asteroids and comets can mobilize soluble minerals in hydrothermal systems or generate new evaporitic materials. Impact hydrothermal systems can remobilize and precipitate salts. The formation of a habitat has been shown within the sheets of recrystallized hydrothermally precipitated selenite (a transparent form of gypsum, $CaSO_4 \cdot 2H_2O$) in the Haughton structure (Parnell *et al.*, 2004). The colonists are primarily cyanobacteria (*Nostoc* and *Gloeocapsa* spp.), which grow within the sheets and presumably obtain nutrients from allochthonous sources leached into the selenite. The selenite is deposited throughout the melt sheet of the Haughton structure and is exposed, allowing access to photosynthetically active radiation. As the selenite is quite clear, however, the organisms are exposed to UV radiation and so they are highly pigmented. The selenite contains lipids associated with present-day microbial inhabitants; uncolonized gypsum also contains lipids which have a very different profile (Bowden and Parnell, 2007).

Microorganisms extensively colonize the impact-rotated and disrupted Ordovician gypsum beds from which the selenite was derived. The original beds can be found within the central uplift underlying the melt rock hills (Fig. 11.9a,b). Along their margins, recrystallization of mobilized gypsum provides a habitat for cyanobacteria and heterotrophs which either feed off organic carbon produced by the phototrophs or from allochthonous organic carbon (Fig. 11.9c,d). Amongst the heterotrophic population are many *Arthrobacter*, a genus of Actinobacteria resistant to desiccation and starvation (Cacciari and Lippi, 1986; Reddy *et al.*, 2002). The desiccated, nutrient-poor conditions within the gypsum probably favour their persistence. They are known to colonize Arctic soils (Nelson and Parkinson, 1978a,b). The organisms are not halophilic and only slightly halotolerant, consistent with the low solubility of gypsum in water (Cockell *et al.*, 2009b). In contrast to the selenite, the precipitated gypsum along the margins of the impact-disrupted evaporite beds is composed of many small crystals which effectively scatter UV radiation and provide a protected microhabitat, although concomitantly the photosynthetically active radiation will also be lower for the phototroph population. Laboratory experiments show that the *Arthrobacter* are capable of causing gypsum neogenesis (Cockell *et al.*, 2009b). The organic material produced by them might either act as an organic site for nucleation or negatively charged moieties on the cell surface might allow for localized supersaturation of gypsum on the cells, thus causing gypsum crystallization. Thus, the bacteria may have a direct role not merely in the passive colonization of the gypsum deposits, but also in the precipitation process. Figure 11.10 shows a phylogenetic tree of *Arthrobacter* found within gypsum deposits at Haughton and the formation of gypsum by two such isolates. Figure 11.11 shows a schematic of the geomicrobiological environment in the impact-disrupted gypsum.

Other impact structures also show evidence of the precipitation of evaporitic deposits. The limnology of the hypersaline lake within the bowl of the simple 1 km diameter Tswaing impact crater in South Africa, formed approximately 50,000 years ago, has been characterized (Ashton and Schoeman, 1983). In this case, the evaporites have no relation to the hydrothermal system. Instead, the crater cavity, by collecting water in the hydrologic depression, creates a hypersaline evaporitic habitat in a climatic region where the rate of evaporation is high. The evaporites around the margins of the crater are inhabited by cyanobacterial microbial mats.

In analogy to the effects of fracturing of target material on habitats for the deep biosphere, which are similar to the surface effects of fracturing, impacts can generate deep subsurface briny environments. The confinement of seawater from shallow marine impacts can generate briny habitats, as is thought to be the case in the subsurface of the Chesapeake Bay impact structure, USA (Sanford, 2003; Sanford *et al.*, 2007). Table 11.1 shows a summary of some of the known geomicrobiological habitats in evaporites that are linked to impact processes or the geological conditions resulting from impact.

## 11.3 Chemical changes

The observations in the previous sections focus on the physical effects of impacts on the geomicrobiological environment for microorganisms. Is it possible that impacts change their availability of elements within rocks? First of all, impacts might enhance some elements. Two mechanisms can result in the enhancement of elements within an impact structure. The localized concentration of metals and other elements within the target area can be caused by mechanisms such as the impact-redistribution of elements present in the pre-impact environment at high concentration or the localized concentration of elements in the impact hydrothermal system (Reimold *et al.*, 2005; Ames *et al.*, 2006). These mobilizations result in economically important metal deposits. The enhancements could be beneficial to a biota or in some environments where heavy metal concentrations exceed several per cent (Ames *et al.*, 2006) they might plausibly be detrimental to some organisms.

Can impacts deplete bioessential elements? Osinski *et al.* (2010) reported data from the sedimentary clasts in the Haughton structure and found no evidence for elemental depletion caused by

**Figure 11.9**   Colonization of disrupted gypsum beds in the meltrock hills of the Haughton impact structure. (a) Gypsum outcrop at 75°2385′N, 89°3237′W embedded within the melt rocks of the Haughton impact structure and harbouring evaporitic communities. Note rotation and disruption of bedding planes (white dashed lines) caused by impact (scale bar 1 m). (b) Image of green zone of phototrophs within precipitated gypsum (arrow; scale bar 1 cm). (c) Micrograph of cyanobacterial morphotypes inhabiting the green zone in the gypsum (scale bar 10 μm). (d) SYBR Green I staining of prokaryotes associated with gypsum crystals and cyanobacteria. Presumptive heterotrophs associated with phototrophs are highlighted by an arrow. Phototrophs are highlighted with a dashed arrow. The SYBR Green I image is superposed with the background of the bright-field image to show location of heterotrophs with respect to phototrophs. (See Colour Plate 27)

impact. However, evidence for depletion of some bioessential cations (Fe, Mg and Ca) in some shocked gneisses (shocked to greater than about 20 GPa) compared with the unshocked material in the Haughton structure was reported by Fike *et al.* (2003). The gneiss is comprised of diverse minerals which yield variations in their bulk composition (Metzler *et al.*, 1988), so that Fike *et al.* (2003) may not have exactly compared like with like in making comparisons between shocked and unshocked gneisses. As rock samples colonized by phototrophs (which were selected in the study by Fike *et al.* (2003)) are exposed on the surface of the melt hills, their highly porous structure may also subject them to enhanced weathering over time, which could be a factor in loss of some cations. However, at high shock pressures, the

crystalline rocks might experience elemental redistributions which may account for the observations of depletions in the colonized rock samples. Further investigations on the effects of impact on the chemical composition of the shocked gneisses are required.

## 11.4   Impact events and weathering

The fracturing of rock or the mixing of clasts of target material, such as, for example, in an impact melt sheet or material deposited after an impact tsunami, would be expected to have consequences for microbe–mineral interactions and the way in which

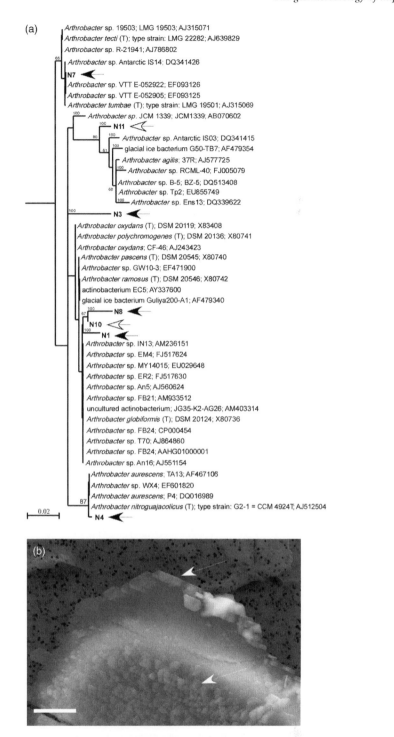

**Figure 11.10**   Gypsum neogenesis by *Arthrobacter* in impact gypsum habitats. (a) A maximum likelihood tree was generated using a neighbour-joining tree (1000 bootstraps) as the input tree and using a mask of 507 nucleotide positions (*Escherichia coli* numbering 210 to 716 covering V2 to V4). *Aquifex* sp. (accession number AB304892) was used as an outgroup (Cockell *et al.,* 2009b). The scale bar represents the number of changes per nucleotide position. The black arrows correspond to *Arthrobacter* isolates from the gypsum. The two white-headed arrows are the isolates used to demonstrate gypsum neogenesis. (b) Formation of gypsum crystals on an organic template in culture N10 (scale bar 10 μm). Arrows indicate gypsum crystals (identified by SEM–EDS) on the organic material.

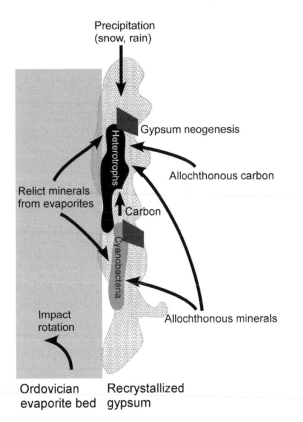

**Figure 11.11**  Colonization of disrupted gypsum beds in the meltrock hills of the Haughton impact structure. Schematic showing major geomicrobiological processes occurring within the material.

**Table 11.1**  Influence of asteroid and comet impacts on evaporitic habitats for microorganisms

| Habitat created by asteroid or comet impact | Example | Reference |
|---|---|---|
| Precipitation of salts in impact hydrothermal system | Colonization of selenite crystals in Haughton impact crater, Nunavut, Canadian High Arctic | Parnell *et al.* (2004) |
| Uplifted, rotated or disrupted (and exposed) ancient evaporitic deposits | Bay Fjord Ordovician evaporite beds, Haughton impact crater, Nunavut, Canadian High Arctic | Cockell *et al.* (2009b) |
| Formation of intra-crater salty or hypersaline lakes and ponds and/or evaporitic deposits at crater cavity margins | Tswaing impact crater lake, South Africa | Schoeman and Ashton (1982), Ashton and Schoeman (1983) |
| Trapped seawater in the deep subsurface | Chesapeake Bay impact crater, USA | Sanford (2003), Sanford *et al.* (2007) |

a microbiota and plants interact with, and weather, the target material. To investigate the potential influence of impacts on the type of weathering that occurs within target material, the Lockne Crater in Sweden was studied. The Lockne impact structure is located approximately 20 km south of Östersund in central Sweden (63°0.33′N, 14°49.5′E; Fig. 11.12; Sturkell, 1998b). It was formed by the collision of an impactor in a shallow sea in the Middle Ordovician at about 455 Ma. The structure consists of an outer crater approximately 13.5 km in diameter and an inner crater approximately 7.5 km in diameter. Following the impact, waves propagated outwards and then washed tsunami material back into the excavation cavity formed by the impact (von Dalwigk and Ormo, 2001; Ormo and Miyamoto, 2002; Lindstrom *et al.*, 2005).

The tsunami deposits consist of two types (Sturkell, 1998b): (a) Lockne Breccia, which consists of large fragments of lithified clasts from centimetres to tens of metres size, and (b) Loftarstone. The Loftarstone has a large percentage of material originating from within the impact excavation cavity and is a lithified mixture of fine gravel- to sand-sized fragments of Ordovician orthoceratite limestone (10–30%), calcitic fossils (mostly 5–10%) and crystalline (mainly granitic) rock (mostly 20–40%). The rock is cemented by a matrix (making up 15–30% of the material) consisting of clay-sized particles of calcium carbonate and variable proportions of clay minerals, most of which is derived from

Ordovician limestone and Cambrian shale that separates the Ordovician from the underlying crystalline basement. The Cambrian shale occasionally occurs as clasts (Simon, 1987). The Loftarstone can contain clasts of granite 1–3 cm in diameter. Similar to many impact-metamorphosed substrates, the material contains shock-melted rock. Up to 20% of the fragments are impact melt glass (Sturkell, 1998a) and many of the clasts are fractured. In the inner crater the Loftarstone has a thickness of at least 47 m and it can be found approximately 40–45 km from the crater centre. The Loftarstone contains an elevated iridium concentration, which has been interpreted as evidence of the impactor (Sturkell, 1998a).

The biological weathering of Loftarstone within the impact crater was examined (Cockell *et al.*, 2007; Fig. 11.12). The Loftarstone at this site consists of predominantly coarse sand with a minor proportion of gravel. At this grain size the composition is about 35% carbonate grains (limestone and fossil fragments), 25% crystalline (mainly granite, but about 3% dolerite and 8% quartz, feldspar and chlorites), 17% melt fragments and 17% matrix (Simon, 1987). The exposed outcrops of Loftarstone are covered in a biological patina. The white patina is composed of crustose lichens. The colonized regions could be easily removed from the rock surface in exfoliation sheets. The sheets were approximately 0.5–1 cm in thickness and, upon removal, fungal hyphae protruded up to approximately 0.5 cm from their

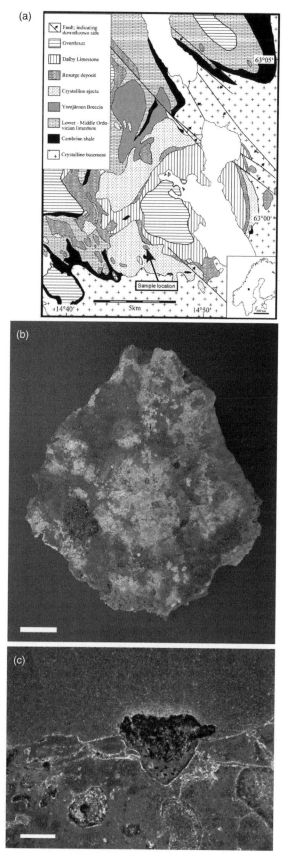

**Figure 11.12** Biological weathering of impact tsunami deposit. (a) Lockne impact crater, Sweden, showing location of tsunami resurge samples studied here (Loftarstone). (b) Photograph of the upper surface of a typical section of weathering crust (scale bar 1 cm). (c) Thin section stained with toluidine-O-blue to show single lichen preferentially colonizing a clast of chlorite in the surface of the Loftarstone (scale bar 100 μm). (From Cockell *et al.* (2007).)

underside. The crust was friable, but sufficiently consolidated for the sheets of rock to remain intact following removal.

The gross biological characteristics showed a layered profile. Thin-section analysis revealed the lichen bodies to be structurally integrated into the rock surface, with fragments of rock cleaved from the underlying substrate and incorporated into the thallus. The lichens were localized to the chlorite component of the Loftarstone. Beneath the lichen surface layer, the crust was penetrated by a well-developed fungal hyphae network. Within the fungal zone there were extensive cavities, identified as areas of calcium carbonate dissolution, since clasts of limestone could not be identified in the weathering crust by energy-dispersive X-ray spectroscopy (EDS), but could be found within unweathered material. The lichen thalli had an elevated calcium concentration compared with their surroundings. Beneath the calcium-rich zone there were regions of amorphous halloysite/kaolinite clay minerals depleted in Fe and Mg, and comprised primarily of Al and Si.

The preferential colonization of the chlorite regions is probably caused by its relative softness compared with the other components of the Loftarstone (feldspar and quartz). Feldspars are known to resist weathering compared with micas (Barker and Banfield, 1998). The lichens and fungi can penetrate the layers of the mineral. Frankel (1977) attributed the rapid biological weathering of biotite to interlayer penetration by microorganisms, and direct observation of the colonization of the interlayer regions of biotite has been reported (Wierzchos and Ascaso, 1994, 1996). The rich nutrient content, particularly abundant Fe and Mg, as shown by scanning electron microscopy (SEM)–EDS and electron microprobe analysis (Cockell *et al.*, 2007), is probably another reason for the preferential colonization of the chlorite.

Lichen colonization of feldspars or quartz on the surface of the rock was not observed. The observation of fragments of quartz and feldspars within thalli by SEM–EDS suggests that the lichens or their mycobionts can cause disaggregation and/or dissolution of these materials once they become established on the Loftarstone. Some of these fragments are probably aeolian and become entrained within the thallus; however, large feldspar and quartz grains deep within the thallus and the eroded appearance of the feldspar/quartz substrate around lichen-colonized zones suggest that dissolution of these components occurs. Organic acids produced by fungi are known to be capable of dissolving aluminosilicates (Prieto *et al.*, 1994; Sterflinger, 2000; Wallander and Hagerberg, 2004).

Apatite may be one important source of phosphorus (Welch *et al.*, 2002). Apatite crystals are found within the feldspar glasses and chlorite within the Loftarstone. The apatite is usually localized to euhedral features in the weathering crust. However, most of the apatite was composed of fine-grained crystals that fill the features, which may be indicative of dissolution and recrystallization of the mineral. The presence of 0.1–0.2% $P_2O_5$ in the sedimentary target rocks of Cambrian and Ordovician age has been previously reported (Karis, 1998; Table 11.1).

The fungal hyphae network under the lichens invades the interface between grain boundaries, but it also filled cavities within the weathering crust. Analogously, intergranular voids in granitic rocks are known to enhance hyphae penetration of rocks (Prieto *et al.*, 1995). The production of organic acids by the fungi and possibly elevated acidity from carbon dioxide produced in

respiration is likely to have been an important mechanism for the dissolution of the calcareous components (Sanders *et al.*, 1994; Gorbushina and Krumbein, 2005). The lichens in the crust were shown to produce lichen acids which might play a role in calcite dissolution.

The observation of poorly ordered aluminosilicate materials at the rock–lichen interface is consistent with work by other workers (Ascaso *et al.*, 1976; Adamo and Violante, 1991), who found the formation of similar materials on lichens growing on basaltic, granitic and gneissic substrates. The extent to which such phases represent weathering of existing material or active neogenesis is still poorly understood. Specifically, Ascaso and Galvan (1976) found biological alteration of biotite and the release of cations to form halloysite, which may be attributable to organic acids produced either by the lichens or the fungal component.

A general scheme for the weathering of the Loftarstone (Fig. 11.13) can be hypothesized. First, lichen propagules become established on the surface of the rock, growing on the nutrient-rich chlorite (Fig. 11.13, A and B). The lichens are likely to aid subsequent development of the fungal hyphae network. The production of photosynthate by the lichen, and physical disaggregation of the substrate, aids the formation of a fungal hyphae network. The fungal hyphae, by secreting acids, dissolve the calcareous matrix and clasts, further enhancing hyphae penetration into the weathering crust. At a depth of approximately 0.5–1 cm,

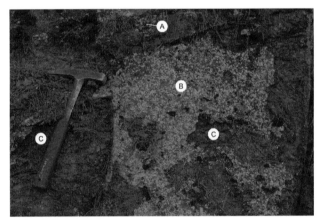

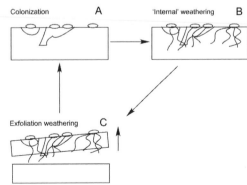

**Figure 11.13**  Summary scheme of weathering process on tsunami resurge (Loftarstone), illustrating exfoliation sequence with photograph showing example regions corresponding to each part of sequence.

under the hyphae network, the leaching of organic acids disaggregates the rock, which is not bound by the biological crust, leading to exfoliation of the rock (Fig. 11.13, C). Colonization can begin again on the new, and now partially degraded, rock surface. The structure of the weathering crust bears gross similarities to the zones of biogeochemical weathering suggested by Barker and Banfield (1998), with the presence of hyphae protruding from the exfoliation crust suggesting that exfoliation occurs at the interface between their Zones 3 and 4 – the biotically and abiotically dominated weathering zones respectively.

The weathering crust can be seen as a model for understanding weathering processes in any mixed materials in addition to impact tsunami deposits. These include tsunamis generated by deep-sea earthquakes (Dawson and Smith, 2000; Fujino *et al.*, 2006; Moore *et al.*, 2006) and sedimentary rocks such as greywacke and shale. The Loftarstone shows how each component of a mineralogically diverse crust, such as material formed in impact, can contribute to the weathering process by providing nutrients and/or improved access into the rock, potentially allowing faster weathering rates compared with more homogeneous mineralogies and establishing a well-defined weathering sequence. In this specific case, the chlorite provides a relatively soft, Fe-rich environment for lichens and other organisms during the initiation of the weathering crust. The chlorite also potentially provides a source of Mg and K, the latter element also being provided by feldspars. The limestone matrix, chlorite and apatite provide a source of Ca and P. The carbonate matrix, on account of its dissolution by organic acids, allows for the formation of cavities within the Loftarstone, which improve microbial access to nutrient-rich clasts.

The weathering crust raises important conjectures concerning impact events and weathering in general. Many of the clasts in the Loftarstone contain fine fracture networks, which are partly attributed to shock (Sturkell, 1998b). Indeed, the shocked features of the quartz within the Loftarstone were used to infer a role for a meteorite impact in the geology of this region of Sweden in the first place (Wickman, 1988). The feldspar and quartz within the weathering crust are derived from heavily shocked and fractured granite basement rock underlying the crater. Fracturing is thought to be one important factor in the determination of weathering rates (Callot *et al.*, 1987), and might account for the differences in weathering rates observed in the laboratory compared with the field (White and Brantley, 2004).

Impacts are known to enhance chemical weathering in target materials. Shocked quartz from the Ries impact crater in Germany has been shown to be susceptible to chemical weathering along its planar deformation features (Leroux, 2005). The fractured clasts within the rock at Lockne might aid microbial penetration. The colonization of impact fractures would accelerate both the dissolution of the clasts and concomitantly contribute to the binding of the clasts by the fungal network, helping to consolidate the exfoliation crust. As it is difficult to directly distinguish between impact-induced fractures and tectonically induced fractures in the weathered crust that was studied, it is not possible to quantify the degree to which impact-induced fractures may have played a role in the rate of weathering of this material.

In addition to weathering at the local scale, the role of impacts in weathering at larger scales should be considered. Weathering rates play an important role in climate control. By regulating $CO_2$ drawdown through the carbonate–silicate cycle, weathering acts as a negative feedback process to regulate planetary surface temperatures. On a heavily bombarded planet, impacts might play a part in increasing weathering rates by increasing either the surface area for chemical weathering or the surface area for the biosphere–geosphere interaction. It is generally thought that the feedback time of weathering on climate control is normally on the order of hundreds of thousands of years (Berner, 1993, 1997). An impact forming a crater of approximately 100 km diameter will form several thousands of cubic kilometres of impact fractured rocks inside the crater (Naumov, 2005). Fractured rocks can extend to at least two radii outwards of the crater centre. If a cubic centimetre of rock was to host fracture networks with a total surface area of approximately $10 \, cm^2$, then the total surface area of fractures in and around a new crater of approximately 100 km might well approach millions of square kilometres, similar to the surface area in, for example, places such as the Congo or the Amazon (Suchet and Probst, 1993), although clearly the effectiveness of weathering would depend upon the exposure of the fractured rocks to circulating water, etc. Such events, which would also excavate and distribute over wide areas vast quantities of small, easily weathered rock fragments, might transiently create conditions more conducive to weathering. On a planetary scale a question is whether such events could transiently contribute to carbon dioxide drawdown and cooling by generating vast areas of new weathering surfaces.

## 11.5   Impoverishment or enrichment?

A question that the data throughout this chapter elicits is whether asteroid and comet impacts, known to be highly destructive to surface-dwelling multicellular life (Alvarez *et al.*, 1980), enrich or impoverish conditions for microorganisms. The answer must be that they simultaneously do both and that the balance of these effects ultimately determines the extent to which particular habitats present new opportunities or deleterious conditions for microorganisms. Both physical and chemical changes can either impoverish or enrich the environment. Physical changes such as fracturing of rocks or impact bulking increase access to rocks and increase the flow of nutrients or redox couples and, therefore, improve conditions. However, impact-induced annealing or pore closure can reduce accessibility of rocks and impoverish the environment, as is seen in some highly shocked sandstones.

In the deep subsurface, fracturing of rocks, although not necessarily providing significant advantages with respect to available space, since this is often not the limiting factor for growth, can improve the flow of nutrients and redox couples, leading to regions of enhanced microbial abundance, as is observed in the deep subsurface of the Chesapeake structure. Chemical changes such as mobilization of evaporites or carbon might enhance the conditions for life; volatilization of elements, such as nitrogen, might impoverish the conditions for microbial life.

There is likely to be a period immediately after an impact, in cases where the impact is of sufficient size and can cause a significant thermal excursion, when the local surface environment is generally impoverished. Many essential nutrients (i.e. the volatile non-metallic nutrients) are volatilized at temperatures well within those expected during an impact event (Melosh, 1989). For example, nitrogen compounds are volatilized at about 200 °C, organic phosphorus at 350 °C and inorganic phosphorus at 750 °C. In this respect, there may be parallels with impoverishment processes associated with fire ecology at the site of high-intensity forest fires where volatilization of essential biological elements occurs (e.g. Agee, 1993). However, this view of early impoverishment cannot be accepted as dogma. At certain temperatures, and depending on the source material, high-intensity fires can mobilize nitrogen and phosphorus, increasing their short-term accessibility by orders of magnitude (Hauer and Spencer, 1998; del Pino *et al.*, 2007), and so it is not inconceivable that in some regions of an impact structure the heating of the target material could mobilize elements such as nitrogen and phosphorus from rocks or organic compounds, enhancing the geomicrobiological environment. As with the recolonization of volcanic substrates, wind-blown microorganisms will be introduced as soon as substrates have cooled to suitable temperatures. It is at this stage that the trade-offs in the longer term physical and chemical effects on the local geology will influence the geomicrobiology at the site of impact.

## 11.6   Astrobiological implications

The observations discussed above raise several astrobiological points that apply to the role of impacts in changing the conditions for planetary habitability. Although the presence of life on other planets remains speculation, these points might equally apply to life on the Archaean Earth. Just some of the implications of the data presented in this chapter are listed below:

- Impact events can generate cryptoendolithic habitats or increase the abundance of chasmoendolithic habitats in crystalline non-sedimentary lithologies. These may be important for the habitability of the surface and subsurface environment of planets where sedimentary rock cover is limited.
- Large impact events may sterilize regions of a planetary surface or, if they are of sufficient size, an entire planetary surface. However, these same impacts, by fracturing rocks in the deep subsurface, will generate refugia for microbial life to escape subsequent impacts.
- Impact events, by creating a hydrological depression in the form of the crater cavity, in which liquid water will gather, may improve conditions for life on planetary surfaces where water may be scarce (Cabrol *et al.*, 2001; Cockell and Lim, 2005).
- By fracturing and mixing rocks, impacts can increase the diversity of available redox couples and geochemical gradients in any given location, which may either directly benefit microbial life or increase the rate of rock weathering, which will itself increase the rate of nutrient release into the local environment. During Late Heavy Bombardment, impacts may have had globally important consequences for enhancing the availability of nutrients and energy from rocks
- The thermal excursion caused by impact can offer warm hydrothermal habitats in which minerals are circulated through a structure or its periphery (Osinski *et al.*, 2001, 2005a; Rathburn and Squyres, 2002; Koeberl and Reimold, 2004; Pope *et al.*, 2006; Hode *et al.*, 2008; Izawa *et al.*, 2011)

or the thermal excursion might create a period of transient melting in permafrost environments that life could take advantage of.

## 11.7   Concluding remarks

Asteroid and comet impacts change the characteristics of target material. Physical changes can include the increase or reduction of porosity, depending on the target lithology and shock levels experienced in a given part of a crater. The subsequent extent of microbial colonization in these physically altered materials will also depend upon the chemical composition or any alterations in the target rocks. These effects have now been observed in surface and subsurface impact environments and they provide a synthetic understanding of the role of impact events in changing geomicrobiological conditions on the Earth through time. In addition to defining the conditions for microorganisms, changes in target lithology and the formation of new substrates, such as tsunami deposits, can change the interaction of microorganisms with minerals, altering the patterns and processes of biogeochemically important processes such as rock weathering. In conclusion, asteroid and comet impacts have had globally important consequences for microorganisms throughout Earth history, changing their habitats, their distribution and their associated biogeochemical cycles.

## References

Abramov, O. and Kring, D.A. (2007) Numerical modeling of impact-induced hydrothermal activity at the Chicxulub Crater. *Meteoritics and Planetary Science*, 42, 93–112.

Abramov, O. and Mojzsis, S.J. (2009) Microbial habitability of the Hadean Earth during the late heavy bombardment. *Nature*, 459, 419–422.

Adamo, P. and Violante, P. (1991) Weathering of volcanic rocks from Mt. Vesuvius associated with the lichen *Stereocaulom vesuvianum*. *Pedobiologia*, 35, 209–217.

Agee, J.K. (1993) *Fire Ecology of Pacific Northwest Forests*, Island Press, Washington, DC.

Alvarez, L.W., Alvarez, W., Asaro, F. and Michel, H.V. (1980) Extraterrestrial cause for the Cretaceous–Tertiary extinction – experimental results and theoretical interpretation. *Science*, 208, 1095–1108.

Ames, D.E., Jonasson, I.R., Gibson, H.L. and Pope, K.O. (2006) Impact-generated hydrothermal system – constraints from large Paleoproterozoic Sudbury Crater, Canada, in *Biological Processes Associated with Impact Craters* (eds C.S. Cockell, C. Koeberl and I. Gilmour), Springer, Heidelberg, pp. 55–100.

Ascaso, C. and Galvan, J. (1976) Studies on the pedogenetic action of lichen acids. *Pedobiologia*, 16, 321–331.

Ascaso, C., Galvan, J. and Rodriguez-Pascal, C. (1976) The pedogenetic action of *Parmelia conspersa*, *Rhizocarpon geographicum* and *Umbilicaria pustulata*. *Lichenologist*, 8, 151–171.

Ashton, P.J. (1999) Limnology of the Pretoria Saltpan Crater-lake, in *Tswaing. Investigations into the Origin, Age and Paleoenvironments of the Pretoria Saltpan* (ed. T.C. Partridge), Geological Survey of South Africa Memoir 85, Council for Geoscience, Pretoria, pp. 72–90.

Ashton, P.J. and Schoeman, F.R. (1983) Limnological studies on the Pretoria Salt Pan, a hypersaline maar lake. 1. Morphometric, physical and chemical features. *Hydrobiologia*, 99, 61–73.

Barker, W.W. and Banfield, J.F. (1998) Zones of chemical and physical interaction at interfaces between microbial communities and minerals: a model. *Geomicrobiology*, 15, 223–244.

Berner, R.A. (1993) Weathering and its effect on atmospheric $CO_2$ over Phanerozoic time. *Chemical Geology*, 107, 373–374.

Berner, R.A. (1997) Paleoclimate – the rise of plants and their effect on weathering and atmospheric $CO_2$. *Science*, 276, 544–546.

Bowden, S.A. and Parnell, J. (2007) Intracrystalline lipids within sulfates from the Haughton impact structure – implications for survival of lipids on Mars. *Icarus*, 187, 422–429.

Büdel, B., Weber, B., Kühl, M. et al. (2004) Reshaping of sandstone surfaces by cryptoendolithic cyanobacteria: bioalkalization causes chemical weathering in arid landscapes. *Geobiology*, 2, 261–268.

Cabrol, N.A., Wynn-Williams, D.D., Crawford, D.A. and Grin, E.A. (2001) Recent aqueous environments in Martian impact craters: an astrobiological perspective. *Icarus*, 154, 98–112.

Cacciari, I. and Lippi, D. (1986) *Arthrobacters*, successful arid soil bacteria. A review. *Arid Soil Research and Rehabilitation*, 1, 1–30.

Callot, G., Maurette, M., Pottier, L. and Dubois, A. (1987) Biogenic etching of microfractures in amorphous and crystalline silicates. *Nature*, 328, 147–149.

Carson, J.L. and Brown, R.M. (1978) Studies of Hawaiian freshwater and soil algae. 2. Algal colonization and succession on a dated volcanic substrate. *Journal of Phycology*, 14, 171–178.

Cockell, C.S. and Lee, P. (2002) The biology of impact craters – a review. *Biological Reviews*, 77, 279–310.

Cockell, C.S. and Lim, D.S.S. (2005) Impact craters, water and microbial life, in *Water and Life on Mars* (ed. T. Tokano), Springer, Heidelberg, pp. 261–275.

Cockell, C.S. and Osinski, G.R. (2007) Impact-induced impoverishment and transformation of a sandstone habitat for lithophytic microorganisms. *Meteoritics and Planetary Science*, 42, 1985–1993.

Cockell, C.S. and Stokes, M.D. (2004) Widespread colonization by polar hypoliths. *Nature*, 431, 414.

Cockell, C.S. and Stokes, M.D. (2006) Hypolithic colonization of opaque rocks in the Arctic and Antarctic polar desert. *Arctic, Antarctic and Alpine Research*, 38, 335–342.

Cockell, C.S., Lee, P., Osinski, G. et al. (2002) Impact-induced microbial endolithic habitats. *Meteoritics and Planetary Science*, 37, 1287–1298.

Cockell, C.S., Osinski, G.R. and Lee, P. (2003a) The impact crater as a habitat: effects of impact-processing of target materials. *Astrobiology*, 3, 181–191.

Cockell, C.S., Rettberg, P., Horneck, G. et al. (2003b) Measurements of microbial protection from ultraviolet radiation in polar terrestrial microhabitats. *Polar Biology*, 26, 62–69.

Cockell, C.S., Lee, P., Broady, P. et al. (2005) Effects of asteroid and comet impacts on habitats for lithophytic organisms – a synthesis. *Meteoritics and Planetary Science*, 40, 1901–1914.

Cockell, C.S., Kennerley, N., Lindstrom, M. et al. (2007) Geomicrobiology of a weathering crust from an impact crater and a hypothesis for its formation. *Geomicrobiology Journal*, 24, 425–440.

Cockell, C.S., Gronstal, A.L., Voytek, M.A. et al. (2009a) Microbial abundance in the deep subsurface of the Chesapeake Bay impact crater: relationship to lithology and impact processes, in *The ICDP–USGS Deep Drilling Project in the Chesapeake Bay Impact Structure: Results from the Eyreville Core Holes* (eds G.S. Gohn, C. Koeberl, K.G. Miller and W.U. Reimold), Geological Society of America Special Paper 458, Geological Society of America, Boulder, CO, pp. 941–950.

Cockell, C.S., Osinski, G., Bannerjee, N.R. et al. (2009b) The microbe–mineral environment and gypsum neogenesis in a weathered polar evaporite. *Geobiology*, 8, 293–308.

Cremer, H. and Wagner, B. (2003) The diatom flora in the ultra-oligotrophic Lake El'gygytgyn, Chukotka. *Polar Biology*, 26, 105–114.

Dawson, S. and Smith, D.E. (2000) The sedimentology of Middle Holocene tsunami facies in northern Sutherland, Scotland, UK. *Marine Geology*, 170, 69–79.

Del Pino, J.S.N., Almenar, I.D., Rivero, F.N. et al. (2007) Temporal evolution of organic carbon and nitrogen forms in volcanic soils under broom scrub affected by a wildfire. *Science of the Total Environment*, 378, 245–252.

Escala, M., Rosell-Melé, A., Fietz, S. and Koeberl, C. (2008) Archaeabacterial lipids in drill core samples from the Bosumtwi impact structure, Ghana. *Meteoritics and Planetary Science*, 43, 1777–1782.

Fermani, P., Mataloni, G. and Van de Vijver, B. (2007) Soil microalgal communities on an Antarctic active volcano (Deception Iceland, South Shetlands). *Polar Biology*, 30, 1381–1393.

Fike, D.A., Cockell, C.S., Pearce, D. and Lee, P. (2003) Heterotrophic microbial colonization of the interior of impact-shocked rocks from Haughton impact structure, Devon Island, Nunavut, Canadian High Arctic. *International Journal of Astrobiology*, 1, 311–323.

Frankel, L. (1977) Microorganism induced weathering of biotite and hornblende grains in estuarine sands. *Journal of Sedimentary Petrology*, 47, 849–854.

French, B.M. (2004) The importance of being cratered: the new role of meteorite impact as a normal geological process. *Meteoritics and Planetary Science*, 39, 169–197.

Friedmann, E.I. (1982) Endolithic microorganisms in the Antarctic cold desert. *Science*, 215, 1045–1053.

Fujino, S., Masuda, F., Tagomori, S. and Matsumoto, D. (2006) Structure and depositional processes of a gravelly tsunami deposit in a shallow marine setting: Lower Cretaceous Miyako Group, Japan. *Sedimentary Geology*, 187, 127–138.

Gohn, G.S., Koeberl, C., Miller, K.G. et al. (2006) Chesapeake Bay impact structure drilled. *EOS, Transactions, American Geophysical Union*, 87, 349–355.

Gohn, G., Koeberl, C., Miller, K.G. et al. (2008) Deep drilling into the Chesapeake Bay impact structure. *Science*, 320, 1740–1745.

Gorbushina, A.A. and Krumbein, W.E. (2005) Role of microorganisms in wear down of rocks and minerals, in *Microorganisms in Soils: Roles in Genesis and Functions* (eds F. Buscot and A. Varma), Springer-Verlag, Berlin, pp. 59–84.

Gronlund, T., Lortie, G., Guilbault, J.P. et al. (1990) Diatoms and arcellaceans from lac du Cratere du Nouveau-Quebec, Ungava, Quebec, Canada. *Canadian Journal of Botany*, 68, 1187–1200.

Gronstal, A., Voytek, M.A., Kirshtein, J.D. et al. (2009) Contamination assessment in microbiological sampling of the Eyreville core, Chesapeake Bay impact structure, in *The ICDP–USGS Deep Drilling Project in the Chesapeake Bay Impact Structure: Results from the Eyreville Core Holes* (eds G.S. Gohn, C. Koeberl, K.G. Miller and W.U. Reimold), Geological Society of America Special Paper 458, Geological Society of America, Boulder, CO, pp 951–964.

Hauer, F.R. and Spencer, C.N. (1998) Phosphorus and nitrogen dynamics in streams associated with wildfire: a study of immediate and long-term effects. *International Journal of Wildland Fire*, 8, 183–198.

Henkel, H. (1992) Geophysical aspects of meteorite impact craters in eroded shield environment, with special emphasis on electric resistivity. *Tectonophysics*, 216, 63–89.

Herrera, A., Cockell, C.S., Self, S. *et al.* (2009) A cryptoendolithic community in volcanic glass. *Astrobiology*, 9, 369–381.

Hode, T., Cady, S.L., von Dalwigk, I. and Kristiansson, P. (2008) Evidence of ancient microbial life in an impact structure and its implications for astrobiology – a case study, in *From Fossils to Astrobiology* (eds J. Seckbach and M. Walsh), Springer, Heidelberg, pp. 249–273.

Horsfield, B. and Kieft, T.L. and the Geobiosphere Group (2007) The geobiosphere, in *Continental Scientific Drilling* (eds U. Harms, C. Koeberl and M.D. Zoback), Springer, Heidelberg, pp. 163–211.

Horton, J.W., Powars, D.S. and Gohn, G.S. (2005) *Studies of the Chesapeake Bay Impact Structure – The USGS–NASA Langley Corehole, Hampton, Virginia, and Related Coreholes and Geophysical Surveys*, USGS Professional Paper 1688, US Geological Survey, Reston, VA.

Horton Jr, J.W., Gibson, R.L., Reimold, W.U. *et al.* (2009) Geologic columns for the ICDP–USGS Eyreville B core, Chesapeake Bay impact structure: impactites and crystalline rocks, 1766 to 1096 m depth, in *The ICDP–USGS Deep Drilling Project in the Chesapeake Bay Impact Structure: Results from the Eyreville Core Holes* (eds G.S. Gohn, C. Koeberl, K.G. Miller and W.U. Reimold), Geological Society of America Special Paper 458, Geological Society of America, Boulder, CO, pp. 21–50.

Izawa, M.R.M., Banerjee, N.R., Osinski, G.R. *et al.* 2011. Weathering of post-impact hydrothermal deposits from the Haughton impact structure: implications for microbial colonization and biosignature preservation. *Astrobiology*, 11, 537–550.

Joshi, A.A., Kanekar, P.P., Kelkar, A.S. *et al.* (2008) Cultivable bacterial diversity of alkaline Lonar Lake, India. *Microbial Ecology*, 55, 163–172.

Karis, L. (1998) Jämtlands östliga fjällberggrund, in *Beskrivning till Berggrundskartan Över Jämtlands Län Del 2. Fjälldelen* (eds L. Karis and A. Strömberg), SGU Serie Ca 53:2, Sveriges Geologiska Undersökning, Uppsala, pp. 4–178.

Kieffer, S.W. (1971) Shock metamorphism of the Coconino sandstone at Meteor Crater, Arizona. *Journal of Geophysical Research*, 76, 5449–5473.

Kieffer, S.W., Phakey, P.P. and Christie, J.M. (1976) Shock processes in porous quartzite: transmission electron microscope observations and theory. *Contributions to Mineralogy and Petrology*, 59, 41–93.

Koeberl, C. and Reimold, W.U. (2004) Post-impact hydrothermal activity in meteorite impact craters and potential opportunities for life. *Bioastronomy 2002: Life among the Stars*, 213, 299–304.

Koeberl, C., Poag, C.W., Reimold, W.U. and Brandt, D. (1996) Impact origin of the Chesapeake Bay structure, and source of the North American tektites. *Science*, 271, 1263–1266.

Kring D.A. (1997) Air blast produced by the Meteor Crater impact event and a reconstruction of the affected environment. *Meteoritics and Planetary Science*, 32, 517–530.

Kring, D.A. (2003) Environmental consequences of impact cratering events as a function of ambient conditions on Earth. *Astrobiology*, 3, 133–152.

Leroux, H. (2005) Weathering features in shocked quartz from the Ries impact crater: Germany. *Meteoritics and Planetary Science*, 40, 1347–1352.

Lindstrom, M., Shuvalov, V. and Ivanov, B. (2005) Lockne Crater as a result of marine-target oblique impact. *Planetary and Space Science*, 53, 803–815.

Maher, K.A. and Stevenson, D.J. (1988) Impact frustration of the origin of life. *Nature*, 331, 612–614.

Malinconico, M.L., Sanford, W.E. and Horton Jr, J.W. (2009) Post-impact heat conduction and compaction-driven fluid flow in the Chesapeake Bay impact structure based on downhole vitrinite reflectance data, ICDP–USGS Eyreville deep core holes and Cape Charles test holes, in *The ICDP–USGS Deep Drilling Project in the Chesapeake Bay Impact Structure: Results from the Eyreville Core Holes* (eds G.S. Gohn, C. Koeberl, K.G. Miller and W.U. Reimold), Geological Society of America Special Paper 458, Geological Society of America, Boulder, CO, pp 905–930.

Maltais, M.J. and Vincent, W.F. (1997) Periphyton community structure and dynamics in a subarctic lake. *Canadian Journal of Botany*, 75, 1556–1569.

Melosh, H.J. (1989) *Impact Cratering: A Geologic Process*, Oxford University Press, Oxford.

Metzler, A., Ostertag, R., Redeker, H.J. and Stoffler, D. (1988) Composition of the crystalline basement and shock metamorphism of crystalline and sedimentary target rocks at the Haughton-impact-crater, Devon Island, Canada. *Meteoritics*, 23, 197–207.

Moore, A., Nishimura, Y., Gelfenbaum, G. *et al.* (2006) Sedimentary deposits of the 26 December 2004 tsunami on the northwest coast of Aceh, Indonesia. *Earth, Planets and Space*, 58, 253–258.

Naumov, M.V. (2005) Principal features of impact-generated hydrothermal circulation systems: mineralogical and geochemical evidence. *Geofluids*, 5, 165–184.

Nelson, L.M. and Parkinson, D. (1978a) Growth characteristics of three bacterial isolates from an arctic soil. *Canadian Journal of Microbiology*, 24, 909–914.

Nelson, L.M. and Parkinson, D. (1978b) Effect of starvation on survival of three bacterial isolates from an arctic soil. *Canadian Journal of Microbiology*, 24, 1460–1467.

Omelon, C.R., Pollard, W.H. and Ferris, F.G. (2006) Chemical and ultrastructural characterization of High Arctic cryptoendolithic habitats. *Geomicrobiology Journal*, 23, 189–200.

Ormo, J. and Miyamoto, H. (2002) Computer modelling of the water resurge at a marine impact: the Lockne Crater, Sweden. *Deep-Sea Research*, 49, 983–994.

Osinski, G.R. (2007) Impact metamorphism of $CaCO_3$-bearing sandstones at the Haughton structure, Canada. *Meteoritics and Planetary Science*, 42, 1945–1960.

Osinski, G.R., Spray, J.G. and Lee, P. (2001) Impact-induced hydrothermal activity within the Haughton impact structure: generation of a transient, warm, wet oasis. *Meteoritics and Planetary Science*, 36, 731–745.

Osinski, G.R., Lee, P., Parnell, J. *et al.* (2005a) A case study of impact-induced hydrothermal activity: the Haughton impact structure, Devon Island, Canadian High Arctic. *Meteoritics and Planetary Science*, 40, 1859–1877.

Osinski, G.R., Lee, P., Spray, J.G. *et al.* (2005b) Geological overview and cratering model of the Haughton impact structure, Devon Island, Canadian High Arctic. *Meteoritics and Planetary Science*, 40, 1759–1776.

Osinski, G.R., Cockell, C.S., Lindgren, P. and Parnell, J. (2010) The effect of meteorite impacts on the elements essential for life. Astrobiology Science Conference, abstract no. 5252.

Parnell, J., Lee, P., Cockell, C.S. and Osinski, G.R. (2004) Microbial colonization in impact-generated hydrothermal sulphate deposits,

Haughton impact structure, and implications for sulphates on Mars. *International Journal of Astrobiology*, 3, 247–256.

Poag, C.W. (1999) *The Chesapeake Invader*, Princeton University Press, Princeton, NJ.

Poag, C.W., Powars, D.S., Poppe, L.J. *et al.* (1992) Deep Sea Drilling Project Site 612 bolide event: new evidence of a late Eocene impact-wave deposit and a possible impact site, U.S. east coast. *Geology*, 10, 771–774.

Poag, C.W., Koeberl, C. and Reimold, W.U. (2004) *The Chesapeake Bay Crater – Geology and Geophysics of a Late Eocene Submarine Impact Structure*, Impact Studies Series, Springer, Heidelberg.

Pope, K.O., Kieffer, S.W. and Ames, D.E. (2006) Impact melt sheet formation on Mars and its implication for hydrothermal systems and exobiology. *Icarus*, 183, 1–9.

Powars, D.S. and Bruce, T.S. (1999) *The Effects of the Chesapeake Bay Impact Crater on the Geological Framework and Correlation of Hydrogeologic Units of the Lower York-James Peninsula, Virginia*, USGS Professional Paper 1612, US Geological Survey, Reston, VA.

Powars, D.S., Poag, C.W. and Mixon, R.B. (1993) The Chesapeake Bay 'impact crater': seismic and stratigraphic evidence. *Geological Society of America Abstracts with Programs*, 33, 417–420.

Prieto, B., Rivas, M.T. and Silva, B.M. (1994) Colonization by lichens of granite dolmens in Galicia (NW Spain). *International Biodeterioration and Biodegradation*, 34, 47–60.

Prieto, B., Rivas, M.T. and Silva, B.M. (1995) Colonization by lichens of granite churches in Galicia (NW Spain). *Science of the Total Environment*, 167, 343–351.

Puskeppeleit, M., Quintern, L.E., ElNaggar, S. *et al.* (1992) Long-term dosimetry of solar UV radiation in Antarctica with spores of *Bacillus subtilis*. *Applied and Environmental Microbiology*, 58, 2355–2359.

Quintern, L.E., Horneck, G., Eschweiler, U. and Bücker, H. (1992) A biofilm used as ultraviolet-dosimeter. *Photochemistry and Photobiology*, 55, 389–395.

Quintern, L.E., Puskeppeleit, M., Rainer, P. *et al.* (1994) Continuous dosimetry of the biologically harmful UV-radiation in Antarctica with the biofilm technique. *Journal of Photochemistry and Photobiology*, 22, 59–66.

Rathbun, J.A. and Squyres, S.W. (2002) Hydrothermal systems associated with Martian impact craters. *Icarus*, 157, 362–372.

Reddy, G.S.N., Prakash, J.S.S., Matsumoto, G.I. *et al.* (2002) *Arthrobacter roseus* sp. nov., a psychrophilic bacterium isolated from an Antarctic cyanobacterial mat sample. *International Journal of Systematic and Evolutionary Microbiology*, 52, 1017–1021.

Reimold, W.U., Koeberl, C., Gibson, R.L. and Dressler, B.O. (2005) Economic mineral deposits in impact structures: a review, in *Impact Tectonics* (eds C. Koeberl and H. Henkel), Springer, Heidelberg, pp. 479–552.

Rettberg, P. and Cockell, C.S. (2004) Biological UV dosimetry using the DLR-biofilm. *Photochemical and Photobiological Sciences*, 3, 781–787.

Rettberg, P., Seif, R. and Horneck, H. (1999) The DLR biofilm as personal UV dosimeter, in *Fundamentals for the Assessment of Risks from Environmental Radiation* (eds C. Baumstark-Khan, S. Kozubek and G. Horneck), NATO Science Series, Series 2: Environmental Security, vol. 55, Kluwer Academic Publishers, Dordrecht, pp. 367–370.

Sanders, W.B., Ascaso, C. and Wierchos, J. (1994) Physical interactions of two rhizomorph-forming lichens with their rock substrate. *Botanica Acta*, 107, 432–439.

Sanford, W.E. (2003) Heat flow and brine generation following the Chesapeake Bay bolide impact. *Journal of Geochemical Exploration*, 78–79, 243–247.

Sanford, W.E., Voytek, M.A., Powars, D.S. *et al.* (2007) Pore-water chemistry from the ICDP–USGS core hole in the Chesapeake Bay impact structure – implications for paleohydrology, microbial habitat, and water resources in *The ICDP–USGS Deep Drilling Project in the Chesapeake Bay Impact Structure: Results from the Eyreville Core Holes* (eds G.S. Gohn, C. Koeberl, K.G. Miller and W.U. Reimold), Geological Society of America Special Paper 458, Geological Society of America, Boulder, CO, pp. 867–890.

Schoeman, F.R. and Ashton, P.J. (1982) The diatom flora of the Pretoria Salt Pan, Transvaal, Republic of South Africa. *Bacillaria*, 5, 63–99.

Sherlock, S.C., Kelley, S.P., Parnell, J. *et al.* (2005) Re-evaluating the age of the Haughton impact event. *Meteoritics and Planetary Science*, 40, 1777–1787.

Simon, S. (1987) *Stratigraphie, Petrographie und Entstehungsbedingungen von Grobklastika in der autochthonen, ordovizischen Schichtenfolge Jämtlands (Schweden)*, SGU Series C 815, Sveriges Geologiska Undersökning.

Sleep, N.H., Zahnle, K.J., Kasting, J.F. and Morowitz, H.J. (1989) Annihilation of ecosystems by large asteroid and comet impacts on the early Earth. *Nature*, 342, 139–142.

Sterflinger, K. (2000) Fungi as geological agents. *Geomicrobiology Journal*, 17, 97–124.

Stuart, J.C. and Binzel R. P. (2004) Bias-corrected population, size distribution, and impact hazard for the near-Earth objects. *Icarus*, 170, 295–311.

Sturkell, E.F.F. (1998a) Impact-related Ir anomaly in the Middle Ordovician Lockne impact structure, Jämtland, Sweden. GFF (*Geologiska Föreningens i Stockholm Förhandlingar*), 120, 333–336.

Sturkell, E.F.F. (1998b) The marine Lockne impact structure, Jämtland, Sweden: a review. *Geologische Rundschau*, 87, 253–267.

Suchet, P.A. and Probst, J.L. (1993) Modelling of atmospheric $CO_2$ consumption by chemical weathering of rocks: application to the Garonne, Congo and Amazon basins. *Chemical Geology*, 107, 205–210.

Surakasi, V.P., Wani, A.A., Shouhe, Y.S. and Ranade, D.R. (2007) Phylogenetic analysis of methanogenic enrichment cultures obtained from Lonar Lake in India: isolation of *Methanocalculus* sp. and *Mathanoculleus* sp. *Microbial Ecology*, 54, 697–704.

Toon, O.W., Zahnle, K., Morrison, D. *et al.* (1997) Environmental perturbations caused by the impacts of asteroids and comets. *Reviews of Geophysics*, 35, 41–78.

Versh, E., Kirsimae, K., Joeleht, A. and Plado, J. (2005) Cooling of the Kardla impact crater: I. The mineral paragenetic sequence observation. *Meteoritics and Planetary Science*, 40, 3–19.

Von Dalwigk, I. and Ormo, J. (2001) Formation of resurge gullies at impacts at sea: the Lockne Crater, Sweden. *Meteoritics and Planetary Science*, 36, 359–369.

Wallander, H. and Hagerberg, D. (2004) Do ectomycorrhizal fungi have a significant role in weathering of minerals in forest soil? *Symbiosis*, 37, 249–257.

Wani, A.A., Surakasi, V.P., Siddharth, J. *et al.* (2006) Molecular analyses of microbial diversity associated with the Lonar soda lake in India: an impact crater in a basalt area. *Research in Microbiology*, 10, 928–937.

Weber, B., Wessels, D.C.J. and Büdel, B. (1996) Biology and ecology of cryptoendolithic cyanobacteria of a sandstone outcrop in the Northern Province, South Africa. *Algological Studies*, 83, 565–579.

Welch, S.A., Taunton, A.E. and Banfield, J.F. (2002) Effect of micro-organisms and microbial metabolites on apatite dissolution. *Geomicrobiology Journal*, 19, 343–367.

White, A.F. and Brantley, S.L. (2003) The effect of time on the weathering of silicate minerals: why do weathering rates differ in the laboratory and field? *Chemical Geology*, 202, 479–506.

Wickman, F.E. (1988) Possible impact structures in Sweden, in *Deep Drilling in Crystalline Bedrock. Volume 1. The Deep Gas Drilling in the Siljan Impact Structure, Sweden and Astroblemes* (eds A. Bodén and K.G. Eriksson), Springer, New York, NY, pp. 298–327.

Wierzchos, J. and Ascaso, C. (1994) Application of back-scattered electron imaging to the study of the lichen–rock interface. *Journal of Microscopy*, 175, 54–59.

Wierzchos, J. and Ascaso, C. (1996) Morphological and chemical features of bioweathered granitic biotite induced by lichen activity. *Clays and Clay Minerals*, 44, 652–657.

# TWELVE

# *Economic deposits at terrestrial impact structures*

## Richard A. F. Grieve

*Earth Sciences Sector, Natural Resources Canada, Ottawa, ON, K1A 0E8, Canada*

## 12.1   Introduction

This chapter represents an update of previous reviews (e.g. Grieve and Masaitis, 1994; Grieve, 2005; Reimold *et al.*, 2005) of natural resources associated with some terrestrial impact structures. This chapter, however, is not totally comprehensive with respect to the full economic value of terrestrial impact structures. A more comprehensive list is given in Reimold *et al.* (2005: Table 1), which lists 56 impact structures with 'economic interests', including structures that provide sites for recreational activities and/or serve as tourist attractions. This chapter also does not consider those structures that have been, or are being, exploited as a source of aggregate, lime or stone for building material (e.g. Ries in Germany, Rouchechouart in France) or are a source of groundwater or serve as reservoirs for hydro-electric power generation (e.g. Manicouagan in Canada, Puchezh-Katunki in Russia). The economic value of such impact structures, however, can be considerable. For example, the electricity generated by the Manicouagan reservoir can reach the order of $5000\,\mathrm{GWh\,yr^{-1}}$, sufficient to supply power to a small city and worth approximately US$300 million per year, at current residential electricity prices in Canada.

Previously, Grieve and Masaitis (1994) and Grieve (2005) considered resource deposits at terrestrial impact structures in the following order: progenetic, syngenetic, epigenetic (Table 12.1). Progenetic economic deposits are those that originated prior to the impact event by purely terrestrial concentration mechanisms. The impact event caused spatial redistribution of these deposits and, in some cases, brought them to a surface or near-surface position, from where they can be exploited. Syngenetic deposits are those that originated during the impact event, or immediately afterwards, as a direct result of impact processes. They owe their origin to energy deposition in the local environment from the impact event, resulting in phase changes and melting. In recent years, there has been greater recognition of the role for post-impact hydrothermal activity at impact structures (e.g.

Ames *et al.*, 1998, 2006; Osinski *et al.*, 2001; Naumov, 2002; Abramov and Kring, 2004, 2007). Post-impact hydrothermal deposits are a logical result of heating due to the impact process and, thus, are considered here as syngenetic deposits. Thus, the remobilization of some progenetic deposits has blurred the separation between progenetic and syngenetic deposits. Epigenetic deposits result from the formation of an enclosed topographic basin, with restricted sedimentation, or the long-term flow of fluids into structural traps formed by impact structures. Commercial accumulations of hydrocarbons at terrestrial impact structures can result from a number of processes related to impact and are considered separately.

## 12.2   Progenetic deposits

Progenetic economic deposits in impact structures include iron, uranium, gold, hydrocarbons and others (Table 12.1). In many cases, the deposits are relatively small. Only the larger and more active deposits are considered here.

### 12.2.1   Iron and uranium at Ternovka

Iron and uranium ores occur in the parautochthonous rocks of the crater floor and in breccias at the Ternovka structure (48°08′N, 33°31′E) in the Krivoj Rog region of the Ukraine, which is well known for its very large iron ore deposits (Nikolsky *et al.*, 1981, 1982). The structure is $375 \pm 25\,\mathrm{Myr}$ old according to Nikolsky (1991) and was formed in a Lower Proterozoic fold belt. It is a complex structure, where erosion has removed most of the allochthonous impact lithologies and exposed the floor of the structure. The central uplift is, in part, brecciated and injected by dikes of impact melt rock up to 20 m wide. The annular trough contains remnants of allochthonous breccia, with patches and lenses of impact melt-bearing breccia (see Chapter 7 for an introduction to impactites). The present diameter of the structure is

*Impact Cratering: Processes and Products*, First Edition. Edited by Gordon R. Osinski and Elisabetta Pierazzo.
© 2013 Blackwell Publishing Ltd. Published 2013 by Blackwell Publishing Ltd.

**Table 12.1** Genetic groups of economic deposits at terrestrial impact structures

| Genesis | Principal mode of origin | Type of economic deposit |
| --- | --- | --- |
| Progenetic | Brecciation | Building stone, silica |
| | Structural displacement | Iron, uranium, gold, oil, natural gas |
| Syngenetic | Phase transitions | Diamond |
| | Melting | Copper, nickel, PGEs,[a] tektite glass |
| | Hydrothermal | Lead, zinc, uranium, gold, PGEs, copper, nickel, zeolite, agate |
| Epigenetic | Confined sedimentation | Oil shale, lignite, calcium phosphate, amber |
| | Fluid flow | Oil, natural gas, fresh and mineralized water |
| | Regional sedimentation | Placer diamonds, tektites |

[a]PGE: platinum group element.

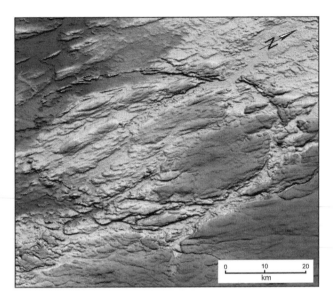

**Figure 12.1** Synoptic image of the topography of the Carswell structure, based on the Shuttle Radar Topographic Mission. Reds and yellows are topographic highs and blues are lows. (See Colour Plate 28)

10–11 km and its original diameter may have been 15–18 km. There is, however, a smaller estimate for the original diameter of approximately 8 km by Krochuk and Sharpton (2003), who also reported a younger, single whole-rock $^{39}$Ar–$^{40}$Ar date of 290 ± 10 Ma for an impact melt rock.

The iron ores at Ternovka have been exploited for more than 50 years through open pit and underground operations down to a depth of about 1 km. The ores are the result of hydrothermal and metasomatic action, which occurred during the Lower Proterozoic, on ferruginous quartzites (jaspilites) and some other lithologies, producing zones of albitites, aegirinites and amphibole–magnetite and carbonate–haematite rocks, along with the uranium mineralization. Post-impact hydrothermal alteration has led to the remobilization of some of the uranium mineralization and the formation of veins of pitchblende. The production of uranium ceased in 1967, but iron ore is still extracted, from two main open pits, Annovsky and Pervomaysk, with the target annual production in 2008 being 14 million tonnes of iron ore concentrate. The total reserves at the Pervomaysk open pit are estimated at 74 million tonnes, with additional reserves of lower grade deposits estimated at 675 million tonnes. The reserves at Annoysky are 450 million tonnes. Owing to impact-related brecciation and megablock displacement, blocks of iron ore are mixed with barren blocks. These blocks are up to hundreds of metres in dimension, having been rotated and displaced from their pre-impact positions. This displacement and mixing of lithologies has caused difficulties in operation and in evaluating the reserves. Conversely, impact-induced fracturing has aided in the physical extraction and processing of the ore.

### 12.2.2 Uranium at Carswell

The Carswell impact structure (58°27′N, 109°30′W) is located in the Athabasca Basin in northern Saskatchewan, Canada, about 120 km south of Uranium City. The Proterozoic-aged Athabasca Basin contains approximately 25% of the world's known uranium

(Jefferson and Delaney, 2007). It is the richest known and second largest uranium-producing region in the world, after the Olympic Dam in Australia. Cumulative uranium production from the basin is approximately 1.5 billion pounds (~0.68 Mt) of uranium oxide. Details of various geological, geophysical and geochemical aspects of the Carswell structure, with heavy emphasis on the uranium ore deposits, can be found in Lainé *et al.* (1985). The Carswell structure, which has been eroded to below the floor of the original crater, is apparent in Shuttle radar topography as two circular ridges, corresponding to the outcrop of the dolomites of the Carswell Formation (Fig. 12.1). The outer ridge is approximately 39 km in diameter and is generally quoted as the diameter of the structure (e.g. Currie, 1969; Harper, 1982). It represents a minimum original diameter. The outer ring of the structure is about 5 km wide and forms cliffs 65 m high of the Douglas and Carswell Formations. Interior to this, there is an annular trough, occupied by sandstones and conglomerates of the William River Subgroup of the Athabasca Group. This trough is also approximately 5 km wide and rises to a core, approximately 20 km in diameter, of metamorphic crystalline basement.

The crystalline basement core consists of mixed feldspathic and mafic gneisses of the Earl River Complex, overlain by the more aluminous Peter River gneiss. Details of their mineralogy and chemistry can be found in Bell *et al.* (1985), Harper (1982), and Pagel and Svab (1985). The basement core is believed to have been structurally uplifted by a minimum of 2 km. From a detailed structural study of the Dominique–Peter uranium deposit near the southern edge of the crystalline core, Baudemont and Fedorowich (1996) estimated that the amount of structural uplift in that area was in the order of 1.2 km. Surrounding the crystalline basement core are units of the unmetamorphosed Athabasca Group of sediments. The inner contact with the basement core is

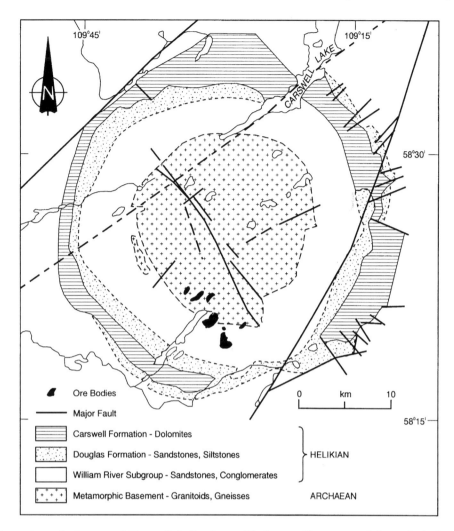

**Figure 12.2** Schematic geologic map of Carswell, indicating uplifted crystalline core and down-faulted annulus of Carswell and Douglas Formations, which appear as the outer topographically high and low annuli respectively in Fig. 12.1. Also shown are the locations of the uranium ore deposits.

faulted and truncated in places and offset by radial faults. The outer contact of the Carswell Formation is also characterized by arcuate faulting, drag folding and local overturning of beds, as well as being offset by radial faults (Fig. 12.2). The outcrop of the Carswell and Douglas Formations, which are the uppermost units of the Athabasca Group, is unique to the area. They owe their preservation to having been down-faulted at least 1 km in the impact to their present position (Harper, 1982; Tona *et al.*, 1985). Brecciation is common at Carswell and affects all lithologies. Currie (1969) used the term 'Cluff Breccias', after exposures near Cluff Lake, to include autochthonous monomict, allochthonous polymict clastic and melt-bearing breccias, as well as clast-rich impact melt rocks (see Chapter 7). The latter three lithologies occur as dike-like bodies and the relationships between the various breccias are locally complex.

Within the Carswell structure, the known commercial uranium deposits (Fig. 12.2) occur in two main settings: at the unconformity between the Athabasca sandstone of the William River Sub-group and the uplifted crystalline basement core, and in mylonites and along faults in the crystalline core. These deposits had grades from 0.3 to over 4% uranium oxide (average 0.6%) and have produced over 45 million pounds (~20.4 kt) of uranium metal (Jefferson and Delaney, 2007), worth over US$2 billion at average 2009 prices. The original uranium mineralization in the Athabasca Basin, and at Carswell, occurred during regolith development in the Precambrian, with later remobilization due to hydrothermal activity in response to thermo-tectonic events (Bell *et al.*, 1985; Lainé, 1986). The original commercial uranium deposit discovered at Carswell, the Cluff Lake D deposit (Fig. 12.3), was a pre-existing or progenetic ore deposit that was brought to its present location by structural uplift in the Carswell impact event and subsequent erosion. The mineralization at Cluff Lake D is also associated with shear zones and faulting (Tona *et al.*, 1985). At the time of mining, it was the richest known uranium ore body in the world (Lainé, 1986). It closed in 2002, having produced 62 million pounds (~28.1 kt) of 'yellow cake', an intermediate milling product consisting mostly of uranium oxide, over its 22 year lifetime.

In their study of Dominique–Peter, the largest basement-hosted uranium deposit, Baudemont and Fedorowich (1996) recognized four episodes of deformation. Two of these were prior to mineralization, the third episode related to mineralization and the final episode related to the Carswell impact event. They noted that Carswell-related deformation reactivated earlier faults,

associated with the main mineralization. 'Cluff Breccia' also occurs commonly in the same fault structures as the uranium mineralization. Baudemont and Fedorowich (1996) found the association 'striking, albeit complexing'. Mineralized material occurs within the 'Cluff Breccias', as veins of coffinite (Lainé, 1986).

It is not clear to what extent the Carswell impact event was involved in remobilizing the ores, beyond physical movement associated with fault reactivation during structural uplift. The basement-hosted ores are all associated with extensive regional alteration, indicative of hydrothermal fluid movement (Lainé, 1986). At present, the last of the known commercial uranium deposits within the Carswell structure is closed down. There are, however, additional active exploration targets, including reactivated faults, with pseudotachylitic breccia and/or 'Cluff Breccias' and uranium mineralization (Baudemont and Fedorowich, 1996). In addition to bringing what were originally 'deep' ore bodies to the surface, Carswell serves as a unique knowledge window into the nature of basement beneath the Athabasca Basin and, as such, serves as a guide to uranium exploration through out the entire basin (C. Jefferson, Geological Survey of Canada, personal communication, 2009).

### 12.2.3 Gold and uranium of Vredefort

The Vredefort structure, South Africa (27°0′S, 27°30′E; Fig. 12.4) consists of a 44 km diameter uplifted central core of predominantly Archaean granitic gneisses surrounded by an 18 km wide collar of steeply dipping to overturned Proterozoic sedimentary

**Figure 12.3** Aerial view in 1999 of DJX open pit and underground mine at Cluff Lake, Carswell, with the Claude pit in the background. Image courtesy of AREVA Resources Canada.

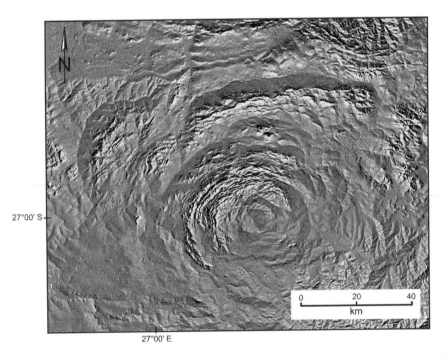

**Figure 12.4** Greyscale synoptic image of the topography of the Vredefort impact structure, based on the Shuttle Radar Topographic Mission. Image has been illuminated from the centre to emphasize topographic features, related to anticlinal and synclinal structures, surrounding the centre of Vredefort. These features are not evident in the SE quadrant, where the impact structure is covered by post-impact Karoo Supergroup sediments and volcanics.

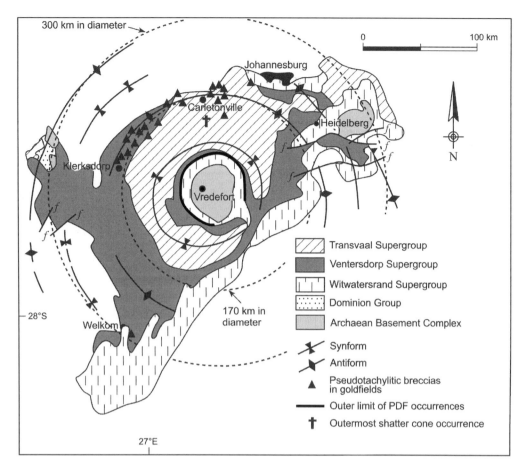

**Figure 12.5** Schematic geologic map of the Vredefort impact structure, with the outcrop of the Karoo cover rocks removed. Also shown are the spatial extent of shatter cones and PDFs in quartz and the various circumscribing structures related to Vredefort and initially identified by McCarthy *et al.* (1986, 1990). The Vredefort goldfields describe the so-called 'golden arch', which passes through or close to Heidelberg, Johannesburg, Carletonville, Klerksdorp and Welkom. Modified from Gibson and Reimold (2008).

and volcanic rocks of the Witwatersrand and Ventersdrop Supergroup and a 28 km wide outer broad synclinorium of gently dipping Proterozoic sedimentary and volcanic rocks of the Transvaal Supergroup. Younger sandstones and shales of the Karoo Supergroup cover the south-eastern portion of the structure. The general circular form with an uplifted central core, the occurrence of stishovite and coesite, as well as planar deformation features (PDFs) in quartz and shatter cones (Chapter 8) have all been presented as evidence that the Vredefort structure is the eroded remnant of a very large, complex impact structure (e.g. Dietz, 1961; Carter, 1965; Manton, 1965; Martini, 1978,1991; Gibson and Reimold, 2000,2005). The most comprehensive and recent guide to the geology of Vredefort can be found in Gibson and Reimold (2008).

The Vredefort impact event occurred at 2023 ± 4 Ma (Kamo *et al.*, 1996). Based on the spatial distribution of impact-related deformation and structural features, Therriault *et al.* (1997) derived a self-consistent, empirical estimate of 225–300 km for the apparent diameter. A similar size estimate was derived by Henkel and Reimold (1998), based on potential field and reflection seismic data, and by Grieve *et al.* (2008), based on comparisons

and scaling between Vredefort, Sudbury and Chicxulub. These estimates effectively equate the impact structure to the entire Witwatersrand Basin. The Witwatersrand Basin is the world's largest goldfield, having supplied some 40–50% of the gold ever mined. Since gold was discovered there in 1886, it has produced 47,000 t of gold and it still contains approximately 45% of currently known global reserves (Robb and Robb, 1998). The annual Witwatersrand gold production for 2002 was approximately 350 t, or approximately 13.5% of the global gold supply, and current reserve estimates are around 20,000 t of gold. Approximately 150,000 t of uranium has been mined, generally as a by-product of gold mining, with estimated reserves of 475,000 t (Reimold *et al.*, 2005).

Independent of impact studies at Vredefort, structural analyses have identified a series of concentric anticlinal and synclinal structures related to Vredefort (e.g. McCarthy *et al.*, 1986, 1990; Fig. 12.5). These Vredefort-related structures have led to the preservation of sediments of the Witwatersrand Supergroup from erosion (McCarthy *et al.*, 1990). The bulk of the gold has been mined from the upper succession of the Witwatersrand Supergroup: the Central Rand Group. The origin of the gold in the basin is still debated. General gold distribution is controlled by sedimentary attributes of the Central Rand, but textures suggest both a detrital and an authigenic hydrothermal origin (e.g. Hayward *et al.*, 2005; Frimmel *et al.*, 2005), with pure detrital and hydrothermal models and combinations of the two having been proposed (e.g. Barnicoat *et al.*, 1997; Minter, 1999; Phillips and Law, 2000). Gold with clear detrital morphological features (Minter *et al.*, 1993) occurs with secondary, remobilized gold. This suggests that detrital gold was introduced into the basin but that some gold was subsequently remobilized by hydrothermal activity (Frimmel and Minter, 2002; Zhao *et al.*, 2006). Grieve and Masaitis (1994) speculated that some remobilization occurred due to the Vredefort event but were equivocal also as to the potential role of thermal activity resulting from the spatially and temporally (2.05–2.06 Ga) close igneous event associated with the Bushveld Complex.

More recent work has clarified the situation, with the Vredefort impact event forming both an important temporal marker and a critical element in the process of gold remobilization. Two thermal or metamorphic events affected the rocks of the basin. A regional amphibolite facies metamorphism predates the Vredefort impact event. A later low-pressure (0.2–0.3 GPa), immediately post-impact event, however, produced peak temperatures of $350 \pm 50\,^{\circ}C$ in the Witwatersrand Supergroup to greater than $700\,^{\circ}C$ in the centre of the crystalline core at Vredefort. This post-impact activity is directly attributed to the combination of post-shock heating and the structural uplift of originally relatively deep-seated parautochthonous rocks during the Vredefort impact event (Gibson *et al.*, 1998). Reimold *et al.* (1999) applied the term 'autometasomatism' to describe the alteration associated with the hydrothermal activity. This activity remobilized the gold (and uranium) within impact-related structures and fractures, which provided channels for fluid migration. It has been recently argued, however, that such a redistribution of the gold was local in scale (Hayward *et al.*, 2005; Boer and Reimold, 2006; Meier *et al.*, 2009). A more detailed discussion of the effect of Vredefort-related hydrothermal activity can be found in Reimold *et al.* (2005). It is now evident the Vredefort impact event played a larger role in the genesis of Witwatersrand Basin gold fields (Frimmel and Minter, 2002; Frimmel *et al.*, 2005; Hayward *et al.*, 2005) than simply preserving them from erosion by structural modification (McCarthy *et al.*, 1990; Grieve and Masaitis, 1994).

### 12.2.4   Other structures

Several other progenetic or pre-impact deposits hosted by impact structures are noted in Grieve and Masaitis (1994) and Reimold *et al.* (2005). Although some have been mined in the past, they are generally small and currently inactive. One exception is the 118 million tonne Century Zn–Pb deposit, within the Lawn Hill impact structure (18°40′S, 138°39′E), Australia. The deposit lies along the Termite Range fault, where it transects the outer reaches of the impact structure. Rhenium–osmium dating of the ores suggests a mineralization age of $1451 \pm 44\,Ma$ (Keays *et al.*, 2006). The estimated age of the impact is around 500–520 Ma (Salisbury *et al.*, 2008). The impact led to some physical redistribution of the ores, the most spectacular occurrence being a 1 million tonne

fragment of ore suspended in brecciated limestone (Broadbent *et al.*, 1998). The impact may have also led to some minor hydrothermal redistribution of ore into overlying limestone and some 17 smaller vein-hosted Zn–Pb deposits situated along faults (Salisbury *et al.*, 2008). Physical redistribution of natural resources also affects mining operations at the Middlesboro structure (36°37′N, 83°44′W) in Kentucky, USA. Faults associated with the structure result in a degree of unpredictability regarding the continuity of coal beds, with the sudden loss of coal, changes in dip and brecciation. Although not an unequivocal conclusion, the local rise in the rank of the coal at Middlesboro by some 20% could be from local post-impact heating (Hower *et al.*, 2009).

### 12.3   Syngenetic deposits

Syngenetic economic natural resources at impact structures include impact diamonds, Cu–Ni sulfides and platinum group and other metals (Table 12.1).

### 12.3.1   Impact diamonds

The first indication of impact diamonds was the discovery in the 1960s of diamond with lonsdaleite, a high-pressure (hexagonal) polymorph of carbon, in placer deposits (e.g. in the Ukraine), although their source was unknown at the time. In the 1970s, diamond with lonsdaleite was discovered in the impact lithologies at the Popigai impact structure (71°30′N, 111°00′E) in Siberia. Since then, impact diamonds have been discovered at a number of structures; for example, Kara (69°05′N, 64°18′E) and Puchezh-Katunki (57°00′N, 43°35′E) in Russia, Lappajärvi (63°12′N, 23°42′E) in Finland, Ries (48°53′N, 10°37′E) in Germany, Sudbury (46°36′N, 81°11′W) in Canada, Ternovka and Zapadnaya (49°44′N, 29°00′E) in Ukraine and others (Masaitis, 1993,1998; Gurov *et al.*, 1996; Langenhorst *et al.*, 1998; Siebenschock *et al.*, 1998). Impact diamonds originate as a result of phase transitions from graphite, or crystallization from coal, and occur when precursor carbonaceous lithologies are subjected to shock pressures greater than 35 GPa (Masaitis, 1998). The diamonds from graphite in crystalline targets usually occur in paramorphs, with inherited crystallographic features (Fig. 12.6) and as microcrystalline aggregates. At Popigai, these aggregates can reach 10 mm in size but most are 0.2–5 mm in size (Masaitis, 1998). They consist of cubic diamond and lonsdaleite, with individual microcrystals of $10^{-4}\,cm$. The diamonds generated from coal, or other carbon in sediments, are generally porous and coloured.

Diamonds are most common as inclusions in impact melt rocks and glass clasts in suevite breccias. For example, at Zapadnaya, impact diamonds occur in dikes of impact melt rock in the central uplift and in suevitic breccias in the peripheral trough (Gurov *et al.*, 1996). Zapadnaya, approximately 3.8 km in diameter and $115 \pm 10\,Myr$ old, was formed in Proterozoic granite containing graphite (Gurov *et al.*, 1996). At Popigai, the allochthonous breccias filling the peripheral trough are capped by diamond-bearing suevitic breccias and coherent bodies of impact melt rocks (Chapters 7 and 9). In the case of Popigai, the original

**Figure 12.6** Impact diamond, showing slight etching along twinning boundaries, after graphite, from the Popigai impact structure. Long dimension is approximately 600 μm. Image from Koeberl *et al.*, 1997.

source of the carbon is graphite in Archaean gneisses. The diamonds at Kara, 65 km in diameter and 67 ± 6 Myr old, are from Permian terrigenous sediments containing coal (Ezerskii, 1982). Impact diamonds can also be found in strongly shocked lithic clasts in suevitic breccias; for example, at Popigai and Ries (Masaitis, 1998; Siebenschock *et al.*, 1998). In impact melt rocks at structures with carbon-bearing lithologies, diamonds occur in relatively minor amounts, with provisional average estimates in the order of 10 ppb, although the cumulative volumes can be enormous. Although diamonds associated with known impact structures are not currently exploited commercially, those produced by shock transformation of graphite tend to be harder and more resistant to breaking than normal cubic diamonds from kimberlites.

Hough *et al.* (1995) have reported the only known occurrence of cubic impact diamond from impact lithologies, again in impact melt-bearing breccias at the Ries structure. They also reported the occurrence of moissanite (SiC) grains. The cubic diamonds are skeletal in appearance and they attributed their origin to chemical vapour deposition of diamond onto SiC in the ejecta plume over the impact site. Based on carbon isotope studies, they considered the source of the carbon and other elements to be from sedimentary rocks, including carbonates, which overlay the crystalline basement at the time of the Ries impact. They calculated that the Ries suevite might contain $7.2 \times 10^4$ t of diamonds and SiC, in the proportion of 3:1. Other workers (e.g. Siebenschock *et al.*, 1998) failed to find either cubic diamond or SiC at the Ries. Silicon carbide has also been reported from the Onaping Formation at the Sudbury impact structure (Masaitis *et al.*, 1999). It is associated with impact diamonds that are a mixture of cubic diamond and hexagonal lonsdaleite from the solid-state transformation of precursor graphite in carbonaceous Huronian target lithologies by shock compression.

## 12.3.2 Cu–Ni sulfides and platinum group metals at Sudbury

The Sudbury structure in central Ontario is the site of world-class Ni–Cu sulfide and PGE metal ores and is Canada's principal mining district. The pre-mining resources associated with the Sudbury Igneous Complex (SIC) are estimated at over $1.5 \times 10^9$ t of 1.2% Ni, 1.1% Cu and 1 g t$^{-1}$ Pd+Pt (Farrow and Lightfoot, 2002). There are also hydrothermal Zn–Pb deposits stratigraphically above the SIC (Ames and Farrow, 2007). Nickel sulfides were first noted at Sudbury in 1856. It was not until they were 'rediscovered' during the building of the trans-Canada railway in 1883, however, that they received attention, with the first production occurring in 1886 (Naldrett, 2003). By 2000, the Sudbury mining camp had produced $9.7 \times 10^6$ t of Ni, $9.6 \times 10^6$ t of Cu, $70 \times 10^3$ t of Co, 116 t of Au, 319 t of Pt, 335 t of Pd, 37.6 t of Rh, 23.3 t of Ru, 11.5 t of Ir, $3.7 \times 10^3$ t of Ag, $3 \times 10^3$ t of Se and 256 t of Te (Lesher and Thurston, 2002). The cumulative total worth of metals produced from Sudbury is estimated at around US$300 billion, based on 2005 metal prices (Ames and Farrow, 2007).

The most prominent feature of the Sudbury structure is the approximately $30 \times 60$ km$^2$ elliptical basin formed by the outcrop of the SIC, the interior of which is known as the Sudbury Basin (Fig. 12.7). Neither the SIC nor the Sudbury Basin is synonymous with the considerably larger Sudbury impact structure. The Sudbury impact structure includes the Sudbury Basin, the SIC and the surrounding brecciated basement rocks and covers a present area of greater than 15,000 km$^2$. From the spatial distribution of shock metamorphic features (e.g. shatter cones; Chapter 8) and other impact-related attributes and by analogy to equivalent characteristics at other large terrestrial impact structures, such as Chicxulub and Vredefort, Grieve *et al.* (2008) estimated that the original crater rim diameter was 150–200 km. Even larger original diameters have been suggested (e.g. Tuchscherer and Spray, 2002; Naldrett, 2003). With the post-impact tectonic deformation and the considerable erosion, estimated to be approximately 10 km (Schwarz and Buchan, 1982), which has taken place at the Sudbury impact structure, it is difficult to constrain its original form. From its estimated original dimensions, it was most likely a peak-ring or a multi-ring basin (Stöffler *et al.*, 1994; Spray and Thompson, 1995; Grieve *et al.*, 2008).

Details of the geology of the Sudbury area can be found in Dressler (1984) and, most recently, Ames *et al.* (2008), while a summary of the salient points of the Sudbury structure and the SIC, in particular, can be found in Grieve (2006). Traditionally, the SIC has been subdivided into a number of phases or units. At the base, there is the Contact Sublayer, which has an igneous-textured matrix with inclusions of locally derived target rocks and cognate and exotic mafic to ultramafic rocks (Lightfoot *et al.*, 1997a,b). Uranium–lead dating of these mafic inclusions indicates an age equivalent to that of the SIC (Corfu and Lightfoot, 1996). Traditionally, included with the Sublayer are the so-called Offset Dikes of the SIC (Lightfoot *et al.* 1997a), which host 25% of the Ni–Cu–PGE resources at Sudbury (Farrow and Lightfoot, 2002). These dikes are most often radial to the SIC, but some are concentric. Some are extensive; for example, the Foy Offset can

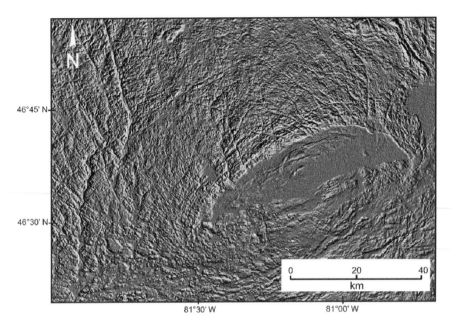

**Figure 12.7**    Greyscale synoptic image of the topography of the Sudbury impact structure, based on the Shuttle Radar Topographic Mission. Image has been illuminated from the centre as in Fig. 12.4. The most prominent feature is the approximately $60 \times 30\,km^2$ elliptical trace of the SIC and the interior Sudbury Basin.

be traced for approximately 30 km from the North Range of the SIC and varies from approximately 400 m in width near the SIC to 50 m at its most distal part (Grant and Bite, 1984; Tuchscherer and Spray, 2002). The Offset Dikes, however, do not have the same composition as the Sublayer and should be considered separately. They have compositions similar to the average SIC (Lightfoot and Farrow, 2002; Keays and Lightfoot, 2004) and are a 'quenched' phase of the SIC, which is the remnant of the impact melt sheet at the Sudbury structure (Grieve et al., 1991).

Stratigraphically above the Sublayer lies the Main Mass of the SIC, which is relatively, but not completely, clast free. For example, there are rare quartz clasts with partially annealed PDFs (Therriault et al., 2002; Chapter 8). Traditionally, the Main Mass has been divided into a number of sub-units: mafic norite, quartz-rich norite, felsic norite, quartz gabbro and granophyre. These terms for these sub-units are, in large part, actually misnomers on the basis of the modal quartz, alkali feldspar and plagioclase proportions, as plotted on a Streckeisen diagram. Only the Sublayer is gabbroic. The others fall into the quartz gabbro, quartz monzogabbro, granodiorite and granite fields (Therriault et al., 2002). The nomenclature confusion arises from the fact that, like most impact melts of granitic/granodioritic composition, the mafic component of the melt crystallized as ortho- and clino-pyroxene, although there are also primary hydrous amphibole and biotite. Therriault et al. (2002), who made their observations on continuous cores rather than discontinuous outcrop, also demonstrated, on the basis of both mineralogy and geochemistry, that the contacts between the various sub-units of the SIC are gradational. The details of the mineralogy and geochemistry support a cogenetic source for the sub-units of the Main Mass of the SIC, produced by fractional crystallization of a single batch

of silicate liquid (e.g. Lightfoot et al., 1997a; Warner et al., 1998; Therriault et al., 2002).

The conclusion that the SIC and its ores are not mantle derived but crustal in composition is borne out by isotopic studies (Faggart et al., 1985; Walker et al., 1991; Dickin et al., 1992,1996,1999). Cohen et al. (2000) analysed the Re–Os isotopes in ultramafic inclusions in the Sublayer of the SIC and concluded that they were also consistent with melting of pre-existing lithologies at 1.85 Ga. They also produced an imprecise crystallization age of $1.97 \pm 0.12\,Ga$ from a number of the inclusions, which, within its uncertainty, is equivalent to the 1.85 Ga crystallization age of the SIC (Krogh et al., 1984). High-precision Os isotope studies of sulfides from several mines have confirmed their crustal origin from a binary mixture of Superior Province and Huronian metasedimentary target rocks (Morgan et al., 2002).

Recently, Farrow and Lightfoot (2002) and Ames and Farrow (2007) reviewed the nature of the ore deposits at Sudbury and placed their formation in an integrated time-sequence model. They recognize: Ni–Cu–Co 'Contact' deposits associated with embayments at the base of the SIC and hosted by Sublayer and footwall breccia; Ni–Cu–Pt–Pd–Au 'Offset' deposits associated with discontinuities and variations in thickness in the offset dikes; and Cu–Pt–Pd–Au-rich 'Footwall' deposits that can occur in the underlying target rock, up to 1 km away from the SIC (Fig. 12.8). They also recognize a fourth deposit environment associated with pseudotachylitic Sudbury Breccia. For example, the Frood–Stobie deposit, which contained some 15% of the entire known Sudbury resources and produced 600 million tonnes of ore, is not hosted in a traditional Offset dike but rather in Sudbury Breccia (Scott and Spray, 2000).

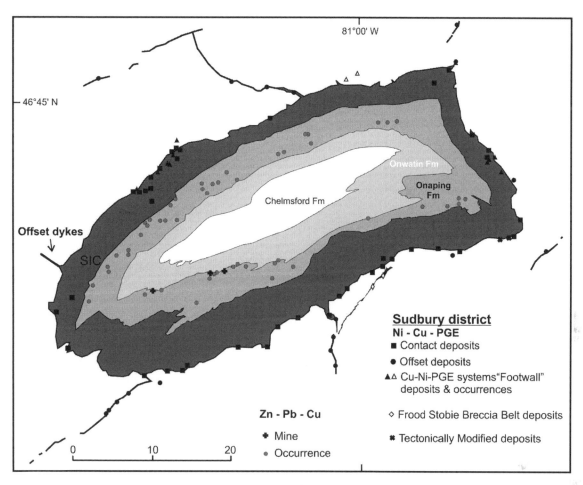

**Figure 12.8** Simplified geologic map of the SIC and overlying post-impact sediments at the Sudbury impact structure. Shown are the locations of the various types of occurrences of Cu–Ni–PGE ores at the base of the SIC (see text for details) and the Zn–Pb–Cu hydrothermal mineralization overlying the SIC. The details of the parautochthonous target rocks (white) are not shown, but are Archaean granite–greenstone terrane and gneisses of the Superior Province in the north and Palaeoproterozoic meta-sediments and meta-volcanics of the Huronian in the south. Image courtesy D. Ames.

The Contact deposits consist of massive sulfides and are volumetrically the largest deposit type, hosting approximately 50% of the known ore resources. They include the Creighton and Murray deposits and North Range deposits of Levack and Coleman. The main economic Offset environments include the Copper Cliff and Worthington Offsets in the South Range, which, along with the Frood–Stobie, contain approximately 40% of the known ores at Sudbury. The Cu–PGE-rich Footwall deposits are volumetrically small (~10% of known ore in 2005) relative to the Contact deposits, but are extremely valuable bodies, as they are relatively enriched in PGEs, in addition to copper. They represent a relatively new ore environment, which is hosted in the brecciated footwall of the SIC and is best known in the North Range; for example, at McCreedy East and West, Strathcona and Levack, where they occur as complex vein networks or sulfide and low sulfide, high precious metal disseminations (Ames and Farrow, 2007). In the last decade, these Footwall deposits have been the focus of much of the exploration activity at Sudbury and were the only deposits mined at Sudbury in the latter half of 2009. There is increasing realization that hydrothermal remobilization played

a role in the genesis of the Footwall deposits (e.g. Farrow and Watkinson, 1997; Marshall *et al.*, 1999; Carter *et al.*, 2001; Molnar *et al.*, 2001; Hanley *et al.*, 2005; Ames and Farrow, 2007). This is consistent with the growing body of knowledge in support of hydrothermal activity driven by the 'local' crustal thermal anomaly that results from large impact events (e.g. Abramov and Kring, 2004; Chapter 6).

The recent work at Sudbury can mostly be fitted into the framework of the formation of an approximately 200 km impact basin at 1.85 Ga, with accompanying massive crustal melting producing a superheated melt of an unusual composition, which gave rise to immiscible sulfides. These sulfides settled gravitationally, resulting ultimately in the present Contact and Offset ore deposits. Complicating factors, but essential components of the evolutionary history of Sudbury, are the creation of a 'localized' but regional-scale impact-related hydrothermal system, which resulted in some ore fractionation and redistribution into the brecciated footwall rocks, and the deformation by the Penokean orogeny that took place shortly after the impact. The Zn–Pb–Cu ores in the post-impact sediments overlying the SIC are the result of the hydrothermal system fuelled by the heat of the SIC (Ames *et al.*, 2006; Ames and Farrow, 2007). This upper hydrothermal system involved sea water, as opposed to the Cl-rich brines that played a role in the origin of the Footwall deposits at the base of the SIC (Ames and Farrow, 2007).

There are a number of small impact-related hydrothermal deposits (mostly Pb–Zn) at other impact structures (Masaitis, 1989; Grieve and Masaitis, 1994; Reimold *et al.*, 2005). The largest are the Pb–Zn deposits associated with the Siljan structure (61°02′N, 14°52′E) in Sweden. These deposits are hosted by Ordovician–Silurian carbonates in the annular trough at this 65 km diameter structure. The largest deposit consists of 0.3 million tonnes of 3% Pb and 1.5% Zn and is located near the town of Boda.

## 12.4   Epigenetic deposits

Significant epigenetic deposits are due mainly to the fact that impact structures can result in isolated topographic basins and can 'locally' influence underground fluid flow. Such deposits may originate almost immediately or over an extended period after the impact event and include reservoirs of liquid and gaseous hydrocarbons. There are also oil shales, various organic and chemical sediments, as well as flows of fresh and mineralized waters (Table 12.1). For example, oil shales are known at Boltysh (48°45′N, 32°10′W), Obolon (49°30′N, 32°55′W) and Rotmistrovka (49°00′N, 32°00′E) in the Ukraine (Masaitis *et al.*, 1980; Gurov and Gurova, 1991). They represent the unmatured equivalent of the hydrocarbon reserves found at some other impact structures, such as Ames in the USA. The most significant reserves are at Boltysh, where there are an estimated 4.5 billion tonnes (Bass *et al.*, 1967). The oil shales are the result of biological activity involving algae in this isolated topographic basin resulting from impact. A more complete listing of such epigenetic deposits can be found in Reimold *et al.* (2005).

## 12.5   Hydrocarbon accumulations

Hydrocarbons occur at a number of impact structures. In North America, approximately 50% of the known impact structures in hydrocarbon-bearing sedimentary basins have commercial oil and/or gas fields. For example, the 25 km diameter Steen River structure (59°31′N, 117°37′W) in Alberta, Canada, has produced 3.5 million barrels of oil and 48.5 billion cubic feet of gas from wells on its rim, with an estimated 1.9 million barrels of oil and 4.5 billion cubic feet of gas as recoverable reserves. Oil and gas are produced from beneath the approximately 13 km diameter Marquez (31°17′N, 96°18′W) and Sierra Madera (30°36′N, 102°55′W) structures in Texas, USA (Donofrio, 1997,1998), which have a combined estimate of reserves of 280 billion cubic feet of gas. Viewfield (49°35′N, 103°04′W; Fig. 12.9) in Saskatchewan, Canada (Sawatzky, 1977), which is a simple bowl-shaped crater (see Chapter 18), produces some 600 barrels of oil and 250 million cubic feet of gas per day. The recoverable reserves associated with Viewfield are estimated to be 10–20 million barrels of oil (Donofrio 1997,1998). More than 500,000 barrels of oil have

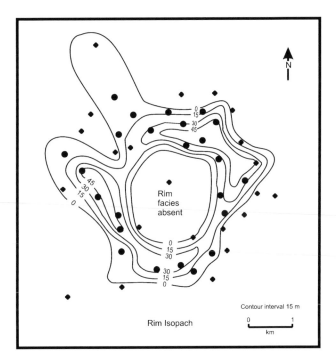

**Figure 12.9**   Thickness of rim facies at Viewfield impact structure. Black dots are producing wells. Diamond-shaped symbols are service or dry wells. Modified from Sawatzky, 1977 with permission from Elsevier.

been produced since 1978 from the Calvin structure (41°51′N, 85°57′W) in Michigan, USA, which is most likely an 8.5 km diameter complex impact structure (Milstein, 1988). Cloud Creek (43°11′N, 106°43′W), an approximately 7 km structure in Wyoming, USA (Stone and Therriault, 2003), has some hydrocarbons in its rim area, but the known amounts are not currently sufficient for commercial production.

Some specific structures in which hydrocarbon have accumulated under differing impact-related circumstances are discussed below.

### 12.5.1   Red Wing Creek

At the Red Wing Creek structure (47°36′N, 103°33′W) in North Dakota, USA, hydrocarbons are recovered from strata of the central uplift. Based on stratigraphy, Red Wing Creek is estimated to be 200 ± 25 Myr old, but the source of the hydrocarbons is in Carboniferous Mississippian strata; that is, 320–360 Myr old strata. It is a complex impact structure, approximately 9 km in diameter, with seismic records and drill-core data indicating a central peak in which strata have been uplifted by up to 1 km, an annular trough containing crater-fill products and a partially eroded rim (Brenan *et al.*, 1975; Sawatzky, 1977). As a result of a pronounced seismic anomaly, Shell Oil drilled the structure in 1965 on the NW flank of the central uplift. The drill hole indicated structurally high and thickened Mississippian and Pennsylvanian sections, compared with drill holes outside the structure. The well, however, was dry. In 1968, Shell drilled another hole in the NW in the annular trough. Here, the Mississippian was found

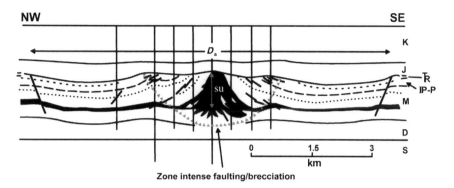

**Figure 12.10** Two-dimensional schematic cross-section (no vertical exaggeration) of the 9 km diameter Red Wing impact structure. The amount of structural uplift is indicated by su. $D_a$ is the apparent diameter. Strata: S, Silurian; D, Devonian; M, Carboniferous (Mississippian); IP-P, Permian; TR, Triassic, J, Jurassic; KL Cretaceous. Area bounded by small triangles corresponds to the volume of most intense deformation. The duplicated, faulted and brecciated Mississippian strata in the structural uplift are the zone of hydrocarbon production. Compare their thickness in the structural uplift with their thickness in the annular trough and outside the structure.

to be structurally low compared with the exterior. It was also dry and the structure, as a whole, was assumed to be dry. True Oil redrilled what was later recognized as the central uplift in 1972 and discovered approximately 820 m of Mississippian oil column, with considerable high-angle structural complexity and brecciation and a net pay of approximately 490 m. This is in contrast to the area outside the structure, which displays gentle dips and approximately 30 m oil columns. In this case, the hydrocarbon resources pre-existed the impact event but were physically displaced in the formation of the impact structure, resulting in enhanced accumulations of reservoir rocks in the central structural uplift.

The large oil column is due to the structural repetition of the Mississippian Mission Canyon Formation in the central uplift (Fig. 12.10; Brenan *et al.*, 1975). The impact-induced porosity and permeability result in relatively high flow rates of more than 1000 barrels per day. Cumulative production since discovery is over 20 million barrels of oil and 30 billion cubic feet of natural gas from 26 wells, of which 22 are still producing. Current production is restricted to about 300,000 barrels per year, to preserve unexploited reserves of natural gas. It is estimated that the brecciated central uplift contains more that 120 million barrels of oil and primary and secondary recoverable reserves may be 60–70 million barrels (Donofrio, 1981,1998; Pickard, 1994). The natural gas reserves are estimated at 100 billion cubic feet. Virtually all the oil has been discovered within a diameter of 3 km, corresponding to the central uplift. Based on net pay and its limited aerial extent, Red Wing is the most prolific oil field in the USA, in terms of producing wells per area, with the wells in the central uplift having the highest cumulative productivity of all wells in North Dakota.

### 12.5.2 Avak

The Avak structure (71°15′N, 156°38′W) is located on the Arctic coastal plain of Alaska, USA. The structure has been known for some time as a 'disturbed zone' in seismic data (Lantz, 1981) and

hydrocarbons were discovered in 1949 (Donofrio, 1998). Evidence of shock metamorphism (see Chapter 8) was discovered in the form of shatter cones and PDFs in quartz (Kirschner *et al.*, 1992; Therriault and Grantz, 1995). The age of the Avak impact is around 90 Ma, based on what is interpreted as Avak ejecta in the sedimentary column in drill cores from 50 km east of the structure (Banet and Fenton, 2008). The structure itself has the form of a 10–12 km diameter complex impact structure (see Chapter 5). It is bounded by listric faults, which define a rim area, and has an annular trough and central uplift. In the central uplift, the Lower–Middle Jurassic Kingak Shale and Barrow Sand are uplifted more than 500 m from their regional levels. The central uplift has been penetrated by the Avak 1 well, which reached a depth of 1225 m. Oil shows occur in Avak 1, but the well is not a commercial producer. Kirschner *et al.* (1992) suggested that pre-Avak hydrocarbon accumulations may have been disrupted and lost due to the formation of the Avak structure.

There are, however, the South Barrow, East Barrow and Sikulik gas fields (Fig. 12.11), which are post-impact and are related to the Avak structure. They occur outside the structure and are due to listric faults in the crater rim, which have truncated the Lower Jurassic Barrow Sand and placed Lower Cretaceous Torok Shales against the Barrow Sand, creating an effective up-dip gas seal. The South and East Barrow fields are currently in production and primary recoverable gas is estimated at 37 billion cubic feet (Lantz, 1981). In this case, the hydrocarbon resource also pre-existed the impact event but, unlike Red Wing Creek, the commercial accumulations occur outside the structure in the rim area. Also unlike Red Wing Creek, the impact did not physically redistribute the resource to form an enhanced accumulation, but rather redistributed lithologies so as to provide a seal to the reservoir rocks.

### 12.5.3 Ames

The Ames structure (36°15′N, 98°10′W) is located in Oklahoma, USA, and is a complex impact structure (Chapter 5) about 14 km

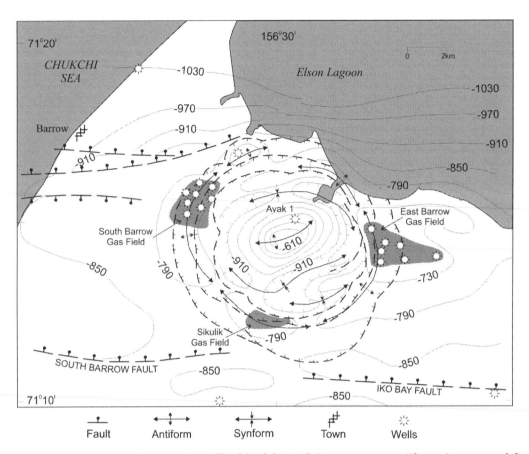

**Figure 12.11**   Bouguer gravity map (contours in milligals) of the Avak impact structure. The main structural features from reflection seismic data are also shown, with the rim locations of the gas fields (see text for details). Modified from Kirschner *et al.* (1992).

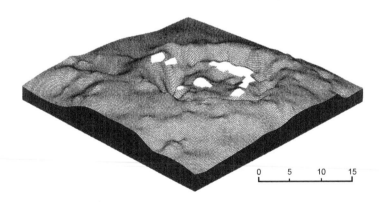

**Figure 12.12**   Three-dimensional mesh diagram of topography on the post-impact Sylvan shale at the 14 km diameter Ames impact structure. View is to the NW at 25° elevation, with 20 times vertical exaggeration. White areas are zones of hydrocarbon production in the underlying impact structure. Original source: Carpenter and Carlson (1997).

in diameter, with a central uplift, an annular trough and slightly uplifted rim (Fig. 12.12). It is buried by up to 3 km of Ordovician to Recent sediments (Carpenter and Carlson, 1992). The structure was discovered in the course of oil exploration in the area (Roberts and Sandridge, 1992) and is the principal subject of a compilation of research papers in Johnson and Campbell (1997). The rim of the structure is defined by the structurally elevated Lower Ordovician Arbuckle dolomite; more than 600 m of Cambrian–Ordovician strata and some underlying basement rocks are missing in the centre of the structure due to excavation. The entire structure is covered by Middle Ordovician Oil Creek shale, which forms both the seal and source for hydrocarbons and may have produced as much as 145 million barrels of oil (Curtiss and Wavrek, 1997).

The first oil and gas discoveries were made in 1990 from an approximately 500 m thick section of Lower Ordovician Arbuckle dolomite in the rim (Fig. 12.12). Owing to impact-induced fracturing and subsequent weathering and karsting, the Arbuckle dolomite in the rim of Ames has considerable economic potential. For example the 27–4 Cecil well, drilled in 1991, had drill stem flow rates of 3440 million cubic feet of gas and 300 barrels of oil per day (Roberts and Sandbridge, 1992). Wells drilled in the centre failed to encounter the Arbuckle dolomite and bottomed in granite breccias of the central uplift and, closer to the rim, in granite–dolomite breccias. These central wells (Fig. 12.9) produce over half the daily production from Ames and include the famous Gregory 1–20, which is the most productive oil well from a single pay zone in Oklahoma at more than 100,000 barrels of oil per year (Carpenter and Carlson, 1997). Gregory 1–20 encountered an approximately 80 m section of granite breccias below the Oil Creek shale, with very effective porosity. A drill-stem test of the zone flowed at approximately 1300 barrels of oil per day, with a conservative estimate of primary recovery in excess of 5 million barrels from this single well (Donofrio, 1998). Approximately 100 wells have been drilled at Ames, with a success rate of 50%. These wells produce more than 2500 barrels of oil and more than 3 million cubic feet of gas per day. Conservative estimates of primary reserves at Ames suggest they will exceed 25–50 million barrels of oil and 15–20 billion cubic feet of gas (Kuyendall *et al.*, 1997; Donofrio, 1998). Hydrocarbon production is from the Arbuckle dolomite, the brecciated granite and granite–dolomite breccias and is largely due to impact induced-fracturing and brecciation, which has resulted in significant porosity and permeability.

In the case of Ames, the impact not only produced the required reservoir rocks, but also the palaeo-environment for the deposition of post-impact shales that provided the source of the oil and gas, upon subsequent burial and maturation (Curtiss and Wavrek, 1997). There are similarities between the Ames crater shale and locally developed Ordovician shale in the Newporte structure (48°58′N, 101°58′W) in North Dakota, USA, an oil-producing (~120,000 barrels per year) 3.2 km diameter simple impact crater in Precambrian basement rocks of the Williston Basin (Clement and Mayhew, 1979; Donofrio, 1998). The Ames and Newporte discoveries have important implications for oil and gas exploration in crystalline rock underlying hydrocarbon-bearing basins. Donofrio (1981, 1997) first proposed the existence of such hydrocarbon-bearing impact craters and that major oil and gas deposits could occur in brecciated basement rocks.

### 12.5.4 Campeche Bank

The Campeche Bank in the SE corner of the Gulf of Mexico is the most productive hydrocarbon-producing area in Mexico. The bulk of the hydrocarbons, from Jurassic source rocks, are recovered from breccia deposits at the Cretaceous–Palaeogene (K–P) boundary. This area includes the world-class Cantarell oil field (Santiago-Acevedo, 1980), which has produced over 11 billion barrels of oil and 3 trillion cubic feet of gas, between discovery in 1976 and 2006. Primary reserves may range as high as 30 billion barrels of oil and 15 trillion cubic feet of gas. Production from the K–Pg boundary rocks is from up to 300 m of dolomitized

limestone breccias, with a porosity of around 10%. Clasts of shocked quartz and plagioclase (Chapter 8) occur in the upper portion of the K–Pg breccias (Grajales-Nishimura *et al.*, 2000). This bentonitic bed, with shocked materials, is considered to be altered ejecta materials from the K–Pg impact structure Chicxulub, which lies some 350–600 km to the NE.

Grajales-Nishimura *et al.* (2000) proposed the following sequence of events for the K–Pg lithologies. The main, hydrocarbon-bearing, breccias resulted from the collapse of the offshore carbonate platform due to seismic energy from the Chicxulub impact. This was followed by the deposition of K–Pg ejecta through slower atmospheric transport. The upper part of the ejecta deposit was later reworked by the action of impact-related tsunamis crossing the Gulf of Mexico. Subsequent dolomitization and Tertiary tectonics served to form the seal and trap respectively for migrating Jurassic hydrocarbons. The net result was the creation of oil fields that produce more than 60% of Mexico's daily production and have reserves in excess of the entire onshore and offshore hydrocarbon reserves of the USA, including Alaska (Donofrio, 1998). The Cantarell oil field is now in decline, with production falling from a peak of around 2 million barrels per day in 2003 to 770,000 barrels per day in January 2009. In 2009, it was replaced by the adjacent Ku–Mallob–Zaap oil field as the most productive oil field in Mexico. Oil from the Campeche Bank fields accounts for the bulk of the over US$50 billion of hydrocarbons (oil at $70 a barrel) produced from North American impact structures per year.

### 12.6 Concluding remarks

Economic deposits associated with terrestrial impact structures range from world class to relatively localized occurrences. The more significant deposits were introduced under the classification progenetic, syngenetic or epigenetic, with respect to the impact event. There is, however, increasing evidence that post-impact hydrothermal systems at large impact structures are of importance, with respect to their potential to redistribute metals. Such hydrothermal systems and metal redistributions can serve to blur the clear distinction between purely progenetic, syngenetic and epigenetic ore deposits related to impact. Examples included the Witwatersrand progenetic gold deposits at the Vredefort impact structure and the syngenetic Contact Ni–Cu–Co ore deposits associated with base of the SIC and the Cu–Pt–Pd–Au-rich Footwall deposits, as well as the Zn–Pb–Cu deposits in the post-impact sediments overlying the SIC, at the Sudbury impact structure.

Although Vredefort and Sudbury are world-class mining districts, hydrocarbon production dominates (in economic terms) the value of natural resource deposits found at impact structures. Commercial hydrocarbon accumulations at impact structures are generally located in the central structural uplift of complex structures and in the rim areas of both complex and simple structures. While spatially localized, such accumulations occur for a variety of reasons, including the physical redistribution of existing reservoir and seal rocks, as well as brecciation to form potential reservoir rocks for migrating hydrocarbons. Many terrestrial impact structures remain to be discovered and, as targets for resource exploration, their relatively invariant morphological and structural properties as a function of diameter, provide an aid to the development of efficient exploration strategies, particularly for hydrocarbons.

## References

Abramov, O. and Kring, D.A. (2004) Numerical modeling of an impact-induced hydrothermal system at the Sudbury crater. *Journal of Geophysical Research*, 109, E10007. DOI: 10.1029/2003JE002213.

Abramov, O. and Kring, D.A. (2007) Numerical modeling of hydrothermal activity at the Chicxulub Crater. *Meteoritics and Planetary Science*, 42, 93–112.

Ames, D.E. and Farrow, C.E.G. (2007) Metallogeny of the Sudbury mining camp, in *Mineral Deposits of Canada*, (ed. W.D. Goodfellow), GAC, Mineral Deposits Division, Special Publication 5, Geological Association of Canada, St Johns, pp. 329–350.

Ames, D.E., Watkinson, D.H. and Parrish, R.R. (1998) Dating of the regional hydrothermal system induced by the 1850 Ma Sudbury impact event. *Geology*, 26, 447–450.

Ames, D.E., Jonasson, I.R., Gibson, H.L. and Pope, K.O. 2006 Impact-generated hydrothermal system from the large Paleoproterozoic Sudbury crater. Canada, in *Biological Process Associated with Impact Events* (eds C. Cockell, I. Gilmour and C. Koeberl), Impact Studies Series, Springer-Verlag, Berlin, pp. 55–100.

Ames, D.E., Davidson, A. and Wodicka, N. (2008) Geology of the giant Sudbury polymetallic mining camp, Ontario, Canada. *Economic Geology*, 103, 1057–1077.

Banet, A.C. and Fenton, J.P.G. (2008) An examination of the Simpson core test wells suggest and age for the Avak impact feature near Barrow, Alaska, in *The Sedimentary Record of Meteorite Impacts* (eds K.R. Evans, J.W. Horton Jr, D.T. King and J.R. Morrow), Geological Society of America Special Paper 437, Geological Society of America, Boulder, CO, pp. 139–145.

Barnicoat, A.C., Henderson, I.H.C., Knipe, R.J. *et al.* (1997) Hydrothermal gold mineralization in the Witwatersrand Basin. *Nature*, 386, 820–824.

Bass Yu.B, Galalka, A.I. and Grabovskly, V.I. (1967) The Boltysh oil shales. *Razvadka i Okhrana Nedr*, 9, 11–15 (in Russian).

Baudemont, D.A. and Fedorowich, J. (1996) Structural control of uranium mineralization at the Dominique Peter deposit, Saskatchewan, Canada. *Economic Geology*, 91, 855–874.

Bell, K., Caccioti, A.D. and Schnessl, J.H. (1985) Petrography and geochemistry of the Earl River Complex, Carswell structure, Saskatchewan – a possible Proterozoic komatiitic succession, in *The Carswell Structure Uranium Deposits, Saskatchewan* (eds R. Lainé, D. Alsonso and M. Scab), GAC Special Paper 29, Geological Association of Canada, St Johns, pp. 71–80.

Boer, R.H. and Reimold, W.U. (2006) Conditions of gold remobilization in the Ventersdrop Contact Reef, Witwatersrand Basin, South Africa, in *Processes on the Early Earth* (eds W.U. Reimold and R.L. Gibson), Geological Society of America Special Paper 405, Geological Society of America, Boulder, CO, pp. 387–402.

Broadbent, G.C., Myers, R.E. and Wright, J.V. (1998) Geology and origin of shale-hosted Zn–Pb–Ag mineralization at the Century Deposit, Northwest Queensland, Australia. *Economic Geology*, 93, 1264–1294.

Brenan, R.L. Peterson, B.I. and Smith, H.J. (1975) The origin or Red Wing Creek Structure, McKenzie County, North Dakota. *Wyoming Geological Association Earth Science Bulletin*, 8, 1–41.

Carpenter, B.N. and Carlson, R. (1992) The Ames impact crater. *Oklahoma Geological Survey*, 52, 208–223.

Carpenter, B.N. and Carlson, R. (1997) The Ames meteorite-impact structure, in *Ames Structure in Northwest Oklahoma and Similar Features: Origin and Petroleum Production* (eds K.S. Johnson and J.A. Campbell), Oklahoma Geological Survey Circular 100, Oklahoma Geological Survey, Norman, OK, pp. 104–119.

Carter, N.L. (1965) Basal quartz deformation lamellae – a criterion for recognition of impactites. *American Journal of Science*, 263, 786–806.

Carter, W.M., Watkinson, D.H. and Jones, P.C. (2001) Post-magmatic remobilisation of platinum-group elements in the Kelly Lake Cu–Ni sulphide deposit, Copper Cliff Offset, Sudbury. *Exploration and Mining Geology*, 10, 95–110.

Clement, J.H. and Mayhew, T.E. (1979) Newporte discovery opens new pay. *Oil and Gas Journal*, 77, 165–172.

Cohen, A.S., Burnham, O.M., Hawkesworth, C.J. and Lightfoot, P.C. (2000) Pre-emplacement Re–Os ages for ultramafic inclusions in the sublayer of the Sudbury igneous complex, Ontario. *Chemical Geology*, 165, 37–46.

Corfu, F. and Lightfoot, P.C. (1996) U–Pb geochronology of the Sublayer environment, Sudbury Igneous Complex, Ontario. *Economic Geology*, 91, 1263–1269.

Currie, K.L. (1969) *Geological Notes on the Carswell Circular Structure, Saskatchewan (74K)*, Canadian Geological Survey of Canada Paper 67–32, Department of Energy, Mines and Resources Ottawa.

Curtiss, D.K. and Wavrek, D.A. (1997) The Oil Creek–Arbuckle (!) petroleum system, Major County, Oklahoma, in *Ames Structure in Northwest Oklahoma and Similar Features: Origin and Petroleum Production* (eds K.S. Johnson and J.A. Campbell), Oklahoma Geological Survey Circular 100, Oklahoma Geological Survey, Norman, OK, pp. 240–258.

Dickin, A.P., Richardson, J.M., Crocket, J.H. *et al.* (1992) Osmium isotope evidence for a crustal origin of platinum group elements in the Sudbury nickel ore, Ontario, Canada. *Geochimica et Cosmochimica Acta*, 56, 3531–2537.

Dickin, A.P., Artan, M.A. and Crocket, J.H. 1996 Isotopic evidence for distinct crustal sources of North and South range ores, Sudbury Igneous Complex. *Geochimica et Cosmochimica Acta*, 60, 1605–1613.

Dickin, A.P., Nguyen, T. and Crocket, J.H. (1999) Isotopic evidence for a single impact melting origin of the Sudbury Igneous Complex, in *Large Meteorite Impacts and Planetary Evolution II* (eds B.O. Dressler and V.L. Sharpton), Geological Society of America Special Paper 339, Geological Society of America, Boulder, CO, pp. 361–371.

Dietz, R.S. (1961) Vredefort ring structure – meteorite impact scar? *Journal of Geology*, 69, 499–516.

Donofrio, R.R. (1981) Impact craters: implications for basement hydrocarbon production. *Journal of Petroleum Geology*, 3, 279–302

Donofrio, R.R. (1997) Survey of hydrocarbon producing impact structures in North America: exploration results to-date and potential for discovery in Precambrian basement rock, in *Ames Structure in Northwest Oklahoma and Similar Features: Origin of Petroleum Production* (eds K.S. Johnson and J.A. Campbell), Oklahoma Geological Survey Circular 100, Oklahoma Geological Survey, Norman, OK, pp. 17–29.

Donofrio, R.R. (1998) North American impact structures hold giant field potential. *Oil and Gas Journal*, 96, 69–83.

Dressler, B.O. (1984) General geology of the Sudbury area, in *The Geology and Ore Deposits of the Sudbury Structure* (eds E.G. Pye,

A.J. Naldrett and P.E. Giblin), Ontario Geological Survey Special Volume 1, Ministry Natural Resources, Toronto, pp. 57–82.

Ezerskii, V.A. (1982) Impact-metamorphosed carbonaceous matter in impactites. *Meteoritika*, 41, 134–140 (in Russian).

Faggart Jr, B.E., Basu, A.R. and Tatsumoto, M. (1985) Origin of the Sudbury complex by meteoritic impact: neodymium isotopic evidence. *Science*, 230, 436–439.

Farrow, C.E.G. and Lightfoot, P.C. (2002) Sudbury PGE revisited: toward and integrated model, in *The Geology, Geochemistry, Mineralogy and Beneficiation of Platinum-Group Elements* (ed. L.I. Cabri), Canadian Institute of Mining and Metallurgy Special Volume 54, Canadian Institute of Mining and Metallurgy, Montreal, pp. 273–297.

Farrow, C.E.G. and Watkinson, G.H. (1997) Diversity of precious-metal mineralization in footwall Cu–Ni–PGE deposits, Sudbury, Ontario. Implications for hydrothermal models of formation. *Canadian Mineralogist*, 35, 817–839.

Frimmel, H.E. and Minter, W.E.I. (2002) Recent developments concerning the geologic history and genesis of the Witwatersrand gold deposits, South Africa, in *Integrated Methods for Discovery: Global Exploration in the Twenty-First Century* (eds R.J. Goldfarb and R.I. Nielsen), Society of Economic Geologists Special Paper 9, Society of Economic Geologists, Lancaster, pp. 17–45.

Frimmel, H.E., Groves, D.I., Kirk, J. et al. (2005) The formation and preservation of the Witwatersrand gold fields, the world's largest gold province. *Economic Geology*, 100, 769–797.

Gibson, R.L. and Reimold, W.U. (2000) Deeply exhumed impact structures; a case study of Vredefort Structure, South Africa, in *Impacts and the Early Earth* (eds I. Gilmour and C. Koeberl), Lecture Notes in Earth Science 91, Springer, Berlin, pp. 249–277.

Gibson, R.L. and Reimold, W.U. (2005) Shock pressure distribution in the Vredefort impact structure, South Africa, in *Large Meteorite Impacts III* (eds T. Kenkmann, F. Hörz and A. Deutsch), Geological Society of America Special Paper 384, Geological Society of America, Boulder, CO, pp. 329–349.

Gibson, R.L. and Reimold, W.U. (2008) *Geology of the Vredefort Impact Structure: A Guide to Sites of Interest*, South African Council for Geoscience Memoir 97, South African Council for Geoscience, Pretoria.

Gibson, R.L., Reimold, W.U. and Stevens, G. (1998) Thermal–metamorphic signature of an impact event in the Vredefort Dome, South Africa. *Geology*, 26, 787–790.

Grajales-Nishimura, J.M., Cedillo-Pardo, E., Rosales-Dominguez, C. et al. (2000) Chicxulub impact. The origin of reservoir and seal facies in the southeastern Mexico oil fields. *Geology*, 28, 307–310.

Grant, R.W. and Bite, A. (1984) Sudbury quartz diorite offset dikes, in *The Geology and Ore Deposits of the Sudbury Structure* (eds E.G. Pye, A.J. Naldrett and P.E. Giblin), Ontario Geological Survey Special Volume 1, Ministry Natural Resources, Toronto, pp. 275–300.

Grieve, R.A.F. (2005) Economic natural resource deposits at terrestrial impact structures, in *Mineral Deposits and Earth Evolution* (eds I. McDonald, A.J. Boyce, I.B. Butler et al.), Geological Society of London Special Publication 248, Geological Society of London, London, pp. 1–29.

Grieve, R.A.F. (2006) *Impact Structures in Canada*, Geological Association of Canada, St Johns.

Grieve, R.A.F. and Masaitis, V.L. (1994) The economic potential of terrestrial impact craters. *International Geology Review*, 36, 105–151.

Grieve, R.A.F., Stöffler, D. and Deutsch, A. (1991) The Sudbury Structure: controversial or misunderstood? *Journal of Geophysical Research*, 96, 22753–22764.

Grieve, R.A.F., Reimold, W.U., Morgan, J. et al. (2008) Observations and interpretations at Vredefort, Sudbury and Chicxulub: toward an empirical model of terrestrial impact basin formation. *Meteoritics and Planetary Science*, 43, 855–882.

Gurov, E.P. and Gurova, E.P. (1991) *Geological Structure and Composition of Rocks in Impact Craters*, Nauk Press, Kiev (in Russian).

Gurov, E.P., Gurova, E.P. and Ratiskaya, R.B. (1996) Impact diamonds of the Zapadnaya crater; phase composition and some properties. *Meteoritics and Planetary Science*, 31 (Supplement), A56 (abstract).

Hanley, J.J., Mungall, J.E., Pettke, T. et al. (2005) Ore metal redistribution by hydrocarbon–brine and hydrocarbon–halide melt phases, North Range footwall of the Sudbury Igneous Complex, Ontario, Canada. *Mineralium Deposita*, 40, 237–256.

Harper, C.T. 1982 Geology of the Carswell structure, central part. *Saskatchewan Geological Survey Report*, 214, 1–6.

Hayward, C.L., Reimold, W.U., Gibson, R.L. and Robb, L.J. (2005) Gold mineralization within the Witwatersrand Basin, South Africa: evidence for a modified placer origin, and the role of the Vredefort impact event, in *Mineral Deposits and Earth Evolution* (eds I. McDonald, A.J. Boyce, I.B. Butler et al.), Geological Society of London Special Publication 248, Geological Society of London London, pp. 31–58.

Henkel, H. and Reimold, W.U. (1998) Integrated geophysical modelling of a giant, complex impact structure: anatomy of the Vredefort Structure, South Africa. *Tectonophysics*, 287, 1–20.

Hough, R.M., Gilmour, I., Pilliger, C.T. et al. (1995) Diamond and silicon carbide in impact melt rock from the Ries impact crater. *Nature*, 378, 41–44.

Hower J.C., Greb, S.F., Kuehen, K.W. and Elbe, C.F. (2009) Did the Middlesboro, Kentucky, bolide impact event influence coal rank? *International Journal of Coal Geology*, 79, 92–96.

Jefferson, C.W. and Delaney, G. (eds) (2007) *EXTECH IV: Geology and Uranium Exploration Technology of the Proterozoic Athabasca Basin, Saskatchewan and Alberta*, Geological Association of Canada, Mineral Deposits Division, Special Publication 4, Geological Association of Canada, St Johns.

Johnson, K.S. and Campbell, J.A. (eds) (1997) *Ames Structure in Northwest Oklahoma and Similar Features: Origin and Petroleum Production*, Oklahoma Geological Survey Circular 100, Oklahoma Geological Survey, Norman, OK.

Kamo, S.L., Reimold, W.U., Krogh, T.E. and Colliston, W.P. (1996) A 2.023 Ga age for the Vredefort impact event and a first report of shock metamorphosed zircons in pseudotachylitic breccias and granophyre. *Earth and Planetary Science Letters*, 144, 69–387.

Keays, R.R. and Lightfoot, P.C. (2004) Formation of Ni–Cu–platinum group element sulfide mineralization in the Sudbury impact melt sheet. *Mineralogy and Petrology*, 82, 217–258.

Keays, R.R., McInnes, B.I.A., Lambert, D.D. and Ihlenfeld, C. (2006) Re–Os geochronology of the Century Pb–Zn–Ag deposit; two stage genesis with mantle input required. *Geochimica et Cosmochimica Acta*, 70 (Supplement), A310 (abstract).

Koeberl, C., Masaitis, V.L., Shafranovsky, G.I. et al. (1997) Diamonds from the Popigai impact structure, Russia. *Geology*, 25, 967–970.

Kirschner, C.E., Grantz, A. and Mullen, W.W. (1992) Impact origin of the Avak Structure and genesis of the Barrow gas fields. *American Association of Petroleum Geologists Bulletin*, 76, 651–679.

Krochuk, R.V. and Sharpton, V.L. (2003) Morphology of the Terny astrobleme based on field observations and sample analysis. 34th Lunar and Planetary Science Conference, CD-ROM, abstract no. 1489.

Krogh, T.E., Davis, D.W. and Corfu, F. (1984) Precise U–Pb zircon and baddeleyite ages for the Sudbury area, in *The Geology and Ore Deposits of the Sudbury Structure* (eds E.G. Pye, A.J. Naldrett and

P.E. Giblin), Ontario Geological Survey Special Volume 1, Ministry Natural Resources, Toronto, pp. 431–446.

Kuyendall, M.D., Johnson, C.L. and Carlson, R.A. (1997) Reservoir characterization of a complex impact structure: Ames impact structure, northern shelf, Anadarko basin, in *Ames Structure in Northwest Oklahoma and Similar Features: Origin and Petroleum Production* (eds K.S. Johnson and J.A. Campbell), Oklahoma Geological Survey Circular 100, Oklahoma Geological Survey, Norman, OK, pp. 199–206.

Lainé, R.T. (1986) Uranium deposits of Carswell structure, in *Uranium Deposits of Canada* (ed. E.L. Evans), Canadian Institute of Mining and Metallurgy Special Volume 33, Canadian Institute of Mining and Metallurgy, Montreal, pp. 155–169.

Lainé, R., Alonso, D. and Svab, M. (eds) (1985) *The Carswell Structure Uranium Deposits, Saskatchewan*, GAC Special Paper 29, Geological Association of Canada, St Johns.

Langenhorst, F., Shafranovsky, G. and Masaitis, V.L. (1998) A comparative study of impact diamonds from the Popigai, Ries, Sudbury and Lappajärvi craters. *Meteoritics and Planetary Science*, 33 (Supplement), A90–A91 (abstract).

Lantz, R. (1981) Barrow gas fields – N. Slope, Alaska. *Oil and Gas Journal*, 79, 197–200.

Lesher, C.M and Thurston, P.C. (eds) (2002) A Special Issue Devoted to Mineral Deposits of the Sudbury Basin. *Economic Geology*, 97, 1373–1606.

Lightfoot, P.C. and Farrow, C.E.G. (2002) Geology, geochemistry and mineralogy of the Worthington offset dike: a genetic model for offset dike mineralization in the Sudbury Igneous Complex. *Economic Geology*, 97, 1419–1446.

Lightfoot, P.C., Keays, R.R., Morrison, G.G. *et al.* (1997a) Geochemical relationships in the Sudbury Igneous Complex: origin of the Main Mass and Offset dikes. *Economic Geology*, 92, 289–307.

Lightfoot, P.C., Keays, R.R., Morrison, G.G. *et al.* (1997b) Geochemical relations in the matrix and inclusions of the Sublayer Norite, Sudbury Igneous Complex: a case study of the Whistle mine embayment. *Economic Geology*, 92, 647–673.

Manton W.I. (1965) The orientation and origin of shatter cones in the Vredefort Ring. *Annals of the New York Academy of Science*, 123, 1017–1049.

Marshall, D., Watkinson D.H., Farrow, C. *et al.* (1999) Multiple fluid generations in the Sudbury Igneous Complex. Fluid inclusions, Ar, O, H, Rb and Sr evidence. *Chemical Geology*, 154, 1–19.

Martini, J.E.J. (1978) Coesite and stishovite in the Vredefort Dome, South Africa. *Nature*, 272, 715–717.

Martini, J.E.J. (1991) The nature, distribution and genesis of the coesite and stishovite associated with the pseudotachylite of the Vredefort Dome, South Africa. *Earth and Planetary Science Letters*, 103, 285–300.

Masaitis, V.L. (1989) The economic geology of impact craters. *International Geology Review*, 31, 922–933.

Masaitis, V.L. (1993) Diamondiferous impactites, their distribution and petrogenesis. *Regional Geology and Metallogeny*, 1, 121–134 (in Russian).

Masaitis, V.L. (1998) Popigai crater: origin and distribution of diamond-bearing impactites. *Meteoritics and Planetary Science*, 33, 349–359.

Masaitis, V.L., Danlin, A.I., Mashchack, M.S. *et al.* (1980) *The Geology of Astroblemes*, Nedra Press, Leningrad (in Russian).

Masaitis, V.L., Shafranovsky, G.I., Grieve, R.A.F. *et al.* (1999) Impact diamonds in suevite breccias of the Onaping Formation, Sudbury Structure, Ontario Canada, in *Large Meteorite Impacts and Planetary Evolution II* (eds B.O. Dressler and V.L. Sharpton), Geological

Society of America Special Paper 339, Geological Society of America, Boulder, CO, pp. 317–321.

McCarthy, T.S., Charlesworth, E.G. and Stanistreet, I.G. (1986) Post-Transvaal structural features of the northern portion of the Witwatersrand Basin. *Transactions of Geological Society of South Africa*, 89, 311–323.

McCarthy, T.S., Stanistreet I.G. and Robb, L.J. (1990) Geological studies related to the origin of the Witwatersrand Basin and its mineralization: an introduction and strategy for research and exploration. *South African Journal of Geology*, 93, 1–4.

Meier, D.L., Heinrich, C.A. and Watts, M.A. (2009) Mafic dikes displacing Witwatersrand gold reefs: evidence against metamorphic–hydrothermal ore formation. *Geology*, 37, 607–610.

Milstein, R.L. (1988) The Calvin 28 structure: evidence for impact origin. *Canadian Journal of Earth Science*, 25, 1524–1530.

Minter, W.E.L. (1999) Irrefutable detrital origin of Witwatersrand gold and evidence of eolian signatures. *Economic Geology*, 94, 655–670.

Minter, W.E.L., Goedhart T.M., Knight, J. and Frimmel, H.E. (1993) Morphology of Witwatersrand gold grains from the Basal Reef: evidence for their detrital origin. *Economic Geology*, 88, 237–248.

Molnar, F., Watkinson, D.H. and Jones, P.C. (2001) Multiple hydrothermal processes in footwall units of the North Range, Sudbury Igneous Complex, Canada and implications for the genesis of vein-type Cu–Ni–PGE deposits. *Economic Geology*, 96, 1645–1670.

Morgan, J.W., Walker, R.J., Horman, M.F. *et al.* (2002) $^{190}Pt–^{186}Os$ and $^{187}Re–^{187}Os$ systematics of the Sudbury Igneous Complex, Ontario. *Geochimica et Cosmochimica Acta*, 66, 273–290.

Naldrett, A.J. (2003) From impact to riches: evolution of geological understanding as seen at Sudbury, Canada. *GSA Today*, 13, 4–9.

Naumov, M.V. (2002) Impact-generated hydrothermal systems: data from Popigai, Kara and Puchez-Katunki impact structures, in *Impacts in Precambrian Shields* (eds J. Plado and L.J. Pesonen), Springer-Verlag, Berlin, pp.117–172.

Nikolsky, A.P. (1991) *Geology of the Pervomaysk Iron-Ore Deposit and Transformation of its Structure by Meteorite Impact*, Nedra Press, Moscow (in Russian).

Nikolsky, A.P., Naumov, V.P. and Korobko, N.I. (1981) Pervomaysk iron deposit, Krivoj Rog and its transformations by shock metamorphism. *Geologia Rudnykh Mestorozhdenii*, 5, 92–105 (in Russian).

Nikolsky, A.P., Naumov, V.P., Mashchak, M.S. and Masaitis, V.L. (1982) Shock metamorphosed rocks and impactites of Ternovka astrobleme. *Transactions of VSEGEI*, 238, 132–142 (in Russian).

Osinski, G.R., Spray, J.G. and Lee, P. (2001) Impact-induced hydrothermal activity within Haughton impact structure, arctic Canada: generation of a transient warm, wet oasis. *Meteoritics and Planetary Science*, 36, 731–745.

Pagel, M. and Svab, M. (1985) Petrographic and geochemical variations within the Carswell structure metamorphic core and their implications with respect to uranium mineralization, in *The Carswell Structure Uranium Deposits, Saskatchewan* (eds R. Lainé, D. Alsonso and M. Svab), GAC Special Paper 29, Geological Association of Canada, St Johns, pp. 55–70.

Phillips G.N. and Law, J.D.M. (2000) Witwatersrand gold fields; geology, genesis and exploration. *Reviews in Economic Geology*, 13, 459–500.

Pickard, C.F. (1994) Twenty years of production from an impact structure, Red Wing Creek Field, McKenzie County, North Dakota. *The Contact*, 41, 2–5.

Reimold, W.U., Koeberl, C., Fletcher, P. *et al.* (1999) Pseudotachylitic breccias from fault zones in the Witwatersrand Basin, South Africa:

evidence of autometasomatism and post-brecciation alteration processes. *Mineralogy and Petrology*, 66, 25–53.

Reimold, W.U., Koeberl, C., Gibson, R.L. and Dressler, B.O. (2005) Economic mineral deposits in impact structures: a review, in *Impact Tectonics* (eds C. Koeberl and H. Henkel), Impact Studies Series 6, Springer, Heidelberg, pp. 479–552.

Robb, L.J. and Robb, V.M. (1998) Gold in the Witwatersrand Basin, in *The Mineral Resources of South Africa* (eds M.G.C. Wilson and C.R. Anhaeusser), Handbook 16, Council for Geoscience, Pretoria, pp. 294–349.

Roberts, C. and Sandridge, B. (1992) The Ames hole. *Shale Shaker*, 42, 118–121.

Salisbury, J.A., Tomkins, A.G. and Schaefer, B.F. (2008) New insights into the size and timing of the Lawn Hill impact structure: relationship to the Century Zn–Pb deposit. *Australian Journal of Earth Sciences*, 55, 587–603.

Santiago-Acevedo, J. (1980) Giant fields of the Southern Zone – Mexico, in *Giant Oil and Gas Fields of the Decade 1968–1978* (ed. M.T. Halbouty), AAPG Memoir 30, American Association of Petroleum Geologists, Tulsa, OK, pp. 339–385.

Sawatzky, H.B. (1977) Buried impact craters in the Williston Basin and adjacent area, in *Impact and Explosion Cratering* (eds D.J. Roddy, R.O. Pepin and R.B. Merrill), Pergamon Press, New York, NY, pp. 461–480.

Schwarz, E.J. and Buchan, K.L. (1982) Uplift deduced from remanent magnetization; Sudbury area since 1250 Ma ago. *Earth and Planetary Science Letters*, 58, 64–74.

Scott, R.G. and Spray, J.G. (2000) The South Range breccia belt of the Sudbury impact structure; a possible terrace collapse feature. *Meteoritics and Planetary Science*, 35, 505–520.

Siebenschock, M., Schmitt, R.T. and Stöffler, D. (1998) Impact diamonds in glass bombs from suevite of the Ries Crater, Germany. *Meteoritics and Planetary Science*, 33 (Supplement), A145 (abstract).

Spray, J.G. and Thompson, L.M. (1995) Friction melt distribution in terrestrial multi-ring impact basins. *Nature*, 373, 130–132.

Stöffler, D., Deutsch, A., Avermann, M. *et al.* (1994) The formation of the Sudbury Structure, Canada: toward a unified impact model, in *Large Meteorite Impacts and Planetary Evolution* (eds, B.O. Dressler, R.A.F. Grieve and V.L. Sharpton), Geological Society of America Special Paper 293, Geological Society of America, Boulder, CO, pp. 303–318.

Stone, D.S. and Therriault, A.M. (2003) Cloud Creek, central Wyoming, USA: impact origin confirmed. *Meteoritics and Planetary Science*, 38, 445–455.

Therriault, A.A. and Grantz, A. (1995) Planar deformation features from mixed breccia of the Avak structure, Alaska. 26th Lunar and Planetary Science Conference, pp. 1403–1404 (abstract).

Therriault, A.M., Grieve, R.A.F. and Reimold, W.U. (1997) Original size of the Vredefort structure: implications for the geological evolution of the Witwatersrand Basin. *Meteoritics and Planetary Science*, 32, 71–77.

Therriault, A.M., Fowler, A.D. and Grieve, R.A.F. (2002) The Sudbury igneous complex: a differentiated impact melt sheet. *Economic Geology*, 97, 1521–1540.

Tona, F., Alonso, D. and Svab, M. (1985) Geology and mineralization in the Carswell structure – a general approach, in *The Carswell Structure Uranium Deposits, Saskatchewan* (eds R. Lainé, D. Alsonso and M. Svab), GAC Special Paper 29, Geological Association of Canada, St Johns, pp. 1–18.

Tuchscherer, M.G. and Spray, J.G. (2002) Geology, mineralisation and emplacement of the Foy offset dike, Sudbury. *Economic Geology*, 97, 1377–1398.

Walker, R.J., Morgan, J.W., Naldrett, A.J. *et al.* (1991) Re–Os isotope systematics of Ni–Cu sulphide ores, Sudbury Igneous Complex, Ontario: evidence for a major crustal component. *Earth and Planetary Science Letters*, 105, 416–429.

Warner, S., Martin, R.F., Abdel-Raham, A.F.M. and Doig, R. (1998) Apatite as a monitor of fractionation, degassing and metamorphism in the Sudbury Igneous Complex, Ontario. *Canadian Mineralogist*, 36, 981–999.

Zhao, B., Robb, L.J., Harris, C. and Jordan, L.J. (2006) Origin of hydrothermal fluids and gold mineralization associated with the Ventersdrop Contact Reef. Witwatersrand Basin, South Africa: constraints from S, O, and H isotopes, in *Processes on the Early Earth* (eds W.U. Reimold and R.L. Gibson), Geological Society of America Special Paper 405, Geological Society of America, Boulder, CO, pp. 333–352.

# THIRTEEN

# *Remote sensing of impact craters*

## Shawn P. Wright*, Livio L. Tornabene[†] and Michael S. Ramsey[‡]

*Institute of Meteoritics, MSC03 2050, University of New Mexico, Albuquerque, NM 87131, USA
[†]Centre for Planetary Science and Exploration, Department of Earth Sciences Western University, 1151 Richmond Street, London, ON, N6A 5B7, Canada
[‡]Department of Geology and Planetary Science, University of Pittsburgh, 4107 O'Hara Street, Pittsburgh, PA 15260-3332, USA

## 13.1 Introduction

As discussed throughout this book, impact cratering is a fundamental geological process that has shaped and modified the surfaces of planets throughout geological time (see summary in Chapter 1). However, the low frequency of meteorite impacts during the recent past and the ongoing active geological processes on Earth offer rare opportunities to examine pristine impact sites, which are common on the surfaces of other terrestrial planets. With the exception of a few robotic and manned space exploration missions, the analysis of orbital data has been the primary method for studying impact craters in a variety of settings and target rocks throughout the Solar System. In this chapter, an overview of remote-sensing datasets and techniques is presented with direct applications to the study of impact craters. Various types of data, as well as methods and techniques for processing and interpreting these datasets, are summarized. For readers who are interested in a more in-depth discussion of the datasets and techniques mentioned in this chapter, refer to Vincent (1997), Sabins (1997), Lillesand and Kiefer (2000) and Jensen (2007), and refer to Garvin et al. (1992), French (1998) and Koeberl (2004) for overviews of remote sensing specifically applied to impact crater studies.

The goal of remote sensing is to provide useful information about planetary surfaces prior to fieldwork, sample collection or *in lieu* of such work in the cases where access may be limited or not possible. The basic principle involves collecting, understanding and using data acquired as a result of the interaction between matter (i.e. rock or surface) and electromagnetic energy (e.g. infrared wavelengths). A variety of remote-sensing datasets and techniques have been used to determine: (1) geology; (2) morphometry/altimetry/topography; (3) composition; and (4) physical properties (e.g. particle/block size, thermal inertia). Different wavelength regions as well as derived datasets and data processing techniques can be utilized to determine one or more of the aforementioned surface properties with applications to the study of impact cratering. This chapter addresses these four themes, followed by two case studies that utilize one or more of these aspects to solve a geological problem about an impact site.

## 13.2 Background

An early and ongoing goal involving remote sensing of terrestrial impact craters has been to simply discover new impact craters solely from remote-sensing data (Garvin et al., 1992; Koeberl, 2004; Folco et al., 2010, 2011) or find spectral differences between shocked rocks and their protoliths (D'Aria and Garvin, 1988; Garvin et al., 1992; Johnson et al., 2002, 2007; Wright et al., 2011). Photogeology, or image analysis, is used to search for the distinctive circular structure of an impact crater, in addition to terraces and central peaks within larger complex craters (see Chapter 5). Several impact structures have been found with this method (e.g. Folco et al., 2010, 2011). Remote sensing alone, however, cannot provide definitive evidence for impact, as many circular structures can be formed by other geological processes (e.g. sinkholes, granite or salt domes, volcanoes). Impact structures can be identified by the recognition of a suite of characteristic impactites (see Chapter 7) and shock metamorphic effects (see Chapter 8); however, this is difficult (albeit not impossible) to accomplish via remote sensing and/or spectroscopy. Impactites, as discussed in Chapters 7 and 9, consist of a variety of impact breccias and melt rocks generally with the same composition as the original target rocks. In some cases, impactites exhibit geochemical enrichments (Chapter 15) from the impactor and/or mineralogical changes due to shock that are difficult to measure from spaceborne instrumentation. Images with very high spatial resolution are commonly required to resolve impact breccias, which must have exposed clasts larger than the spatial resolution of the imaging instrument. The current highest pixel dimensions of modern imaging systems is on the order of tens of centimetres, and thus megabreccias are most easily identified (e.g. Grant et al., 2008; Marzo et al., 2010; Fig. 13.1). Breccias with smaller exposed clasts will be difficult to identify.

*Impact Cratering: Processes and Products*, First Edition. Edited by Gordon R. Osinski and Elisabetta Pierazzo.
© 2013 Blackwell Publishing Ltd. Published 2013 by Blackwell Publishing Ltd.

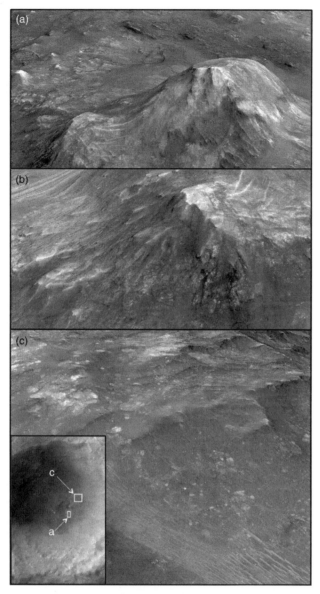

**Figure 13.1**    Various portions of a HiRISE red mosaic image (PSP_005842_1970) covering the eastern portion of the central uplift complex of the approximately 40 km Toro crater, Mars, are shown to demonstrate the use of topographic data. The image is draped on a digital terrain model (DTM) derived using SOCET set v5.4.1 (BAE systems) and the PSP_005842_1970 and ESP_011538_1970 stereo pair (credit: NASA/JPL/UA); resolution is 2 m/pixel with a vertical exaggeration of approximately 5×). (a) A portion of the HiRISE red mosaic image (PSP_005842_1970; ~1 km across from the base to the summit of the uplifted and rotated megablock) covering the southeastern portion of Toro's central uplift complex and crater floor. The perspective view is slightly to the northeast. The uplifted megablock, although covered with fine-grained debris and talus, shows evidence of the underlying light-toned, fractured and brecciated bedrock. A smaller megablock on the lower left shows what appear to be breccia dikes consistent with field observations of central uplift bedrock within terrestrial impact structures. (b) A rotated and close-up view of the smaller megablock exhibiting the dike-like structures in (a). These could be pseudotachylites or possible impact melt-bearing breccia dikes that were injected into the crater floor during the formation and expansion of the transient cavity, and subsequent formation of the central uplift. (c) A perspective view to the south-southwest of Toro's crater-fill deposits. These deposits are observed to drape the uplifted bedrock of the central uplift complex and are continuous with the crater-fill deposits on the crater floor. These materials are relatively smooth, dark toned and possess abundant light-toned megaclasts. The unit is interpreted to be an impact melt-bearing breccia deposit formed by materials that once lined the transient cavity of Toro during the excavation phase of crater formation. Reprinted from Marzo *et al.*, 2010, with permission from Elsevier.

## 13.3   Photogeology

Photogeology is the use of aerial or orbital data to view and interpret geological and geomorphologic features. Geologists have used aerial photographs for decades to: (1) determine rock exposures from surfaces with and without vegetation; (2) study the expression of landforms for insight into their origins (geomorphology); (3) determine the structural arrangements of disturbed bedrock (i.e. folds and faults); (4) evaluate dynamic changes from natural events (e.g. floods; volcanic eruptions); and (5) use as a visual base map, commonly in conjunction with topographic maps, on which a more detailed map can be derived (e.g. a geological map).

The advantage of a large, synoptic view of a study region allows for the examination of structures or features of interest in their entirety, whereas a field geologist likely could not map a large area in a short time. The ability to merge different types of remote-sensing data discussed in this chapter can further facilitate the recognition of relationships between various surface components to be determined with specific applications towards impact cratering studies. Figure 13.2 illustrate the types of images used for photogeologic analyses of three craters and describe various spatial resolutions and examples of using false-colour images and more than one type of remote-sensing analysis of an impact site.

## 13.4   Morphometry, altimetry, topography

Elevation can be extracted using photometric stereo data processing, as well as from active techniques such as radar and laser altimetry. The latter can produce high-resolution DTMs that provide three-dimensional (3D) geospatial visualization and characterization of impact-related features. Elevation data can also be combined with other datasets to create 3D visualizations that facilitate correlations with spectral units and/or high-resolution morphologic units, and also for interpretation of structural/stratigraphic relationships. The terms 'laser altimetry' or formerly 'laser-induced swath mapping' has been replaced by the acronym LiDAR (light detection and ranging). Depending on the instrument, either ultraviolet, visible or near-infrared energy is backscattered or reflected from the surface and the distance from the instrument to the ground is calculated by measuring the time delay between the transmission of the laser pulse and the detection of the reflected signal. This creates high-resolution topographic data from millions of data points from a scanning swath width instrument to dozens of gridded points from an orbital system where the laser pulse is backscattered every several metres.

One of the primary uses of topographic data, particularly with respect to planetary applications for impact craters, is the measure of morphometric aspects of these features, such as depth, diameter, volume of crater-fill and various aspects of crater ejecta (e.g. Stewart and Valiant, 2006). In most cases, visible laser altimeters have been employed to derive gridded global DTMs of planetary surfaces (Zuber *et al.*, 1992; Smith *et al.*, 2001, 2010). Furthermore, for craters on Earth, these data can be used by field geologists to correlate features mapped on the ground (e.g. Zumsprekel and Bischoff, 2005). Another method for deriving morphometry and altimetry is the use of individual or paired stereo images (i.e. obtained from at least two viewing angles and having a proper separation to reproduce the third dimension). These stereo images are typically processed into a standard qualitative image anaglyph (red–blue images showing relative elevation) or into a more quantitative DTM (e.g. Fig. 13.1, Fig. 13.3 and Fig. 13.4). For the 3D analysis of terrestrial craters, numerous options exist that include datasets such as the Shuttle Radar Topography Mission (SRTM; e.g. van Zyl, 2001; Farr *et al.*, 2007) and the global digital elevation model (GDEM) product created by the stereo imaging capability of the Advanced Spaceborne Thermal Emission and Reflection Radiometer (ASTER) sensor (e.g. ASTER, 2010) to high-resolution airborne LiDAR flown over specific targets. The global datasets have DEM resolutions on the order of 30 m/pixel, whereas airborne LiDAR can have very high resolutions (on the order of centimetres).

Scanning LiDAR is a powerful tool that can also acquire data of the surface topography (Bufton *et al.*, 1991) independent of local vegetation and other ground cover. With enough data points, a statistically significant number of laser pulses will miss vegetation cover and reflect off the ground surface rather than (tall) vegetation. The laser pulses that reflect off vegetation can then be statistically filtered to create a 'bald earth' model, which can be used to trace faults and other subtle surface features (Webster *et al.*, 2006; Prentice *et al.*, 2009). As an example of its high accuracy, LiDAR has also been used for change detection (Woolard and Colby, 2002). This can be especially useful for assessing the structure and morphology of terrestrial impact structures covered by dense vegetation (after filtering). The example shown (Fig. 13.5) is not from an impact crater, but displays the utility of using LiDAR.

## 13.5   Composition derived from remote sensing

Measurements of the electromagnetic spectrum are only practical within certain wavelength regions, as they provide specific diagnostic information ranging from atomic/molecular interactions to physical properties of surface materials. Some wavelength regions useful for diagnostic compositional analysis of the surface using remotely sensed data may be obscured by absorption or scatter in the atmosphere of the planetary body being observed. In order to quantitatively extract the position, shape and magnitude of surface spectral features, one must take into account: (1) the interaction of that energy with materials on the surface; (2) the passage of that energy through the intervening atmosphere; and (3) the subsequent capture of that energy and response by the instrument detectors. All of these interactions need to be understood and well constrained in order to extract and separate the atmospheric and surface spectra. Absorption and scattering by water vapour, carbon dioxide and stratospheric ozone in the Earth's atmosphere, for example, cause all of the gamma-ray region, most of the ultraviolet and thermal infrared (TIR) region, and some of the visible to near-infrared (VNIR) and shortwave infrared (SWIR) regions to be obscured (Hudson, 1969). This is

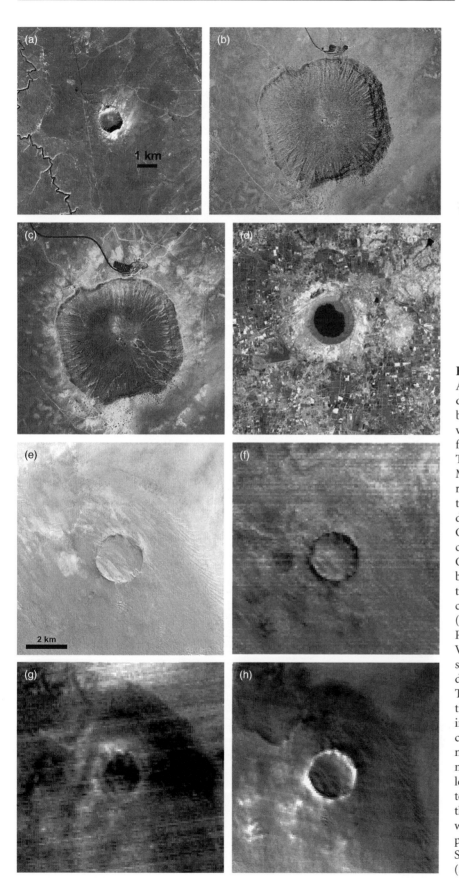

**Figure 13.2** (a) 15 m resolution ASTER VNIR of Meteor Crater, Arizona displayed as bands 3–2–1 in red–green–blue (refer to Colour Plate 29), showing very little vegetation save for a small farm with trees east of the crater. (b) True-colour aerial photograph of Meteor Crater with approximately 1 m resolution. (c) False-colour aerial photograph of Meteor Crater stretched to display ejecta lobes of bright-white Coconino and Kaibab ejecta. (d) False-colour ASTER VNIR image of Lonar Crater, India (Chapter 18), displayed as bands 3–2–1 in red–green–blue; vegetation (as red) can be seen circum the crater lake but inside the crater rim. (e)–(h) ASTER VNIR and TIR data of Roter Kamm, Namibia; (e) daytime VNIR colour image showing the pervasive mantling of aeolian sand; (f) daytime TIR decorrelation stretch of TIR bands 14, 12, 10 in R, G, B (respectively) indicating significant variability in the sand composition (Fe-rich silicate sands shown in blue/purple and non-Fe-rich silicates shown in red); (g) night-time temperature image showing less than 4 °C variation from crater rim to the near-rim ejecta; (h) apparent thermal inertia (ATI) image (see text) with high ATI values (bright) indicating potential blocky and less mantled ejecta. See text for explanations of acronyms. (See Colour Plate 29)

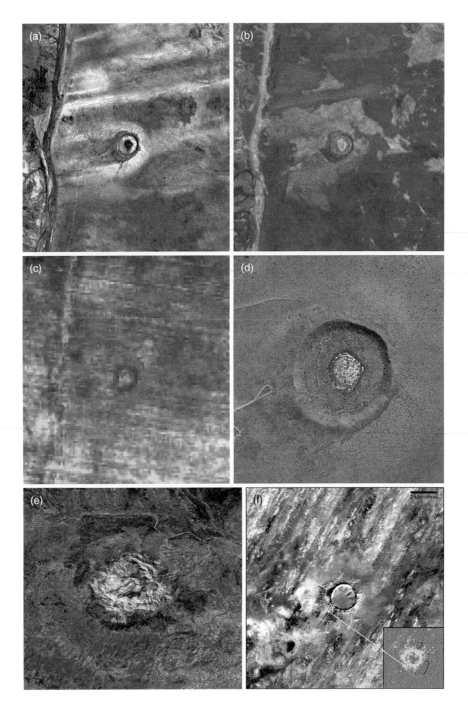

**Figure 13.3** Various remote images of the Wolfe Creek (a–d) and Gosses Bluff (e) craters in Australia. (a) ASTER VNIR image with bands 3, 2 and 1 displayed as red, green and blue (see Colour Plate 30), with a 2% stretch. Image taken 02-02-2003. (b) Landsat band ratio image with Fe ratio, normalized difference vegetation index and OH-bearing ratio (see text for details) displayed as red, green and blue. Data are stretched 2% to bring out differences. Image from 09-03-1999. (c) ASTER TIR decorrelation stretch with bands 14, 12 and 10 displayed as red, green and blue. Image from 02-02-2003. (d) High-resolution image derived from © Google Earth, 2012. (e) View to northwest of the 23 km diameter Gosses Bluff Crater, with a hue saturation value transformed VNIR with SWIR minimum noise fraction (MNF) bands applied to a VNIR image (15 m resolution) and draped on a DTM with a vertical exaggeration of 3. Green represents bedrock from several Mesozoic sandstone lithologies (Barlow, 1979; Prinz, 1996) uplifted above alluvium shown as blue and purple. (f) ASTER stereo-derived DEM and VNIR data covering the 1.9 km terrestrial impact structure Tenoumer located at 22°55′7″N, 10°24′29″W, in Mauritania, Africa. The 15 m/pixel colour infrared image of Tenoumer reveals what appears to be a two-facies, circumferential deposit around the crater rim reminiscent of ejecta around fresh impact craters on other planetary surfaces. The colour-coded DEM in the inset shows elevated terrain consistent with the inner most facies of these deposits, which strongly supports the idea that some ejecta was preserved around this infilled simple crater. The DEM shows similar to results from Meteor Crater (Garvin *et al.*, 1989), that the wind regime that has dominated this region since the time of crater formation has preferentially muted and eroded the windward (southwest) side. (Note: the two areas outlined in black and in solid purple are areas of spurious data values where difficulties due to surface or atmospheric properties confounded attempts to derive digital elevation for these areas.)

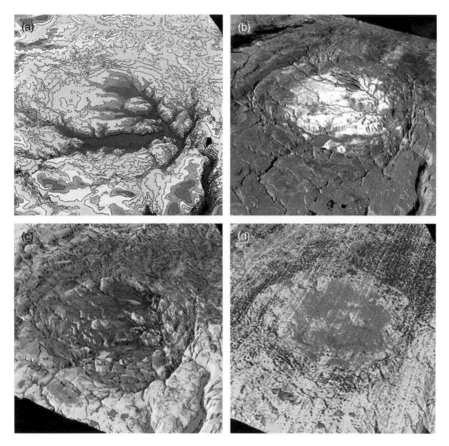

**Figure 13.4** Morphometric and spectral maps (VNIR, SWIR and TIR) of the Haughton impact structure, Devon Island, Nunavut territory, Canada. North is to the bottom right of each subimage. (a) A 25 m colourized DTM of the Haughton impact structure (courtesy of the Natural Resources of Canada http://www.nrcan-rncan.gc.ca/com/index-eng.php), with the vertical exaggeration set to 10. (b) VNIR image from *Landsat 7* ETM+ (30 m/pixel). The impact melt rock of the Haughton Formation is the most apparent lithological unit. This is manifested as a white unit that predominately lines the crater floor, but smaller occurrences can also be observed in the vicinity of the wall/terraces and rim area. (c) ASTER SWIR (1.656–2.400 μm) colour composite of MNF-transformed bands using principle components 2, 1 and 6 in RGB (30 m/pixel). Lithological units are as follows: magenta, dolomite; cyan, limestone; red, gypsum; blue, impact melt rock. (d) ASTER TIR colour-composite using DCS bands 14, 12 and 10 (90 m/pixel; Tornabene *et al.*, 2005). Lithological units are as follows: dark blue–green is dolomite; cyan–light green is limestone; red is gypsum; and magenta–pink is impact melt rock. (See Colour Plate 31)

less of an issue for the microwave portion; however, even within a relatively clear atmospheric window, scattering/emission can still occur, which can further complicate an accurate analysis of the surface spectra for specific mineral features. Because the various wavelength regions can provide information as varied as elemental, mineralogical and thermal details of the surface, each is summarized below by wavelength region.

### 13.5.1   Gamma and ultraviolet

As all gamma rays and virtually all ultraviolet light is absorbed by the Earth's ozone layer, these data are not typically analysed for terrestrial applications and, hence, are not discussed in detail here. On other planets lacking atmospheric ozone (or atmospheres for that matter), such as the Moon and Mars, gamma ray and neutron remote sensing can be used to place constraints on the elemental

composition of the uppermost approximately 1 m of the surface, including the presence of hydrogen inferred to be subsurface water ice. However, very large pixel sizes common in this wavelength region generally prevent detailed analyses of all but the largest impact craters and basins. For example, the Mars Odyssey Gamma Ray Spectrometer has a spatial resolution of approximately 300 km per pixel (Boynton *et al.*, 2004).

### 13.5.2   Visible to near-infrared (VNIR) to shortwave infrared (SWIR)

The most familiar type of remote sensing involves detecting energy in the 0.4–0.67 μm wavelength region. This includes aerial and satellite-based images of Earth and other planets (Fig. 13.2 and Fig. 13.3a). In general, this wavelength region is the most easily understood, and has had the longest period of technological

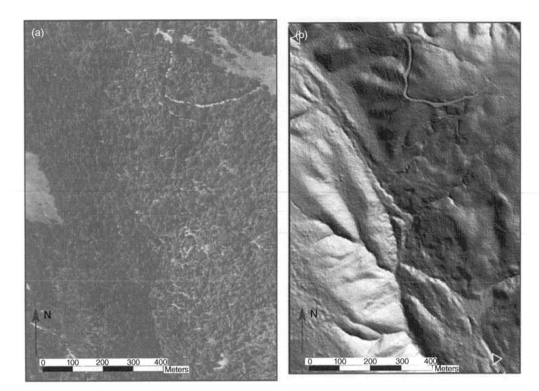

**Figure 13.5**   Analysis of the San Andreas fault from Prentice *et al.* (2009). (a) Aerial photograph of the fault obscured by a dense vegetation canopy. (b) Hillshade image of same area shown in (a) derived from LiDAR after removal of the vertical vegetation structure. The San Andreas fault (arrow) and several minor topographic features can be seen in much more detail. Similar data of terrestrial impact sites may permit the identification of previously unmapped faults and fractures.

development through extensive use by the military, commercial and scientific communities. Compositional properties, in addition to morphology, morphometry and altimetry (via stereo observations), can be gleaned using visible data. The VNIR 0.4–2.6 μm wavelength region is particularly sensitive to changes in valence and the energy state of the outer electrons of transition metals (e.g. $Fe^{3+}$ or $Fe^{2+}$) and vibrations between atoms chemically bonded in mineral structures (e.g. $OH^-$ and $H_2O$). The electrons and vibrating molecules absorb or reflect radiation at characteristic frequencies, leading to distinctive spectral features that are diagnostic of characteristic metal cation–anion bonds. Generally, electronic transition absorptions are broad and occur at wavelengths shorter than approximately 1.5 μm, whereas sensitivity to bond vibrational energy generally occurs at wavelengths longer than this. Depending upon band positions, multispectral data (generally <10 bands) can be used to distinguish several mineral groups, but hyperspectral data (tens to hundreds of bands) is commonly required in order to accurately identify specific minerals based on their image-derived spectra. These spectra can be compared with a spectral library of laboratory-measured rocks and minerals. Most clays, sulfates and iron-bearing minerals, such as pyroxene and olivine, all have diagnostic VNIR absorptions, whereas unhydrated silica phases, chlorides, glasses and some sulfates and oxides do not, and thereby are difficult if not impossible to detect using this particular wavelength range.

### 13.5.3   Thermal infrared (TIR), including mid-infrared to far infrared

Vibrational spectroscopy (also known as mid infrared (MIR) and TIR) involves the measurement of emitted energy of planetary surfaces over approximately the 3–50 μm region and has been used as a tool to address a variety of geological problems requiring major rock-forming mineral identification and mapping. This is made possible because many of these materials display prominent spectral absorption features (emissivity lows/reflectance highs) within this wavelength region. In the context of geological studies, spectral features arise from the bending/stretching frequencies of the Al–Si–O and Ca–O bonds, for example, making TIR remote sensing excellent for the study of silicate and carbonate rocks on planetary surfaces (Lyon, 1965; Hunt, 1980; Salisbury and Walter, 1989). In order to compare emitted spectra of planetary surfaces with those from the laboratory, the surface radiance must be separated into the temperature and emissivity components to facilitate comparisons between them. The emitted radiance is described by the Planck equation, which states that perfectly emitting surfaces (i.e. blackbodies) do not have temperature–wavelength-dependent variations (Realmuto, 1990; Gillespie *et al.*, 1998). Fortunately, most natural materials and surfaces are not blackbodies and have spectral features at characteristic wavelengths as a result of the asymmetric bonds of the minerals they contain (Salisbury and D'Aria, 1992; Salisbury and Wald, 1992; Salisbury, 1993).

For the TIR emissivity of most rocks and soils, the photon–matter interaction of the component minerals can be modelled as linear combinations because of the high absorption coefficients (Gillespie, 1992; Adams *et al.*, 1993; Ramsey and Christensen, 1998). This property results in simple surface (Fresnel) reflections of the emitted photons. These photons have a much shorter path length relative to the particle size and, therefore, generally interact only once after being emitted/reflected from particles. These photons either reach the detector containing only information about that particular particle, or they are quickly absorbed after being scattered and never reach the detector. As a result, the majority of the energy detected remotely has interacted with only one surface particle and the spectral features from the surface particles are retained in proportion to their areal extent.

Mixing of emitted energy from spectral end-members occurs at all scales. What defines a component or end-member depends on the scientific goals of the study (e.g. specific minerals, rock units, sand populations, micrometre-scale surface textures). For cases such as these, a more complex data extraction model is required in order to positively identify the presence and amount of a particular end-member within a TIR scene. If the pure mineral spectra (end-members) are known in advance, this allows TIR spectra to be linearly deconvolved (i.e. unmixed) in order to ascertain the areal percentages of end-member constituents (i.e. mineral or rock components) within the field of view of the instruments and several micrometres into the surface (Thomson and Salisbury, 1993; Ramsey and Christensen, 1998; Ramsey and Fink, 1999; Hamilton, 2003; Wright and Ramsey, 2006; Carter *et al.*, 2009). However, non-isothermal surfaces or very fine grained surface materials (grains with sizes smaller than or on the order of the wavelength) behave non-linearly with respect to their components (Gillespie, 1992; Ramsey and Christensen, 1998).

## 13.6 Physical properties derived from remote sensing

### 13.6.1 Thermophysical properties

In addition to the spectral information derived from TIR data, the radiant energy also contains temperature information that can be used to determine information deeper than the uppermost surface layer (i.e. thermophysical properties). Thermophysical properties include brightness temperature, thermal inertia and estimates of rock abundance (as described in Fergason *et al.* (2006) and Nowicki and Christensen (2007)). Thermal inertia is a measure of the resistance of a material to change in temperature and can be calculated accurately if a material's thermal conductivity, density and specific heat are known. Unfortunately, these variables cannot be obtained directly from remote sensing, and thus a thermal model of the surface is needed that relies on several assumptions in conjunction with infrared radiance data to derive both temperature and thermal inertia (Cracknell and Xue, 1996; Fergason *et al.*, 2006). For the Earth, the Moon and Mars, thermal inertia is an excellent proxy for particle size of surface materials, and is used to discriminate bedrock and various grain-sized unconsolidated deposits (coarse sands down to 'dust'-sized

particles). Similarly, both brightness temperature and thermal inertia have been used for a more accurate understanding of impact sites on Mars (e.g. Pelkey *et al.*, 2001; Christensen *et al.*, 2003; Tornabene *et al.*, 2005; Wright and Ramsey, 2006), and more recently for the Moon (Bandfield *et al.*, 2011). Changes in the particle size and/or thermophysical properties can also help to derive soil moisture, detect crater rims buried by thick sand deposits and distinguish ejecta from the surrounding terrain, as shown with high-resolution TIR data of Mars (Christensen *et al.*, 2003; Wright and Ramsey, 2006) and Roter Kamm Crater in Namibia (Fig. 13.2h). Cracknell and Xue (1996) reviewed how to calculate thermal inertia values from remote-sensing data.

### 13.6.2 Surface roughness from microwave imaging (i.e. radar)

The basic principle of radar (former acronym for 'radio detection and ranging') remote sensing involves the acquisition and processing of microwaves that have interacted with materials (~1 cm to several metres in wavelength). In side-looking radar, radar pulses are transmitted at an angle (the depression angle) by the airborne or spaceborne instrument after which the energy, timing and polarization of the reflected energy are captured. Surface roughness, on the scale of the incident wavelength, is the primary attribute measured. 'Rough' and coarse-grained surfaces (centimetres to metres) generally reflect the radar wave back to the detector, whereas low-return surfaces with fewer reflections are interpreted as 'smoother' or fine grained. A surface with higher dielectric constants (i.e. soil moisture, metals and buried ground ice) will also increase the radar return.

The best example of radar remote sensing used for analysis of craters on planetary surfaces is the global mosaic of Venus collected by the *Magellan* spacecraft. The surface of Venus is obscured to most remote-sensing techniques by a thick, $CO_2$-rich, approximately 90 bar atmosphere; however, radar waves can penetrate thick planetary atmospheres (e.g. Venus and Titan). Therefore, radar is commonly the remote-sensing dataset chosen to examine such bodies, although some infrared datasets are also useful (e.g. Brown *et al.*, 2004). After the *Pioneer Venus Orbiter* spacecraft's radar altimeter allowed for the earliest topography of Venus, albeit with a 150 km surface resolution (Masursky *et al.*, 1980; Pettengill *et al.*, 1980), *Magellan's* 12.6 cm radar wavelength data gave better views of the impact craters on the Venusian surface (Saunders *et al.*, 1990). It also revealed a surprising paucity of craters on Venus, indicating a younger surface than previously suspected. The *Cassini* spacecraft's Titan Radar Mapper (1.88 cm wavelength) acquired radar images of Titan that showed craters. Figure 13.6 highlights examples of complex craters with bright and dark interior and exterior deposits imaged on Venus and Titan. Radar-bright areas are generally coarse grained or 'rougher' than radar-dark areas. Hummocky crater-fill deposits and ejecta blankets are commonly radar bright, whereas radar-dark surfaces are finer grained or 'smoother'. The radar images of complex craters in Fig. 13.6 exhibit several attributes of fresh complex craters: ejecta, rim, wall/terraces and a central peak. Several workers have focused on radar data of impact sites on Venus and Titan (Phillips *et al.*, 1991; Schaber *et al.*, 1992; Campbell

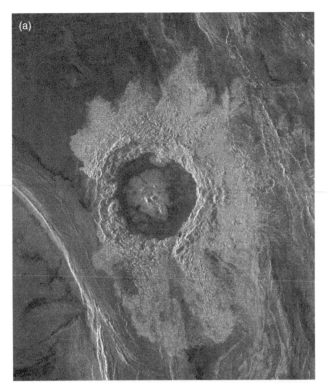

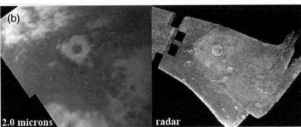

**Figure 13.6**   (a) *Magellan* radar image (12.6 cm wavelength) of Dickinson Crater on Venus, an approximately 69 km diameter complex crater exhibiting a radar-bright crater rim and wall/terrace region, and both 'smooth' and 'rough' interior deposits. The ejecta deposits are both very bright (rough) in the northern ejecta and show some less bright to darker portions (smoother) in the southern ejecta, whereas the crater interior shows a 'rough' central peak surrounded by a smoother crater floor. (Image credit: NASA/JPL; Image PIA00479: http://photojournal.jpl.nasa.gov/.) (b) *Cassini* orbiter images of one of Titan's few impact craters imaged in near infrared (left) and radar (right). The impact crater is an approximately 80 km diameter complex crater (note the 'darker' central peak feature at 2.0 μm and its outline in radar). (Image credit: http://photojournal.jpl.nasa.gov/catalog/PIA07868.)

*et al.*, 1992; Chadwick and Schaber, 1993; Lorenz *et al.*, 2007; Le Mouélic *et al.*, 2008; Wood *et al.*, 2010). Recent work has showed the utility of using a particular type of radar called circular polarization ratio radar to identify secondary craters on the Moon (Wells *et al.*, 2010).

## 13.7   General spectral enhancement and mapping techniques

Various methods of enhancing spectral features on remote images are briefly described here. For more information, consult chapters concerning this topic in textbooks (Vincent, 1997; Sabins, 1997; Lillesand and Kiefer, 2000; Jensen, 2007). With an understanding of the properties discussed above, it is the scientist's task to visually display a feature or characteristic of the crater or geological scene under study using remote data. Several of the more common approaches are described here, with further detail on certain methods provided in the case studies in Section 13.8.

With multispectral data, there are limited colour combinations one can make to emphasize mineral and lithological spectral variability in a scene. Individual inspection of bands is typically necessary in order to determine which areas are affected by absorptions at a particular wavelength. By noting which bands show the most variability, one can simply make a first-order spectral map by placing three of these bands into red, green and blue (R–G–B). Spectral units are contiguous areas of similar spectral properties (i.e. colour) contained in the scene. Related to this approach is creating band ratios (e.g. dividing a wavelength channel positioned on the continuum of a spectrum by a channel positioned within the deepest portion of an absorption feature) for the purpose of mapping the presence of diagnostic absorption features. Band ratioing is most useful for multispectral imagery that lacks the spectral resolution needed for direct matching to specific library spectra (e.g. ETM+ or ASTER data for Earth or Thermal Emission Imaging System (THEMIS) data for Mars), and has been used extensively to map specific mineral groups in terrestrial remote-sensing studies (e.g. Rowan and Mars, 2003; Fig. 13.3b). In the event that colour differences are subtle where the band-to-band data are highly correlated, a decorrelation stretch (DCS) can be applied (Fig. 13.2f, Fig. 13.3c and Fig. 13.4d). The DCS identifies the major axes of correlation, stretches data perpendicular to these axes and, thus, fills the entire dynamic range of the colour space while retaining the original hues of the units in the original R–G–B composite (Gillespie *et al.*, 1986). Additional mapping techniques include the MNF transform noise (Boardman and Kruse, 1994), a type of principal component (PC) analysis, which is designed to decrease the dimensionality of hyperspectral datasets (Fig. 13.4c). Such techniques are necessary for large (hyperspectral) datasets where it would not be practical to inspect all bands individually. These techniques reduce the spectral variability in a data cube containing hundreds of bands to as few as a dozen bands with each one being individually influenced by an important surface attribute, such as albedo, illumination, temperature and individual spectral components (i.e. compositional variability). Another means of analysing hyperspectral variability is to identify the image-derived spectral endmembers (IEMs) in a scene, which typically represent the purest spectral components that can be resolved within the spatial and spectral limits of the dataset (i.e. purest mineral and lithological compositions). The significance of such a classification is that, ideally, every pixel within a scene can be potentially described as a mixture of these purest end-member spectra, and the distribution and concentration of each IEM can be displayed. Generally, displaying the abundance of each IEM is more easily understood than displaying individual bands or values.

## 13.8  Case studies

Two examples are described below that demonstrate the usefulness of remote sensing in determining the distribution of lithologies at impact craters. The composition of the Haughton impact structure (HIS), Canada, and Meteor Crater, Arizona, USA, are related to the pre-impact stratigraphy and erosion since impact.

### 13.8.1  VNIR, TIR and DEM analyses of the HIS, Canadian Arctic

*In lieu* of, or in addition to, *a priori* knowledge from the field, remote-sensing techniques can be used to construct lithological maps of terrestrial impact structures that are well exposed and well preserved in areas with low vegetation coverage. Such maps are reasonably accurate with respect to features on the order of about three or four pixels and can be used prior to mapping and sampling in the field, or generally augment observations from the field. As a specific example, we present a summary of the work of Tornabene *et al.* (2005) on the lithological mapping, via spectral analyses of multiple spaceborne datasets, of the HIS – an approximately 23 km diameter complex crater centred at 75°22′N, 89°41′W on the western part of Devon Island, in the Canadian Arctic Archipelago, Nunavut. The HIS is particularly well suited for study by remote-sensing spectral and mapping techniques as: (1) the structure is sizable enough for lithological units to be resolved spectrally using the spatial resolution of the available orbital datasets; (2) the structure is moderately to well preserved with clearly discernible complex crater morphologic features; (3) the crater is well exposed and covered by minimal to sparse vegetation; and (4) the structure has been well characterized, sampled and geologically mapped, and also offers reasonable access to the undisturbed target sequence stratigraphy for comparison with mapping both in the field and via remote sensing. Such factors are important to note here, as they are particularly relevant to remote-sensing studies of impact structures elsewhere on Earth.

Although the primary objective of this study was a proof of concept for spectral mapping of impact-exposed mineral and lithological compositions on Mars, this study highlights various aspects of terrestrial complex craters that are complementary to the second case study of a simple crater and, therefore, are relevant to this chapter. For a complete summary of the history and geology of Haughton, see Chapter 19.

#### 13.8.1.1  Spectral and lithological characterization

A spectral unit map of the Haughton structure was created using standard image contrast stretches, band ratios and both MNF and PC algorithms on the ETM+ and ASTER scenes (Tornabene *et al.*, 2005). Maps generated using the 30 m/pixel Landsat Enhanced Thematic Mapper+ (ETM+) and ASTER SWIR data (Fig. 13.4) were very useful, as the VNIR wavelengths covered by these instruments are particularly sensitive to the carbonates and hydrated sulfates that make up a major portion of Haughton's pre-impact target sequence (Osinski *et al.*, 2005). Spectral units (i.e. units of contiguous spectral similarity) defined using SWIR and TIR datasets were lithologically characterized in a similar fashion to the case study presented on Meteor Crater (Section 13.8.2). TIR-derived IEM spectra were matched to rock and mineral spectra from a library compiled and described by Christensen *et al.* (2000). These results were then ground-truthed with laboratory-collected spectra of samples obtained in the field, and via a direct comparison of our spectral/lithological maps with the lithostratigraphic map constructed from previous workers (Thorsteinsson and Mayr, 1987; Osinski *et al.*, 2005).

#### 13.8.1.2  Results and discussion

An understanding of fresh complex crater structure and morphology observed on other planetary surfaces enables recognition of similar features at Haughton. Despite the estimated approximately 200 m of erosion at the HIS (Grieve, 1988; Osinski *et al.*, 2005), it is well preserved compared with many complex impact structures on Earth. The preservation and morphology of the HIS are readily observed in a colourized DTM of the structure (Fig. 13.4a). Elevation profiles derived from the DTM compare favourably to an ideal cross-section of a complex crater despite the level of degradation, infilling and other post-impact modification of the structure (see Chapter 19). The most readily recognizable feature in the colourized DTM is the presence of a circular depression bounded by a zone of circumferential listric faults. The concentric faults are indicative of a complex crater's rim and wall-terrace region (Fig. 13.4a), which is formed during the modification stage of complex crater formation (Chapter 5). On the northeast side of the structure, we can observe that one of these arcuate listric faults has been clearly exploited by ongoing fluvial erosion. Raised topography in the central region is consistent with a structural uplift, typical of complex craters; however, this interpretation cannot be based on topography/morphology alone. In addition, it is important to note that wall terraces can form in volcanic craters/calderas, so their presence is not necessarily diagnostic of an impact origin either. However, by linking these features observed in the DTM and in the visible images along with the spectral/lithological units from complementary spectral datasets (below), the relationships between the observed lithologies and morphologic expression of the structure, especially where compared with the lithological succession observed in the undisturbed target sequence, provides the best evidence for an impact (i.e. exogenic) origin versus an endogenic one (e.g. calderas, sinkholes, glacial features, salt domes). Unfortunately, erosion at the HIS has left no evidence of an ejecta blanket, which, if present, can be recognized in DTMs (e.g. Fig. 13.3e and Fig. 13.4a) and aerial/satellite imagery of other impact structures.

Combing our spectral results with the digital elevation from the DTM provides an important synthesis that allows the HIS to be recognized as a complex crater and the pre-impact target stratigraphy to be deciphered and reconstructed. In the case of stratified targets, complex craters preserve the relative succession and increasing depth of stratigraphic units in the pre-impact target within their features (e.g. wall terraces, central uplift).

These subsurface materials are exposed at the surface from a depth that is directly scaled to the size of the crater (Housen *et al.*, 1983; Schmidt and Housen, 1987; Melosh, 1989). These estimates provide a means to approximate quantitative depths of excavation and uplift of the lithologies exposed. Complex craters are of particular interest for reconstructing subsurface and shallow crustal stratigraphy, because they excavate rocks and minerals from both the immediate subsurface (e.g. within crater walls, terraces and ejecta) and from the shallow crust via stratigraphic uplift in a central feature (see Chapter 5).

All of the major units sampled and exposed by the Haughton impact event can be identified and mapped by remote-sensing techniques alone. The gentle regional tilt (~3–5° to the WSW) of the stratigraphy in the vicinity of Haughton conveniently offers the means to compare the lithologies exposed by the crater with the undisturbed target sequence just to the west of the crater (Fig. 13.4) – both from orbit and in the field. The matched compositions summarized below correlate and agree with the lithological units and the pre-impact stratigraphic sequence as mapped during recent field studies of the HIS (Chapter 19; Tornabene *et al.*, 2005). Units undisturbed by the HIS and the position of their equivalents exposed by the HIS are consistent with a complex crater and an impact origin. The summarized progression from the shallowest subsurface unit exposed within the rim to the deepest exposed unit within the central uplift reveals: dolomitic limestone in the 'rim area' and wall terraces; limestone in the wall/terraces; gypsum-rich carbonate (eastern floor exposure); and limestone in the central uplift.

Dolomitic limestone occurs in the eroded crater 'rim area' (i.e. the elevated circumferential peaks surrounding the HIS) and represents the shallowest unit in the target sequence exposed by the HIS. Alternating dolomite and limestone outcrops that repeat on the eastern side of the crater are the clearest example of a set of terraces exposing the uppermost units of the pre-impact target sequence (Fig. 13.3c,d). The presence of two, or possibly three, crater terraces can be observed as a repetition of these two units, which correlates with the zone of concentric faults observed in the profiles derived from the HIS DTM. Together, these two lithological units, including the dolomitic limestone in the 'rim area', are consistent with the dolomites and limestones of the Middle and Lower Members of the Allen Bay Formation and Thumb Mountain Formation (Osinski *et al.*, 2005). The rim and wall-terrace units are stratigraphically followed by gypsum-rich unit (Fig. 13.4) observed exposed on or near the 'crater floor', which correlates with the gypsum-rich portions of the Thumb Mountain Formation (Osinski *et al.*, 2005). These occurrences of Thumb Mountain Formation appear to be part of the lower portions of the wall terraces in the eastern portion of the HIS, but rather may represent shallow portions of the structural uplift exposed from the erosion of the structure. Limestone occurs again associated with small distinctive topographic peaks in the central region of the crater, interpreted to be limestones of the Eleanor River Formation (Osinski *et al.*, 2005; Tornabene *et al.*, 2005; Chapter 19) that were stratigraphically uplifted from approximately 1.1 to 1.5 km beneath the pre-impact surface (Osinski *et al.*, 2005). Uplifted blocks of limestone from the Eleanor River Formation are embayed by a crater moat unit (Fig. 13.4, Fig. 13.7), which was interpreted to be a mix of all major lithotypes and is

consistent with clast-rich impact melt rocks at Haughton (Osinski *et al.*, 2005; Tornabene *et al.*, 2005).

## 13.8.2   High-resolution airborne TIR of Meteor Crater, Arizona

Meteor Crater (also known as Barringer Crater) is located 69 km east of Flagstaff, Arizona, near the city of Winslow, Arizona. It is arguably the most well known impact crater in the world, despite the initial misinterpretation as being volcanic in origin due to its slightly squarish shape and proximity to the cinder cones and maar craters of the San Francisco volcanic field to the west (Gilbert, 1896). Furthermore, the lack of large buried iron mass below the crater, which was the prevailing assumption at the time for meteorite impacts, further perpetuated the belief of a volcanic origin for many decades (Barringer, 1905). One of the first scientific investigations of the crater was in 1891 by G.K. Gilbert, who concluded that the crater was hydrovolcanic in origin based on the shape and lack of erupted lava. It was not until the early 1960s, when the dynamics of hypervelocity impact crater formation were better understood and high-pressure silica (stishovite) was discovered, that the impact origins of the crater were confirmed (Shoemaker, 1960). Since Shoemaker's early pioneering work at Meteor Crater, numerous studies have examined the geology (Shoemaker and Kieffer, 1974; Roddy, 1978), post-crater formation weathering/erosion (Pilon *et al.*, 1991; Grant and Schultz, 1993), as well as the current surface conditions derived from remote sensing (Ramsey, 2002; Wright and Ramsey, 2006) (see Chapter 18 for further details on Meteor Crater).

The crater has provided an ideal test locale for numerous studies, particularly in the TIR region, because of its well-preserved state, semi-arid environment, paucity of vegetation, lithological exposures and low relief. The crater has been the target of ground-based, airborne and satellite-based TIR data collections, with the primary focus on the exposed crater ejecta and the aeolian reworking of those sediments into the distinctive northeast-trending wind-streak. The analysis of the aeolian processes using remote sensing was either focused on trying to confirm the amount of erosion that had taken place, and thus the palaeoclimatic conditions (Ramsey, 2002), or using the data as an analogue for similar studies on Mars (Wright and Ramsey, 2006). Because the amount and style of erosion of a crater's ejecta have a direct bearing on the past climatic and geological history, the possibility of further constraining these rates with remote sensing becomes important for exploration of different craters on Earth or other planetary bodies.

### 13.8.2.1   *Geology and crater formation*

Meteor Crater is located in north-central Arizona east of Canyon Diablo (Fig. 13.2a–c). The current age estimate places the crater at approximately 50 ka (Nishiizumi *et al.*, 1991), making it one of the most recent and well-preserved impact sites on Earth (Chapter 18). The impact of the iron meteorite produced a simple crater that is approximately 180 m deep and 1200 m in diameter with a 30–60 m high rim. The region contains NW-trending normal faults and orthogonal joint sets that are responsible for the unusual squarish appearance of the crater (Shoemaker and

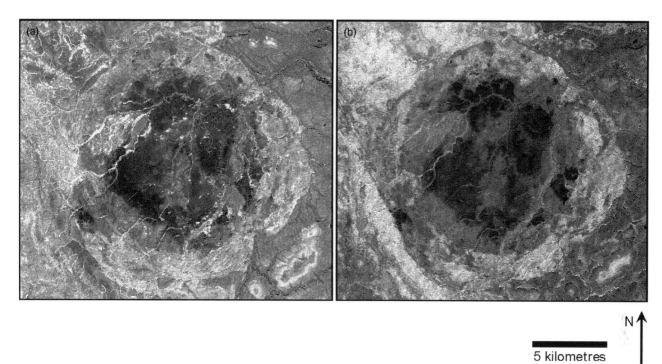

N↑

5 kilometres

**Figure 13.7** *Landsat 7* ETM+ band ratio images of the Haughton impact structure. (a) Band ratio image consisting of bands 5/7 (1.65 μm data divided by 2.215 μm data) highlights areas within the scene that may contain carbonates, phyllosilicates and various hydrothermal alteration minerals, which typically have absorption features in the VNIR in the vicinity of approximately 2.1–2.3 μm (band 7) and generally higher reflectance in the vicinity of 1.65 μm (band 5). In the case of Haughton, discrete high ratio-value areas on the eastern part of the crater floor correlate with gypsum of the Bay Fiord Formation and clays within the Haughton Formation in the vicinity of the Haughton River on the western end of the structure. Some sensitivity to carbonates is also exhibited in this band ratio combination. (b) Ratio of bands 3/2 (0.660 μm/0.560 μm) highlights areas bearing minerals rich (white) or poor (dark) in ferric iron. $Fe^{2+}$ substitution for $Mg^{2+}$ in the dolomites of the Allen Bay Formation is particularly prevalent in this image, whereas limestones and the impact melt-bearing breccias are particularly $Fe^{2+}$ poor. The $Fe^{2+}$-poor nature of the crater floor 'melt' sheet may be a consequence of pervasive hydrothermal alteration (e.g. oxidation) of any $Fe^{2+}$-bearing materials that may have been incorporated into these deposits during and just shortly after impact (LandSat 7/NASA).

Kieffer, 1974). The principal flat-lying stratigraphic units (Permian Coconino Sandstone, Permian Kaibab Limestone and Triassic Moenkopi Formation) are preserved in inverted order as primary ejecta up to two crater radii (~1.2 km) away from the rim. These hummocky, near-rim ejecta deposits consist of a continuous blanket formed by larger blocks ranging in size from 0.5–30 m. The western, northern and eastern sides of the crater are dominated by Kaibab Formation ejecta with several lobate deposits of Coconino Sandstone interspersed. On the southern rim, the vast majority of the deposit consists of Coconino Sandstone, which is easily discernible due to its high albedo. Because of its friability and high porosity, this deposit has been the source for the majority of material that now comprises the northeast-trending patchy wind-streak (Fig. 13.8). To the west and southwest, the character of the hummocky ejecta is considerably more muted than the rest of the crater rim, likely due to mantling by aeolian fines transported by the dominant winds from the southwest. Within the crater, the uppermost floor deposits are formed from trapped airborne dust and basaltic ash from the nearby volcanic field. Below these surficial deposits, drill cores reveal 30 m of interfingered Quaternary lake beds and alluvium

overlying approximately 10 m of mixed impact breccia (Shoemaker and Kieffer, 1974).

### 13.8.2.2 Previous research results

On the basis of the apparent lack of mapped ejecta from units such as the Coconino Sandstone, previous studies by Shoemaker (1960) and Roddy (1978) reported 20–30 m of vertical erosion had occurred in the near-rim vicinity (~600 m). However, subsequent investigations concentrated on the distal ejecta field using ground-penetrating radar (Pilon *et al.*, 1991) and trenching/sieve analysis (Grant and Schultz, 1991) and found the total erosion was significantly lower (1–2 m). Based on these results, Grant and Schultz (1993) reinforced their argument of low erosion by citing the apparent lack of any appreciable amount of reworked Coconino Sandstone in the surrounding depositional traps. It was hypothesized that detectable volumes of reworked Coconino Sandstone should be present in these traps if the crater rim had been eroded by more than 20 m, unless this eroded material had been entirely removed from the region. Extensive transport and removal was not deemed likely based on the palaeoclimate

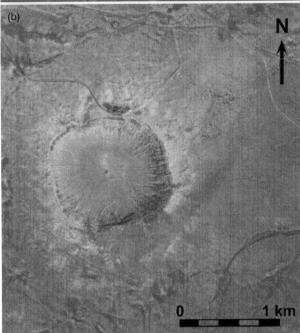

**Figure 13.8** (a) Meteor Crater as seen from the east. (b) TIMS image with Coconino Sandstone spectral end-member displayed as red, Kaibab Limestone spectral end-member displayed as green and Moenkopi Formation (siltstone/mudstone in this region) displayed as blue. (see Colour Plate 32)

conditions in the region since the crater was formed, which suggest 1–2 cm of erosion per 1000 years (Greeley and Iverson, 1985) and is consistent with the lower estimate.

Several remote-sensing-based studies took place during the 1980s and 1990s at Meteor Crater. Garvin *et al.* (1989) examined airborne laser altimetry and characterized the nature and morphology of the near-rim hummocky ejecta. They found the ejecta was considerably less pronounced to the west and southwest and attributed it to aeolian burial by fines transported from the southwest. Later studies used TIR data to characterize the processes that produced the lithological composition of the ejecta deposits and wind streak.

Ramsey (2002) sought to validate the degree of erosion at Meteor Crater through the application of airborne TIR remote sensing. This study also served as an applied test of the linear deconvolution approach previously developed for unmixing laboratory-collected spectral mixtures data (Ramsey and Christensen, 1998). The results were also used to simulate data from Mars orbit to be acquired from the THEMIS instrument (Christensen *et al.*, 2003), similar to the other case study herein. Thermal Infrared Multispectral Scanner (TIMS) airborne data were acquired in 1994 at several different altitudes, which resulted in spatial resolutions from 3.2 to 10.9 m/pixel. The TIMS instrument had six spectral bands from 8 to 12 µm and was flown by NASA from 1982 to 1996 (Palluconi and Meeks, 1985). The highest spatial resolution data were used to extract spectral end-members from regions high in the three primary statigraphic units. The location of each of the end-member pixels was based on *a priori* knowledge of rock outcrops from field mapping and the strength of the spectral features observed in the processed emissivity data. The spectral end-members used with the linear deconvolution model produced images showing the relative spectral concentration of each IEM representative of the three lithologies (Fig. 13.8). These were compared with geological maps of the crater region (Shoemaker, 1960).

The TIMS intermediate flight line was chosen because it had good areal coverage as well as moderate spatial resolution to determine the areal extent and maximum traceable distance of the Coconino Sandstone. The Coconino comprised the uppermost layer of the wind-streak and was remotely detected up to five crater radii (3 km) to the northeast, with strong (>30%) concentrations as far out as 1 km. This is more extensive than the previously reported 0.5 crater radii from the crater rim (Grant and Schultz, 1993). Ramsey (2002) postulated that the reworked ejecta comprising the wind-streak is not strongly concentrated and consists of grains less than 300 µm, making it indistinguishable from the sediments derived from the other units and, therefore, easily overlooked in the field. In order to extend these lithological end-member maps from areal extent to eroded volume, an estimate of the sandstone ejecta thickness was made with the knowledge of the pixel size and an assumption of the volume of Coconino ejected based on theoretical modelling. This type of calculation is only accurate for the uppermost tens of micrometres because emitted TIR energy can only be detected to that depth. However, in conjunction with other geophysical data sets, it was extended to greater depths. Only pixels directly comprising the primary ejecta and wind-streak were counted. This end-member percentage summation was multiplied by the pixel resolution to produce the total areal coverage. Using this value and the theoretical maximum excavated volume proposed by Roddy (1978), an average thickness of 1.47 m was calculated. This estimate is consistent with the field and theoretical data and agrees with the values of Grant and Schultz (1993), but is far less than the original 20–30 m amount of Shoemaker (1960). It was the conclusion of this study that the lack of Coconino ejecta in the region surrounding Meteor Crater was caused by the differential uplift and subsequent ejection of the sedimentary units rather than a large amount of erosion post-impact.

Wright and Ramsey (2006) expanded upon the previous work by examining newly-acquired ASTER TIR data of the crater. ASTER was the Earth's first spaceborne multispectral TIR instrument to obtain high spatial resolution data. It has five bands in the TIR region from 8.125 to 11.65 μm at 90 m/pixel spatial resolution. Unlike the previous study, end-members were statistically chosen in order to identify the most spectrally distinct pixel clusters in the ASTER emissivity data. These pixels were selected as IEMs, thereby requiring no prior knowledge of the surface geology. However, the lower spatial resolution of ASTER compared with that of TIMS produced clear spectral mixing of the IEMs. Wright and Ramsey (2006) found that the areal abundances derived from the deconvolution of the ASTER data were within 6% of those derived from the higher spatial and spectral resolution TIMS data. Furthermore, the areal extent of the wind-streak was found to be 2.7 km from the crater rim compared with 2.4 km calculated by Ramsey (2002). Therefore, it was determined that, despite a resolution degradation of nearly 800% from the TIMS data, the ASTER data produced similar results. This indicated the spatial scale of lithological mixing in the TIR pixels likely became homogeneous at scales larger than approximately 10–15 m and, therefore, validated the use of TIR data on the order of 100 m/pixel for similar studies of Earth and Mars.

### 13.8.2.3  *Applicability to other craters*

For craters that are significantly older and more modified/mantled, TIR remote sensing can also be used to determine other surface parameters. Roter Kamm Crater, Namibia (Fig. 13.2), is dated to about 3.5 Ma, is approximately 2.5 km in diameter and is located in the arid Namib Desert. It has been well mantled by active, broad sand dunes emanating from a larger sand sheet to the north (Fig. 13.2e). The target includes primarily Precambrian crystalline rocks and outcrops of impact melt breccias are found exclusively on the crater rim (Grant, 1999). Compositional analysis of the TIR data indicates that the region is dominated by two slightly different compositions of sand, which completely mantle the crater, the rim and the ejecta. However, the thermophysical properties of mantled surfaces become more obvious if the ATI image is examined. For Earth, we have a limited understanding of the relationships between the thermophysical data and natural surfaces. Thermal inertia can be approximated using the ATI, which is a comparison of the day/night temperature difference to the complement of one minus the daytime VNIR–SWIR-derived albedo. The ATI image of Roter Kamm Crater shows the crater rim and regions to the southwest to have higher values (Fig. 13.2g,h). These locations also agree well with spaceborne radar data of the region, which show radar-rough surfaces in the same locations. These higher ATI values are approximately half of those extracted from rock outcrops in the mountains 10 km to the east, indicating that even these surfaces are likely mantled to some degree. ATI data can give a good indication of the degree of mantling and presence of blocky ejecta and the crater rim rocks up to several metres below the exposed surface (Putzig and Mellon, 2007).

## 13.9  Concluding remarks

The above case studies demonstrate the use of airborne and orbital data to provide compositional details for preserved, well-exposed and non-vegetated structures and how craters can be used as a means to map subsurface lithologies for constraints on stratigraphy. By using ASTER and ETM+ datasets as analogues for Mars remote-sensing instruments, the study of the Haughton structure serves as a terrestrial proof of concept that subsurface geology of Mars may be successfully mapped via moderately to well-preserved impact craters that are not obscured by dust or other surface deposits. Had it not been for the regional tilt of the stratigraphic target sequence, the HIS would be completely surrounded by the dolomite with the only subsurface exposures of lower stratigraphic units occurring within nearby valleys and within the excavated and uplifted target materials outcropping within the impact structure as a natural 'window' into the crust (Tornabene *et al.*, 2005). Because tectonic processes are less pervasive on Mars and other terrestrial bodies than on Earth, impact craters provide the most convenient means to locate, map and identify the diverse mineral and lithological compositions within planetary crusts.

For younger craters where the primary ejecta is still present, remote sensing provides an ideal way to characterize the amount of erosion, the distribution and its thickness. At Meteor Crater, these parameters were constrained using TIR remote sensing. This work not only demonstrated the utility of using remote sensing for insight into the crater formation and subsequent erosional processes, but also the geological history of a region. Using the unique spectral characteristics of a lithology not otherwise observable at the surface if not for impact, the areal extent and volume of that ejected lithology were calculated. Comparable investigations of the surfaces of other planets and moons are warranted where only remote data exist.

Whether searching for new impact sites or through detailed analyses of known impact structures, remote sensing provides an approach to extract quantitative aspects or details that would not otherwise be known from the field. This ranges from identifying terrestrial impact structures via morphology and topography to analyses of the composition or physical properties of the materials affected by impact. This chapter represents a summary of what properties can be determined from the analyses of remote sensing.

## References

Adams, J.B., Smith, M.O. and Gillespie, A.R. (1993) Imaging spectroscopy: interpretation based on spectral mixture analysis, in *Remote Geochemical Analysis: Elemental and Mineralogical Composition* (eds C.M. Pieters and P.A. Englert), Cambridge University Press, New York, NY, 145–166.

ASTER (2010) ASTER global digital elevation model, http://www.ersdac.or.jp/GDEM/E/4.html.

Bandfield, J., Ghent, R., Vasavada, A. *et al.* (2011) Lunar surface rock abundance and regolith fines temperatures derived from LRO Diviner Radiometer data. *Journal of Geophysical Research*, 116, E00H02. DOI: 10.1029/2011JE003866.

Barlow, B.C. (1979) Gravity investigations of the Gosses Bluff impact structure, central Australia. *Journal of Australian Geology and Geophysics*, 4, 323–329.

Barringer, D.M. (1905) Coon Mountain and its crater. *Proceedings of the Academy of Natural Sciences of Philadelphia*, 66, 861–886.

Boardman J.W. and Kruse F.A. (1994) Automated spectral analysis: a geological example using AVIRIS data, north Grapevine Mountains, Nevada. Proceedings, ERIM 10th Thematic Conference on Geologic Remote Sensing, pp. I-407–I-418.

Boynton, W.V., Feldman, W.C., Mitrofanov, I.G. *et al.*, (2004) The Mars Odyssey Gamma-Ray Spectrometer instrument suite. *Space Science Reviews*, 110, 37–83. DOI: 10.1023/B:SPAC.0000021007.76126.15.

Brown, R.H., and 21 colleagues, (2004) The *Cassini* Visual and Infrared Mapping Spectrometer (VIMS) investigation. *Space Science Reviews*, 115, 111–168. DOI: 10.1007/s11214-004-1453-x.

Bufton, J.L., Garvin, J.B., Cavanaugh, J.F. *et al.* (1991) Airborne lidar for profiling of surface topography. *Optical Engineering*, 30, 72–78.

Campbell, D.B., Stacy, N.J.S., Newman, W.I. *et al.* (1992) Magellan observations of extended impact crater related features on the surface of Venus. *Journal of Geophysical Research*, 97, 16249–16278.

Carter, A.J., Ramsey, M.S., Durant, A.J. *et al.* (2009) Micron-scale roughness of volcanic surfaces from thermal infrared spectroscopy and scanning electron microscopy. *Journal of Geophysical Research*, 114, B02213. DOI: 10.1029/2008JB005632.

Chadwick, D.J. and Schaber, G.G. (1993) Impact crater outflows on Venus: morphology and emplacement mechanisms. *Journal of Geophysical Research*, 98 (E11), 20891–20902.

Christensen, P.R., Bandfield, J.L., Hamilton, V.E. *et al.* (2000) A thermal emission spectral library of rock forming minerals. *Journal of Geophysical Research*, 105, 9735–9739.

Christensen, P.R., Bandfield, J.L., Bell III, J.F. *et al.* (2003) Morphology and composition of the surface of Mars: *Mars Odyssey* THEMIS results. *Science*, 300, 2056–2061.

Cracknell, A.P. and Xue, Y. (1996) Thermal inertia determination from space – a tutorial review. *International Journal of Remote Sensing*, 17 (3), 431–461. DOI: 10.1080/01431169608949020.

D'Aria, D. and Garvin, J.B. (1988) Thermal infrared reflectance spectroscopy of impact-related rocks: implications for geologic remote sensing of Mars and Earth. Lunar and Planetary Science Conference XIX, pp. 243–244

Farr, T.G., Rosen, P.A., Caro, E. *et al.* (2007) The Shuttle Radar Topography Mission. *Reviews of Geophysics*, 45, RG2004. DOI: 10.1029/2005RG000183.

Fergason, R.L., Christensen, P.R. and Kieffer, H.H. (2006) High resolution thermal inertia derived from THEMIS: thermal model and applications. *Journal of Geophysical Research*, 111, E12004. DOI: 12010.11029/12006JE002735.

Folco, L., Di Martino, M., El Barkooky, A. *et al.* (2010) The Kamil Crater in Egypt. *Science*, 329, 804. DOI: 10.1126/science.1190990.

Folco, L., Di Martino, M., El Barkooky, A. *et al.* (2011) Kamil Crater (Egypt): ground truth for small-scale meteorite impacts on Earth. *Geology*, 39, 179–182. DOI: 10.1130/G31624.1

French, B.M. (1998) *Traces of Catastrophe: A Handbook of Shock-Metamorphic Effects in Terrestrial Impact Structures*, LPI Contribution No. 954, Lunar and Planetary Institute, Houston, TX.

Garvin, J.B., Bufton, J.L., Campell, B.A. and Zisk, S.H. (1989) Terrain analysis of Meteor Crater ejecta blanket. Lunar and Planetary Science Conference XX, pp. 333–334.

Garvin, J.B., Schnetzler, C.C. and Grieve, R.F. (1992) Characteristics of large terrestrial impact structures as revealed by remote sensing studies. *Tectonophysics*, 216, 45–62.

Gilbert, G.K. (1896) The origin of hypotheses, illustrated by the discussion of a topographical problem. *Science*, 3, 1–13.

Gillespie, A.R. (1992) Spectral mixture analysis of multispectral thermal infrared images. *Remote Sensing of Environment*, 42, 137–145.

Gillespie A.R., Kahle A.B. and Walker, R.E. (1986) Color enhancement of highly correlated images. 1. Decorrelation and HSI contrast stretches. *Remote Sensing of Environment*, 20, 209–235.

Gillespie, A.R., Matsunaga, T., Rokugawa, S. and Hook, S.J. (1998) Temperature and emissivity separation from Advanced Spaceborne Thermal Emission and Reflection Radiometer (ASTER) images. *IEEE Transactions on Geoscience Remote Sensing*, 36, 1113–1126.

Grant J.A. (1999) Evaluating the evolution of process specific degradation signatures around impact craters. *International Journal of Impact Engineering*, 23, 331–340.

Grant, J.A. and Schultz, P.H. (1991) Characteristics of ejecta and alluvial deposits at Meteor Crater, Arizona and Odessa Craters, Texas: results from ground penetrating radar. Lunar and Planetary Science Conference XXII, pp. 481–482.

Grant, J.A. and Schultz, P.H. (1993) Erosion of ejecta at Meteor Crater, Arizona. *Journal of Geophysical Research*, 98, 15033–15047.

Grant, J.A., Irwin, R.P., Grotzinger, J.P. *et al.* (2008) HiRISE imaging of impact megabreccia and sub-meter aqueous strata in Holden Crater, Mars. *Geology*, 36, 195–198.

Greeley, R. and Iverson, J.D. (1985) *Wind as a Geological Process on Earth, Mars, Venus, and Titan*, Cambridge University Press, New York, NY.

Grieve, R.A.F. (1988) The Haughton impact structure: summary and synthesis of the results of the HISS project. *Meteoritics*, 23, 249–254.

Hamilton, V.E. (2003) Thermal infrared emission spectroscopy of titanium-enriched pyroxenes. *Journal of Geophysical Research*, 108, 5095. DOI: 10.1029/2003JE002052.

Housen K.R., Schmidt R.M. and Holsapple K.A. (1983) Crater ejecta scaling laws – fundamental forms based on dimensional analysis. *Journal of Geophysical Research*, 88, 2485–2499.

Hudson Jr, R.D. (1969) *Infrared System Engineering*, John Wiley and Sons, Inc., New York, NY.

Hunt, G.R. (1980) Electromagnetic radiation: the communication link in remote sensing, in *Remote Sensing in Geology* (eds B.S. Siegel and A.R. Gillespie), John Wiley and Sons, Inc., New York, NY, pp. 5–45.

Jensen, J.R. (2007) *Remote Sensing of the Environment: An Earth Resource Perspective*, 2nd edn, Prentice Hall (ISBN: 0-13-188950-8).

Johnson, J.R., Hörz, F., Lucey, P.G. and Christensen, P.R. (2002) Thermal infrared spectroscopy of experimentally shocked anorthosite and pyroxenite: implications for remote sensing of Mars. *Journal of Geophysical Research*, 107 (E10), 5073. DOI: 10.1029/2001JE001517.

Johnson, J.R., Staid, M.I. and Kraft, M.D. (2007) Thermal infrared spectroscopy and modeling of experimentally shocked basalts. *American Mineralogist*, 92, 1148–1157.

Koeberl, C. (2004) Remote sensing studies of impact craters: how to be sure? *Comptes Rendus Geoscience*, 336, 959–961.

Le Mouélic, S., Paillou, P., Janssen, M.A. *et al.*, (2008) Mapping and interpretation of Sinlap crater on Titan using *Cassini* VIMS and RADAR data. *Journal of Geophysical Research*, 113, E04003. DOI: 10.1029/2007JE002965.

Lillesand, T.M. and Kiefer, R.W. (2000) *Remote Sensing and Image Interpretation*, John Wiley and Sons, Inc., New York, NY.

Lorenz, R.D., Wood, C.A., Lune, J.I. *et al.* (2007) Titan's young surface: initial impact crater survey by *Cassini* RADAR and model comparison. *Geophysical Research Letters*, 34, L07204. DOI: 10.1029/2006GL028971.

Lyon, R.J.P. (1965) Analysis of rocks by spectral infrared emission (8 to 25 microns). *Economic Geology*, 60, 715–736.

Marzo, G.A., Davila, A.F., Tornabene, L.L. *et al.* (2010) Evidence for Hesperian impact-induced hydrothermalism on Mars. *Icarus*, 208, 667–683. DOI: 10.1016/j.icarus.2010.03.013.

Masursky, H., Eliason, E., Ford, P.G. *et al.* (1980) Pioneer Venus radar results: geology from images and altimetry. *Journal of Geophysical Research*, 85 (A13), 8232–8260. DOI: 10.1029/JA085iA13p08232.

Melosh, H.J. (1989) *Impact Cratering: A Geologic Process*, Oxford University Press.

Nishiizumi, K., Kohl, C.P., Shoemaker, E.M. *et al.* (1991) *In situ* $^{10}$Be–$^{26}$Al exposure ages at Meteor Crater, Arizona. *Geochimica et Cosmochimica Acta*, 55, 2699–2703.

Nowicki, S.A. and Christensen, P.R. (2007) Rock abundance on Mars from the Thermal Emission Spectrometer. *Journal of Geophysical Research (Planets)*, 112, 5007. DOI: 10.1029/2006JE002798.

Osinski, G.R., Lee, P., Spray, J.G. *et al.* (2005) Geological overview and cratering model for the Haughton impact structure, Devon Island, Canadian High Arctic. *Meteoritics and Planetary Science*, 40, 1759–1776. DOI: 10.1111/j.1945-5100.2005.tb00145.x.

Palluconi, F.D. and Meeks, G.R. (1985) Thermal infrared multispectral scanner (TIMS): an investigators guide to TIMS data. JPL Publication 85-32, Jet Propulsion Laboratory, Pasadena, CA.

Pelkey, S.M., Jakosky, B.M. and Mellon, M.T. (2001) Thermal inertia of crater-related wind streaks on Mars. *Journal of Geophysical Research*, 106, 23909–23920.

Pettengill, G.H., Eliason, E., Ford, P.G. *et al.* (1980) Pioneer Venus radar results altimetry and surface properties. *Journal of Geophysical Research*, 85 (A13), 8261–8270. DOI: 10.1029/JA085iA13p08261.

Phillips, R.J., Arvidson, R.E., Boyce, J.M. *et al.* (1991) Impact craters on Venus: initial analysis from *Magellan. Science*, 252, 288–297. DOI: 10.1126/science.252.5003.288.

Pilon, J.A., Grieve, R.A.F. and Sharpton, V.L. (1991) The subsurface character of Meteor Crater, Arizona, as determined by ground-probing radar. *Journal of Geophysical Research*, 96, 15563–15576.

Prentice, C.S., Crosby, C.J., Whitehill, C.S. *et al.*, (2009) Illuminating northern California's active faults. *EOS, Transactions, American Geophysical Union*, 90 (7), 55–56.

Prinz, T. (1996) Multispectral remote sensing of the Gosses Bluff impact crater, central Australia (N.T.) by using Landsat-TM and ERS-1 data. *Journal of Photogrammetry and Remote Sensing*, 51, 137–149.

Putzig, N.E. and Mellon, M.T. (2007) Apparent thermal inertia and the surface heterogeneity of Mars. *Icarus*, 191, 68–94.

Ramsey, M.S. (2002) Ejecta distribution patterns at Meteor Crater, Arizona: on the applicability of lithologic end-member deconvolution for spaceborne thermal infrared data of Earth and Mars. *Journal of Geophysical Research*, 107, 5059. DOI: 10.1029/2001JE001827.

Ramsey, M.S. and Christensen, P.R. (1998) Mineral abundance determination: quantitative deconvolution of thermal emission spectra. *Journal of Geophysical Research*, 103, 577–596.

Ramsey, M.S. and Fink, J.H. (1999) Estimating silicic lava vesicularity with thermal remote sensing: a new technique for volcanic mapping and monitoring. *Bulletin of Volcanology*, 61, 32–39.

Realmuto, V. (1990) Separating the effects of temperature and emissivity: emissivity spectrum normalization, in *Proceedings of the Second Annual Airborne Earth Science Workshop*, vol. 2 (ed. E.A. Abbott), JPL Publication 90-55, Jet Propulsion Laboratory, Pasadena, CA, pp. 31–35.

Roddy, D.J. (1978) Pre-impact geologic conditions, physical properties, energy calculations, meteorite and initial crater dimensions and orientations of joints, faults and walls at Meteor Crater, Arizona. *Proceedings of the Lunar and Planetary Science Conference*, 9, 3891–3980.

Rowan L.C. and Mars J.C. (2003) Lithologic mapping in the Mountain Pass, California area using Advanced Spaceborne Thermal Emission and Reflection Radiometer (ASTER) data. *Remote Sensing of the Environment*, 84, 350–366.

Sabins, F.F. (1997) *Remote Sensing: Principles and Interpretation*, W.H. Freeman and Company, New York, NY.

Salisbury, J.W. (1993) Mid-infrared spectroscopy: laboratory data, in *Remote Geochemical Analysis: Elemental and Mineralogical Composition* (eds C.M. Pieters and P.A. Englert), Cambridge University Press, New York, NY, pp. 79–98.

Salisbury, J.W. and D'Aria, D.M. (1992) Emissivity of terrestrial materials in the 8–14 μm atmospheric window. *Remote Sensing of Environment*, 42, 83–106.

Salisbury, J.W. and Wald, A. (1992) The role of volume scattering in reducing spectral contrast of reststrahlen bands in spectra of powdered minerals. *Icarus*, 96, 121–128.

Salisbury, J.W. and Walter, L.S. (1989) Thermal infrared (2.5–13.5 μm) spectroscopic remote sensing of igneous rock types on particulate planetary surfaces. *Journal of Geophysical Research*, 94, 9192–9202.

Saunders, R.S., Pettengill, G.H., Arvidson, R.E. *et al.* (1990) The *Magellan* Venus radar mapping mission. *Journal of Geophysical Research*, 95 (B6), 8339–8355. DOI: 10.1029/JB095iB06p08339.

Schaber, G.G., Strom, G.H., Moore, H.J. *et al.* (1992) Geology and distribution of impact craters on Venus: what are they telling us? *Journal of Geophysical Research*, 97 (E8), 13257–13301.

Schmidt, R.M. and Housen, K.R. (1987) Some recent advances in the scaling of impact and explosion cratering. *International Journal of Impact Engineering*, 5, 543–560.

Shoemaker, E.M. (1960) Impact mechanics at Meteor Crater, Arizona. PhD dissertation, Princeton University.

Shoemaker, E.M. and Kieffer, S.W. (1974) *Guidebook to the Geology of Meteor Crater, Arizona*, Arizona State University Center for Meteorite Studies Publication 17, Center for Meteorite Studies, Arizona State University, Tempe, AZ.

Smith, D.E., Zuber, M.T., Frey, H.V. *et al.* (2001) Mars Orbiter Laser Altimeter: experiment summary after the first year of global mapping of Mars. *Journal of Geophysical Research*, 106, 23689–23722. DOI: 10.1029/2000JE001364.

Smith, D.E., Zuber, M.T., Neumann, G.A. *et al.* (2010) Initial observations from the Lunar Orbiter Laser Altimeter (LOLA). *Geophysical Research Letters*, 37, L18204. DOI: 10.1029/2010GL043751.

Stewart, S.T. and Valiant, G.J. (2006) Martian subsurface properties and crater formation process inferred from fresh impact crater geometries. *Meteoritics*, 41, 1509–1537. DOI: 10.1111/j.1945-5100.2006.tb00433.x.

Thomson, J.L. and Salisbury, J.W. (1993) The mid-infrared reflectance of mineral mixtures (7–14 μm). *Remote Sensing of Environment*, 45, 1–13.

Thorsteinsson R. and Mayr U. (1987) *The Sedimentary Rocks of Devon Island, Canadian Arctic Archipelago*, Geological Survey of Canada Memoir 411, Geological Survey of Canada, Ottawa.

Tornabene, L.T. Moersch, J.E., Osinski, G.R. *et al.* (2005) Spaceborne visible and thermal infrared lithologic mapping of impact-exposed subsurface lithologies at the Haughton impact structure, Devon Island, Canadian High Arctic: applications to Mars. *Meteoritics*, 40 (12), 1835–1858.

Van Zyl, J.J. (2001) The Shuttle Radar Topography Mission (SRTM): a breakthrough in remote sensing of topography. *Acta Astronautica*, 48, 559–565.

Vincent, R.K. (1997) *Fundamentals of Geological and Environmental Remote Sensing*, Prentice Hall.

Webster, T.L., Murphy, J.B. and Gosse, J.C. (2006) Mapping subtle structures with light detection and ranging (LIDAR): flow units and phreatomagmatic rootless cones in the North Mountain Basalt, Nova Scotia. *Canadian Journal of Earth Sciences*, 43, 157–176.

Wood, C.A., Lorenz, R., Kirk, R. *et al.* (2010) Impact craters on Titan. *Icarus*, 206, 334–344. DOI: 10.1016/j.icarus.2009.08.021.

Woolard, J.W. and Colby, J.D. (2002) Spatial characterization, resolution, and volumetric change of coastal dunes using airborne LIDAR: Cape Hatteras, North Carolina. *Geomorphology*, 48, 269–287.

Wells, K.S., Campbell, D.B., Campbell, B.A. and Carter, L.M. (2010) Detection of small lunar secondary craters in circular polarization ratio radar images. *Journal of Geophysical Research*, 115, E06008. DOI: 10.1029/2009JE003491.

Wright, S.P. and Ramsey, M.S. (2006) Thermal infrared data analyses of Meteor Crater, Arizona: implications for Mars spaceborne data from the Thermal Emission Imaging System. *Journal of Geophysical Research*, 111, E02004. DOI: 10.1029/2005JE002472.

Wright, S.P., Christensen, P.R. and Sharp, T.G. (2011) Laboratory thermal emission spectroscopy of shocked basalt from Lonar Crater, India and implications for Mars orbital and sample data. *Journal of Geophysical Research*, 116, E09006. DOI: 10.1029/2010JE003785.

Zuber, M.T., Smith, D.E., Solomon, S.C. *et al.* (1992) The Mars Observer Laser Altimeter investigation. *Journal of Geophysical Research*, 97 (E5), 7781–7797. DOI: 10.1029/92JE00341.

Zumsprekel, H. and Bischoff, L. (2005) Remote sensing and GIS analyses of the Strangways impact structure, Northern Territory. *Australian Journal of Earth Science*, 52, 621–630.

# FOURTEEN

# Geophysical studies of impact craters

**Joanna Morgan* and Mario Rebolledo-Vieyra[†]**

*Department of Earth Science and Engineering, Imperial College London, South Kensington Campus, London, SW7 2AZ, UK
[†]Centro de Investigación Científica de Yucatán, A.C., CICY, Calle 43 No 130, 97200 Mérida, Mexico

## 14.1 Introduction

Our knowledge and understanding of terrestrial impact craters is obtained from geological outcrop, observations on other planetary bodies, drill holes, dynamic models of crater formation, small-scale laboratory experiments and geophysical data. We use geophysical methods to image beneath the Earth's surface and to obtain models of subsurface physical properties (e.g. seismic velocity, density, resistivity and magnetization). Changes in near-surface physical properties are good indicators of lithological changes and impact craters often have a clear geophysical signature. A thorough review of the geophysical anomalies commonly associated with impact craters can be found in Pilkington and Grieve (1992), Grieve and Pilkington (1996) and Grieve (2006). Table 14.1 summarizes their principal findings, includes some additional observations and provides information on the resolution of each geophysical method. Below, we describe the common geophysical signatures associated with impact craters, examine resolution for each technique, explain the basic principles involved in modelling geophysical data and present case histories from three craters: Araguainha, Chicxulub and Silverpit. We include the Araguainha Crater as novel resistivity images have recently been obtained across this crater (Fig. 14.1), which have implications for central uplift formation. Chicxulub was chosen as it has been the subject of a wide range of geophysical studies and illustrates that the availability of a range of different datasets, and higher resolution data in particular (Fig. 14.2a), can improve the imaging of crater features. Although the Silverpit structure is not a confirmed impact crater, it was selected owing to the unparalleled images that have been obtained across this structure using three-dimensional (3D) seismic reflection data. These data show us just how powerful geophysical imaging can be – albeit quite costly.

## 14.2 Geophysical signature of terrestrial impacts

Geophysical surveys have played a crucial role in the discovery of number of impact craters, including, for example, Chicxulub and Montagnais (Jansa et al., 1989; Hildebrand et al., 1991). They are successful because meteorite impacts cause changes within the target rocks that can be remotely sensed at surface (Table 14.1). The two most commonly used exploration techniques are gravity and magnetic (potential field) methods. Gravity data are sensitive to changes in near-surface density, with positive and negative gravity anomalies signifying higher and lower densities respectively, relative to a background value. Impact craters are associated with negative gravity anomalies that are typically circular and extend out to, or slightly beyond, the crater rim (e.g. Grieve, 2006). The density reduction is principally caused by fracturing and brecciation of the target rocks, but contributions from low-density impact breccias and sedimentary infill are also possible (Grieve and Pilkington, 1996). For example, at Chicxulub, the deep-water marls and carbonates that gradually filled the impact basin have a much lower density than the adjacent platform carbonates and are responsible for about half the total gravity anomaly (Hildebrand et al., 1998). Large craters may have a central gravity high, which is produced by the structural uplift of denser material at the centre of the crater (e.g. at Manicouagan; Sweeney, 1978). At Chicxulub, there is also a circular gravity low that correlates with an inward-dipping zone of low-velocity rocks that form the topographic peak ring (Vermeesch and Morgan, 2004; Morgan et al., 2011; Fig. 14.2a). The total gravity anomaly produced by any target structure is independent of source depth (from Gauss's theorem); hence, integrals of the total gravity anomaly can be used for mass excess/deficit calculations. This means that gravity data can also be used to estimate the total excess or missing mass below surface

*Impact Cratering: Processes and Products*, First Edition. Edited by Gordon R. Osinski and Elisabetta Pierazzo.

**Table 14.1**   Geophysical anomalies associated with impact craters

| Method | Physical properties | Resolution[a] | Signature[b] | |
|---|---|---|---|---|
| Gravity | Density | Low (undefined) | Gravity low (negative) | Impact fracturing, basin infill, impact breccias, peak ring |
| | Mass excess/deficit | | Central high ($D > 30$ km) | Central uplift |
| Magnetics | Magnetic susceptibility | Low (undefined) | Magnetic low ($D < 40$ km) | Demagnetized target rocks |
| | Remanent magnetism | | Short $\lambda$ anomalies ($D > 10$ km) | Shock metamorphism, hydrothermal processes, thermoremanent magnetization (TRM) |
| | | | Long $\lambda$ anomaly ($D > 40$ km) | Central uplift |
| Direct resistivity | Resistivity | Low (undefined) | Low resistivity | Impact breccias, basin infill, fractured rocks |
| | | | High resistivity | Central uplift |
| Magnetotelluric | Resistivity | Med (~skin depth) | Low resistivity | Impact breccias, basin infill, fractured rocks |
| | | | High resistivity | Central uplift |
| Seismic refraction | Velocity | Med–high (~Fresnel zone) | Low velocity | Impact breccias/melts, basin infill, fractured rocks, peak ring |
| | | | High velocity | Central uplift, impact melts |
| Full wavefield refraction | Velocity, attenuation | High (~wavelength) | As above | As above |
| Seismic reflection | Acoustic impedance (velocity × density) | High (~wavelength) | Reflectivity | Target rocks, megablock zone, faulting, crater rim, basin infill, central peak, peak ring, central uplift, melt sheet, max. depth of disruption |
| | | | Loss of coherent reflectivity | Impact breccias, fractured/brecciated rocks |
| Ground-penetrating radar (GPR) | Electric and magnetic properties | High (~wavelength) | Reflectivity | Crater floor (small craters) |

[a]The dimensions of the smallest body that can be resolved with each geophysical method are shown in parentheses. Resolution can also be adversely affected by survey design, noise and modelling approach adopted.

[b]$D$ is crater diameter. A magnetic low means there is only a small magnetic anomaly (positive or negative), whereas a gravity low means that the gravity anomaly is negative.

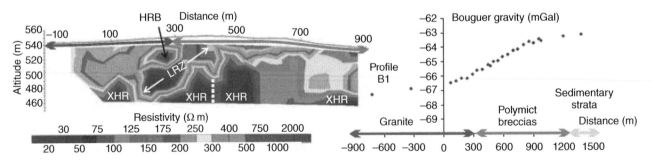

**Figure 14.1**   Left-hand figure is a tomographic model of electrical resistivity at Araguainha Crater (also see Plate 33); right-hand figure is Bouguer gravity and its relationship with surface outcrop (from Tong *et al.* (2010)). A low-resistivity zone (LRZ), which corresponds to polymict breccias at surface, dips beneath the granitic core of the central uplift. (See Colour Plate 33)

– see calculations in Campos-Enriquez *et al.* (1998) for the Chicxulub Crater.

Magnetic anomalies are observed when magnetic rocks are close to the Earth's surface. The anomalies are dipolar (have a positive and negative component) except near the poles and equator, and for any particular magnetic body the recorded anomaly changes with latitude. Magnetic properties in rocks can vary by orders of magnitude over short distances and may comprise both a permanent and induced magnetic component. These factors mean that producing good subsurface models of magnetic

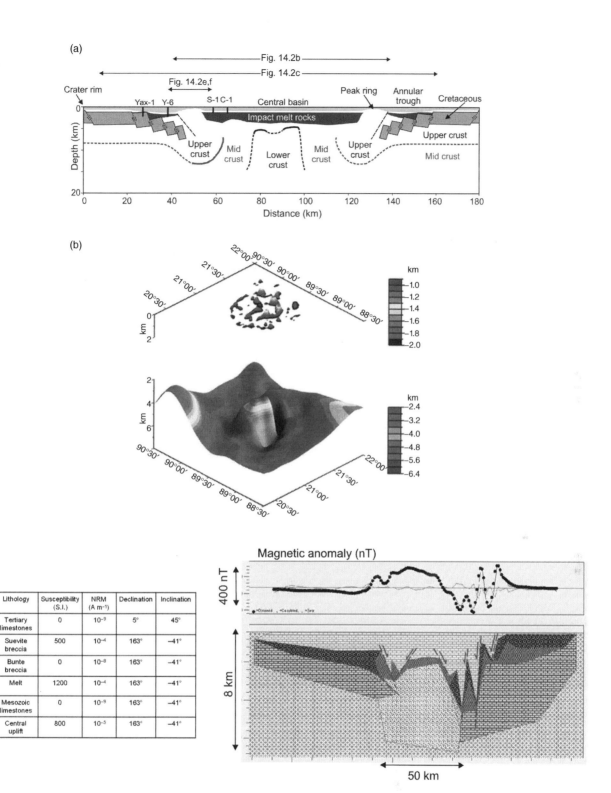

The table in part (c) reads:

| Legend | Lithology | Susceptibility (S.I.) | NRM (A m⁻¹) | Declination | Inclination |
|---|---|---|---|---|---|
| | Tertiary limestones | 0 | $10^{-9}$ | 5° | 45° |
| | Suevite breccia | 500 | $10^{-4}$ | 163° | −41° |
| | Bunte breccia | 0 | $10^{-8}$ | 163° | −41° |
| | Melt | 1200 | $10^{-4}$ | 163° | −41° |
| | Mesozoic limestones | 0 | $10^{-9}$ | 163° | −41° |
| | Central uplift | 800 | $10^{-5}$ | 163° | −41° |

**Figure 14.2** (a) Model of the Chicxulub Crater, which is consistent with drill core and geophysical data (from Vermeesch and Morgan (2008)). Seismic reflection data are used to locate the Cretaceous sediments, post-impact sediments and peak ring (see (e) and (f)). The location of the lower crust, mid crust and upper crust is derived from 3D tomographic velocity models obtained by jointly inverting gravity and travel-time data; solid lines show contours that are well constrained. S-1, C-1, Y-6 and Yax-1 are onshore drill holes. The nature of the rocks that form the peak ring are unknown. The total thickness and lateral extent of the melt sheet, as well as the structure of the zone of central uplift, are not well constrained. (b) Result from the inversion of magnetic data across Chicxulub (from Pilkington and Hildebrand (2000)). The inversion is constrained by imposing the assumption that the long-wavelength magnetic anomaly is produced by magnetized basement rocks and the short-wavelength features by magnetic bodies in the allogenic melt breccias. (c) A model of the Chicxulub Crater derived from modelling the magnetic field (from Rebolledo-Vieyra *et al.* (2010)). Profile runs from west to east through the crater centre. Rapid changes in magnetic anomaly are modelled by faulting in the central uplift.

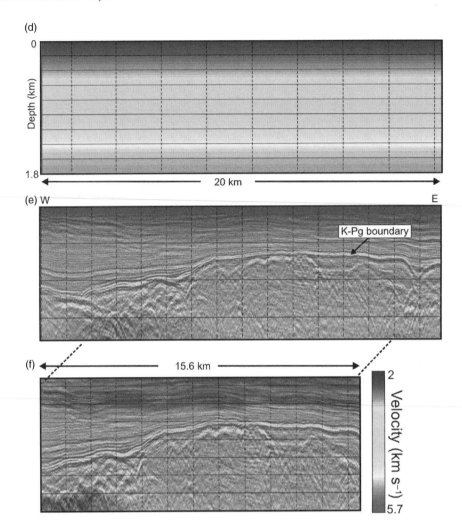

**Figure 14.2** (*Continued*) (d) Starting velocity model for travel-time inversions across the peak ring at Chicxulub (from Morgan *et al.* (2011)). (e) Tomographic velocity model (colour), obtained by inverting travel-time data along seismic profile 10 (acquired in 2005) using the FAST program (Zelt and Barton, 1998). Background is 2D seismic reflection data for the same profile, converted to depth using the tomographic velocity model (Morgan *et al.*, 2011). (f) Tomographic velocity model (colour), obtained by inverting for the full-seismic wavefield along profile 10 (Morgan *et al.*, 2011). Background is 2D seismic reflection data for the same profile, converted to depth using the tomographic velocity model. Model edges and depths below 1.6 km have been omitted as the model is considered to be poorly constrained in these areas. (See Colour Plate 34 for 14.2 b–f)

properties is more challenging than is the case for models of density. Impact craters are often characterized by a magnetic low (a magnetic anomaly close to zero; e.g. Clark, 1983) and this indicates that the rocks at the impact site have been demagnetized by the impact process. Larger craters may have short-wavelength magnetic anomalies in the centre of the crater (Table 14.1), and if the zone of central uplift is formed from magnetic basement material it will produce a magnetic anomaly with a longer wavelength. The cause of the short-wavelength anomalies is varied and can be produced by shock metamorphism, hydrothermal processes and/or post-impact cooling (TRM) of melts and impact breccias (e.g. Grieve and Pilkington, 1996).

Magnetic surveys measure the Earth's magnetic field over a region; this field is a vector quantity having both amplitude and direction. The orientation of the magnetic field is the result of the interaction between the geometry of the source and the orientation of the effective magnetic vector (Ugalde *et al.*, 2007). Commercial sensors (caesium vapour based) are incapable or detecting the orientation of the magnetic vector (Nabighian and Asten, 2002); therefore, only the amplitude of the magnetic vector is recorded, restricting the resolution of these data (Table 14.1). As only the scalar magnitude of the Earth's magnetic field is recorded, interpretation of magnetic data in areas of low magnetic latitudes is not straightforward (Ugalde *et al.*, 2007); however, the magnetic anomaly across impact craters yields information on the horizontal location, size and depth to magnetic sources. The wavelength of gravity and magnetic anomalies increases with increasing depth to source; hence, wavelength is a direct indicator of the maximum depth of burial (it is a maximum depth because many models with shallower changes in physical properties can also explain the same potential field data). High gradients (rapid changes) in potential field data are observed when changes in density or magnetic properties are close to surface and can be used to locate sub-vertical boundaries when the geology is relatively simple (i.e. there are not multiple sources for the recorded anomalies at the surface). At the Bosumtwi structure, two aeromagnetic surveys were conducted: a medium-resolution survey with a flight altitude of 300 m and line spacing of 500 m; and a second, high-resolution survey flown at low altitude (70 m) (Ugalde *et al.*, 2007). In the high-resolution surveys, short wavelength anomalies (<1 km) were detected and used to locate the position and depth of magnetic sources in the rocks that form the crater floor.

Resistivity and magnetotelluric surveys are sensitive to the electrical properties of rocks and are used to model subsurface resistivity. Resistivity usually decreases with increasing porosity, as electric current flows more easily in the saline water that typically fills the pore space and cracks. Thus, at impact craters, resistivity is usually lower in fractured or brecciated rocks, allocthonous deposits and porous sedimentary infill and higher in less-fractured rocks and deeper parts of the central uplift (Table 14.1). For example, measurements in the Nördlingen borehole at the Ries Crater show resistivity values of approximately 5 Ω m in the post-impact sediments, rising to approximately 100 Ω m in the impact breccias and fractured basement (Pohl *et al.*, 1977), whereas dry rocks typically have resistivities of greater than 1000 Ω m.

Seismic refraction surveys are used to obtain models of subsurface velocity. Seismic velocity is often reduced at impact craters (e.g. Pilkington and Grieve, 1992) and usually correlates with zones of increasing porosity, decreasing density and decreasing resistivity – as observed in well logs in the Nördlingen borehole at the Ries Crater (Pohl *et al.*, 1977). In larger craters, high velocities may be observed within the zone of central uplift (e.g. at Vredefort; Green and Chetty, 1990), although the velocity of these uplifted rocks might be slightly reduced by impact-induced fracturing and brecciation (e.g. Grieve and Pilkington, 1996). Impact melt may have velocities that are higher (e.g. at Chicxulub; Barton *et al.*, 2010) or lower than the surroundings rocks, depending on the target rocks at the impact site. Within the differentiated melt sheet at the Sudbury impact crater, the granophyre and quartz gabbro have relatively low average velocities of 6.1–6.2 km s$^{-1}$, whereas, at the base of the melt sheet, the norite has a velocity of 6.6–6.7 km s$^{-1}$, which is comparable to velocities within the footwall rocks (Salisbury *et al.*, 1994).

Seismic reflection surveys are used to provide detailed maps of subsurface structure. Seismic waves are reflected from subsurface boundaries (reflectors) across which there is a change in acoustic impedance (velocity times density). A variety of crater features have been imaged with seismic reflection data (Table 14.1), including coherent reflections in targets rocks that become increasingly disturbed when tracked towards the crater centre, faulted blocks of target rocks that are downthrown and/or rotated in the terrace or megablock zone (e.g. at Chicxulub; Morgan *et al.*, 1997) and uplifted rocks in the crater centre (e.g. at Mjolnir; Tsikalas *et al.*, 1998). Seismic reflection surveys across craters that form lakes and craters on continental shelves may be characterized by layered coherent reflections in the post-impact sediments above a more chaotic or reflector-free zone in the fractured, brecciated or allocthonous crater deposits. For example, the bases of the reflective post-impact sediments at the Bosumtwi and Chicxulub Craters show clear images of the morphology of a central peak and peak ring respectively (Bell *et al.*, 2004; Scholtz *et al.*, 2007). Other features imaged by seismic surveys are the crater rim (e.g. at Chesapeake; Poag *et al.*, 1994) and the maximum depth of disruption beneath the crater (e.g. at Red Wing; Brenan *et al.*, 1975).

Ground-penetrating radar (GPR) surveys are also used to map subsurface structure, although their shallow penetration means this technique is not widely used in studies of impact craters. In GPR surveys, electromagnetic waves are reflected from subsurface boundaries across which there is a change in electrical and magnetic properties. GPR surveys were recently used to locate the floor of the Kamil Crater in Egypt (Folco *et al.*, 2010).

## 14.3 The resolution of geophysical data

Resolution is dependent upon the background physics that underpins the geophysical method. When we use the word resolution, we usually mean 'What is the smallest body that we can resolve with a particular geophysical technique?' The resolution for different geophysical methods is shown in Table 14.1, with the size of the smallest body that can be resolved shown in parentheses. Resolution is also dependent on the quality of data acquired and may be adversely affected by noise and poor data coverage, as well as any assumptions and approximations that have been adopted when modelling and inverting these geophysical data. In all cases, resolution gets worse with burial depth, which means that finer details can be resolved when the crater features of interest are closer to surface.

Resistivity, gravity and magnetic methods are all governed by Laplace's equation and their resolution is undefined. This means that it is not possible to resolve a subsurface body of any size with these data and all geophysical models produced using these data suffer severely from non-uniqueness – many models will fit the data equally as well. Hence, although potential field data are excellent reconnaissance techniques for discovering impact craters, any subsurface models derived from these data alone can only be considered as first-order approximations.

Magnetotelluric techniques are controlled by the diffusion equation, and the resolution is approximately equal to the skin depth: 500 × (resistivity/frequency)$^{1/2}$ (Sharma, 1997). This typically leads to resolutions of several hundred metres to a few kilometres in the upper crust. The depth of penetration is increased by decreasing the source frequency, although lowering the frequency has the disadvantage of leading to a poorer resolution. Thus, these data can be used to identify broad features; for example, magnetotelluric data were able to detect the thickening of the Tertiary basin and zone of central uplift at Chicxulub (Unsworth *et al.*, 2002).

In seismic refraction surveys, denser sampling leads to significant improvements in resolution. For velocity models derived from inverting travel times (travel-time tomography) the resolution is approximately equal to the Fresnel zone: ~(depth × wavelength/2)$^{1/2}$; this typically varies from a few hundred metres near the surface to several kilometres at the base of the crust. Resolution is also controlled, however, by the shot–receiver spacing, with resolution being approximately equal to half the shot or receiver spacing – whichever is the largest (Zelt, 1998). This means that, if either the shots or receivers are a long distance apart, the resolution may be worse than estimated using the Fresnel zone calculation. For example, much of the Ries Crater refraction data were acquired with a single shot into multiple receivers or as an expanding shot–receiver spread with a common midpoint (Pohl and Will, 1974; Angenheister and Pohl, 1969) and thus have no easily defined resolution. Line 11, acquired across the central crater region, had shots located at each end of an approximately 10 km long line of receivers – hence the resolution is approximately 5 km. The limited resolution means that these seismic data are consistent with either a high- or low-velocity zone beneath the central crater, leading to contrasting interpretations on whether there is a zone of central uplift at the Ries Crater (Ernstson and Pohl, 1977; Wünnemann *et al.*, 2005). Resolution tests show that 3D refraction datasets with regular grid spacing lead to the most reliable subsurface velocity models (Zelt, 1998).

The techniques with the best resolution are those that are governed by the wave equation; for example, seismic refraction (when the wavefield is modelled and not just the travel time), seismic reflection and GPR. Full wavefield inversions have a potential resolution close to the seismic wavelength (velocity that the wave is travelling at divided by its frequency), which is an improvement of around an order of magnitude over travel-time tomography (Pratt, 1999). Compare, for example, the velocity models obtained from inverting travel times (Fig. 14.2e) and the full-wavefield (Fig. 14.2f) across the peak ring at Chicxulub. In the model derived from travel times, velocities are smoothed over several hundred metres, whereas the full wavefield model contains finer detail – for example, the 100–200 m thick low-velocity zone at the top of the peak ring. Within the peak ring, the velocity model obtained from travel-time tomography has a resolution of approximately 400 m, whereas the model derived using full-wavefield tomography has a resolution of approximately 100 m (Morgan *et al.*, 2011).

In ideal conditions seismic reflection and GPR methods can resolve subsurface structures that are of the order of the seismic or electromagnetic wavelength. In GPR surveys we use source frequency to control depth of penetration – as we decrease source frequency, the signal strength remains high at deeper depths below surface and we can detect more deeply buried reflectors. The resolution, however, gets worse with decreasing frequency. If we wish to image the uppermost metre or so we can use a high source frequency of a few hundred megahertz and obtain a resolution of a few millimetres. When we address resolution in seismic reflection data, we usually mean vertical resolution. Vertical resolution is the smallest distance two reflectors can be apart, but still be seen as two separate reflectors: approximately wavelength/4 (e.g. Berkhout, 1984). In industry-standard 3D reflection data, this corresponds to approximately 10 m in near-surface sediments and approximately 40 m in basement rocks. The fidelity of seismic reflection data is best when the target structure is relatively shallow, the overburden is geologically simple, the acquisition is 3D (adjacent profiles are ≤100 m apart) and the acquired data have a high frequency content (>40 Hz). Grids of two-dimensional (2D) seismic reflection profiles spaced a few hundred metres apart can provide good images of the subsurface, but may be adversely affected by spatial aliasing and lack the absolute accuracy in reflector positioning than can be achieved by migrating the data in 3D (Yilmaz, 2001). Marine seismic reflection data are typically of better quality than land data are, which is often adversely affected by rapid changes in near-surface properties, source variations or through the difficulties of obtaining well-sampled 3D data in an urban setting, or regions of high vegetation or rough topography.

## 14.4    Modelling geophysical data

Given the limitations in resolution and problems with non-uniqueness when modelling geophysical data, the recommended approach is to: (1) explore the parameter space to find a range of models that fit these geophysical data equally as well; (2) carry out resolution tests to determine which parts of the model are well constrained; (3) model or jointly invert more than one dataset to obtain a subsurface model that is consistent with two or more independent datasets; and (4) use physical property measurements from wells or local outcrop to more narrowly constrain the range of models that fit the data. When modelling geophysical data we try to find a model that is consistent with geological observations and expectations. In the case that many models are consistent with the data, best practice is to choose the simplest subsurface model that fits the data reasonably well (to within the level of uncertainty in the data). In accordance with Occam's principle, we should try to avoid introducing subsurface structure unless justified by other *a priori* constraints.

In geophysical modelling we would ideally like to use a method that is as objective as possible. This can only really be achieved through using some sort of inversion program, and works best for geophysical techniques with good resolution. In a forward modelling approach the user updates the model of the subsurface physical properties until the fit between the observed and calculated data is good. Such a trial-and-error approach is unavoidably subjective and it is extremely unlikely that a user will find the entire range of subsurface models that can reproduce the data. One way of dealing with non-uniqueness is to use a method that performs a wide search of the parameter space; for example, a Monte Carlo approach (Sambridge and Mosegaard, 2002). Many thousands of subsurface models are tested and only those that fit the data reasonably well are retained. It is possible to invert geophysical data that are inherently non-unique by imposing constraints on the derived models. For example, in Pilkington and Hildebrand's (2000) model of the Chicxulub Crater, the inversion is constrained by imposing the assumption that the long-wavelength magnetic anomaly is produced by magnetized basement rocks and the short-wavelength features by magnetic bodies in the allogenic melt breccias (Fig. 14.2b). In such a case, the accuracy of the subsurface model is dependent on the validity of the assumptions imposed.

A useful tool before modelling, or even acquiring, geophysical data is to perform some sort of sensitivity test. In such tests, a model of the predicted subsurface structure is used to generate a synthetic geophysical dataset, and then the model is perturbed to see how much the perturbation affects the generated data. These tests allow the user to discover whether they can resolve the target structure with the acquired data and see how sensitive the data are to the subsurface properties that are of interest. The more sensitive the data to subsurface changes in physical properties, the better the data are able to resolve subsurface structure. Synthetic tests are also useful for assessing the reliability of derived models. For example, Christeson *et al.* (2001) investigated the resolution of 2D velocity models across the central crater at Chicxulub to see how well a zone of central uplift could be resolved with the acquired data. Chequerboard testing is another common method for assessing geophysical models, although they only work for geophysical methods with good resolution (e.g. the 3D tomographic velocity model at Chicxulub; Morgan *et al.*, 2002). A chequerboard is added to the final model and a tomographic inversion run to recover the chequerboard. Regions of the model where the chequerboard is recovered are considered well resolved. The size of the chequerboard is gradually reduced until it cannot be recovered – this tells us the minimum size of a body than can be resolved with the acquired data.

Where multiple geophysical datasets have been acquired across the target structure, the aim is to find a subsurface model that fits all data. Ideally, this would be carried out using a joint inversion; that is, having one inversion program search for a model that fits both datasets. For example, seismic travel-time and gravity data were jointly inverted across the central uplift at Chicxulub (Vermeesch *et al.*, 2009). Whereas density and velocity are typically correlated, and the relationship between density and velocity can be estimated for particular rock types, this is less true for other physical properties, such as resistivity and magnetization. Hence, joint inversions remain a challenge, with the best approach being perhaps a statistical one; for example, as used by the oil industry in developing rock physics models (Avseth *et al.*, 2005). A common approach, where multiple datasets are involved, is to use the higher resolution dataset to guide the construction of subsurface models for the lower resolution dataset. For example, Karp *et al.* (2002) use the reflection data across the central crater at Bosumtwi to locate subsurface boundaries between the water, sediments and breccias when they model their 2D refraction data.

Whenever possible, direct measurements of petrophysical properties (magnetic susceptibility, natural remanent magnetization (NRM), velocity, density, etc.) should be used to guide the development of geophysical models. Parameter variation can be restricted so that the physical properties within each layer are consistent with the measured petrophysical properties. Geological observations in boreholes can be used to constrain the depth and thickness of particular geological layers in the derived geophysical models.

## 14.5 Case studies

### 14.5.1 Araguainha

The Araguainha impact structure in central Brazil is an approximately 245 Ma complex crater with a central uplift of approximately 12 km in diameter. Geological mapping of the crater has shown that the central uplift is composed of a granitic core surrounded by younger, steeply dipping sediments and that the boundary between the two is discordant (Lana *et al.*, 2007). The

contact between the core and sediment is occasionally covered by polymict and monomict impact breccias. Some of the sediments are overturned, and duplications within the sedimentary layer are responsible for the thickening of these strata (Lana *et al.*, 2008). A suite of resistivity and gravity data were acquired across the central crater with the aim of obtaining a geological model of the structure below the surface (Tong *et al.*, 2010). These data revealed some intriguing resistivity anomalies, most notably a low-resistivity zone (LRZ) dipping underneath the granitic core which, at the surface, corresponds to the impact breccias (Fig. 14.1). If impact breccias do lie beneath the central uplift at Araguainha, then this is the first such observation at a terrestrial crater, and has implications for central crater formation. Tong *et al.* (2010) postulated that there was a resurge of impact breccias during crater formation and that the resurge was influenced by crater floor topography and the geometry of the sedimentary and granitic units. The Araguainha impact structure has a crater diameter of 40 km and, hence, might be expected to have a central gravity high (Table 14.1). The observation that the granitic core is associated with a gravity low (Fig. 14.1) indicates that the central uplift has been strongly fractured (Tong *et al.*, 2010).

### 14.5.2 Chicxulub

Potential field data were central to the discovery of the Chicxulub Crater (Penfield and Camargo-Zanoguera, 1981; Hildebrand *et al.*, 1991). At Chicxulub there is an approximately 180 km circular gravity low, an approximately 50 km wide central gravity high and an approximately 100 km wide magnetic anomaly that has a semicircular shape with a large-amplitude central anomaly surrounded by smaller reversed dipolar anomalies (e.g. Pilkington *et al.*, 1994; Ortiz-Aleman *et al.*, 2001). The crater is buried within a platform carbonate sequence in the northwestern Yucatán peninsula, Mexico; half the crater is onshore and the other half is offshore in shallow water. Initial stratigraphic information came mainly from PeMex (Petróleos Mexicanos) boreholes and regional models (e.g. Cornejo-Toledo and Hernandez-Osuna, 1950; López-Ramos, 1973, 1983; Weidie, 1985). Potential field and PeMex borehole data were used to derive models of Chicxulub that were distinctly different in diameter and structure (e.g. Hildebrand *et al.*, 1991; Sharpton *et al.*, 1996; Urrutia-Fucugauchi *et al.*, 1996). Thus, Chicxulub is one example that illustrates how useful potential field data are for reconnaissance, as well as the problem of inherent non-uniqueness when modelling these data.

Since its discovery, a wealth of geophysical data has been acquired across Chicxulub, including additional magnetic and gravity data, a grid of marine 2D seismic reflection profiles, 3D wide-angle refraction and magnetotelluric data (Camargo-Zanoguera and Suárez-Reynoso, 1994; Hildebrand *et al.*, 1998; Pilkington and Hildebrand, 2000; Unsworth *et al.*, 2002; Gulick *et al.*, 2008; Vermeesch and Morgan, 2008; Rebolledo-Vieyra *et al.*, 2010). In addition, scientific drilling by UNAM (Universidad National Autónoma de México) and ICDP (International Continental Drilling Program) has provided us with more information on stratigraphy, subsurface lithology and the physical properties of each rock type (e.g. Urrutia-Fucugauchi *et al.*, 1996, 2004; Vermeesch and Morgan, 2004; Rebolledo-Vieyra and

Urrutia-Fucugauchi, 2006; Mayr *et al.*, 2008). These data greatly aid the development of geophysical models because, as mentioned above, they allow realistic physical properties to be assigned to subsurface geological layers in models of the crater.

Over the last approximately 20 years geophysical models of the subsurface have been developed, refined and updated. The use of higher resolution 2D seismic reflection and 3D seismic refraction techniques have allowed some aspects of crater structure to be reasonably well agreed; for example, that the crater is 180–200 km in diameter and that it possesses a peak ring, melt sheet and a zone of central uplift (Fig. 14.2a). The detailed structure of some crater features remains unclear, and interpretations of the total thickness and lateral extent of the melt sheet, as well as its relationship with the zone of central uplift, differs between competing geophysical models of the crater (e.g. Sharpton *et al.*, 1996; Campos-Enriquez *et al.*, 1998; Hildebrand *et al.*, 1998; Pilkington and Hildebrand, 2000; Ortiz-Aleman *et al.*, 2001; Vermeesch and Morgan, 2008; Rebolledo-Vieyra *et al.*, 2010). One outstanding question has arisen from these geophysical studies: What is the nature (lithology, depth of origin and physical state) of the rocks that form the peak ring? This question will be addressed if the proposed joint IODP–ICDP drilling at Chicxulub goes ahead as planned in 2014.

The magnetic data have been particularly useful for investigating the structure of the central uplift and its interaction with the allochthonous impact melts and breccias. At Chicxulub, the magnetic gradient between the target rocks (Mesozoic limestones with magnetic susceptibilities in the diamagnetic range, $\kappa \approx 0$–30 SI) and the impact rocks ($\kappa \approx 155$–4820 SI) (Rebolledo-Vieyra and Urrutia-Fucugauchi, 2004) leads to a very clear magnetic anomaly that can be used to estimate the horizontal location, size and depth of the magnetic sources.

Pilkington and Hildebrand (2000) modelled the magnetic data by separating out the long- and short-wavelength anomalies, on the assumption that they have different sources: the central uplift and impact breccias respectively. The inversion was constrained by imposing the condition that short-wavelength anomalies are produced by shallow magnetic sources with an average layer depth of 2 km and the long-wavelength anomalies by topography on a magnetic basement with an average layer depth of 5 km. In their selected inverted model (Fig. 14.2b), short-wavelength magnetic anomalies were modelled by small bodies within the allochthonous breccias and the long-wavelength magnetic anomaly is reproduced by a zone of central uplift in the middle of the crater. An increase/decrease in assigned magnetization or average layer depth would change the relief on these features.

Additional measurements of NRM and magnetic susceptibility were performed on Chicxulub core samples, and new aeromagnetic data were acquired at a flight altitude of 450 m and digitized on a $1 \times 1$ km$^2$ grid (Rebolledo-Vieyra *et al.*, 2010). Suevitic breccias had a mean magnetic susceptibility of $229 \times 10^{-5}$ SI, with values as high as $1200 \times 10^{-5}$ SI, while individual clasts of impact melt and crystalline basement had susceptibilities of $500 \times 10^{-5}$ SI and $400 \times 10^{-5}$ SI respectively. All other crater lithologies had low susceptibility values and were found to be within the diamagnetic range. A radially averaged power spectrum was used to estimate the depth to magnetic source, and found to be approximately 6000 m within the centre of the structure and approximately

500 m towards the edge of the anomaly. These new data were used to develop a model of Chicxulub, with a system of sub-vertical faults across the central uplift and terrace zone (Fig. 14.2c; Rebolledo-Vieyra *et al.*, 2010). Faulting is suggested by lineaments in the magnetic data and reinforced by rapid changes in the magnetic anomaly – which are apparent in the plot of the first derivative of the magnetic anomaly. The model shown in Fig. 14.2c utilizes the most extensive dataset of magnetic properties of crater lithologies to date, incorporates *a priori* data, such as depth to source, and considers previous models of the magnetic data (Pilkington *et al.*, 1994; Pilkington and Hildebrand, 2000; Ortiz-Aleman *et al.*, 2001).

In 2005, a second suite of 2D seismic reflection and 3D refraction data was acquired across the crater and 3D tomographic inversions of the travel time and gravity data provided improved velocity models across the central crater (Vermeesch and Morgan, 2008; Vermeesch *et al.*, 2009; Christeson *et al.*, 2009; Barton *et al.*, 2010). As discussed in Section 14.3, higher resolution imaging techniques allow us to map finer scale crater features. High-resolution velocity models across the peak ring were obtained using 2D full-wavefield inversions of seismic data recorded on a 6 km multichannel streamer (Fig. 14.2f; Morgan *et al.*, 2011). These models confirm that the peak ring is formed from low-velocity rocks and show, for the first time, that there is a strong velocity inversion between the lowermost Palaeocene and the top of the peak ring. The uppermost peak ring is formed from a 100–200 m thick layer of low-velocity rocks (Fig. 14.2f). The rest of the peak-ring is formed from rocks with velocities of between 3900 and 4500 m s$^{-1}$. If the material that forms the bulk of the peak ring originates from overturned sediments and basement, as predicted by some numerical models (e.g. Collins *et al.*, 2002; Ivanov, 2005; Senft and Stewart, 2009), then they have a much lower velocity than expected (see Chapter 4). Both the Cretaceous sediment and basement rocks in this region have average velocities of greater than 5 km s$^{-1}$ (Christeson *et al.*, 2001; Vermeesch and Morgan, 2004; Mayr *et al.*, 2008).

### 14.5.3   Silverpit, North Sea

In terms of financial cost, more funds are spent on the acquisition and processing of 3D seismic reflection data in the search for hydrocarbons than on any other geophysical method. This technique is unparalleled in its ability to obtain a high-fidelity structural image of the near subsurface. Silverpit was discovered by chance, using 3D seismic reflection data acquired by Western Geophysical on behalf of Arco International Oil and Gas Company (Stewart and Allen, 2002). The structure is located approximately 130 km off the east coast of England in the North Sea, is about 20 km in diameter and buried approximately 600 m below surface. Silverpit is probably an impact crater, as other explanations for the structure, such as salt withdrawal (Thomson, 2004; Underhill, 2004), are inconsistent with the observation of structural uplift at the centre of the feature, the age of the structure and the circular deformation pattern (Stewart and Allen, 2004; Wall *et al.*, 2008). Industry-quality 3D seismic reflection data have exceptional spatial resolution, and this means that 3D geometries and structures can be mapped across the crater and placed in a structural and stratigraphic context. Drill holes through the Silverpit

(a)

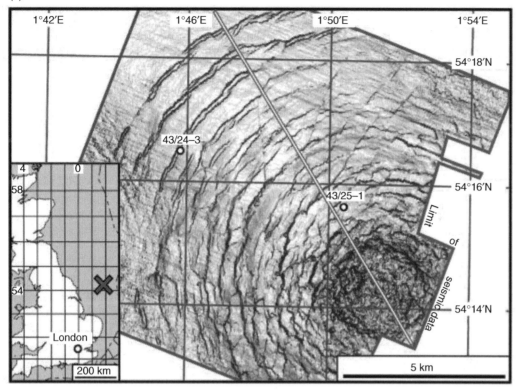

(b)

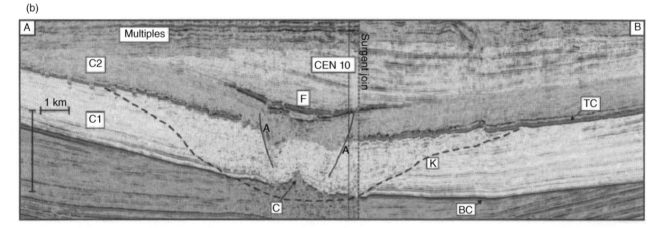

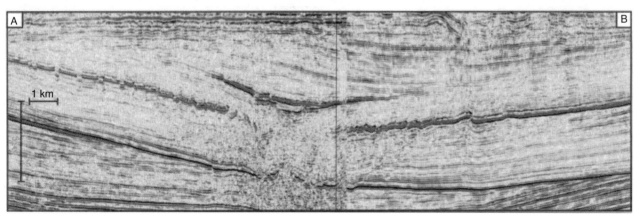

**Figure 14.3** (a) The topography of a single horizon at Silverpit (the reflection labelled TC in (b)), that lies a few hundred metres below the seabed in the North Sea, UK – see inset for location (from Stewart and Allen (2002)). Local drill holes (open circles) are used to interpret reflection TC as the top layer of Cretaceous chalk. Offsets in the horizon are observed as continuous rings that are shown to be extensional fault-bounded grabens in (b). (b) A 2D vertical time slice through the 3D seismic reflection volume that crosses through the centre of the Silverpit structure (from Wall *et al.* (2008)). The horizontal bar is 1 km and the vertical bar is 0.5 s two-way travel-time (~1 km). The Silverpit structure forms a depression in Cretaceous and Palaeogene reflections (packages C1 and C2); reflections exhibit a parallel onlap fill onto the bowl-shaped crater floor (package F). (See Colour Plate 35)

structure have been used to determine the lithology and age of individual reflections (horizons) and these horizons can be tracked across the crater. Figure 14.3a (also see Plate 35) is a plot of a single horizon – the reflection identified as top of Cretaceous chalk. Offsets in this horizon form rings that are traceable all around the crater and are shown to be fault-bounded grabens on 2D vertical time slices (Fig. 14.3b). Stewart and Allen (2002) proposed that these grabens were produced by extension as the chalk slumped inwards towards the excavation cavity.

Two factors made Silverpit quite special: this was potentially the first terrestrial impact crater that appeared to have the same morphology as multi-ring basins on icy bodies in our Solar System and the proposed age of between 60 and 65 Ma (Stewart and Allen, 2002) raised the possibility of multiple impacts at the K–Pg boundary. Individual rings in multi-ring impact basins on Jupiter's moons (e.g. Valhalla, Callisto) are closely spaced ridges and grabens, with inward- and outward-facing asymmetric scarps (see Chapter 1). Silverpit has since been interpreted as mid-Eocene in age (Wall *et al.*, 2008), which, if correct, may mean that rings were never a topographic feature of this crater. Grabens are absent or less topographically pronounced in the suite of reflections between the top Cretaceous reflector and the mid-Eocene reflection that is interpreted as the post-impact surface by Wall *et al.* (2008) (Fig. 14.3b).

The Silverpit seismic data contain a number of features that are common in impact craters. Reflections become increasingly disturbed towards the crater centre and are down-dropped to form a terrace zone – see offsets in package C1 (yellow) in Figure 14.3b. The middle of the crater shows a central uplift – although the reflectors are more disturbed here, they do show a decrease in travel time beneath the centre of the crater (see horizon BC in Figure 14.3b). Finally, there appears to be a maximum zone of disturbance beneath the crater – the deepest Triassic reflectors are sub-horizontal and continuous beneath the crater.

The spectacular high-fidelity images of the Silverpit structure illustrate the fine structural detail that can be obtained across an impact crater, if industry-quality 3D reflection data are available.

# References

Angenheister, G. and Pohl, J. (1969) Die seismischen Messungen im Ries von 1948–1969. *Geologica Bavarica*, 61, 304–326.

Avseth, P., Mukerji, T. and Mavko, G. (2005) *Quantitative Seismic Interpretation: Applying Rock Physics Tools to Reduce Interpretation Risk*, Cambridge University Press, Cambridge.

Barton, P.J., Grieve, R.A.F., Morgan, J.V. *et al.* (2010) Seismic images of Chicxulub impact melt sheet and comparison with the Sudbury structure, in *Large Meteorite Impacts and Planetary Evolution IV* (eds R.L. Gibson and W.L. Reimold), Geological Society of America Special Paper 465, Geological Society of America, Boulder, CO, pp. 103–113.

Bell, C., Morgan, J.V., Hampson, G.J. and Trudgill, B. (2004) Stratigraphic and sedimentological observations from seismic data across the Chicxulub impact basin. *Meteoritics and Planetary Science*, 39, 1089–1098.

Berkhout, A.J. (1984) *Seismic Resolution: A Quantitative Analysis of Resolving Power of Acoustical Echo Techniques*, Handbook of Geophysical Exploration vol. 12, Geophysical Press, London.

Brenan, R.L., Petreson, B.L. and Smith, H.J. (1975) The origin of Red Wing Creek structure: McKenzie county, North Dakota. *Earth Science Bulletin (Wyoming)*, 8, 1–41.

Camargo-Zanoguera, A. and Suárez-Reynoso, G. (1994) Evidencia sísmica del cráter impacto de Chicxulub. *Boletín de la Asociación Mexicana de Geofísicos de Exploración*, 34, 1–28.

Campos-Enriquez, J.O., Morales-Rodriguez, H.F., Dominguez-Mendez, F. and Birch, F.S. (1998) Gauss's theorem, mass deficiency at Chicxulub Crater (Yucatán, Mexico), and the extinction of the dinosaurs. *Geophysics*, 63, 1585–1594.

Christeson, G.L., Nakamura, Y., Buffler, R.T. *et al.* (2001) Deep crustal structure of the Chicxulub impact crater. *Journal Geophysical Research*, 106, 21751–21769.

Christeson, G., Collins, G., Morgan, J. *et al.* (2009) Mantle topography beneath the Chicxulub impact crater. *Earth and Planetary Science Letters*, 284, 249–257.

Clark, J.F. (1983) Magnetic survey data at meteoritic impact sites in North America. Geomagnetic Service of Canada, Earth Physics Branch open file 83-5, pp. 1–32.

Collins, G.S., Melosh, H.J., Morgan, J.V. and Warner, M.R. (2002) Hydrocode simulations of Chicxulub Crater collapse and peak-ring formation. *Icarus*, 157, 24–33.

Cornejo-Toledo, A. and Hernandez-Osuna, A. (1950) Las anomalias gravimetricas en la Cuenca salina del istmo, planicie costera de Tabasco, Campeche y Peninsula de Yucatán. *Boletín del la Asociación Mexicana de Geológos Petroleros*, 2, 453–460.

Ernston, K. and Pohl, J. (1977) Neue Modelle zur Verteilung der Dichte und Geschwindigkeit im Ries-Krater. *Geologica Bavarica*, 75, 355–371.

Folco, L., Di Martino, M., El Barkooky, A. *et al.* (2010) The Kamil Crater in Egypt. *Science*, 329, 804.

Green, R.W. and Chetty, P. (1990) Seismic refraction studies in the basement of the Vredefort impact structure. *Tectonophysics*, 171, 105–113.

Grieve, R.A.F. (2006) *Impact Structures in Canada*, Geological Association of Canada, St John's.

Grieve, R.A.F. and Pilkington, M. (1996) The signature of terrestrial impacts. *AGSO Journal of Australian Geology and Geophysics*, 16, 399–420.

Gulick, S.P.S., Barton, P.J., Christeson, G.L. *et al.* (2008) Importance of pre-impact crustal structure for the asymmetry of the Chicxulub impact crater. *Nature Geoscience*, 1, 131–135.

Hildebrand, A.R., Penfield, G.T., Kring, D.A. *et al.* (1991) A possible Cretaceous–Tertiary boundary impact crater on the Yucatán Peninsula, Mexico. *Geology*, 19, 867–871.

Hildebrand, A.R., Pilkington, M., Ortiz-Aleman C. *et al.* (1998) Mapping Chicxulub Crater structure with gravity and seismic reflection data, in *Meteorites: Flux with Time and Impact Effects* (eds M.M. Grady, R. Hutchinson, G.J.H. McCall and D.A. Rothery), Geological Society of London Special Publication 140, The Geological Society, London, pp. 153–173.

Ivanov, B.A. (2005) Numerical modeling of the largest terrestrial meteorite craters. *Solar System Research*, 39, 381–409.

Jansa, L.F., Pe-Piper, G., Robertson, P.B. and Freidenreich, O. (1989) Montagnais: a submarine impact structure on the Scotian Shelf, eastern Canada. *Geological Society of America Bulletin*, 101, 450–463.

Karp, T., Milkereit, B., Janle, P. *et al.* (2002) Seismic investigation of the Lake Bosumtwi impact crater: preliminary results. *Planetary and Space Science*, 50, 735–742.

Lana, C., Filho, C.R.S., Marangoni, Y.R. *et al.* (2007) Insights into the morphology, geometry, and post-impact erosion of the Araguainha peak-ring structure, central Brazil. *Geological Society of America Bulletin*, 119, 1135–1150.

Lana, C., Filho, C.R.S., Marangoni, Y.R. *et al.* (2008) Structural evolution of the 40-km wide Araguainha impact structure, central Brazil. *Meteoritics and Planetary Science*, 43, 701–716.

López-Ramos, E. (1973) Estudio geológico de la Península de Yucatán. *Boletín Asociación Mexicana de Geólogos Petroleros*, 25, 23–76.

López-Ramos, E. (1983) *Geología de México*, vol. 3, 3a edición, Universidad Nacional Autónoma de México, México D.F.

Mayr, S.I., Wittmann, A., Burkhardt, H. *et al.* (2008) Integrated interpretation of physical properties of rocks of the borehole Yaxcopoil-1 (Chicxulub impact structure). *Journal of Geophysical Research*, 113, B07201. DOI:10.1029/2007JB005420.

Morgan, J.V., Warner, M.R. and Chicxulub Working Group (1997) Size and morphology of the Chicxulub impact crater. *Nature*, 390, 472–476.

Morgan, J.V., Christeson, G. and Zelt, C. (2002) 3D velocity tomogram across the Chicxulub Crater: testing the resolution. *Tectonophysics*, 355, 217–228.

Morgan, J.V, Warner, M.R., Collins, G.S. *et al.* (2011) Full waveform tomographic images of the peak ring at the Chicxulub impact crater. *Journal of Geophysical Research*, 116, B06303. DOI: 10.1029/2010JB008015.

Nabighian, M.N. and Asten, M. (2002) Metalliferous mining geophysics: state of the art in the last decade of the 20th century and the beginning of the new millennium. *Geophysics*, 67, 964–978.

Ortiz-Aleman, C., Urrutia-Fucugauchi, J. and Pilkington, M. (2001) Three-dimensional modeling of aeromagnetic anomalies over the Chicxulub Crater. 32nd Lunar and Planetary Science Conference, CD-ROM, abstract no. 1962.

Penfield, G.T. and Camargo-Zanoguera, A. (1981) Definition of a major igneous zone in the central Yucatán Platform with aeromagnetics and gravity (abstract). Society Exploration Geophysics Annual Meeting, 51, p. 37.

Pilkington, M. and Grieve, R.A.F. (1992) The geophysical signature of terrestrial impact craters. *Reviews of Geophysics*, 30, 161–181.

Pilkington, M. and Hildebrand, A.R. (2000) Three-dimensional magnetic imaging of the Chicxulub Crater. *Journal of Geophysical Research*, 105, 23479–23491.

Pilkington, M., Hildebrand, A.R. and Ortiz-Aleman, C. (1994) Gravity and magnetic field modeling and structure of the Chicxulub Crater, Mexico. *Journal of Geophysical Research*, 99, 13147–13162.

Poag, C.W., Powars, D.S., Poppe, L.J. and Mixon, R.B. (1994) Meteoroid mayhem in Ole Virginny: source of the North American tektite strewn field. *Geology*, 22 (8), 691–694.

Pohl, J. and Will, M. (1974) Vergleich der Geschwindigkeitsmessungen im Bohrloch der Forschungsbohrung Nördlingen 1973 mit seismischen Tiefensondierungen innerhalb und außerhalb des Ries. *Geologica Bavarica*, 72, 75–80.

Pohl, J., Stoeffler, D., Gall, H. and Ernstson, K. (1977) The Ries impact crater, in *Impact and Explosion Cratering: Planetary and Terrestrial Implications; Proceedings of the Symposium on Planetary Cratering Mechanics, Flagstaff, Ariz., September 13–17, 1976* (eds D.J. Roddy, R.O. Pepin and R.B. Merrill), Pergamon Press, New York, pp. 343–404.

Pratt, R.G. (1999) Seismic waveform inversion in the frequency domain. Part I: theory and verification in a physical scale model. *Geophysics*, 64, 888–901.

Rebolledo-Vieyra, M. and Urrutia-Fucugauchi, J. (2004) Magnetostratigraphy of the impact breccias and post-impact carbonates from borehole Yaxcopoil-1, Chicxulub impact crater, Yucatan, Mexico. *Meteoritics and Planetary Sciences*, 39, 821–830.

Rebolledo-Vieyra, M. and Urrutia-Fucugauchi, J. (2006) Magnetostratigraphy of the Cretaceous/Tertiary boundary and Early Paleocene sedimentary sequence from the Chicxulub impact crater. *Earth, Planets and Space*, 58, 1309–1314.

Rebolledo-Vieyra, M., Urrutia-Fucugauchi, J. and Lopez-Loera, H. (2010) Aeromagnetic anomalies and structural model of the Chicxulub impact crater, Yucatán, Mexico. *Revistas Mexicana de Ciencias Geológicas*, 27, 185–195.

Salisbury, M.H., Iuliucci, R. and Long, C. (1994) Velocity and reflection structure of the Sudbury structure from laboratory measurements. *Geophysical Research Letters*, 21, 923–926.

Sambridge, M. and Mosegaard, K. (2002) Monte Carlo methods in geophysical inverse problems. *Reviews of Geophysics*, 40, 1009. DOI: 10.1029/2000RG000089.

Scholtz, C.A., Karp, T. and Lyons, R.P. (2007) Structure and morphology of the Bosumtwi impact structure from seismic reflection data. *Meteoritics and Planetary Science*, 42, 549–560.

Senft, L.E. and Stewart, S.T. (2009) Dynamic fault weakening and the formation of large impact craters. *Earth and Planetary Science Letters*, 287, 471–482.

Sharma, P.V. (1997) *Environmental and Engineering Geophysics*, Cambridge University Press, Cambridge.

Sharpton, V.L., Marín, L.E., Carney, J.L. *et al.* (1996) A model of the Chicxulub impact basin based on evaluation of geophysical data, well logs, and drill core samples, in *The Cretaceous–Tertiary Event and other Catastrophes in Earth History* (eds G. Ryder, D. Fastovsky and S. Gartner), Geological Society of America Special Paper 307, Geological Society of America, Boulder, CO, pp. 55–74.

Stewart, S.A. and Allen, P.J. (2002) A 20-km diameter multi-ringed impact structure in the North Sea. *Nature*, 418, 520–523.

Stewart, S.A. and Allen, P.J. (2004) An alternative origin for the Silverpit Crater (reply). *Nature*, 428, 280. DOI: 10.1038/nature02480.

Sweeney, J.F. (1978) Gravity study of a great impact. *Journal of Geophysical Research*, 83, 2809–2815.

Thomson, K. (2004) Overburden deformation associated with halokinesis in the southern North Sea: implications for the origin of the Silverpit Crater. *Visual Geosciences*, 9, 39–47. DOI: 10.1007/s10069-004-0019-0.

Tong, C.H., Lana, C., Marangoni, Y.R. and Elis, V.R. (2010) Geolectric evidence for centripetal resurge of impact melt and breccias over central uplift of Araguainha impact structure. *Geology*, 38, 91–94.

Tsikalas, F., Gudlaugsson, S.T. and Faleide, J.I. (1998) The anatomy of a buried complex impact structure: the Mjølnir structure, Barents Sea. *Journal of Geophysical Research*, 103, 30469–30484.

Ugalde, H., Morris, W.A., Pesonen, L.J. and Danuor, S.K. (2007) The Lake Bosumtwi meteorite impact structure, Ghana – where is the magnetic source? *Meteoritics and Planetary Science*, 42, 867–882.

Underhill, J.R. (2004) An alternative origin for the Silverpit Crater. *Nature*, 428, 280. DOI: 10.1038/ nature02476.

Unsworth, M., Campos-Enriquez, O., Belmonte, S. *et al.* (2002) Crustal structure of the Chicxulub impact imaged with magnetotelluric exploration. *Geophysical Research Letters*, 29, 1788. DOI: 10.1029/2002GL014998.

Urrutia-Fucugauchi, J., Marin, L. and Trejo-Garcia, A. (1996) UNAM Scientific drilling program of Chicxulub impact structure. Evidence for a 300 kilometer crater diameter. *Geophysical Research Letters*, 23, 1565–1568.

Urrutia-Fucugauchi, J., Soler-Arechalde, A.M., Rebolledo-Vieyra, M. and Vera-Sánchez, P. (2004) Paleomagnetic and rock magnetic study of the Yaxcopoil-1 impact breccia sequence, Chicxulub impact crater, Mexico. *Meteoritics and Planetary Science*, 39, 843–856.

Vermeesch, P.M. and Morgan, J.V. (2004) Structure of Chicxulub: where do we stand? *Meteoritics and Planetary Science*, 39, 1019–1034.

Vermeesch, P.M. and Morgan, J.V. (2008) Structural uplift beneath the Chicxulub impact crater. *Journal of Geophysical Research*, 113, B07103. DOI: 10.1029/2007JB005393.

Vermeesch, P., Morgan, J., Christeson, G. *et al.* (2009) 3D joint inversion of travel time and gravity data across the Chicxulub impact crater. *Journal of Geophysical Research*, 114, B02105. DOI: 10.1029/2008JB005776.

Wall, M.L.T., Cartwright, J. and Davies, R.J. (2008) An Eocene age for the proposed Silverpit impact crater. *Journal of the Geological Society of London*, 165, 781–794.

Weidie, A.E. (1985) Geology of the Yucatan Platform, in *Geology and Hydrology of the Yucatan and Quaternary Geology of Northeastern Yucatán Peninsula* (eds W.C. Ward, A.E. Weidie and

W. Back), New Orleans Geological Society, New Orleans, LA, pp. 1–19.

Wünnemann, K., Morgan, J.V. and Jödicke, H. (2005) Is Ries a typical example of a middle-sized terrestrial crater? in *Large Meteorite Impacts and Planetary Evolution III* (eds T. Kenkmann, F. Hörz and A. Deutsch), Geological Society of America Special Paper 384, Geological Society of America, Boulder, CO, pp. 67–83.

Yilmaz, Ö. (2001) *Seismic Data Analysis: Processing, Inversion and Interpretation of Seismic Data*, Society of Exploration Geophysicists, Tulsa, OK.

Zelt, C.A. (1998) Lateral velocity resolution from three-dimensional seismic refraction data. *Geophysical Journal International*, 135, 1101–1112.

Zelt, C.A. and Barton, P.J. (1998) Three-dimensional seismic refraction tomography: a comparison of two methods applied to data from the Faeroe Basin. *Journal of Geophysical Research*, 103, 7187–7210.

# FIFTEEN

# Projectile identification in terrestrial impact structures and ejecta material

**Steven Goderis\*,†, François Paquay‡ and Philippe Claeys\***

*\*Earth System Science, Vrije Universiteit Brussel, Pleinlaan 2, BE-1050 Brussels, Belgium*
*†Department of Analytical Chemistry, Universiteit Gent, Krijgslaan 281 – S12, BE-9000 Ghent, Belgium*
*‡Department of Geology and Geophysics, University of Hawaii at Manoa, Honolulu, HI, USA*

## 15.1 Introduction

During crater formation (Chapters 3, 4 and 5), the impactites incorporate a small amount of meteoritic material, resulting in a distinct chemical signature compared with the local or average continental crust values (see Chapter 7 for an overview of impactites). The extraterrestrial contribution (ETC) in impactites generally amounts to less than 1 wt% of meteoritic material, although much higher values occur, such as, for example, 5.7 wt% in the Morokweng Crater (McDonald *et al.*, 2001). Two geochemical approaches pinpoint this cosmic contribution in terrestrial rocks: elevated concentrations of specific siderophile elements and atypical isotope ratios. Generally, the projectile contribution occurs in impact melt rock (e.g. East Clearwater, Morokweng, Popigai; Palme *et al.*, 1978,1981; McDonald *et al.*, 2001; McDonald, 2002; Tagle and Claeys, 2005). However, other impactites also contain enrichments in projectile components, as documented for the distal ejecta distributed worldwide at the Cretaceous–Tertiary (K–Pg) boundary (Alvarez *et al.*, 1980; Smit and Hertogen, 1980; Claeys *et al.*, 2002). Owing to its diluted concentration, meteoritic material is often heterogeneously distributed in impact melt rock compared with the more uniform composition in major (and some trace) elements (Grieve *et al.*, 1977; Chapter 9).

Several classes of meteorites (and their asteroid or comet parent-body precursors) display significant enrichments in moderately and strongly siderophile elements, compared with crustal lithologies (Taylor and McLennan, 1985). The first projectile identification studies carried out on terrestrial impactites indicated that siderophile elements, such as the platinum group metals (PGEs: ruthenium (Ru), rhodium (Rh), palladium (Pd), osmium (Os), iridium (Ir), platinum (Pt)) and nickel (Ni), along with the more moderate siderophiles such as chromium (Cr) and cobalt (Co), were effective as cosmic indicators (Morgan *et al.*, 1975; Palme *et al.*, 1978,1979; Wolf *et al.*, 1980). These studies relied on neutron activation methods that allowed the identification of the Ir peaks with relative ease compared with the other

PGEs. As a result, analysis for Ir became the tool of predilection to recognize ETC in sedimentary sequences and impactites. Since 1980 and the K–Pg boundary hypothesis, the term 'positive Ir anomaly' is widely used in the literature to designate a meteoritic contribution. During the last 20 years, the rapid development of inductively coupled plasma mass spectrometry (ICP-MS) rendered possible the simultaneous determination of the concentrations of all the highly siderophile PGEs and gold (Au) (e.g. Schmidt *et al.*, 1997; McDonald *et al.*, 2001; Norman *et al.*, 2002; Tagle and Claeys, 2005; Lee *et al.*, 2006).

Extraterrestrial materials also differ from terrestrial crustal values in terms of the isotope proportions of specific elements. The development of isotope geology, ion-exchange chromatographic separation, and thermal ionization mass spectrometry (TIMS) and multi-collector (MC)-ICP-MS now make it possible to detect small isotope variations induced by the addition of almost indiscernible ETCs into impactites. The Os and Cr isotope systematics are commonly used. The determination of the $^{187}Os/^{188}Os$ ratio is by far the most sensitive method to detect a minute projectile contribution in terrestrial lithologies (unless significant ultramafic components occur within the target). However, it does not distinguish between different types of meteorites. The $^{53}Cr/^{52}Cr$ and $^{54}Cr/^{52}Cr$ isotope ratios can discriminate between carbonaceous and other chondrites and, in favourable cases, even within different types of chondrites. Other isotope systematics, such as tungsten (W), lead (Pb), neodymium (Nd) and strontium (Sr), have been proposed as potential impact indicators but remain controversial because of their variable degrees of success (e.g. Schoenberg *et al.*, 2002; Trinquier *et al.*, 2006, Moynier *et al.*, 2009).

## 15.2 Current situation: projectile identification at impact craters and ejecta layers

Of the known terrestrial impact structures, approximately 50 have been investigated using one or more of the projectile-tracing

**Table 15.1** Type of impactor determined for terrestrial impact structures and ejecta deposits. This list is an updated assemblage of previous compilations (Grieve and Shoemaker, 1994; Koeberl, 1998,2007; Tagle and Hecht, 2006). Only craters larger than 0.1 km in diameter are listed. Size (longest dimension of largest crater) and age of the impact structures were obtained from the Earth Impact Database (2010)

| Crater name | Location | Age (Ma) | Diameter (km) | Impactor type[a] | Evidence[b] | Reference |
|---|---|---|---|---|---|---|
| Morasko | Poland | <0.01 | 0.10 | IIIC – iron | M | Koblitz (2000) |
| Kaalijärvi | Estonia | 0.004 ± 0.001 | 0.11 | IA – iron | M | Buchwald (1975), Koblitz (2000) |
| Wabar | Saudi Arabia | 0.006 ± 0.002 | 0.12 | IIIA – iron | M, S | Morgan et al. (1975), Mittlefehldt et al. (1992b) |
| Henbury | Australia | <0.005 | 0.16 | IIIA – iron | M, S | Taylor (1967), Koblitz (2000) |
| Odessa | USA | <0.05 | 0.17 | IA – iron | M | Buchwald (1975), Koblitz (2000) |
| Boxhole | Australia | 0.0300 ± 0.0005 | 0.17 | IIIA – iron | M | Buchwald (1975), Koblitz (2000) |
| Macha | Russia | <0.007 | 0.30 | Iron | M, S | Gurov (1996) |
| Aouelloul | Mauretania | 3.1 ± 0.3 | 0.39 | Iron | S, Os* | Morgan et al. (1975), Koeberl and Auer (1991) |
| Monturaqui | Chile | <1 | 0.46 | IA? – iron | M, S | Bunch and Cassidy (1972), Buchwald (1975) |
| Kalkkop | South Africa | 0.250 ± 0.050 | 0.64 | Chondrite? | S, Os* | Koeberl et al. (1994b), Reimold et al. (1998) |
| Wolfe Creek | Australia | <0.3 | 0.88 | IIIB – iron | M, S | Attrep et al. (1991), Koblitz (2000) |
| Tswaing (formerly Pretoria Saltpan) | South Africa | 0.220 ± 0.052 | 1.13 | Chondrite | S, Os* | Koeberl et al. (1994a) |
| Barringer | USA | 0.049 ± 0.003 | 1.19 | IA – iron | M, S | Morgan et al. (1975), Mittlefehldt et al. (1992a) |
| Roter Kamm | Namibia | 3.7 ± 0.3 | 2.5 | Non CC/chondrite? | S | Hecht et al. (2008) |
| New Quebec | Canada | 1.4 ± 0.1 | 3.4 | OC; type L? | S | Grieve et al. (1991), Evans et al. (1993) |
| Brent | Canada | 450 ± 30 | 3.8 | OC; type L or LL | S | Palme et al. (1981), Evans et al. (1993) |
| Gow | Canada | <250 | 4.0 | Iron?/no contamination? | S | Wolf et al. (1980) |
| Rio Cuarto | Argentina | <0.1 | 4.5 | Chondrite (H?) | M, S, Os* | Schultz et al. (1994), Koeberl (2007) |
| Gardnos | Norway | 500 ± 10 | 5.0 | Chondrite/NMI | S, Os* | French et al. (1997), Goderis et al. (2009) |
| Sääksjärvi | Finland | ~560 | 6.0 | Stony-iron, iron, chondrite?/NMI | S | Palme et al. (1980), Schmidt et al. (1997), Tagle et al. (2009) |
| Wanapitei | Canada | 37.2 ± 1.2 | 7.5 | OC; type L or LL | S | Wolf et al. (1980), Evans et al. (1993), Tagle et al. (2006) |
| Ilyinets | Ukraine | 378 ± 5 | 8.5 | Iron? | S | Grieve and Shoemaker (1994) |
| Mien | Sweden | 121.0 ± 2.3 | 9.0 | Stone? | S | Palme et al. (1980), Schmidt et al. (1997) |
| Bosumtwi | Ghana | 1.03 ± 0.02 | 11 | Chondrite? | S, Os*, Cr* | Koeberl et al. (2007), Goderis et al. (2007), McDonald et al. (2007) |
| Ternovka | Ukraine | 280 ± 10 | 11 | Chondrite? | S | Grieve and Shoemaker (1994) |
| Nicholson | Canada | <400 | 12.5 | Achondrite (olivine rich) | S | Wolf et al. (1980) |
| Zhamanshin | Kazakhstan | 0.9 ± 0.1 | 14 | Chondrite or iron? | S | Palme et al. (1978), Glass et al. (1983) |
| El'gygytgyn | Russia | 3.5 ± 0.5 | 18 | Achondrite? | S | Grieve and Shoemaker (1994) |
| Dellen | Sweden | 89.0 ± 2.7 | 19 | Stone? Chondrite | S | Palme et al. (1980), Schmidt et al. (1997), Tagle et al. (2008b) |
| Obolon | Ukraine | 169 ± 7 | 20 | Iron? | S | Grieve and Shoemaker (1994) |
| Lappajärvi | Finland | 77.3 ± 0.4 | 23 | OC; type H | S, Cr* | Tagle et al. (2007), Koeberl et al. (2007) |
| Rochechouart | France | 218 ± 8 | 23 | Stony-iron, chondrite or iron (IIA or NMI) | S, Cr* | Janssens et al. (1977), Wolf et al. (1980), Koeberl et al. (2007), Tagle et al. (2009) |

| Name | Location | Age (Ma) | Diameter (km) | Type | Method | References |
|---|---|---|---|---|---|---|
| Haughton | Canada | 39 | 23 | No contamination | S | Morgan et al. (1975), Palme et al. (1978) |
| Ries | Germany | 15 ± 1 | 24 | No contamination or achondrite | S | Morgan et al. (1979), Schmidt and Pernicka (1994) |
| Boltysch | Ukraine | 65.17 ± 0.64 | 24 | Chondrite? | S | Grieve and Shoemaker (1994) |
| Strangways | Australia | 646 ± 42 | 25 | Achondrite (olivine rich) | S | Morgan and Wandless (1983) |
| East Clearwater | Canada | 290 ± 20 | 26 | OC; type LL (H?) | S, Cr* | Palme et al. (1979), McDonald (2002), Koeberl et al. (2007) |
| Mistastin | Canada | 36.4 ± 4 | 28 | Iron?/no contamination?/achondrite | S | Morgan et al. (1975), Wolf et al. (1980), Palme et al. (1978, 1981) |
| Manson | USA | 74.1 ± 0.1 | 35 | Chondrite | S, Os* | Pernicka et al. (1996), Koeberl and Shirey (1996) |
| West Clearwater | Canada | 290 ± 20 | 36 | No contamination | S | Palme et al. (1978) |
| Mjølnir | Norway | 142.0 ± 2.6 | 40 | Iron? | S | Dypvik and Attrep (1999) |
| Saint Martin | Canada | 220 ± 32 | 40 | No contamination | S | Palme (1982) |
| Charlevoix | Canada | 342 ± 15 | 54 | Chondrite? | S | Tagle et al. (2008b) |
| Morokweng | South Africa | 145.0 ± 0.8 | 70 | OC; type LL | M, S, Os*, Cr* | McDonald et al. (2001), Koeberl et al. (2002), Maier et al. (2006) |
| Chesapeake Bay | USA | 35.3 ± 0.1 | 90 | Chondrite or achondrite? No contamination? | S, Os* | Lee et al. (2006), McDonald et al. (2009), Goderis et al. (2010) |
| Acraman[c] | Australia | ~590 | 90 | Chondrite | S | Gostin et al. (1989), Wallace et al. (1990) |
| Kara | Russia | 70.3 ± 2.2 | 65 | Chondrite? | S | Grieve and Shoemaker (1994) |
| Popigai | Russia | 35 ± 5 | 100 | OC; type L | S | Tagle and Claeys (2005) |
| Manicouagan | Canada | 214 ± 1 | 100 | No contamination/chondrite? | S | Palme et al. (1978, 1981) |
| Chicxulub[c] | Mexico | 64.98 ± 0.05 | 170 | CC (CM2) | M, S, Os*, Cr* | Shukolyukov and Lugmair (1998), Kyte (1998), Trinquier et al. (2006), Quitté et al. (2007) |
| Vredefort | South Africa | 2023 ± 4 | 300 | Chondrite? | S, Os* | Koeberl et al. (1996, 2002) |
| Eltanin | South Pacific | 2.15 | No crater | Mesosiderite | S | Kyte (2002) |
| Stac Fada | Scotland | 1199 ± 70 | No crater | Chondrite | S, Cr* | Amor et al. (2009) |
| Spherule layers in Dale Gorge Member, Wittenoom Fm, Jeerinah Fm and Carawine Dolomite | Hamersley Basin, Australia | 2630–2490 | No crater | OC? | S, Cr* | Simonson et al. (2009) |
| Spherule layers in Kuruman BIF, Reivilo Fm and Monteville Fm | Griqualand West Basin, South Africa | 2630–2490 | No crater | OC? | S, Cr* | Simonson et al. (2009) |
| Barberton (S3, S4) | South Africa | 3100–3500 | No crater | CC | S, Cr* | Kyte et al. (2003) |

[a]OC: ordinary chondrite; EC: enstatite chondrite; CC: carbonaceous chondrite; NMI: non-magmatic iron; ?: questionable.
[b]S = siderophile element abundances (PGE, Cr, Co, Ni, Au); Cr*: chromium isotopes; Os*: osmium isotopes; M: projectile fragment.
[c]Enrichment in ejecta layer.

methods described above (Table 15.1). A major issue for projectile characterization studies is the variety of possible impactors – stony, iron and stony–iron meteorites – subdivided in numerous types and subtypes, with widely varying siderophile element compositions. Distinctly higher siderophile element concentrations in impactites, compared with target rock abundances, indicate the presence of either a chondritic or an iron meteoritic component. It is more complicated to discern an achondritic projectile, as it has significantly lower abundances of the key siderophile elements and, consequently, requires the application of different methodologies, e.g. Cr isotopes. With these difficulties, only in a small fraction of the known impact structures (approximately 10) has the projectile composition determined down to the meteorite-type (Table 15.1). In the majority of cases, the identification of the impactor remains approximated at the level of a stone versus iron meteorite. Nevertheless, over the last few years the situation has improved significantly (e.g. McDonald *et al.*, 2001; McDonald 2002; Tagle and Claeys, 2005; Koeberl *et al.*, 2007; Tagle *et al.*, 2007; Goderis *et al.*, 2009; Simonson *et al.*, 2009).

## 15.3    Methodology

While under favourable circumstances chondrites, primitive achondrites, iron and stony-iron meteorites possess a clearly detectable elemental or isotopic signature, some differentiated achondrites, such as HEDs, do not. To complicate matters, not all achondrites are depleted in PGEs (e.g. ureilites, aubrites, brachinites, lodranites, acapulcoites and winonaites; Morgan *et al.*, 1979; Rankenburg *et al.*, 2008). The impact of a siderophile-poor bolide results in PGE signals that are low or even indiscernible from background crustal values, especially if a mafic or ultramafic component is present in the target rock. Eventually, the application of isotope methods (e.g. $^{53}Cr/^{52}Cr$ and $^{54}Cr/^{52}Cr$ ratios) can bring out the ETC.

### 15.3.1    Identification of impactor fragments

In rare cases, meteorite fragments occur within terrestrial impact craters (Table 15.1). However, weathering quickly affects any exposed meteorite debris and generally destroys them after a few thousand years. Impactor fragments prevail in the vicinity of the youngest terrestrial impact structures, which, as a result of the size–frequency distribution, also happen to be the smallest (i.e. less than 1.5 km; Table 15.1). In this small impact crater size range, iron meteorites probably form the only type of body capable of surviving atmospheric passage relatively intact and to collide with the terrestrial crust with enough kinetic energy to create a hyper-velocity impact structure (Melosh, 1989). However, exceptions do occur, such as the H4-5 Carancas chondrite that impacted near Lake Titicaca (Peru) in September 2007 (Kenkmann *et al.*, 2009) (see Chapter 2 for further discussion). Protected from alteration, impactor fragments can survive for extended periods within the impactites. The recovery of a 2.5 mm diameter carbonaceous chondritic inclusion from the 65 Ma K–Pg boundary sediments in Deep Sea Drilling Project (DSDP) drill core 576 (Kyte, 1998) most likely originated from the Chicxulub impactor. Similarly, a drill core within the Morokweng impact melt rocks revealed a large (25 cm), unaltered, chondritic meteorite fragment, as well as

several smaller fragments (Maier *et al.*, 2006). This type of discovery is more by chance than the rule, and the geochemical tools, described below, must be used to characterize the precise type of meteorite.

### 15.3.2    Siderophile elements

A distinction is generally made between the highly siderophile PGEs and more commonly used moderately siderophile elements (Cr, Co, Ni; McDonough and Sun, 1995). This denotes the geochemical tendency of these elements to partition into metal phases, quantitatively expressed by the ratio of the concentration of an element in liquid or solid metal to that in silicate melt at given pressure, temperature and oxygen fugacity, and assuming equilibrium between metal and silicate (Palme, 2008). Elements with partition coefficients above approximately 10,000 are called highly siderophile elements (HSEs) and include the PGEs, Re and Au. With partition coefficients one to two orders of magnitude lower, Ni and Co exhibit a moderately siderophile behaviour (e.g. Kramers, 1998). Although it can be considered a siderophile element, Cr additionally behaves as a lithophile (i.e. rock loving; McDonough and Sun, 1995).

Early on in Earth history, much of the HSEs partitioned into the core, as the result of the differentiation of the Earth into a metallic core, mantle and crust (Lorand *et al.*, 2008), whereas Cr remained mostly concentrated in the mantle (Drake and Righter, 2002). Based on the abundances of the HSEs in the Earth's mantle (e.g. Morgan *et al.*, 1981) compared with values predicted by core–mantle equilibrium partitioning models (e.g. Borisov *et al.*, 1994), the addition of approximately 0.8% chondritic material to the mantle after the Earth's initial differentiation has been suggested (e.g. Kimura *et al.*, 1974; O'Neill, 1991). Additionally, substantiated by the chondritic Os isotopic composition inferred for the primitive upper mantle (Meisel *et al.*, 1996), this 'late-veneer hypothesis' was interpreted as an intense meteoritic bombardment event that hit the primitive Earth shortly after formation of the siderophile-rich core (e.g. Morgan, 1986; O'Neill, 1991; Halliday, 2004). However, the most recent models that account for the observed HSE characteristics of the terrestrial, lunar and Martian mantles are complex combinations of inefficient core formation, HSE partitioning at elevated temperatures and pressures, and late accretion (Walker, 2009).

In projectile identification studies, the use of both the moderately siderophile elements and HSEs relies on localized enrichments in these elements compared with the low background crustal values; that is, if an (ultra-)mafic mantle-derived component can be excluded. Unlike the PGEs, for which background values are generally low, the target lithologies can, on top of the meteoritic contribution, already contain elevated concentrations of Cr, Co and Ni. Their presence disturbs the inter-element correlations and blurs the characteristic projectile signature (e.g. Palme, 1980). Nevertheless, these moderately siderophile elements act as a proxy for the presence of PGE enrichments. They can be considered as the initial indicators for the presence of an ETC, and, in some cases, even allow an initial characterization (Koeberl *et al.*, 2007).

Often Cr, Co and Ni concentrations as well as inter-element ratios are used, sometimes in combination with the HSE Ir, to discriminate between chondrites, iron-rich meteorites, and

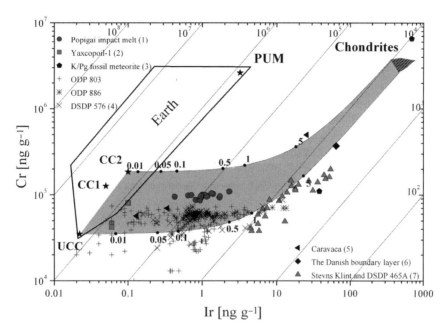

**Figure 15.1** Cr versus Ir concentrations of terrestrial target rocks compared with the composition of the Popigai impact melt rocks and selected K–Pg boundary clays and surrounding sediments (modified after Tagle and Hecht (2006)). The grey field indicates the most likely mixing trajectories between chondritic projectiles and common terrestrial targets. Numbers represent weight per cent chondritic material in the mixing trajectories. PUM: primitive upper mantle; MORB: mid-ocean ridge basalt; UCC: upper continental crust; CC: continental crust. Popigai impact melt rock data from (1) Tagle and Claeys (2005); Chicxulub (from the Yaxcopoil-1 drill core) melt data from (2) Tagle *et al.* (2004); and K–Pg boundary data from (3) Kyte (1998), (4) Kyte *et al.* (1995), (5) Smit and Hertogen (1980), (6) Alvarez *et al.* (1980) and (7) Kyte *et al.* (1980).

achondrites (e.g. Palme *et al.*, 1978,1979,1981; Wolf *et al.*, 1980; Morgan and Wandless, 1983; Evans *et al.*, 1993). A strong enrichment of these siderophile elements (e.g. more than 0.1 wt% meteoritic contribution) in impactites indicates either a chondritic or iron meteoritic impactor. Chondrites show elevated Cr abundances (2575–3810 µg g$^{-1}$) and Ni/Cr ratios that vary between roughly 2 and 7 (Tagle and Berlin, 2008). Chromium varies widely in iron meteorites, but its concentration is generally approximately 100 times lower than in chondrites (Buchwald, 1975). Consequently, iron meteorites display much higher Ni/Cr or Co/Cr ratios (up to three orders of magnitude). However, exceptions occur; for example, non-magmatic iron meteorites that can acquire highly variable ratios by mixing of the different constituents of which they are composed (Mittlefehldt *et al.*, 1998; Tagle *et al.*, 2009). The combination of the presence of a significant Cr enrichment and chondritic refractory metal ratios (e.g. Ni/Cr, Ni/Co, Co/Cr) has proven effective in the identification of a chondritic impactor (e.g. Palme *et al.*, 1979,1981; Wolf *et al.*, 1980). High Ni and Co concentrations (and high Ni/Cr and Co/Cr ratios), without elevated Cr concentrations, point towards iron meteorite projectiles (e.g. Janssens *et al.*, 1977). The absence of clear Ni and Co (as well as HSE) enrichment in impactites, accompanied by elevated Cr values (and low Ni/Cr and Co/Cr ratios) can be interpreted as the result of the impact of an achondritic meteorite (Palme, 1980), although it is often difficult to completely rule out an indigenous (ultra)mafic origin for the Cr. The only impact structures verifiably produced by achondrites,

Lake Nicholson and Strangways, were formed by olivine-rich achondrites, as attested by the presence of their PGE enrichments (Wolf *et al.*, 1980; Morgan and Wandless, 1983).

Combining the moderately siderophile elements with Ir (using elemental ratios Cr/Ir, Co/Ir and Ni/Ir) also discriminates between iron and chondrite groups. Additionally, a strong correlation between these moderately and strongly siderophile elements in the impactites supports a common origin and likely excludes the presence of mafic or ultramafic contributions, as well as significant post-impact fractionation and remobilization (Palme, 1980; Morgan and Wandless, 1983). Tagle and Hecht (2006) illustrate the use of a Cr versus Ir plot on a double-logarithmic scale. Impactites from craters made by chondritic projectiles plot along a path that tracks the admixture of meteoritic material with the continental crust. Most terrestrial rocks have Cr/Ir ratios higher than approximately 10$^6$ and plot within a specific zone of Fig. 15.1, designated as 'Earth'. Most materials impacting Earth contain much lower Cr/Ir ratios, with chondrites typically exhibiting Cr/Ir ratios between two and more than three orders of magnitude lower. The mixing of a chondritic projectile with an original crustal composition results in a decrease of the Cr/Ir ratio, depending on the degree of projectile contamination (as illustrated in Fig. 15.1 for the Popigai and Chicxulub impact melts and selected K–Pg boundary samples; Alvarez *et al.*, 1980; Kyte *et al.*, 1980,1995; Smit and Hertogen, 1980; Kyte, 1998; Tagle and Claeys, 2005; Tagle and Hecht 2006; Tagle *et al.*, 2004).

**Table 15.2** Projectile contribution reported for some impact craters and deposits. Although generally the bulk projectile components are less than 1 wt%, exceptions occur (East Clearwater and Morokweng). Even in the case of a relatively limited ETC (0.1–1 wt%), a precise characterization of the type of projectile is possible following the procedures described in the text. The calculated ETC depends strongly on the type of projectile and the PGE content reported for the impactor (varies between databases)

| Impact structure | Highest Ir content[a] (ng g$^{-1}$) | Impactite type | Impactor type | Ir in impactor type[b] | Bulk projectile component (wt%) | Reference for Ir data |
|---|---|---|---|---|---|---|
| East Clearwater | 40.1 | Melt rock | Possibly type L | 501 | 8.0 | McDonald (2002) |
| Morokweng | 19.1 | Sulfide-poor melt rock | LL | 336 | 5.7 | McDonald *et al.* (2001) |
| Popigai | 2.31 | Melt rock | L | 501 | 0.46 | Tagle and Claeys (2005) |
| Wanapitei | 1.89 | Melt rock | Possibly type L | 501 | 0.38 | Tagle *et al.* (2006) |
| Lappajärvi | 5.36 | Melt rock | H | 749 | 0.72 | Tagle *et al.* (2007) |
| K–Pg boundary | 64.3 | Ejecta | CM2 | 605 | 11 | Alvarez *et al.* (1980) |
|  | 25.5 | Ejecta |  |  | 4.2 | Smit and Hertogen (1980) |

[a]As reported in referred study.
[b]Tagle and Berlin (2008).

### 15.3.3    Platinum group elements

The PGE group of metals Ru, Rh, Pd, Os, Ir, Pt have similar physical and chemical properties (Brenan, 2008; Palme, 2008). Owing to comparable properties, Re and Au are commonly associated with them. Among the PGEs, determination of the Ir concentration is the most commonly used technique to search for chemical impact markers and, coupled with the moderate siderophiles, identify the type of impacted projectile. The choice of this particular element derives from the ease of analysing it using instrumental neutron activation analysis compared with the other PGEs.

Over the last 10 years, the PGE determination methodology has shifted towards the use of ICP-MS, in combination with isotope dilution and/or nickel sulfide (NiS) fire assay pre-concentration. This ICP-MS technique possesses the advantage of determining the concentrations of PGEs at the same time and on the same aliquot (although mono-isotopic Rh cannot be determined with isotope dilution). However, the challenges in PGE analysis are not restricted to reaching low detection limits and simplifying chemical procedures. The inhomogeneous distribution of PGEs, most likely in the form of micrometre-size nuggets, makes PGE concentrations hard to reproduce between aliquots of the same sample and requires large enough sample sizes (of several gram) to dilute possible nugget effects (Hall and Pelchat, 1994; Plessen and Erzinger, 1998; Tagle and Claeys, 2005).

Iridium abundance in chondrites (in the range of 472–1066 ng g$^{-1}$; Wasson and Kallemeyn, 1988; Tagle and Berlin, 2008) is commonly a factor of 100 higher than those of the terrestrial mantle (estimated 3.5 ± 0.4 ng g$^{-1}$ for the primitive upper mantle; Becker *et al.*, 2006) and four orders of magnitude higher than crustal abundances (0.022 ng g$^{-1}$ Ir for the upper continental crust; Peucker-Ehrenbrink and Jahn, 2001). These significant concentration differences apply to all the PGEs, rendering them ideally suited for detecting even minute amounts of meteoritic material (except for PGE-poor meteorites, such as some achondrites) in impactites. Terrestrial impactites usually contain less

than 1 wt% of bulk meteoritic component (Table 15.2). Based on Ir excesses, around 8 wt% of a chondritic component is dissolved in the melt rocks of the East Clearwater impact structure (Palme *et al.*, 1978; McDonald, 2002). As such, the East Clearwater melt rocks contain the highest fraction of ETC of any terrestrial impact structure, followed by the Morokweng melt rocks with up to 5.7 wt% (McDonald *et al.*, 2001; Maier *et al.*, 2006; Table 15.2).

For rapid preliminary examination, the PGE concentrations detected in terrestrial impactites are plotted on a logarithmic-scale CI-normalized diagram (Fig. 15.2), with the PGE along the *X*-axis, most commonly in order of decreasing condensation temperature. A relatively flat pattern points towards a chondritic projectile (Palme, 1982; Evans *et al.*, 1993; Schmidt *et al.*, 1997). However, the target rock contribution influences this pattern, in some cases, obscuring or even masking completely the projectile signature (e.g. for Bosumtwi; Goderis *et al.*, 2007; McDonald *et al.*, 2007). A greatly uneven PGE contribution originating from various target rock components, with very low to relatively high PGE crustal contents, also blurs the flat chondritic pattern. In some cases, it is possible to subtract the inherited PGE target-rock composition to clarify the meteoritic PGE pattern. However, this approach requires a good knowledge of the precise nature and relative abundances of the different target-rock components (e.g. Schmidt and Pernicka 1994; Pernicka *et al.*, 1996). It is a difficult to sometimes impossible task, especially if the target rock composition is poorly known or not easily sampled. Plotting elemental patterns offers a first-hand classification of the projectile type, but it does not permit a precise projectile identification down to the sub-family level.

More recently, a different approach has achieved the precise identification of the projectile without accounting for the target component (e.g. McDonald *et al.*, 2001; McDonald 2002; Tagle and Claeys, 2005; Tagle and Hecht, 2006; Goderis *et al.*, 2009; Tagle *et al.*, 2009). Linear regression analysis is carried out on the obtained PGE concentrations by plotting two PGEs (e.g. Rh versus Ir) on an *X*–*Y* diagram. The slopes of the mixing lines obtained by plotting all the PGE abundances against each other

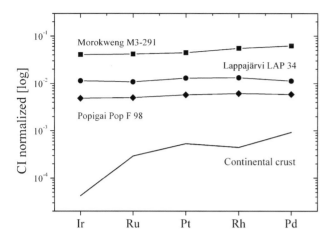

**Figure 15.2** CI-normalized logarithmic plot of the highest PGE abundances (in order of increasing volatility) in the Morokweng sulfide-poor impact melt rocks (McDonald *et al.*, 2001) and in the Lappajärvi and Popigai impact melt rock (Tagle and Claeys, 2005; Tagle *et al.*, 2007). The average composition of the continental crust is shown for comparison (Ru, Pd, Ir and Pt values from Peucker-Ehrenbrink and Jahn (2001); Rh from Wedepohl (1995)). CI values from Tagle and Berlin (2008). Osmium was not reported in these studies, as it volatilizes during nickel sulfide fire assay. The impact melt rocks are enriched in PGE compared with the continental crust and show a flat (chondritic) PGE pattern.

produce a series of individual PGE ratios characteristic of the meteoritic contribution in the impactites analysed. Owing to the sheer concentration, the PGE ratio of the projectile dominates the slope obtained from the linear regression. The slope deviation from the original projectile values depends on the PGE composition of the target rock. To illustrate the dominance of the projectile on the slope of the mixing line, the improbable and extreme case of a CM carbonaceous projectile impacting a pure komatiite target is demonstrated in Fig. 15.3a,b. In komatiites, Pd concentrations are high relative to Ir abundances, even compared with the projectile; consequently, the Pd/Ir illustrated in this figure is most divergent from the original projectile ratio. Nevertheless, the difference of the determined Pd/Ir ratio from the original projectile ratio remains relatively small (less than 1%) and possibly below analytical resolution. Impact on more typical crustal lithologies, with limited amounts of ultramafic contributions, displays little disruption. A similar approach cannot be used for Cr/Ir, Co/Ir or Ni/Ir, due to the much higher Cr, Co and Ni background of the target rocks compared with the PGEs.

In impact melt rocks, considered 'relatively' homogeneous in composition (e.g. Grieve *et al.*, 1977; Dressler and Reimold, 2001; Chapter 9), the projectile component is usually unevenly distributed. The use of siderophile element ratios determined by linear regression analysis cancels out sampling effects in heterogeneous materials, provides a more robust picture of the siderophile element signature (e.g. Tagle and Berlin, 2008) and takes into account PGE-rich target components (Fig. 15.3b).

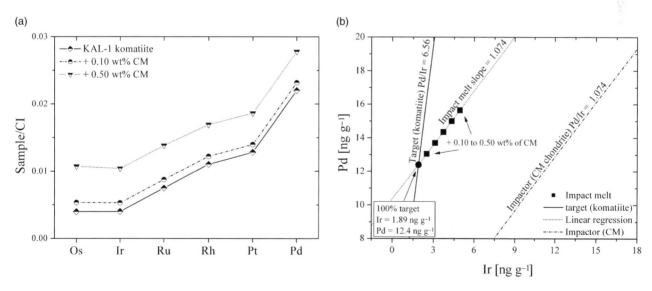

**Figure 15.3** (a) Calculated CI-normalized PGE patterns (in order of decreasing melting temperatures) of the impactites that would result from the addition of 0.10 and 0.50 wt% of projectile contribution after the hypothetical impact of a CM carbonaceous chondrite on a komatiite target (data from Meisel and Moser (2004) and Tagle and Berlin (2008)). The PGE patterns are strongly disturbed by the composition of the target. No clear identification of the projectile is possible. (b) Different amounts of admixed projectile in the modelled impactites (the black squares indicate 0.10, 0.20, 0.30, 0.40 and 0.50 wt% of CM material added to the indigenous komatiite PGE content) allow the determination of the Pd/Ir ratio of the projectile through linear regression. Even though the Pd/Ir ratio of the target (6.56) differs strongly from that of the impactor (1.074), the slope of the mixing line in the impactites is almost identical to the original Pd/Ir ratio of the projectile.

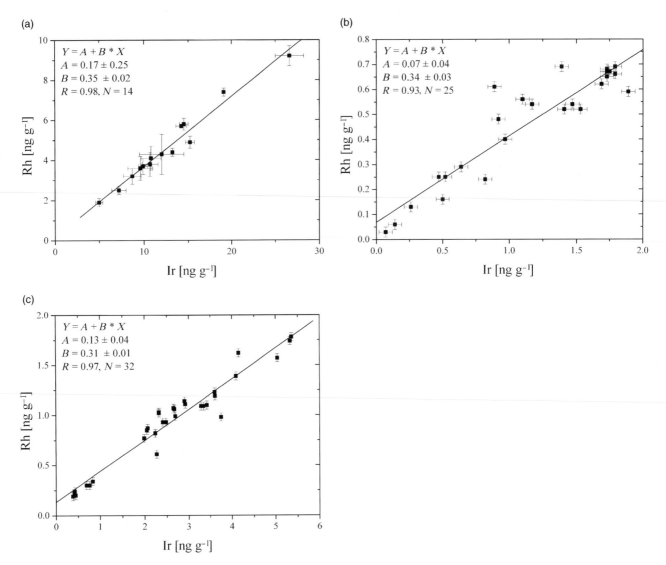

**Figure 15.4**   Linear regression analysis of Rh versus Ir ($Y = A + B \times C$; $A$: intercept; $B$: slope; $R$: regression factor; $N$: number of included samples) in the (a) Morokweng (McDonald *et al.*, 2001), (b) Popigai (Tagle and Claeys, 2005) and (c) Lappajärvi (Tagle *et al.*, 2007) impact melt rocks. The error bars represent reported analytical error.

Although the chondritic projectile cannot be recognized in the normalized PGE pattern (Fig. 15.3a), combining the different PGE ratios (sometimes together with moderately siderophile element ratios) allows a highly precise identification of the impactor.

The series of PGE elemental ratios and eventually the other siderophile elements obtained by regression analyses for the Morokweng, Popigai and Lappajärvi impact structures are given in Fig. 15.4a–c (McDonald *et al.*, 2001; Tagle and Claeys, 2005; Tagle *et al.*, 2007) and compared with the mean ratios of the different types of meteorites, compiled in siderophile element databases (e.g. McDonald *et al.*, 2001; Tagle and Berlin, 2008; Tagle *et al.*, 2009; Fischer-Gödde *et al.*, 2010). Where absolute concentrations of siderophile elements show large variations in the proportion of metals in a given aliquot (both in impactites and meteorites), element ratios are more reproducible (e.g. Morgan

*et al.*, 1985; Kallemeyn and Wasson 1986; McDonald *et al.*, 2001; Horan *et al.*, 2003).

Not all possible element combinations have the same cogency for the identification of projectile types (Tagle and Hecht, 2006; Tagle and Berlin, 2008). Generally between 7 and 14 ratios are used, supplemented with four moderately siderophile element ratios (Ni/Ir Cr/Ir, Co/Ir and Ni/Cr) and Au/Ir (e.g. McDonald, 2002; McDonald *et al.*, 2001,2009; Tagle and Claeys, 2004,2005; Tagle *et al.*, 2007,2009; Goderis *et al.*, 2009). Figure 15.5a–c illustrates the discriminative value of the different ratios according to Tagle and Berlin (2008). Condensation processes in the solar nebula mainly control the fractionation of PGEs in chondrites (e.g. Horan *et al.*, 2003). Element ratios of low condensation temperature elements (Rh or Pd) combined with those with higher condensation temperatures (such as Os, Ir and Ru) offer the best discrimination between different chondrite types; that is,

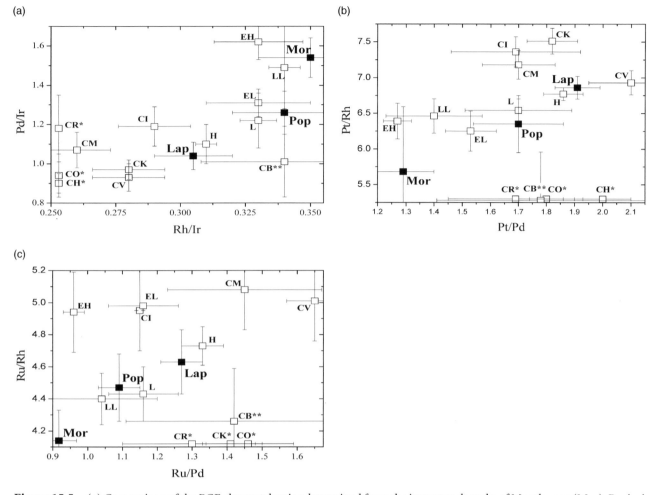

**Figure 15.5**    (a) Comparison of the PGE elemental ratios determined from the impact melt rocks of Morokweng (Mor), Popigai (Pop) and Lappajärvi (Lap), with the elemental ratios of chondrites. These ratios were calculated for each type of chondrite, after compilation of literature and new data in the database of Tagle and Berlin (2008). The uncertainty bars for the meteorites represent 1σ standard deviations. *Only one of the two ratios plotted is reported. **Only data for the metal fraction. The Pd/Ir versus Rh/Ir (A) and Pt/Ir versus Pt/Pd (b) plots separate ordinary and enstatite chondrites from carbonaceous chondrites, as the former have higher Rh/Ir and lower Pt/Rh and Pt/Pd ratios than the latter. The Ru/Rh versus Ru/Pd ratios help to distinguish ordinary from enstatite chondrite groups (c).

different associations of Rh/Ir, Pd/Ru, Rh/Os, etc. provide the most efficient way to distinguish between different types of chondritic projectiles (Fig. 15.5a–c). In undecided cases, it has proved helpful to combine Pt with both sets of elements, because of its intermediate condensation temperature between Ru and Rh (Tagle and Berlin, 2008). On a Pd/Ir versus Rh/Ir and a Pt/Rh versus Pt/Pd plot (Fig. 15.5a,b), OC and EC are clearly separated from CC by higher Rh/Ir ratios and lower Pt/Rh and Pt/Pd ratios. The differences in the Rh/Ir ratio appear to be most indicative for distinguishing between CC and OC/EC (Fig. 15.5a). Unfortunately, the relatively low number of Rh analyses available for all chondrite groups hampers the most detailed projectile identification in some cases (Tagle and Berlin, 2008). The Ru/Rh versus Ru/Pd ratios separate the different OC groups from the EC groups (Fig. 15.5c).

This method has been applied to identify the projectiles responsible for the formation of the Morokweng, East Clearwater, Popigai, Lappajärvi, Wanapitei, Rochechouart, Sääksjärvi and Gardnos impact structures, as well as the 2.63–2.49 Ga impact spherule layers in Western Australia and South Africa (e.g. McDonald *et al.*, 2001; McDonald, 2002; Tagle and Claeys, 2005; Tagle *et al.*, 2006,2007,2009; Goderis *et al.*, 2009; Simonson *et al.*, 2009). The identification of the projectiles of the Morokweng, Popigai and Lappajärvi impact structures are shown in Fig. 15.5. The fields defined by the slope of the different PGEs overlap with the fields of discrete ordinary chondrites, allowing a precise identification of the projectile; for example, an L or, most likely, LL ordinary chondrite for the Morokweng structure (South Africa; McDonald *et al.*, 2001), an ordinary chondrite, possibly L-type for the Popigai crater (Siberia; Tagle and Claeys 2005) and an H chondrite for Lappajärvi (Finland; Tagle *et al.*, 2007). The 25 cm ordinary chondrite fragment discovered within the impact melt sheet of the Morokweng impact structure (Maier *et al.*, 2006) conclusively supports the PGE-based projectile identification carried out by McDonald *et al.* (2001). The ordinary chondrite composition for the Morokweng and Lappajärvi impact structures was also confirmed by the measurement of positive Cr isotopic ratios (Table 15.1 and references therein).

### 15.3.4    Os isotope system

As discussed above, Os is a refractory and highly siderophile PGE. It consists of seven stable isotopes, with each isotope produced by different stellar nucleosynthetic processes. The relative abundance of these isotopes is as follows: $^{184}$Os, 0.024%; $^{186}$Os, 1.600%; $^{187}$Os, 1.510%; $^{188}$Os, 13.286%; $^{189}$Os, 16.252%; $^{190}$Os, 26.369%; and $^{192}$Os, 40.958% (Faure, 1986). Of these isotopes, two are radiogenic: $^{186}$Os and $^{187}$Os. As such, Os is the daughter element in two decay systems, the very long-lived $^{190}$Pt to $^{186}$Os alpha decay (half-life: $4.5 \times 10^{11}$ years) and the beta decay of $^{187}$Re to $^{187}$Os (half-life: $4.16 \times 10^{10}$ years) (Walker *et al.*, 1997,2002). The previously reported $^{187}$Os/$^{186}$Os ratios have been changed to the more appropriate $^{187}$Os/$^{188}$Os notation in most recent publications.

The discovery of the Ir anomaly at the K–Pg boundary triggered the use of Os isotopes as a meteorite impact marker. Osmium isotopes ($^{187}$Os/$^{188}$Os) confirmed the meteoritic nature of the PGE anomalies occurring in the K–Pg clay layers (Turekian, 1982; Luck and Turekian, 1983). The application of Os isotopes in impact studies is the result of the geochemical nature of Os and Re (both are refractory and siderophile; McDonough and Sun, 1995). As Os partitions into metal phases rather than silicate phases, the range of Os concentrations found in metal alloys of differentiated planetary bodies is in the range 400–800 ng g$^{-1}$ (Anders and Grevesse, 1989; Wasson and Kallemeyn, 1988) contrary to $10^{-12}$ g g$^{-1}$ concentrations most commonly measured in silicates (i.e. PGE-poor achondrites) or Earth's crustal material. Earth's mantle peridotites, oceanic and continental basalts, are extremely poor in Os relative to the core of planets. Rhenium and Os behave differently during magmatic processes. This yields very high Re/Os ratios in continental and crustal rocks and low Re/Os ratios in the Fe–Ni core of terrestrial planets. Consequently, owing to large Re/Os ratios in crustal rocks, radioactive decay of $^{187}$Re leads to elevated $^{187}$Os crustal values. In terms of isotopic ratios, the commonly used $^{187}$Os/$^{188}$Os average for the upper continental crust is 1.4 (Peucker-Ehrenbrink and Jahn, 2001). However, this ratio varies significantly as a function of the type and age of the rock. Chondritic materials yield lower unradiogenic $^{187}$Os/$^{188}$Os ratios of approximately 0.13 with little variability between the different types of chondrites. As a result, Os isotope ratios cannot determine the detailed chemical nature of the chondritic meteorite material in either marine sections or impact melts.

The large disparity between crustal and meteoritic end-members renders the application of Os isotopes the most sensitive technique to identify a meteoritic contribution in impactites; as long as the impacted projectile is chondritic and no significant ultramafic rocks exist within the target lithology. Although the abundance of meteoritic Os in impact melts is generally small, the Os contribution induces a significant decrease in the $^{187}$Os/$^{188}$Os compared with crustal values. Older impactites usually require

a Re correction to obtain the exact initial $^{187}$Os/$^{188}$Os values. The Re–Os isotope system has been applied successfully at several terrestrial impact craters, ejecta layers and in tektites, mainly to confirm a chondritic or iron meteorite signature (Koeberl and Shirey, 1993,1997). In other cases, such as the Bosumtwi crater in Ghana, the radiogenic $^{187}$Os/$^{188}$Os signatures measured in the impactites indicate a lack of meteoritic contribution or a strong dominance of the sulfide mineralizations present in the target rock (McDonald *et al.*, 2007).

Unfortunately, the use of the Re–Os isotope system in continental crustal material does not allow the detection of differentiated basaltic achondrites, as their PGE abundances are significantly lower than chondritic concentrations and approximate abundances of the silicate Earth. Nevertheless, in most cases, the very high sensitivity of the Os isotope method detects even minute meteoritic contributions in sediments and impactites. Recently, $^{187}$Os/$^{188}$Os ratios refuted extraordinary claims that several major impact events, other than the K–Pg boundary, were responsible for sudden mass extinction. Highly radiogenic $^{187}$Os/$^{188}$Os initial ratios from sections in Italy and Austria occur at the Permo-Triassic boundary, discrediting claims of an impact event (Koeberl *et al.*, 2004). Similarly, high $^{187}$Os/$^{188}$Os ratios are measured in Eifelian/Givetian sections (Schmitz *et al.*, 2006) and in marine and terrestrial sections of the Bølling–Allerød/Younger Dryas transition (Paquay *et al.*, 2009).

### 15.3.5    Cr isotope system

The measurements of Cr isotope ($^{53}$Cr/$^{52}$Cr and $^{54}$Cr/$^{52}$Cr) ratios provide not only an unambiguous detection of meteoritic material, but also discriminate between different types of projectiles (Table 15.1). Applications of the method to crater impactites and ejecta layers are described in detail by Shukolyukov and Lugmair (1998), Shukolyukov *et al.* (2000), Koeberl *et al.* (2002, 2007), Kyte *et al.* (2003), Trinquier *et al.* (2006), Quitté *et al.* (2007), and Simonson *et al.* (2009). The conventional methodology relies on the measurement of the relative abundances of radiogenic isotope $^{53}$Cr, the daughter product of the long since extinct radionuclide $^{53}$Mn (half life of 3.7 Myr). The parent isotope $^{53}$Mn existed during the formation of the first solids and the subsequent formation of large planetesimals. So far, all the meteorite classes show characteristic relative $^{53}$Cr abundances that differ clearly from the terrestrial–Moon value (e.g. Lugmair and Shukolyukov 1998; Shukolyukov *et al.*, 2003; Shukolyukov and Lugmair, 2004,2006). No terrestrial variation of $^{53}$Cr/$^{52}$Cr ratios is expected because the Earth–Moon system homogenized long after all $^{53}$Mn had decayed. Consequently, deviation of the $^{53}$Cr/$^{52}$Cr ratio in a sample relative to the standard terrestrial value (by definition zero and usually expressed in $\varepsilon$-units; $1\,\varepsilon = 1$ part in $10^4$, or 0.01%) implies the presence of an ETC. Unlike the PGEs, Cr was not completely scavenged from the crust during the differentiation of the Earth; a result of its partially lithophile character. Consequently, the identification of the type of impacted meteorite using Cr isotopes requires an elevated proportion of projectile contribution in the impactites, in the order of a few weight per cent. This requirement limits the use of the Cr isotope method.

The determination of a Cr isotope signature depends on high-precision analytical techniques. Earlier studies used the $^{54}$Cr/$^{52}$Cr ratio to implement a correction for a 'second-order' mass fractionation resulting from the instrumental equipment (TIMS) used to achieve the required precision level of several parts per million in the measurement of $^{53}Cr/^{52}Cr$ ratios (e.g. Lugmair and Shukolyukov, 1998). The presence of a pre-solar component enriched in $^{54}Cr$ in carbonaceous chondrites explains their elevated $^{54}Cr/^{52}Cr$ ratios (e.g. Shukolyukov and Lugmair, 2006; Trinquier *et al.*, 2006,2007). Using the second-order normalization, the elevated $^{54}Cr/^{52}Cr$ ratios translate into apparent deficits in $^{53}Cr$ (i.e. negative $\varepsilon(^{53}Cr)$ values), while unnormalized, 'raw' values of $\varepsilon(^{53}Cr)$ are positive and similar to those of other undifferentiated meteorites. Normalized $^{53}Cr/^{52}Cr$ ratios for meteorites vary between +0.1 and +1.3 $\varepsilon$, except for carbonaceous chondrites, which show values around −0.4 $\varepsilon$. The original normalization procedure made possible the first isotopic recognition of the presence of a carbonaceous chondritic projectile component at the K–Pg boundary layer (Shukolyukov and Lugmair, 1998; Quitté *et al.*, 2007). The 'raw' $\varepsilon(^{53}Cr)$ and $\varepsilon(^{54}Cr)$ ratios (although less precise than the normalized equivalent) can distinguish between different classes of carbonaceous chondrites and refined the type of impactor at the K–Pg boundary down to a CM2 carbonaceous meteorite (Trinquier *et al.*, 2006). This identification agrees with the type of meteorite that was recovered, as a presumed projectile fragment, from the K–Pg boundary layer (Kyte, 1998).

The Cr isotope ratios measured in samples from the S2, S3 and S4 Barberton spherule layers, from the Archaean of South Africa, also clearly indicate carbonaceous chondrite impactors (possibly CV-type for S3 and S4, based on the raw ratios; Shukolyukov *et al.*, 2000; Kyte *et al.*, 2003). The late Eocene clinopyroxene spherule marine deposit (Kyte *et al.*, 2004), impact melt rocks from the Morokweng, East Clearwater, Lappajärvi and Rochechouart impact structures (Shukolyukov and Lugmair, 2000; Koeberl *et al.*, 2002, 2007), an Ivory Coast tektite from the Bosumtwi impact crater (Koeberl *et al.*, 2007) and seven spherule layers occurring in the formations deposited between about 2.49 and 2.63 Ga around the Archaean–Proterozoic boundary in Western Australia and South Africa (Simonson *et al.*, 2009) all exhibit Cr isotope signatures consistent with an ordinary chondrite projectile (Fig. 15.6). However, although carbonaceous and enstatite chondrites show distinguishable Cr isotope ratios, all other types of chondrites and differentiated meteorites share a rather similar Cr isotope composition (Kyte *et al.*, 2003).

The relatively high background of Cr in terrestrial rocks not only restricts the detection of an ETC in impactites and distal ejecta layers, but also requires precise estimates on the amount of meteoritic material present. As the moderately siderophile elements and HSEs are carried by other phases than the partly lithophile Cr, these assessments could be precarious. The limit of detection of an ETC is proportional to the indigenous Cr concentrations and the proportion of meteoritic material admixed in the impactites (Koeberl *et al.*, 2002; Frei and Rosing, 2005).

### 15.3.6 Other potential projectile indicators

Several attempts have been made to resolve the origin of geochemical anomalies at impact structures and ejecta layers using other isotope systems (e.g. Frei and Frei, 2002). The uranium–lead (U–Pb), rubidium–strontium (Rb–Sr) and samarium–neodymium (Sm–Nd) isotope systems can be used as tracers of

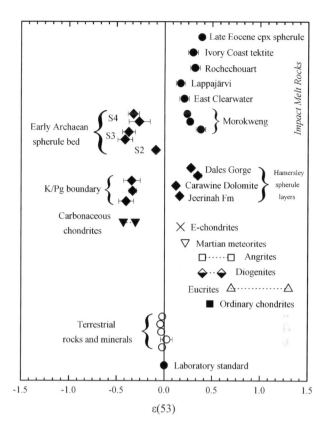

**Figure 15.6** Plot of second-order corrected, normalized $\varepsilon(^{53}Cr)$ of various impact-related materials (impact melt rock and impact ejecta deposits) compared with terrestrial and meteoritic samples. The Cr isotope signature of the Early Archaean spherule beds and the K–Pg boundary clays is non-terrestrial and similar to that in carbonaceous chondrites. The Morokweng, East Clearwater, Lappajärvi and Rochehouart impact melt rocks, an Ivory Coast tektite, a late Eocene clinopyroxene-bearing spherule layer and the 2.63–2.49 Ga Hamersley Basin spherule layers show close affinities to ordinary chondrites. Error bars represent $2\sigma$ mean uncertainties. Data from Lugmair and Shukolyukov (1998), Shukolyukov and Lugmair (1998, 2000, 2004, 2006), Shukolyukov *et al.* (2000, 2003), Koeberl *et al.* (2002, 2007), Kyte *et al.* (2003) and Simonson *et al.* (2009), with permission from Elsevier.

source rocks, and by extension perhaps as a clue to the impacted projectile type. Similarly, provenance studies can be carried out to trace the source crater of tektites or other forms of ejected material. However, the terrestrial values dominate and mask the possible meteoritic abundances and isotope signatures, hampering their use for projectile identification.

Tungsten ($^{182/184}W$) isotope ratios (based on the decay of $^{182}Hf$ to $^{182}W$ with a half-life of 9 Myr) have been proposed to evaluate the degree of mixing that occurred between meteoritic material and the surrounding terrestrial sediments (Quitté *et al.*, 2007). Similar to Cr isotope systematics, the terrestrial W isotope composition is constant ($\varepsilon(^{182}W)=0$ by definition; Lee and Halliday, 1996; Schoenberg *et al.*, 2002), while different classes of meteorites show a high variability in W isotope ratios. Iron meteorites have the lowest $^{182/184}$W ratios ($-4.4$ to $-2.5$ $\varepsilon_W$), reflecting the early removal of metal-loving W from silicate into the core of planetesimals, when most $^{182}$Hf was still present. Enstatite, ordinary chondrites and carbonaceous chondrites yield $\varepsilon_W$ of $-2.2$, $-2$ to $-0.64$ and $-1.9$ respectively, while angrites and eucrites show positive values, of $+4.6$ and $+2$ to $+39$ $\varepsilon_W$ respectively (see Quitté et al. (2007) and references therein). So far, very few impactite samples have been analysed for W isotope ratios. The W isotope compositions of five K–Pg boundaries and surrounding sediments (Stevns Klint, Caravaca, Bidart, Berwind Canyon and Gams) as well as selected impactites from the Gardnos, Vredefort, Morokweng and Ries impact structures are indistinguishable from the terrestrial value, within uncertainties (Quitté et al., 2007; Moynier et al 2009). This results from the extreme dilution of the isotope signature, because of the rather low W concentration in meteorites. However, all of the W could also be terrestrial. The only controversial evidence of W isotope ratios as an impact signature comes from metamorphosed sedimentary rocks from the 3.7–3.8 Ga Isua greenstone belt of West Greenland and closely related rocks from northern Labrador, Canada. These samples show meteoritic W isotope anomalies, interpreted as the only trace of the Late Heavy Bombardment on Earth (Schoenberg et al., 2002). However, Cr isotope analyses fail to verify these observations (Frei and Rosing, 2005).

### 15.3.7   Multi-tracer approach

Multi-tracer approaches provide the most complete and coherent identification of the impacted projectile (e.g. Frei and Frei, 2002; Kyte et al., 2004; Trinquier et al., 2006). They also allow comparison of different techniques, and testing of possible projectile types and components. For example, the Cr isotope data complements the inter-element correlations between siderophiles. Combined application of these methods can determine a specific type of ordinary chondrite impactor. In the case of the Lappajärvi impact structure, the combination of siderophile element and Cr isotope data (Koeberl et al., 2007) agrees with PGE ratios (Tagle et al., 2007), both clearly indicating an H-chondrite as the most likely projectile.

Once the type of projectile has been recognized, the Ir values present in the impactites also refine the estimation of the meteoritic contribution incorporated during the cratering event (Table 15.2). For the Popigai impact structure that likely resulted from the impact of an ordinary chondrite, the meteoritic contamination does not exceed 0.5 wt% (Tagle and Claeys, 2005). In the case of the Morokweng crater, the impactor contribution amounts up to an exceptional 5.7 wt% of an LL chondrite projectile (McDonald et al., 2001; Maier et al., 2006). In addition, excursions in the marine $^{187}$Os/$^{188}$Os isotope ratio record have provided an estimation of the size of the impactors for the K–Pg boundary and the Late Eocene impact events, independent of the iridium data (Paquay et al., 2008).

### 15.4   Review of identified projectiles

Systematic studies of projectile identification document the origin of the objects that have collided with the Earth over the last 3 billion years. So far, the record is rather incomplete (Table 15.1). The scarcity of projectile identification studies is in part the result of a lack of projectile-enriched material. For the terrestrial craters larger than 1 km, most of the projectiles are only characterized down to the level of chondrite or iron meteorite, without further details. Some of the classifications remain ambiguous; for example, for Zhamanshin, both a chondrite and an iron meteorite have been proposed (Palme et al., 1978; Glass et al., 1983).

Only for a small fraction of the structures investigated is the proposed projectile type known down to a specific type of chondrite. For a substantial amount of structures, no clear meteoritic contribution occurs, in some cases leading to the suggestion that a differentiated, PGE-poor, possibly achondritic projectile formed this particular structure (e.g. Ries, Manicouagan, West Clearwater and Saint Martin; Morgan et al., 1979; Palme et al., 1980; Wolf et al., 1980; Palme, 1982). However, differentiated achondrites represent only around 2% of the total present meteoritic population (Grady, 2000). Alternatively, unknown types of meteorite (such as PGE-poor cometary bodies) could be advocated.

Some impact melt rocks seem deprived of any meteoritic component (e.g. Lake Saint Martin, Ries; Morgan et al., 1979; Palme et al., 1980) or the meteoritic component can be overshadowed by the presence of mantle-derived mafic or ultramafic target rock components (Bosumtwi; Goderis et al., 2007; McDonald et al., 2007). The percentage of meteoritic material that remains in the crater after the impact also depends strongly on the impact velocity and angle of the impacting meteorites (Pierazzo and Melosh, 2000; Artemieva et al., 2004).

Although the distal K–Pg boundary ejecta of the Chicxulub impact structure are marked by high enrichments in impactor-derived elements and anomalous isotope ratios (e.g. Alvarez et al., 1980; Smit and Hertogen, 1980; Shukolyukov and Lugmair, 1998), the impactites of the Chicxulub structure have yielded no detectable meteoritic contamination to date (Gelinas et al., 2004; Tagle et al., 2004). The apparent lack of a projectile signature within an impact structure, therefore, is not a strict argument for the absence of a PGE-rich projectile. The meteoritic signature can also be modified by fractionation, alteration and post-impact remobilization, as observed at a number of terrestrial impact structures (e.g. Evans et al., 1993; Colodner et al., 1992; McDonald et al., 2001), the Archaean spherule layers in Western Australia and South Africa (Simonson et al., 1998,2009; Reimold et al., 2000), the ~590 Ma Acraman impact ejecta and host shales (Wallace et al., 1990) and the 65 Ma K–Pg boundary clay (Evans et al., 1995). However, these processes are unlikely to account for the complete absence of a meteoritic contribution. Clearly, projectile identification in impact structures is not always a straightforward task.

Based on the existing data, chondrites dominate the terrestrial impactor population (Table 15.1). Seven of the Phanerozoic structures (>1 km) for which the projectiles are known down to the level of a specific class or type were produced by ordinary chondrites (OC). Only the projectile of the largest known impact in the last 1 billion years, the Chicxulub structure, stands out as a carbonaceous chondrite (CM2-type; Kyte, 1998; Shukolyukov and Lugmair, 1998; Trinquier et al., 2006; Quitté et al., 2007). Stony-iron projectiles prevail in the non-chondritic fraction of the projectiles identified, including the mesosiderite fragments found at the Eltanin impact site (Kyte, 2002). Non-magmatic iron (NMI) meteorites created the 6 km Sääksjärvi, 23 km Rochechouart, 5 km Gardnos and 7.5 km Lockne impact structures (Tagle *et al.*, 2008a, 2009; Goderis *et al.*, 2009). For the Nicholson Lake and Strangways impact structures, olivine-rich achondrites have been proposed (Morgan and Wandless, 1983; Wolf *et al.*, 1980). Further back in time, both PGE ratios and Cr isotopic anomalies indicate that most, if not all, of the impactors that produced the spherule layers deposited between circa 2.63 and 2.49 Ga in Western Australia and South Africa were ordinary chondrites in composition (Simonson *et al.*, 2009). In contrast, carbonaceous chondritic impactors produced all known terrestrial spherule layers older than approximately 3.0 Ga (Kyte *et al.*, 2003; Lowe *et al.*, 2003).

The first projectile-identification studies preferred differentiated bodies over chondrites (e.g. Palme *et al.*, 1978; Palme 1982). However, the view on the impactor population hitting Earth through time has changed over the last 30 years in favour of ordinary chondrites (Chapter 2). With the systematic application of a wide range of analytical techniques in the impactites of the currently recognized and yet undiscovered terrestrial impact craters, the projectile population is perhaps likely to evolve even further.

## 15.5   Concluding remarks

Of the impact structures recognized on Earth today, approximately 50 have been characterized, with various degrees of precision, for the type of impactor responsible for its formation. A variety of impactors, compositionally similar to the samples in the present meteorite population and the corresponding parent bodies in the Main Asteroid Belt, have been found. Ordinary chondritic meteorites dominate the impactor population, at least for the Phanerozoic. Only the Chixculub impact structure and the Precambrian spherule layers of the Barberton Mountain Land older than circa 3.0 Ga were determined as carbonaceous chondrites.

Although our present knowledge on the projectile population distribution remains fragmentary, terrestrial and lunar projectile identification studies have unravelled a few trends that need to be confirmed. Systematic application of one or, preferably, more of the methodologies (determination of siderophile element abundances, and Os and Cr isotope ratios) in all available impact material samples will further characterize the nature and the frequency of the projectiles that have bombarded the Earth for the last 4.5 billion years. Eventually, changes in impactor-types will be highlighted, perhaps reflecting modification in the dynamics of small bodies in the Solar System.

## References

Alvarez, L.W., Alvarez, W., Asaro, F. and Michel, H.V. (1980) Extraterrestrial cause for the Cretaceous–Tertiary extinction. *Science*, 208, 1095–1108.

Amor, K., Hesselbo, S.P., Porcelli, D. *et al.* (2009) A Precambrian proximal ejecta blanket from Scotland. *Geology*, 36, 303–306.

Anders, E. and Grevesse, N. (1989) Abundances of the elements: meteoritic and solar. *Geochimica et Cosmochimica Acta*, 53, 197–214.

Artemieva, N., Karp, T. and Milkereit, B. (2004) Investigating the Lake Bosumtwi impact structure: insight from numerical modeling. *Geochemistry, Geophysics, Geosystems*, 5, 1–20.

Attrep, M., Orth, C.J., Quintana, L.R. *et al.* (1991) Chemical fractionation of siderophile elements in impactites from Australian meteorite craters, in *Lunar and Planetary Science XXII*, The Lunar and Planetary Institute, Houston, TX, pp. 39–40.

Becker, H., Horan, M.F., Walker, R.J. *et al.* (2006) Highly siderophile element composition of the Earth's primitive upper mantle: constraints from new data on peridotite massifs and xenoliths. *Geochimica et Cosmochimica Acta*, 70, 4528–4550.

Borisov, A., Palme, H. and Spettel, B. (1994) Solubility of palladium in silicate melts: implications for core formation in the Earth. *Geochimica et Cosmochimica Acta*, 58, 705–716.

Brenan, J.M. (2008) The platinum-group elements: 'admirably adapted' for science and industry. *Elements*, 4, 227–232.

Buchwald, V.F. (1975) *Handbook of Iron Meteorites*, University of California Press, Berkeley, CA.

Bunch, T.E. and Cassidy, W.A. (1972) Petrographic and electron microprobe study of the Monturaqui impactite. *Contributions to Mineralogy and Petrology*, 36, 95–112.

Claeys, P., Kiessling, W. and Alvarez, W. (2002) Distribution of Chicxulub ejecta at the Cretaceous–Tertiary boundary, in *Catastrophic Events and Mass Extinctions: Impacts and Beyond* (eds C. Koeberl and K.G. MacLeod), Geological Society of America, Boulder, CO, pp. 55–68.

Colodner, D.C., Boyle, E.A., Edmond, J.M. and Thomson, J. (1992) Post-deposital mobility of platinium, iridium and rhenium in marine sediments. *Nature*, 358, 402–404.

Drake, M.J. and Righter, K. (2002) Determining the composition of the Earth. *Nature*, 416, 39–44.

Dressler, B.O. and Reimold, W.U. (2001) Terrestrial impact melt rocks and glasses. *Earth-Science Reviews*, 56, 205–284.

Dypvik, H. and Attrep Jr, M. (1999) Geochemical signals of the Late Jurassic, marine Mjølnir impact. *Meteoritics and Planetary Science*, 34, 393–406.

Earth Impact Database (2010) http://www.unb.ca/passc/Impact Database (last accessed 31 May 2010).

Evans, N.J., Gregoire, D.C., Grieve, R.A.F. *et al.* (1993) Use of platinum-group elements for impactor identification – terrestrial impact craters and Cretaceous–Tertiary boundary. *Geochimica et Cosmochimica Acta*, 57, 3737–3748.

Evans, N.J., Ahrens, T.J. and Gregoire, D.C. (1995) Fractionation of ruthenium from iridium at the Cretaceous–Tertiairy boundary. *Earth and Planetary Science Letters*, 134, 141–153.

Faure, G. (1986) *Principles of Isotope Geology*, 2nd edn, John Wiley and Sons, Inc., New York, NY.

Fischer-Gödde, M., Becker, H. and Wombacher, F. (2010) Rhodium, gold and other siderophile abundances in chondritic meteorites. *Geochimica et Cosmochimica Acta*, 74, 356–379.

Frei, R. and Frei, K.M. (2002) A multi-isotopic and trace element investigation of the Cretaceous–Tertiary boundary layer at Stevens Klint, Denmark – inferences for the origin and nature of siderophile and lithophile geochemical anomalies. *Earth and Planetary Science Letters*, 203, 691–708.

Frei, R. and Rosing, M.T. (2005) Search for traces of the late heavy bombardment on Earth – results from high precision chromium isotopes. *Earth and Planetary Science Letters*, 236, 28–40.

French, B.M., Koeberl, C., Gilmour, I. *et al.* (1997) The Gardnos impact structure, Norway: petrology and geochemistry of target rocks and impactites. *Geochimica et Cosmochimica Acta*, 61, 873–904.

Gelinas, A., Kring, D.A., Zurcher, L. *et al.* (2004) Osmium isotope constraints on the proportion of bolide component in Chicxulub impact melt rocks. *Meteoritics and Planetary Science*, 39, 1003–1008.

Glass, B.P., Fredricksson, K. and Florensky, P.V. (1983) Microirghizites recovered from a sediment sample from the Zhamanshin impact structure. *Journal of Geophysical Research*, 88, B319–B330.

Goderis, S., Tagle, R., Schmitt, R.T. *et al.* (2007) Platinum group elements provide no indication of a meteoritic component in ICDP cores from the Bosumtwi crater, Ghana. *Meteoritics and Planetary Science*, 42, 731–741.

Goderis, S., Kalleson, E., Tagle, R. *et al.* (2009) A non-magmatic iron projectile for the Gardnos impact event. *Chemical Geology*, 258, 145–156.

Goderis, S., Hertogen, J., Vanhaecke, F. and Claeys, Ph. (2010) Siderophile elements from the Eyreville drill cores of the Chesapeake Bay impact structure do not constrain the nature of the projectile, in *Large Meteorite Impacts and Planetary Evolution IV* (eds R.L. Gibson and W.U. Reimold), Geological Society of America, Boulder, CO, pp. 395–409.

Gostin, V.A., Keays, R.R. and Wallace, M.W. (1989) Iridium anomaly from the Acraman impact ejecta horizon: impacts can produce sedimentary iridium peaks. *Nature*, 340, 542–544.

Grady, M.M. (2000) *Catalogue of Meteorites*, 5th edn, Cambridge University Press, Cambridge.

Grieve, R.A.F. and Shoemaker, E.M. (1994) The record of past impacts on Earth, in *Hazards Due to Comets and Asteroids* (ed. T. Gehrels), University of Arizona Press, Tucson, AZ, pp. 417–462.

Grieve, R.A.F., Dence, M.R. and Robertson, P.B. (1977) Cratering processes: as interpreted from the occurrence of impact melts, in *Impact and Explosion Cratering* (eds D.J. Roddy, R.O. Pepi and R.B. Merrill), Pergamon Press, New York, NY, pp. 791–814.

Grieve, R.A.F., Bottomley, R.B., Bouchard, M.A. *et al.* (1991) Impact melt rocks from New Quebec Crater, Quebec, Canada. *Meteoritics*, 26, 31–39.

Gurov, E.P. (1996) The group of Macha craters in Western Yakutia, in *Lunar and Planetary Science XXVII*, The Lunar and Planetary Institute, Houston, TX, pp. 473–474.

Hall, G.E.M. and Pelchat, J.C. (1994) Analysis of geochemical materials for gold, platinum and palladium at low ppb levels by fire assay-ICP mass spectrometry. *Chemical Geology*, 115, 61–72.

Halliday, A.N. (2004) Mixing, volatile loss and compositional change during impact-driven accretion of the Earth. *Nature*, 427, 505–509.

Hecht, L., Reimold, U.W., Sherlock, S. *et al.* (2008) New impact-melt rock from the Roter Kamm impact structure, Namibia: further constraints on impact age, melt rock chemistry, and projectile composition. *Meteoritics and Planetary Science*, 43, 1201–1218.

Horan, M.F., Walker, R.J., Morgan, J.W. *et al.* (2003) Highly siderophile elements in the Earth and meteorites. *Chemical Geology*, 196, 5–20.

Janssens, M.-J., Hertogen, J., Takahashi, H. *et al.* (1977) Rochechouart Meteorite Crater: identification of projectile. *Journal of Geophysical Research*, 82, 750–758.

Kallemeyn, G.W. and Wasson, J.T. (1986) Composition of enstatite (EH3, EH4,5 and EL6) chondrites: implication regarding their formation. *Geochimica et Cosmochimica Acta*, 50, 2153–2164.

Kenkmann, T., Artemieva, N.A., Wünnemann, K. *et al.* (2009) The Carancas meteorite impact crater, Peru: geologic surveying and modeling of crater formation and atmospheric passage. *Meteoritics and Planetary Science*, 44, 985–1000.

Kimura, K., Lewis, R.S. and Anders, E. (1974) Distribution of gold and rhenium between nickel–iron and silicate melts. *Geochimica et Cosmochimica Acta*, 38, 683–781.

Koblitz, J. (2000) MetBase 5.0. CD-ROM.

Koeberl, C. (1998) Identification of meteoritic component in impactites, in *Meteorites: Flux with Time and Impact Effects* (eds M.M. Grady, R. Hutchinson, G.J.H. McCall and R.A. Rothery), The Geological Society, London, pp. 133–153.

Koeberl, C. (2007) The geochemistry and cosmochemistry of impacts, in *Treatise of Geochemistry* (eds A. Davis), Elsevier, Amsterdam, pp. 1.28.1–1.28.52 (online edition).

Koeberl, C. and Auer, P. (1991) Geochemistry of impact glass from the Aouelloul crater, Mauretania, in *Lunar and Planetary Science XXII*, The Lunar and Planetary Institute, Houston, pp. 731–732.

Koeberl, C. and Shirey, S.B. (1993) Detection of a meteoritic component in Ivory Coast tektites with rhenium–osmium isotopes. *Science*, 261, 595–598.

Koeberl, C. and Shirey, S.B. (1996) Re–Os isotope study of rocks from the Manson impact structure, in *The Manson Impact Structure, Iowa: Anatomy of an Impact Crater* (eds C. Koeberl and R.R. Anderson), Geological Society of America, Boulder, CO, pp. 331–339.

Koeberl, C. and Shirey, S.B. (1997) Re–Os isotope systematics as a diagnostic tool for the study of impact craters and distal ejecta. *Palaeogeography, Palaeoclimatology, Palaeoecology*, 132, 25–46.

Koeberl, C., Reimold, W.U. and Shirey, S.B. (1994a) Saltpan impact crater, South Africa: geochemistry of target rocks, breccias, and impact glasses, and osmium isotope systematics. *Geochimica et Cosmochimica Acta*, 58, 2893–2910.

Koeberl, C., Reimold, W.U., Shirey, S.B. and Le Roux, F.G. (1994b) Kalkkop crater, Cape Province, South Africa: confirmation of impact origin using osmium isotope systematics. *Geochimica et Cosmochimica Acta*, 58, 1229–1234.

Koeberl, C., Reimold, W.U. and Shirey, S.B. (1996) A Re–Os isotope and geochemical study of the Vredefort Granophyre: clues to the origin of the Vredefort structure, South Africa. *Geology*, 24, 913–916.

Koeberl, C., Peucker-Ehrenbrink, B., Reimold, W.U. *et al.* (2002) Comparison of Os and Cr isotopic methods for the detection of meteoritic components in impactites: examples from the Morokweng and Vredefort impact structures, South Africa, in *Catastrophic Events & Mass Extinctions: Impacts and Beyond* (eds C. Koeberl and K.G. MacLeod), Geological Society of America, Boulder, CO, pp. 607–617.

Koeberl, C., Farley, K.A., Peucker-Ehrenbrink, B. and Sephton, M.A. (2004) Geochemistry of the end-Permian extinction event in Austria and Italy: no evidence for an extraterrestrial component. *Geology*, 32, 1053–1056.

Koeberl, C., Shukolyukov, A. and Lugmair, G.W. (2007) Chromium isotopic studies of terrestrial impact craters: identification of meteoritic components at Bosumtwi, Clearwater East, Lappajärvi, and Rochechouart. *Earth and Planetary Science Letters*, 256, 534–546.

Kramers, J.D. (1998) Reconciling siderophile element data in the Earth and Moon, W isotopes and the upper lunar age limit in a simple model of homogeneous accretion. *Chemical Geology*, 145, 461–478.

Kyte, F.T. (1998) A meteorite from the Cretaceous/Tertiary boundary. *Nature*, 396, 237–239.

Kyte, F.T. (2002) Unmelted meteoric debris collected from Eltanin ejecta in Polarstern cores from expedition ANT XII/4. *Deep-Sea Research II*, 49, 1063–1071.

Kyte, F.T., Zhou, Z. and Wasson, J.T. (1980) Siderophile-enriched sediments from the Cretaceous–Tertiary boundary. *Nature*, 288, 651–656.

Kyte, F.T., Bostwick, J.A. and Zhou, L. (1995) Identification of the Cretaceous–Tertiary boundary at ODP Site 886, ODP Site 803, and DSDP Site 576. *Proceedings, Ocean Drilling Program, Scientific Results*, 145, 427–434.

Kyte, F.T., Shukolyukov, A., Lugmair, G.W. *et al.* (2003) Early Archean spherule beds: chromium isotopes confirm origin through multiple impacts of projectiles of carbonaceous chondrite type. *Geology*, 31, 283–286.

Kyte, F.T., Shukolyukov, A., Hildebrand, A.R. *et al.* (2004) Initial Cr-isotopic and iridium measurements of concentrates from late Eocene CPX-spherule deposits. 35th Lunar and Planetary Science Conference, CD-ROM, abstract no. 1824.

Lee, D.C. and Halliday, A.N. (1996) Hf–W isotopic evidence for rapid accretion and differentiation in the early Solar System. *Science*, 274, 1876–1879.

Lee, S.R., Horton Jr, J.W. and Walker, R.J. (2006) Confirmation of a meteoritic component in impact melt rocks of the Chesapeake Bay impact structure, Virginia, USA – evidence from osmium isotopic and PGE systematics. *Meteoritics and Planetary Science*, 41, 819–833.

Lorand, J.-P., Luguet, A. and Alard, O. (2008) Platinum-group elements: a new set of key tracers for the Earth's interior. *Elements*, 4, 247–252.

Lowe, D.R., Byerly, G.R., Kyte, F.T. *et al.* (2003) Spherule beds 3.47–3.24 billion years old in the Barberton Greenstone Belt, South Africa: a record of large meteorite impacts and their influence on early crustal and biological evolution. *Astrobiology*, 3, 7–48.

Luck, J.M. and Turekian, K.K. (1983) Osmium-187/osmium-186 in manganese nodules and the Cretaceous–Tertiary boundary. *Science*, 222, 613–615.

Lugmair, G.W. and Shukolyukov, A. (1998) Early Solar System timescales according to $^{53}$Mn–$^{53}$Cr systematics. *Geochimica et Cosmochimica*, 62, 2863–2886.

Maier, W.D., Andreoli, M.A.G., McDonald, I. *et al.* (2006) Discovery of a 25-cm asteroid clast in the giant Morokweng impact crater, South Africa. *Nature*, 411, 203–206.

McDonald, I. (2002) Clearwater East impact structure: a reinterpretation of the projectile type using new platinum-group element data. *Meteoritics and Planetary Science*, 37, 459–464.

McDonald, I., Andreoli, M.A.G., Hart, R.J. and Tredoux, M. (2001) Platinum-group elements in the Morokweng impact structure, South Africa: evidence for the impact of a large ordinary chondrite projectile at the Jurassic–Cretaceous boundary. *Geochimica et Cosmochimica Acta*, 65, 299–309.

McDonald, I., Peucher-Ehrenbrink, B., Coney, L. *et al.* (2007) Search for a meteoritic component in drill cores from the Bosumtwi impact structure, Ghana: platinum group element contents and osmium isotopic characteristics. *Meteoritics and Planetary Science*, 42, 743–753.

McDonald, I., Bartosova, K. and Koeberl, C. (2009) Search for a meteoritic component in impact breccia from the Eyreville core, Chesapeake Bay impact structure: considerations from platinum group element contents, in *The ICDP–USGS Deep Drilling Project in the Chesapeake Bay Impact Structure: Results from the Eyreville Core Hole* (eds G.S. Gohn, C. Koeberl, K.G. Miller and W.U. Reimold), Geological Society of America, Boulder, CO, pp. 469–479.

McDonough, W.F. and Sun, S.S. (1995) The composition of the Earth. *Chemical Geology*, 120, 223–253.

Meisel, T. and Moser, J. (2004) Reference materials for geochemical PGE analysis: new analytical data for Ru, Rh, Pd, Os, Ir, Pt and Re by isotope dilution ICP-MS in 11 geological reference materials. *Chemical Geology*, 208, 319–338.

Meisel, T., Walker, R.J. and Morgan, J.W. (1996) The osmium isotopic composition of the primitive upper mantle. *Nature*, 383, 517–520.

Melosh, H.J. (1989) *Impact Cratering: A Geologic Process*, Oxford University Press, New York, NY.

Mittlefehldt, D.W., See, T.H. and Hörz, F. (1992a) Projectile dissemination in impact melts from Meteor crater, Arizona, in *Lunar and Planetary Science XXIII*, The Lunar and Planetary Institute, Houston, TX, pp. 919–920.

Mittlefehldt, D.W., See, T.H. and Hörz, F. (1992b) Dissemination and fractionation of projectile materials in the impact melts from Wabar crater, Saudi Arabia. *Meteoritics*, 27, 361–370.

Mittlefehldt, D.W., McCoy, T., Goodrich, C.A. and Kracher, A. (1998) Non-chondritic meteorites from the asteroidal bodies, in *Planetary Materials* (ed. J.J. Papike), Mineralogical Society of America, Washington, DC, pp. 1–195.

Morgan, J.W. (1986) Osmium isotope constraints on Earth's late accretionary history. *Nature*, 317, 703–705.

Morgan, J.W. and Wandless, G.A. (1983) Strangways Crater, northern territory, Australia: siderophile element enrichment and lithophile element fractionation. *Journal of Geophysical Research*, 88, A819–A829.

Morgan, J.W., Ganapathy, R. and Anders, E. (1975) Meteoritic material in four terrestrial meteorite craters. Proceedings, 6th Lunar Science Conference, pp. 1609–1623.

Morgan, J.W., Janssens, M.-J., Hertogen, J. *et al.* (1979) Ries impact crater, southern Germany: search for meteoritic material. *Geochimica et Cosmochimica Acta*, 43, 803–815.

Morgan, J.W., Wandless, G.A., Petrie, R.K. and Irving, A.J. (1981) Composition of the Earth's upper mantle – 1. Siderophile trace elements in ultramafic nodules. *Tectonophysics*, 75, 47–67.

Morgan, J.W., Janssens, M., Takahashi, H. *et al.* (1985) H-chondrites: trace element clues to their origin. *Geochimica et Cosmochimica Acta*, 49, 247–259.

Moynier, F., Koeberl, C., Quitté, G. and Telouk, P. (2009) A tungsten isotope approach to search for meteoritic components in terrestrial impact rocks. *Earth and Planetary Science Letters*, 286, 35–40.

Norman, M.D., Bennett, V.C. and Ryder, G. (2002) Targeting the impactors: siderophile element signatures of lunar impact melts from Serenitatis. *Earth and Planetary Science Letters*, 202, 217–228.

O'Neill, H.S.C. (1991) The origin of the Moon and the early history of the Earth – a chemical model. Part, 2, the Earth. *Geochimica et Cosmochimica Acta*, 55, 1159–1172.

Palme, H. (1980) The meteoritic contamination of terrestrial and lunar impact melts and the problem of indigenous siderophiles in the lunar highland. *Proceedings of the Lunar and Planetary Science Conference*, 11, 481–506.

Palme, H. (1982) Identification of projectiles of large terrestrial impact craters and some implications for the interpretation of Ir-rich Cretaceous/Tertiary boundary layers, in *Geological Implication of Impacts of Large Asteroids and Comets on Earth* (eds L.T. Silver and P.H. Schultz), Geological Society of America, Boulder, CO, pp. 223–233.

Palme, H. (2008) Platinum-group elements in cosmochemistry. *Elements*, 4, 233–238.

Palme, H., Janssens, M.J., Takahashi, H. *et al.* (1978) Meteoritic material at five large impact craters. *Geochimica et Cosmochimica Acta*, 42, 313–323.

Palme, H., Göbel, E. and Grieve, R.A.F. (1979) The distribution of volatile and siderophile elements in the impact melt of East Clearwater (Quebec). *Proceedings of the Lunar and Planetary Science Conference*, 10, 2465–2492.

Palme, H., Rammensee, W. and Reimold, W.U. (1980) The meteoritic component of impact melts from European impact craters. *Proceedings of the Lunar and Planetary Science Conference*, 11, 848–851.

Palme, H., Grieve, R.A.F. and Wolf, R. (1981) Identification of the projectile at the Brent crater, and further considerations of projectile types at terrestrial craters. *Geochimica et Cosmochimica Acta*, 45, 2417–2424.

Paquay, F.S., Ravizza, G.E., Dalai, T.K. and Peucker-Ehrenbrink, B. (2008) Determining chondritic impactor size from the marine osmium isotope record. *Science*, 320, 214–218.

Paquay, F.S., Goderis, S., Ravizza, G. et al. (2009) Absence of geochemical evidence for an impact event at the Bolling, Allerod/Younger Dryas transition. *Proceedings of the National Academy of Sciences of the United States of America*, 106, 21505–21510.

Pernicka, E., Kaether, D. and Koeberl, C. (1996) Siderophile element concentrations in drill core samples from the Manson crater, in *The Manson Impact Structure, Iowa: Anatomy of an Impact Crater* (eds C. Koeberl and R.R. Anderson), Geological Society of America, Boulder, CO, pp. 325–330.

Peucker-Ehrenbrink, B. and Jahn, B.-M. (2001) Rhenium–osmium isotope systematics and platinum group element concentrations: loess and the upper continental crust. *Geochemistry Geophysics Geosystems*, 2 (10), 1061.

Pierazzo, E. and Melosh, H.J. (2000) Hydrocode modeling of oblique impacts: the fate of the projectile. *Meteoritics and Planetary Science*, 35, 117–130.

Plessen, H.-G. and Erzinger, J. (1998) Determination of the platinum-group elements and gold in twenty rock reference material by inductively coupled plasma-mass spectrometry (ICP-MS) after pre-concentration by nickel fire assay. *Geostandards Newsletter*, 22, 187–194.

Quitté, G., Levasseur, S., Capmas, F. et al. (2007) Osmium, tungsten, and chromium isotopes in sediments and in Ni-rich spinel at the K–T boundary: signature of a chondritic impactor. *Meteoritics and Planetary Science*, 42, 1567–1580.

Rankenburg, K., Humayun, M., Brandon, A.D. and Herrin, J.S. (2008) Highly siderophile elements in ureilites. *Geochimica et Cosmochimica Acta*, 72, 4642–4659.

Reimold, W.U., Koeberl, C. and Reddering, J.S. V. (1998) The 1992 drill core from the Kalkkop impact crater, Eastern Cape Province, South Africa: stratigraphy, petrography, geochemistry and age. *Journal of African Earth Sciences*, 26, 573–592.

Reimold, W.U., Koeberl, C., Johnson, S. and McDonald, I. (2000) Early Archean spherule beds in the Barberton Mountain Land, South Africa: impact or terrestrial origin? in *Impacts and the Early Earth* (eds I. Gilmour and, C. Koeberl), Springer, Heidelberg, pp. 117–180.

Schmidt, G. and Pernicka, E. (1994) The determination of platinum group elements (PGE) in target rocks and fall-back material of the Nördlinger Ries impact crater (Germany). *Geochimica et Cosmochimica Acta*, 58, 5083–5090.

Schmidt, G., Palme, H. and Kratz, K.L. (1997) Highly siderophile elements (Re Os, Ir, Ru, Rh, Pd, Au) in impact melts from three European impact craters (Sääksjärvi, Mien and Dellen): clues to the nature of the impacting bodies. *Geochimica et Cosmochimica Acta*, 61, 2977–2987.

Schmitz, B., Ellwood, B.B., Peucker-Ehrenbrink, B. et al. (2006) Platinum group elements and Os-187/Os-188 in a purported impact ejecta layer near the Eifelian–Givetian stage boundary, Middle Devonian. *Earth and Planetary Science Letters*, 249, 162–172.

Schoenberg, R., Kamber, B.S., Collerson, K.D. and Moorbath, S. (2002) Tungsten isotope evidence from ~3.8-Gyr metamorphosed

sediments for early meteorite bombardment of the Earth. *Nature*, 418, 403–405.

Schultz, P.H., Koeberl, C., Bunch, T. et al. (1994) Ground truth for oblique impact processes: new insight from the Rio Cuarto, Argentina, crater field *Geology*, 22, 889–892.

Shukolyukov, A. and Lugmair, G.W. (1998) Isotopic evidence for the Cretaceous–Tertiary impactor and its type. *Science*, 282, 927–929.

Shukolyukov A. and Lugmair G.W. (2000) Extraterrestrial matter on Earth: evidence from the Cr isotopes. International Conference on Catastrophic Events and Mass Extinctions: Impacts and Beyond, 9–12 July, Vienna, Austria, pp. 197–198 (abstract no. 3041).

Shukolyukov, A. and Lugmair, G.W. (2004) Manganese–chromium isotope systematics of enstatite meteorites. *Geochimica et Cosmochimica Acta*, 68, 2875–2888.

Shukolyukov, A. and Lugmair, G.W. (2006) Manganese–chromium isotope systematics of carbonaceous chondrites. *Earth and Planetary Science Letters*, 250, 200–213.

Shukolyukov, A., Kyte, F.T., Lugmair, G.W. et al. (2000) The oldest impact deposits on Earth – first confirmation of an extraterrestrial component, in *Impacts and the Early Earth* (eds I. Gilmour and, C. Koeberl), Springer, Heidelberg, pp. 99–115.

Shukolyukov, A., Lugmair, G.W. and Bogdanovski, O. (2003) Manganese–chromium isotope systematics of Ivuna, Kainsaz and other carbonaceous chondrites. 34th Lunar and Planetary Science Conference, CD-ROM, abstract no. 1279.

Simonson, B.M., Davies, D., Wallace, M. et al. (1998) Iridium anomaly but no so shocked quartz from Late Archean microkrystite layer: oceanic impact ejecta? *Geology*, 26, 195–198.

Simonson, B.M., McDonald, I., Shukolyukov, A. et al. (2009) Geochemistry of 2.63–2.49 Ga impact spherule layers and implications for stratigraphic correlations and impact processes. *Precambrian Research*, 175, 51–76.

Smit, J. and Hertogen, J. (1980) An extraterrestrial event at the Cretaceous–Tertiary boundary. *Nature*, 285, 198–200.

Tagle, R. and Berlin, L. (2008) A database of chondrite analyses including platinum group elements, Ni, Co, Au, and Cr: implications for the identification of chondritic projectiles. *Meteoritics and Planetary Science*, 43, 541–559.

Tagle, R. and Claeys, P. (2004) Comet or asteroid shower in the late Eocene? *Science*, 305, 492.

Tagle, R. and Claeys, P. (2005) An ordinary chondrite impactor for the Popigai crater, Siberia. *Geochimica et Cosmochimica Acta*, 69, 2877–2889.

Tagle, R. and Hecht, L. (2006) Geochemical identification of projectiles in impact rocks. *Meteoritics and Planetary Science*, 41, 1721–1735.

Tagle, R., Erzinger, J., Hecht, L. et al. (2004) Platinum group elements in impactites of the ICDP Chicxulub drill core Yaxcopoil-1: are there traces of the projectile? *Meteoritics and Planetary Science*, 39, 1009–1016.

Tagle, R., Claeys, P., Grieve, R.A.F. et al. (2006) Evidence for a second L chondrite impact in the late Eocene: preliminary results from the Wanapitei Crater, Canada. 37th Lunar and Planetary Science Conference, CD-ROM, abstract no. 1278.

Tagle, R., Ohman, T., Schmitt, R.T. et al. (2007) Traces of an H chondrite in the impact-melt rocks from the Lappajärvi impact structure, Finland. *Meteoritics and Planetary Science*, 42, 1841–1854.

Tagle, R., Schmitt, R.T. and Erzinger, J. (2008a) The Lockne impact is not related to the Ordovician L-chondrite shower. 39th Lunar and Planetary Science Conference, CD-ROM, abstract no. 1418.

Tagle, R., Spray, J.G. and Schmitt, R.T. (2008b) Search for projectile traces in melt rocks of the Charlevoix and Dellen impact structures.

39th Lunar and Planetary Science Conference, CD-ROM, abstract no. 1391.

Tagle, R., Schmitt, R.T. and Erzinger, J. (2009) Identification of the projectile component in the impact structures Rochechouart France and Sääksjärvi, Finland: implications for the impactor population for the Earth. *Geochimica et Cosmochimica Acta*, 73, 4891–4906.

Taylor, S.R. (1967) Composition of meteorite impact glass across the Henbury strewn field. *Geochimica et Cosmochimica Acta*, 31, 961–968.

Taylor, S.R. and McLennan, S.M. (1985) *The Continental Crust: Its Composition and Evolution*, Blackwell, Oxford.

Trinquier, A., Birck, J.-L. and Allègre, C.J. (2006) The nature of the KT impactor. A $^{54}$Cr reappraisal. *Earth and Planetary Science Letters*, 241, 780–788.

Trinquier, A., Birck, J.-L. and Allègre, C.J. (2007) Widespread $^{54}$Cr heterogeneity in the inner Solar System. *The Astrophysical Journal*, 655, 1179–1185.

Turekian, K.K. (1982) Potential of $^{187}$Os/$^{186}$Os as a cosmic versus terrestrial indicator in high iridium layers of sedimentary strata. *Geological Society of America Bulletin*, 190, 243–249.

Walker, R.J. (2009) Highly siderophile elements in the Earth, Moon and Mars: update and implications for planetary accretion and differentiation. *Chemie der Erde – Geochemistry*, 69, 101–125.

Walker, R.J., Morgan, J.W., Beary, E.S. *et al.* (1997) Applications of the Pt-190–Os-186 isotope system to geochemistry and cosmochemistry. *Geochimica et Cosmochimica Acta*, 61, 4799–4807.

Walker, R.J., Horan, M.F., Morgan, J.W. *et al.* (2002) Comparative $^{187}$Re–$^{187}$Os systematics of chondrites: implications regarding early Solar System processes. *Geochimica et Cosmochimica Acta*, 66, 4187–4201.

Wallace, M.W., Gostin, V.A. and Keays, R.R. (1990) Acraman impact ejecta and host shales: evidence for low-temperature mobilization of iridium and other platinoids. *Geology*, 18, 132–135.

Wasson, J.T. and Kallemeyn, G.W. (1988) Composition of chondrites. *Philosophical Transactions of the Royal Society of London A*, 325, 535–544.

Wedepohl K.H. (1995) The composition of the continental-crust. *Geochimica et Cosmochimica Acta*, 59, 1217–1232.

Wolf, R., Woodrow, A.B. and Grieve, R.A.F. (1980) Meteoritic material at four Canadian impact craters. *Geochimica et Cosmochimica Acta*, 44, 1015–1022.

# SIXTEEN

# The geochronology of impact craters

## Simon P. Kelley and Sarah C. Sherlock

*Centre for Earth, Planetary, Space and Astronomical Research, Department of Earth and Environmental Sciences, Open University, Milton Keynes, MK7 6AA, UK*

## 16.1  Introduction

This chapter is concerned with the ages of terrestrial meteorite impact structures, whether determined by relative techniques (such as bracketing between the age of the impacted rocks and sediments filling the crater) or by absolute techniques (such as radioisotope dating). Radioisotope dating of impact craters uses the radioactive decay of naturally occurring elements such as potassium, uranium and rubidium, and has led to several important advances in our understanding of their distribution in the geological record. There are several high-quality critical reviews of the different radioisotope dating techniques applied to meteorite impacts (Deutsch and Schärer, 1994), and the present state of dating craters (Jourdan *et al.*, 2009). Here, we will discuss some of the technical aspects, but also consider the dating of large impacts close to the K–Pg boundary, correlations between the ages of large igneous provinces, large impacts and mass extinctions, and finally the contribution made by radioisotope dating towards our understanding of meteorite impact clusters.

There are currently over 180 confirmed craters on Earth, but their ages are poorly constrained, and only around half are dated with a precision of less than 10 million years. Thus, the history of terrestrial cratering remains an important but poorly constrained aspect of Earth history. There are many more craters on the Moon and other Solar System bodies, but the only ones for which it is possible to obtain samples and directly constrain ages are the terrestrial craters and some areas of the Moon visited by the Apollo missions. The ages of a small number of lunar craters have been measured directly (using impact melts). In addition, impact-generated glass spherules from lunar soils have yielded a spectrum of ages reflecting crater-forming events (e.g. Culler *et al.*, 2000). An alternative method of dating lunar craters, crater counting, determines the relative ages of craters by counting the craters formed on the surface of known age and within the confines of earlier craters, and matches that record to estimates of cratering rates through time. Age estimates based on crater counting are

thus model dependent and less accurate than radioisotope ages, but it remains the most commonly used technique for dating craters on the Moon and other Solar System bodies (Hartmann and Neukum, 2001; Neukum *et al.*, 2001). Crater counting cannot be used on the Earth with its constantly reworked surface. In fact, only two craters on Earth can be dated relative to one another. The 2.7 km diameter Rotmistrovka Crater in Ukraine must predate the Boltysh Crater also in Ukraine, since the former contains an ejecta layer from the latter. Radioisotope dating has yielded an age of 65.17 ± 0.64 Ma for Boltysh and stratigraphic constraints yield an age of 120 ± 10 Ma for Rotmistrovka, thus confirming the relationship.

Thus, the Earth cratering record remains an important source of information regarding changing populations of small bodies in the Solar System through time. Determining the ages of craters contributes to our understanding of Earth history: constraining the overall cratering rate on the continents (and thus the likely present-day threat from impact by asteroids and comets); precisely dating large impacts of impacts in relation to known events in the geological record (such as the end-Cretaceous mass extinction); and constraining the ages of showers that caused crater clusters in the geological record.

For many years the very concept of impact craters on Earth, let alone their formation ages, was controversial, and many older geology textbooks list only one crater, the Barringer Crater (also known as Meteor Crater) in Arizona, USA (e.g. Holmes and Holmes, 1978). Early catalogues of terrestrial impact craters included a range of impact and non-impact craters (Classen, 1977), but the first comprehensive tabulation of crater dates appears in Grieve and Robertson (1979), when just 78 craters had been confirmed on Earth. In this early publication, many of the ages were listed as constrained only by the stratigraphy of their target rocks, or bracketed by the age of crater filling or cover sediments. Later compilations show the numbers of craters increasing, and they have reached 178 at the time of writing this chapter. The ages range from the oldest well-constrained crater, Vredefort

*Impact Cratering: Processes and Products*, First Edition. Edited by Gordon R. Osinski and Elisabetta Pierazzo.
© 2013 Blackwell Publishing Ltd. Published 2013 by Blackwell Publishing Ltd.

in South Africa, with an age of 2023 ± 4 Ma, to Wabar Crater in Saudi Arabia, which formed around 1860, and a cluster of small craters at Sikhote Alin in Russia which were formed by a fall of iron meteorites in 1947. There is no written record of a large meteorite crater forming unless we include the Tunguska event in 1908, which is thought to have been an air burst (see Chapter 10) and left no crater, although it caused extensive damage at the surface.

The questions over crater ages and their relationship, if any, to other geological events has led to a surge in radioisotope dating research to determine the precise age of large meteorite impact craters since a hypothesis linked a large impact to mass extinction at the end of the Cretaceous period (Alvarez *et al.*, 1980).

## 16.2 Techniques used for dating terrestrial impact craters

There are two fundamentally different ways to view dating of events in the deep past that are preserved only as signals in rocks or as features on the Earth.

- For many geologists, the concept of relative dating is useful to describe rocks in the field where there may not be any lithologies suitable for dating. Sediments, for example, are often difficult to date in absolute terms; but it is easy to determine their relative ages, since the younger beds overlie earlier beds. For example, rocks of the Maastrichtian stage lie above Campanian rocks in the upper Cretaceous period but below the earliest rocks of the Palaeogene period. While we can now place quite good absolute age constraints upon the Maastrichtian stage and sometimes key fossils can lead to very tight constraints (particularly widespread and rapidly evolving fossils such as ammonites, or microfossils).
- An absolute date is how many years ago the event happened. For example, the Chicxulub meteorite impact crater formed 65.6 million years ago. It is a simple concept but often difficult to achieve if there are no suitable rocks to measure a direct age. Igneous rocks, such as a lava flow, form very quickly in geological terms and their age can be measured using the decay of radioactive elements, and large meteorites often involve melting of country rocks, thus providing potentially dateable samples. We will discuss some of the techniques below.

The ages of many large meteorite impact craters are poorly constrained both in absolute and relative terms, since they are just holes in the ground or scars in the rocks. We will outline examples of both relative and absolute dating below.

### 16.2.1 Stratigraphic (relative) dating

Stratigraphy of terrestrial and marine craters sometimes provides the only available dating method when melted rocks are absent, or when the structure is highly eroded, or altered. However, this technique sometimes only provides a broad bracket of ages. For example, the Dhala impact crater, which is the remnant of a large Proterozoic crater, formed sometime between 2.5 Ga (the age of

the target rock) and 1.7 Ga (the age of local overlying sediments), a potential range of 0.8 Ga (Pati *et al.*, 2010). In many cases, crater formation can only be constrained by the age of the target rock; for example, Decaturville, in the USA, is younger than the Permian rocks which it excavated (at roughly 300 Ma), but its actual age is unknown. In contrast, stratigraphy has also been used to great effect in situations where ejecta layers occur within well-dated sequences: for example, Manson Crater, which was correlated with a local ejecta deposit (Izett *et al.*, 1993, 1998), or Acraman Crater in Australia, linked to a more distal ejecta deposit which has been placed within a stratigraphic sequence providing some age constraints (Gostin *et al.*, 1986). Perhaps the best example of a widely occurring narrow ejecta layer is the Ir and shocked-quartz-rich layer which documents the globally synchronous impact event at the end of the Cretaceous period (Smit, 1999).

Marine craters, although rarer than continental craters (since they are more difficult to find), are more amenable to stratigraphic dating. For example, the Mjolnir Crater (Smelrør *et al.*, 2001; Dypvik *et al.*, 2004, 2006) formed in ocean sediments that have been precisely dated using biostratigraphy. Moreover, the ocean filled the crater immediately after impact, depositing more microfossil-rich sediments. In the case of the Mjolnir Crater, an impact age at 144 ± 5 Ma can be assigned, on the basis of different microfossils in the underlying and overlying layers, using the newer stratigraphic ages (Gradstein *et al.*, 2005) and propagating the errors in the ages of the boundaries (Dypvik *et al.*, 2006).

### 16.2.2 Radioisotope (absolute) dating

A wide range of radioisotope techniques have been used to directly constrain the ages of impact craters by dating the formation of impact melts, and friction melts which sometimes evolve to form thicker melt layers, bodies and veins in the crater and its surroundings (see Chapter 9). While there are now many radioisotope age determinations in the literature, impact crater dating seems to have suffered more than other areas of geochronology from overinterpretation with too poor regard paid to the associated analytical errors, which are often precise, but inaccurate. All radioisotope dates reported here are quoted at the 95% confidence level unless stated otherwise.

In this regard, the review of Deutsch and Schärer (1994) sets out the techniques used to date impact crater deposits, and Jourdan *et al.* (2009) reviewed the current state of impact crater dating, emphasizing the pitfalls of interpretation of a set of apparently well-constrained craters. Both reviews are regarded as essential reading alongside any interpretation of a new set of radioisotope age determinations. No attempt will be made here to reproduce these reviews; instead, we will focus on some key time intervals and the relationships between different events which are illuminated by precise radioisotope dates.

All of the radioisotope techniques that have been used to determine the ages of large meteorite impacts date the instantaneous melting and crystallization of target rocks, processes that reset the radioisotope clock by mixing and homogenizing the elements or, in the case of Ar, by degassing them. While this is very effective when the rocks are totally melted, many of the issues surrounding the accuracy of such ages arise from incomplete melting and mixing of partially melted material with fully melted glasses and their subsequent alteration. The largest terrestrial craters, such as Sudbury, Vredefort and Manicouagan, have been dated very precisely since the large igneous bodies resulting from melting cooled and crystallized sufficiently slowly, forming crystals of accessory minerals, including zircon, which are commonly used for dating. Techniques for dating such crystals have advanced significantly in recent years (Dickin, 2005) as a result of efforts to measure the absolute ages of volcanic ash horizons in order to improve the precision and accuracy of the geological timescale (Gradstein *et al.*, 2005). Radioisotope techniques such as $^{40}Ar$–$^{39}Ar$, (U,Th)–He and Rb–Sr have been applied to glasses or finely crystalline rocks to yield formation ages for smaller craters where there was too little melt to coalesce and form larger bodies with the potential to crystallize zircon. While such dating techniques are providing important new constraints on intermediate-size craters, there are sometimes complications resulting from partial melting, contaminating unmelted material and later alteration.

### 16.2.2.1   U–Pb dating

The most precise and accurate method for dating geological events, U–Pb dating, records the age at which minerals crystallize, but it is only possible to achieve high-precision ages for a specific set of accessory minerals, such as zircon and baddeleyite, that accommodate the trace element uranium but do not incorporate lead (a second series of minerals that incorporate $^{232}Th$, including allanite and monazite, can be dated via a similar decay chain to $^{208}Pb$). These minerals do not occur in large quantities in a rock but are present in many gneisses and granitic rocks, and they are generally easy to separate using density and magnetic techniques. The U–Pb dating method relies on two separate decay chains via a series of unstable isotopes, both terminating in an isotope of Pb: the decay of $^{235}U$ to $^{207}Pb$ (which has a half-life of 704 million years) and $^{238}U$ to $^{206}Pb$ (which has a half-life of 4468 million years). The combination of these two well-calibrated decay schemes pinpoints the age to very high precision and accuracy (Dickin, 2005). Such ages are robust and resistant to resetting by later heating events; but this resistance is also the main barrier to their use in smaller scale melts. Accessory minerals, particularly zircon, are resistant to melting and partially reset fragments of grains remain in many impact melts, causing them to yield anomalously old ages known as inherited ages (Dickin, 2005). The ages are commonly measured by two laboratory methods: thermal ionization mass spectrometry (TIMS) and secondary ion mass spectrometry (SIMS). In the former technique, a small aliquot of the purified minerals is first dissolved, then U and Pb are separated using an ion-exchange technique, and their isotopes are measured by TIMS in a high-resolution mass spectrometer. In SIMS, the grains are directly mounted in resin blocks and polished flat so that the internal zoning can be imaged, and spots around 30 μm in diameter are ablated using a focused beam of ions. The beam simultaneously ablates and ionizes the material, which is accelerated into a magnetic sector mass spectrometer. TIMS uses larger samples and produces more precise ages, but SIMS is undertaken on ablated spots and can thus be used to measure ages of cores and rims of partially altered or overgrown grains. Both techniques have been used for dating terrestrial craters.

The U–Pb technique has most commonly been used to date the largest terrestrial impact craters, which result in slowly cooled igneous bodies, including the Vredefort impact crater in South Africa in which a medium-grained norite-composition igneous body formed. Zircon from the norite yielded precise age determinations using the *in-situ* SIMS technique to analyse individual grains of 2017 ± 5 Ma (Gibson *et al.*, 1997) and a more precise age of 2023 ± 4 Ma by separating the grains and analysing them using the TIMS technique (Kamo *et al.*, 1996). Ar–Ar ages from friction-generated pseudotachylite melt veins yielded a less precise age of 2014 ± 14 Ma (Spray *et al.*, 1995) but can be more closely correlated with the impact events. More recently, the correlation between U–Pb ages in deformed zircon grains and deformation in rocks surrounding the Vredefort impact crater has demonstrated the potential of zircons to detect the ages of rock fabrics caused by exogenic as well as the more common tectonic processes. The possibility of recognizing exogenic events without direct evidence of a crater may be possible in the future to determine their ages. Other large impacts closely associated with large impact melt igneous bodies have also been assigned very precise ages using the U–Pb technique, including the Sudbury impact 1849.5 ± 0.3 Ma (Krogh *et al.*, 1982; Ostermann *et al.*, 1994; Jourdan *et al.*, 2009) and Manicouagan 214 ± 1 Ma (Hodych and Dunning, 1992).

U–Pb dating has also been used to measure the age of zircons in a distal ejecta deposit of the Chicxulub Crater, demonstrating capability to measure both the age of the impact and, importantly, the age of the rocks in which the crater formed with just a few grains from a thin sedimentary deposit (Krogh *et al.*, 1993a,b; Kamo and Krogh, 1995).

### 16.2.2.2   Rb–Sr dating

The Rb–Sr dating technique has also been used to date meteorite impact craters and has the potential to yield ages but does not generally produce ages as precise as the U–Pb technique. The technique is based on the decay of $^{87}Rb$ to $^{87}Sr$ with the release of a $\beta^-$ particle, with a half-life of 48,800 million years. Unlike the U–Pb technique, Rb–Sr dating is strongly dependent upon measuring more than one phase or sample from the rock. In order to record an age, at least two phases must have been in equilibrium with one another at the time the rock was melted and cooled (Rb–Sr dating has only been used on rocks which actually melted). The method relies on either minerals which incorporate Rb upon crystallization or suites of minerals and rocks with varying Rb/Sr ratios. Since all minerals and rocks contain $^{87}Sr$, the measurements have to be corrected for the $^{87}Sr/^{86}Sr$ ratio of the rock at the time of crystallization using an isochron diagram (Dickin, 2005). The most common minerals used are at least one Rb-rich–Sr-poor phase (such as potassium feldspar or mica) and one Rb-poor–Sr-rich phase (such as amphibole or plagioclase feldspar). Rb–Sr analysis involves obtaining pure mineral separate, dissolving and obtaining clean aliquots of Sr using an ion-exchange technique and high-precision measurements using the TIMS technique.

For example Hart *et al.* (1981) obtained an Rb–Sr age for the Sudbury impact-related intrusion of 1950 ± 60 Ma (1σ). However, in rocks that have not fully melted and homogenized, the presence of Sr-rich plagioclase fragments and altered glass may lead to isochron ages related to the host rock age or an alteration age, rather than the age of the meteorite impact and melting event. For example, Rb–Sr determinations of heterogeneous melt rocks at the Rochechouart impact yielded an age of 186 ± 8 Ma, significantly younger than more recent Ar–Ar age determinations of 201 ± 2 Ma (Schmieder *et al.*, 2010); an earlier Ar–Ar determination of 214 ± 8 Ma (Kelley and Spray, 1997) may reflect excess Ar. The significance of the accuracy of absolute crater ages is well illustrated by this example: the Rb–Sr age places the impact event in the Lower Jurassic period close to the age of a well-known mass extinction, whereas the Ar–Ar ages place Rochechouart in the Late Triassic.

The issue with using the Rb–Sr system to date impact and friction melts is that, in order to yield a true age, the melt must have caused complete homogenization of the $^{87}Sr/^{86}Sr$ ratio within the samples (Dickin, 2005). In large-scale melts this criterion is sometimes met following the impact event, but the precision of the U–Pb dating method is generally better, so the Rb–Sr technique has become less commonly used. In contrast, small-scale impact melts are rarely completely homogenized and many preserve plagioclase fragments, and they are commonly the main Sr reservoir of the rock. In such small-scale melts the Sr concentration of the melt is dominated by the extent of plagioclase melting and Rb–Sr dating is thus not commonly used for dating small-scale meteorite impact events. However, while the Rb–Sr dating technique had been in decline, the recent introduction of microsampling techniques (Charlier *et al.*, 2006) is enabling new developments and has the potential to make significant contributions in the future.

### 16.2.2.3 K–Ar and Ar–Ar dating

Both K–Ar and Ar–Ar dating techniques have been used to determine the ages of small and intermediate-size impact craters. While K–Ar was used in early attempts to date impact melts, in recent years it has been largely replaced by Ar–Ar dating. Both techniques measure the age using the radioactive decay of $^{40}K$ to $^{40}Ar$. Natural $^{40}K$ decay occurs via two routes: 89% decays to the most common isotope of calcium, $^{40}Ca$, with the release of a β particle; only 11% of $^{40}K$ decay is to $^{40}Ar$ via electron capture decay, but Ar is a rare element and provides the best radioisotope clock. The total decay of $^{40}K$ (via both routes) has a half life of 1250 million years (McDougall and Harrison, 1999).

The essential difference between K–Ar and Ar–Ar analyses lies in the measurement of K. In K–Ar dating, K is measured on one aliquot of the mineral or rock sample and Ar isotope measurements are performed on a separate aliquot. Thus, like Rb–Sr dating, K–Ar dating is reliant upon separating a homogeneous mineral phase or whole-rock sample so the K and Ar measurements can be combined to yield an age, an assumption sometimes compromised in heterogeneous impact melt rocks. In contrast, in the Ar–Ar dating technique a small proportion of K is transmuted from $^{39}K$ to $^{39}Ar$ by neutron bombardment. The Ar–Ar age is calculated on the basis of the ratio of $^{40}Ar$ to $^{39}Ar$, measured in a gas-source mass spectrometer specifically used for measuring noble gases. Such mass spectrometers measure in static mode (in other words, isolated from any pumps) and the noble gas isotope ratios are measured on very small (circa $10^{-10} cm^3$ of Ar). The capability of measuring an age on very small amounts and hetero-

geneous samples is provided by this technique because it relies upon the ratio of two isotopes of the same element. Experiment design is an important factor in achieving high-precision ages using the Ar–Ar technique; the aim is to irradiate the sample to produce the amount of $^{39}Ar$ required to achieve a $^{40}Ar/^{39}Ar$ ratio between 1 and 100 while minimizing the interfering reactions, such as production of $^{36}Ar$, $^{37}Ar$ and $^{39}Ar$ from Ca (McDougall and Harrison, 1999; Kelley, 2002).

The fact that the Ar–Ar dating technique does not require sample homogeneity and the ubiquitous presence of potassium in most rocks mean the technique is ideally suited to measuring the ages of small, heterogeneous and partially melted rocks. Indeed, almost the first application of the Ar–Ar dating technique was to determine the ages of meteorites and lunar rocks returned by the Apollo missions (Turner, 1970, 1971). The stepped heating technique has provided useful information on partial resetting, and heterogeneous samples and more recent innovations (such as the use of lasers to reduce blank levels and thus sample sizes) have led to this being the most commonly used technique for dating terrestrial impact craters. The Ar–Ar dating technique can be used on mineral separates, whole-rock samples, melts and even glasses, but the utility of the technique derives from the fact that it can retrieve more than just a single age. The Ar–Ar dating techniques can be used to discriminate against Ar loss by post-impact heating, retention of Ar through the impact melting event (yielding a pre-crater age) and the state of alteration of samples.

Several different extraction techniques have been developed for Ar–Ar dating, including vacuum furnace heating, laser spot dating and laser stepped heating (Kelley, 2002). The laser or furnace stepped heating technique is perhaps the most commonly used, in which the sample is heated to sequentially higher temperatures and the Ar released at each step is measured in the noble gas mass spectrometer. The furnace is generally an externally heated narrow tube under ultra-high vacuum, and reaches temperatures of over 1600 °C (McDougall and Harrison, 1999). Lasers are also used as a heat source for stepped heating, generally infrared wavelength systems such as $CO_2$ (10.6 μm) and Nd–YAG (1064 nm). While furnaces produce higher blank levels (atmospheric Ar released from the hot metal), lasers have the advantage of lower blanks, since only the sample grains are heated; but they also introduce greater thermal gradients within the sample and are limited to a maximum sample size of around 100 mg, whereas furnaces are capable of step heating over a gram.

In order to portray Ar–Ar data, two different plots are commonly used: isochron diagrams and stepped heating release diagrams. The use of both stepped heating diagrams and isochron diagrams is covered in detail elsewhere (McDougall and Harrison, 1999; Kelley, 2002). An isochron diagram plots points of a single age to achieve high precision and check the goodness of fit, similar to plots used for other radioisotope techniques. A stepped heating release diagram is a plot of the ages for individual steps plotted against the percentage $^{39}Ar$ released; if all steps yield the same age, then the age is very likely to be geologically meaningful. However, step heating experiments of real samples commonly show variations in measured ages, including low ages in the initial low-temperature steps, anomalously old ages in the later high-temperature steps or yet more variable patterns which cannot be interpreted as geologically meaningful. Thus, in order for an

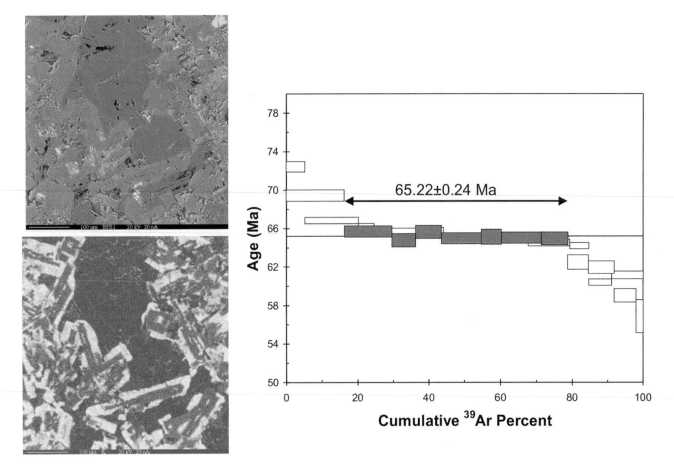

**Figure 16.1**    Scanning electron microscope images of an impact melt rock from the Boltysh impact crater showing a back-scattered electron image and an X-ray map showing the potassium content. The images show a large area of glass in the centre and zoned crystals with a potassium-rich rim a few tens of micrometres thick which contains the majority of the potassium for the rock. Two Ar–Ar stepped heating release patterns for this sample exhibit a reproducible plateau age, and also illustrate some of the complexity that can be induced (higher ages in the early release and low age for the later high-temperature release).

Ar–Ar step heating age to be interpreted as meaningful, there is a convention that steps from at least 50% of the $^{39}$Ar released comprising at least three consecutive steps should yield the same age within analytical errors. This is known as a 'plateau' age (Fig. 16.1).

An alternative approach to stepped heating entails using a laser to measure spot ages by producing instantaneous melting or by ablation. Spots produce sample areas of around 25–100 μm diameter and can resolve spatial variations and test for heterogeneity within complex samples, identify altered areas and even relict ages or excess Ar. However, the laser spot technique is limited by the size of the laser spot and heating outside the visible spot except where UV lasers are used to ablate samples, producing even higher spatial resolution and smaller sample sizes (Kelley et al., 1994; Kelley, 2002).

The age measured by the Ar–Ar technique reflects the time since the rock cooled trapping radiogenic Ar produced by decay of $^{40}$K (McDougall and Harrison, 1999; Kelley, 2002). This may be the 'age' of the impact melt, given that cooling is often very rapid, or alternatively the age of the most recent cooling event if

the rock has been reheated or remained deep in the crust for long periods. The strength of the technique is that it can reveal meaningful ages even when it has been partially compromised by Ar loss or gain. Several impact craters have yielded melted rocks that reveal a simple history. For example, samples of melt from the Logoisk Crater yielded near-concordant step heating plateau ages with a weighted mean age for the impact events of 29.7 ± 0.5 Ma (Sherlock et al., 2009). However, this high level of reproducibility between samples from meteorite impacts is not universal (Fig. 16.2): impact melt samples are often affected by either Ar loss subsequent to the cratering event, anomalously old ages caused by retention of fragments of unmelted rock or significant concentrations of $^{40}$Ar dissolved in the melt (Kelley, 2002). Even in cases where the rocks yield relatively homogeneous ages, there are minor effects which can affect precision. For example, a series of melt sheet samples from the Siljan impact in Sweden yielded laser stepped heating and laser spot ages with a tightly constrained mean of 378 ± 4 Ma (Reimold et al., 2005, Jourdan et al., 2009), whereas a sample exhibiting slightly more alteration, and thus higher contaminating atmospheric Ar contents, yielded an age of

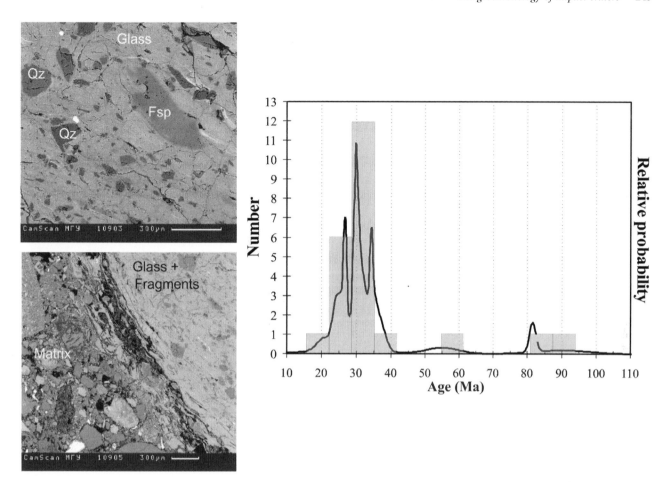

**Figure 16.2** Scanning electron microscope images of an impact melt rock from the Logoisk impact crater illustrating the incorporation of crystals of the country rock and a complex margin between melt and country rock. The histogram of laser spot measurements illustrates the range of ages that can arise from such a complex rock. The range of ages reflects the melt age, but also older partially reset clasts.

366.3 ± 9.0 Ma (Reimold *et al.*, 2005). Samples of older friction melts associated with both Vredefort (Spray *et al.*, 1995) and Sudbury (Thompson *et al.*, 1998) impact craters have lost Ar as a result of alteration and during subsequent burial. In particular, the melt samples from Sudbury were heated in later orogenic disturbances. The main potassium-bearing minerals in fine-grained friction and impact melt rocks such as pseudotachylites are alkali feldspar, biotite, amphibole or plagioclase. However, the grain sizes are as little as 1–5 μm and, thus, they have lower closure temperatures than coarse-grained minerals found in many metamorphic rocks (McDougall and Harrison, 1999). Thompson *et al.* (1998) showed that the pseudotachylites lost Ar at temperatures of 160–180 °C while buried to depths of just 5–6 km for long periods of time.

The Ar–Ar technique has also been used to measure the age of shocked grains in a Late Triassic distal ejecta deposit in the UK. Thackrey *et al.* (2009) measured Ar–Ar ages of individual biotite grains (analysed by fusing individual grains using a laser), which they combined with garnet chemistry from the same deposit, to link it to the rocks in an area of the Grenville Shield adjacent to the Manicouagan Crater in Canada.

In summary, Ar–Ar dating provides several techniques, including laser spot analysis at high spatial resolution and stepped heating of carefully selected samples, in order to surmount some of the issues in dating heterogeneous melt samples. Ar–Ar dating can yield useful chronological information in samples with some excess Ar or Ar loss, which is a particular advantage when very little sample is available if craters are inaccessible or material is only available from drill core. This review reports all Ar–Ar ages relative to the conventional decay constant (Steiger and Jäger, 1977) and standards (Renne *et al.*, 1998).

### 16.2.3 Other dating techniques

Several other dating techniques have been used to date meteorite impacts craters, though none has been sufficiently successful to be applied widely. Measurements of palaeomagnetism, thermoluminescence, cosmogenic nuclides and fission track analyses of glasses have been used occasionally and were reviewed in detail by Deutsch and Schärer (1994). More recently, the (U,Th)–He dating technique has been applied successfully (van Soest *et al.*, 2009; Wartho *et al.*, 2010). The latter technique takes advantage of the resistant nature of accessory minerals such as zircon and apatite, which commonly survive impact and even partial melting. However, while the mineral grains survive, radiogenic $^4$He produced by the decay of U and Th is lost by diffusion. Early experiments indicate that $^4$He measured in the grains today has built up since the heating event and can thus be used to measure the crater formation age. The capability of measuring ages from shocked zircon and apatite in impact ejecta offers the prospect of dating small and intermediate-size impact craters where glasses that might be used for Ar–Ar dating have been altered or contain excess Ar. (U,Th)–He ages are likely to be less precise than Ar–Ar, but will be important in dating smaller craters.

## 16.3    Impact craters at the K–Pg boundary

Several large terrestrial craters have been linked to sudden environmental change and mass extinctions, but there is only one example of a large terrestrial impact crater confirmed as synchronous with a very sudden change to within a few thousand years, and that is the K–Pg boundary (Hallam and Wignall, 1997). In fact, the main reason why scientists are so certain that this impact was synchronous with sudden extinctions of many genera in the oceans is a thin layer of impact ejecta in ocean sediments around the world. The high concentrations of Ir (Alvarez *et al.*, 1980) and shocked quartz (Bohor *et al.*, 1987) found in this layer demonstrated that it had an extraterrestrial cause, but the crater from which the ejecta originated was only confirmed later (Hildebrand *et al.*, 1991), at the same time as a series of dates from glass-rich ejecta deposits on Haiti, determined using the $^{40}$Ar–$^{39}$Ar dating technique, appeared to confirm its status as a crater at the K–Pg boundary (Izett *et al.*, 1991). Swisher *et al.* (1992) measured radioisotope dates on both melted rocks from the Chicxulub Crater and tektites from K–Pg boundary-age deposits in Haiti and Mexico using the $^{40}$Ar–$^{39}$Ar dating technique.

It is important to note that Swisher *et al.* (1992) reported the age as $64.98 \pm 0.05$ Ma based on three Ar–Ar plateau ages obtained on impact melt rocks and compared them with step heating and total fusion ages of tektites, which have been correlated with the impact event from various locations. However, this is a good example of how absolute dating can lead to some confusion. The age was quoted in the paper in 1992 using the known ages of international standards and the geological timescale at that time. As standard calibrations have moved on, the age can be recalculated to $65.6 \pm 0.05$ Ma (e.g. Knight *et al.*, 2003) and, more recently, independent calibrations appear to indicate an age of $65.81 \pm 0.14$ Ma (e.g. Renne *et al.*, 1998; Kuiper *et al.*, 2008; Jourdan *et al.*, 2009). These changes to the calibration are part of a research community effort to improve the absolute calibration of the geological timescale. For the purposes of comparing this age with others in this chapter, the age of 65.6 Ma is most appropriate.

### 16.3.1    Other K–Pg craters, including Boltysh

While the identification of the source of the K–Pg ejecta layer as the Chicxulub Crater is generally accepted, it was not clear cut for several years. Several other craters were considered and attempts were made to determine radioisotope ages for craters thought to have ages close to the K–Pg boundary. The Manson Crater, for example, although too small to have been responsible for the global catastrophe, initially yielded an age of $65.4 \pm 0.3$ Ma (Kunk *et al.*, 1989) based on partially outgassed samples. Later work on Manson linked it to a local impact deposit, and a combination of stratigraphy (relative) and radioisotope (absolute) dating of potassium feldspar (sanidine) from the impact deposits (Izett *et al.*, 1993) appears to show that the Manson Crater formed around $73.8 \pm 0.3$ Ma, 9 million years before the end of the Cretaceous (Izett *et al.*, 1998). This constant updating and refinement of ages has been a feature of impact crater dating, and is a consequence of advancing analytical techniques.

Radioisotope dating has also played an important role in the recent discovery of a second confirmed meteorite impact crater with an age close to the K–Pg boundary. The Boltysh Crater in Ukraine had been assigned an age of $88 \pm 3$ Ma based on K–Ar dating, but the rapid formation and cooling of impact glasses in this 24 km diameter crater caused partial retention of radiogenic Ar because there simply was not time for the Ar to escape while the rock was molten; thus, the K–Ar age was anomalously old. At Boltysh, the target rocks were Proterozoic gneisses with ages around 1500 Ma (Kelley and Gurov, 2002), and thus contained high concentrations of radiogenic Ar when the impact occurred. Melting in the Boltysh Crater did not create a thick melt sheet, but layers of melt several metres thick formed in the ejecta, filling the crater. In some cases, rapid cooling of the melts produced glassy layers, whereas other similar layers' cooled rocks showed that the glassy samples yielded very variable ages as old as 95 Ma (this is probably why the earlier determined K–Ar age for the crater was $88 \pm 3$ Ma). However, samples of Boltysh impact melts, which had cooled more slowly and crystallized, yielded reproducible ages with a weighted mean of $65.17 \pm 0.64$ Ma and represent the true age of the impact crater (see Fig. 16.1).

The formation age of the Boltysh impact crater is within analytical errors of the ages measured for Chicxulub melt rocks and tektites from Haiti and Mexico. Radioisotope dating has thus been able to confirm that two large impact craters formed very close to the K–Pg boundary. However, this conclusion should not be confused with proof that the two craters formed simultaneously, since all radioisotope ages have an associated measurement error, meaning that there is a limit to the precision to which ages can be measured and thus what the technique reveals. It is thus not possible to use radioisotope dating to prove beyond doubt that two impacts were simultaneous. In fact, recent research on sediments from the Boltysh Crater shows that it formed as little as 2000 years before Chicxulub (Jolley *et al.*, 2010). The conclusion is derived from an estimate based upon the flora in sediments recovered from a drill core of the crater fill which indicate that the local area of the ejecta blanket became colonized by early-sequence plants after the impact, but that this ecosystem was in turn suddenly devastated by a second event, believed to be the result of the Chicxulub impact. It is thus a relative age, quantified by comparison with other ecosystems.

Cratering rates on stable continental areas of the Earth show that, on average, one 25 km crater forms on the continental crust every million years (Shoemaker, 1998; Hughes, 2000). Thus, the radioisotope dating of Chicxulub and Boltysh shows that it is unlikely that two large craters would form within a few thousand years. The Boltysh locality is too far from Mexico for a double ejecta layer to have been detected, but the only multiple K–Pg ejecta layer recorded outside Mexico is at Tetritskaro in Georgia (Smit, 1999), which is around 1000 km from Boltysh. Controversial interpretation of ejecta layers in Mexico (at a locality close to Chicxulub Crater), have been cited as evidence of multiple impacts over timescales of a few thousand years (Keller *et al.*, 2004a,b; Smit *et al.*, 2004), but only one multiple layers have been detected outside Mexico.

While it is unlikely that radioisotope dating will solve all the debates over multiple impacts at the K–Pg boundary, it can be used to identify repeated patterns and constrain relationships between impacts, mass extinctions and large igneous provinces at other boundaries, which continues to be an area of vigorous debate.

Despite the restrictions of analytical precision, radioisotope dating has played a central role in the debate over causal relationships between volcanism, impacts and mass extinctions. A case has been made for repeated correlation of mass extinctions with both massive volcanism (Wignall, 2001; Courtillot and Renne, 2003) and large meteorite impacts (Rampino and Stothers, 1986; Stothers, 1989, 2006; Napier, 2006) based on their ages determined by radioisotope dating (note the earlier discussion on use of radioisotope ages and reviews by Deutsch and Schärer (1994) and Jourdan *et al.* (2009)). In fact, radioisotope dating offers a practical and precise method for checking repeated coincidence, and thus potentially discriminating between hypotheses. If they did indeed cause global catastrophes, large meteorite impacts and large igneous provinces should coincide repeatedly with sudden climate change and/or mass extinctions. However, in order for this to be successfully applied, we need to know the age of many of the Earth's impact craters to a level of precision suitable for correlation, and the ages of many medium-sized and small craters are currently very poorly constrained.

## 16.4  Geochronology of impacts, flood basalts and mass extinctions

Perhaps one of the more continuously active debates in Earth science over the last few years concerns the relationship between large volcanic provinces, large meteorite impacts and sudden environmental change or mass extinctions of life on Earth (cf. Hallam and Wignall, 1997). In particular, the debate over the importance of a single large asteroid impact on the Yucatán Peninsula synchronously with the K–Pg boundary remains controversial (Archibald *et al.*, 2010; Courtillot and Fluteau, 2010; Keller *et al.*, 2010; Schulte *et al.*, 2010a,b).

The key question addressed is whether there are any links between catastrophic global events such as mass extinctions and events such as large meteorite impact craters (bear in mind that another group of scientists contend that mass extinctions may have multiple causes; e.g. Macleod, 2005). Here, we will focus on what geochronology can tell us about the best known mass extinction at the end of the Cretaceous period, for which both a large igneous province and a large meteorite impact have been advocated as causal killing mechanisms (Alvarez *et al.*, 1980;

Courtillot *et al.*, 1988). Other mass extinctions have been explained as sudden catastrophes mainly by analogy with the K–Pg event (e.g. Courtillot, 1999; Becker *et al.*, 2001; Wignall, 2001; Basu *et al.*, 2003), but causality in individual cases is extremely difficult to demonstrate. If global catastrophic events did drive sudden environmental changes or mass extinctions, then the cause may be related to the rapidity of events or their global scale which overwhelmed the Earth system. Refinement of the ages and peak flow periods of large volcanic provinces have now been constrained using high-precision Ar–Ar dating, showing that most of them exhibit sudden high-flux events lasting less than 1 million years (see review by Kelley (2007)), but even this pace of eruptions may hide more rapid eruption rates which cause globally important environmental damage (Self, 2006).

So how can geochronology of craters advance the debate concerning the relationship between large igneous provinces, meteorite impacts and mass extinctions if the geochronology has a precision of ±1%? The answer is that, if there is a general relationship between such events, we would expect a repeated association between either large igneous provinces or large meteorite impacts and stratigraphic boundaries associated with extinction in the geological record. One way to demonstrate this association for the K–Pg events is by comparison of the Ir anomaly associated with the impact and sediment accumulation rates measured by extraterrestrial dust input (Mukhopadhyay *et al.*, 2001b), but this is not possible for many of the claimed associations in the geological record.

Ar–Ar dating shows that the Deccan large volcanic province initiated around 68 Ma and peaked close to 65.5 Ma, subsiding rapidly but continuing until around 62 Ma (Basu *et al.*, 1993; Hofmann *et al.*, 2000; Widdowson *et al.*, 2000). K–Pg tektites from Haiti yield Ar–Ar ages for the crater at Chicxulub of 65.6 ± 0.1 Ma (recalculated to calibrated international mineral standards; Izett *et al.*, 1991; Renne *et al.*, 1998; Knight *et al.*, 2003). Note that the absolute ages of the peak of Deccan eruptions and the Chicxulub tektites have been constrained by Ar–Ar dating to lie within 0.1 Ma of each other, but the absolute age of the events is known only to within 0.5 million years because both sets of ages rely upon the precision to which the potassium decay constant is known (±1.25%; see discussion in Renne *et al.* (1998)). The issue of precision of the decay constant can be overcome when the ages being compared are reported relative to the same decay constant and standards. The larger error must be used when such ages are compared with other decay schemes and relative ages based on stratigraphic correlations which commonly rely on U–Pb dates. The most important consequence of this is that the current estimate of the K–Pg boundary is 65.6 Ma (Knight *et al.*, 2003) rather than 65 Ma as commonly reported (note, however, that the most recent timescale shows the K–Pg boundary at 65.5 ± 0.3 Ma; Gradstein *et al.*, 2005).

Despite the problems of using radioisotope dating to monitor the pace of individual extinction events, it offers a practical and precise method for checking repeated coincidence, and thus potentially discriminating between hypotheses advocating large volcanic provinces (Wignall, 2001; Courtillot and Renne, 2003) and large meteorite impacts (Rampino and Stothers, 1986; Stothers, 1989, 2006; Napier, 2006) as causal mechanisms in mass extinction events. If either of the two types of global catastrophes

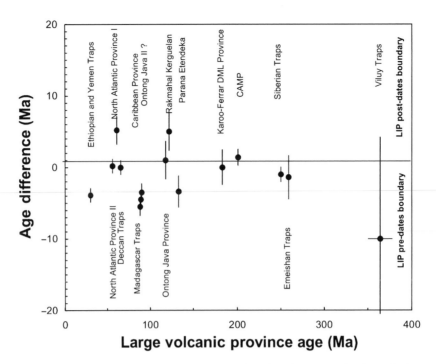

**Figure 16.3**  A plot of radioisotope ages for large igneous provinces (LIPs) and stratigraphic boundaries associated with sudden environmental change. The plot is normalized to allow clear comparison of the differences between flood basalt peak age and stratigraphy. Where points lie above the line the flood basalt pre-dates the potential extinction boundary; where they lie below the line the potential extinction boundary pre-dates the flood basalt.

are indeed causal, such events should coincide repeatedly with sudden climate change and/or mass extinctions. Kelley (2007) reviewed the current state of dating craters, large igneous provinces and mass extinctions and showed that there is a repeatedly close association between large igneous provinces and sudden environmental change (which is, in turn, sometimes associated with mass extinctions) while the association between large impacts and sudden environmental change is limited to two cases (Fig. 16.3 and Fig. 16.4).

Kelley (2007) concluded that up to six of the 15 known large volcanic provinces have peak eruption ages coinciding to within 1–2 million years with sudden environmental change (Fig. 16.3). The repeated coincidence has been confirmed by improvement in precision and accuracy of dates on volcano provinces, but also demonstrated that there have been large volcanic provinces for which no environmental effects have yet been detected. There is a good case, based on geochronology, for a link between large igneous provinces such as Deccan and sudden environmental change, though the precise mechanism remains unclear.

The dating of large meteorite impact craters is far less mature and as much as 90% of all impact craters larger than 20 km diameter that have formed on Earth have been covered or eroded, or are yet to be discovered. Thus, the correlation between known craters and times of sudden environmental change is much less evident. Of the 12 largest meteorite impact craters formed in the last 400 million years, four have radioisotope ages within error of a boundary associated with sudden environmental change, but three of those are associated with uncertainties of greater than 4

million years (Fig. 16.4). The case for repeated correlation of large meteorite impact craters with sudden global environmental change is unproven.

## 16.5  Using geochronology to identify clusters of impacts in the geological record

There are several temporally related clusters of impacts on Earth that have been shown to be significant by determining their ages (see Table 16.1 for a more complete list), notably a short burst of impacts during the Ordovician period (Schmitz *et al.*, 1997), another cluster during the Late Devonian period (Claeys *et al.*, 1992; Leroux *et al.*, 1995); a potential cluster during the Late Triassic period (Spray *et al.*, 1995), at least two craters at the K–Pg boundary (see discussion above) and, finally, several craters and heightened extraterrestrial dust infall during the Late Eocene (Farley *et al.*, 1998; Koeberl, 2009).

Two hypotheses have emerged to explain clusters of craters on Earth: comet showers and asteroid showers. The two scenarios, comets and asteroids, yield different patterns and should be amenable to testing by the age distribution of craters on Earth. Showers of comets can be caused by disturbance in the Oort cloud (Hut *et al.*, 1987), which modelling indicated would potentially cause a cluster of impacts on Earth lasting 1–3 million years. The second hypothesis, a shower of asteroids from the asteroid belt between Mars and Jupiter, could cause showers of impacts lasting from a few thousand to several million years. Asteroid showers

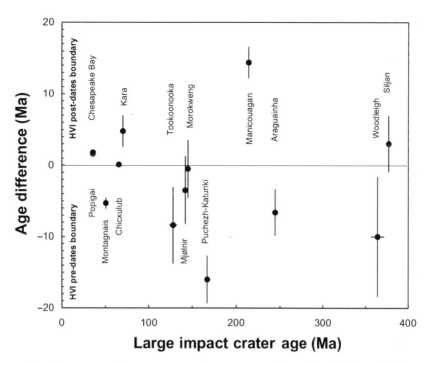

**Figure 16.4** A plot of radioisotope ages for large meteorite impacts and stratigraphic boundaries associated with sudden environmental change (HVI: hypervelocity impact). The plot is normalized to allow clear comparison of the differences between meteorite crater age and stratigraphy. Where points lie above the line the meteorite crater pre-dates the potential extinction boundary; where they lie below the line the potential extinction boundary pre-dates the meteorite crater.

**Table 16.1** Clusters of meteorite impacts on Earth[a]

| Cluster age (Ma) | ± | Stratigraphic boundary | ± | Crater name | Age (Ma) | Diameter (km) |
|---|---|---|---|---|---|---|
| 2 | 2 | 3 | 0.1 | Zhamanshin | 0.9 ± 0.1 | 14 |
| | | | | Bosumptwi | 1.1 ± 0.0 | 10 |
| | | | | Kara-Kul | 2.5 ± 2.5 | 52 |
| | | | | El'gygytyn | 3.5 ± 0.5 | 18 |
| | | | | Bigach | 5.0 ± 3.0 | 8 |
| 36 | 1 | 33.9 | 0.1 | Chesapeake Bay | 35.5 ± 0.3 | 90 |
| | | | | Popigai | 35.7 ± 0.2 | 100 |
| | | | | Mistastin | 36 ± 4 | 28 |
| | | | | Wanapetei | 37.2 ± 1.2 | 8 |
| 49 | 3 | 55.8 | 0.2 | Ragozinka | 46 ± 3 | 9 |
| | | | | Gusev | 49.0 ± 0.2 | 3 |
| | | | | Kamensk | 49.0 ± 0.2 | 25 |
| | | | | Montagnais | 50.5 ± 0.8 | 45 |
| 65.5 | 1 | 65.5 | 0.3 | Chicxulub | 65.6 ± 0.1 | 170 |
| | | | | Boltysh | 65.2 ± 0.6 | 25 |
| 72 | 3 | 70.6 | 0.6 | Ust-Kara, Kara | 70.3 ± 2.2 | 25, 65 |
| | | | | Lappajarvi | 73.3 ± 5.3 | 23 |
| | | | | Manson | 73.8 ± 0.3 | 35 |
| 144 | 2 | 145.5 | 4 | Mjolnir | 142.0 ± 2.6 | 40 |
| | | | | Gosses Bluff | 142.5 ± 0.8 | 22 |
| | | | | Morokweng | 145.0 ± 0.8 | 70 |
| 214 | 1 | 216.5 | 2 | Manicougan | 214 ± 1 | 100 |
| | | | | Rochechouart | 214 ± 8 | 23 |
| | | | | St Martin | 220 ± 32 | 40 |
| 370 | 5 | 374.5 | 2.4 | Woodleigh | 364 ± 8 | 40 |
| | | | | Siljan | 376.8 ± 1.7 | 52 |
| | | | | Ilyinets | 378 ± 5 | 8.5 |
| | | | | Kaluga | 380 ± 5 | 15 |

[a]Crater ages and diameters are from the PASSC impact database, stratigraphic boundary ages from Gradstein *et al.* (2005).

are caused as asteroids in the belt between Mars and Jupiter are introduced into resonance bands (known as Kirkwood gaps) by collisions. The resonance bands are zones where the orbital periods of asteroids are in resonance with Jupiter (the important resonances lie at orbital period ratios of 3:1, 5:2, 7:3 and 2:1 with Jupiter). Gladman *et al.* (1997) and Morbidelli and Gladman (1998) showed that many asteroids ejected from resonance bands fall into the inner Solar System and either fall into the Sun or impact one of the inner planets. Zappala *et al.* (1998) modelled the result of large collisions and break-ups using some of the known asteroid families and showed that a small percentage would hit the Earth in a shower of meteorites between 2 and 30 million years after the break-up event, with the shower lasting variable times depending upon the resonance band involved.

Perhaps the best-documented example of a cluster of meteorite impacts on Earth in the past is documented not by radioisotope dating but by relative dating or, in other words, by superposition of sediments. The example is not a set of large impacts, but an Ordovician limestone sequence in Sweden that contains a multitude of fossil meteorites. Notably, all those meteorites found in the limestone are of the type 'L-chrondrite' (Schmitz *et al.*, 1997), indicating a period of extreme meteorite flux for just over 1 million years (Schmitz *et al.*, 2001) when the estimated flux of small meteorites was around two orders of magnitude greater than the present day. Dating of the L-chondrite class of meteorites which have recently fallen to Earth also yields dates around 465 Ma (McConville *et al.*, 1988; Korochantseva *et al.*, 2007), implying a major collision in the asteroid belt involving a parent body of L-chondrite composition. In addition, Nesvorny *et al.* (2002) showed that the 'Flora' family of asteroids in the main asteroid belt, thought to be the source of L-chondrites, initiated around the same time. There are also several small impact craters with radioisotope and relative ages similar to 465 Ma – many of them in Scandinavia. Although it is tempting to cite these impact craters as evidence for a cluster, the ages themselves are scattered (Schmitz *et al.*, 2001) and are not sufficiently precise to confirm synchronicity with the heightened meteorite flux (Table 16.1). The number of known large craters is not significantly greater than the predicted steady state, but is certainly worthy of further study.

Perhaps the best-documented impact crater cluster is the coincidence of two of the largest craters to form on Earth during the last 500 Ma: Popigai (100 km diameter, 35.7 ± 0.2 Ma; Bottomley *et al.*, 1997) and Chesapeake Bay (90 km diameter, 35.5 ± 0.3 Ma; Obradovich and Snee, 1989), which fell within a few thousand years of each other (Whitehead *et al.*, 2000; Koeberl, 2009). Accompanying the impacts was a heightened flux of extraterrestrial dust detected by monitoring the $^3$He content of the non-carbonate fraction of limestones of the same age at Gubbio in Italy. The variation of $^3$He showed a slow rise and fall over a period of 2–3 million years surrounding the two impacts. The combination of large meteorite impact craters (two of the largest in the Phanerozoic) and heightened extraterrestrial dust has been interpreted as indicating a comet shower (Farley *et al.*, 1998; Mukhopadhyay *et al.*, 2001a). Whitehead *et al.* (2000) also suggested that other smaller craters including, Wanapetei (37.2 ± 1.2 Ma) and Mistastin (36 ± 4 Ma) (Table 16.1), were part of the same cluster as the larger impacts, though it is very difficult to prove

the association since the analytical errors on individual ages are sufficiently large that the cluster does not necessarily exceed the present-day large-body impact rate. Conversely, given an approximately 10% preservation rate for impact craters, it is unlikely that four such craters would be preserved with ages so closely spaced had there not been several more craters now lost through erosion or burial. Recent work on the composition of the bodies, which impacted during the late Eocene, based on trace element chemistry of the impact rocks (Tagle and Claeys, 2005; Tagle *et al.*, 2006), appears to show that the impact craters may not have been the result of a comet shower, but in fact asteroid bodies with L-chondrite composition. In the light of new trace element evidence for the impacting bodies, higher precision measurements of the ages of smaller craters might not help to constrain the cluster because, as we see from the above discussion, ejections from the main asteroid belt can give rise to impact clusters lasting over 10 million years. Such showers are probably common in the geological record. Indeed, a recent shower of meteorites was identified in a study of $^3$He in Miocene ocean floor sediments. A peak in $^3$He was identified in sediments beginning around 8.2 Ma and coinciding with modelled ages for the origin of the Veritas family of asteroids, and a peak in cosmic ray exposure ages determined on H-chondrites, a common class of meteorite found on Earth (Trieloff *et al.*, 2003; Farley *et al.*, 2006).

Table 16.1 identifies potential meteorite impact crater clusters in the terrestrial record (from Kelley (2007)). In most cases three or more craters have been identified, though in one case only two have been identified; and in the oldest case, the presence of a series of fossilized meteorites is the key evidence. Other clusters have been advocated, but only eight can be sustained on the basis of absolute or relative dating showing evidence of impact cratering rate approaching or exceeding the modern steady state. In fact, the youngest cluster is almost certainly an artefact of preservation, with five craters formed in the last 5 million years. Only one impact crater in the recent population, the Kara-Kul Crater in Tajikistan (52 km), is greater than 20 km in diameter, which may indicate that there are several undiscovered craters greater than 20 km which formed in the last 5 million years. However, note that, despite its young age and size, the absolute age of Kara-Kul is poorly constrained (no reported radioisotope age).

In summary, while the geochronology of impact crater clusters is relatively new, radioisotope dating shows that clusters do exist in the terrestrial record but, like large volcanic provinces and the very largest individual impact craters, there is not a one-to-one correlation with sudden environmental change or mass extinctions.

## 16.6   Concluding remarks

The geochronology of terrestrial impact craters remains an area of active research, with new and improved dates for craters appearing every year, improving our understanding of their effects and their celestial mechanical cause. Craters have been dated by relative and absolute techniques, and the recent work has focused primarily on radioisotope techniques; but the record is still poorly constrained, and some craters that have been assigned precise ages are, in fact, not well constrained. There is still a lot of work required to bring the resolution of crater dating up to the level seen in dating large volcanic provinces. The main techniques being used currently are the Ar–Ar and U–Pb techniques, although other techniques such as (U,Th)–He dating may prove important in dating smaller craters.

Geochronology has been able to demonstrate the close relationship between some craters and sudden environmental change, but the evidence for a relationship between large igneous provinces and such changes is much stronger on the basis of geochronology of Phanerozoic events. While most attention has focused on larger impacts, the possibility that the record reflects clusters of impacts rather than a continuum has become better recognized, and future work may help to resolve just how many of the terrestrial craters formed in short periods of time.

## References

Alvarez, L.W., Alvarez, W., Asaro, F. and Michel, H.V. (1980) Extraterrestrial cause for the Cretaceous–Tertiary extinction. *Science*, 208, 1095–1108.

Archibald, J.D., Clemens, W.A., Padian, K. *et al.* (2010) Cretaceous extinctions: multiple causes. *Science*, 328, 973–973.

Basu, A.R., Renne, P.R., Dasgupta, D.K. *et al.* (1993) Early and late alkali igneous pulses and a high-$^3$He plume origin for the Deccan flood basalts. *Science*, 261, 902–906.

Basu, A.R., Michail, I.P., Poreda, R.J. *et al.* (2003) Chondritic meteorite fragments associated with the Permian–Triassic boundary in Antarctica. *Science*, 302, 1388–1391.

Becker, L., Poreda, R.J., Hunt, A.G. *et al.* (2001) Impact event at the Permian–Triassic boundary: evidence from extraterrestrial noble gases in fullerenes. *Science*, 291, 1530–1533.

Bohor, B.F., Modreski, P.J. and Foord, E.E. (1987) Shocked quartz in the Cretaceous–Tertiary boundary clays – evidence for a global distribution. *Science*, 236, 705–709.

Bottomley, R., Grieve, R., York, D. and Masaitis, V. (1997) The age of the Popigai impact event and its relation to events at the Eocene/Oligocene boundary. *Nature*, 388, 365–368.

Charlier, B.L.A., Ginibre, C., Morgan, D. *et al.* (2006) Methods for the microsampling and high-precision analysis of strontium and rubidium isotopes at single crystal scale for petrological and geochronological applications. *Chemical Geology*, 232, 114–133.

Claeys, P., Casier, J.G. and Margolis, S.V. (1992) Microtektites and mass extinctions – evidence for a Late Devonian asteroid impact. *Science*, 257, 1102–1104.

Classen, J. (1977) Catalogue of 230 certain, probable, possible and doubtful impact structures. *Meteoritics*, 12, 61–78.

Courtillot, V. (1999) *Evolutionary Catastrophes – The Science of Mass Extinction*, Cambridge University Press, Cambridge.

Courtillot, V. and Fluteau, F. (2010) Cretaceous extinctions: the volcanic hypothesis. *Science*, 328, 973–974.

Courtillot, V.E. and Renne, P.R. (2003) On the ages of flood basalt events. *Comptes Rendus Geoscience*, 335, 113–140.

Courtillot, V., Feraud, G., Maluski, H. *et al.* (1988) Deccan flood basalts and the Cretaceous/Tertiary boundary. *Nature*, 333, 843–846.

Culler, T.S., Becker, T.A., Muller, R.A. and Renne, P.R. (2000) Lunar impact history from $^{40}$Ar/$^{39}$Ar dating of glass spherules. *Science*, 287, 1785–1788.

Deutsch, A. and Schärer, U. (1994) Dating terrestrial impact events. *Meteoritics*, 29, 301–322.

Dickin, A.P. (2005) *Radiogenic Isotope Geology*, Cambridge University Press.

Dypvik, H., Sandbakken, P.T., Postma, G. and Mork, A. (2004) Early post-impact sedimentation around the central high of the Mjolnir impact crater (Barents Sea, Late Jurassic). *Sedimentary Geology*, 168, 227–247.

Dypvik, H., Smelrør, M., Sandbakken, P.T. *et al.* (2006) Traces of the marine Mjolnir impact event. *Palaeogeography, Palaeoclimatology, Palaeoecology*, 241, 621–636.

Farley, K.A., Montanari, A., Shoemaker, E.M. and Shoemaker, C.S. (1998) Geochemical evidence for a comet shower in the Late Eocene. *Science*, 280, 1250–1253.

Farley, K.A., Vokrouhlicky, D., Bottke, W.F. and Nesvorny, D. (2006) A Late Miocene dust shower from the break-up of an asteroid in the main belt. *Nature*, 439, 295–297.

Gibson, R.L., Armstrong, R.A. and Reimold, W.U. (1997) The age and thermal evolution of the Vredefort impact structure: a single-grain U–Pb zircon study. *Geochimica et Cosmochimica Acta*, 61, 1531–1540.

Gladman, B.J., Migliorini, F., Morbidelli, A. *et al.* (1997) Dynamical lifetimes of objects injected into asteroid belt resonances. *Science*, 277, 197–201.

Gostin, V.A., Haines, P.W., Jenkins, R.J.F. *et al.* (1986) Impact ejecta horizon within late Precambrian shales, Adelaide geosyncline, South Australia. *Science*, 233, 198–200.

Gradstein, F.M., Ogg, J.G. and Smith, A.G. (2005) *The Geological Timescale 2004*, Cambridge University Press.

Grieve, R.A.F. and Robertson, P.B. (1979) Terrestrial cratering record: 1. Current status of observations. *Icarus*, 38, 212–229.

Hallam, A. and Wignall, P.B. (1997) *Mass Extinctions and their Aftermath*, Oxford University Press, Oxford.

Hart, R.J., Welke, H.J. and Nicolaysen, L.O. (1981) Geochronology of the deep profile through Achaean basement at Vredefort with implications for early crustal evolution. *Journal of Geophysical Research*, 86, 663–680.

Hartmann, W.K. and Neukum, G. (2001) Cratering chronology and the evolution of Mars. *Space Science Reviews*, 96, 165–194.

Hildebrand, A.R., Penfield, G.T., Kring, D.A. *et al.* (1991) Chicxulub Crater – a possible Cretaceous/Tertiary boundary impact crater on the Yucatan Peninsula, Mexico. *Geology*, 19, 867–871.

Hodych, J.P. and Dunning, G.R. (1992) Did the Manicouagan impact trigger end-of-Triassic mass extinction? *Geology*, 20, 51–54.

Hofmann, C., Feraud, G. and Courtillot, V. (2000) $^{40}$Ar/$^{39}$Ar dating of mineral separates and whole rocks from the Western Ghats lava pile: further constraints on duration and age of the Deccan Traps. *Earth and Planetary Science Letters*, 180, 13–27.

Holmes, A. and Holmes, D.C. (1978) *Holmes Principles of Physical Geology*, Thomas Nelson, Sunbury-on-Thames.

Hughes, D.W. (2000) A new approach to the calculation of the cratering record of the Earth over the last 125 ± 20 Myr. *Monthly Notices of the Royal Astronomical Society*, 317, 429–437.

Hut, P., Alvarez, W., Elder, W.P. *et al.* (1987) Comet showers as a cause of mass extinction. *Nature*, 329, 118–126.

Izett, G.A., Dalrymple, G.B. and Snee, L.W. (1991) $^{40}$Ar/$^{39}$Ar age of Cretaceous–Tertiary boundary tektites from Haiti. *Science*, 252, 1539–1542.

Izett, G.A., Cobban, W.A., Obradovich, J.D. and Kunk, M.J. (1993) The Manson impact structure – $^{40}$Ar/$^{39}$Ar age and its distal impact ejecta in the Pierre Shale in southeastern South Dakota. *Science*, 262, 729–732.

Izett, G.A., Cobban, W.A., Dalrymple, G.B. and Obradovich, J.D. (1998) $^{40}$Ar/$^{39}$Ar age of the Manson impact structure, Iowa and correlative impact ejecta in the Crow Creek member of the Pierre Shale (Upper Cretaceous), South Dakota and Nebraska. *Geological Society of America Bulletin*, 110, 361–376.

Jolley, D.W., Gilmour, I., Kelley, S.P. *et al.* (2010) Two large meteorite impacts at the K/Pg boundary. *Geology*, 38, 835–838.

Jourdan, F., Renne, P.R. and Reimold, W.U. (2009) An appraisal of the ages of terrestrial impact structures. *Earth and Planetary Science Letters*, 286, 1–13.

Kamo, S.L. and Krogh, T.E. (1995) Chicxulub Crater source for shocked zircon crystals from the Cretaceous–Tertiary boundary layer, Saskatchewan – evidence from new U–Pb data. *Geology*, 23, 281–284.

Kamo, S.L., Reimold, W.U., Krogh, T.E. and Colliston, W.P. (1996) A 2.023 Ga age for the Vredefort impact event and a first report of shock metamorphosed zircons in pseudotachylitic breccias and Granophyre. *Earth and Planetary Science Letters*, 144, 369–387.

Keller, G., Adatte, T., Stinnesbeck, W. *et al.* (2004a) Chicxulub impact predates the K–T boundary mass extinction. *Proceedings of the National Academy of Sciences of the United States of America*, 101, 3753–3758.

Keller, G., Adatte, T., Stinnesbeck, W. *et al.* (2004b) More evidence that the Chicxulub impact predates the K/T mass extinction. *Meteoritics and Planetary Science*, 39, 1127–1144.

Keller, G., Adatte, T., Pardo, A. *et al.* (2010) Cretaceous extinctions: evidence overlooked. *Science*, 328, 974–975.

Kelley, S.P. (2002) K–Ar and Ar–Ar dating, in *Noble Gases in Geochemistry and Cosmochemistry* (eds D. Porcelli, C.J. Ballentine and R. Wieler), Reviews in Mineralogy and Geochemistry, Volume 47, Mineralogical Society of America, Washington, DC, pp. 785–818.

Kelley, S.P. (2007) The geochronology of large igneous provinces, terrestrial impact craters, and their relationship to mass extinctions on Earth. *Journal of the Geological Society*, 164, 923–936.

Kelley, S.P. and Gurov, E. (2002) Boltysh, another impact at the KT boundary. *Meteoritics and Planetary Science*, 37, 1031–1043.

Kelley, S.P. and Spray, J.G. (1997) A late Triassic age for the Rochechouart impact structure, France. *Meteoritics and Planetary Science*, 32, 629–636.

Kelley, S.P., Arnaud, N.O. and Turner, S.P. (1994) An ultraviolet laser probe $^{40}$Ar–$^{39}$Ar extraction technique. *Geochimica et Cosmochimica Acta*, 58, 3519–3525.

Knight, K.B., Renne, P.R., Halkett, A. and White, N. (2003) Erratum to '$^{40}$Ar/$^{39}$Ar dating of the Rajahmundry Traps, eastern India and their relationship to the Deccan Traps'. *Earth and Planetary Science Letters*, 209, 257.

Koeberl, C. (2009) Late Eocene impact craters and impactoclastic layers – an overview, in *Late Eocene Earth: Hothouse Icehouse, and Impacts* (eds C. Koeberl and A. Montanari), Geological Society of America Special Paper 452, Geological Society of America, Boulder, CO, pp. 17–26.

Korochantseva, E.V., Trieloff, M., Lorenz, C.A. *et al.* (2007) L-chondrite asteroid breakup tied to Ordovician meteorite shower by multiple isochron Ar-40-Ar-39 dating. *Meteoritics and Planetary Science*, 42, 113–130.

Krogh, T.E., McNutt, R.H. and Davis, G.L. (1982) High-precision U–Pb zircon ages for the Sudbury nickel irruptive. *Canadian Journal of Earth Sciences*, 19, 723–728.

Krogh, T.E., Kamo, S.L. and Bohor, B.F. (1993a) Fingerprinting the K/T impact site and determining the time of impact by U–Pb dating of single shocked zircons from distal ejecta. *Earth and Planetary Science Letters*, 119, 425–429.

Krogh, T.E., Kamo, S.L., Sharpton, V.L. *et al.* (1993b) U–Pb ages of single shocked zircons linking distal K/T ejecta to the Chicxulub Crater. *Nature*, 366, 731–734.

Kuiper, K.F., Deino, A., Hilgen, F.J. *et al.* (2008) Synchronizing rock clocks of Earth history. *Science*, 320, 500–504.

Kunk, M.J., Izett, G.A., Haugerud, R.A. and Sutter, J.F. (1989) $^{40}$Ar–$^{39}$Ar dating of the Manson impact structure: a Cretaceous–Tertiary boundary crater candidate. *Science*, 244, 1565–1568.

Leroux, H., Warme, J.E. and Doukhan, J.-C. (1995) Shocked quartz in the Alamo breccia, southern Nevada: evidence for a Devonian Impact event. *Geology*, 23, 1003–1006.

MacLeod, N. (2005) Mass extinction causality: statistical assessment of multiple-cause scenarios. *Russian Geology and Geophysics*, 46, 979–987.

McConville, P., Kelley, S.P. and Turner, G. (1988) Laser probe $^{40}$Ar/$^{39}$Ar studies of the Peace River shocked L6 chondrite. *Geochimica et Cosmochimica Acta*, 52, 2487–2499.

McDougall, I. and Harrison, T.M. (1999) *Geochronology and Thermochronology by the $^{40}$Ar/$^{39}$Ar Method*, Oxford University Press, New York, NY.

Morbidelli, A. and Gladman, B. (1998) Orbital and temporal distributions of meteorites originating in the asteroid belt. *Meteoritics and Planetary Science*, 33, 999–1016.

Mukhopadhyay, S., Farley, K.A. and Montanari, A. (2001a) A 35 Myr record of helium in pelagic limestones from Italy: implications for interplanetary dust accretion from the early Maastrichtian to the middle Eocene. *Geochimica et Cosmochimica Acta*, 65, 653–669.

Mukhopadhyay, S., Farley, K.A. and Montanari, A. (2001b) A short duration of the Cretaceous–Tertiary boundary event: evidence from extraterrestrial helium-3. *Science*, 291, 1952–1955.

Napier, W.M. (2006) Evidence for cometary bombardment episodes. *Monthly Notices of the Royal Astronomical Society*, 366, 977–982.

Nesvorny, D., Morbidelli, A., Vokrouhlicky, D. *et al.* (2002) The Flora family: a case of the dynamically dispersed collisional swarm? *Icarus*, 157, 155–172.

Neukum, G., Ivanov, B.A. and Hartmann, W.K. (2001) Cratering records in the inner Solar System in relation to the lunar reference system. *Space Science Reviews*, 96, 55–86.

Obradovich, J.D. and Snee, L.W. (1989) Is there more than one glassy impact layer in the Late Eocene? *Geological Society of America Abstracts with Programs*, 21, A134.

Ostermann, M., Scharer, U. and Deutsch, A. (1994) Impact melting and 1850 Ma offset dikes emplacement in the Sudbury Impact Structure – constraints from zircon and baddeleyite U–Pb ages. *Meteoritics*, 29, 513–513.

Pati, J.K., Jourdan, F., Armstrong, R.A. *et al.* (2010) First SHRIMP U–Pb and $^{40}$Ar/$^{39}$Ar chronological results for the impact melt breccia from the Paleoproterozoic Dhala impact structure, India, in *Large Meteorite Impacts and Planetary Evolution IV* (eds R. Gibson and W. Reimold), Geological Society of America Special Paper 465, Geological Society of America, Boulder, CO, pp. 571–591.

Rampino, M.R. and Stothers, R.B. (1986) Geologic periodicities in the galaxy, in *The Galaxy and the Solar System* (eds R. Smoluchowski, J.N. Bahcall and M.S. Mathews), University of Arizona Press, Tucson, AZ, pp. 241–259.

Reimold, W.U., Kelley, S.P., Sherlock, S.C. *et al.* (2005) Laser argon dating of melt breccias from the Siljan impact structure, Sweden: implications for a possible relationship to Late Devonian extinction events. *Meteoritics and Planetary Science*, 40, 591–607.

Renne, P.R., Swisher, C.C., Deino, A.L. *et al.* (1998) Intercalibration of standards, absolute ages and uncertainties in $^{40}$Ar/$^{39}$Ar dating. *Chemical Geology*, 145, 117–152.

Schmieder, M., Buchner, E., Schwarz, W.H. *et al.* (2010) A Rhaetian $^{40}$Ar/$^{39}$Ar age for the Rochechouart impact structure (France) and implications for the latest Triassic sedimentary record. *Meteoritics and Planetary Science*, 45, 1225–1242.

Schmitz, B., Peucker-Ehrenbrink, E., Lindstrom, M. and Tassinari, M. (1997) Accretion rates of meteorites and cosmic dust in the early Ordovician. *Science*, 278, 88–90.

Schmitz, B., Tassinari, M. and Peucker-Ehrenbrink, B. (2001) A rain of ordinary chondritic meteorites in the early Ordovician. *Earth and Planetary Science Letters*, 194, 1–15.

Schulte, P., Alegret, L., Arenillas, I. *et al.* (2010a) The Chicxulub asteroid impact and mass extinction at the Cretaceous–Paleogene boundary. *Science*, 327, 1214–1218.

Schulte, P., Alegret, L., Arenillas, I. *et al.* (2010b) Response – Cretaceous extinctions. *Science*, 328, 975–976.

Self, S. (2006) The effects and consequences of very large explosive volcanic eruptions. *Philosophical Transactions of the Royal Society A – Mathematical Physical and Engineering Sciences*, 364, 2073–2097.

Sherlock, S.C., Kelley, S.P., Glazovskaya, L. and Peate, I.U. (2009) The significance of the contemporaneous Logoisk impact structure (Belarus) and Afro-Arabian flood volcanism. *Journal of the Geological Society*, 166, 5–8.

Shoemaker, E.M. (1998) Long-term variations in the impact cratering rate on Earth, in *Meteorites: Flux with Time and Impact Effects* (eds M.M. Grady, R. Hutchison, G.J.H. McCall and D.A. Rothery), Geological Society Special Publication 140, Geological Society of London, London, pp. 7–10.

Smelrør, M., Kelly, S.R.A., Dypvik, H. *et al.* (2001) Mjolnir (Barents Sea) meteorite impact ejecta offers a Volgian–Ryazanian boundary marker. *Newsletters on Stratigraphy*, 38, 129–140.

Smit, J. (1999) The global stratigraphy of the Cretaceous–Tertiary boundary impact ejecta. *Annual Reviews of Earth and Planetary Sciences*, 27, 75–113.

Smit, J., Van Der Gaast, S. and Lustenhouwer, W. (2004) Is the transition impact to post-impact rock complete? Some remarks based on XRF scanning, electron microprobe, and thin section analyses of the Yaxcopoil-1 core in the Chicxulub Crater. *Meteoritics and Planetary Science*, 39, 1113–1126.

Spray, J.G., Kelley, S.P. and Reimold, W.U. (1995) Laser probe argon-40/argon-39 dating of coesite- and stishovite-bearing pseudotachylytes and the age fo the Vredefort impact event. *Meteoritics*, 30, 335–343.

Steiger, R.J. and Jäger, E. (1977) Subcommission on geochronology: convention on the use of decay constants in geo- and cosmochronology. *Earth and Planetary Science Letters*, 36, 359–362.

Stothers, R.B. (1989) Structure and dating errors in thegeologic time scale and periodicity in mass extinctions. *Geophysical Research Letters*, 16, 119–122.

Stothers, R.B. (2006) The period dichotomy in terrestrial impact crater ages. *Monthly Notices of the Royal Astronomical Society*, 365, 178–180.

Swisher, C.C., Grajalesnishimura, J.M., Montanari, A. *et al.* (1992) Coeval $^{40}$Ar/$^{39}$Ar ages of 65.0 million years ago from Chicxulub Crater melt rock and Cretaceous–Tertiary boundary tektites. *Science*, 257, 954–958.

Tagle, R. and Claeys, P. (2005) An ordinary chondrite impactor for the Popigai Crater, Siberia. *Geochimica et Cosmochimica Acta*, 69, 2877–2889.

Tagle, R., Claeys, P., Grieve, R.A.F. *et al.* (2006) Evidence for a second L chondrite impact in the Late Eocene: preliminary results from the Wanapitei Crater, Canada. 37th Annual Lunar and Planetary Science Conference, League City, TX, abstract no. 1278.

Thackrey, S., Walkden, G., Indares, A. *et al.* (2009) The use of heavy mineral correlation for determining the source of impact ejecta: a Manicouagan distal ejecta case study. *Earth and Planetary Science Letters*, 285, 163–172.

Thompson, L.M., Spray, J.G. and Kelley, S.P. (1998) Laser probe argon-40/argon-39 dating of pseudotachylyte from the Sudbury structure: evidence for postimpact thermal overprinting in the North Range. *Meteoritics and Planetary Science*, 33, 1259–1269.

Trieloff, M., Jessberger, E.K., Herrwerth, I. *et al.* (2003) Structure and thermal history of the H-chondrite parent asteroid revealed by thermochronometry. *Nature*, 422, 502–506.

Turner, G. (1970) Argon-40/argon-39 dating of lunar rock samples. *Science*, 167, 466–468.

Turner, G. (1971) $^{40}$Ar–$^{39}$Ar ages from the lunar maria. *Earth and Planetary Science Letters*, 11, 169–191.

Van Soest, M.C., Wartho, J.-A., Monteleone, B.D. *et al.* (2009) (U–Th)/He dating of single zircon and apatite crystals – a new tool for dating terrestrial impact structures. 40th Lunar and Planetary Science Conference, Houston, TX, 2041.pdf.

Wartho, J.-A., van Soest, M.C., Cooper, F.J. *et al.* (2010) Updated (U–Th)/He zircon ages for the Lake Saint Martin impact structure (Manitoba, Canada) and implications for the Late Triassic multiple impact theory. 41st Lunar and Planetary Science Conference, Houston, TX, 1930.pdf.

Whitehead, J., Papanastassiou, D.A., Spray, J.G. *et al.* (2000) Late Eocene impact ejecta: geochemical and isotopic connections with the Popigai impact structure. *Earth and Planetary Science Letters*, 181, 473–487.

Widdowson, M., Pringle, M.S. and Fernandez, O.A. (2000) A post K–T boundary (Early Palaeocene) age for Deccan-type feeder dykes, Goa, India. *Journal of Petrology*, 41, 1177–1194.

Wignall, P.B. (2001) Large igneous province and mass extinctions. *Earth-Science Reviews*, 53, 1–33.

Zappala, V., Cellino, A., Gladman, B.J. *et al.* (1998) Asteroid showers on Earth after family breakup events. *Icarus*, 134, 176–179.

# SEVENTEEN

# Numerical modelling of impact processes

## Gareth S. Collins*, Kai Wünnemann†, Natalia Artemieva‡ and Elisabetta Pierazzo§

*IARC, Department of Earth Science and Engineering, Imperial College London, London, SW7 2AZ, UK
†Museum für Naturkunde, Leibniz-Institut an der Humboldt-Universität Berlin, Invalidenstrasse 43, 10115 Berlin, Germany
‡Institute for Dynamics of Geospheres, Russian Academy of Science, Leninsky pr. 38, blg.1, Moscow 119334, Russia
§Planetary Science Institute, 1700 E. Fort Lowell Road, Suite 106, Tucson, AZ 85719, USA

## 17.1  Introduction

Numerical modelling is now a well-established and important component of impact research. When used to complement small-scale impact experiments, it provides a powerful tool for exposing the physical processes in a cratering event and for investigating the effect of individual physical parameters that are not otherwise under the experimenter's control. Moreover, once adequately validated against small-scale laboratory experiments, numerical models allow us to simulate impacts much larger and more energetic than those that can be studied experimentally.

The first numerical impact simulation was published in 1961 by R.L. Bjork (Fig. 17.1). That calculation of the Meteor Crater impact simulated only the first 60 ms of the collision and did not account for gravity or the strength of the target rocks (Bjork, 1961). Nevertheless, that simulation, and those that followed in the next decades, provided great insight into the physics of hypervelocity impact, including the formation and attenuation of the detached shock wave, the fate of the impactor, and the scaling of crater dimensions (e.g. Ahrens and O'Keefe, 1977; Orphal, 1977; O'Keefe and Ahrens, 1993).

Nearly half a century on, and thanks to advances in numerical impact models and much improved computational resources, numerical models that include gravity and complex material models are used routinely to simulate large terrestrial impacts, aiding interpretation of geological and geophysical observations and quantifying the energy of the impact and impact-related consequences. For example, several numerical impact models of the Chicxulub impact have been performed and compared with observational data to constrain the impact energy, the fate of the projectile, the volume of climate forcing gases released into the atmosphere, the transport and re-entry of ejected material and the regional seismicity (e.g. Pierazzo et al., 1998; Pierazzo and Melosh, 1999, 2000a; O'Keefe and Ahrens, 1999; Collins et al., 2002, 2008; Ivanov and Artemieva, 2002; Ivanov, 2005; Artemieva and Morgan, 2009; Goldin and Melosh, 2009).

Constraints imposed by computer hardware meant that most early impact models assumed vertical impact, because the axial symmetry of the process allows the simplification of the model to two dimensions. In non-vertical impacts, the axial symmetry is broken, and more complex and computationally intense three-dimensional (3D) models are required. The step from two-dimensional (2D) to 3D is an important advance in realism, because natural impacts in which the projectile strikes the target vertically are virtually nonexistent: the angle of impact of maximum frequency is 45°, whereas the probability of vertical as well as grazing impacts is negligible (Pierazzo and Melosh, 2000c). Furthermore, impact angle (especially low impact angles) appears to have played a major role in various problems of geological interest, like the ejection of matter from planetary surfaces into interplanetary space, the giant impact theory for the origin of the Moon, the environmental effects of large planetary impacts and the spin rate of asteroids. Systematic high-resolution modelling of oblique impacts was first described in Pierazzo and Melosh (1999, 2000a,b). Since that work, advances in computer hardware have allowed more widespread simulation of oblique impacts (Artemieva et al., 2002, 2004; Shuvalov et al., 2002; Artemieva and Ivanov, 2004; Gisler et al., 2004; Shuvalov and Dypvik, 2004; Elbeshausen et al., 2009).

Numerical impact models are particularly useful for investigating impact-related processes in planetary science that cannot be simulated in the laboratory, and have not been (or are seldom) observed, such as the formation of the Moon (e.g. Benz et al., 1989; Cameron, 2000; Canup and Asphaug, 2001; Canup, 2004, 2008) and the ejection of material from planetary surfaces (e.g. Head et al., 2002; Artemieva and Ivanov, 2004). An excellent illustration of the power of numerical modelling was the prediction made by Crawford et al. (1994), based on numerical simulations of kilometre-scale ice fragments impacting the Jovian atmosphere, that the impact of comet Shoemaker–Levy 9 with

Impact Cratering: Processes and Products, First Edition. Edited by Gordon R. Osinski and Elisabetta Pierazzo.

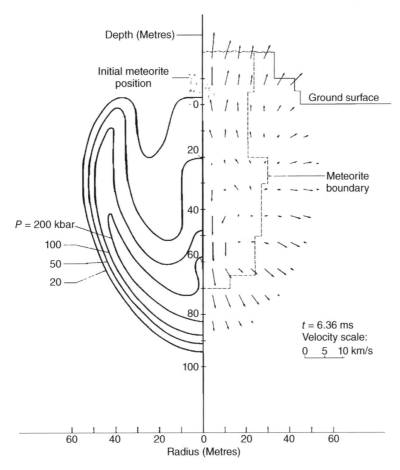

**Figure 17.1** This is the first numerical simulation of an impact event (Bjork, 1961). Contours of pressure (left) and velocity vectors (right) 6.36 ms after the vertical impact of a 12,000-ton cylindrical iron projectile, resolved by 12 cells across its radius, striking a half-space target of tuff at 30 km s$^{-1}$.

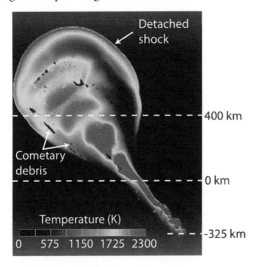

**Figure 17.2** Impact plume (fireball) formed by the impact of a 3 km fragment of comet Shoemaker–Levy 9 on Jupiter (simulated using the 3D CTH hydrocode) approximately 69 s after entering Jupiter's atmosphere taken from Crawford *et al.* (1994). This is a cross-sectional view with temperature represented by colour (see Plate 36); the vertical scale is altitude above 1 bar (0.1 MPa) atmospheric pressure level. The computational mesh consists of 6.3 million 5-km-sized cubical cells. (See Colour Plate 36)

Jupiter would produce an impact plume visible from Earth (see Fig. 17.2). The results of these calculations made testable predictions that were later confirmed by the spectacular events in July 1994 and provided constraints on the mass of the cometary fragments.

Since the earliest numerical simulations of meteorite impacts in the 1960s and 1970s the field of numerical impact modelling has grown into a research field in its own right. A single chapter cannot do justice to the depth and breadth of research in numerical impact modelling, either in terms of the progress in numerical impact model development or the broad application of numerical impact models in planetary science. This chapter is intended to provide a brief overview of how numerical models are used to study impact processes. The questions we address are: How do impact models work in principle? What are their strengths and weaknesses? and What can they tell us about impact processes? Readers searching for a deeper, more mathematical, and more detailed introduction to the numerical techniques and methodologies commonly employed by modern impact models are directed to the relevant literature in the discussion below; however, excellent reviews of the theory and implementation of impact models are found in Anderson (1987) and Benson (1992).

## 17.2    Fundamentals of impact models

### 17.2.1    Governing equations

Most numerical impact models are based on the foundations of continuum mechanics, which supposes that the region of interest is a continuous medium, without any gaps or empty spaces. Such an approach is valid for impacts at scales large enough that the molecular structure of matter can be ignored. The dynamics of a continuous medium can be described by a set of differential equations that represent the principles of the conservation of mass, momentum and energy. Derivations of these equations can be found in any text on continuum mechanics (e.g. Malvern, 1969).

The conservation of momentum equation relates the change in velocity of the material to the sum of the forces acting on the material per unit mass. In the frame of reference of the moving material, it may be written as

$$\frac{Du_i}{Dt} = g_i + \frac{1}{\rho}\frac{\partial \sigma_{ji}}{\partial x_j} \tag{17.1}$$

where $u$ is the velocity, $g$ is the acceleration due to gravity, but could be any other external acceleration (force per unit mass), $\rho$ is the material density, $\sigma$ is the stress, $x$ is distance and $t$ is time; $Du_i/Dt$ is the total time derivative, the index $i$ (and $j$) refers to the coordinate directions ($i, j = 1 \ldots M$, where $M$ is the number of dimensions) and summation over the indices is implied.

In numerical impact models the full stress tensor $\sigma_{ij}$ is often separated into two parts: an isotropic scalar part $p$ (the pressure) and a deviatoric part $s_{ij}$ (the deviatoric stress tensor; where $s_{ij} = \sigma_{ij} - p$ for $i = j$ and $s_{ij} = \sigma_{ij}$ for $i \neq j$). In this case, eqn (17.1) becomes

$$\frac{Du_i}{Dt} = g_i + \frac{1}{\rho}\left(\frac{\partial s_{ji}}{\partial x_j} - \frac{\partial p}{\partial x_j}\right) \tag{17.2}$$

The advantage of this separation is that the forces acting on the material due to changes in volume (compression or expansion, i.e. changes in $p$) can be calculated independently of those forces acting on the material due to changes in shape (deformation, i.e. changes in $s_{ij}$).

In the moving material frame of reference, the differential equations representing the conservation of mass and energy may be written as

$$\frac{D\rho}{Dt} = -\rho \frac{\partial u_i}{\partial x_i} \tag{17.3}$$

and

$$\frac{DI}{Dt} = -\frac{p}{\rho}\frac{\partial u_i}{\partial x_i} + \frac{1}{\rho}s_{ij}\dot{\varepsilon}_{ij} \tag{17.4}$$

where $I$ is the specific internal energy and $\dot{\varepsilon}_{ij}$ is the deviatoric strain rate. In physical terms, eqn (17.3) relates the change in density to the divergence of the flow field; that is, the flux of material away from or toward the material's position. Equation (17.4) relates the change in internal energy to the rate at which heat is generated by volume work (compression or expansion; first term on the right-hand side) and viscous dissipation (shearing; second term).

Two more relationships are required to complete the set of equations that govern the dynamics of continuous media: the equation of state (EoS) and the deviatoric stress model, which are described in greater detail in Sections 17.3.1 and 17.3.3 respectively. The *EoS* of a given material relates pressure, density and either temperature or internal energy. It describes compressibility, thermal expansion, wave speeds and other thermodynamic properties, and may also include descriptions of the material's phase changes such as solid–solid, melt and vaporization. As internal energy is computed directly by most impact models, the most convenient form for the EoS is typically

$$p = p(\rho, I) \tag{17.5}$$

However, entropy and temperature are often computed in addition.

The *deviatoric stress model* (sometimes termed the constitutive model or strength model) relates deviatoric stress to a combination of strain, strain rate, pressure and internal energy or temperature:

$$s_{ij} = f(\varepsilon_{ij}, \dot{\varepsilon}_{ij}, p, I) \tag{17.6}$$

It describes the response of a material to stresses that induce deviatoric deformations – changes of shape.

The EoS and the deviatoric stress model together define a material's response to stress, which is often referred to as the material model. Specific material properties govern the response of materials to stress, implying that different materials behave differently for nominally the same impact conditions. Hence, the ability of a numerical model to simulate collision between two objects depends in large part on the sophistication and accuracy of the material models employed to describe the impactor and the target. The difficulty of building accurate material models remains the major challenge for numerical modelling of impacts, and is discussed in more detail in Section 17.3.

### 17.2.2    Discretization

A computer has a finite memory allocation and, therefore, can only represent a continuous medium by dividing it into a manageable number of separate elements – a process known as discretization. Likewise, the continuous differential equations (eqns (17.2)–(17.6)) can only be evaluated or solved over a finite number of points in space and time. Apart from differences in material models, the many numerical impact models in use today differ primarily in the techniques they employ to discretize the domain of interest and discretize and solve eqns (17.2)–(17.6).

Numerical impact models can be broadly classified according to whether they use a mesh to represent the continuous medium of interest or not. Mesh-based methods divide up the entire domain into a number of small cells (or elements) with no gaps or overlaps. Physical quantities of interest (density, pressure,

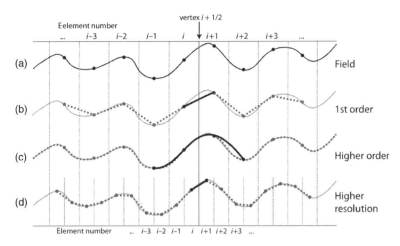

**Figure 17.3**    Schematic illustration of the principle of the finite-difference approximation. A smooth continuous field is represented at a discrete number of cell centres (a). The derivative of the discrete field at a vertex between two cells approximates the derivative of the continuous field (b). The approximation is improved by using a wider stencil to compute the derivative of the discrete field (c) or by using more cells, with closer separation, to represent the field (d).

velocity, etc.) may be associated with the cell itself, with the vertices (or nodes) that define the corners of the cell or with the edges/faces that define the boundary of the cells between vertices. The cells may be arranged in a regular grid pattern, like a chessboard, in the case of a structured mesh, or in an irregular or arbitrary pattern, in the case of an unstructured mesh. Each cell is physically connected to its neighbours (or the edge of the domain, for boundary cells) so that force, momentum and mass are easily transferred from one cell to the next, as required, and boundary conditions are straightforward to apply.

Mesh-free (or meshless) methods, such as the smooth particle hydrodynamics (SPH; Monaghan, 1988), represent physical quantities at a number of unconnected points. In this case, force must be transferred from one point to another by locating two points close enough to influence each other's behaviour and computing the interaction between the materials at each location. SPH codes use interpolation nodes (i.e. smooth particles) from which weighted functions ('kernels') representing parameters of interest may be calculated. Mesh-free approaches are valuable methods for problems where material expands into a huge volume, since only the part of space containing material is modelled, whereas a mesh-based code must model the total space. In other cases, however, mesh-free methods tend to be computationally more expensive due to the large number of particles required to give an accurate solution. In addition, boundary conditions are more complicated to treat in mesh-free codes because there are no natural boundaries in the domain.

Mesh-based codes may be classified according to how the governing equations are discretized and solved over the mesh of cells. Three common techniques are the finite-difference method, the finite-element method and the finite-volume method. A description of all these methods is well beyond the scope of this contribution; the interested reader is referred to the review by Benson (1992), or one of the various textbooks on the subject (e.g. Richtmyer and Morton, 1967; LeVeque, 2002, 2007; Zienkiewicz *et al.*, 2005). However, it is instructive to briefly consider one

such technique, the finite-difference method, as it exposes some important underlying principles that almost all numerical models share.

The principle of the finite-difference method is to replace the derivatives in the differential equation with differences computed between points separated in space or time. Variants of the finite-difference method differ in how the spatial derivatives (e.g. the pressure gradient in eqn (17.2)) are approximated. Figure 17.3 illustrates the spatial discretization procedure graphically. A smoothly varying field (pressure, for example) is represented at a finite number of cell centres (Fig. 17.3a). The simplest estimate of the derivative of this field at the highlighted vertex is given by the slope of the straight line connecting adjacent cells (Fig. 17.3b). The error in this approximation (the mismatch between the actual derivative of the field and the finite-difference approximation to this derivative) is known as the truncation error of the finite-difference scheme. This error can be reduced, and the accuracy of the solution improved, either by increasing the 'order' of the finite-difference approximation – that is, using information from more surrounding cells to approximate the derivative at a point (Fig. 17.3c) – or by increasing the number of sampling points (and reducing the separation between them; Fig. 17.3d). Both of these options increase the number of mathematical operations and the amount of data storage involved in one timestep; hence, there is an obvious trade-off between accuracy and both calculation time and computer memory.

The finite-element and finite-volume methods are alternatives to the finite-difference method for solving the governing equations. Although they differ significantly in their implementation, the same types of error and stability considerations exist, as well as the same fundamental trade-off between solution accuracy and the cost of the solution in time and memory.

Most impact models use an explicit finite-difference approximation to advance the solution in time, meaning that the solution (e.g. the velocity at each node) is updated using the state of the system at the current time level only. Explicit methods suffer from the disadvantage that they are only stable (i.e. they only work) if the time interval over which the update is computed (the timestep) is sufficiently small. In other words, the solution cannot be extrapolated too far forward in time.

Several factors may limit the stable duration of an explicit timestep, the most important of which is the requirement that information may not propagate entirely across any cell in the mesh in one timestep. Formally, this is known as the Courant–Friedrichs–Levy (CFL) criterion, after the three scientists who first described the limit (Courant, 1928). The CFL criterion limits the timestep according to the rule

$$\Delta t \leq \frac{\Delta \chi}{c} \qquad (17.7)$$

where $\Delta \chi$ is the minimum cell dimension and $c$ is the signal speed, which is given by the sum of the sound speed and the speed that material moves. The latter, the material velocity, may be neglected if the reference frame (discrete grid of cells) moves along with the material flow (see Lagrangian reference frame in Section 17.2.3). In practice, to ensure stability, the timestep is often chosen to be a small fraction (e.g. 20–50%) of that defined by the CFL criterion.

In impact simulations, initial material velocities are often several times greater (typically $20\,\mathrm{km\,s^{-1}}$ on Earth) than the material sound speed (on average $\sim 5\,\mathrm{km\,s^{-1}}$). To calculate the development of the crater, which might last several minutes (see Chapters 3–5), tens to hundreds of thousands of computational timesteps are required, depending on the minimum discrete cell size.

### 17.2.3  Frames of reference

The conservation equations (eqns (17.2)–(17.4)) are written with respect to the material; they describe the motion (displacement, velocity and acceleration) with respect to a reference frame that moves with the continuous medium itself. This type of description is known as a Lagrangian (or material) description. The equations may also be written with respect to a fixed reference frame; a formulation termed an Eulerian (or spatial) description. Numerical impact models exist that employ either or both formulations.

To obtain a Lagrangian description the computational mesh is defined to represent the geometry of the simulated materials; points within the mesh (vertices) are attached to the material and move with it. As time progresses, cells defined by adjacent vertices become deformed in shape and size due to the forces acting on them and the constitutive relations between force and deformation. Mass, momentum and energy are transported by moving and deforming the entire mesh in space. Mass within a cell is invariant; changes in density are exclusively due to changes in a cell's volume.

In contrast, the Eulerian description relies on material flowing through a fixed mesh, which must, therefore, define the entire space of interest. As time progresses, the model variables are calculated at the fixed points of the grid. Thus, mass, momentum and energy must flow across cell boundaries. The amount of flow between cells is used to compute the new variables within each cell. In this formulation it is the volume of the cell that is invariant and changes in density are due to changes in the mass within a particular cell.

Figure 17.4a shows a snapshot of a Lagrangian impact simulation; in this case, only the space occupied by material of interest is divided into cells. The free surface (the boundary between material and vacuum) is defined naturally by the edges of the cells (unlike the Eulerian description). An additional advantage of the Lagrangian description is that, because each cell represents the same material throughout the simulation, the history of this material is naturally recorded, and processes that depend on the material's history (strain softening, strain hardening, etc.) are straightforward to model. The major weakness with the Lagrangian description is the inaccuracy of the numerical approximation when the cells become distorted significantly. Extreme cell deformation can lead to the unphysical inversion of the cell or a vanishingly small timestep, either of which will terminate the calculation. Strategies exist to prolong Lagrangian calculations that involve severe deformation; for example, by deleting (eroding) those cells that exceed some user-specified level of distortion (see Fig. 17.4a). However, in most cases impacts involve such extreme deformation that purely Lagrangian mesh-based calculations are impractical.

Figure 17.4b illustrates the same impact simulation as in Fig. 17.4a performed using an Eulerian description. In this formulation the mesh must define the entire space in which the material movement is contained. Consequently, to achieve the same spatial resolution as the Lagrangian description in Fig. 17.4a, the number of cells required is significantly greater. During the calculation certain cells are completely filled with material, other cells are partially filled and the remainder are empty. This is illustrated by the shading in Fig. 17.4b; cells containing a greater concentration of mass are shaded darker. Hence, unlike the Lagrangian description, where the free surface and material boundaries are precisely defined, these interfaces are not tracked exactly in the Eulerian description and in practice they are often computed approximately based on the fraction of material in each cell. Furthermore, in cells that contain more than one material, mesh variables such as pressure must be computed for the material mixture in the cell. On the other hand, Eulerian simulations are not limited by cell deformations and the associated solution inaccuracies or mesh erosion. As a result, practical impact simulations often employ this method.

A hybrid method, referred to as arbitrary Lagrangian–Eulerian (ALE), exists in which mesh movement occurs, but not necessarily to follow the exact movement of the material throughout the domain. The advantages of the Lagrangian formulation when cell deformations are not extreme mean that the ALE mesh is often chosen to be close to that of the Lagrangian mesh, and hence the advection step is only performed near zones of high deformation.

Meshless methods are by definition purely Lagrangian, but they do not suffer the limitations of mesh-based Lagrangian methods as there is no grid to distort. Figure 17.4c shows the same impact simulated using an SPH solution algorithm. In this case, the deformation around the crater is visualized by plotting the location of the centres of the 'smoothed particles'.

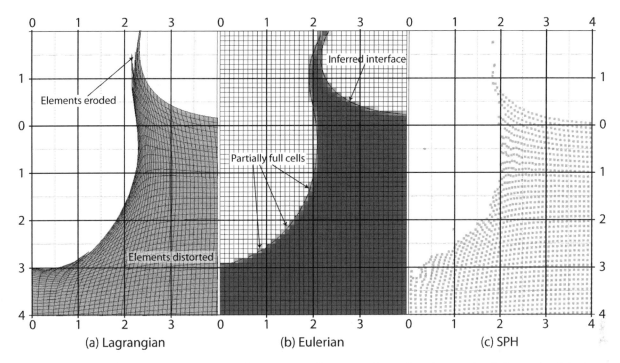

**Figure 17.4**  Schemes for numerical impact simulations. Time-frames are shown from impact simulations using (a) a Lagrangian mesh method (with mesh erosion), (b) an Eulerian mesh method and (c) an SPH meshless method. (a) and (c) were generated using AUTODYN by M. Price and used by kind permission; (b) generated using iSALE. Courtesy of Mark Price.

### 17.2.4  Resolution

As discussed above, all numerical impact models have an intrinsic error associated with representing a continuous domain at a finite number of points. As the number of points used to represent the domain (resolution) is increased, the error decreases. At sufficiently high resolution the error may become negligible, in which case the solution is said to have converged on the 'true' solution. However, increasing resolution also incurs costs in terms of computer memory and calculation time. The computer memory required to store all mesh information is roughly proportional to the total number of cells in the mesh. If $N$ is the number of cells per dimension of the domain, the required memory allocation is proportional to $N^M$, where $M$ is the number of dimensions. Moreover, the total calculation time is proportional to $N^{M+1}$. Hence, doubling the resolution of a 2D impact model increases the memory overhead by a factor of 4 and the calculation time by a factor of 8. For 3D simulations, the cost of doubling the resolution is twice as expensive as it is for 2D simulations.

The preceding discussion illustrates that the choice of resolution must be parsimonious: the need for accuracy must be balanced by the time and computer power available. In practical impact simulations a converged solution is often not possible. In such cases, it is important to quantify the error in the solution due to underresolution. In general, this is not possible to determine analytically (Hirt, 1968); instead, a 'brute force' approach is used in which a spectrum of computer runs, with different resolutions, is performed to determine the sensitivity of the solution to the number of mesh cells.

The number of computational cells per projectile radius (CPPR) is commonly used to define the resolution of an impact model. With this notation, the higher the resolution (CPPR) is, the smaller is the size of individual cells. The required number of CPPR in an impact simulation depends on the model and also on the question addressed by the model. For example, several modelling studies of impact melt production have demonstrated that the volume of material that experiences a certain peak pressure is quite sensitive to resolution (Pierazzo *et al.*, 1997; Wünnemann *et al.*, 2008), whereas crater depth and diameter tend to be relatively insensitive to resolution. In a comparison of eight impact models, Pierazzo *et al.* (2008) quantified the dependence of peak shock pressure at a given distance from the point of impact and crater dimensions on the number of cells used to resolve the projectile. In most cases, a resolution of 20–40 CPPR was sufficient to calculate peak shock pressure to within 10% of the true (high-resolution) value. In contrast, the same study showed that 10–20 CPPR were sufficient to compute crater size (depth and diameter) to within 10% of the true value.

Resolution or convergence testing is an important component of proper numerical modelling work. It provides verification that the model is working correctly and, where the use of very high resolution is not feasible, an estimate of the uncertainty in the solution due to the use of lower resolution. However, this uncertainty should not be confused with the real 'error' in the solution; that is, the difference between the numerical solution and reality. The fact that a numerical model converges on a solution does not mean that the solution is a true representation of reality. If important physics (such as material strength or an important phase change) is missing from the model, for example, the model may well converge on a solution that is very different from reality. For this reason, numerical models must also be rigorously tested against reality. This process, known as validation, is discussed further in Section 17.4.

### 17.2.5  Adaptive resolution

The previous discussion of the computational cost of increasing resolution applies to numerical models that employ uniform spatial resolution; that is, where the cell size is constant across the mesh or spatial domain and remains the same size throughout the simulation. This is the case for many impact models in current use, although such models often use uniform resolution only in a zone proximal to the impact site within which the crater grows. External to this, the mesh is extended efficiently by coarsening the spatial resolution outward so that the boundaries are far from the crater rim despite the relatively low number of cells in this region. However, technological advances have led to the growing use of adaptive resolution, where the number and size of mesh cells change during the simulation to increase resolution in zones of dynamic significance or interest and decrease resolution in zones of low significance or interest. The most widely employed adaptive resolution strategy in numerical impact models is adaptive mesh refinement (AMR; Berger and Oliger, 1984; Berger and Colella, 1989), where cells are subdivided into a number of finer 'daughter' cells, or combined with neighbouring sister cells to form a coarser 'parent' cell, to achieve a uniform spatial truncation error. Typically, the mesh is refined, increasing the resolution, in parts of the mesh where there are large changes in properties such as pressure, velocity, density, etc., over small distances and hence large truncation errors. Conversely, the mesh is coarsened in zones where such variables are relatively constant and the truncation error is low. In impact simulations, these techniques enable resolution to track and focus on the propagation of shock waves and the movement of material interfaces, in particular (see Fig. 17.5 for example).

Adaptive mesh methods can provide large efficiency gains because they achieve the same effective spatial resolution as a uniform mesh with a smaller total number of mesh cells. Hence, these methods can either (a) enable higher resolution to be employed in regions of importance than are affordable with a uniform-mesh simulation or (b) decrease the calculation time compared with a uniform-mesh simulation for the same effective spatial resolution. However, these methods are not without their drawbacks. The mesh refinement process can be a time-consuming procedure and efficiency savings may be reduced or lost entirely for certain problems. Moreover, the refinement criteria (i.e. the variables and tolerances that drive the refinement) are often user-defined options, which provide more opportunities for the user to affect the solution. Finally, the mesh refinement process can generate or amplify instabilities itself by encouraging or

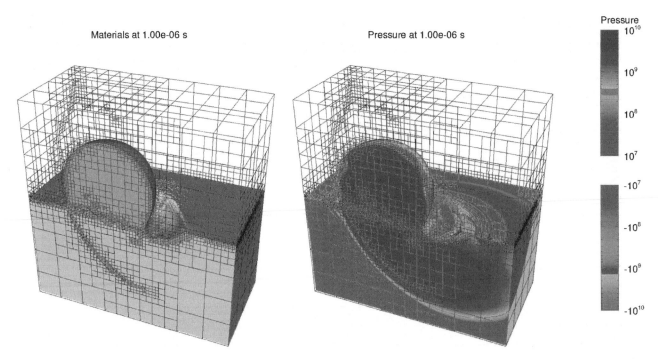

**Figure 17.5**  Contour plots of material (left) and pressure (Pa, right) from a high-resolution CTH simulation with AMR of a ¼-inch aluminium sphere impacting an aluminium target at 5 km s$^{-1}$ and at an angle of 15° to the horizontal. AMR is a mechanism for dynamically modifying the resolution during a simulation to ensure that computational effort is expended only where necessary. Observe in this example how resolution is focused along material interfaces and around the asymmetric shock wave, where it is most needed for an accurate solution. Note that the smallest mesh-block outlines (shown in black) represent 8 × 8 × 8 computational cells. This image is courtesy of Dave Crawford. (See Colour Plate 37)

suppressing physical processes that are sensitive to resolution, such as fracturing. Despite these weaknesses, when used appropriately and with care, adaptive mesh methods can provide huge efficiency savings, allowing otherwise intractable problems to be tackled (e.g. Pierazzo *et al.*, 2008; Crawford, 2010).

### 17.2.6   Shock modelling

A hypervelocity impact generates a shock wave that propagates through the target and projectile, irreversibly modifying their physical state and initiating the cratering process (see Chapter 3). Consequently, an essential feature of numerical impact models is the ability to simulate accurately the formation and propagation of shock waves. Mathematically, a shock wave is a discontinuity – an instantaneous jump – in physical properties, which is obviously problematic to represent on a mesh with a finite separation between mesh points. To address this problem, von Neumann and Richtmyer (1950) introduced a concept, termed artificial viscosity, inspired by the smearing of naturally occurring shockwaves due to viscous dissipation (in reality, the shock width can be millimetres in metals and metres in rocks due to pore collapse and fracture growth, which are usually not resolved in simulations by the computational grid). Their idea, which was later refined by several workers (e.g. Landshoff, 1955; Richtmyer and Morton, 1967; Wilkins, 1980), was to supplement the actual pressure in the governing equations with a fictitious pressure-like term, often referred to as the artificial viscosity *q*, that has the effect of smoothing abrupt increases in pressure over a few mesh cells (see Anderson (1987) for more details). Von Neumann and Richtmyer (1950) showed that the Hugoniot equations of conservation of mass, momentum and energy across a shock are still satisfied despite the use of artificial viscosity, provided that the pre- and post-shock states are compared away from the smoothed shock front.

The use of artificial viscosity is not, however, without its drawbacks (e.g. Benson, 1992; Caramana *et al.*, 1998). The most obvious of these is that in large impacts, where mesh cells can be on the order of a kilometer across, the simulated shock width of a few kilometres is significantly larger than in reality. Other weaknesses are more technical; perhaps the most important of which is that artificial viscosity introduces one or two problem/material-specific artificial parameters that are usually either chosen in an ad-hoc manner or not modified at all from one problem to the next. Improved shock capturing schemes now exist (e.g. Schulz, 1964; Benson and Schoenfeld, 1993; Caramana *et al.*, 1998); however, many impact models in current use employ artificial viscosity schemes similar to the original von Neumann and Richtmyer method. The use of such codes, therefore, requires care in the selection of model parameters and in the interpretation and analysis of results, particularly where those results are sensitive to the details of the shock front.

### 17.2.7   Multi-material and multiphase modelling

As discussed in Section 17.2.3, the most common approach in numerical modelling of impact processes is an *Eulerian* approach where material flows (advects) through a mesh fixed in space. This approach naturally leads to cells that contain more than one material; so-called mixed cells (see Fig. 17.4). These cells are problematic because variables such as pressure, distension, stress, strength and so on must be defined for the cell as a whole, not for each material, which requires some kind of averaging process and assumes that the mixture is uniform over the cell. The mixture will clearly not be an entirely accurate representation of the state and behaviour of the materials in the cell.

A good example is the determination of pressure in a mixed cell. Each material in the cell may have a different density and internal energy and, hence, a different pressure. However, the cell must have a single pressure; the pressure of each material must be in equilibrium. A simple approach is to use the mean pressure averaged over the volume fraction of each material. A more accurate, but also computationally more time-consuming, method is to adjust iteratively the volume fraction of each material and, thus, the density and pressure, until an equilibrium pressure for the cell is found. However, if the impedance contrast between different material fractions in a cell is high then the latter procedure may fail and the average pressure has to be used.

In many circumstances, mixed cells are unphysical because the materials in the cell are not uniformly mixed; the cell represents a mixture that is only just unresolved. For example, the upper third of the cell may represent one material and the lower two-thirds another material with a sharp interface separating the materials. In such cases, it is important for the numerical model to maintain sharp interfaces between different materials or material and void space – in other words, to minimize the width of zones of mixed cells. Unfortunately, the advection of material through a fixed computational mesh will inherently introduce numerical diffusion (artificial mixing), which tends to smear material interfaces over several cells, increasing the number of mixed cells (see Fig. 17.6). To solve this problem and preserve sharp material boundaries an 'interface tracking' or 'interface reconstruction' procedure is required (Fig. 17.6; e.g. Benson, 1992, 2002), the most popular of which is known as the volume of fluid (VOF) method (Hirt and Nichols, 1981). A detailed description of the various VOF algorithms to determine the location of the material interface is beyond the scope of this chapter and we refer interested readers to Benson (2002).

The numerical procedure of constructing a sharp interface between materials is not always physically appropriate. For example, if one of the materials is in a gaseous state and the other represents fragments of rock or melt droplets with a length scale much smaller than the cell width, it may be appropriate to allow mixing between the materials. If the mass of solid/molten particles greatly exceeds the mass of the surrounding gas (or liquid) then the gas fraction can be neglected and the motion of the particles described ballistically. Similarly, if the mass of particles is substantially lower than the mass of gas, the particles can be treated as a passive component, transported with the same velocity as the gas. In all intermediate cases, an alternative approach, known as multiphase hydrodynamics, allows the interpenetration and mixing of particles and gas. It is important to distinguish between the meaning of 'multiphase' and 'multi-material' in this context. The standard *multi-material* hydrodynamic approach assumes that the two (or more) materials in a single cell are transported with the same velocity; the *multiphase* approach, on

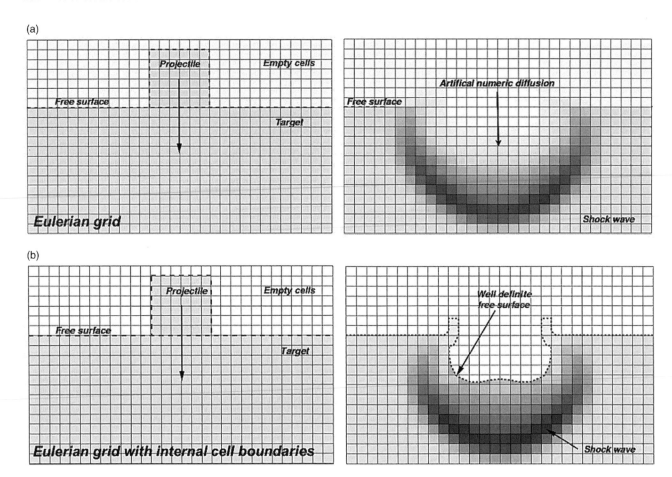

**Figure 17.6**   Eulerian simulations of a block impacting a half-space of the same material. (a) Without interface construction the free surface becomes diffused. (b) Interface construction maintains a sharp free surface and specifies its location. The same algorithm can be used to track interfaces between different materials within the target.

the other hand, allows the different materials (in this context called phases) to move with different velocities so that they can pass through one another.

Multiphase modelling of impact processes is generally only important for proper treatment of the impact plume and ejecta, where particles and fragments ejected from the crater interact with vaporized material and the surrounding atmosphere. In the standard two-phase hydrodynamics method (e.g. Harlow and Amsden, 1975; Valentine and Wohletz, 1989), the two 'phases' (one representing the particles and the other the surrounding gas or liquid) are treated as separate fluids, with their own velocity and EoS. This approach, widely used in volcanology (e.g. Valentine and Wohletz, 1989; Neri *et al.*, 2003) and recently modified to include physics for impact applications (Goldin and Melosh, 2009), allows exchange between phases of heat, mass (via evaporation/condensation) and momentum (via drag). An alternative multiphase approach is the so-called 'dusty flow' model (Boothroyd, 1971; Shuvalov, 1999), where representative particles (describing the collective behaviour of many real particles with the same size, velocity and trajectory) move through the surrounding fluid, as governed by buoyancy and drag forces. More details may be found in Shuvalov (1999). The advantage of this approach is that it easily allows particles of various sizes to be modelled. On the other hand, mass and heat exchange between

particles and gas are not treated, neglecting important physical processes in higher concentration particle flows.

### 17.2.8   Tracer particles

To permit large deformations, impact models often employ an Eulerian description, where the computational mesh is fixed and material is advected (moved) through it. As a consequence, the provenance of material contained in a given mesh cell changes during the simulation, and the state (pressure, temperature, etc.) of each mesh cell at a given time represents only the state of that spatial location, and the material within it, at that time. To track the movement and thermodynamic history of 'parcels' of material, information that would otherwise be lost, such codes often use Lagrangian tracer particles (tracers). These are massless particles that move with material flow without interacting with it and record the position and changing state of the material during the impact event.

### 17.3   Material models

To accurately simulate impact of one substance into another requires an adequate description of each material's response to stress, known as the material model. Constructing material models accurate and sophisticated enough to represent the complexity of geological materials remains the major challenge for numerical modelling of impacts. An accurate EoS, which describes the materials' responses to changes in pressure, density and temperature, is critical for modelling the generation and propagation of the shock wave in the early stages of the impact.

On the other hand, an accurate deviatoric (strength) model is critical for modelling the late stages of impact cratering, where the competition between gravity and material strength determines the final shape and characteristics of the crater. This section summarizes the equations of state and strength models in wide use for simulating impact processes.

### 17.3.1 Equations of state

In the context of numerical impact simulations, the EoS describes the unique relationship between pressure, internal energy and density (or specific volume) for a given material. It may take the form of anything from a single equation to a complex subprogram or look-up table. The simplest EoS models parameterize a material's thermodynamic behaviour using empirical laws; more complex EoS models attempt to represent the complexities of the material's atomic, molecular and crystalline structure using a combination of physical and empirical laws. Melosh (1989: appendix II) provides a general description of many EoSs widely used in impact studies.

Two of the simplest, yet most popular, EoSs for solids used in impact codes are the Mie–Grüneisen and the Tillotson EoS (Tillotson, 1962). Both models separate the pressure into a thermal and non-thermal (cold) part. The thermal pressure is defined in a manner similar to that for a perfect gas, while the non-thermal pressure is related to the change in density (compression or expansion) relative to a reference state. In both cases the model parameters must be found by fitting the model to experimentally derived thermodynamic data (e.g. Melosh, 1989: appendix II). As the Tillotson EoS involves more parameters than the Mie–Grüneisen EoS and has a separate form for hot expanded states, it usually provides a better approximation of material behaviour over a larger parameter space. However, major weaknesses of both the Mie–Grüneisen and the Tillotson EoS are that neither directly provides the temperature (although approximate temperature can be inferred indirectly; e.g. Ivanov *et al.*, 2002) or the entropy of the material. Nor do they account for melting and vaporization. A crude estimate of melt production can be made using the Tillotson EoS by dividing the internal energy by a constant known as the melt energy, but this approach is not reliable; for instance, the Tillotson EoS does not account for the latent heat of fusion.

More recent EoSs use increasingly complex descriptions that rely on different physical approximations and equations in different regions of phase space where those approximations are valid. The best-known example of these EoSs is ANEOS (Thompson and Lauson, 1972), a semi-analytical model now used in a number of hydrocodes. In ANEOS, pressures, temperatures and densities are derived from the Helmholtz free energy and are, hence, thermodynamically consistent. Explicit treatment of melt and vapour is included. Although clearly superior to prior analytical EoSs,

the original ANEOS has several limitations, such as the treatment of gases as monatomic species, which causes it to overestimate the liquid–vapour phase curve and critical point of most complex materials. Complex materials are still difficult to model, especially when they involve several solid–solid phase transitions (Ivanov, 2005; Melosh, 2007). An updated version of ANEOS is now available, which includes, among other things, the treatment of bi- and tri-atomic molecular gases (Melosh, 2007). However, it does not address the problem of different complex molecules present in the vapour phase. Moreover, ANEOS can only account for a single solid–solid phase transition or the melt transition, not both. ANEOS input arrays for several materials of geologic interest have been developed in recent years (e.g. Dunite: Benz *et al.*, 1989; Granite: Pierazzo *et al.*, 1997; Basalt: Pierazzo *et al.*, 2005; Quartz: Melosh, 2007), but much more work is still needed.

An important disadvantage in directly coupling complex codes such as ANEOS to impact codes is that they can significantly increase computation time as the EoS must be evaluated at least once per cell that contains material each timestep. A more efficient approach is to use tabular EoSs, which permits the fast look-up of predetermined EoS data. A classic example of this approach is SESAME, a tabular EoS database developed by the Los Alamos National Laboratory (e.g. Lyon and Johnson, 1992). Figure 17.7 illustrates an example of a SESAME EoS table developed for $H_2O$ (Senft and Stewart, 2008). At each point on the density–temperature grid a pressure, energy and entropy are

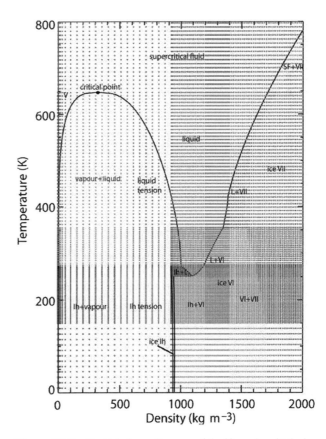

**Figure 17.7** Partial phase diagram of the 'five-phase' tabular $H_2O$ EoS, showing phases and phase boundaries, from Senft and Stewart (2008).

stored; intermediate states are calculated by interpolating between the nearest points in the table. These tabular forms are often built (at least in part) using semi-analytical models, such as ANEOS, and thus can suffer the same limitations, including being restricted to limited ranges of thermodynamic space. In addition, interpolation schemes used to define values between those stored in the table often 'smooth' away discontinuities in the EoS, which can be a problem in particular at phase boundaries (e.g. Swesty, 1996). However, EoS tables do provide a straightforward and efficient way to describe the thermodynamic behaviour of very complex materials.

The EoS is an essential part of all numerical impact models; however, the sophistication and accuracy required of the EoS depends on the impact conditions and the application. An accurate EoS is paramount for simulating the early stages of the impact, when the stresses are very large compared with the yield strength (i.e. deviatoric stresses are small compared with the pressure and can be neglected; see Section 17.3.3). During this stage, the EoS determines the peak and post-shock states of the target and impactor, as well as the attenuation of the shock wave.

For most materials, complex phase transitions, such as melting and vaporization, occur only when the impact velocity is greater than several kilometres per second. As a result, simple analytical equations of state (e.g. Mie–Grüneisen or Tillotson) are often adequate descriptions of material response in low-velocity impacts. However, if the impact velocity is high enough to induce important phase changes in the material, then more sophisticated EoS models that account for these must be used, particularly if the objective is to calculate melt or vapour production (e.g. Pierazzo *et al.*, 1997; Wünnemann *et al.*, 2008).

EoSs for geological materials have grown in sophistication in recent years and continue to be developed; however, detailed differences between different geological minerals and mixtures of minerals are still outside our ability to model.

### 17.3.2 Porous and mixed-material EoSs

EoS models, as discussed in the previous section, describe homogeneous, isotropic materials. They do not account for textural or compositional variations in the rocks that are common for geological materials. Rocks, in particular sedimentary material, are often porous and contain water or ice, or they consist of large clasts embedded in a fine-grained matrix of different modal composition and petrophysical properties. Heterogeneities within geological materials are often small in comparison with the scales involved in an impact, such as the diameter of the impactor or the width of the shock wave. However, they are of crucial importance for material behaviour and cause different phenomena on different scales (e.g. see Wünnemann *et al.* (2006, 2008)). It is important, therefore, for continuum impact models to account for the effect of material heterogeneity on a scale smaller than one computational cell by parameterizing the bulk (macroscale) behaviour of the heterogeneous material.

#### 17.3.2.1 Modelling porous materials

The added complication to material modelling introduced by porosity is that changes in a material's bulk density during shock compression are due to both the closing of pore space (compaction) and compression of the solid component (Zel'dovich and Raizer, 1966). A popular method for modelling porous materials is to separate the two processes of pore compaction and solid matrix compression. To do this, porosity is often parameterized by distension $\alpha$:

$$\alpha = \frac{1}{1-\phi} = \frac{V}{V_{S}} = \frac{\rho_{S}}{\rho} \tag{17.8}$$

where $V$ is the specific volume of the porous material, $V_{S}$ is the specific volume of the solid component, $\rho$ is density of the porous material, $\rho_{S}$ is the density of the solid component and $\phi$ is porosity. Using this definition, the pressure in a porous material $p$ can be described by the thermodynamically consistent relationship

$$p = f(\rho, I, \alpha) = \frac{1}{\alpha} p_{S}(\alpha\rho, I) = \frac{1}{\alpha} p_{S}(\rho_{S}, I) \tag{17.9}$$

after Herrmann (1969), Carroll and Holt (1972) and Kerley (1992). In this equation, $p_{S}$ denotes the pressure in the nonporous matrix material, which can be calculated using the EoS for the pure nonporous material from the specific internal energy $I$ and the solid matrix density ($\rho_{S} = \alpha\rho$). In this case, the only additional requirement to describe the pressure in a porous material is an equation relating $\alpha$ to another state variable (e.g. pressure or volume strain), which is often referred to as the compaction (crush) curve or compaction function. For further details on different compaction models, see Hermann (1969), Kerley (1992), Wünnemann *et al.* (2006), Jutzi *et al.* (2008) and Collins *et al.* (2011).

#### 17.3.2.2 Material mixtures

Porosity models are based on the assumption that pore space is filled with air or void. In this case, the effect of the pore filling on compaction is usually negligible; crushing of pores depends only on the strength of the surrounding solid rock matrix or grains. However, porous rocks are often saturated with water or ice that inhibits pore closure, resulting in a more complex behaviour than described by the simple compaction functions discussed.

Material mixtures are difficult to describe thermodynamically and may be modelled in different ways. One approach is to treat cells containing the material mixture as 'mixed cells' (see Section 17.2.7); another approach is to develop a separate EoS for the mixture and treat the mixture as a pure material. In either case, the bulk behaviour of the mixture should be some sort of average of the behaviour of the two individual materials, based on the relative volume fractions and assumptions of mechanical and/or thermal equilibrium.

A mixed material EoS may be constructed by combining the EoSs of two (or more) pure materials prior to any simulation. An example of the application of this approach for a basalt–water mixture is described in Pierazzo *et al.* (2005: appendix B). This approach assumes total equilibrium (pressure and temperature) between the two (or more) materials and a constant mass ratio of the components, which may not always be appropriate. This

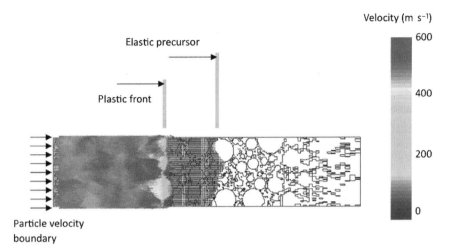

**Figure 17.8** A 3D meso-mechanical simulation of shock wave propagation in concrete (Riedel, 2000). The computational domain comprises zones of cement and zones of gravel, for which the EoSs are known, to represent a concrete mixture. A velocity boundary condition is used to generate a plastic wave and an elastic precursor that propagate through the mixture. The calculated shock wave velocity and average material velocity behind the shock can be used to define the Hugoniot curve for the concrete mixture and hence an EoS. (See Colour Plate 38)

approach is not suitable if the mass ratio of the component materials changes during the simulation; for example, if vaporized water escapes solid basalt, changing the mass ratio.

An alternative approach to develop a macroscopic EoS (table or equation) for a two- or multi-material mixture of a given mass ratio is to examine its thermodynamic behaviour explicitly in so-called meso-scale models. In such models, the components of a mixture are simulated using the standard continuum mechanical approach described above. The spatial distribution of each material is resolved by the model and the thermodynamic interaction at material boundaries (e.g. plastic flow localized on interfaces of particles or around pores and enclosures) is exactly computed. By a series of meso-scale numerical experiments of shock loading and unloading, Hugoniot curves and release adiabats for the material mixture can be derived (see Fig. 17.8). For more details about this approach see, for example, Benson (1994), Riedel (2000) and Nesterenko (2001).

### 17.3.3 Deviatoric stress (strength) model

The deviatoric stress model (also known as the constitutive or strength model) is a set of equations that define the non-hydrostatic part of the stress tensor as a function of a combination of strain (displacement), strain rate (velocity), pressure and internal energy or temperature, for a given material. In so doing it describes the resistance of a material to changes in shape.

The simplest form of this model assumes that the material is inviscid and has no strength; in this 'hydrodynamic' case, no deviatoric stress is supported: $s_{ij}=0$. Although all materials offer some resistance to shear deformation, the hydrodynamic approximation is useful for simulating the early stages of impact, in which pressures are so large that the deviatoric part of the stress tensor is insignificant and can reasonably be ignored. It is for this reason

that the earliest impact models did not include strength effects and became known as 'hydrocodes'.

Most geological materials (soils, rocks, ices and metals) are solid, rather than fluid; in other words, they possess a 'yield strength' that allows them to support a certain stress without permanently deforming. At stresses lower than the yield strength, deformation is reversible (elastic) and the material returns to its initial state when stresses are removed. When the yield strength is exceeded, permanent and irreversible deformation occurs, which may take the form of ductile flow, where strain is uniformly distributed, or brittle fracturing and breaking, where strain is localized, depending on the properties of the material and its physical state. Deformation can become arbitrarily large and is often approximated by so-called plasticity or failure models, in which stress is capped at a certain threshold stress state that is itself dependent on state variables.

Stress at any point is measured by a symmetric tensor, which has six independent components in three dimensions. Equivalently, stress in three dimensions can be represented by three principal stresses, aligned in three principal directions. For isotropic materials the direction in which stress is applied does not affect whether permanent deformation occurs and, hence, stress is adequately quantified by the three principal stresses alone. A common strength in numerical impact models tests the square root of the second invariant of the deviatoric stress tensor $J_2$ in each cell to test whether it exceeds the yield strength $Y$ in the cell. $J_2$ is a function of the principal stresses:

$$J_2 = \frac{1}{6}[(\sigma_1 - \sigma_2)^2 + (\sigma_2 - \sigma_3)^2 + (\sigma_3 - \sigma_1)^2] \qquad (7.10)$$

If $\sqrt{J_2}$ exceeds the yield strength $Y$, each term in the stress tensor may be reduced by a constant factor $(Y/\sqrt{J_2})$ to ensure that the stress state does not lie above the yield envelope (Wilkins, 1964).

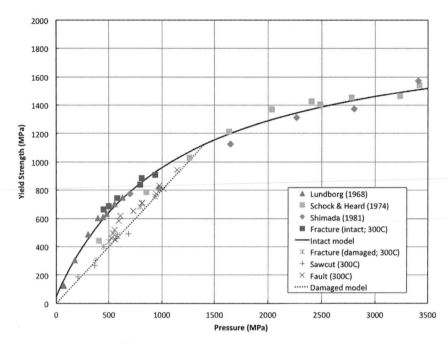

**Figure 17.9**    Yield strength (defined as $\sqrt{J_2}$) as a function of pressure for intact (closed symbols) and damaged (open symbols) samples of Westerly granite. The data labelled 'Fracture', 'Sawcut' and 'Fault' are from Stesky *et al.* (1974) and refer to different sample preparation; see that reference for more details. Note the increase in strength with increasing pressure, characteristic of all rock-like materials. At low pressures the strength of damaged rock samples is considerably lower than that of intact samples, which results in brittle failure and strain localization. Other data from Lundborg (1968), Schock and Heard (1974) and Shimada (1981).

The complexity in $J_2$-type strength models stems from how the yield strength $Y$ is defined. A common approach to use (semi-) empirical relationships for the behaviour of geological materials under stress and to choose the value of model parameters based on data from laboratory rock mechanics experiments. Such experiments show that a rock's yield strength is a function of confining pressure, temperature, strain, strain rate, porosity and sample size (for a recent review, see Lockner (1995); or any text-book on rock mechanics, e.g. Jaeger *et al.*, 2008). Collins *et al.* (2004) discussed the elements of rock strength most important for impact simulations and described one way to define the yield envelope for rock materials. Holsapple (2009) reviewed this and many alternative approaches for defining the yield envelope used in modern impact codes. Here, we summarize the key important elements of rock strength.

The most important characteristic of rock strength is that it increases with increasing confining pressure as a consequence of its granular structure – as the grains are squeezed more tightly together they are harder to move over one another. Figure 17.9 shows yield strength data (defined as the square root of $J_2$ at failure) as a function of pressure from a number of laboratory rock mechanics experiments that tested both intact and pre-fractured granite samples. The observed trends are typical for granular rock materials. In the absence of overburden pressure, fragmented rock materials have little or no strength and do not support tensile stress. For intact samples, on the other hand, there is a substantial yield strength at zero pressure, termed the cohesion, and a tensile strength. As pressure increases, the yield

strength for pre-fractured samples is approximately linearly proportional to the confining pressure (Stesky *et al.*, 1974; Byerlee, 1978), where the constant of proportionality is known as the coefficient of friction (slope of the dashed line in Fig. 17.9). For intact rock the yield strength rises smoothly, but not as a straight line; the local slope of the strength versus pressure curve (the coefficient of internal friction) decreases with increasing pressure. A number of approximations have been used to describe intact rock strength as a function of pressure (e.g. Lundborg, 1968; Hoek and Brown, 1980).

Figure 17.9 illustrates that at low pressure the yield strength of damaged rock is less than that of intact rock. In this regime, once a narrow zone of damaged rock is formed, it offers less resistance to sliding than the surrounding intact rocks do and further deformation of the rock mass occurs preferentially along these zones (Rice, 1976). This style of localized (brittle) deformation is often modelled by introducing a scalar 'damage' field, which varies from 0 for intact rocks to 1 for completely damaged rocks. Damage may be defined simply as an increasing function of permanent plastic strain (e.g. Johnson and Holmquist, 1994; Ivanov *et al.*, 1997) or it may be related to the parameterized growth of tensile flaws (e.g. Melosh *et al.*, 1992) or it may be defined using a combination (Collins *et al.*, 2004). The damage parameter is used to smoothly modify the yield envelope from that of intact rock to that of damaged rock.

The strength of rock materials, both intact and fragmented, decreases with increasing temperature. As temperature approaches the melting temperature, the resistance to shear reduces, reaching zero at the melting point. This behaviour can be described by any of a number of simple functions (e.g. Johnson and Cook, 1983; Ohnaka, 1995).

As pressure (and temperature) increases, rocks respond to applied stress more and more homogeneously; such behaviour is described as ductile. For ductile rocks the concept of damage loses its meaning and relevance, as the difference between the strength of intact and damaged rocks is negligible (see Fig. 17.9). In addition, many ductile metals exhibit strain hardening, where yield strength increases with increasing strain. The Johnson and Cook (1983) and the Steinberg–Guinan (Steinberg *et al.*, 1980) strength models both include strain hardening terms.

A further complexity associated with deformation of geological materials is the effect of porosity. Porous materials, when compressed, initially just compact with no associated rise in strength. Furthermore, when brittle materials fail there is an associated increase in volume, as new pores are created and rock fragments rearrange themselves to move over each other, known as shear bulking or dilatancy (Reynolds, 1885). The volume of a rock mass undergoing bulking rises monotonically with increasing strain until the fragmented rock mass achieves a fully dilatant state and the bulk volume reaches a maximum. Compaction and bulking are difficult to implement into hydrocodes for simulating impact events, because they affect both the constitutive model and the EoS, which are usually treated separately. Preliminary efforts to include the effects of bulking in an impact code were described by O'Keefe *et al.* (2001).

A weakness of basing the deviatoric stress model on (semi-) empirical trends from laboratory strength data is that conditions in many impact scenarios will be very different to the small-scale, quasi-static conditions typical of rock mechanics experiments. In particular, there is much evidence that natural rock materials are weaker on scales of tens to hundreds of metres than laboratory strength measurements of centimetre-scale rock samples would suggest (Hoek and Brown, 1980; Schmidt and Montgomery, 1995; Hoek *et al.*, 2002). Moreover, results of numerical impact models demonstrate that strength models based solely on laboratory strength measurements do not permit the gravitational collapse of large-scale impact craters required to form complex craters (see Chapter 5). The physics of rock failure and deformation during large crater collapse is still not understood (Melosh and Ivanov, 1999). However, simulating large-scale impacts requires modification of the strength model parameters or the incorporation of additional physics to explain the apparent weakening of target materials during impact. Weakening processes that have been investigated by numerical models include acoustic fluidization (Melosh and Ivanov, 1999; Wunnemann and Ivanov, 2003), enhanced thermal softening (O'Keefe and Ahrens, 1993, 1999), comminution (O'Keefe *et al.*, 2001) and dynamic fault weakening (Senft and Stewart, 2009) and are discussed in Chapter 5.

## 17.4   Validation, verification and benchmarking

Verification and validation are two important, but quite different, ways that all numerical models are tested. Verification is the process of demonstrating that the model is correctly solving the governing equations (e.g. eqs (17.2)–(17.6)). Validation, on the other hand, is the process of testing the model's ability to reproduce known real-world situations: a laboratory experiment or a naturally occurring impact of known initial conditions, for example. In other words, verification asks 'Have we built the model correctly?' whereas validation asks 'Have we built the correct model?'

Validation and verification might appear to be synonymous – a verified code should produce valid physical results – but a well-verified code will not produce accurate or meaningful results for a problem that it is not designed to simulate or not capable of simulating. As discussed above, different impact models have different capabilities, strengths and weaknesses, which make them more suitable to particular problems and make them less suitable to others. Hence, to evaluate and reliably apply an impact model to a specific problem it is essential to validate the model as rigorously as possible.

Benchmarking is the process of comparing the results of one numerical model with those from another. This type of testing exposes differences arising from both the underlying numerical treatments and the material models used in impact models. Benchmarking is particularly useful because comparing several different model simulations of a given scenario is often the only available metric for confidence in hydrocode results, since the scenarios of most interest often explore new regimes beyond the realm of laboratory experiments.

Pierazzo *et al.* (2008) reported on results of one very successful benchmarking and validation project, involving eight impact models from commercial and US National Laboratory codes, to those developed in the academic community. This study showed that the 'error' between different codes and between model results and experiments was approximately 10% for a range of diagnostics, such as crater diameter and peak shock pressure.

## 17.5   Concluding remarks

Numerical modelling is an important facet of impact cratering research. Impact models offer a means for studying various aspects of the impact process that cannot be investigated by other methods. Moreover, they provide detailed information regarding all variables of interest for the entire simulation; in a sense they may be regarded as the best instrumented experiment (Anderson, 1987). When combined with experimental and observational ground truth, impact models can provide unparalleled insight into the cratering process.

Impact modelling has advanced understanding of the impact process. In particular, the amount of melt and vapour produced during the early stages of an impact and the fate of the impactor (see Chapter 3), the development and evolution of ejecta and the impact plume (see Chapter 4) and formation of the complex morphologies observed in large craters (see Chapter 5). As computers become more powerful and high-performance computing becomes more accessible, the number of impact modelling studies is likely to increase dramatically. It is imperative, therefore, for all students of impact cratering to understand the limitations of impact models and 'best practice' approach for their application, as described in this chapter.

Apart from hardware advances, impact models are continually increasing in sophistication, both in terms of efficiency – the amount of data that can be processed in a reasonable time-frame – and the complexity of the physics that is implemented in them. For example, 3D impact models using AMR are becoming a matter of course (e.g. Crawford, 2010). There is no doubt that future advances in impact modelling will rely as much on the implementation of new physics and the development of intelligent technology, such as AMR, as they will on the inexorable rise of computer power.

Impact cratering is a geological process. Hence, impact models must account for the complex behaviour of heterogeneous geological materials. Material models employed in impact simulations are improving in realism all the time, exemplified by recent advances to the EoS for ice (Senft and Stewart, 2008) and methods for incorporating pore space collapse (Wünnemann et al., 2006). However, in many respects the material models in use today are inadequate; important questions concerning the behaviour of material under the extreme conditions associated with an impact event remain unanswered. It is clear that improving material models will continue to be the biggest challenge at the frontier of impact modelling. but also that many scientific fruits lie in wait to be discovered along the way.

## References

Ahrens, T.J. and O'Keefe, J.D. (1977) Equations of state and impact-induced shock-wave attenuation on the Moon, in *Impact and Explosion Cratering* (eds D.J. Roddy, R.O. Pepin and R.B. Merrill), Pergamon Press, New York, NY, pp. 639–656.

Anderson, C.E. (1987) An overview of the theory of hydrocodes. *International Journal of Impact Engineering*, 5, 33–59.

Artemieva, N.A. and Ivanov, B.A. (2004) Launch of Martian meteorites in oblique impacts. *Icarus*, 171, 84–101.

Artemieva, N. and Morgan, J. (2009) Modeling the formation of the K–P boundary layer, *Icarus*, 201, 768–780.

Artemieva, N.A., Pierazzo, E. and Stöffler, D. (2002). Numerical modeling of tektite origin in oblique impacts: implication to Ries–Moldavite strewn field. *Bulletin of the Czech Geological Survey*, 77 (4), 31–39.

Artemieva, N., Karp, T. and Milkereit, B. (2004) Investigating the Lake Bosumtwi impact structure: insight from numerical modeling. *Geochemistry Geophysics Geosystems*, 5, Q11016.

Benson, D.J. (1992) Computational methods in Lagrangian and Eulerian hydrocodes. *Computer Methods in Applied Mechanics and Engineering*, 99 (2–3), 235.

Benson, D.J. (1994) An analysis by direct numerical simulation of the effects of particle morphology on the shock compaction of copper powder. *Modelling and Simulation in Materials Science and Engineering*, 2, 535–550.

Benson, D.J. (2002) Volume of fluid interface reconstruction methods for multi-material problems. *Applied Mechanics Reviews*, 55 (2), 151–165.

Benson, D.J. and Schoenfeld, S. (1993) A total variation diminishing shock viscosity. *Computational Mechanics*, 11 (2), 107–121.

Benz, W., Cameron, A.G.W. and Melosh, H.J. (1989) The origin of the Moon and the single-impact hypothesis III. *Icarus*, 81, 113–131.

Berger, M.J. and Colella, P. (1989) Local adaptive mesh refinement for shock hydrodynamics. *Journal of Computational Physics*, 82, 64–84.

Berger, M.J. and Oliger, J. (1984) Adaptive mesh refinement for hyperbolic partial differential equations. *Journal of Computational Physics*, 53, 484–512.

Bjork, R.J. (1961) Analysis of the formation of Meteor Crater, Arizona. A preliminary report. *Journal of Geophysical Research*, 66, 3379–3387.

Boothroyd, R.G. (1971) *Flowing Gas–Solids Suspensions*, Chapman and Hall, London.

Byerlee, J. (1978) Friction of rocks. *Pure and Applied Geophysics*, 116, 615–626.

Cameron, A.G.W. (2000) Higher-resolution simulations of the giant impact, in *Origin of the Earth and Moon* (eds R.M. Canup and K. Righter), University of Arizona Press, Tucson, AZ, pp. 133–144.

Canup, R.M. (2004) Simulations of a late lunar-forming impact. *Icarus*, 168, 433–456.

Canup, R.M. (2008) Lunar-forming collisions with pre-impact rotation. *Icarus*, 192, 518–538. DOI: 10.1016/j.icarus.2008.03.011.

Canup, R.M. and Asphaug, E. (2001). Origin of the Moon in a giant impact near the end of the Earth's formation. *Nature*, 412, 708–712.

Caramana, E.J., Shashkov, M.J. and Whalen, P.P. (1998) Formulations of artificial viscosity for multi-dimensional shock wave computations. *Journal of Computational Physics*, 144 (1), 70–97.

Carroll, M.M. and Holt, A.C. (1972) Static and dynamic pore-collapse relations for ductile porous materials. *Journal of Applied Physics*, 43, 1626–1636.

Collins, G.S., Melosh, H.J., Morgan, J.V. and Warner, M.R. (2002) Hydrocode simulations of Chicxulub Crater collapse and peak-ring formation. *Icarus*, 157, 24–33.

Collins, G.S., Melosh, H.J. and Ivanov, B.A. (2004) Modeling damage and deformation in impact simulations. *Meteoritics and Planetary Science*, 39 (2), 217–231.

Collins, G.S., Morgan, J.V., Barton, P. et al. (2008) Dynamic modeling suggests terrace zone asymmetry in the Chicxulub Crater is caused by target heterogeneity. *Earth and Planetary Science Letters*, 270, 221–230.

Collins, G.S., Melosh, H.J. and Wünnemann, K. (2011) Improvements to the epsilon–alpha compaction model for simulating impacts into high-porosity Solar System objects. *International Journal of Impact Engineering*, 38, 434–439.

Courant, R. (1928) Über die partiellen Differenzengleichungen der mathematischen Physik. *Mathematische Annalen*, 100 (1), 32.

Crawford, D. (2010) Parallel *N*-body gravitational interaction in CTH for planetary defense and large impact simulations. Proceedings of 11th Hypervelocity Impact Symposium, Freiburg, Germany, abstract no. 155.

Crawford, D.A., Boslough, M.B., Trucano, T.G. and Robinson, A.C. (1994) The impact of comet Shoemaker–Levy 9 on Jupiter. *Shock Waves*, 4, 47–50.

Elbeshausen, D., Wünnemann, K. and Collins, G.S. (2009) Scaling of oblique impacts in frictional targets: implications for crater size and formation mechanisms. *Icarus*, 204, 716–731.

Gisler, G., Weaver, R., Mader, C. and Gittings, M. (2004) Two- and three-dimensional asteroid impact simulations. *Computing in Science and Engineering*, 6 (3), 46–55.

Goldin, T.J. and Melosh, H.J. (2009) Self-shielding of thermal radiation by Chicxulub impact ejecta: firestorm or fizzle? *Geology*, 37, 1135–1138. DOI: 10.1130/G30433A.1.

Harlow, F.H. and Amsden, A.A. (1975) Numerical calculations of multiphase fluid flow. *Journal of Computational Physics*, 17, 19–52.

Head, J.N., Melosh, H.J. and Ivanov, B.A. (2002) Martian meteorite launch: high-speed ejecta from small craters. *Science*, 298, 1752–1756.

Herrmann, W. (1969) Constitutive equation for the dynamic compaction of ductile porous material. *Journal of Applied Physics*, 40, 2490–2499.

Hirt, C.W. (1968) Heuristic stability theory for finite difference equations. *Journal of Computational Physics*, 2, 339.

Hirt, C.W. and Nichols, B.D. (1981) Volume of fluid (VOF) method for the dynamics of free boundaries. *Journal of Computational Physics*, 39 (1), 201–225.

Hoek, E. and Brown, E.T. (1980) *Underground Excavations in Rock*, Institution of Mining and Metallurgy, London.

Hoek, E., Carranza-Torres, C. and Corkum, B. (2002) Hoek–Brown criterion 2002 edition. Proceedings of the NARMS-TAC Conference, Toronto, vol. 1, pp. 267–273.

Holsapple, K.A. (2009) On the 'strength' of the small bodies of the Solar System: a review of strength theories and their implementation for analyses of impact disruptions. *Planetary and Space Science*, 57, 127–141. DOI: 10.1016/j.pss.2008.05.015.

Ivanov, B.A. (2005) Numerical modeling of the largest terrestrial meteorite craters. *Solar System Research*, 39 (5), 381–409.

Ivanov, B.A. and Artemieva, N.A. (2002) Numerical modeling of the formation of large impact craters, in *Catastrophic Events and Mass Extinctions: Impacts and Beyond* (eds C. Koeberl and K.G. MacLeod), Geological Society of America Special Paper 356, Geological Society of America, Boulder, CO, pp. 619–630.

Ivanov, B.A., Deniem, D. and Neukum, G. (1997) Implementation of dynamic strength models into 2D hydrocodes: applications for atmospheric breakup and impact cratering. *International Journal of Impact Engineering*, 20, 411–430.

Ivanov, B.A., Langenhorst, F., Deutsch, A. and Hornamann, U. (2002) How strong was impact-induced $CO_2$ degassing in the Cretaceous–Tertiary event? Numerical modeling of shock recovery experiments, in *Catastrophic Events and Mass Extinctions: Impacts and Beyond* (eds C. Koeberl and K.G. MacLeod), Geological Society of America Special Paper 356, Geological Society of America, Boulder, CO, pp. 587–594.

Jaeger, J.C., Cook, N.G.W. and Zimmermann, R.W. (2008) *Fundamentals of Rock Mechanics*, fourth edition, Chapman and Hall, London.

Johnson, G.R. and Cook, W.H. (1983) A constitutive model and data for metals subjected to large strains, high strain rates and high temperatures. Proceedings, 7th International Symposium on Ballistics, Hague, The Netherlands.

Johnson, G.R. and Holmquist, T.J. (1994) An improved computational model for brittle materials, in *High-Pressure Science and Technology – 1993* (eds S.C. Schmidt, J.W. Shaner, G.A. Samara and M. Ross), AIP Press, Woodbury, NY, pp. 981–984.

Jutzi, M., Benz, W. and Michel, P. (2008) Numerical simulations of impacts involving porous bodies: I. Implementing sub-resolution porosity in a 3D SPH hydrocode. *Icarus*, 198 (1), 242–255.

Kerley, G.I. (1992) CTH equation of state package: porosity and reactive burn models. Sandia National Lab Report.

Landshoff, R. (1955) A numerical method for treating fluid flow in the presence of shocks. Report LA-1930, Los Alamos National Laboratory, Los Alamos, NM.

LeVeque, R.J. (2002) *Finite Volume Methods for Hyperbolic Problems*, Cambridge University Press, Cambridge.

LeVeque, R. (2007) *Finite Difference Methods for Ordinary and Partial Differential Equations: Steady-State and Time-Dependent Problems*, Classics in Applied Mathematics, Cambridge University Press, Cambridge.

Lockner, D.A. (1995) Rock failure, in *Rock Physics and Phase Relations: A Handbook of Physical Constants* (ed. T.J. Ahrens), American Geophysical Union, Washington, DC, pp. 127–147.

Lundborg, N. (1968) Strength of rock-like materials. *International Journal of Rock Mechanics and Mineral Science*, 5, 424–454.

Lyon, S.P. and Johnson, J.D. (1992) SESAME: the LANL equation of state database. Report LA-UR-92-3407, Los Alamos National Laboratories, Los Alamos, NM.

Malvern, L.E. (1969) *An Introduction to the Mechanics of a Continuous Medium*, Prentice-Hall, Englewood Cliffs, NJ.

Melosh, H.J. (1989) *Impact Cratering: A Geologic Process*, Oxford University Press, New York, NY.

Melosh, H.J. (2007) A hydrocode equation of state for $SiO_2$. *Meteoritics and Planetary Science*, 42, 2079–2098.

Melosh, H.J. and Ivanov, B.A. (1999) Impact crater collapse. *Annual Reviews in Earth and Planetary Science*, 27, 385–415.

Melosh, H.J., Ryan, E.V. and Asphaug, E. (1992) Dynamic fragmentation in impacts: hydrocode simulation of laboratory impacts. *Journal of Geophysical Research*, 97 (E9), 14735–14759.

Monaghan, J.J. (1988) An introduction to SPH. *Computer Physics Communications*, 48, 89–96.

Neri, A., Ongaro, T.E., Macedonio, G. and Gidaspow, D. (2003) Multiparticle simulations of collapsing volcanic columns and pyroclastic flow. *Journal of Geophysical Research*, 108 (B4), 2202. DOI: 10.1029/2001 JB000508.

Nesterenko, V.F. (2001) *Dynamics of Heterogeneous Materials*, Springer-Verlag New York Inc., New York, NY.

Ohnaka, M. (1995) A shear failure strength law of rock in the brittle–plastic transition regime. *Geophysical Research Letters*, 22, 25–28.

O'Keefe, J.D. and Ahrens, T.J. (1993) Planetary cratering mechanics. *Journal of Geophysical Research*, 98, 17001–17028.

O'Keefe, J.D. and Ahrens, T.J. (1999) Complex craters: relationship of stratigraphy and rings to impact conditions. *Journal of Geophysical Research*, 104 (E11), 27091–27104.

O'Keefe, J.D., Stewart, S.T., Lainhart, M.E. and Ahrens, T.J. (2001) Damage and rock–volatile mixture effects on impact crater formation. *International Journal of Impact Engineering*, 26, 543–553.

Orphal, D.L. (1977) Calculations of explosion cratering – II. Cratering mechanics and phenomenology, in *Impact and Explosion Cratering* (eds D.J. Roddy, R.O. Pepin and R.B. Merrill), Pergamon Press, New York, NY, pp. 907–917.

Pierazzo, E. and Melosh, H.J. (1999) Hydrocode modeling of Chicxulub as an oblique impact event. *Earth and Planetary Science Letters*, 165, 163–176.

Pierazzo, E. and Melosh, H.J. (2000a) Hydrocode modeling of oblique impacts: the fate of the projectile. *Meteoritics and Planetary Science*, 35 (1), 117–130.

Pierazzo, E. and Melosh, H.J. (2000b) Melt production in oblique impacts. *Icarus*, 145, 252–261.

Pierazzo, E. and Melosh, H.J. (2000c) Understanding oblique impacts from experiments, observations, and modeling. *Annual Reviews in Earth and Planetary Science*, 28, 141–167.

Pierazzo, E., Vickery, A.M. and Melosh, H.J. (1997) A reevaluation of impact melt production. *Icarus*, 127, 408–423.

Pierazzo, E., Kring, D.A. and Melosh, H.J. (1998) Hydrocode simulation of the Chicxulub impact event and the production of climatically active gases. *Journal of Geophysical Research*, 103, 28607–28625.

Pierazzo, E., Artemieva, N. and Ivanov, B. (2005) Starting conditions for hydrothermal systems underneath Martian craters: numerical modelling, in *Large Meteorite Impacts III* (eds T. Kenkmann, F. Horz and A. Deutsch), Geological Society of America Special Paper 384, Geological Society of America, Boulder, CO, pp. 443–457.

Pierazzo, E., Artemieva, N., Asphaug, E. *et al.* (2008) Validation of numerical codes for impact and explosion cratering. *Meteoritics and Planetary Science*, 43 (12), 1917–1938.

Reynolds, O. (1885) On the dilatancy of media composed of rigid particles in contact. With experimental illustrations. *Philosophical Magazine, Series 5*, 20, 469–481.

Rice, J.R. (1976) The localization of plastic deformation. Proceedings, 14th International Congress of Theoretical and Applied Mechanics, pp. 207–220.

Richtmyer, R.D. and Morton, K.W. (1967) *Difference Methods for Initial-Value Problems*, 2nd edition, Interscience Publishers, New York, NY.

Riedel, W. (2000) Beton unter dynamischen Lasten. Meso- und makromechanische Modelle und ihre Parameter, in *Forschungsergebnisse aus der Kurzzeitdynamik*, Fraunhofer Institute for High Speed Dynamics, Ernst Mach Institute.

Schmidt, K.M. and Montgomery, D.R. (1995) Limits to relief. *Science*, 270 (5236), 617–620.

Schock, R.N. and Heard, H.C. (1974) Static mechanical properties and shock loading response of granite. *Journal of Geophysical Research*, 79, 1662–1666.

Schulz, W.D. (1964) Tensor artificial viscosity for numerical hydrodynamics. *Journal of Mathematical Physics*, 5, 133–138.

Senft, L.E. and Stewart, S.T. (2008) Impact crater formation in icy layered terrains on Mars. *Meteoritics and Planetary Science*, 43 (12), 1993–2013.

Senft, L.E. and Stewart, S.T. (2009) Dynamic fault weakening and the formation of large impact craters. *Earth and Planetary Science Letters*, 287, 471–482. DOI: 10.1016/j.epsl.2009.08.033.

Shimada, M. (1981) The method of compression test under high pressures in a cubic press and the strength of granite. *Tectonophysics*, 72 (3–4), 343–357.

Shuvalov, V.V. (1999) Multidimensional hydrodynamic code SOVA for interfacial flows: application to the thermal layer effect. *Shock Waves*, 9, 381–390.

Shuvalov, V. and Dypvik, H. (2004) Ejecta formation and crater development of the Mjølnir impact. *Meteoritics and Planetary Science*, 39, 467–479.

Shuvalov, V., Dypvik, H. and Tsikalas, F. (2002) Numerical simulations of the Mjølnir marine impact crater. *Journal of Geophysical Research*, 107, 5047. DOI: 10.1029/2001JE001698.

Steinberg, D.J., Cochran, S.G. and Guinan, M.W. (1980) A constitutive model for metals applicable at high strain rates. *Journal of Applied Physics*, 51, 1498–1504.

Stesky, R.M., Brace, W.F., Riley, D.K. and Robin, P.Y.F. (1974) Friction in faulted rock at high temperature and pressure. *Tectonophysics*, 23, 177–203.

Swesty (1996) Thermodynamically consistent interpolation of EoS tables. *Journal of Computational Physics*, 127, 118–127.

Thompson, S.L. and Lauson, H.S. (1972) Improvements in the chart-D radiation hydrodynamic code III: revised analytical equation of state. Report SC-RR-710714, Sandia National Laboratories, Albuquerque, NM.

Tillotson, J.H. (1962) Metallic equations of state for hypervelocity impact. General Atomics Report GA-3216, General Atomics Division, General Dynamics San Diego, CA.

Valentine, G.A. and Wohletz, K.H. (1989) Numerical models of Plinian eruption columns and pyroclastic flows. *Journal of Geophysical Research*, 94, 1867–1887.

Von Neumann, J. and Richtmyer, R.D. (1950) A method for the numerical calculation of hydrodynamic shocks. *Journal of Applied Physics*, 21 (3), 232–237.

Wilkins, M.L. (1964) Calculation of elastic–plastic flow, in *Methods of Computational Physics*, Vol. 3 (eds B. Alder, S. Fernback and M. Rotenberg), Academic Press, New York, NY.

Wilkins, M.L. (1980) Use of artificial viscosity in multidimensional fluid dynamic calculations. *Journal of Computational Physics*, 36 (3), 281–303.

Wünnemann, K. and Ivanov, B.A. (2003) Numerical modelling of the impact crater depth–diameter dependence in an acoustically fluidized target. *Solar System Research*, 51 (13), 831–845.

Wünnemann, K., Collins, G.S. and Melosh, H.J. (2006) A strain-based porosity model for use in hydrocode simulations of impacts and implications for transient crater growth in porous targets. *Icarus*, 180 (2), 514–527.

Wünnemann, K., Collins, G.S. and Osinski, G.R. (2008) Numerical modelling of impact melt production in porous rocks. *Earth and Planetary Science Letters*, 269 (3–4), 530–539.

Zel'dovich, Y.B. and Raizer, Y.P. (1966) *Physics of Shock Waves and High-Temperature Hydrodynamic Phenomena*, Academic Press, New York, NY.

Zienkiewicz, O.C., Taylor, R.L. and Nithiarasu, P. (2005) *The Finite Element Method for Fluid Dynamics*, Butterworth-Heinemann, Oxford.

# EIGHTEEN

# Comparison of simple impact craters: a case study of Meteor and Lonar Craters

## Horton E. Newsom[*], Shawn P. Wright[*], Saumitra Misra[†] and Justin J. Hagerty[‡]

[*]Institute of Meteoritics and Department of Earth and Planetary Sciences MSC03 2050, University of New Mexico, Albuquerque, NM 87131, USA
[†]School of Geological Sciences, University of Kwazulu-Natal, Private Bag X54001, Durban 4000, South Africa
[‡]United States Geological Survey, Astrogeology Science Center, 2255 N. Gemini Drive, Flagstaff, AZ 86001, USA

## 18.1 Introduction

The 50,000-year-old Meteor Crater, Arizona, USA is approximately 1.2 km in diameter and was formed when an iron meteorite (Canyon Diablo) impacted the sedimentary target rocks of the Colorado Plateau. The Lonar crater, recently dated at 570,000 years old (Jourdan et al., 2011), is located in Maharashtra, India. The crater is 1.8 km in diameter (Fig. 18.1) and it was formed in basaltic target rocks of the Deccan Traps. The Lonar target basalts are Fe rich and similar to Martian basalts in their mineralogy, geochemistry and thermal infrared spectroscopy (Hagerty and Newsom, 2003; Wright et al., 2011). Both Meteor Crater and Lonar Crater (Fig. 18.1, Table 18.1) have played important roles in the history of impact cratering (Hoyt, 1987; Mark, 1987; Kring, 2007). G.K. Gilbert in his paper about Meteor Crater (Gilbert, 1896) used the absence of meteorite material at Lonar Crater as one of his arguments against the impact origin of Meteor Crater. Nevertheless, Gilbert's enthusiasm about these structures is seen in his statement that, 'The thought of examining the scar produced on the Earth by the collision of a star was so attractive that I desired to visit the Arizona crater'. Only recently, after more than a hundred years, have traces of a chondritic impactor been found in Lonar materials by Misra et al. (2009).

This chapter provides a comparison and discussion of the materials and properties of these simple craters and the resulting implications for impact crater processes. There are 35 terrestrial craters between 1 and 4 km in diameter (see Chapter 1; Earth Impact Database, 2011), but none as well known and well preserved as the Meteor and Lonar Craters. The significance of these impact craters for planetary science will be summarized. For example, Mars may have approximately 250,000 impact craters in this size range, in both sedimentary and basaltic targets. The ejecta blankets of these craters contain large blocks of ejected target rocks, and shocked and melted ejecta materials that may be similar to those observed on Mars from orbit and from rovers. As planetary analogues, understanding the nature of these craters will improve our understanding of similar structures on the planets, and of impact-related processes that formed planetary regolith and soils.

## 18.2 Meteor Crater, Arizona

Meteor Crater (Fig. 18.1) is located in north-central Arizona near the Canyon Diablo in the southern part of the Colorado Plateau. The impact produced a simple, bowl-shaped crater that has an average diameter of 1186 m, with a rim crest 30–60 m above the surrounding plain, and a rim crest to floor depth of 167 m. The official name is 'Meteor Crater', according to the USGS Geographic Names Information System (GNIS), and was suggested by D.M. Barringer (Barringer, 1993). The crater is also informally called the 'Barringer Meteorite Crater' in honour of Daniel M. Barringer, who first suggested that it was produced by meteorite impact (Kring, 2007). The most comprehensive geological map, based on work by Shoemaker (1960), is published in colour in Kring (2007).

### 18.2.1 Age

Nishiizumi et al. (1991) used $^{10}$Be–$^{26}$Al exposure ages of impact-derived samples at Meteor Crater to derive an age of 49,200 ± 1700 years. Phillips et al. (1991) used cosmogenic $^{36}$Cl accumulation in exhumed rocks from the normally buried Kaibab Formation, as well as $^{14}$C in rock varnish on rocks ejected by the impact event, yielding a mean exposure age of 49,700 ± 850 years. Sutton (1985) obtained an age of 49,000 ± 3000 years, based on thermoluminescence of quartz in shocked rocks from the floor of the crater.

Impact Cratering: Processes and Products, First Edition. Edited by Gordon R. Osinski and Elisabetta Pierazzo.
© 2013 Blackwell Publishing Ltd. Published 2013 by Blackwell Publishing Ltd.

### 18.2.2    Nature of impactor

The fragments of the impactor that formed Meteor Crater are the coarse octahedrite iron meteorites officially named 'Canyon Diablo', after an arroyo near the crater. The mass of meteorite fragments located near the crater is estimated to be about 30 tons.

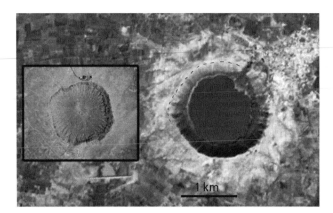

**Figure 18.1**    Meteor Crater and Lonar Crater at the same scale. The larger size of the rim of Lonar Crater is clearly evident (Meteor Crater image, Chapter 15, Lonar Crater image courtesy NASA – ASTER instrument). Image: NASA.

The meteorite fragments largely consist of the Fe–Ni alloys kamacite and taenite, forming a distinct Widmanstätten pattern.

### 18.2.3    Target rocks and structural geology

In the region surrounding the crater, the surface of the Colorado Plateau consists of relatively flat-lying Permian and Triassic beds. Three lithologies compose over 99% of the target rocks and subsequent ejecta. The Coconino Sandstone has a thickness of 210–240 m, but only the upper 80 m was sampled by impact (Roddy, 1978). Conformably overlying the Coconino are the Toroweap Formation and the Permian Kaibab Formation. The Toroweap is only approximately 1–1.5 m thick and consists of interbedded thin layers of sandstones and limestones, virtually identical to the underlying Coconino Sandstone and overlying Kaibab Limestone and is indistinguishable from them in ejecta. The Kaibab Formation is a fossiliferous dolomite and dolomitic limestone 79.5–81 m thick. Unconformably overlying the Kaibab is a thin, patchy veneer of Moenkopi Formation, a dark, reddish-brown, 3–6 m thick Triassic fissile siltstone. The near-horizontal strata that were located at or near the point of impact have been uplifted and folded back, with the uppermost layer of Moenkopi being folded back onto itself, creating an inverted stratigraphy (Shoemaker, 1960; Roddy *et al.*, 1975). Faulting at the time of impact along two mutually perpendicular joint sets caused uplift to various degrees in the crater wall. As a result, the exposures of each of the

**Table 18.1**    Comparison of some major characteristics between Meteor Crater and Lonar Crater

| Characteristic | Meteor Crater | Lonar Crater |
|---|---|---|
| Age of impact | ~49 ± 1 ka | ~570 ± 47 ka |
| Diameter | Avg. 1186 m (1070–1220 m) | 1880 ± 50 m |
| Latitude, Longitude | N 35°2′, W 111°1′ | N 19°58′, E 76°31′ |
| Probable original diameter | Only ~10–15 m less than current | ~1700 m |
| Highest elevation of rim above surroundings | 47 m | 20–30 m |
| Depth from rim to current floor | 200 m | 120 m |
| Depth from rim to top of inner crater breccias | 205 m | 220 m |
| Depth from rim to base of inner crater breccias | 360 m | >520 m |
| Sedimentary fill thickness | ~155 m | 100 m |
| Breccia lens thickness | 150 m | >300 m |
| Depth to extent of disturbed bedrock/country rock | 1040 m | ~700 m |
| Average ejecta radius from rim | 950 m | 1350 m |
| Proximal ejecta thickness | ~15 m at rim; ~5 m away from rim | ~9 m at rim and ejecta |
| Target rock type and age | Sedimentary: Permian sandstone and limestone, Triassic siltstone/mudstone, ~230–210 Ma | Igneous: five or six Late Cretaceous Deccan basalt flows, ~67–65 Ma |
| Impactor type | Iron meteorite (Canyon Diablo) | Probably chondritic |
| Ejecta block size | ~30 m maximum | ~5–10 m maximum |
| Ejecta matrix | Target lithologies pulverized to 5–10 μm | Fine-grained basalt fragments; suevite contains localized lapilli |
| Impact melts | Millimetre- to centimetre-sized irregular fragments, splash-form spheres and dumbbells, melt-coated rocks | Decimeter-sized glass bombs, millimetre- and sub-millimetre-size splash-type spherules, scoriacious clasts |
| Hydrothermal alteration | Not reported | Breccia lens only |
| Lapilli | Lapilli reported | Armoured and accretionary lapilli present |
| Topography references | Roddy (1978) | Fudali *et al.* (1980), Maloof *et al.* (2010) |
| Geophysics references | See Kring (2007) for refs to radar, gravity, and magnetic studies | Rajasekhar and Mishra (2005) |

three primary lithologies within the inner-crater walls and rim are not uniform. For example, Coconino Sandstone is observed on the eastern, northern and southern walls, but is not found on the western wall.

### 18.2.4   Ejecta and crater fill deposits

A principal feature of Meteor Crater is the layer of ejecta exposed on the crater walls and forming a continuous ejecta blanket outside the crater (Fig. 18.2). The thickness of the continuous ejecta blanket from a drilling study ranges from less than 3 m to approximately 25 m at the rim (Roddy *et al.*, 1975). However, re-examination of the drill cuttings (now curated at the USGS Flagstaff) is leading to a more accurate determination of the ejecta depth (Gaither *et al.*, 2011). Studies by ground-penetrating radar did find substantial local variations in thickness (Pilon *et al.*, 1991; Grant and Schultz, 1991). The continuous ejecta blanket is hummocky and consists of fragments of the three principal stratigraphic units, usually inverted, and is preserved up to two crater radii from the rim. Boulder- to cobble-size blocks of the three main lithologies are deposited on the crater rim. With the exception of the southern ejecta flap, the majority of the ejecta consist of blocks of Kaibab ranging in size from 0.5 to 30 m in diameter, including the famous 'House Rock' on the east rim. Unconsolidated ejecta lobes of Coconino Sandstone are also evident on the northern and northeastern ejecta blanket. To the south and southeast of the crater, the ejecta blanket is composed almost entirely of Coconino Sandstone that shows varying degrees of shock metamorphism (Shoemaker and Kieffer, 1974). A more detailed discussion of the effects of shock can be found in Section 18.3.

Shoemaker (1960) and Kring (2007) noted the presence on the interior crater walls of impact-melt-bearing breccias (suevite) present as inner-crater wall dikes, patchy deposits of allogenic

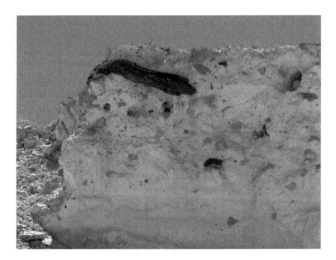

**Figure 18.2**   A large outcrop (3 m in width) of Meteor Crater ejecta, containing a large clast of Moenkopi Formation (dark) and several smaller clasts of Kaibab Limestone (light) can be seen in a fine-grained matrix of primarily pulverized Coconino Sandstone (image S.P. Wright).

breccia, plastered against the crater wall, all covered by fall-out ejecta, that is also present in the crater floor shafts. These overlying fall-out materials, as recorded by Shoemaker and Kieffer (1974) at station 13 in the crater, consist of rare fragments of strongly shocked rock scattered through mixed debris. The material includes fragments of Coconino, showing all five stages of shock metamorphism, and also lapilli of shock-melted Kaibab dolomite. Of particular note is the presence of frothy white glasses, or lechatelierite, from the Coconino Sandstone. In modern terms, the presence of glass requires that this fall-out material be classified as suevite. Shoemaker (1960) suggested that the dikes were injected into the walls during formation of the transient cavity. The fine-grain matrix consists of pulverized silica (from the Coconino Sandstone) and/or dolomite/calcite (from the Kaibab Limestone) with local tiny (<5 mm) ballistic melt fragments and spherules (Hörz *et al.*, 2002). Highly shocked clasts are not found in the lithic breccia that makes up the majority of the Meteor Crater ejecta blanket. However, remnant clasts and fines from the suevite breccia that originally made up the upper layer of ejecta can be found in the ejecta blanket in topographic lows (Osinski *et al.*, 2006).

The nature of the deposits beneath the crater floor is revealed in the shafts and associated dumps from the early mining, described in some detail by Shoemaker (1960; Shoemaker and Kieffer, 1974). They found that beneath 30 m of lake sediments is 10.3 m of fall-out breccia with an average grain size of about 2 cm. The unit is generally massive, but there is a subtle grading upwards from coarse debris at the bottom to finer grain debris at the top of the unit. The basal 1.3 m is particularly coarse, although fragments at the base seldom exceed 0.3 m in diameter. The allogenic breccia in the shafts beneath the fall-out deposit is composed entirely of Coconino Sandstone. Some of the blocks are more than a metre in size. The breccias below the mixed layer material extend to a depth of at least 100–200 m based on the early drilling in the centre of the crater floor. Cuttings examined by Shoemaker from a depth of 180 m contained impact melt glass with spherules of meteoritic metal (Shoemaker and Kieffer, 1974).

### 18.2.5   Shocked minerals and rocks and impact melts

The impact melts at Meteor Crater include spherules, melted sandstones and rocks covered with 5–6 cm of melt (Nininger, 1956; Kring, 2007). The ubiquitous Ni-bearing metallic spherules in the melts are similar to those found at the Henbury and Wabar Craters (Nininger, 1956). The silicate melts are heterogeneous in composition (Greenwood and Morrison, 1969; Hörz *et al.*, 2002; See *et al.*, 2002; Osinski *et al.*, 2006; Kring, 2007). Vesicular mafic silicate melts are common and were produced by devolatilization of the target rocks and variable incorporation of material from the impactor (Hörz *et al.*, 2002; Mittlefehldt *et al.*, 2005). As pointed out by Kring (2007), the melt volume is very hard to evaluate due to erosion and the limited information on the melts beneath the floor of the crater. Recently, vesicular impact melt particles in the 5 mm size range have been found in the drill cuttings from the ejecta blanket (Gaither *et al.*, 2011). Hagerty *et al.* (2011) have identified two new types of impact glass based on studies of the USGS Meteor Crater sample collection.

### 18.2.6    Geophysics and palaeomagnetism

Most of the drill core and geophysical data for the breccia lens in the crater indicate a reasonably symmetrical breccia lens that extends to 190 m beneath the crater floor, but rapidly thins toward the walls of the crater (Kring, 2007). Hargraves and Perkins (1969) investigated 32 cores at four sites around the rim of Meteor Crater and at two flat-lying sites outside of the crater, and found no evidence of any palaeomagnetic effect that could be attributed to the shock wave generated by the impact. However, Cisowski and Fuller (1978) found that the Coconino Sandstone shows changes in magnetic properties with increasing degrees of shock metamorphism (from unshocked up to Class 5). The principle effect of the impact event was to demagnetize the pre-existing remanence of the target materials. A secondary effect was the acquisition of a component of remanence whose direction was related to the ambient field direction at the time of impact (Cisowski and Fuller, 1978).

### 18.2.7    Lake and lake deposits

Approximately 20–30 m of lake sediments were described by Barringer (1905) and Shoemaker and Kieffer (1974) from the crater floor drill cores. Shoemaker and Kieffer (1974) noted that lechatelierite, which is lighter than water, was concentrated at the top of the lake sediments, indicating that they were floating. The lake sediments in the shafts on the crater floor are in contact with the underlying fall-out breccias lens, with no alluvium in-between, indicating that the lake began to form immediately after impact (Shoemaker and Kieffer, 1974). Kring (1997) and Kumar *et al.* (2010) suggested the lake dried up at approximately 11 ka, coeval with the end of the last ice age and the beginning of the current, arid climate.

### 18.2.8    State of erosion

Shoemaker (1960) and Roddy (1978) estimated that over 20 m of erosion of the ejecta blanket has occurred, while Grant and Schultz (1993) used trenching and sieve analyses to estimate no more than 1–2 m of vertical erosion. Grant and Schultz (1993) suggested that, to the north and northeast, drainages along radial gullies have transported some of the original near-rim continuous ejecta blanket to mantle the outer ejecta blanket. Aeolian erosion is indicated by a northeastern-trending erosional wind streak formed from the deflation of Coconino Sandstone, ejecta from the southern crater rim, as well as the erosion of Coconino and Kaibab from smaller ejecta lobes on the northern and eastern sides of the crater (Ramsey, 2002; Wright and Ramsey, 2006). Ramsey (2002) mapped the distribution of Coconino ejecta using thermal infrared remote sensing and calculated that only approximately 1.5 m could have been eroded, consistent with terrestrial erosion rates (1–2 cm per thousand years) for arid regions (Greeley and Iverson, 1985). The recent work by Kumar *et al.* (2010) on the erosion of the crater walls showed a well-developed centripetal drainage pattern consisting of individual alcoves, channels and fans (e.g. Fig. 18.3), and evidence of selective discharge through caves and fractures on the crater wall. They

**Figure 18.3**    The walls of Meteor Crater with enhanced gully formation on the lower slopes of the crater walls below the outcrops of the Kaibab limestone (image S. Misra).

concluded that, in general, the gully locations are the locales where the surface runoff from rain precipitation and snow melting can preferentially flow, causing erosion and crater degradation, although an erosion rate was not estimated.

### 18.3    Lonar Crater

Lonar Crater (19°58′N, 76°31′E) in the state of Maharashtra, India (Fig. 18.1 and Fig. 18.4) is a simple, bowl-shaped, near-circular impact crater with a N–S axis of approximately 1830 m and an E–W axis of approximately 1790 m (Fudali *et al.*, 1980; Kumar, 2005; Maloof *et al.*, 2010; Misra *et al.*, 2010). Except for a small sector in the NE, there is a continuous rim raised approximately 15–30 m above the adjacent plains. The estimated depth of the crater from its rim crest is approximately 222 m, whereas the crater floor lies approximately 90 m below the pre-impact surface (Fredriksson *et al.*, 1973a; Fudali *et al.*, 1980). A remote image of Lonar is shown in Figure 13.2d and Figure 18.4. The most recent comprehensive geological map is published in Maloof *et al.* (2010).

Evidence in support of an impact origin of this crater included the relative youthfulness of the crater in comparison with the age of the Deccan target rocks, the presence of glassy spherules and the presence of melt breccias with shocked minerals (e.g. La Fond and Dietz, 1964; Nayak, 1972, 1993; Fredriksson *et al.*, 1973b, 1979; Fudali *et al.*, 1980). The history of the earlier studies is summarized in La Fond and Dietz (1964) and Fudali (1999) and included early work by Newbold (1844), Blandford (1868) and Medicott and Blanford (1879), who considered a cryptovolcanic origin (e.g. Bucher, 1963) caused by a 'rush of steam' as a possible origin.

About 700 m north of the rim of Lonar Crater there is another relatively shallow depression known as 'Ambar Lake', or 'Little Lonar', approximately 300 m in diameter and surrounded by a raised rim along its southern and western margins (Fig. 18.4). Although Master (1999) suggested that Little Lonar might have formed as a double impact with Lonar Crater, no evidence for this

**Figure 18.4**    Portion of a four-band (RGB+NIR) pan-sharpened (0.6 m resolution) Quickbird image showing Lonar Crater (Maloof *et al.*, 2010). The Ambar Lake ('Little Lonar') is a blue patch surrounded by a circular feature consisting of distal ejecta beneath the label. Distal ejecta are also well exposed in outcrops near the Kalapani Dam and in hand dug water wells southeast of the crater. Image provided courtesy of Dr. Sarah Stewart. (http://www.fas.harvard.edu/~planets/sstewart/resources/Lonar-quickbird-RGB.jpg) (See Colour Plate 39)

has been found during recent mapping (Maloof *et al.*, 2010), or during early drilling into the structure (Fredriksson, personal communication, 2003).

### 18.3.1    Age

Early attempts at fission track dating by Fredriksson *et al.* (1979) led to an upper limit of 50,000 years, and thermoluminescence studies by Sengupta *et al.* (1997) on glasses led to an age of 52,000 ± 6000 years. The latest information on the age of the crater comes from $^{40}$Ar–$^{39}$Ar dating of the crater by Jourdan (2010; Jourdan *et al.*, 2011), who found an age of approximately 570,000 years for glass bombs from the crater ejecta. This age has great implications for the state of preservation and extent of erosion of the crater and deposits, especially in comparison with the younger Meteor Crater as discussed below.

### 18.3.2    Erosion

The motion of the Indian plate and the geography of India suggest that Lonar Crater has been located in a semi-arid region during the last half a million years similar to that of Meteor Crater (Misra *et al.*, 2010). The crater walls at Lonar (Fig. 18.4 and Fig. 18.5) are uniformly sloped into the crater and consist of the strongly resistant basalts of the target rocks and weak interflow palaeosols and sediments. The gullies on the crater wall originate generally at the top of the rim and extend all the way to the base of the wall (Fig. 18.5). Observations in the field (HEN, SPW, SM) and topographic

**Figure 18.5**    The eastern wall of Lonar crater exhibits gullies that extend from the rim to the lake. The white structure is the two-story tourist hotel, and in the distance the small temple on top of the pre-impact hill east of the crater is visible (image S.P. Wright).

maps indicate no evidence for slumps or debris flow scars down to the lake level, suggesting that fluvial erosion has been the primary agent of erosion. In addition, Fudali *et al.* (1980) suggested that approximately 50 m of the rim has been backwasted around the crater; thus, the 'hinge line' where ejecta is emplaced as inverted strata on top of the original bedrock is not as well preserved as at Meteor Crater. Further investigation of the rim is partly complicated by human-caused erosion around the entire rim of the crater.

Some characteristics of the ejecta blanket at Lonar, such as its hummocky nature, surface lag of larger basalt blocks and lack of deep incision by gullies (Fudali *et al.*, 1980; Maloof *et al.*, 2010), led to suggestions that the state of erosion is similar to Meteor Crater (Grant, 1999). The young gullies in the continuous ejecta blanket are generally 1–2 m deep and approximately 2 m across, with a large, approximately 3–4 m deep gully on the eastern ejecta being an exception. However, there is a strong possibility that the current state of the crater's ejecta blanket reflects a greater amount of erosion than previously suggested, based on the absence of the upper melt-bearing layer on most of the ejecta surface (Wright and Newsom, 2011).

### 18.3.3  Lonar Lake and associated deposits

Beneath the shallow saline lake, a sequence of unconsolidated sediments up to 100 m thick overlies highly weathered Deccan Trap basalts (Nandy and Dey, 1961; Fudali *et al.*, 1980). Lonar Lake is alkaline (pH around 10) and its chemistry is dominated by sodium cations (95%) and chloride, carbonate and bicarbonate anions (Chowdhury and Handa, 1978). Although Nayak (1985) suggested a connection between impact-generated hydrothermal processes and the lake chemistry, simple evaporation in a closed basin with absorption of potassium by the lake sediments can explain the water chemistry (Chowdhury and Handa, 1978). The proposed age of 570,000 years makes the evaporation model more plausible, and similar alkaline lakes are known in basaltic environments (Rosen *et al.*, 2004).

### 18.3.4  Nature of impactor

No evidence of any meteorite fragments has ever been reported in the field (Nayak, 1972; Fredriksson *et al.*, 1973a,b; Fudali *et al.*, 1980) in contrast to Meteor Crater (e.g. Nininger, 1956). Bulk geochemical analysis of numerous melt rocks has also failed to turn up enrichments of nickel consistent with a meteorite component (Nayak, 1972; Fredriksson *et al.*, 1973a,b; Kieffer *et al.*, 1976; Stroube *et al.*, 1978; Morgan, 1978; Ghosh and Bhaduri, 2003; Osae *et al.*, 2005). An additional study of over 100 Lonar impact glasses, soils and ejecta was recently conducted using a portable X-ray fluorescence detector and no evidence of any Ni-rich material was found (Newsom *et al.*, 2010b).

The only chemical evidence for the impactor at Lonar is from a microprobe study of separated spherules found in the ejecta (Misra *et al.*, 2009). The impact spherules, which are approximately 0.3–1.0 mm in diameter, including spheres, teardrops, cylinders, dumbbells and spindles, are found in the proximal ejecta. A chondritic meteorite component is indicated by very high Ni contents, up to 14 times (2500 ppm) the average content of Lonar basalt, and high Cr (450–550 ppm) in a few of the spherules (Fig. 18.6). Mixing models between chondrites and target basalt suggest a maximum meteorite component in the spherules of 12–20% by weight, but the uncertainties do not allow a specific chondrite type to be identified. Fragments of a chondritic impactor would not be very resistant to erosion, which might bear on the lack of

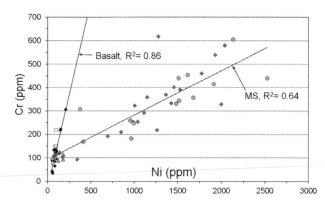

**Figure 18.6**    Cr and Ni abundances in Lonar samples (Misra *et al.*, 2009). Symbols: open square, impact melt (IM); light grey-coloured circle, sub-millimetre-sized spherule with melt-dominated portion (SMS-ML); dark grey-coloured rhombus, sub-millimetre-sized spherule with magnetite dominated portion (SMS-MT); black solid dot, target basalt; light grey-coloured triangle, millimetre-sized spherule (MS). Sub-millimetre-sized spherules show enrichments in the siderophile element Ni up 2500 ppm and Cr up to 600 ppm, consistent with addition of a chondritic impactor to the basaltic target rock.

meteorite fragments at Lonar. Jull *et al.* (1993), for example, estimated a half-life for the destruction of H and L chondritic meteorites by weathering of only 10,000–15,000 years in the high plains of the USA, in a climate similar to the semi-arid climate at Lonar.

### 18.3.5  Target rocks

Lonar Crater is located in the Deccan Trap flood basalt, which erupted at approximately 65 Ma (e.g. Chernet *et al.*, 2007). Six flows ranging from approximately 8 to 40 m thick are exposed in the vicinity of Lonar. Exposures of the four lower flows are only found in the crater itself. Ghosh and Bhaduri (2003) described the flows as *a'a* type except the flow at the bottom, which is *pahoehoe* type; although Maloof *et al.* (2010) argue that all flows are pahoehoe type. Lonar basalts are generally vesicular, aphanitic basalts with sporadic occurrences of plagioclase phenocrysts. Labradorite, as phenocrysts (1.5 mm to 1 cm in length) and in the groundmass, comprises approximately 45–50% of the bulk basalt (Kieffer *et al.*, 1976). Clinopyroxenes (primarily augite, but some pigeonite) comprise approximately 30–35% of the mineralogy. The remaining groundmass is composed of oxides such as ilmenite and titanomagnetite, limited (<1%) sulfides and alteration products such as clays and palagonite (Kieffer *et al.*, 1976; Fredriksson *et al.*, 1973a, 1979). Fresh, dense basalts occur only in the upper approximately 50 m of the crater wall, whereas the flows below this level are heavily weathered and friable with serpentine, chlorite and celadonite in the groundmass (Fudali *et al.*, 1980). A connection between the alteration of the lower flows and the impact is unlikely, as the ejecta contains fragments of both the altered and unaltered flows, and basalt flows several kilometres from the crater also exhibit similar alteration (unpublished data,

HEN). The lava flows are separated by discontinuous red and green palaeosols, chilled and vesicular margins, and amygdaloidal vugs filled with secondary chlorite, zeolite, quartz and limonite (Kieffer *et al.*, 1976). Intertrappean sediments of limited areal extent consist of chert, sandstone and impure limestone of varied thickness (Jhingran and Rao, 1958; Venkatesh, 1967). Ghosh and Bhaduri (2003) observed pre-impact black, sticky, humus-rich soil of approximately 5–90 cm thickness at some places between flows and below ejecta.

### 18.3.6 Structure and topography

The basalt flows around most of the rim dip shallowly away from the crater at angles from 17° to 35°, due to uplift from the formation of the crater (Misra *et al.*, 2010). The western rim of the crater from the WSW to the WNW, shows steep reverse dips averaging from 46° to 56°, indicating that the basalt flows are overturned (Maloof *et al.*, 2010). The deep vertical notches that transect the northeastern and western rim (Fig. 18.4) correspond to a normal fault (Kumar, 2005). Remote-sensing images together with field observations suggest that the fault was part of a NE–SW trending lineament that is radial to the crater but crosscuts the crater rim both in the northeast and west. According to eye witnesses (S. Bugdani, personal communication, 2010), movement along this lineament took place in October 1998. There is another less prominent E–W lineament that also radially crosscuts the crater rim.

### 18.3.7 Geophysics

Gravity and magnetic anomalies over Lonar Crater were measured by Subrahmanyam (1985) and Rajasekhar and Mishra (2005). The circular/semicircular-shaped gravity and magnetic anomalies are approximately 2.5 mGal and 550 nT respectively, similar to those reported over other impact craters. Analysis and modelling of these anomalies and comparison with drill core lithologies suggest that the impact has modified the density of the target rock to a depth of 500–600 m below the present surface. A 135 m thick brecciated layer has a bulk density of 2.60 g cm$^{-3}$ and a 150 m thick fragmented layer has a bulk density of 2.7 g cm$^{-3}$.

The depth of the affected rocks is approximately 0.3–0.35 times the diameter of the crater.

Several studies conclude that any shock remanent magnetization acquired during the impact was replaced by a low-coercivity and/or low-temperature post-impact viscous and/or chemical remanent magnetization (Nishioka, 2007; Nishioka *et al.*, 2007; Nishioka and Funaki, 2008; Louzada *et al.*, 2008; Arif *et al.*, 2011). The magnetic data also indicate that the crater walls did not exceed approximately 200 °C due to shock heating during the impact (Louzada *et al.*, 2008). Recent studies of the anisotropic magnetic susceptibility (AMS) of target basalt around Lonar Crater suggest that a magnetic signature associated with the impact has been preserved, but this work is at an early stage (Nishioka, 2007; Arif *et al.*, 2009; Misra *et al.*, 2010).

### 18.3.8 Crater fill deposits, breccias and drill core studies

The floor of Lonar Crater was sampled by a number of drill cores obtained by the Geological Survey of India and personnel from the Smithsonian Institution (Fredriksson *et al.*, 1973a, 1979). The drill holes penetrated approximately 90–100 m of lake sediments before reaching a lens of shocked breccia described by Fredriksson *et al.* (1973a) as 'micro breccia' and 'coarse breccia'. In terms of the current classification terminology (Stöffler and Grieve, 2007; see Chapter 7), the breccias would be classified as impact melt-bearing breccia or 'suevite' (having melt particles). The deposit would be further characterized as 'allochthonous monomict breccia'. Early studies of the drill cores resulted in construction of a cross-section of the crater floor and deposits (Fig. 18.7). The drill holes encountered solid basalt at approximately 450 m below the rim, and some rock flour layers were encountered at greater depths up to 520 m. The breccia lens is therefore roughly 250–280 m thick, which is consistent with the gravity data (Fudali *et al.*, 1980; Subrahmanyam, 1985). The alteration minerals and the chemistry of the breccias and basement in samples from the drill cores have been studied by Fredriksson *et al.* (1973a, 1979), Hagerty and Newsom (2003) and Chaklader *et al.* (2004).

Hagerty and Newsom (2003) observed evidence for aqueous and hydrothermal alteration samples from drill cores into the

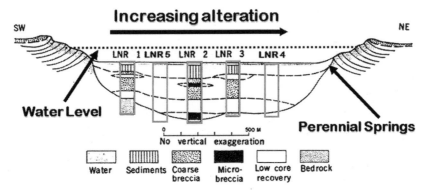

**Figure 18.7** Schematic cross-section of the Lonar crater (Hagerty and Newsom, 2003) modified from Fredriksson *et al.* (1973a). This diagram shows the location of the five drill cores obtained in the early 1970s.

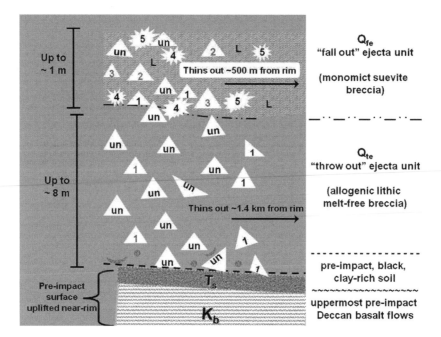

**Figure 18.8**   Schematic diagram of the two-layer structure of the ejecta blanket at Lonar Crater. The ejecta overlie the pre-impact palaeosol on top of the target basalts. The shocked rock classification scale of Kieffer *et al.* (1976) is shown on the triangles that represent clasts in each breccia unit. The lower layer contains a matrix of pulverized basalt with unshocked clasts (labelled as 'un') and limited amounts of Class 1 shocked basalt. The upper layer contains finely comminuted basalt in the matrix but also contains materials shocked to all levels, including impact melts (Kieffer *et al.*, 1976). The upper layer also contains spherules and armoured lapilli represented by the label 'L' (Fredriksson *et al.*, 1973b; Beal *et al.*, 2011). Figure after Wright, S.P. (2008).

crater floor, including textural evidence from backscattered electron images of replacement textures, ubiquitous pockets of alteration, alteration of impact glasses and multiple generations of hydrothermal carbonate growth. The degree of alteration in the cores increases in abundance and severity from west to east across the crater (Fig. 18.7). Alteration minerals include Fe-rich saponite (core LNR-3) and celadonite (core LNR-1), determined by X-ray diffraction signatures and confirmed by microprobe analyses. Geochemical modelling and data from terrestrial geothermal systems indicate that saponite and minor celadonite are produced during basalt alteration at temperatures of 130–200 °C (Hagerty and Newsom, 2003). Although evidence for alteration is present, ion-probe analysis of drill core samples (Newsom *et al.*, 2004; Chaklader *et al.*, 2004) provides little evidence for vertical or horizontal transport of Li, Be, B and Ba.

### 18.3.9   Ejecta

The continuous ejecta deposit extends outward 1300–1400 m from the crater rim with a very gentle slope of 2–6°, covering an area of approximately 6.7 km$^2$ (Fredriksson *et al.*, 1973a; Fudali *et al.*, 1980; Kumar, 2005, Maloof *et al.*, 2010; Misra *et al.*, 2010). The reflectivity of the ejecta in satellite images near its outer fringe in the west is low due to vegetation cover and the patchy distribution of ejecta. The high-reflectivity areas to the southeast, east and

northeast of the crater are hillocks formed by basalt that do not contain any ejecta (Fig. 18.4).

The ejecta consists of two layers, shown schematically in Fig. 18.8, with an upper layer containing an abundance of shocked and melted target rocks and a lower layer with little to no evidence of shock. This structure is reminiscent of the classic ejecta blanket at the Ries Crater, Germany, where shocked and melted suevite deposits overlie the relatively unshocked Bunte Breccia (Hörz *et al.*, 1983; Chapter 4). As the ejecta at Lonar Crater consists of two discrete layers of ejecta, Lonar probably appeared originally as a double-layer ejecta (DLE) structure in plan form, similar to many Martian craters (Barlow *et al.*, 2000). Furthermore, Maloof *et al.* (2010) reported a thickening of the distal edge of the ejecta similar to fluidized ejecta around craters on Mars.

The uppermost melt-bearing ejecta layer meets the definition of 'suevite' (Koeberl *et al.*, 2004), or impact melt-bearing breccia (see Chapter 7) and corresponds to the 'fallout' layer described by Shoemaker (1960) at Meteor Crater. At Lonar, this allogenic suevite unit is present in discontinuous patches up to approximately 1 m in thickness and is found up to 500–600 m away from the rim. The suevite has not been identified in near-rim (<0.1 km) ejecta deposits, presumably due to erosion. The basalt clasts in the suevite breccia have shock levels that range from unshocked through intermediate (Classes 2/3) and up to highly shocked basalts (Classes 4/5) (Kieffer *et al.*, 1976). Highly shocked and melted basalt (>60 GPa) is present as clasts in the

Figure 18.9 (a) Proximal ejecta on the northeast rim of Lonar crater with the characteristically small fraction of matrix. (b) Distal ejecta at the outcrop near the Kalapani dam with underlying palaeosol and a 40 cm hammer for scale. Note the high amount of matrix, but there is no chemical evidence for incorporation of the palaeosol in the matrix. Note 40 cm long hammer for scale in both images (images H.E. Newsom).

top approximately 1 m of the uppermost basalt ejecta unit. The fine-grained matrix of the upper layer contains 'splash form' glass spherules, dumbbells (Fredriksson *et al.*, 1973b) and 'glassy ash' (Kieffer *et al.*, 1976). Recently, accretionary lapilli and armoured lapilli have also been found associated with the upper layer of ejecta (Beal *et al.*, 2011). These observations are consistent with the upper layer of ejecta forming from fall-out material.

The lower layer of ejecta consists of a lithic monomict melt-free breccia unit containing clasts of unshocked basalt fragments in a matrix of crushed, sand-size, basalt fragments. The high abundance of basalt clasts in the proximal ejecta results in a block-supported breccia (Fig. 18.9a). The distal ejecta contains a small fraction of blocks in addition to coherent lumps of palaeosol in a fine-grain matrix (Fig. 18.9b). Recent work has presented field evidence of a ground-hugging ejecta flow beyond the rim of Lonar Crater (Maloof *et al.*, 2010). The unaltered blocks from the upper flows are distinctive and often form a lag deposit on the surface. The largest coherent ejected basalt clasts are 2–5 m in size, and are found in the ejecta blanket within approximately 500 m of the northern, western and southern crater rim (Fig. 18.10). There is also evidence in the eastern and southern ejecta blanket for large fragments, up to 5 m or even larger, of the lower, altered basalt flows (unpublished observations HEN; Maloof *et al.*, 2010).

At Meteor Crater, Shoemaker (1960) described the lower unit as a 'throw-out' unit, with the original stratigraphy still preserved in distal ejecta deposits (Shoemaker, 1960; Roddy, 1978). The stratigraphy is not as evident at Lonar Crater, because the ejecta is all basalt, though the upper three flows and bottom three flows can be discriminated in patches in this unit and as clasts in the upper ejecta unit (Kieffer *et al.*, 1976). As the rim is backwasted as much as approximately 50 m, there is no equivalent to the exposures of the overturned flap at Meteor Crater. However,

Figure 18.10 Large 5 m ejected block on the western portion of the continuous ejecta blanket. S. Wright (1.8 m tall) for scale (image H.E. Newsom).

the rim and proximal ejecta do show evidence of upturned and inverted lava flows and ejecta blocks, suggesting some inversion of the stratigraphy did occur.

### 18.3.10 Ejecta blanket and palaeosol alteration

Palaeosol samples from beneath the ejecta blanket were collected during several field campaigns by H.E. Newsom, S.P. Wright and S. Misra. In the distal ejecta, lumps of torn up palaeosol are present as macroscopic clasts within the ejecta. However, chemical data do not support incorporation of the palaeosol into the distal ejecta matrix as a well-mixed component (Newsom *et al.*, 2010a). Based on X-ray diffraction methods, palaeosol samples contain plagioclase, pyroxene and phyllosilicate (smectite) in variable amounts, as well as iron–titanium oxide minerals and alkaline zeolite heulandites (N. Muttik, personal communication, 2011). Smectite is the principal product of alteration in the Lonar palaeosol and ejecta blanket material, as indicated by dominant peak at 13 Å in the sodium saturated air-dry state that expands to 16 Å after ethylene glycol salvation in all samples. The smectite composition does not show any significant variance and fully expandable smectite is characterized by random (R0) ordering (Moore and Reynolds, 1997). While hydrothermal processes of dissolution and precipitation may have occurred beneath the floor of the crater (Hagerty and Newsom, 2003), the mineralogical data suggest alteration of the ejecta blanket took place basically under ambient weathering conditions, with no indication of hydrothermal processes in the ejecta blanket.

Newsom *et al.* (2010a) reported major and trace element data for a large suite of samples from Lonar (e.g. Fig. 18.11). The ternary plots of Nesbitt and Wilson (1992) provide a powerful tool for understanding the chemical fractionations during basalt weathering on Earth and Mars (e.g. Ming *et al.*, 2008). The unaltered Lonar basalts in this diagram (Fig. 18.11) plot in the centre of the field between the FeO + MgO corner and the feldspar composition. Many of the samples of the proximal and distal ejecta and the palaeosol have compositions similar to the unaltered basalts. However, the most weathered or altered samples,

including some of the palaeosol samples, are distinctly depleted in alkali elements and some trace elements like Sr and P (not shown). In contrast, some of the proximal and distal ejecta samples are enriched in alkali elements, suggesting some transport by fluids. Extensive caliche deposits, suggesting ambient alteration and fluid transport, are associated with the exposed base of the ejecta in gullies on the western sides of the crater and as a calcrete soil horizon excavated during cultivation of fields on the southern flank of the ejecta blanket (Maloof *et al.*, 2010; Newsom *et al.*, 2010a).

### 18.3.11    Impact melts and shocked rocks

Shocked Lonar Crater basalt can be compared with basaltic Martian meteorites, achondrite meteorites, Apollo samples and samples like 'Bounce Rock' observed by the Mars Exploration Rovers (Zipfel *et al.*, 2011). Shocked rocks are found mainly in the upper layer of the ejecta and within drill cores from the crater floor. The dark impact-melt clasts are reasonably easy to identify, but the immediately-shocked basalts are very similar in colour to unshocked basalt. Textural characteristics of shocked rocks can include a distinctive weathering style that looks like foliation or layering, but is due to 'decompression cracks' (Fig. 18.12; Newsom *et al.*, 2011; Wright and Newsom, 2011). The effects of shock on minerals and textures can be divided into five classes (e.g. Kieffer *et al.*, 1976; Fig. 18.13):

- *Class 1.* At low (<20 GPa) shock pressures basalt exhibits slight fracturing of plagioclase and clinopyroxene grains. Planar fractures and planar deformation features can form in labradorite grains.
- *Class 2.* At intermediate (20–40 GPa) shock pressures, plagioclase grains are converted via solid-state transformation (low heat, no melting) to a diaplectic glass (maskelynite). Clinopyroxene is fractured but not melted.
- *Class 3.* At 40–60 GPa the plagioclase glass begins to melt and flow. Clinopyroxenes are more extensively fractured than in Class 2 shocked basalt.
- *Class 4.* At 60–80 GPa, plagioclase grains are melted, flowed freely and vesiculated. Clinopyroxene grains are heavily fractured.
- *Class 5.* Above 80 GPa, all minerals are melted into an isotropic brown or black glass.

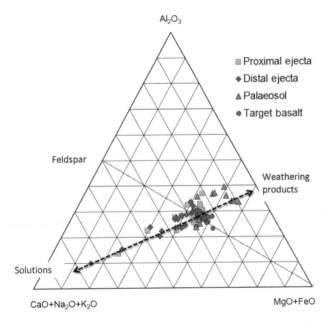

**Figure 18.11**    Major element composition of Lonar materials including proximal ejecta within 100 m of the rim of the crater (*n* = 27), distal ejecta (*n* = 23), palaeosol samples (*n* = 14) and unaltered target basalts (*n* = 17) (Newsom *et al.*, 2010a). Some palaeosol samples are fractionated from the target basalt composition due to loss of alkali elements due to weathering. Some of the ejecta samples show an enrichment of alkali elements, mainly CaO, due to mobile element transport associated with alteration or weathering. Figure this study.

The effect of shock on the chemistry of Lonar samples has been the subject of a few studies (Chaklader *et al.*, 2004; Misra *et al.*, 2007; Misra and Newsom, 2010). The shock transformation during formation of maskelynite did not result in any substantial chemical fractionations. However, the characteristic Na loss of Lonar impact melts and spherules from the proximal ejecta blanket, which were shocked to a higher temperature, was likely caused by the impact process (Osae *et al.*, 2005). Chaklader *et al.* (2004) provided some preliminary ion-probe data on the effects of shock on trace elements in pyroxene from Lonar samples. The shocked Lonar sample, which also contained maskelynite, showed a distinct depletion of both Li and B in the Fe-enriched rims of the shocked pyroxene grains which was not observed in the mineral grains from the unshocked sample. This observation is

**Figure 18.12**    Natural and cut face of a Class 2 (20–40 GPa) shocked maskelynite-bearing basalt (Newsom *et al.*, 2011). The sample is 7 cm wide and exhibits 'decompression cracks' resulting from the release of the shock wave by the rarefraction wave. (a) The weathered surface shows a distinct weathering style likely controlled by the decompression cracks. Figure this study.

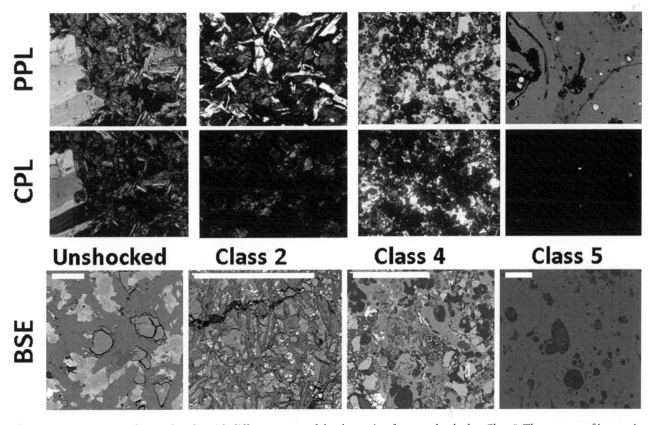

**Figure 18.13**    Images of Lonar basalts with different stages of shock ranging from unshocked to Class 5. The top row of images is in plane polarized light (PPL), the second row shows the same areas of the thin sections with crossed polarized light (CPL) (optical images are 5× magnification). The third row shows a series of backscattered electron (BSE) images for the different shock levels. Optical images approximately 1.5 mm across; scale bars for the BSE images are 1 mm long. Figure this study. (See Colour Plate 40)

important because Li and B variations in Martian meteorites are usually ascribed to fluid processes and not to shock.

The impact glasses occurring in and around Lonar Crater are classified into two fundamental petrographic classes (Osae *et al.*, 2005). The type 'c' glasses include centimetre-sized (≤1–30 cm)

impact-melt bombs that contain unmelted fragments of basalt (Fig. 18.14) and the type 'a' glasses consist of millimetre-sized impact-spherules or 'splash-form' spherules. Both types of glasses are rare within the ejecta blanket. Similar impact melt particles are also be present at Meteor Crater (see above). The large impact melt clasts (type c) appear partly layered, almost like Muong Nong tektites. These clasts are black or brown, and exhibit schlieren and flow structures outlined by magnetite crystallites. At Lonar, care is needed to distinguish ~1 ka manmade bricks manmade bricks from impact-derived products. Ancient builders produced bricks from the local materials and small-scale low-fired brick manufacturing continues adjacent to the crater today (Misra, 2006; Deenadayalan *et al.*, 2009).

Chemical fractionations between the target basalt and impact melts (type 'a' impactites, after Osae *et al.* (2005)) were examined in detail by Osae *et al.* (2005) and Son and Koeberl (2007), with additional work by Misra *et al.* (2009). Minor volatile element depletions in the melts include $H_2O$, $Na_2O$, $K_2O$ and $P_2O_5$ when compared with unshocked Deccan basalt. These depletions are probably due to volatility during the formation of the different types of impact melts. The possibility that the impact melts contain traces from the Archaean basement underneath the Deccan basalt has also been considered in a series of papers (Chakrabarti and Basu, 2006; Chakrabarti *et al.*, 2006; Misra, 2006).

### 18.3.12    Spherules and lapilli in Lonar ejecta

A wide range of droplets (i.e. splash-form spherules) are found in the ejecta blanket. In addition, armoured lapilli have recently been discovered in the ejecta (Beal *et al.*, 2011). The impact spherules consist of at least two different categories: millimetre-sized spherules and sub-millimetre-sized spherules. The spherules are highly variable in terms of vesicularity and show variable geometric shapes, such as teardrop, dumbbell, elongate and bended, horseshoe and spherical shapes. Schlieren, flow-banding and partly melted mineral inclusions have also been observed (Fredriksson *et al.*, 1973b). Weiss *et al.* (2010) found that the small 'splash-form' impact spherule samples (<0.5 g) have a weak and unstable natural remnant magnetization (NRM), indicating that they cooled during flight. Some of the spherules also contain significant proportions (~12–20%) of the impactor, as described above (Misra *et al.*, 2009).

Investigation of sub-millimetre-size spherules (Beal *et al.*, 2011) led to the discovery of spherules and other particles coated with fine-grained mineral fragments (Fig. 18.15), similar to volcanic and impact lapilli (Graup, 1981; Schumacher and Schmincke, 1994; Warme *et al.*, 2002; Koeberl *et al.*, 2007). The fine-grained

**Figure 18.14**    A typical large impact-melt clast, 20 cm in length, is shown after excavation from the eastern ejecta blanket (image H.E. Newsom). This clast is flattened with evidence for plastic deformation. The samples often fracture when removed from their matrix, providing evidence that the samples were found undisturbed in place. The shape is suggestive of the flädle from the Ries Crater suevite (Hörz, 1965). The matrix in which the melts are found contains spherules and armoured lapilli.

(a)                                          (b)

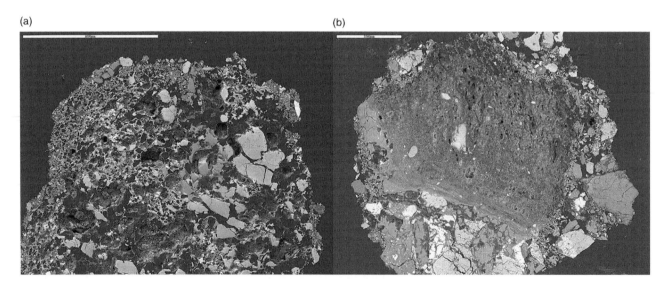

**Figure 18.15**    Backscattered electron images of lapilli from the matrix of the upper ejecta layer (Beal *et al.*, 2011). (a) An ultra-vesicular spherule from the Lonar ejecta blanket. This example contains relict mineral grains embedded in the glass and a thin accretionary rim of particles, making it an armoured lapillus. Scale bar is 500 μm. (b) An armoured lapilli particle consisting of the partially melted core with adhering fragments of basaltic mineral grains. Scale bar 200 μm.

mantles adhering to the cores are resistant to cleaning with alcohol and the grain size of the mantling material is often finer than the loose ejecta in which the particles are found. Scanning electron microscope (SEM) analyses of the nearly continuous coatings of fine-grain material are consistent with pyroxene, plagioclase and iron oxides derived from the Lonar basalt (Beal *et al.*, 2011). The most common type of particle has a central core of glass or highly shocked basalt surrounded by a fine-grained rim of mineral fragments. A second type is similar to accretionary lapilli and contains only ash-size particles. The Lonar particles have characteristics of both the accretionary and armoured types, and usually have a single, fine-grained mantling layer. In some cases, lapilli show evidence of minor post-accretionary devitrification and alteration.

## 18.4 Comparisons and planetary implications

### 18.4.1 Nature of the excavated cavity, crater-fill deposits and ejecta

The external profiles of Meteor and Lonar Craters are similar, with hummocky ejecta sloping up to the uplifted rims. Numerical models predict transient cavities with vertical walls that collapse into the crater forming breccia lenses (Artemieva and Pierazzo, 2011). A significant difference between the two craters is the larger diameter, but similar absolute depth of Lonar Crater compared with Meteor Crater, which is clearly visible in cross-sections of the two craters reproduced at the same scales (Fig. 18.16). The

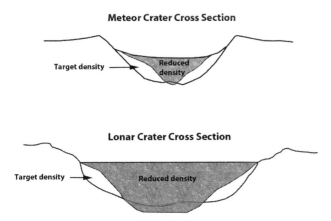

**Figure 18.16** Cross-sections of Meteor Crater and Lonar Crater (no vertical exaggeration). The outline of the craters beneath the crater floors are based on drill holes and seismic data for Meteor Crater (Shoemaker data presented in Kring (2007)) and drill holes for Lonar Crater (Fredriksson *et al.*, 1973a). The shaded areas of reduced density are those used to fit gravity models from Regan and Hinze (1975) for Meteor Crater and Rajasekhar and Mishra (2005) for Lonar Crater. Notice that the reduced density for both craters is restricted to the centre portion of the mapped breccia deposits. The portions of the craters labelled 'Target density' must contain materials with a density similar to the unshocked basement rocks.

origin of these differences could be due to a combination of the different target materials, the nature of the impact process and a possible older age of Lonar. Possible differences in the impact processes between the craters might also include variables such as impact angle and amount of fragmentation of the impactor in the atmosphere before impact.

Gravity models for both craters require reduced densities only in the central portions of the breccia deposits in the craters. The edges of the inferred breccia deposits are somehow similar in density to the target, either because of induration or less brecciation due to the collapse or slump of portions of the craters walls, but there is no evidence of major slumps in either impact crater. In the case of Meteor Crater, the seismic results of Ackermann *et al.* (1975) suggest extensive fracturing beneath the crater as reflected in the geological cross-section (Kring, 2007). Seismic data for Lonar Crater are needed to confirm the nature of the geological cross-section.

The similarities between the ejecta blankets include their hummocky nature and the presence of large blocks in the ejecta. The size distribution of the largest blocks of ejecta may be comparable to ejecta blocks from possible impacts into basalts, which may cover portions of the Vastitas Borealis Formation in the northern plains of Mars (Catling *et al.*, 2011). There is also a similarity in the presence of two ejecta layers, with an upper impact melt-rich layer (see Chapter 4). A difference may be found in the thicknesses of the ejecta as a function of distance from the crater. The ejecta at Meteor Crater becomes systematically thinner as a function of distance from the rim (Roddy *et al.*, 1975), while Maloof *et al.* (2010) find that the ejecta at Lonar may show a thickening near the distal edge indicative of processes also seen in Martian craters.

### 18.4.2 Impactor fragmentation and impact plume properties

In the case of Meteor Crater, Melosh and Collins (2005) suggested that the breakup of the iron impactor in the atmosphere led to a relatively low impact velocity of 12 km s$^{-1}$. The low impact velocity was used to explain the presence of unmelted meteorite fragments near the crater and the scarcity of impact melt deposits. Further studies by Artemieva and Pierazzo (2009, 2011) also argue for impact of a fragmented projectile, but at slightly higher velocities based on the fate of the projectile. The atmospheric breakup of impactors results in the formation of an airburst involving the impingement on the surface of a high-temperature atmospheric jet of vaporized impactor material (Boslough and Crawford, 2008; Newsom and Boslough, 2008; Newsom *et al.*, 2010b). At Meteor Crater, there are some materials, such as melt-coated rocks (e.g. Kring, 2007), that could be a result of this process, and Artemieva and Pierazzo (2011) suggest this could explain the relatively small amount of solid projectile material around the crater. SEM analyses of impact glasses from Meteor Crater indicate the pervasive occurrence of small spherules of meteoritic fragments in all types of impact glasses. As was suggested by Mittlefehldt *et al.* (2005), the occurrence of these materials likely indicates that fragmentation of meteoritic material occurred at the earliest stages of crater formation (likely prior to impact). The impact melts may also then represent a significant sink for meteoritic materials. At Lonar Crater, minute amounts of a chondritic impactor have been identified within impact melt spherules (e.g. Misra *et al.*, 2009). The absence of any meteorites may also suggest that an atmospheric breakup was likely, although other explanations are likely, including human removal and especially simple erosion.

The presence of an airburst implies that the total energy of the impact, including the energy involved in the ejecta plume above the crater, can be much greater than would be predicted from the diameter of the crater, with the usual assumption (e.g. Artemieva and Pierazzo, 2011) that most of the energy is deposited into the target rocks. The high temperatures from the air blast coupled with the heat associated with the crater formation can explain the high-temperature formation of the armoured lapilli cores, including glassy melt spherules and partially melted basalt fragments seen at Lonar Crater (Beal *et al.*, 2011). The report by Shoemaker and Kieffer (1974) of the presence of lapilli in the upper fall-out layer suggests a similar environment at Meteor Crater.

### 18.4.3   Erosion

To a first order, the hillslope processes on the walls of the two craters and the erosion of the ejecta blankets appear similar, reflecting similar climates and impacts into flat terrains with mechanically strong target rocks. Both craters also contain gullies on the crater walls that are partly controlled by faults. The ejecta blankets of the two craters exhibit removal of some of the upper surface, but retention of the hummocky morphology. In detail, there are some significant differences, with evidence for greater erosion at Lonar Crater. At Meteor Crater, around the inside of the rim there is an obvious contact between the in-place beds and the overturned flap of ejecta. However, the erosion of the crater rim at Lonar has removed the comparable exposures, although further away from the rim there is evidence for overturned and inverted basalt ejecta layers (Maloof *et al.*, 2010; Misra *et al.*, 2010). In addition, the remnants of the uppermost melt-bearing ejecta layer at Lonar Crater seem to be very limited in extent compared with Meteor Crater, probably due to erosion, even though Lonar Crater is significantly larger. The climatic factors controlling erosion at Lonar Crater and Meteor Crater relative to processes on Mars remain to be fully determined. Kumar *et al.* (2010) argued that, at Meteor Crater, a warmer climate in the past, indicated by the presence of lake sediments, was the most important driver for erosion. Erosion in arid climates on the Earth reflects a complex interplay between formation of weathered debris layers and episodic rainfall from single, intense storms to variations in rainfall on a decadal time-scale (e.g. McAuliffe *et al.*, 2006). On Mars, the presence of near-surface ice (Lanza *et al.*, 2010), possible brines (McEwen *et al.*, 2011) and aeolian erosion may be of major importance.

### 18.4.4   Hydrothermal alteration

The presence of hydrothermal alteration clays in the inner-crater breccias at the Lonar Crater contrasts with the absence of reported hydrothermal products at Meteor Crater. However, curated drill core samples from beneath the floor of Meteor Crater are not available for modern studies (drill core samples from the ejecta blanket are available; Gaither *et al.*, 2011). The data for Lonar and other small terrestrial impact craters show that the lower size limit for impact craters that can produce and sustain a significant hydrothermal system resulting in the formation of alteration clays is near the 1.8 km diameter size of Lonar Crater (Hagerty and Newsom, 2003; Naumov, 2005). Smaller craters like Pretoria Salt Pan (1.13 km diameter) do not have any evidence for hydrothermal alteration (Reimold *et al.*, 1999). However, according to the compilation by Naumov (2005), somewhat larger craters do have evidence of hydrothermal alteration; for example, the 2.5 km diameter Roter Kamm and Mischina Gora Craters, the 3.2 km diameter Zapadnaya Crater and the 3.8 km diameter Brent Crater.

The location of hydrothermal processes in terrestrial impact craters in the size range of Lonar Crater seem to be limited to the breccias filling the crater (Hagerty and Newsom, 2003; Naumov, 2005) (see Chapter 6). On the ejecta blankets of craters, the hot material is quite thin and will be deposited in the upper layers, isolated from groundwater. On the other hand, small impacts do produce shocked and fragmented rock that can be more readily altered under ambient conditions. There is also a substantial amount of fine-grained dust produced during impacts that is thrown up and dispersed by winds. The formation of accretionary lapilli provides evidence for this dust, which can include shocked and melted material. Small craters can, therefore, contribute to the formation of fine-grained and mobile surficial materials on Mars and other planetary bodies. However, the limited potential for the formation of lakes and hydrothermal deposits, for craters in this size range and smaller, greatly limits the astrobiology potential of such deposits, in contrast to larger impact craters (Chapters 6 and 11; Newsom, 2010; Schwenzer *et al.*, 2010).

### 18.4.5   Comparisons with lunar and Martian impact craters

The nature of fresh planetary impact craters compared with terrestrial craters provides constraints on the effects of different target rocks, the role of an atmosphere and the presence of water and aqueous alteration products in both sedimentary and basaltic target rocks. The crater Linné (2.2 km diameter, 27.7°N, 11.8°E) on the Moon provides an excellent example of a fresh crater the size of Lonar Crater, formed in the absence of an atmosphere in flat-laying basalt flows (Garvin *et al.*, 2011, Fig. 18.17). Of particular note are the exposed layers of the uplifted basalt on the crater walls (like the layers in Meteor Crater and Lonar Crater) and the smooth bowl-shaped crater form without any evidence for collapse of the crater walls. The ratio of rim height to diameter for Lonar (0.17) is shallower than is seen in other terrestrial craters (Fudali *et al.*, 1980), but is consistent with images of Linné. Another interesting feature is the incipient narrow gully-forms on the walls of the crater, possibly due to dry mass wasting of loose material.

**Figure 18.17** The Lunar Reconnaissance Orbiter Camera (LROC)-based digital elevation model of lunar crater Linné (2.2 km diameter) combined with LROC narrow-angle camera images (m046 and m161) at 2 m resolution (Garvin *et al.*, 2011). This crater represents a fresh impact on a basaltic target in NW Mare Serenitatis. Among the interesting features are the incipient narrow gully-forms on the crater wall. NASA/Goddard Space Flight Center/Arizona State University source: lroc.sese.asu.edu.

## 18.5  Summary and concluding remarks

Meteor Crater and Lonar Crater continue to provide important natural laboratories for impact crater research. The science merit of Meteor Crater includes being able to study: (1) the nature of impact processes in a layered sedimentary target; (2) the effects of erosion on crater walls and ejecta blankets; and (3) the role of airbursts and ejecta plumes. The engineering merit of the crater includes easy access to relatively young ejecta blanket material with some properties similar to planetary surfaces, such as the loose unconsolidated material on the surface.

The science merit of Lonar Crater includes being able to study: (1) signatures of habitable environments in proximal and distal ejecta; (2) crater-floor breccias with aqueous or hydrothermal alteration phases such as phyllosilicates and chemical precipitates; (3) shocked and melted rocks in the layered ejecta; and (4) fluidized ejecta features and the effects of multiple lava flows and inter-flow deposits during an impact event. The engineering merit of Lonar Crater includes the presence of an ejecta blanket with material properties expected for craters in basaltic terrains, similar to those on the Moon and Mars. The site can support tests of drilling and tests of analytical instrumentation, including contact and non-contact devices.

The amounts of erosion at Meteor Crater and Lonar Crater were stated by some past studies to be similar (e.g. Grant, 1999; Maloof *et al.*, 2010), when the ages for the two craters were also thought to be similar at approximately 50,000 years. This impression needs further examination given the publication of the new age for Lonar Crater of approximately 570,000 years. Several factors are probably relevant to the discussion. The early work of Grant (1999) utilized low-resolution Landsat data, and later work has not actually made a quantitative comparison of features, such

as depth of gullies on ejecta blankets, which are much deeper on the Lonar Crater. A further issue is the larger size of Lonar Crater (e.g. Fig. 18.1). When images of Lonar Crater are scaled to the same diameter as Meteor Crater, the differences seem less significant. Finally, only a few geologists have done extensive field work at both craters and co-author SPW, who has spent weeks at both craters, does believe that Lonar Crater is substantially more eroded. In addition to the need for more quantitative analysis of the erosion of the craters, future studies of the sedimentation record in the crater floors may provide important information on their erosional history. The long-standing mystery about the absence of meteorite fragments at Lonar Crater, in contrast to Meteor Crater, may be explained by the much greater time span for meteorite fragments to disintegrate and the evidence that the impactor at Lonar Crater was chondritic and not an iron meteorite (Misra *et al.*, 2009).

Information about the formation of the two craters continues to be discovered. The thickening of the distal edge of the ejecta blanket at Lonar Crater (Maloof *et al.*, 2010) is not seen at Meteor Crater (Roddy *et al.*, 1975), raising questions about the respective emplacement processes. Recent advances in impact crater studies involving these craters includes the discovery of the role of atmospheric breakup of impactors, based on the distribution of meteorites at Meteor Crater (Melosh and Collins, 2005; Artemieva and Pierazzo, 2009, 2011). This discovery is leading to an understanding of the enhancement of atmospheric effects due to a possible airburst from the breakup of the incoming projectiles (Boslough and Crawford, 2008). At Lonar Crater, the discovery of lapilli (Beal *et al.*, 2011) and evidence for a chondritic impactor (Misra *et al.*, 2009) could also be consistent with formation of an airburst. At Meteor Crater, the pervasive occurrence of meteoritic fragments in all impact melt types may also be consistent with an airburst event (Hagerty *et al.*, 2011). On Mars, the effects of an airburst mechanism on the surface are expected to be more common (Boslough and Crawford, 2008).

Additionally, new information on the macroscopic nature and spectroscopic effects of shock in basalts from Lonar Crater (Glotch *et al.*, 2011; Wright *et al.*, 2011) may have important utility for studies of extraterrestrial impacts, as the early studies of sandstone from Meteor Crater did for terrestrial impacts (Kieffer, 1971, 1974). Both Meteor Crater and Lonar Crater are near the lower limit in size required for the formation of significant hydrothermal deposits. Excavation, shock and comminution enhance the later ambient weathering of the ejecta material at both craters. Thus, in the presence of water, alteration of shocked and disrupted materials from impact craters the size of Lonar Crater (1.8 km diameter) or larger may contribute to substantial transport of mobile elements on planetary surfaces. This process can produce a chemically fractionated regolith and possibly contribute to the formation of brines in the near-surface environment of Mars.

### Acknowledgements

Research support: NASA Planetary Geology and Geophysics grants NNH07DA001N (H.E. Newsom) and NNH09AK43I (J.J.

Hagerty). Planex grant, Indian Space Research Organization (S. Misra). Previous support for research at Lonar includes the Barringer Family Fund (S.P. Wright), and by a Senior Research Associateship (grant B-10220), Council of Scientific and Industrial Research, New Delhi, India (S. Misra).

## References

Ackermann, H.D., Godson, R.H. and Watkins, J.S. (1975) A seismic refraction technique used for subsurface investigations at Meteor crater, Arizona. *Journal of Geophysical Research*, 80, 765–775.

Arif, Md., Misra, S., Basavaiah, N. and Newsom, H.E. (2009) Distribution of impact-induced stress around Lonar Crater, India. Meteoritical Society Meeting 2009, Nancy, France, Lunar and Planetary Science Institute, no. 5397.

Arif, Md., Deenadayalan, K., Basavaiah, N. and Misra, S. (2011) Variation of primary magnetization of basaltic target rocks due to asteroid impact: example from Lonar Crater, India. 42nd Lunar and Planetary Science Conference, 1383.pdf.

Artemieva, N. and Pierazzo, E (2009) The Canyon Diablo impact event: projectile motion through the atmosphere. *Meteoritics*, 44, 25–42.

Artemieva, N. and Pierazzo, E (2011) The Canyon Diablo impact event: 2. Projectile fate and target melting upon impact. *Meteoritics*, 46, 805–829.

Barlow, N.G., Boyce, J.M., Costard, F.M. *et al.* (2000) Standardizing the nomenclature of Martian impact crater ejecta morphologies. *Journal of Geophysical Research*, 105, 26733–26738.

Barringer, D.M. (1905) Coon Mountain and its crater. *Proceedings of the Academy of Natural Sciences of Philadelphia*, 66, 861–886.

Barringer, J.P. (1993) Acceptance speech. *Meteoritics*, 28, 9–11.

Beal, R.A., Newsom, H.E., Wright, S.P. and Misra, S. (2011) Discovery of mantled sub-millimeter lapilli from the Lonar Crater, India. 42nd Lunar and Planetary Science Conference, 1509.

Blandford, W.T. (1868) *Records of the Geological Survey of India*, 1, 62.

Boslough, M.B.E. and Crawford, D.A. (2008) Low-altitude airbursts and the impact threat. *International Journal of Impact Engineering*, 35, 1441–1448.

Bucher, W.H. (1963) Cryptoexplosions structures caused from without or within the Earth? ('Astroblemes' or 'Geoblemes?'). *American Journal of Science*, 261, 597–649.

Catling, D.C., Leovy, C.B., Wood, S.E. and Day, M. (2011) A lava sea in the Northern Lowlands of Mars: circumpolar oceans reconsidered. 42nd Lunar and Planetary Science Conference, no. 2529.

Chaklader, J., Shearer, C.K., Hörz, F. and Newsom, H.E. (2004) Volatile behavior in lunar and terrestrial basalts during shock: implications for Martian magmas. Lunar and Planetary Science XXXV: Martian Meteorites: Petrology, LPI Contribution 1197, no. 1397.

Chakrabarti, R. and Basu, A.R. (2006) Trace element and isotopic evidence for Archean basement in the Lonar Crater impact breccia, Deccan Volcanic Province. *Earth and Planetary Science Letters*, 247, 197–211.

Chakrabarti, R., Basu, A.R. and Ghatak, A. (2006) Reply to: Comments on 'Trace element and isotopic evidence for Archean basement in the Lonar crater impact breccia, Deccan volcanic province' by Ramananda Chakrabarti and Asish R. Basu. *Earth and Planetary Science Letters*, 263, 669–670.

Chernet, A.-L., Quidelleur, X., Fluteau, F. *et al.* (2007) $^{40}$K–$^{40}$Ar dating of the Main Deccan large igneous province: further evidence of KTB age and short duration. *Earth and Planetary Science Letters*, 263, 1–15.

Chowdhury, A.N. and Handa, B.K. (1978) Some aspects of the geochemistry of Lonar lake water. *Indian Journal of Earth Sciences*, 5, 111–118.

Cisowski, S.M. and Fuller, M. (1978) The effect of shock on the magnetism of terrestrial rocks. *Journal of Geophysical Research*, 83 (B7), 3441–3458.

Deenadayalan, K., Basavaiah, N., Misra, S. and Newsom, H.E. (2009) Absence of Archean basement in the genesis of Lonar Crater, India. Meteoritical Society Meeting 2009, Nancy, France, no. 5388. http://www.lpi.usra.edu/meetings/metsoc2009/pdf/5388.pdf.

Earth Impact Database (2011) http://www.passc.net/EarthImpact Database/index.html.

Fredriksson, K., Dube, A., Milton, D.J. and Balasundaram, M.S. (1973a) Lonar Lake, India: an impact crater in basalt. *Science*, 180, 862–864.

Fredriksson, K., Noonan, A. and Nelen, J. (1973b) Meteoritic, lunar and Lonar impact chondrules. *The Moon*, 7, 475–482.

Fredriksson, K., Brenner, P., Dube, A. *et al.* (1979) Petrology, mineralogy and distribution of Lonar (India) and lunar impact breccias and glasses. *Smithsonian Contributions to the Earth Sciences*, 22, 1–12.

Fudali, R.F. (1999) Commentary, in *Deccan Volcanic Province* (ed. K.V. Subbarao), Geological Society of India, Bangalore, pp. 911–914.

Fudali, R.F., Milton, D.J., Fredriksson, K. and Dube, A. (1980) Morphology of Lonar Crater, India: comparisons and implication. *Moon and the Planets*, 23, 493–515.

Gaither, T.A., Hagerty, J.J., Clark, S.E. *et al.* (2011) Multi-dimensional characterization of impact ejecta deposits from Meteor Crater, AZ. 42nd Lunar and Planetary Science Conference, no. 1474.

Garvin, J.B., Robinson, M.S., Frawley, J. *et al.* (2011) Linné: simple lunar mare crater geometry from LRO observations. 42nd Lunar and Planetary Science Conference, no. 2063.

Ghosh, S. and Bhaduri, S.K. (2003) Petrography and petrochemistry of impact melts from Lonar Crater, Buldana District, Maharashtra, India. *Indian Minerals*, 57, 1–26.

Gilbert, G.K. (1896) The origin of hypotheses, illustrated by the discussion of a topographic problem. *Science*, 3, 1–13.

Glotch, T.D., Wright, S.P., McKeeby, B.E. and Ferrari, M.J. (2011) Micro-FTIR and micro-Raman spectroscopy of shocked basalts from Lonar Crater, India. 42nd Lunar and Planetary Science Conference, no. 1566.

Grant, J.A. (1999) Evaluating the evolution of process specific degradation signatures around impact craters. *International Journal of Impact Engineering*, 23, 331–340.

Grant, J.A. and Schultz, P.H. (1991) Characteristics of ejecta and alluvial deposits at Meteor Crater, Arizona and Odessa Craters, Texas: results from ground penetrating radar. 22nd Lunar and Planetary Science Conference, pp. 481–482.

Grant, J.A. and Schultz, P.H. (1993) Erosion of ejecta at Meteor Crater, Arizona. *Journal of Geophysical Research*, 98, 15033–15047.

Graup, G. (1981) Terrestrial chondrules, glass spherules, and accretionary lapilli from the suevite, Ries Crater, Germany. *Earth and Planetary Science Letters*, 55, 407–418.

Greeley, R. and Iverson, J.D. (1985) *Wind as a Geological Process on Earth, Mars, Venus, and Titan*, Cambridge University Press, New York.

Greenwood, W.R. and Morrison, D.A. (1969) Genetic significance of the morphology of some impact bombs from Meteor crater, Arizona. *Meteoritics*, 4, 182–183.

Hagerty, J.J. and Newsom, H.E. (2003) Evidence for impact-induced hydrothermal alteration at the Lonar crater, India. *Meteoritics and Planetary Science*, 38, 365–381.

Hagerty, J.J., Gaither, T.A., McHone, J.F. and Sauer, K. (2011) SEM characterization of impact ejecta deposits from Meteor Crater, Arizona. 2nd Planetary Crater Consortium Meeting, abstract no. 1109.

Hargraves, R.B. and Perkins, W.E. (1969) Investigations of the effect of shock on natural remanent magnetism. *Journal of Geophysical Research*, 74 (10), 2576–2589.

Hörz, F. (1965) Untersuchungen an Riesgläsern. *Beiträge zur Mineralogie und Petrographie*, 11, 621–661. DOI: 10.1007/BF01128707.

Hörz, F., Ostertag, R. and Rainey, D.A. (1983), Bunte Breccia of the Ries: continuous deposits of large impact craters. *Reviews of Geophysics*, 21 (8), 1667–1725.

Hörz, F., Mittlefehldt, D.W., See, T.H. and Galindo, C. (2002) Petrographic studies of the impact melts from Meteor Crater, Arizona, USA. *Meteoritics and Planetary Science*, 37, 501–531.

Hoyt, W.G. (1987) *Coon Mountain Controversies: Meteor Crater and the Development of the Impact Theory*, The University of Arizona Press, Tucson, AZ.

Jhingran, A.G. and Rao, K.V. (1958) Lonar lake and its salinity. *Records of Geological Survey of India*, 85 (3), 313–334.

Jourdan, F. (2010) First $^{40}Ar/^{39}Ar$ age of the Lonar Crater: a ~0.65 Ma impact event? 41st Lunar and Planetary Science Conference, 1661.pdf.

Jourdan, F., Moynier, F., Koeberl, C. and Eroglu, S. (2011) $^{40}Ar/^{39}Ar$ age of the Lonar Crater and consequence for the geochronology of planetary impacts. *Geology*, 39, 671–674.

Jull, A.J.T., Donahue, D.J., Cielaszyk, E. and Wlotzka, F. (1993) Carbon-14 terrestrial ages and weathering of 27 meteorites from the southern high plains and adjacent areas (USA). *Meteoritics*, 28, 188–195.

Kieffer, S.W. (1971), Shock metamorphism of the Coconino Sandstone at Meteor Crater, Arizona. *Journal of Geophysical Research*, 76, 5449–5473.

Kieffer, S.W. (1974) Shock metamorphism of the Coconino sandstone at Meteor Crater, in *Guidebook to the Geology of Meteor Crater, Arizona* (eds E.M. Shoemaker and S.W. Kieffer), Center for Meteorite Studies Publication No. 17, Arizona State University, Phoenix, AZ, pp. 12–19.

Kieffer, S.W., Schaal, R.B., Gibbons, R. *et al.* (1976) Shocked basalt from Lonar impact crater, India, and experimental analogues. Proceedings of 7th Lunar Science Conference, pp. 1391–1412.

Koeberl, C., Bhandari, N., Dhingra, D. *et al.* (2004) Lonar impact crater, India: occurrence of a basaltic suevite? 35th Lunar and Planetary Science Conference, no. 1751.

Koeberl, C., Brandstätter, F., Glass, B.P. *et al.* (2007) Uppermost impact fallback layer in the Bosumtwi crater (Ghana): mineralogy, geochemistry, and comparison with Ivory Coast tektites. *Meteoritics and Planetary Science*, 42 (4), 709–729.

Kring, D.A. (1997) Air blast produced by the Meteor Crater impact event and a reconstruction of the affected environment. *Meteoritics and Planetary Science*, 32, 517–530.

Kring, D.A. (2007) *Guidebook to the Geology of Barringer Meteorite Crater, Arizona (a k a Meteor Crater)* LPI Contribution no. 1355.

Kumar, P.S. (2005) Structural effects of meteorite impact on basalt: evidence from Lonar Crater, India. *Journal of Geophysical Research*, 110, B12402.

Kumar, P.S., Head, J.W. and Kring, D.A. (2010) Erosional modification and gully formation at Meteor Crater, Arizona: insights into crater degradation processes on Mars. *Icarus*, 208, 608–620.

La Fond, E.C. and Dietz, R.S. (1964) Lonar Crater, India, a meteorite crater? *Meteoritics*, 2 (2), 111–116.

Lanza, N.L., Meyer, G.A., Okubo, C. *et al.* (2010) Evidence for debris flow and shallow subsurface flow on Mars. *Icarus*, 205, 103–112. DOI: 10.1016/j.icarus.2009.04.014.

Louzada, K.L., Weiss, B.P., Maloof, A.C. *et al.* (2008) Paleomagnetism of Lonar impact crater, India. *Earth and Planetary Science Letters*, 275, 308–319.

Maloof, A.C., Stewart, S.T., Weiss, B.P. *et al.* (2010) Geology of Lonar Crater, India. *GSA Bulletin*, 122, 109–126.

Mark, K. (1987) *Meteorite Craters*, University of Arizona Press, Tucson, AZ.

Master, S. (1999) Evidence for an impact origin of the Amber lake structure: a smaller companion crater to the Lonar impact crater, Maharashtra, India. *Meteoritics and Planetary Science*, 34, A78.

Medicott, H.B. and Blanford, W.T. (1879) *A Manual of the Geology of India*, Pt. 1, Geological Survey Office, Calcutta, p. 379.

McAuliffe, J.R., Scuderi, L.A. and McFadden, L.D. (2006) Tree-ring record of hillslope erosion and valley floor dynamics: landscape responses to climate variation during the last 400yr in the Colorado Plateau, northeastern Arizona. *Global and Planetary Change*, 50, 184–201.

McEwen, A.S., Ojha, L., Dundas, C.M. *et al.* (2011) Seasonal flows on warm Martian slopes. *Science*, 333 (6043), 740–743.

Melosh, H.J. and Collins, G.S. (2005) Meteor Crater formed by low-velocity impact. *Nature*, 434, 157.

Ming, D.W., Morris, R.V. and Clark, B.C. (2008) Aqueous alteration on Mars, in *The Martian Surface* (ed. J. Bell), Cambridge University Press, pp. 519–540.

Misra, S. (2006) Comments on 'Trace element and isotopic evidence for Archean basement in the Lonar Crater impact breccia, Deccan volcanic province' by R. Chakrabarti and A. R. Basu in *Earth and Planetary Science Letters* 247, 197–211. *Earth and Planetary Science Letters*, 250, 667–668.

Misra, S. and Newsom, H.E. (2010) Geochemistry and petrology of maskelynite in NWA1195 shergottites and its comparison with maskelynite from Lonar Crater, India, South Africa. 41st Lunar and Planetary Science Conference, no. 1806.

Misra, S., Newsom, H., Mukherjee, T. *et al.* (2007) No evidence of impact induced volatile loss from maskelynite of Lonar Crater, India. 38th Lunar and Planetary Science Conference, no. 1672.

Misra, S., Newsom, H.E., Shyam Prasad, M. *et al.* (2009) Geochemical evidence of the impactor for Lonar Crater, India: solution to a century-old mystery. *Meteoritics and Planetary Science*, 44, 1001–1018.

Misra, S., Arif, M.D., Basavaiah, N. *et al.* (2010) Structural and anisotropy of magnetic susceptibility (AMS) evidence for oblique impact on terrestrial basalt flows: Lonar crater, India. *GSA Bulletin*, 122, 563–574.

Mittlefehldt, D.W., Hörz, F., See, T.H. *et al.* (2005) Geochemistry of target rocks, impact-melt particles and metallic spherules from Meteor Crater, Arizona: empirical evidence on the impact process, in *Large Meteorite Impacts III* (eds T. Kenkmann, F. Hörz, and A. Deutsch), Geological Society of America Special Paper 384, Geological Society of America, Boulder, CO, pp. 367–390.

Moore D.E. and Reynolds Jr, R.C. (1997) *X-Ray Diffraction and the Identification and Analysis of Clay Minerals*, second edn., Oxford University Press, Oxford, p. 378.

Morgan, J.W. (1978) Lonar Crater glasses and high-magnesium australites: trace element volatilization and meteoritic contamination. Proceedings of 9th Lunar Science Conference, pp. 2713–2730.

Nandy, N.C. and Dey, V.B. (1961) Origin of the Lonar lake and its alkalinity. *Technical Journal of the Tata Iron and Steel Company, Ltd., India*, 8 (3), 144–155.

Naumov, M.V. (2005) Principal features of impact-generated hydrothermal circulation systems: mineralogical and geochemical evidence. *Geofluids*, 5, 165–184.

Nayak, V.K. (1972) Glassy objects (impactite glasses?): a possible new evidence for meteoritic origin of the Lonar Crater, India. *Earth and Planetary Science Letters*, 14, 1–6.

Nayak, V.K. (1985) Trona in evaporite from the Lonar impact crater, Maharashtra. *Indian Journal of Earth Sciences*, 12, 221–222.

Nayak, V.K. (1993) Maskelynite from the Indian impact crater at Lonar. *Journal of the Geological Society of India*, 41, 307–312.

Nesbitt, H.W. and Wilson, R.E. (1992) Recent chemical weathering of basalts. *American Journal of Science*, 292, 740–777.

Newbold, T.J. (1844) Summary of the geology of southern India. Art. II, Part X, newer or overlying trap. *Journal of the Royal Asiatic Society*, 12, 20–42.

Newsom, H.E. (2010) Heated lakes on Mars, in *Lakes on Mars* (eds N. Cabrol and E. Grin), Elsevier, Amsterdam, pp. 91–110 (ISBN 978-0-444-52854-4).

Newsom, H.E. and Boslough, M.B.E. (2008) Impact melt formation by low-altitude airburst processes, evidence from small terrestrial craters and numerical modeling. 39th Lunar and Planetary Science Conference, no. 1460.

Newsom, H.E., Nelson, M.J., Shearer, C.K. *et al.* (2004) Major and trace element variations in impact crater clays from Chicxulub, Lonar, and Mistastin, implications for the Martian soil. 35th Lunar and Planetary Science Conference, no. 1087.

Newsom, H.E., Misra, S., Wright, S.P. and Muttik, N. (2010a) Contrasting alteration and enrichment of mobile elements during weathering of basaltic ejecta and ancient soils at Lonar Crater, India. 41st Lunar and Planetary Science Conference, no. 2210.

Newsom, H.E., Wright, S.P. and Boslough, M.B.E. (2010b) Layered melts and melt-covered rocks from small impact craters – evidence for melting by an airburst. 2010 GSA Denver Annual Meeting, 31 October–3 November 2010, paper no. 69-13.

Newsom, H.E., Wright, S.P., Misra, S. *et al.* (2011) Role of impact craters in the origin of phyllosilicates and surficial materials on Mars – new understanding from earth analog studies. 42nd Lunar and Planetary Science Conference, 7–11 March, The Woodlands, TX, LPI Contribution No. 1608, p. 1298.

Nininger, H.H. (1956) *Arizona's Meteorite Crater*, American Meteorite Museum, Sedona, AZ.

Nishioka, I. (2007) Rock magnetic study of basalt at Lonar impact crater in India: Effects of stress waves on rock magnetic properties. PhD thesis, Graduate University for Advanced Studies, Japan.

Nishioka, I. and Funaki, M. (2008) Irreversible changes in anisotropy of magnetic susceptibility: study of basalts from Lonar Crater and experimentally impacted basaltic andesite. 71st Annual Meteoritical Society Meeting no. 5207, p. A116.

Nishioka, I., Funaki, M. and Sekine, T. (2007) Shock-induced anisotropy of magnetic susceptibility: impact experiment on basaltic andesite. *Earth, Planets, and Space*, 59, e45–e48.

Nishiizumi, K., Kohl, C.P., Shoemaker, E.M. *et al.* (1991) In situ $^{10}$Be–$^{26}$Al exposure ages at Meteor Crater, Arizona. *Geochimica et Cosmochimica Acta*, 55, 2699–2703.

Osae, S., Misra, S., Koeberl, C. *et al.* (2005) Target rocks, impact glasses and melt rocks from the Lonar Crater, India: petrography and geochemistry. *Meteoritics and Planetary Science*, 40, 1473–1492.

Osinski, G.R., Bunch, T.E. and Wittke, J. (2006) Proximal ejecta at Meteor Crater, Arizona: discovery of impact-melt bearing breccias. 37th Lunar and Planetary Science Conference, no. 1005.

Phillips, F.M., Zreda, M.G., Smith, S.S. *et al.* (1991) Age and geomorphic history of Meteor Crater, Arizona from cosmogenic $^{36}$Cl and $^{14}$C in rock varnish. *Geochimica et Cosmochimica Acta*, 55, 2695–2698.

Pilon, J.A., Grieve, R.A.F. and Sharpton, V.L. (1991) The subsurface character of Meteor Crater, Arizona, as determined by ground-probing radar. *Journal of Geophysical Research*, 96, 15563–15576.

Rajasekhar, R.P. and Mishra, D.C. (2005) Analysis of gravity and magnetic anomalies over Lonar Lake, India: an impact crater in a basalt province. *Current Science*, 88, 1836–1840.

Ramsey, M.S. (2002) Ejecta distribution patterns at Meteor Crater, Arizona: on the applicability of lithologic end-member deconvolution for spaceborne thermal infrared data of Earth and Mars. *Journal of Geophysical Research*, 107 (E8), 3–14.

Regan, R.D. and Hinze, W.J. (1975) Gravity and magnetic investigations of Meteor crater, Arizona. *Journal of Geophysical Research*, 80, 776–778.

Reimold, W.U., Koeberl, C. and Brandt, D. (1999) The origin of the Pretoria Saltpan crater. *Geological Survey of South Africa*, 85, 35–54.

Roddy, D.J. (1978) Pre-impact geologic conditions, physical properties, energy calculations, meteorite and initial crater dimensions and orientations of joints, faults and walls at Meteor Crater, Arizona. *Proceedings of the 9th Lunar and Planetary Science Conference*, vol. 3, Pergamon Press, New York, NY, pp. 3891–3930.

Roddy, D.J., Boyce, J.M., Colton, G.W. and Dial Jr, A.L. (1975) Meteor crater, Arizona, rim drilling with thickness, structural uplift, diameter, depth, volume, and mass-balance calculations. *Proceedings of 6th Lunar Science Conference*, vol. 3, Pergamon Press, New York, NY, pp. 2621–2644.

Rosen, M.R., Arehart, G.B. and Lico, M.S. (2004) Exceptionally fast growth rate of 100-yr-old tufa, Big Soda Lake, Nevada: implications for using tufa as a paleoclimate proxy. *Geology*, 32, 409–412.

Sengupta, D., Bhandari, N. and Watanabe, S. (1997) Formation age of Lonar Meteor crater, India. *Revista de Fisica Aplicada e Instrumentacao*, 12, 1–7.

Schumacher, R. and Schmincke, H.-U. (1994) Models for the origin of accretionary lapilli. *Bulletin of Volcanology*, 56, 626–639.

Schwenzer, S.P., Abramov, O., Allen, C.C. *et al.* (2010) Exploring Martian impact craters: what they can reveal about the surface and why they are important in the search for life. 41st Lunar and Planetary Science Conference (Lunar and Planetary Science XLI), Houston, TX, abstract no. 1589.

See, T.H., Hörz, F., Mittlefehldt, D.W. *et al.* (2002) Major element analyses of the target rocks at Meteor Crater, Arizona. NASA Technical Memorandum (TM)-2002-210787.

Shoemaker, E.M. (1960) Impact mechanics at Meteor Crater, Arizona. PhD dissertation, Princeton University.

Shoemaker, E.M. and Kieffer, S.W. (eds) (1974) *Guidebook to the Geology of Meteor Crater, Arizona*, Center for Meteorite Studies Publication No. 17, Arizona State University, Phoenix, AZ.

Son, T.H. and Koeberl, C. (2007) Chemical variation in Lonar impact glasses and impactites: GFF. *The Geological Society of Sweden*, 129, 161–176.

Stöffler, D. and Grieve, R.A.F. (2007) Impactites, in *Metamorphic Rocks: A Classification and Glossary of Terms, Recommendations of the International Union of Geological Sciences*, (eds D. Fettes and J. Desmons), Cambridge University Press, Cambridge, UK, pp. 82–92, 111–125, 126–242.

Stroube Jr, W.B., Garg, A.N., Ali, M.Z. and Ehmann, W.D. (1978) A chemical study of the impact glasses and basalts from Lonar Crater, India. *Meteoritics*, 13, 201–208.

Subrahmanyam, B. (1985) Lonar Crater, India: a crypto-volcanic origin. *Journal of the Geological Society of India*, 26, 326–335.

Sutton, S.R. (1985) Thermoluminescence measurements on shock-metamorphosed sandstone and dolomite from Meteor Crater, Arizona; shock dependence of thermoluminescence properties. *Journal of Geophysical Research*, 90 (B5), 3683–3689.

Venkatesh, V. (1967), Geology and origin of the Lonar Crater, Maharashtra. *Record of Geological Survey of India*, 97, 30–45.

Warme, J.E., Morgan, M. and Kuehner, H.-U. (2002) Impact-generated carbonate accretionary lapilli in the Late Devonian Alamo Breccia, in *Catastrophic Events and Mass Extinctions: Impacts and Beyond* (eds C. Koeberl and K.G. MacLeod), Geological Society of America Special Paper 356, Geological Society of America, Boulder, CO, pp. 489–504.

Weiss, B.P., Pedersen, S., Garrick-Bethell, I. *et al.* (2010) Paleomagnetism of impact spherules from Lonar Crater, India and a test for impact-generated fields. *Earth and Planetary Science Letters*, 298, 66–76.

Wright, S.P. (2008) Intermediate (20–40 GPa) shocked basalt from Lonar Crater, India: ejecta locality and spectroscopy of a sher-

gottite analog. 39th Lunar and Planetary Science Conference, no. 2330.

Wright, S.P. and Newsom, H.E. (2011) Analyses of a large sample collection of shocked Deccan basalt reveal a range of shock pressures, protoliths, and pre- and post-impact alteration products. 42nd Lunar and Planetary Science Conference, no. 1619.

Wright, S.P. and Ramsey, M.S. (2006) Thermal infrared data analyses of Meteor Crater, Arizona: implications for Mars spaceborne data from the Thermal Emission Imaging System. *Journal of Geophysical Research, Planets*, 111, E02004. DOI: 10.1029/2005JE002472.

Wright, S.P., Christensen, P.R. and Sharp, T.G. (2011) Laboratory thermal emission spectroscopy of shocked basalt from Lonar Crater, India: Implications for Mars orbital and sample data. *Journal of Geophysical Research, Planets*, 116, E09006. DOI: 10.1029/2010JE003785.

Zipfel, J., Schröder, C., Jolliff, B.L. *et al.* (2011) Bounce Rock – a shergottite-like basalt encountered at Meridiani Planum, Mars. *Meteoritics and Planetary Science*, 46, 1–20.

# NINETEEN

# Comparison of mid-size terrestrial complex impact structures: a case study

## Gordon R. Osinski[*,†] and Richard A. F. Grieve[*]

*Department of Earth Sciences, Western University, 1151 Richmond Street, London, ON, N6A 5B7, Canada
†Department of Physics and Astronomy, Western University, 1151 Richmond Street, London, ON, N6A 5B7, Canada

## 19.1  Introduction

Despite the ubiquitous nature of impact craters in the Solar System, some important second-order aspects of the processes and products of their formation remain incompletely understood (see Chapter 1). One such aspect is the parameters and processes controlling the final morphology and morphometry of complex impact craters (see Chapter 5). The generation of impact melt (see Chapter 9), particularly in different target rocks, and the origin and emplacement of impact ejecta deposits (see Chapter 4) also remain topics of debate within the impact community. For these reasons, the systematic study of mid-size impact structures was one of the main recommendations for future research resulting from the first 'Bridging the Gap' conference in 2003 (Herrick and Pierazzo, 2003) and provides the motivation for this contribution. Mid-size structures in the 20–40 km size range were highlighted as being important to study as they are large enough that they possess all the characteristics of complex impact craters (e.g. central uplift, terraced crater rim), yet they are small enough that detailed studies are possible within a handful of field expeditions. In addition, and perhaps most importantly, there are enough examples present in different target rocks that comparisons become meaningful. This is not the case for large complex impact structures in the 150 km size range and above, where only three such structures exist on Earth (Grieve *et al.*, 2008).

We present, here, a comparative study of several mid-size terrestrial impact structures. The Haughton, Ries and Mistastin impact structures are the main focus (see Table 19.1 for a list of the important parameters for these impact sites), as these structures are similar in size, are exposed and are relatively well studied and preserved. We also consider the Rochechouart structure, which is considerably eroded, and the Boltysh structure, which is almost completely buried. An important variable for these structures is that, with respect to target composition and the proportion of sedimentary versus crystalline rocks, Haughton and Boltysh–Mistastin–Rochechouart can be viewed as end-members, with Ries being an 'intermediate' case (Fig. 19.1; Table 19.1). Thus, a comparison of these structures has the potential to yield valuable information as to the effect of target lithology on the processes and products of impact cratering.

## 19.2  Overview of craters

### 19.2.1  Boltysh impact structure, Ukraine

The Boltysh impact structure is located in Ukraine (48°54′N, 32°15′E) in the basin of the Tyasmin River. Like many impact structures, it was originally thought to be volcanic in origin, but the discovery of shocked quartz confirmed its impact origin (Masaitis, 1974). The structure is largely buried by approximately 30 m of Quaternary sediments, such that the diameter of 24–25 km for Boltysh is based on drilling and geophysical surveys (Masaitis, 1999). Boltysh formed in the Proterozoic porphyroblastic granites and biotite gneisses of the Ukrainian Shield and the impact event is dated at $65.17 \pm 0.64$ Ma by Ar–Ar methods (Kelley and Gurov, 2002). Despite being buried, a large amount of information exists based on numerous drill cores and geophysical investigations that were undertaken in the 1960s and 1980s in the search for hydrocarbons; although, unfortunately, the drill cores have since been largely lost and/or destroyed.

Boltysh is clearly a complex crater, with a topographic central peak that emerges from the surrounding crater-fill impactites (Grieve *et al.*, 1987; Fig. 19.2). The depth to the crater floor is approximately 1 km around this central uplift. Boltysh is associated with an approximately 22–23 km diameter negative Bouguer gravity anomaly of −190 g.u. The interior of the crater (~12 km in diameter) is filled with impact melt rocks, impact melt-bearing breccias ('suevites') and lithic breccias (Gurov *et al.*, 2003; Fig. 19.2). The impact melt rocks form an annular sheet approximately 12 km in diameter and up to 220 m thick, surrounding the central uplift (Grieve *et al.*, 1987). The lowermost impact melt

*Impact Cratering: Processes and Products*, First Edition. Edited by Gordon R. Osinski and Elisabetta Pierazzo.
© 2013 Blackwell Publishing Ltd. Published 2013 by Blackwell Publishing Ltd.

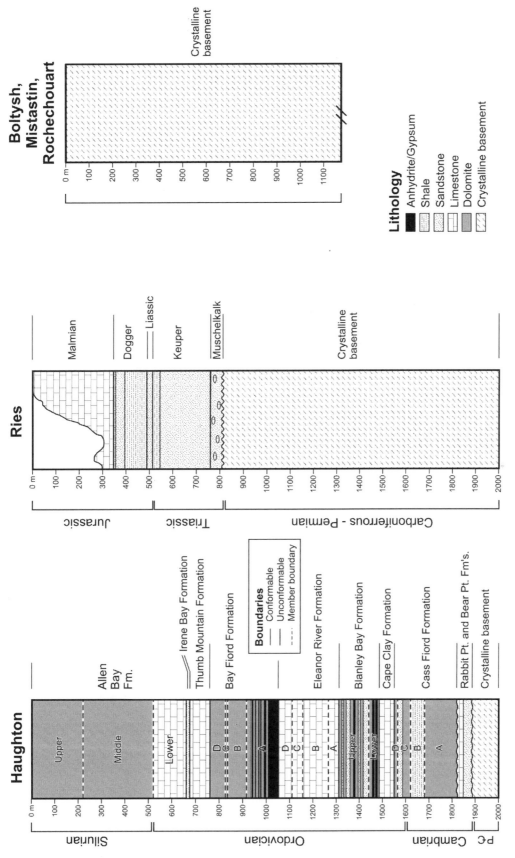

**Figure 19.1** Comparison of the target stratigraphy at the Haughton, Ries, and Mistastin impact structures. Modified from Osinski *et al.* (2008a).

**Table 19.1**  Summary of the important statistics and parameters of the Boltysh, Haughton, Mistastin, Ries and Rochechouart impact structures[a]

| Parameter | Boltysh | Notes and reference(s) | Haughton | Notes and reference(s) | Mistastin | Notes and reference(s) | Ries | Notes and reference(s) | Rochechouart | Notes and reference(s) |
|---|---|---|---|---|---|---|---|---|---|---|
| Age | 65.17±0.64 Ma | Kelley and Gurov (2002) | 39±2 Ma | Sherlock et al. (2005) | 36.4±2 Ma | Mak et al. (1976) | 14.6 | Buchner et al. (2010) | 201±2 Ma | Most recent age estimate (Schmieder et al., 2010); 214±8 Ma also reported (Kelley and Spray, 1997) |
| Target | Cryst. | Porphyroblastic granites and biotite gneisses of the Ukrainian Shield (Kelley and Gurov, 2002) | Mixed | Predominantly lst and dst, with minor evap and sst, overlying Precambrian gn (Thorsteinsson and Mayr, 1987; Osinski et al., 2005b) | Cryst. | Anorthosite, mangerite and granodiorite intrusive rocks (Currie, 1971) | Mixed | Predominantly sandstone, siltstone, marl and limestone overlying Hercynian crystalline basement (Schmidt-Kaler, 1978) | Cryst. | Granitic intrusive and metamorphic rocks (Kraut and French, 1971) |
| Thickness of sedimentary cover | 0 m | – | 1880 m | Osinski et al. (2005b) | 0 m | – | 470–820 m | Schmidt-Kaler (1978) | 0 m | – |
| Apparent crater diameter | 24 km | Based on drilling and geophysical surveys (Masaitis, 1999) | 23 km | Outermost ring of concentric normal faults at present-day erosion level (Osinski and Spray, 2005) | 28 km | Based on topography surrounding Lake Mistastin (Grieve, 1975) | 24–26 km | Pohl et al. (1977) | 24 km | Based on a shock zoning study (Lambert, 1977) |

| | | | | | | | | | | |
|---|---|---|---|---|---|---|---|---|---|---|
| Rim (final crater) diameter | ? | — | ~16 km | Location of large displacement normal fault (Osinski and Spray, 2005) | 18 km | Estimated based on fieldwork (Marion, 2008) | ? | — | ? | — |
| Central uplift | Yes | Central peak | Yes | Subdued; no central peak or peak ring | Yes | Eroded; unknown original morphology | Yes | No central peak; uplifted so-called inner ring of crystalline rocks (Pohl et al, 1977) | ? | — |
| Crater-fill impactites | Yes | Clast-rich to clast-poor silicate impact melt rocks overlain by impact melt-bearing breccias (Grieve et al., 1987) | Yes | Clast-rich impact melt rocks, comprising calcite, sulfate and Mg-rich glasses (Osinski et al., 2005c) | Yes | Clast-rich to clast-poor silicate impact melt rocks (Grieve, 1975) | Yes | ~400 m of impact melt-bearing breccia (suevite) | Yes | Lithic breccias overlain by series of different impact melt-bearing impactites (Lambert, 1977) |
| Ejecta | Yes | Clast-rich impact melt rocks overlying lithic impact breccias (Gurov et al., 2003) | Yes | Clast-rich impact melt rocks overlying impact breccias | Yes | Impact melt rocks overlying impact melt-bearing and -free impact breccias (Mader et al., 2011) | Yes | Impact melt-bearing breccia (suevite) overlying lithic breccia (Bunte Breccia) (von Engelhardt, 1990) | ? | — |

[a]Abbreviations: cryst, crystalline; lst, limestone; dst, dolomite; sst, sandstone; gn, gneiss.

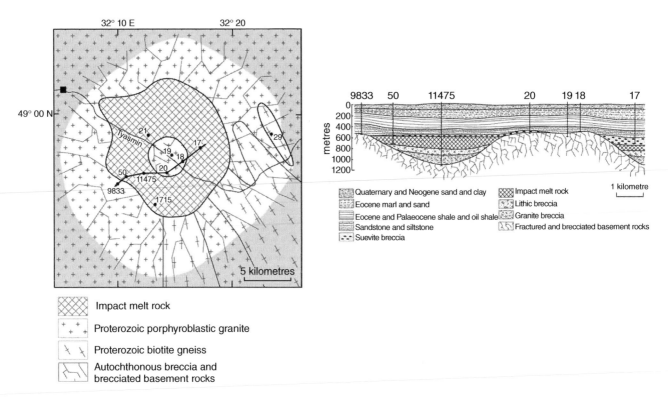

**Figure 19.2**    Simplified geological map (left) and cross-section (right) of the Boltysh impact structure. Modified from Gurov *et al.* (2006).

rocks possess a glassy matrix, whereas the upper parts of the sheet are composed of microcrystalline impact melt rock (Grieve *et al.*, 1987). Impact melt-bearing breccias ('suevite') of variable thickness (~12–22 m) overlie the impact melt rocks (Fig. 19.2). The crater-fill impactites are overlain by post-impact lacustrine sediments up to approximately 575 m thick, comprising argillites, siltstones, sandstones, sands and oil shales (Gurov *et al.*, 2006). Impact ejecta deposits are present at Boltysh but they are very poorly exposed and eroded in close proximity to the crater rim (Gurov *et al.*, 2003). In the Tyasmin River valley, approximately 6–8 km outside the crater rim, low-shock, melt-free 'monomict' breccias are overlain by polymict breccias (Gurov *et al.*, 2003). The polymict breccias contain more highly shocked materials and are described as 'lithic breccias' (Gurov *et al.*, 2003); however, they also are reported as containing altered melt particles, in which case they should be termed impact melt-bearing breccias (see Chapters 7 and 9).

### 19.2.2    Haughton impact structure, Canada

Haughton is a well-preserved complex impact structure situated on Devon Island in the Canadian Arctic Archipelago (75°22′N, 89°41′W; Fig. 19.3). It was originally thought to be a salt dome (Greiner *et al.*, 1963); however, the isolation of the structure from these other salt tectonic features prompted Dence (1972) to include it in a list of 'possible impact structures'. An impact origin for Haughton was subsequently confirmed by the discovery of shatter cones (Robertson and Grieve, 1978) and coesite-bearing gneiss clasts from impact breccias (Frisch and Thorsteinsson,

1978). Field studies carried out in 1984, under the auspices of the multidisciplinary Haughton Impact Structure Studies (HISS) project, resulted in a substantial increase in our knowledge of Haughton (summarized in Grieve (1988)). Then, beginning in 1998, initially under the auspices of the Haughton–Mars Project, fieldwork was carried out for a dozen consecutive years (see overview in Osinski *et al.* (2005b)), providing new insights into this unique structure. It was during this time that Haughton was mapped in detail for the first time, and this structure remains the only crater of its size to have a detailed published geological map (Osinski, 2005b).

Originally thought to be Miocene in age (Jessberger, 1988), more recent high-precision $^{40}$Ar–$^{39}$Ar laser-probe dating of potassic glasses contained in highly shocked basement clasts yields an age of $39 \pm 2$ Ma for Haughton (Sherlock *et al.*, 2005). The pre-impact target sequence at Haughton comprised an approximately 1880 m thick series of Lower Palaeozoic sedimentary rocks of the Arctic Platform, overlying Precambrian metamorphic basement rocks of the Canadian Shield (Fig. 19.1). Allochthonous crater-fill deposits form a virtually continuous 54 km² unit covering the central area of the structure (Fig. 19.3, Fig. 19.4 and Fig. 19.5a,b). Originally thought to be clastic or fragmental breccias (Redeker and Stöffler, 1988), field and analytical scanning electron microscopy studies indicate that these rocks are clast-rich carbonate melt-bearing impact melt rocks (Osinski and Spray, 2001,2003; Osinski *et al.*, 2005c). Ejecta deposits are preserved over an approximately 5 km² area in the south of the structure and comprise a two-layer stratigraphy (Fig. 19.3 and Fig. 19.4; Osinski *et al.*, 2005c).

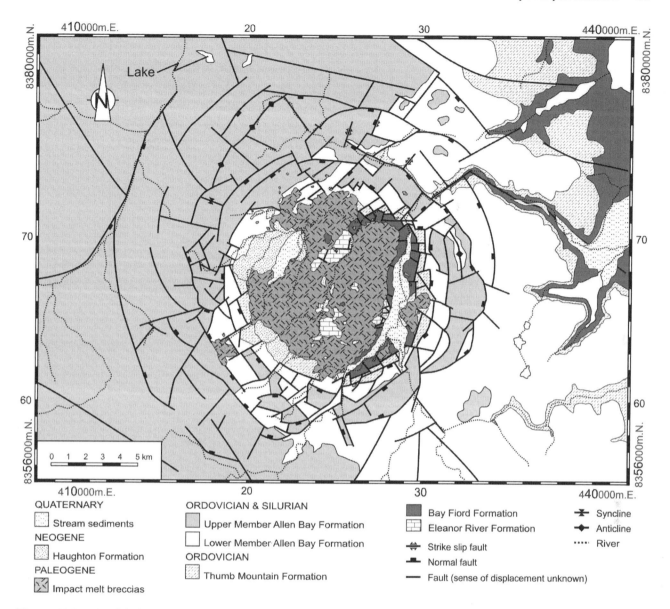

**Figure 19.3** Simplified geological map of the Haughton impact structure, Devon Island, Nunavut. Modified from Osinski *et al.* (2005b).

The crater-fill impactites and, in places, the Palaeozoic target rocks, are unconformably overlain by the Haughton Formation, a series of post-impact lacustrine sediments (Fig. 19.3). These sediments consist of dolomitic silts and muds, with subordinate fine-grained dolomitic sands (Hickey *et al.*, 1988). Field studies suggest that the Haughton Formation was laid down $10^5$–$10^6$ years following the impact event, after a substantial amount of erosion of impact melt rocks and pre-impact target rocks (Osinski and Lee, 2005).

Haughton has an apparent crater diameter of approximately 23 km, with an estimated rim (final crater) diameter of approximately 16 km (Fig. 19.4; Osinski and Spray, 2005). The structure lacks a central topographic peak or peak ring, which is unusual for complex craters of this size. Haughton is associated with a

approximately 24 km diameter negative Bouguer gravity anomaly (Pohl *et al.*, 1988). The centre of the Haughton structure is associated with a minimum of approximately 30 g.u. with very steep gradients surrounded by weaker relative maxima at approximately 6–7 km radial distance (Pohl *et al.*, 1988). The overall negative anomaly has been explained as a bowl-shaped zone with reduced densities, as compared to the undisturbed surroundings (Pohl *et al.*, 1988). A ground survey revealed a 300 nT positive magnetic anomaly at the geometric centre of Haughton (Pohl *et al.*, 1988), which has been confirmed by recent airborne magnetic surveys (Glass *et al.*, 2002). The positive magnetic anomaly coincides with the central negative gravity anomaly and may be equated with the presence of a core of very low-density material (Pohl *et al.*, 1988).

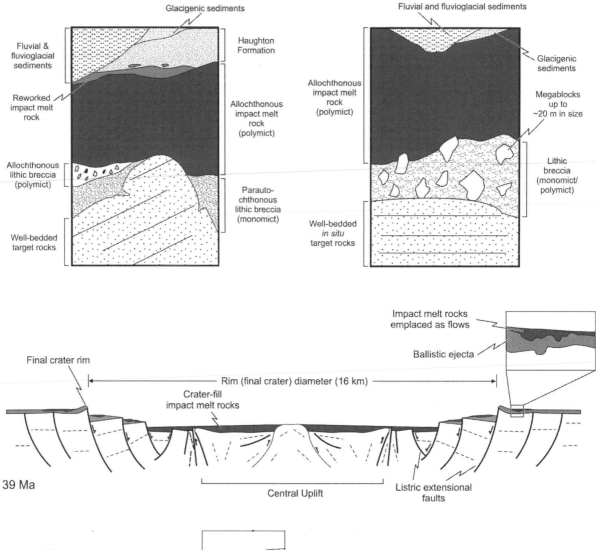

CRATER INTERIOR IMPACTITE SEQUENCE:

RIM IMPACTITE SEQUENCE:

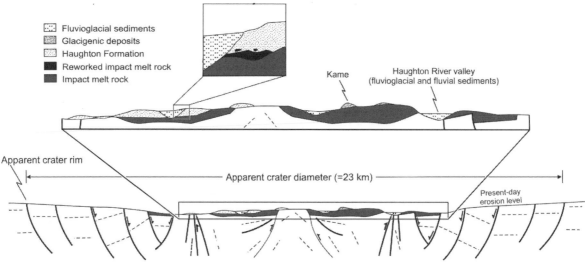

**Figure. 19.4** *Top*: Schematic cross-sections showing the different types of impactites and their stratigraphic sequence in the crater interior and near-surface crater rim regions of the Haughton impact structure. *Bottom*: Schematic cross-sections depicting the newly formed Haughton impact crater with an uplifted rim approximately 16 km in diameter, a terraced zone and a buried central uplift. The present-day Haughton impact structure bears the signs of erosion and sedimentary infilling over the past 39 Myr. The crater rim has been eroded away, along with much of the ejecta deposits, exposing other concentric normal faults not visible in the newly formed Haughton crater.

**Figure 19.5** Field photographs of impactites from impact craters developed in sedimentary (Haughton; a,b), mixed sedimentary–crystalline (Ries; b,c), and crystalline (Mistastin; e,f) targets. (a) Field photograph of a well-exposed section of the crater-fill impact melt rocks at the Haughton structure. (b) Close-up view of impact melt rocks showing the clast-rich character and fine-grained microscopic nature of the pale grey groundmass. (c) Surficial impact melt-bearing breccia ('suevite') (light grey) overlying Bunte Breccia (dark grey) at the Aumühle quarry. (d) Close-up view of the impact melt-bearing breccia showing the fine-grained groundmass and macroscopic irregular silicate glass bodies. (e) Oblique aerial view of the approximately 80 m high cliffs of impact melt rock at the Discovery Hill locality, Mistastin impact structure, Labrador. Photograph courtesy of Derek Wilton. (f) Close-up view of massive fine-grained (aphanitic) impact melt from the Discovery Hill locality. Camera case for scale. (See Colour Plate 41)

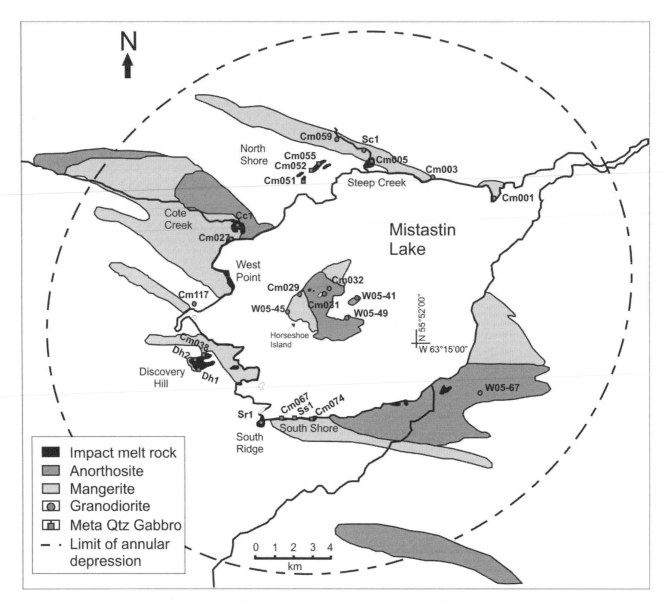

**Figure 19.6** Simplified geological map of the Mistastin Lake impact structure. Modified from Marion (2008).

### 19.2.3  Mistastin Lake impact structure, Canada

The approximately 36 Ma Mistastin Lake impact structure, Labrador (55°53′N, 63°18′W), was formed entirely within crystalline rocks of the approximately 1.4 Ga Mistastin Lake Batholith, comprising mangerite, anorthosite, and minor granodiorite and granitic gneiss (Currie, 1971; Grieve, 1975; Mak *et al.*, 1976; Marchand and Crocket, 1977; Marion *et al.*, 2007) (Fig. 19.6; Table 19.1). The Mistastin Lake structure (locally known as Kamestastin) was originally identified as an impact structure by Taylor and Dence (1969). An oval-shaped lake occupies the inner approximately 16 km diameter portion of the crater. Two islands, the approximately 3 km wide Horseshoe Island and the small Bullseye Island, are located in the centre of the lake, and represent a central uplift (Fig. 19.6). The topographic rim of the structure is eroded, but is still visible in airborne images, yielding an apparent crater

diameter of approximately 28 km. To date, no geophysical surveys or modelling specific to the Mistastin structure have been completed. Regional Bouguer gravity and aeromagnetic surveys indicate a negative anomaly centred over the southeast portion of the lake. The occurrence of anorthosite, a relatively low-density rock, in that area may explain why the anomaly is not centred in the middle of the structure (Grieve, 2006).

Relatively little work has been conducted at Mistastin in the past. Most recently, the impact melt rocks have been the focus of renewed study (Marion, 2008; Marion and Sylvester, 2010). Fieldwork conducted over the past three summers, together with ongoing laboratory work, has revealed important new insights into Mistastin, including the discovery of ejecta deposits (Mader *et al.*, 2011). One of the characteristic features of Mistastin is the presence of widespread impact melt rocks, including an approximately 60–80 m sequence at the Discovery Hill locality, displaying classic columnar jointing, typical of volcanic and intrusive igneous rocks (Fig. 19.5e,f). This impact melt rock layer has been subdivided into four gradational subunits (Fig. 19.6; Grieve, 1975). A variety of impactites are present below the coherent impact melt rocks, ranging from melt-free lithic breccias to melt-bearing breccias (Fig. 19.5e,f, Fig. 19.6). The original impact melt sheet at Mistastin has been estimated at 200 m thick, yielding an original melt volume estimate of approximately 20 km³ (Grieve, 1975).

### 19.2.4  Ries impact structure, Germany

The Ries impact structure is one of the best-preserved terrestrial complex impact structures on Earth (Pohl *et al.*, 1977). Located in Bavaria, southern Germany (48°53′N, 10°37′E), it formed approximately 14.6 Ma (Buchner *et al.*, 2010) in an approximately 470–820 m thick flat-lying sequence of predominantly Mesozoic sedimentary rocks that unconformably overlay Hercynian crystalline basement (Fig. 19.1; Pohl *et al.*, 1977; Schmidt-Kaler, 1978). The Ries structure consists of a central crater cavity approximately 12 km in diameter (so-called 'inner ring') that is filled by a series of crater-fill impactites (Fig. 19.7; Table 19.1), overlain by an approximately 400 m thick series of post-impact lacustrine sedimentary rocks (Fig. 19.7; Pohl *et al.*, 1977). No surface exposures exist of the crater-fill or so-called 'crater suevites'; however, this unit has been sampled by three drill holes (Deiningen, Nördlingen 1973, and Hole 1001; Fig. 19.7). All impact melt products in this unit have been completely hydrothermally altered. Magnetic field measurements reveal a negative anomaly (−300 nT) coincident with the central cavity, which has been attributed to the presence of reversely magnetized crater-fill suevites (Pohl and Angenheister, 1969). The central basin is bounded by a prominent 'inner ring' composed of weakly shocked rocks from the uppermost part of the crystalline basement, together with sedimentary lithologies. No central peak is present at the surface; however, geophysical studies suggest the presence of an uplifted ring of basement rocks with a diameter of approximately 4–5 km (Fig. 19.7; Pohl *et al.*, 1977). The region between the inner ring and the tectonic rim (~24 km diameter) is characterized by a chaotic mixture of large (metre to 100 m scale) blocks of both crystalline and sedimentary rocks ('megablock zone'; Pohl *et al.*, 1977), overlain by various different types of proximal impactites, which also extend out beyond the crater rim (Fig. 19.5c,d, Fig. 19.7 and Table 19.1): (1) Bunte Breccia and megablocks; (2) polymict crystalline breccias; (3) surficial suevites; and (4) coherent impact melt rocks.

The Ries structure is most notable as being the type locality for 'suevite'. It was first recognized at the Ries and takes its name from the Roman name for the region: 'Provincia Suevia' (Sauer, 1920). Isolated, surficial outcrops of suevite, from a few metres to approximately 25–30 m thick, occur inside the morphological rim, and up to radial distances of approximately 14 km beyond the rim to the south-southwest and east-northeast (Fig. 19.7; von Engelhardt, 1990). However, approximately 80 m of suevite was penetrated in the Wörnitzostheim drillhole (Förstner, 1967). The origin and emplacement of the Ries surficial suevites has been debated for decades. Early workers, based on optical microscopy studies, suggested that the groundmass is essentially clastic – with

melt present as glass fragments – with a generally airborne mode of emplacement (Stöffler, 1977; von Engelhardt and Graup, 1984; von Engelhardt, 1990,1997; von Engelhardt *et al.*, 1995). More recent work, utilizing scanning electron microscopy, provided evidence for melt products in the groundmass, with the interpretation of flow during and after emplacement and consistent with an origin as melt-rich flows with entrained clasts that emanated from different regions of the growing crater (Osinski *et al.*, 2004).

It is notable that no coherent impact melt sheet has been documented at the Ries structure to date; however, given that impact melt rocks derived from the crystalline basement are preserved around the crater rim (von Engelhardt *et al.*, 1969; Osinski, 2004) and that the drill hole may not be representative of the entire crater-fill impactites sequence (Stöffler, 1977), it was suggested that a small coherent melt body(s) may be present within the interior of the Ries structure (Stöffler, 1977; Dressler and Reimold, 2001; Osinski, 2004). The discovery of approximately 13 m of impact melt rock in the newly acquired Enkingen drill core provides a tantalizing hint at such a possible body (Pohl *et al.*, 2010).

### 19.2.5  Rochechouart impact structure, France

The Rochechouart impact structure was formed in Hercynian age (400–300 Ma) granitic intrusive and metamorphic rocks in the northwestern edge of the French Massive Central (45°50′N, 0°56′E; Kraut and French, 1971). The generally accepted age of the Rochechouart structure is 214 ± 8 Ma, based on ⁴⁰Ar–³⁹Ar laser spot fusion dating of pseudotachylite generated during transient crater collapse (Kelley and Spray, 1997), although a younger age of 201 ± 2 Ma has also recently been reported (Schmieder *et al.*, 2010). The Rochechouart structure has been deeply eroded, such that no topographic expression of the crater remains (Kraut and French, 1971; Fig. 19.1). Accepted estimates of the crater diameter range from 20 to 25 km, based on a shock zoning study (Lambert, 1977), to 18 to 26 km, based on correlation with a negative gravity anomaly centred on the structure (Pohl *et al.*, 1978).

Despite the erosion level, scattered outcrops of impact breccias and impact melt-bearing rocks occur over an area of 150 km² (Lambert, 1977). The Rochechouart impactites are complex and heterogeneous at all scales. The allochthonous Rochechouart impactites form a gradational continuous spectrum from melt-free lithic breccias to aphanitic crystalline melt rocks comprising 99 vol.% melt (Lambert, 1977). Five main units from the base of the structure upwards have been described as follows, with many of the units having names associated with their location of discovery: unit 1, shocked parautochthonous basement rock; unit 2, finely crystalline melt rock ('Babadous melt'), unit 3, red 'welded' breccia/suevite ('Montoume breccia'); unit 4, lithic breccia ('Rochechouart breccia'); and unit 5, suevitic breccia ('Chassenon suevite') (Kraut and French, 1971). Recent work based on scanning electron microscopy shows that the so-called Montoume breccia is in fact a particulate clast-rich impact melt rock (Sapers *et al.*, 2009). Impact-generated hydrothermal alteration is pervasive in places, with all impactite units having elevated $K_2O/Na_2O$ ratios indicative of pervasive K-metasomatism (Lambert, 1977).

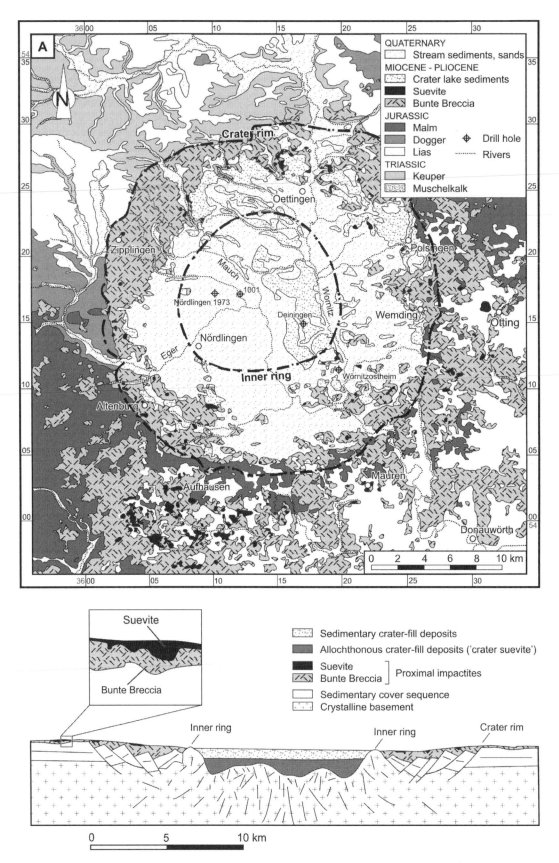

**Figure 19.7**  Simplified geological map (top) and cross-section (bottom) of the Ries impact structure. Modified from Schmidt-Kaler (1978).

## 19.3 Comparisons and implications

### 19.3.1 Size

A fundamental question is what the various listed diameters for these craters actually represent. Do they represent the rim (or final crater) diameter or apparent crater diameter (see discussion in Chapter 1 and Turtle *et al.* (2005))? It is notable that in the Earth Impact Database (2012) no distinction is made and care should be used when using these diameter estimates for scaling and calculating variables such as stratigraphic uplift. For most craters on Earth it is often not possible to relate rim and apparent crater diameter. In terms of the craters compared in this contribution, it has been suggested for Haughton that the commonly quoted diameter of 23 km is actually the apparent diameter and a robust estimate of 16 km has been suggested for the rim diameter (Osinski and Spray, 2005). This rim diameter represents the outer limit of a topographic depression marked by a semi-continuous line of concentric normal faults that record large-scale (>100–400 m) displacements of slump blocks in towards the crater centre (Osinski and Spray, 2005), consistent with the definition by Turtle *et al.* (2005). At the more eroded craters (Mistastin and Rochech-ouart), the listed diameter is clearly an apparent diameter. It is not so clear, however, for Ries and Boltysh – which both retain much of their original morphology – what the listed diameter represents. Cross-sections of the Boltysh structure do not show a topographic feature in the rim region (Masaitis, 1999), consistent with the largely eroded ejecta blanket, so this most likely represents the apparent crater diameter.

### 19.3.2 Central uplift formation

It has long been recognized that lithologies in the centre of complex impact structures are structurally uplifted above their pre-impact stratigraphic position (Dence *et al.*, 1968). The formation mechanism(s) of central uplifts, however, is still not fully understood (see Chapter 5). There are notable differences in the surface expressions of the central uplifts at the similarly sized complex impact structures under consideration here. Rochechouart and Mistastin are too eroded to make any definitive affirmations regarding their original uplift morphology. Boltysh possesses a central uplift that is emergent through the crater-fill deposits (Masaitis, 1999) and can, therefore, be termed a *central peak* crater (see discussion in Chapters 1 and 5). At Haughton, there can be no argument that it has a *central uplift*, as large, kilometre-size fault-bounded blocks of the Eleanor River Formation and smaller blocks (up to ~50–150 m across) of the Blanley Bay Formation have been uplifted greater than 1050 m to less than 1300 m and greater than 1300 m to less than 1450 m respectively above their pre-impact stratigraphic positions (Osinski and Spray, 2005). Field evidence, however, strongly suggests that the central uplift was completely covered with impact melt rocks in the newly formed Haughton crater. There is, therefore, no central peak or peak ring at Haughton, as the original definitions of these features require that they emerge through the crater-fill deposits (Grieve and Therriault, 2004). Like Haughton, the Ries structure also lacks a central emergent topographic peak.

Instead, an inner ring of uplifted basement material is present (Pohl *et al.*, 1977).

These observations suggest that the lack of a central peak could be due to the presence of a thick sedimentary cover sequence at Haughton (~1880 m) and the Ries (~500–850 m), whereas Boltysh formed in an entirely crystalline target. Importantly, the outer edge of the central uplift at Haughton has been interpreted as representing an interference zone between the outward collapsing central uplift and inward collapsing crater walls (Osinski and Spray, 2005) (cf. numerical models of Collins *et al.* (2002)). If this interpretation is correct, it may help to explain the lack of a central peak or peak ring at Haughton. That is, a 'peak' may have formed early on during the modification stage, but subsequently collapsed during the final stages of crater formation. This collapse, presumably, was aided by relatively lower bulk strength of the collapsing material, due to the presence of layering in the sedimentary target rocks. However, other impact characteristics (e.g. projectile density, impact velocity) may also be involved (see discussion in Grieve and Therriault (2004)).

### 19.3.3 Generation of impact melt

One of the most notable differences between these various craters is in their allochthonous impactites. Figure 19.5 clearly show that these impactites are very different in terms of their visual appearance. However, early suggestions that impacts into sedimentary targets do not produce impact melt-bearing lithologies have now largely been superseded by the realization that the volumes of melt produced within craters formed within different target lithologies are similar (Osinski *et al.*, 2008a,b; Wünnemann *et al.*, 2008). It is apparent, however, that even in craters developed purely in crystalline targets, there are substantial differences in the characteristics of the allochthonous crater-fill materials. At Mistastin, a large coherent melt layer was generated, with characteristic columnar jointing and igneous textures, but impact melt-bearing breccias (suevites) also underlie the melt sheet and intrude into the crater floor (Grieve, 1975). At Boltysh, suevites underlie and overlie the melt sheet (Masaitis, 1999). The presence of these impact melt-bearing breccias beneath these coherent impact melt sheets is clearly not explainable by an airborne mode of emplacement.

At Rochechouart, for a long time it appeared that no coherent clast-poor melt sheet was preserved and/or formed, but rather a series of more heterogeneous melt-bearing breccias. Recent work, however, suggests that the greater than 80 m thick Montoume impactites are, in fact, clast-rich impact melt rocks and may represent the base of an originally much larger and thicker coherent impact melt sheet (Sapers *et al.*, 2009).

### 19.3.4 Ejecta emplacement

Impact ejecta deposits are preserved at four of the craters considered here. Despite the poor preservation of ejecta at Boltysh, field and drill core evidence indicates the presence in the proximal ejecta consisting of a low-shock, lithic breccia overlain by a melt-bearing deposit (Table 19.1). A similar situation exists at the Haughton, Ries and Mistastin structures. The range of target rocks, at these structures, suggests, therefore, that this trait is not due to the effect of volatiles, layering or other effects of target lithology on the impact cratering process.

Some of the best-preserved and exposed ejecta deposits on Earth occur at the Ries structure, Germany (von Engelhardt, 1990). The Ries clearly displays a distinctive two-layer ejecta configuration (Fig. 19.5c and Fig. 19.7), with a series of impact melt-bearing breccias (suevites) and minor impact-melt rocks overlying the continuous ejecta blanket (Bunte Breccia). Studies of the Bunte Breccia strongly support the concept of ballistic sedimentation (Hörz *et al.*, 1983). The sharp contact between the Ries ejecta layers (Fig. 19.5c) indicates that there is a temporal hiatus between emplacement of the ballistic Bunte Breccia and the overlying suevites/impact melt deposits (Hörz, 1982). Early workers suggested that the overlying surficial 'suevite' ejecta at the Ries were deposited sub-aerially from an ejecta plume and they were described as 'fallout' suevites (von Engelhardt and Graup, 1984; von Engelhardt, 1990). This is despite the fact that the surficial suevites are unsorted to poorly sorted, which is not predicted by sub-aerial deposition (i.e. fallout from an ejecta plume). Rather, such deposits are typically sorted and display normal grading, as is the case in pyroclastic fall deposits (Fisher and Schmincke, 1984).

A contrasting hypothesis is that the proximal surficial suevites were emplaced as surface flow(s), either comparable to pyroclastic flows or as ground-hugging volatile- and melt-rich flows (Newsom *et al.*, 1986; Bringemeier, 1994; Osinski *et al.*, 2004). A flow origin also has been suggested for the relatively rare coherent impact melt rocks that reside atop the Bunte breccia, at the Ries (Osinski, 2004). Osinski *et al.* (2011) also suggested that the most intransigent argument used to support an airborne origin for suevites at the Ries – namely, melt glass clasts, similar in shape to 'fladen' (German for 'round, flat dough cake'), and 'gneiss-cored glass bombs' – can also be formed during transport in a confining flow. Indeed, melt glass coatings on lithic clasts and so-called aerodynamically shaped 'fladen' can be found in impact melt-bearing breccias that lie beneath coherent impact melt rocks at the Mistastin and Manicouagan structures. As with the respective coherent melt rocks, these breccias with melt 'fladen' below the melt rocks were never airborne, with their melt glass morphologies the result of flow.

Observations at these four mid-size impact structures were recently used to provide a new working hypothesis for the origin and emplacement of ejecta on the terrestrial planets, in which the ejecta are emplaced in a multistage process (Osinski *et al.*, 2011). In this model, the generation of the continuous ejecta blanket occurs during the excavation stage of cratering, via the conventional ballistic sedimentation and radial flow model (Oberbeck, 1975). This is followed by the emplacement of more melt-rich, ground-hugging flows – the 'surface melt flow' phase – during the terminal stages of crater excavation and the modification stage of crater formation. Fallback is minor and occurs during the final stages of crater formation.

### 19.3.5    Hydrothermal alteration

Over the past several years, evidence has been mounting that the generation of hydrothermal systems is commonplace after impact into $H_2O$-bearing planetary objects (see Chapter 6). A recent review suggests that evidence for impact-induced hydrothermal activity can be recognized at over 70 of the approximately 180 terrestrial craters (Osinski *et al.*, 2012). Based on this and earlier reviews (e.g. Naumov, 2005), it seems highly probable that any hypervelocity impact capable of forming a complex crater (>2–4 km and >5–10 km diameter on Earth and Mars respectively) will generate a hydrothermal system. Out of the craters studied here, hydrothermal alteration is by far better studied at the Haughton and Ries structures. Interestingly, there are major differences between these two craters. At the Ries, hydrothermal alteration was pervasive throughout the crater-fill impact melt-bearing breccias, completely altering all impact melt phases (Osinski, 2005a). In contrast, at the Haughton impact structure, evidence for hydrothermal alteration is widespread, but is typically restricted to isolated vugs and veins within the crater-fill impactites and pristine impact glasses are common (Osinski *et al.*, 2005a). The difference in the intensity of hydrothermal alteration of crater-fill impactites between the Haughton and Ries impact structures is notable given their very similar size (23 km and 24 km respectively) and the fact that they both occurred in a continental setting. It has been observed previously that craters in marine targets are typically more hydrothermally altered (Naumov, 2005). Critically, the crater-fill suevites at the Ries are overlain by approximately 400 m of lacustrine crater-fill sediments and sedimentation appears to have commenced immediately following impact (Arp, 1995). In contrast, at Haughton, there is no evidence preserved of a crater lake immediately post-impact (Osinski and Lee, 2005). This has been used to infer that the presence/absence of an overlying crater lake may play a critical role in determining the level of hydrothermal alteration of crater-fill impactites (Osinski *et al.*, 2005a; Chapter 6).

### 19.4    Comparisons with lunar and Martian impact craters

Observations of planetary craters are critical for providing information regarding the original pristine morphology and morphometry of complex impact structures. Impact craters on Earth, in contrast, are typically eroded to varying degrees, as is the case for all the structures studied here, but they provide the only ground truth data on the three-dimensional structure of complex craters and on the various types of impactites produced. Comparison of craters on the Earth with other planetary bodies must be approached with some care. In particular, it is clear from the above discussion that the diameter of craters on Earth is typically difficult to define. For most craters on Earth, it is the apparent crater diameter that is used, whereas on other planetary bodies the diameter will typically mean the rim or final crater diameter. The difference between the apparent and rim diameter can be several kilometres, if observations and interpretations at the Haughton structure are correct.

One of the key conclusions of this comparative case study of mid-size impact structures is that it is apparent that the target lithology plays a major role in defining crater morphology, in particular of central uplifts (e.g. compare the lack of an emergent central uplift at Haughton with a well-formed central peak at Boltysh). Differences in central uplift characteristics have also been noted on the Moon and also ascribed to target differences. For example, Cintala *et al.* (1977) showed that there was a greater abundance in central peaks in mare versus highland craters, which they ascribed to the layered but coherent nature of the mare substrate, in contrast to the mega-regolith of the highlands. This is consistent with the observation that central uplifts form at smaller diameters (~2 km) in sedimentary (weaker) rocks than in crystalline (stronger) rocks (~4 km) on Earth (Dence, 1968). This case study suggests also that central uplifts formed in sedimentary rocks then become unstable as diameter increases, such that subdued central uplifts appear to be the norm in mid-size craters in such targets on Earth.

In terms of impact ejecta, continuous ejecta blankets around lunar craters are typically blocky and of high albedo, when fresh. 'Ponds' of impact melt are common around many lunar craters and overlie the continuous ejecta blanket (Hawke and Head, 1977). In contrast, there is a wide variation in impact ejecta morphology around Martian impact craters, with so-called layered ejecta that display single (SLE; 86%), double (DLE; 9%) or multiple (MLE; 5%) layer morphologies being commonplace (Barlow, 2005). It is outside the scope of this chapter to discuss in detail impact ejecta emplacement, but a commonality between the mid-size impact structures considered here, and many lunar and Martian structures, is the apparent presence of impact melt-rich deposits (which may be patchy or continuous) overlying the continuous ejecta blanket. As a result of these observations, a new multistage impact ejecta emplacement model has been proposed (Osinski *et al.*, 2011), which is discussed in Chapter 4.

## 19.5 Concluding remarks

This comparative study suggests that the fundamental processes of impact cratering are basically the same in different target rocks. Despite the major differences in the visual appearance of impactites in craters developed in sedimentary, crystalline and mixed sedimentary–crystalline targets (Fig. 19.5), the use of advanced micro-beam analytical techniques suggests that the basic products are genetically equivalent and that similar volumes of melt are produced. It is likely, however, that target differences played a role in the noted morphological differences between craters formed in various types of target rocks, in particular the formation of central uplifts. Further comparative studies are required and encouraged, as it is not yet clear how these different parameters pertain to achieving these differences.

## References

Arp, G. (1995) Lacustrine bioherms, spring mounds, and marginal carbonates of the Ries impact crater (Miocene, southern Germany). *Facies*, 33, 35–89.

Barlow, N.G. (2005) A review of Martian impact crater ejecta structrues and their implications for target properties, in *Large Meteorite Impacts III* (eds T. Kenkmann, F. Hörz and A. Deutsch), Geological Society of America Special Paper 384, Geological Society of America, Boulder, CO, pp. 433–442.

Bringemeier, D. (1994) Petrofabric examination of the main suevite of the Otting Quarry, Nordlinger Ries, Germany. *Meteoritics and Planetary Science*, 29, 417–422.

Buchner, E., Schwarz, W.H., Schmieder, M. and Trieloff, M. (2010) Establishing a 14.6±0.2 Ma age for the Norrdlinger Ries impact (Germany) – a prime example for concordant isotopic ages from various dating materials. *Meteoritics and Planetary Science*, 45, 662–674.

Cintala, M.J., Wood, C.A. and Head, J.W. (1977) The effects of target characteristics on fresh crater morphology – preliminary results for the moon and Mercury. *Proceedings of the Lunar and Planetary Science Conference*, 8, 3409–3425.

Collins, G.S., Melosh, H.J., Morgan, J.V. and Warner, M.R. (2002) Hydrocode simulations of Chicxulub Crater collapse and peak-ring formation. *Icarus*, 157, 24–33.

Currie, K.L. (1971) *Geology of the Resurgent Cryptoexplosion Crater at Mistastin Lake, Labrador*, Bulletin of the Geological Survey of Canada 207, Geological Survey of Canada, Ottawa.

Dence, M.R. (1968) Shock zoning at Canadian craters: petrography and structural implications, in *Shock Metamorphism of Natural Materials* (eds B.M. French and N.M. Short), Mono Book Corporation, Baltimore, MD, pp. 169–184.

Dence, M.R. (1972) The nature and significance of terrestrial impact structures. 24th International Geological Congress Proceedings, pp. 77–89.

Dence, M.R., Innes, M.J. S., Robertson, P.B. *et al.* (1968) Recent geological and geophysical studies of Canadian craters, in *Shock Metamorphism of Natural Materials* (eds B.M. French and N.M. Short), Mono Book Corporation, Baltimore, MD.

Dressler, B.O. and Reimold, W.U. (2001) Terrestrial impact melt rocks and glasses. *Earth Science Reviews*, 56, 205–284.

Earth Impact Database (2012) Earth Impact Database, http://www.unb.ca/passc/ImpactDatabase/.

Fisher, R.V. and Schmincke, H.U. (1984) *Pyroclastic Rocks*, Springer-Verlag, Berlin.

Förstner, U. (1967) Petrographische untersuchungen des suevit aus den bohrungen Deiningen und Wörnitzostheim im Ries von Nördlingen. *Contributions to Mineralogy and Petrology*, 15, 281–308.

Frisch, T. and Thorsteinsson, R. (1978) Haughton astrobleme: a Mid-Cenozoic impact crater, Devon Island, Canadian Arctic Archipelago. *Arctic*, 31, 108–124.

Glass, B.J., Lee, P. and Osinski, G.R. (2002) Airborne geomagnetic investigations at the Haughton impact structure, Devon Island, Nunavut, Canada: new results. 33rd Lunar and Planetary Science Conference, CD-ROM, abstract no. 2008.

Greiner, H.R., Fortier, Y.O., Blackadar, R.G. *et al.* (1963) Haughton Dome and area southwest of Thomas Lee Inlet, in *Geology of the North-Central Part of the Arctic Archipelago, Northwest Territories (Operation Franklin)*, Geological Survey of Canada Memoir 320, Geological Survey of Canada, Ottawa.

Grieve, R.A.F. (1975) Petrology and chemistry of impact melt at Mistastin Lake Crater, Labrador. *Geological Society of America Bulletin*, 86, 1617–1629.

Grieve, R.A.F. (1988) The Haughton impact structure: summary and synthesis of the results of the HISS project. *Meteoritics*, 23, 249–254.

Grieve, R.A.F. (2006) *Impact Structures in Canada*, Geological Association of Canada, Ottawa.

Grieve, R.A.F. and Therriault, A. (2004) Observations at terrestrial impact structures: their utility in constraining crater formation. *Meteoritics and Planetary Science*, 39, 199–216.

Grieve, R.A.F., Reny, G., Gurov, E.P. and Ryabenko, V.A. (1987) The melt rocks of the Boltysh impact crater, Ukraine, USSR. *Contributions to Mineralogy and Petrology*, 96, 56–62.

Grieve, R.A.F., Reimold, W.U., Morgan, J.V. *et al.* (2008) Observations and interpretations at Vredefort, Sudbury and Chicxulub: towards an empirical model of terrestrial impact basin formation. *Meteoritics and Planetary Science*, 43, 855–882.

Gurov, E.P., Kelley, S.P. and Koeberl, C. (2003) Ejecta of the Boltysh impact crater in the Ukrainian Shield, in *Impact Markers in the Stratigraphic Record, Impact Studies*, vol. 3 (eds C. Koeberl and F.C. Martinez-Ruiz), Springer-Verlag, Heidelberg, pp. 179–202.

Gurov, E.P., Kelley, S.P., Koeberl, C. and Dykan, N.I. (2006) Sediments and impact rocks filling the Boltysh impact crater, in *Biological Processes Associated with Impact Events* (eds C.S. Cockell, C. Koeberl and I. Gilmour), Springer-Verlag, Berlin, pp. 335–358.

Hawke, B.R. and Head, J.W. (1977) Impact melt on lunar crater rims, in *Impact and Explosion Cratering* (eds D.J. Roddy, R.O. Pepin and R.B. Merrill), Pergamon Press, New York, NY, pp. 815–841.

Herrick, R.R. and Pierazzo, E. (2003) *Results of the Workshop on Impact Cratering: Bridging the Gap Between Modeling and Observations*, LPI Contribution No. 1162, Lunar and Planetary Institute, Houston, TX.

Hickey, L.J., Johnson, K.R. and Dawson, M.R. (1988) The stratigraphy, sedimentology, and fossils of the Haughton Formation: a post-impact crater-fill, Devon Island, N.W.T., Canada. *Meteoritics*, 23, 221–231.

Hörz, F. (1982) Ejecta of the Ries Crater, Germany, in *Geological Implications of Impacts of Large Asteroids and Comets on the Earth* (eds L.T. Silver and P.H. Schultz), Geological Society of America Special Paper 190, Geological Society of America, Boulder, CO, pp. 39–55.

Hörz, F., Ostertag, R. and Rainey, D.A. (1983) Bunte Breccia of the Ries: continuous deposits of large impact craters. *Reviews of Geophysics and Space Physics*, 21, 1667–1725.

Jessberger, E.K. (1988) $^{40}$Ar–$^{39}$Ar dating of the Haughton impact structure. *Meteoritics*, 23, 233–234.

Kelley, S.P. and Gurov, E.P. (2002) Boltysh, another end-Cretaceous impact. *Meteoritics and Planetary Science*, 37, 1031–1043.

Kelley, S.P. and Spray, J.G. (1997) A Late Triassic age for the Roche-chouart impact structure, France. *Meteoritics and Planetary Science*, 32, 629–636.

Kraut, F. and French, B.M. (1971) The Rochechouart impact structure, France. *Journal of Geophysical Research*, 76, 5407–5413.

Lambert, P. (1977) The Rochechouart Crater: shock zoning study. *Earth and Planetary Science Letters*, 35, 258–268.

Mader, M., Osinski, G.R. and Marion, C. (2011) Impact ejecta at the Mistastin Lake impact structure, Labrador. 41st Lunar and Planetary Science Conference, no. 2505.

Mak, E.K., York, D., Grieve, R.A.F. and Dence, M.R. (1976) The age of the Mistastin Lake Crater, Labrador, Canada. *Earth and Planetary Science Letters*, 31, 345–357.

Marchand, M. and Crocket, J.H. (1977) Sr isotopes and trace element geochemistry of the impact melt and target rocks at the Mistastin Lake Crater, Labrador. *Geochimica et Cosmochimica Acta*, 41, 1487–1495.

Marion, C.L. (2008) Petrology of impact melts at the Mistastin impact structure, Labrador. MSc thesis, Memorial University, St Johns.

Marion, C.L. and Sylvester, P.J. (2010) Composition and heterogeneity of anorthositic impact melt at Mistastin Lake crater, Labrador. *Planetary and Space Science*, 58, 552–573.

Marion, C.L., Sylvester, P.J., Tubrett, M. and Shaffer, K. (2007) Microbeam-based analytical approach to define melt composition and identify target rocks in impact melt sheets: an analogue study from the Mistastin Lake impact crater, Labrador. *Geological Association of Canada – Mineralogical Association of Canada Annual Meeting, Program with Abstracts*, 32, 53.

Masaitis, V.L. (1974) Some ancient meteorite craters in the territory of USSR. *Meteoritica*, 33, 64–68.

Masaitis, V.L. (1999) Impact structures of northeastern Eurasia: the territories of Russia and adjacent countries. *Meteoritics and Planetary Science*, 34, 691–711.

Naumov, M.V. (2005) Principal features of impact-generated hydro-thermal circulation systems: mineralogical and geochemical evidence. *Geofluids*, 5, 165–184.

Newsom, H.E., Graup, G., Sewards, T. and Keil, K. (1986) Fluidization and hydrothermal alteration of the suevite deposit in the Ries Crater, West Germany, and implications for Mars. *Journal of Geophysical Research, B, Solid Earth and Planets*, 91, 239–251.

Oberbeck, V.R. (1975) The role of ballistic erosion and sedimentation in lunar stratigraphy. *Reviews of Geophysics and Space Physics*, 13, 337–362.

Osinski, G.R. (2004) Impact melt rocks from the Ries impact structure, Germany: an origin as impact melt flows? *Earth and Planetary Science Letters*, 226, 529–543.

Osinski, G.R. (2005a) Hydrothermal activity associated with the Ries impact event, Germany. *Geofluids*, 5, 202–220.

Osinski, G.R. (2005b) Geological map, Haughton impact structure, Devon Island, Nunavut, Canada. *Meteoritics and Planetary Science*, 40 (12).

Osinski, G.R. and Lee, P. (2005) Intra-crater sedimentary deposits at the Haughton impact structure, Devon Island, Canadian High Arctic. *Meteoritics and Planetary Science*, 40, 1887–1900.

Osinski, G.R. and Spray, J.G. (2001) Impact-generated carbonate melts: evidence from the Haughton Structure, Canada. *Earth and Planetary Science Letters*, 194, 17–29.

Osinski, G.R. and Spray, J.G. (2003) Evidence for the shock melting of sulfates from the Haughton impact structure, Arctic Canada. *Earth and Planetary Science Letters*, 215, 357–370.

Osinski, G.R. and Spray, J.G. (2005) Tectonics of complex crater formation as revealed by the Haughton impact structure, Devon Island, Canadian High Arctic. *Meteoritics and Planetary Science*, 40, 1813–1834.

Osinski, G.R., Grieve, R.A.F. and Spray, J.G. (2004) The nature of the groundmass of surficial suevites from the Ries impact structure, Germany, and constraints on its origin. *Meteoritics and Planetary Science*, 39, 1655–1684.

Osinski, G.R., Lee, P., Parnell, J. *et al.* (2005a) A case study of impact-induced hydrothermal activity: the Haughton impact structure, Devon Island, Canadian High Arctic. *Meteoritics and Planetary Science*, 40, 1859–1878.

Osinski, G.R., Lee, P., Spray, J.G. *et al.* (2005b) Geological overview and cratering model for the Haughton impact structure, Devon Island, Canadian High Arctic. *Meteoritics and Planetary Science*, 40, 1759–1776.

Osinski, G.R., Spray, J.G. and Lee, P. (2005c) Impactites of the Haughton impact structure, Devon Island, Canadian High Arctic. *Meteoritics and Planetary Science*, 40, 1789–1812.

Osinski, G.R., Grieve, R.A.F., Collins, G.S. *et al.* (2008a) The effect of target lithology on the products of impact melting. *Meteoritics and Planetary Science*, 43, 1939–1954.

Osinski, G.R., Grieve, R.A.F. and Spray, J.G. (2008b) Impact melting in sedimentary target rocks: an assessment, in *The Sedimentary Record of Meteorite Impacts* (eds K.R. Evans, W. Horton, D.K. King Jr *et al.*), Geological Society of America Special Publication 437, Geological Society of America, Boulder, CO, pp. 1–18.

Osinski, G.R., Tornabene, L.L. and Grieve, R.A.F. (2011) Impact ejecta emplacement on the terrestrial planets. *Earth and Planetary Science Letters*, 310, 167–181.

Osinski, G.R., Tornabene, L.L., Banerjee, N.R. *et al.* (2012) Impact-generated hydrothermal systems on Earth and Mars. *Icarus*, in press.

Pohl, J. and Angenheister, G. (1969) Anomalien des Erdmagnetfeldes und Magnetisierung der Gesteine im Nördlinger Ries. *Geologica Bavarica*, 61, 327.

Pohl, J., Stöffler, D., Gall, H. and Ernstson, K. (1977) The Ries impact crater, in *Impact and Explosion Cratering* (eds D.J. Roddy, R.O. Pepin and R.B. Merrill), Pergamon Press, New York, NY pp. 343–404.

Pohl, J., Ernstson, K. and Lambert, P. (1978) Gravity measurements in the Rochechouart impact structure (France). *Meteoritics*, 13, 601–604.

Pohl, J., Eckstaller, A. and Robertson, P.B. (1988) Gravity and magnetic investigations in the Haughton impact structure, Devon Island, Canada. *Meteoritics*, 23, 235–238.

Pohl, J., Poschlod, K., Reimold, W.U. *et al.* (2010) Ries crater, Germany: the Enkingen magnetic anomaly and associated drill core SUBO 18, in *Large Meteorite Impacts and Planetary Evolution IV* (eds R.L. Gibson and W.U. Reimold), Geological Society of America Special Papers 465, Geological Society of America, Boulder, CO, pp. 141–163.

Redeker, H.J. and Stöffler, D. (1988) The allochthonous polymict breccia layer of the Haughton impact crater, Devon Island, Canada. *Meteoritics*, 23, 185–196.

Robertson, P.B. and Grieve, R.A.F. (1978) The Haughton impact structure. *Meteoritics*, 13, 615–619.

Sapers, H.M., Osinski, G.R. and Banerjee, N.R. (2009) Re-evaluating the Rochechouart impactites: petrographic classification, hydrothermal alteration and evidence for carbonate-bearing target rocks. 40th Lunar and Planetary Science Conference, 1284.pdf.

Sauer, A. (1920) Erläuterungen zur geologischen, Karte Württemberg, 1 : 50 000, Blatt 20, Bopfingen.

Schmidt-Kaler, H. (1978) Geological setting and history, in *Principle Exposures of the Ries Meteorite Crater in Southern Germany* (eds E.C.T. Chao, R. Hüttner and H. Schmidt-Kaler), Verlag Bayerisches Geologisches Landesamt, Munich, pp. 8–11.

Schmieder, M., Buchner, E., Schwarz, W.H. *et al.* (2010) A Rhaetian $^{40}Ar/^{39}Ar$ age for the Rochechouart impact structure (France) and implications for the latest Triassic sedimentary record. *Meteoritics and Planetary Science*, 45, 1225–1242.

Sherlock, S.C., Kelley, S.P., Parnell, J. *et al.* (2005) Re-evaluating the age of Haughton impact event. *Meteoritics and Planetary Science*, 40, 1777–1787.

Stöffler, D. (1977) Research drilling Nördlingen 1973: polymict breccias, crater basement, and cratering model of the Ries impact structure. *Geologica Bavarica*, 75, 443–458.

Taylor, F.C. and Dence, M.R. (1969) A probable meteorite origin for Mistastin Lake, Labrador. *Canadian Journal of Earth Sciences*, 6, 39–45.

Thorsteinsson, R. and Mayr, U. (1987) *The Sedimentary Rocks of Devon Island, Canadian Arctic Archipelago*, Geological Survey of Canada Memoir 411, Geological Survey of Canada, Ottawa.

Turtle, E.P., Pierazzo, E., Collins, G.S. *et al.* (2005) Impact structures: what does crater diameter mean? in *Large Meteorite Impacts III* (eds T. Kenkmann, F. Hörz and A. Deutsch), Geological Society of America Special Paper 384, Geological Society of America, Boulder, CO, pp. 1–24.

Von Engelhardt, W. (1990) Distribution, petrography and shock metamorphism of the ejecta of the Ries Crater in Germany – a review. *Tectonophysics*, 171, 259–273.

Von Engelhardt, W. (1997) Suevite breccia of the Ries impact crater, Germany: petrography, chemistry and shock metamorphism of crystalline rock clasts. *Meteoritics and Planetary Science*, 32, 545–554.

Von Engelhardt, W. and Graup, G. (1984) Suevite of the Ries Crater, Germany: source rocks and implications for cratering mechanics. *Geologische Rundschau*, 73, 447–481.

Von Engelhardt, W., Stöffler, D. and Schnieder, W. (1969) Petrologische Untersuchungen im Ries. *Geologica Bavarica*, 61, 229–295.

Von Engelhardt, W., Arndt, J., Fecker, B. and Pankau, H.G. (1995) Suevite breccia from the Ries Crater, Germany: origin, cooling history and devitrification of impact glasses. *Meteoritics*, 30, 279–293.

Wünnemann, K., Collins, G.S. and Osinski, G.R. (2008) Numerical modelling of impact melt production on porous rocks. *Earth and Planetary Science Letters*, 269, 529–538.

# TWENTY

# Processes and products of impact cratering: glossary and definitions

## Gordon R. Osinski

*Departments of Earth Sciences/Physics and Astronomy, Western University, 1151 Richmond Street, London, ON, N6A 5B7, Canada*

### 20.1 Introduction

The goal of this chapter is to provide a compendium of useful definitions and formulae for calculating various attributes of meteorite impact craters. Definitions have been complied largely from Grieve and Cintala (1981), Grieve *et al.* (1981), Melosh (1989), Grieve and Pilkington (1996) and Turtle *et al.* (2005). Other references are provided, where applicable.

### 20.2 General definitions

**Equation of state:** Thermodynamic equation describing a shock transition from an uncompressed state, $P_0, \rho_0, E_0$, to a compressed state, $P_1, \rho_1, E_1$, in terms of pressure $P$, density $\rho$ and internal energy $E$.

**Hugoniot curve:** Locus of shock states that can be achieved by shock wave compression of variable intensity in any specific material.

**Hugoniot elastic limit:** The shock pressure below which the shock-compressed material behaves elastically.

**Hypervelocity impact:** Hypervelocity impact occurs when a cosmic projectile is large enough (typically >50 m for a stony object and >20 m for an iron body) to pass through the atmosphere on Earth with little or no deceleration and so strike at virtually its original cosmic velocity (>11 km s$^{-1}$; French, 1998). This produces high-pressure shock waves in the target. Smaller projectiles lose most of their original kinetic energy in the atmosphere and produce small metre-size 'penetration craters', without the production of shock waves.

**Impact ejecta:** Target materials, regardless of their physical state, that are transported beyond the rim of the transient cavity (see Chapter 4). In complex craters, therefore, ejecta deposits occur in the crater rim region interior to the final crater rim (Osinski *et al.*, 2011). This term does not carry any genetic connotation.

**Impact ejecta, ballistic:** Solid, liquid and vapourized rock, or any combination thereof, ejected ballistically from an impact crater.

**Impact ejecta, distal:** Impact ejecta deposited beyond five crater radii of the impact point.

**Impact ejecta, proximal:** Impact ejecta deposited within five crater radii of the impact point.

**Impact ejecta blanket:** The continuous ejecta deposit around an impact crater.

**Impact metamorphism:** This is essentially the same as shock metamorphism (see below), except that it also encompasses the melting and vapourization of target rocks.

**Shock deformation:** Deformation by shock wave compression at shock pressures above the Hugoniot elastic limit, leading to either transient or residual shock effects after pressure release.

**Shock effects:** Also known as shock metamorphic effects. These form due to deformation and/or transformation of minerals and rocks induced by the passage of a shock wave. Shock effects may be transient during shock compression or permanent after pressure release (see Chapter 8).

**Shock melting:** Melting of solid matter by shock wave compression resulting from high post-shock temperature after pressure release (see Chapter 9).

**Shock metamorphism:** The metamorphism of rocks and minerals caused by shock wave compression and decompression due to impact of a solid body or due to the detonation of high-energy chemical or nuclear explosives (see Chapter 8).

**Shock wave:** A step-like discontinuity in pressure, density, particle velocity and internal energy which propagates in gaseous, liquid or solid matter with supersonic velocity.

**Shock vapourization:** Vapourization of solid or liquid matter by shock wave compression and resulting from high post-shock temperature after pressure release.

## 20.3   Morphometric definitions and equations

**Central uplift:** Target rocks found approximately in the centre of a complex impact crater that have been uplifted above their pre-impact stratigraphic level. For most craters on Earth, this is the preferred terminology to use, as it does not reflect the surface morphology in a fresh crater.

**Central peak:** If the central uplift is observed to be emergent from surrounding allochthonous crater-fill impactites (e.g. a coherent impact melt sheet) and takes the form of a single peak or tightly clustered group of peaks, then the term central peak can be applied.

**Central-peak basin:** If the central uplift is observed to be emergent from surrounding allochthonous crater-fill impactites (e.g. a coherent impact melt sheet) and takes the form of a fragmentary ring of peaks surrounding a central peak, then the term central-peak basin can be applied.

**Complex impact crater:** An impact crater with relatively low depth-to-diameter ratio and with central uplift, annular trough and down-faulted, terraced rim structure.

**Depth, apparent crater:** Apparent depth measured from the present ground surface to the top of the allochthonous crater fill (Grieve *et al.*, 1981). Related to the rim (or final crater) diameter $D$ by the empirical relationship $d_a=0.13D^{1.06}$ for terrestrial simple craters (Grieve and Pilkington, 1996).

**Depth, true crater:** Depth to the base of the allochthonous breccia lens. Related to the rim (or final crater) diameter $D$ by the empirical relationship $d_t=0.28D^{1.02}$ for terrestrial simple craters (Grieve and Pilkington, 1996).

**Depth, excavation:** Various estimates for the depth of excavation $d_e$ are in the literature. Based on stratigraphic considerations, $d_e>0.08D$ at Barringer Crater (Shoemaker, 1963), where $D$ is the final rim diameter. The maximum $d_e$ of material in the ballistic ejecta deposits of the Haughton and Ries structures is $0.035D_a$ (Osinski *et al.*, 2011), where $D_a$ is the apparent crater diameter. If the final rim diameter $D$ is used, which is the parameter measured in planetary craters, a value $0.05D$ is obtained for Haughton. No other data are available for $d_e$ in terrestrial complex impact craters.

**Diameter, apparent crater $D_a$:** Represents the diameter of the outermost ring of (semi-) continuous concentric normal faults, measured with respect to the pre-impact surface (i.e. accounting for the amount of erosion that has occurred). For the majority of impact structures on Earth this will be the only measurable diameter.

**Diameter, transient cavity:** Several estimates for relating the diameter of the transient cavity $D_{tc}$ to the final rim diameter in complex craters have been proposed: $D_{tc}=0.5D–0.65D$ (Grieve *et al.*, 1981), $D_{tc}=1.23D^{0.85}$ (Croft, 1985), $D_{tc}=0.57D$ (Lakomy, 1990).

**Impact crater:** A generally circular crater formed either by impact of an interplanetary body (projectile) on a planetary surface or by an experimental hypervelocity impact of a projectile into solid matter. Applicable for fresh craters; for eroded craters, the term *impact structure* should be used (see below).

**Impact structure:** A roughly circular geological structure caused by impact, irrespective of its state of preservation.

**Multi-ring basin:** An impact crater with relatively low depth-to-diameter ratio and with at least two concentric rings inside the crater.

**Peak-ring basin:** If the central uplift is observed to be emergent from surrounding allochthonous crater-fill impactites (e.g. a coherent impact melt sheet) and takes the form of a well-developed ring of peaks but no central peak, then the term peak-ring basin can be applied.

**Rim (or final crater) diameter $D$:** Defined as the diameter of the topographic rim that rises above the surface for simple craters, or above the outermost slump block not concealed by ejecta for complex craters. This is relatively easy to measure on most planetary bodies, where the topographic rim is usually preserved due to low rates of erosion. On Earth, however, such pristine craters are rare and the rim region is typically eroded away.

**Simple impact crater:** A relatively simple, bowl-shaped impact crater, with relatively high depth-to-diameter ratio.

**Structural uplift:** The observed amount of uplift undergone by the deepest marker horizon now exposed in the centre of a complex impact structure (Grieve *et al.*, 1981). Based on observations at 24 impact craters developed in sedimentary rocks on Earth, the structural uplift (SU) in the centre of the crater is generally taken as $0.086D^{1.03}$, where $D$ is the crater diameter (Grieve and Pilkington, 1996).

**Transient cavity:** The transient, roughly bowl-shaped cavity that is thought to form directly by excavation and displacement of the target rocks by the cratering flow-field, during the formation of an impact crater.

## 20.4 Impactites

**Diaplectic glass:** Glass formed through the solid-state transformation from an original mineral; original grain boundaries are preserved.

**Fragmental impact breccia:** Synonymous to lithic impact breccia. Note: it has been recommended that this term should no longer be used (Stöffler and Grieve, 2007).

**Impact breccia:** A monomict or polymict breccia, which occurs around, inside and below impact craters.

**Impactite:** Rock affected by impact metamorphism (includes shocked rocks, impact breccias and impact melt rocks).

**Impactoclastic deposit:** Consolidated or unconsolidated sediment resulting from ballistic excavation, transport and deposition of rocks at impact craters; may contain particles of impact melt rock.

**Impactoclastic air fall bed:** A sedimentary layer containing a certain fraction of shock-metamorphosed material (e.g. shocked minerals and melt particles) which has been ejected from an impact crater and deposited over large regions of a planet or globally.

**Impact regolith:** Fine-grained impactoclastic deposit formed by multiple impacts on the surface of planetary bodies lacking an atmosphere, such as the Moon, Mercury or asteroids. The lunar regolith contains unshocked and shocked lithic and mineral clasts, glass fragments, glass bodies of revolution and agglutinate glass.

**Impact melt:** Melt formed by shock melting of target rocks in impact craters (see Chapter 9).

**Impact melt rock:** Crystalline or glassy rock solidified from impact melt containing variable amounts of target rock clasts. Can be clast-rich, -poor or -free.

**Impact melt rock, particulate:** Impact melt rock that comprises a heterogeneous mixture of impact melt phases and where there is evidence for these groundmass phases being fluid during and after transport (Osinski *et al.*, 2008). This distinguishes such rocks from coherent crystalline silicate melt rocks and impact melt-bearing breccias that possess a clastic groundmass.

**Impact melt-bearing breccia:** Impact breccia with a *clastic* matrix containing lithic and mineral clasts in various stages of shock metamorphism and including clasts of impact melt that are in a glassy or crystallized state.

**Impact melt breccia:** Impact melt rock containing lithic and mineral clasts displaying variable degrees of shock metamorphism in a crystalline or glassy matrix. Note: it has been

recommended that this term should no longer be used as it is synonymous with a clast-rich impact melt rock (Stöffler and Grieve, 2007).

**Impact glass:** Impact melt quenched to glass; may contain crystallites and target rock clasts.

**Lithic impact breccia:** Impact breccia with clastic matrix containing shocked and unshocked mineral and lithic clasts but containing no melt, either as clasts or matrix material.

**Maskelynite:** Diaplectic plagioclase glass.

**Planar microstructures:** A collective term comprising shock-induced planar fractures and planar deformation features.

**Planar fractures:** Fractures occurring in shocked minerals as multiple sets of planar fissures parallel to rational crystallographic planes, which are usually not observed as cleavage planes under normal geological (non-shock) conditions.

**Planar deformation features:** Submicroscopic amorphous lamellae occurring in shocked minerals as multiple sets of planar lamellae (optical discontinuities under the petrographic microscope) parallel to rational crystallographic planes; indicative of shock metamorphism.

**Pseudotachylite:** Also spelled pseudotachylite, these typically dike-like impactites are thought to be produced by either frictional melting and/or shock melting in the basement of impact craters (see Chapter 9). They may contain unshocked and shocked mineral and lithic clasts in a very fine grained crystalline or glassy melt matrix.

**Suevite:** Originally defined for impactites at the Ries impact structure, Germany, 'suevites' are polymict impact breccia with *clastic* matrix containing lithic and mineral clasts in various stages of shock metamorphism, including cogenetic impact melt particles which are in a glassy or crystallized state (Stöffler, 1977; Stöffler *et al.*, 1977). It is recommended that the term not be used, but instead replaced with *impact melt-bearing breccia*.

**Tektite:** Impact glass formed from near-surface melt ejected ballistically and deposited, often as aerodynamically shaped bodies, in a strewn field outside the continuous ejecta blanket at some terrestrial impact structures.

## References

Croft, S.K. (1985) The scaling of complex craters. Proceedings of the 15th Lunar and Planetary Science Conference, pp. 828–842.

French, B.M. (1998) *Traces of a Catastrophe: A Handbook of Shock Metamorphic Effects in Terrestrial Meteorite Impact Structures*, LPI Contribution No. 954, Lunar and Planetary Institute, Houston, TX.

Grieve, R.A.F. and Cintala, M.J. (1981) A method for estimating the initial impact conditions of terrestrial cratering events, exemplified by its application to Brent Crater, Ontario. *Proceedings of the Lunar and Planetary Science Conference*, 12B, 1607–1621.

Grieve, R.A.F. and Pilkington, M. (1996) The signature of terrestrial impacts. *Journal of Australian Geology and Geophysics*, 16, 399–420.

Grieve, R.A.F., Robertson, P.B. and Dence, M.R. (1981) Constraints on the formation of ring impact structures, based on terrestrial data, in *Proceedings of the Conference on Multi-Ring Basins: Formation and Evolution* (eds P.H. Schultz and R.B. Merrill), Pergamon Press, New York, NY, pp. 37–57.

Lakomy, R. (1990) Distribution of impact induced phenomena in complex terrestrial impact structures: implications for transient cavity dimensions. *Proceedings of the Lunar and Planetary Science Conference*, 21, 676–677.

Melosh, H.J. (1989) *Impact Cratering: A Geologic Process*, Oxford University Press, New York, NY.

Osinski, G.R., Grieve, R.A.F., Collins, G.S. *et al.* (2008) The effect of target lithology on the products of impact melting. *Meteoritics and Planetary Science*, 43, 1939–1954.

Osinski, G.R., Tornabene, L.L. and Grieve, R.A.F. (2011) Impact ejecta emplacement on the terrestrial planets. *Earth and Planetary Science Letters*, 310, 167–181.

Shoemaker, E.M. (1963) Impact mechanics at Meteor Crater, Arizona, in *The Moon, Meteorites and Comets* (eds B.M. Middlehurst and G.P. Kuiper), University of Chicago Press, Chicago, IL, pp. 301–336.

Stöffler, D. (1977) Research drilling Nördlingen 1973: polymict breccias, crater basement, and cratering model of the Ries impact structure. *Geologica Bavarica*, 75, 443–458.

Stöffler, D. and Grieve, R.A.F. (2007) Impactites, in *Metamorphic Rocks* (eds D. Fettes and J. Desmons), Cambridge University Press, Cambridge, pp. 82–92.

Stöffler, D., Ewald, U., Ostertag, R. and Reimold, W.U. (1977) Research drilling Nördlingen 1973 (Ries): composition and texture of polymict impact breccias. *Geologica Bavarica*, 75, 163–189.

Turtle, E.P., Pierazzo, E., Collins, G.S. *et al.* (2005) Impact structures: what does crater diameter mean? in *Large Meteorite Impacts III* (eds T. Kenkmann, F. Hörz and A. Deutsch), Geological Society of America Special Paper 384, Geological Society of America, Boulder, CO, pp. 1–24.

# *Index*

Note: Page numbers referring to figures are in *italics* and those referring to tables are in **bold**. Plate numbers are in brackets.

---

*Impact Cratering: Processes and Products*, First Edition. Edited by Gordon R. Osinski and Elisabetta Pierazzo.
© 2013 Blackwell Publishing Ltd. Published 2013 by Blackwell Publishing Ltd.